**Genetics :
Questions
and
Problems**

D1319019

Genetics: Questions and Problems

John Kuspira
Professor of Genetics
University of Alberta

G. W. Walker
Professor of Genetics
University of Alberta

McGraw-Hill Book Company
New York St. Louis San Francisco Düsseldorf Johannesburg Kuala Lumpur London
Mexico Montreal New Delhi Panama Rio de Janeiro Singapore Sydney Toronto

Genetics:
Questions
and
Problems

Copyright © 1973 by McGraw-Hill, Inc. All rights reserved.
Printed in the United States of America. No part of this
publication may be reproduced, stored in a retrieval system,
or transmitted, in any form or by any means, electronic,
mechanical, photocopying, recording, or otherwise, without
the prior written permission of the publisher.

1234567890 KPKP 79876543

This book was set in Times New Roman.
The editors were James R. Young, Jr. and David Damstra;
the designer was Edward A. Butler; and the production supervisor was Thomas J. Lo Pinto.
The drawings were done by Reproduction Drawings Ltd.
The printer and binder was Kingsport Press, Inc.

Library of Congress Cataloging in Publication Data

Kuspira, John, 1928–
    Genetics.

    Includes bibliographies.
    1. Genetics—Problems, exercises, etc.
I. Walker, George William, 1913–   joint author.
II. Title. [DNLM: 1. Genetics. QH 431 K97ge 1973]
QH431.K89   575.1'076   72-6855
ISBN 0-07-035672-6

Contents

Preface

The bases for a well-founded course of instruction in genetics have been clearly laid out by Sinnott and Dunn:

> The principles of genetics have developed out of arduous study of scores of investigators, and understanding of principles can best be gained by the student through a process similar to that employed in their original discovery. This process begins with, and is continually stimulated by, curiosity as to the methods and the mechanism of inheritance; it proceeds by the collection and study of facts, and by a critical discrimination between those which are true and relevant and those which are untrue and irrelevant; and finally it involves a considerable practice of the reasoning faculty by which deductions are made, and applied or tested on many similar cases. It is only in this way that the process of inheritance can be *understood*. The learning of facts alone cannot accomplish this.

The authors of "Genetics: Questions and Problems" have sought to facilitate some of these aspects of instruction that are rarely found in texts, viz. "the collection and study of facts" and "practice of the reasoning faculty by which deductions are made and applied or tested on many similar cases." The authors have attempted to cover all areas of genetics, including as wide a range of organisms as possible and emphasizing current trends of thought and technique. "Genetics: Questions and Problems" should therefore complement any current text in genetics. It is hoped also that instructors will find the data useful for teaching purposes.

The problems in each chapter are placed in three categories: (1) questions, (2) problems, and (3) problems for further reading. The questions are intended to assist students in their study and review of lecture topics. The problems, included for student assignment, are roughly graded in difficulty, being in general more difficult toward the end of the chapters and in category 3. The great majority of the problems are based on actual experimental data, so that the student can gain a sense of true exploration in the science. To emphasize this feature we have included references at the end of each chapter to papers and texts used or referred to in setting the problems. We have tried to include all the classic and significant papers in genetics, and it is hoped that students, especially upper undergraduates and graduates, will use them to explore the science beyond the confines of texts and this book. Instructors are advised to review questions chosen for assignments carefully to make sure that their students are adequately prepared to handle them. Some interchapter redundancy has been permitted in simpler problems that have relevance in different areas so that problem assignments will involve as little cross-referencing as possible.

In this book taxonomy is concerned only with clear identification of the organisms used in each experimental study, together with their genetic variants. Some laxity has therefore

---

[1] E. W. Sinnott and L. C. Dunn, "Principles of Genetics," p. xiii, McGraw-Hill Book Company, New York, 1939.

been permitted in the current binomial nomenclature, so that outdated names are retained (along with their current equivalents) to facilitate reference to the original papers and scientific names are frequently omitted in favor of their colloquial equivalents. Explanations of biochemical abbreviations and other necessary tools will be found in the Appendix.

The following texts have been used to provide models or guidelines for some of the questions: F. B. Hutt, "Animal Genetics," The Ronald Press Company, New York, 1964; R. C. King, "Genetics," Oxford University Press, New York, 1965; E. W. Sinnott, L. C. Dunn, and T. Dobzhansky, "Principles of Genetics," 5th ed., McGraw-Hill Book Company, New York, 1958; L. H. Snyder and P. R. David, "The Principles of Heredity," 5th ed., D. C. Heath and Company, Boston, 1957; A. M. Srb and R. D. Owen, "General Genetics," W. H. Freeman and Company, San Francisco, 1952; and M. W. Strickberger, "Genetics," The Macmillan Company, New York, 1968.

We gratefully acknowledge the patience and understanding of our wives, Joanne Marlene and Minnie Jane, the valuable assistance of Drs. Asad Ahmed and Dave Cameron in reading and commenting on the molecular and population genetics sections, and the reviewers of the manuscript. We wish also to thank Margaret Brennan and Kay Baert for their expert secretarial assistance and Nick Muntjewerff for the painstaking checking of details.

John Kuspira
G. W. Walker

1
Mitosis and Meiosis

These questions are concerned with those aspects of mitosis and meiosis which are necessary for a full comprehension of transmission genetics and chromosome aberrations. Questions regarding the chemical nature and structure of chromosomes are included in Chap. 25.

QUESTIONS

1 a What are the main organelles of the nucleus? What is the function of each?
  b Why are cytogeneticists more interested in the nucleus than in other parts of the cell?

2 a According to the geneticist's point of view, what is the major function of mitosis and how is it accomplished?
  b How does mitosis (understood here to include both nuclear and cytoplasmic division) in animals differ from mitosis in plants?
  c Is the process of mitosis basically the same or different in diploid and haploid cells of the same organism? In different organisms of (1) the same species, (2) different species? Explain.
  d Mitosis is essential to the asexual reproduction of lower organisms and the growth of higher ones. Using specific examples, explain and illustrate this statement.
  e Explain how you would demonstrate experimentally that mitosis achieves a duplication rather than a halving of the chromosomes during each division.
  f It is often stated that telophase is prophase in reverse. Explain.

3 Rarely do all the somatic cells of a multicellular individual have the same chromosome number; e.g., in man, although most of the body cells have a $2n = 46$ chromosome number, at least some of the cells of the liver have 92 ($4n$) chromosomes. How can a $4n$ cell arise from a $2n$ one?

4 A zoospore (asexual cell) in *Chlamydomonas*, an alga, divides mitotically, as do the daughter zoospores during each of the subsequent five divisions. All cells remain viable.
  a How many cells do you expect in the clone after the sixth mitosis?
  b Develop a formula to indicate the number of cells expected in a clone after:
    (1) One mitotic division.
    (2) Two such divisions.
    (3) n mitotic divisions.

5 The chromosome constitution of a somatic *Drosophila melanogaster* cell and its daughter cells as seen at mitotic prometaphase is illustrated. Count the chromosomes in all three cells and suggest one way in which mitosis could be modified to produce daughter cells with the aberrant chromosome numbers they possess.

Parent cell   Daughter cells

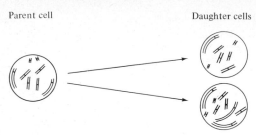

Figure 5

6 At interphase chromosomes are not visible with the light microscope. Nevertheless, we know they are there, having maintained their integrity from the previous division. Suggest ways of determining whether a particular interphase nucleus is *n*, 2*n*, or polyploid (3*n*, 4*n*, etc.).

7 **a** Which, if either, of the two cells illustrated is haploid? Why?

Figure 7

  b Which of the mitotic stages can these cells not be at?

8 **a** What in your opinion is the main function of meiosis? How is this accomplished? What makes such a process necessary?
  b What genetic functions does meiosis perform? Explain.

9 **a** What conditions are necessary to prove that the segregation of any one chromosome pair is independent of that of any other?
  b Diagram what you would expect to see at anaphase I under these conditions.
  c Show with an idiogram (chromosome diagram) how these conditions might be achieved in an experimental organism such as corn or the fruit fly *Drosophila*.

10 What would be the consequences to future generations of a diploid organism if meiosis were omitted during gamete formation?

11 **a** In what main features does the first division of meiosis differ from mitosis?
  b How does the second division of meiosis differ from mitosis?

12 State whether the following statements are true or false and why:
  a Any chromosome may synapse with any other chromosome in the same cell at zygotene.

b During mitosis chromosomes divide, and their chromatids separate at anaphase to form two nuclei, each with the same number of chromosomes as the mother cell.

c A secondary meiocyte has half as many chromosomes as a meiocyte.

d In a primary spermatocyte containing 18 chromosomes, 15 of these chromosomes may be paternal.

e The chromosomes in a somatic cell, regardless of the organism, are all morphologically alike.

f A microspore may contain more paternal chromosomes than the somatic cells of the same plant.

13 Is it possible for meiosis to occur in haploid species? Haploid individuals? Explain.

14 Which of the two processes, mitosis or meiosis, has greater significance in genetic studies and why?

15 In somatic cells of the fruit fly *Drosophila colorata*, at mitotic anaphase there are 12 chromosomes at each pole (Wharton, 1943). Five are straight and rod-shaped, two are small and V-shaped, three are large and V-shaped, and two are J-shaped. What does this tell you about the position of the centromere in each of these types?

16 If half of the chromosomes of a primary oocyte segregate into the first polar body, why aren't some chromosomes of the complement absent from the egg?

17 The somatic cells of man (*Homo sapiens*) contain 23 pairs of chromosomes (Tjio and Levan, 1956). What relation do the members of each pair bear to the parents? Explain, using one pair of chromosomes.

18 When do the chromosomal strands become double in (1) a somatic cell and (2) a pollen mother cell (microsporocyte)? Cite and discuss the evidence for your answer.

19 In *Haplopappus gracilis* ($2n = 4$), one of the Compositae, a zygote receives the chromosomes A and B from the male parent and their homologs A^1 and B^1 from the female parent. Which of the following chromosome complements would you expect to find in the somatic cells of the plant arising from this zygote: A^1A^1BB, $AA^1B^1B^1$, $AABB$, AA^1BB^1, $A^1A^1B^1B^1$? Why?

20 Chromosomes or chromosome regions that are heterochromatic are usually very compact and stain densely during interphase and early prophase, whereas euchromatic chromosomes or chromosome regions stain lightly or not at all at these stages. What does this suggest about the coiling cycle of the two kinds of chromatin?

21 **a** Explain, with the aid of diagrams, how meiosis reduces the chromosome number to one-half the somatic number whereas mitosis produces nuclei with the same chromosome number as the parent nucleus.

b Is it theoretically possible to simplify meiosis so that only one nuclear division would be required to obtain cells containing the haploid complement of chromosomes and genes? Explain your answer with the aid of diagrams.

22 Distinguish between cytokinesis and karyokinesis and cite examples of tissues in which the two processes do not occur concurrently.

23 A plant in the diploid species *Arabidopsis thaliana*, a crucifer, has five homologous pairs of chromosomes, which we may label as KK, LL, MM, NN, and OO. When it is self-fertilized, which of the following chromosome complements would you expect to find in the somatic cells of its offspring?
KLMO   MMLLNN   KKLLMMOO
KNO   KKLL   KKLLMMNNOO
MNO   KKKKLLLLMMMMNNNNOOOO

24 Suggest a reason why three of the four products of meiosis in females are nonfunctional.

25 In *Euschistus variolarius*, a hemipteran insect, there are 14 chromosomes in the somatic cells. Seven bivalent associations were always found at prophase I of meiosis (Montgomery, 1901).
a Argue that this is evidence that bivalent associations at prophase I are always made up of one paternal and one homologous maternal chromosome.
b Show what you would expect if pairing were between chromosomes of a given parent.

26 What bearing do the following findings have on (1) chromosome continuity from cell to cell and organism to organism and (2) qualitative differences between chromosomes?

1 In the horse threadworm (*Ascaris megalocephala*) it is found that chromosomes occupy the same relative positions at anaphase of one division and prophase of the next.
2 If two members of an identical homologous chromosome pair are rendered heteromorphic in a given cell, the heteromorphic difference is retained in successive mitotic divisions of this cell.
3 Products of the first division of 3n and 4n zygotes in the sea urchin, with few exceptions, develop abnormally. However, they are not all alike in abnormal development. Moreover, the chromosome numbers are different in most cases.

27 A rare type of *parthenogenesis*, called *thelytoky*, occurs in certain species of insects, rotifers, crustaceans, and nematodes. Unfertilized eggs give rise to females which are identical in all respects. Males are either absent or sexually nonfunctional.
a Outline one method of reproduction that may lead to such progeny.
b Discuss the advantages and disadvantages of such a form of reproduction with regard to (1) reproductive efficiency and (2) genetic variability.

28 Answer the following questions with respect to each of the processes illustrated in Figs. 28A and 28B.
   a What is the process? Give reasons for your decisions.
   b Does it occur in a plant or animal or both? Explain.
   c Label the various stages.

A   B   C   D

E   F   G   H

Figure 28A

A   B   C

D   E   F

G   H   I

J   K   L

Figure 28B

PROBLEMS

29 *Crepis neglecta*, a member of the Compositae, has four pairs of chromosomes (Babcock, 1942). Assume that a plant receives the chromosomes A, B, C, and D from the male parent and the homologs A^1, B^1, C^1, and D^1 from the female parent.

a If no crossing-over occurs, what proportion of the gametes of this plant will be expected to carry all the chromosomes of paternal origin?

b Would the proportion carrying only chromosomes from the maternal parent be the same? Explain.

c What proportion of the gametes would carry chromosomes from both the male and the female parent?

d Derive a formula that indicates the number of different kinds of gametic chromosomal combinations an individual with any number of pairs of chromosomes can produce if the members of all pairs are qualitatively different.

30 The American bullfrog has 13 pairs of chromosomes ($2n = 26$; Makino, 1951). Considering only three of these pairs and designating the chromosomes from the male parent with solid lines and those from the female parent with broken ones, show diagrammatically all the possible alignments of these pairs at metaphase I. Assuming no crossing-over, how many different kinds of meiotic products (spermatids, ootids) are possible from such an individual?

31 Sutton (1902) showed that in male *Brachystola magna* (a grasshopper), all the chromosomes could be individually recognized. All somatic cells had 23 chromosomes, which were classified by size into 11 pairs and 1 other chromosome. At prophase I of meiosis, 11 bivalents and 1 univalent were observed in each nucleus, and chromosomes showed the same size differences as at mitosis. What conclusions can you draw from these observations regarding (1) maintenance of morphological individuality and (2) meiotic pairing?

32 Carothers (1913) in a cytological study of *Brachystola magna*, a grasshopper, examined 300 cells showing metaphase I or anaphase I of meiosis in males with 23 chromosomes. She found 11 bivalents, 1 of them heteromorphic (one long, one short), and an unpaired chromosome. This unpaired chromosome (the X) moved in its entirety to one pole or the other at anaphase I. In 154 cells the large chromosome of the heteromorphic pair went to the same pole as the X chromosome, and in 146 cells the shorter one segregated with the X chromosome. What conclusions can you draw regarding the manner in which chromosome pairs orient and segregate in relation to each other?

33 The majority of the somatic cells of multicellular organisms, whether diploid or haploid, have chromosome complements that are quantitatively and qualitatively (viz. morphologically) identical. Using a plant, outline how you would obtain evidence for this statement.

34 In the males of certain insects such as *Protenor belfragei* the somatic cells and meiocytes possess an odd number of chromosomes. *Protenor* males have thirteen chromosomes (six morphologically different types in pairs and one without a pairing partner; Wilson, 1906). Two types of gametes are produced, with six and seven chromosomes, respectively.

a What would you expect to happen to the univalent (chromosome without a partner) at pachytene, metaphase I, anaphase I, and metaphase II?

b What kinds of spermatids, with respect to their chromosome number, would you expect to be formed?   What fraction of the sperms would you expect to be of each type?

35 In a certain plant of the species *Haplopappus gracilis* ($2n = 4$), one of the Compositae (Darlington and Ammal, 1945), a part of chromosome 1 became attached to chromosome 2, so that one could easily distinguish cytologically between the members of each of the chromosome pairs.   The karyotype of this plant is shown in Fig. 35*A*.   Cytological analysis of 1,000 microspores

Figure 35*A*

showed that four haploid complements were present in the proportions given in Fig. 35*B*.   Discuss the relation of the data (1) to the manner of segregation

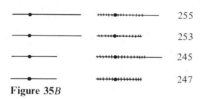

	255
	253
	245
	247

Figure 35*B*

of the chromosomes of each pair and (2) to the segregation of one chromosome pair in relation to that of the other.

36 In most organisms cytokinesis accompanies karyokinesis.   *Arabidopsis thaliana*, a crucifer, has a diploid chromosome number of $2n = 10$ (Darlington and Ammal, 1945).

a What will happen to the nuclear content of a cell if the replication of chromosomes at late interphase is not followed by a mitotic division?

b If this happens in an embryo and subsequent mitosis proceeds normally, what will be the chromosome content of the plant that develops from it?

37 A plant has a karyotype consisting of rod-shaped and V-shaped chromosomes.

Figure 37

What proportion of the progeny will have only rod-shaped chromosomes after four generations of self-fertilization, and what proportion of the progeny will have the original karyotype?   Explain.

38 In the tomato (*Lycopersicon* [=*Lycopersicum*] *esculentum*) chromosome 2 has a satellite (S) on the short arm (Brown, 1949).   Suppose you discover a plant that has a satellite missing from one of the second chromosomes and a satellite (S) attached to one of the third chromosomes, so that the karyotype of the plant with respect to these chromosomes is as shown.

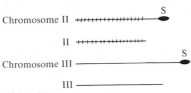

Figure 38

a What types of meiotic products (with respect to chromosome number and kind) would you expect on the male side (in the nuclei of pollen grains) and in what proportions would you expect to find them?

b Would the same results be obtained on the female side?   Explain.

c The plant is self-fertilized (pollen of this plant is used to fertilize its eggs). Show the types of chromosome complements you would expect in a large population of offspring and the proportions of each.

39 In armadillos the pattern of armored bands around the body frequently varies within, as well as between, families.   However, the pattern is always the same in all the offspring (normally 4) of a litter (Newman, 1913).   Explain why this is so.

40 In some of the coccids, meiocytes and gametes are $2n$ and n in chromosome number, respectively, as in eukaryotes with the common pattern of meiosis. However, the secondary meiocytes in these insects are $2n$, whereas those in eukaryotes with the common pattern of meiosis are n.   Suggest how the events at meiosis in the coccids might differ from those in eukaryotes with the common pattern to account for both the similarities and the difference between the two patterns.   (Assume $2n = 2$; each meiocyte produces four haploid meiotic products.)

41 a The housefly (*Musca domestica*) has 12 chromosomes in its somatic cells: three long pairs, one metacentric, one acrocentric, and one telocentric; three short pairs, one metacentric, one acrocentric, and one telocentric.   What proportion of the spermatids of such a male would you expect to possess:
   (1) A long metacentric pair, a long acrocentric pair, and a short telocentric pair?
   (2) Two telocentric pairs and one acrocentric pair?

(3) Two metacentrics, two acrocentric, and two telocentric chromosomes?

(4) Three long chromosomes (one metacentric, one acrocentric, and one telocentric) and three short chromosomes (one telocentric, one acrocentric, and one metacentric)?

b What kinds of chromosomes should the progeny possess if this male is mated with a female with the same number and kinds of chromosomes?

c A spermatid is found with the same number and kinds of chromosomes as in the body cells.   How could such a cell arise?   Illustrate.

42 Prior to 1900 Boveri had shown that n and $3n$ individuals of the sea urchin *Paracentrotus lividus* ($2n = 36$) developed normally.   He later (1902) studied the developmental behavior of eggs that had been fertilized by two sperms, each contributing one centriole.   One or both of these reproduced once before the first division in the embryo, so that at the first cleavage, the embryos ($3n = 54$) divided simultaneously into three or four, instead of two, cells.   The distribution of the chromosomes to the three or four poles at anaphase was irregular.   Most of the first-division embryos isolated from these eggs were found to be abnormal in development (although not alike in their abnormalities).   Boveri noted that only 1 out of 1,500 embryos from four-centriole zygotes developed normally whereas 58 out of 719 from three-centriole zygotes developed normally.

a Discuss the effect of chromosomal imbalance (the occurrence of chromosome numbers other than n, $2n$, $3n$, etc.) on embryonic viability.

b What conclusions can you draw from these data regarding differences (in a genetic sense) between chromosomes?

43 The males in the Hymenoptera (bees, wasps), *Icerya* coccids, and a few other groups develop from unfertilized eggs and are haploid (Hughes-Schrader, 1948; White, 1954).   Nevertheless they produce haploid gametes that fertilize the haploid gametes of the female to give rise, with few exceptions, to $2n$ females. In what way might meiosis be modified in these males to produce haploid gametes?

44 In certain organisms the chromosomes have been found to have diffuse centromeres that extend along their entire length, e.g., in homopteran and hemipteran insects (Brown and Nelson-Rees, 1961; Hughes-Schrader, 1948) and *Luzula*, a member of the plant family Juncaceae (Castro et al., 1949).   It is probable that certain algae and fungi also have this feature (Godward, 1954).

a In what respects does their mitotic chromosome behavior differ from that in organisms with localized centromeres?

b With the aid of diagrams compare and contrast meiosis in organisms with localized centromeres with that of organisms with diffuse centromeres, with respect to:

(1) Synapsis at prophase I.

(2) Type of orientation (auto- or co-) at metaphase I.

(3) Type of disjunction and segregation (for a particular locus) at anaphase I.

(4) The chromosome number of secondary meiocytes.

(5) Second division (type of orientation, disjunction, and segregation).

(6) Chromosome number of the meiotic products.

45 In the majority of cases where two species can be successfully crossed (e.g., common wheat, $2n = 42$; and rye, $2n = 14$) the hybrids are completely or almost completely sterile.   Show how this can be explained by chromosome behavior at meiosis.

46 The male and female hybrids from crosses between the horse ($2n = 64$) and the donkey ($2n = 62$) have 63 chromosomes.   If there is no pairing between the chromosomes of the two species at meiosis in the hybrids, would you expect the hybrids to be fertile or sterile?   Why?

QUESTIONS INVOLVING FURTHER READING

47 Argue for the hypothesis: Of all the cellular organelles the chromosome is the structure most likely to serve as the genetic material or as its carrier. (Herskowitz, 1965; see also Sutton, 1903.)

48 In certain species it is easy to demonstrate cytologically whether non-homologous chromosomes are morphologically different.   To show whether they are qualitatively different in the genetic sense (viz. carry different sets of genes) is much more difficult.

a Outline how you might proceed to prove that they are genetically different.

b Read the classic paper by T. Boveri (1902) for the first demonstration of qualitative differences between nonhomologous chromosomes.

49 The exact time of synapsis of homologous chromosomes in eukaryotes has not been conclusively established.   Speculate as to the exact time the process may occur, realizing that almost all genetic material (DNA) is synthesized in pre-meiotic interphase.   See Comings and Okada (1970), Craig-Cameron (1970), Henderson (1966), Moens (1970), Parchman and Stern (1969), Stern and Hotta (1969), and Walters (1970).

REFERENCES

Anderson, N. G. (1956), *Q. Rev. Biol.*, **31**:169, 243.
Babcock, E. B. (1942), *Bot. Rev.*, **8**:139.
Beadle, G. W. (1932), *Cytologia*, **3**:142.
Belling, J. (1929), *Univ. Calif. Publ. Bot.*, **14**:379.
Benedin, E. van (1883), *Arch. Biol.*, **4**:265; excerpts trans. and reprinted in Gabriel and Fogel (1955), pp. 245–248.
Boveri, T. (1887), *Jena. Z.*, **21**:423.
——— (1888), *Jena. Z.*, **22**:685.
——— (1902), *Verh. Physiol. Med. Ges. Wuerzb.*, N.F., **35**:165.
Brachet, J. (1961), *Sci. Am.*, **205**(September):50.
——— and A. E. Mirsky (eds.) (1961), "The Cell," Academic, New York.

Brown, S. W. (1949), *Genetics,* **34:**437.

———— and W. A. Nelson-Rees (1961), *Genetics,* **46:**983.

Callan, H. G. (1963), *Int. Rev. Cytol.,* **15:**1.

Carothers, E. E. (1913), *J. Morphol.,* **24:**487.

———— (1921), *J. Morphol.,* **35:**457.

Castro, D., A. Camara, and N. Malheiros (1949), *Genet. Iber.,* **1:**48.

Chandra, H. S. (1962), *Genetics,* **47:**1441.

Coleman, W. (1965), *Proc. Am. Philos. Soc.,* **109:**124.

Comings, D. E., and T. A. Okada (1970), *Nature,* **227:**451.

Craig-Cameron, T. A. (1970), *Chromosoma,* **30:**169.

Croes, A. F. (1969), *Planta,* **76:**227.

Darlington, C. D. (1929), *J. Genet.,* **21:**17.

———— and E. K. J. Ammal (1945), "Chromosome Atlas of Cultivated Plants," G. Allen, London.

Farmer, J. B., and J. E. S. Moore (1905), *Q. J. Microsc. Sci.,* **48:**489.

Flemming, W. (1879), *Arch. Mikrosk. Anat.,* 16:302; reprinted in abridged and translated form in Gabriel and Fogel (1955), pp. 127–131.

———— (1887), *Arch. Mikrosk. Anat.,* **29:**389.

Gabriel, M. L., and S. Fogel (eds.) (1955), "Great Experiments in Biology," Prentice-Hall, Englewood Cliffs, N.J.

Gall, J. G. (1956), *Brookhaven Symp. Biol.,* **8:**17.

———— (1961), *J. Biophys. Biochem. Cytol.,* **10:**163.

Godward, M. B. E. (1954), *Ann. Bot.,* **18:**144.

Gregoire, V. (1904), *Cellule,* **21:**297.

———— (1907), *Cellule,* **24:**369.

Gustavsson, I., and C. O. Sundt (1965), *Hereditas,* **54:**249.

Henderson, S. A. (1966), *Nature,* **211:**1043.

Herskowitz, I. H. (1965), "Genetics," 2d ed., Little, Brown, Boston.

Hertig, O. (1876), *Morphol. Jahrb.,* **1:**347.

Hooke, R. (1665), "Micrographia," London; reprinted in Gabriel and Fogel (1955), pp. 3–6.

Hughes-Schrader, S. (1948), *Adv. Genet.,* **2:**127.

———— and H. Ris (1941), *J. Exp. Zool.,* **87:**429.

Jagiello, G., J. Karnicki, and R. J. Ryan (1968), *Lancet,* **1:**178.

John, B., and K. R. Lewis (1965), "The Meiotic System," *Protoplasmologia,* **6**(FI):1–335.

Kitchin, R. M. (1970), *Chromosoma,* **31:**165.

LaCour, L. F. (1953), *Heredity Suppl.,* **6:**77.

Makino, S. (1951), "An Atlas of the Chromosome Number in Animals," Iowa State College Press, Ames.

Mazia, D. (1961*a*), *Sci. Am.,* **205**(September):100.

———— (1961*b*), in Brachet and Mirsky (1961), vol. I, pp. 77–142.

McLeish, John, and B. Snoad (1958), "Looking at Chromosomes," St. Martin's, New York.

Moens, P. B. (1970), *Proc. Natl. Acad. Sci.,* **66:**94.

Montgomery, T. H. (1901), *Trans. Am. Philos. Soc.,* **20:**154.

Moses, M. J. (1968), *Annu. Rev. Genet.,* **2:**363.

Newman, H. H. (1913), *Am. Nat.,* **47:**513.

Ohno, S., H. P. Klinger, and N. B. Atkins (1962), *Cytogenetics,* **1:**42.

Parchman, L. G., and H. Stern (1969), *Chromosoma,* **26:**298.

Patterson, J. T., and W. S. Stone (1952), "Evolution in the Genus *Drosophila,*" Macmillan, New York.

Rhoades, M. M. (1961), in Brachet and Mirsky (1961), vol. III, pp. 1–75.

Riley, R. (1966), *Sci. Prog.,* **54:**193.

———— and V. Chapman (1958), *Nature,* **182:**713.

Ruddle, F. H. (1962), *J. Natl. Cancer Inst.,* **29:**1247.

Sasaki, M. S., and S. Makino (1962), *J. Hered.,* **53:**157.

Schrader, F. (1953), "Mitosis: The Movement of Chromosomes in Cell Division," 2d ed., Columbia, New York.

Sharp, L. W. (1943), "Fundamentals of Cytology," McGraw-Hill, New York.

Sci. Am. (1961), "The Living Cell." **205**(September).

Steffensen, D. (1953), *Proc. Natl. Acad. Sci.*, **39**:613.

Stern, H., and Y. Hotta (1969), *Genetics Suppl.*, **61**:27.

Strasburger, E. (1879), "Die Angiospermen und die Gymnospermen," Gustav Fischer, Jena.

—— (1884), Die Controversen der indirecten Kerntheilung, *Arch. Mikrosk. Anat.*, **23**:246.

—— (1888), "Histologische Beiträge," Heft I, Gustav Fischer, Jena.

Sutton, W. S. (1902), *Biol. Bull.*, **4**:24.

—— (1903), *Biol. Bull.*, **4**:231.

Swanson, C. P. (1957), "Cytology and Cytogenetics," Prentice-Hall, Englewood Cliffs, N.J.

—— (1960), "The Cell," Prentice-Hall, Englewood Cliffs, N.J.

Swift, H. (1950), *Proc. Natl. Acad. Sci.*, **36**:643.

Tjio, J. H., and A. Levan (1956), *Hereditas*, **42**:1.

Waldeyer, W. (1888), *Arch. Mikrosk. Anat.*, **32**:1; English trans. (1889), *Q. J. Microsc. Sci.*, **30**:159.

Walters, M. S. (1970), *Chromosoma*, **29**:375.

Weismann, A. (1889), English trans. of a paper published in 1887, in "Essays upon Heredity and Kindred Biological Problems," vol. I, Clarendon Press, Oxford.

Wenrich, D. H. (1916), *Bull. Mus. Comp. Zool. Harv.*, **60**:55.

Wettstein, D. von (1971), *Proc. Natl. Acad. Sci.*, **68**:851.

Wharton, L. T. (1943), *Univ. Tex. Publ.* 4313, p. 282.

White, M. J. D. (1954), "Animal Cytology and Evolution," Macmillan, New York.

Wilson, E. B. (1906), *J. Exp. Zool.*, **3**:1.

—— (1925), "The Cell in Development and Inheritance," 3d ed., Macmillan, New York.

Winiwarter, H. von (1900), *Arch. Biol. (Paris)*, **17**:33.

Yuncken, C. (1968), *Cytogenetics*, **7**:234.

2
Life (Chromosome) Cycles of Eukaryotes

2
Life
(Chromosome) Cycles
of Eukaryotes

Questions on reproduction and life cycles of bacteria and viruses are included in Chaps. 16 and 17.

QUESTIONS

50 **a** What is the fundamental cytological feature of all forms of asexual reproduction?

b Of what use is it for genetic studies?
Briefly describe the common form it takes in any three of:

Lily	*Neurospora*	*Spirogyra*
Hydra	Aphids	*Paramecium*

51 **a** What are the main differences between the life cycles of higher plants and animals?

b Illustrate with a sketch the salient features of the life cycles of:
  (1) Man (*Homo sapiens*), typical of multicellular diploid animal eukaryotes.
  (2) *Neurospora crassa*, a haploid fungus typical of haploid plant eukaryotes.
  (3) Corn (*Zea mays*) or some other plant typical of diploid plant eukaryotes.

c Compare these life cycles with respect to:
  (1) The conspicuous phase.
  (2) The relationship of meiosis and syngamy to the haplontic and diplontic phases.
  (3) Gamete formation (whether direct or indirect).
  (4) Mitosis: its place of occurrence and its function.

52 **a** Describe the origin and development of the endosperm. In what group of organisms does it occur, and what is its function?

b In corn (*Zea mays*), a monoecious species, the $3n$ endosperm may be *flinty*, F, or *floury*, F^1. Reciprocal crosses between true-breeding *flinty* and true-breeding *floury* strains give the tabulated results.

Table 52

Parent		Endosperm type in seeds
♀	♂	
Flinty	*Floury*	*Flinty*
Floury	*Flinty*	*Floury*

  (1) Sketch as accurately as possible a pollen grain (male gametophyte) and an embryo sac (female gametophyte) from each strain. Identify (label) all nuclei and show the genotypes of each in both structures.
  (2) Explain genetically why reciprocal crosses give different results.

53 What is the function of each of the two sperms injected from the pollen tube into the embryo sac? Apart from its role as a means of sperm transport, what other function may be ascribed to the pollen tube in sexual reproduction?

54 In what ways does syngamy in higher animals differ from that in higher plants?

55 Distinguish between cross-fertilization and self-fertilization and state what type or types can occur in:
a A unisexual (dioecious) organism.
b A bisexual (monoecious) organism.

56 In *Neurospora crassa* (*n*), an mt^+ strain with *orange* conidia is mated with an mt^- strain with *white* conidia. Do you expect all the progeny from this cross to be phenotypically identical if *orange* vs. *white* are due to alleles W and W^1 of a gene on chromosome 3? Explain. Show the types of progeny expected.

57 Discuss the statement: The human egg is much larger than the human sperm, yet a child inherits equally from both parents.

58 In a number of organisms it is known that some characters are controlled by hereditary determiners in the cytoplasm. Would children more often resemble their mothers or their fathers with respect to characters of this nature? Explain.

59 **a** Which form of reproduction, asexual or sexual, would you expect to give greater phenotypic differences between parents and offspring and among offspring, and why?
b Would you consider genetic variability an advantage or disadvantage to the survival of a species under natural conditions?
c What is one possible advantage of sexual reproduction to the maintenance of a species?

60 You find a very small plant species in which some individuals are *green* and others *bluish* in color. Only one of the two phases of the life cycle, the diplont or the haplont, is conspicuous (large enough to be seen and studied). After crossing the two, you find that progeny plants are *green* and *bluish* in equal numbers.
a Is the species diploid or haploid and why?
b Outline one type of study you might undertake to verify or refute your answer to (a). Indicate the results expected.

61 The life cycle of all sexually reproducing organisms follows the sequence shown. However, three variations of this basic pattern are known, depending on the time and place of meiosis and whether the haplont (*n*), diplont (*2n*), or both reproduce by mitosis.
a Describe and illustrate each of these variations and name two organisms or groups of organisms in which each occurs.
b Which type of life cycle is probably primitive? Suggest how the other types could have evolved from it.

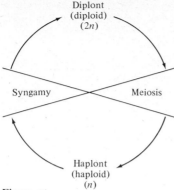

Figure 61

62 In the sea lettuce (*Ulva ulva*), a multicellular alga, both the gametophytic (haploid) and sporophytic (diploid) phases of the life cycle are equally large and conspicuous.   A plant with *crinkly* leaves is crossed with a plant with *smooth* leaves.   All the progeny have *smooth* leaves.   Is it possible to determine whether the parents were diploid or haploid?   Explain with the aid of diagrams.

63 With respect to the process or principle concerned, state which alternatives are incorrect and give reasons for your selection.
  a Syngamy:
    (1) Is essential to the maintenance of a constant chromosome number from generation to generation.
    (2) Provides for the doubled chromosome number in the diploid phase.
    (3) Occurs between any two cells of an organism.
    (4) Occurs in asexually reproducing organisms.
  b The multicellular phase of an organism:
    (1) May be diplontic or haplontic.
    (2) Consists of genetically and chromosomally identical cells.
    (3) May increase in size by mitosis or meiosis.
    (4) Begins life as a zygote.
  c Asexual reproduction:
    (1) Involves both mitotic and meiotic divisions.
    (2) Preserves the genotype without variation.
    (3) Occurs only when male and female gametes unite to form a new individual.

64 Organisms are frequently described as *diploid* or *haploid*.   What is meant when these terms are used?

65 The gene T for leg thickness in *Drosophila melanogaster* ($2n$) is on chromosome 2.   A true-breeding line in which all flies have the genotype TT and possess *fat* legs is crossed with a true-breeding line in which all flies are T^1T^1 and have *thin* legs.
  a Would you expect all the progeny from this cross to be phenotypically identical?   Explain.

b What phenotype(s) might the progeny express?   Explain.

66 What are some of the advantages of the following organisms for genetic research: *Neurospora, Paramecium, Zea, Drosophila*?

67 *Crepis capillaris* is a sexually reproducing bisexual plant species. The karyotype of the somatic cells of the diplont phase of the life cycle of all individuals in all generations is identical; all possess a *2n* chromosome

Figure 67

number of 6 with the morphology sketched.   After showing the karyotype of a cell of the haplont phase, briefly explain, with the aid of illustrations, why the chromosome number in *Crepis* remains constant (1) within the diplont and haplont phases and (2) from one diplont and haplont phase to the next.

68 Label each of the various stages and answer the following questions regarding the life cycle of corn illustrated in Fig. 68.

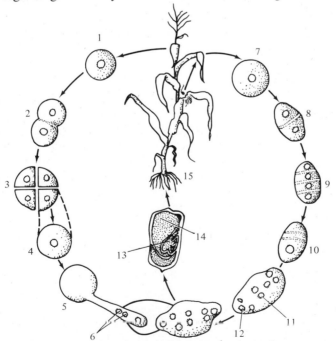

Figure 68   (After Adrian M. Srb, Ray D. Owen, and Robert S. Edgar, "General Genetics," 2d ed., W. H. Freeman and Company. Copyright © 1965.)

a What stages constitute the diplont and haplont generations?
b Which stages belong to the male and which to the female part of the reproductive cycle?
c Where do meiosis and syngamy occur?
d Which of the haplontic stages are missing from the life cycle of animals?
e What event occurs here but is absent in all animals? What is its purpose in plants?

69 State whether each of the following statements is true or false. Give a reason or example supporting your decision in each case.
a Syngamy is a process involving two cells; it occurs only in diploid organisms.
b In dioecious organisms the terms male and female are always applicable.
c Organisms that are hermaphroditic cannot cross-fertilize.
d Sex is always necessary for biological multiplication.
e Cytokinesis and karyokinesis always occur concurrently.
f In certain isogamous species either gametic type may develop parthenogenetically into an adult, while in oogamous species only the eggs are capable of developing in this manner.
g Individuals produced by parthenogenesis in diploid organisms are always diploid.

70 Figure 70 illustrates four kinds of gametogenesis.

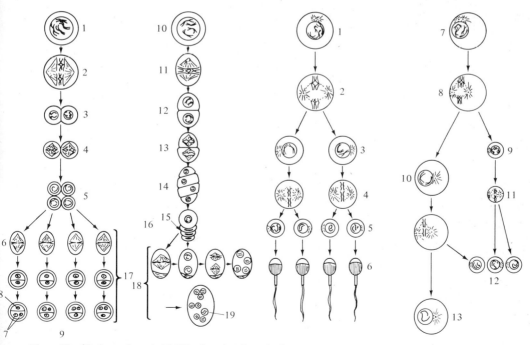

Figure 70   (Redrawn from A. M. Winchester, "Genetics," 2d ed., Houghton Mifflin Company, Boston, 1958.)

a For each of these four types state (1) the sex in which it occurs and (2) whether the process occurs in a plant or animal.   State why in each case.

b Label the numbered stages or cell types and give the general chromosome number (n or $2n$) of each.

71 If the chromosome carries genetic material, all the body cells derived by mitosis should possess an identical genotype.   Describe how you would proceed to test this hypothesis using a plant like the carrot or the geranium.

72 Eggs of many species of higher plants and animals can develop parthenogenetically into mature individuals.   Only in rare instances is the sperm, regardless of the species, believed capable of this.   Suggest a reason for this difference.

PROBLEMS

73 The diplontic sporophyte and the haplontic gametophyte of the brown algae are both large plants.   A botanist discovers a new species of this group and identifies two forms A and B, which he suspects represent the two phases of the life cycle.   He distinguishes two classes of form B, *elongate* vs. *ovata*, and only one of form A, *ovate*.   Experiments show the following:

1 Isolated plants of form A produce young form B plants of both classes.

2 Isolated plants of form B produce no young.

3 A pair of plants of form B, one *elongate*, the other *ovata*, give rise to young plants of form A.

a State which form represents the haplont generation and why.

b Explain the genetic data, showing the genotypes for form A and the two types of form B.

c Is the species self- or cross-fertilizing?

74 In *Neurospora crassa*, a haploid fungus, one strain has a normal metacentric chromosome 2, the other a mutant homolog 2 which is deficient for most of one arm.   How would you determine whether the first meiotic division in the zygote (meiocyte) of a cross between the two strains was reductional for the short arm?   Equational for the short arm?   Show the results expected in each case.

Strain A   Strain B

Figure 74

75 a In common wheat (*Triticum aestivum*) the haploid number of chromosomes is 21.   How many chromosomes would you expect to find in the following:

(1) The tube nucleus   (2) A microsporocyte
(3) One of the polar nuclei   (4) The nucleus of an aleurone cell
(5) A root-tip nucleus   (6) An egg nucleus
(7) A leaf cell

b In man (*Homo sapiens*) the diploid chromosome number is 46.   How many chromosomes would you find in the following?
(1) An egg
(2) A brain cell
(3) A red blood cell (erythrocyte)
(4) A white blood cell (leukocyte)
(5) A primary spermatocyte
(6) A spermatid
(7) A polar body
(8) An oogonium
(9) A cell in the cornea of the eye

76 **a** How many sperms will be obtained from:
(1) 100 primary spermatocytes?
(2) 100 spermatids?
b How many eggs will be obtained from:
(1) 100 primary oocytes?
(2) 100 ootids?
c Is there a difference in the number of meiotic products in the female and in the male?   If so, offer an explanation for it.

77 The members of a pair of homologous chromosomes are heteromorphic, one long and one short.   At anaphase I the chromosome with the two long chromatids always segregates from the chromosome with the two short chromatids.   Show diagrammatically where the centromere is located.

78 **a** The common red fox (*Vulpes vulpes* [= *fulva*]) has 38 chromosomes in its somatic cells; the Arctic fox (*Alopex lagopus*) has 50, which are smaller than those in the red fox (Gustavsson and Sundt, 1965).   Hybrids from crosses between the two species are sterile.   Studies of meiosis in these hybrids by Wipf and Shackleford (1949) have shown that both bivalents and univalents are present at diakinesis and metaphase I.
(1) How many chromosomes would you expect to find in the somatic cells of the hybrids?
(2) Account for the observed cytological behavior.
(3) Do you think there is any relationship between the cytological behavior and the sterility of the hybrids?   If so, explain.
b A cytotaxonomist collects bugs belonging to the genus *Protenor* ($2n = 4$) from two regions about $\frac{1}{4}$ mile apart.   The four chromosomes in all offspring of matings between males from the one region and females from the other behave as shown at metaphase I of meiosis.   Are the bugs in the two regions likely to be members of the same or different species?   Why?

Figure 78

QUESTIONS INVOLVING FURTHER READING

79 **a** Explain how *Paramecium* reproduces (1) asexually and (2) sexually.

b Compare or contrast conjugation with autogamy in *Paramecium aurelia*.

c Do you think autogamy facilitates genetic analysis?   Explain.

See Sonneborn (1947) for a detailed and authoritative review of the life cycle and genetics of *Paramecium*.

80 **a** Explain how the fungus *Aspergillus nidulans* reproduces (1) asexually and (2) sexually.

b Compare reproduction in *A. nidulans* and *Neurospora crassa*.

See Pontecorvo (1953) to obtain an understanding of the life cycle and methods of genetic analysis in this organism.

REFERENCES

Allen, C. E. (1935), *Bot. Rev.*, **1**:269.
—— (1937), *Am. Nat.*, **71**:193.
Beale, G. H. (1954), "The Genetics of *Paramecium aurelia*," Cambridge University Press, Cambridge.
Berrill, N. F. (1953), "Sex and the Nature of Things," Dodd, Mead, New York.
Demerec, M. (ed.) (1950), "The Biology of *Drosophila*," Wiley, New York.
—— and B. P. Kaufmann (1961), "Drosophila Guide," 7th rev. ed., Carnegie Institution, Washington, D.C.
Ephrussi, B. (1953), "Nucleo-cytoplasmic Relations in Micro-organisms," Oxford University Press, London.
Fincham, J. R. S., and P. R. Day (1963), "Fungal Genetics," Blackwell, Oxford.
Grant, V. (1951), *Sci. Am.*, **184**(June):6.
Gustavsson, I., and C. O. Sundt (1965), *Hereditas*, **54**:249.
Haskell, G. (1961), "Practical Heredity with *Drosophila*," Oliver & Boyd, Edinburgh.
Kiesselbach, T. A. (1949), *Univ. Nebr. Coll. Agric., Agric. Exp. Stn. Res. Bull.* 161.
Loeb, J. (1899), *Am. J. Physiol.*, **31**:135.
Maheshwari, P. (1950), "An Introduction to the Embryology of the Angiosperms," McGraw-Hill, New York.
Pincus, G. (1951), *Sci. Am.*, **184**(March):44.
Pontecorvo, G. (1953), *Adv. Genet.*, **5**:141.
Pool, R. J. (1941), "Flowers and Flowering Plants," 2d ed., McGraw-Hill, New York.
Randolph, L. F. (1936), *J. Agric. Res.*, **53**:881.
Runnstrom, J., B. E. Hagstrom, and P. Perlmann (1959), in J. Brachet and A. E. Mirsky (eds.), "The Cell," vol. I, pp. 327–397, Academic, New York.
Sharp, L. W. (1943), "Fundamentals of Cytology," McGraw-Hill, New York.
Smith, G. M. (1955), "Cryptogamic Botany," vol. II, 2d ed., pp. 1–3, McGraw-Hill, New York.
Sonneborn, J. M. (1947), *Adv. Genet.*, **1**:263.
Sprague, G. F. (1955), "Corn and Corn Improvement," Academic, New York.
Srb, A. M., R. D. Owen, and R. S. Edgar (1965), "General Genetics," 2d ed., Freeman, San Francisco.
Strickberger, M. W. (1962), "Experiments in Genetics with *Drosophila*," Wiley, New York.
Swanson, C. P. (1964), "The Cell," 2d ed., Prentice-Hall, Englewood Cliffs, N.J.
Tyler, A. (1954), *Sci. Am.*, **190**(June):70.
Wagner, R. P., and H. K. Mitchell (1964), "Genetics and Metabolism," 2d ed., Wiley, New York.
Wilson, E. B. (1925), "The Cell in Development and Heredity," 3d ed., Macmillan, New York.
Wipf, L., and R. M. Shackleford (1949), *Proc. Natl. Acad. Sci.*, **35**:468.

3
Monohybrid Inheritance

3
Monohybrid Inheritance

The First Law of Genetics:
Mendel's Law of Segregation

QUESTIONS

81 The members of a population of mature tobacco (*Nicotiana tabacum*) plants from the same farm vary greatly in size. How would you demonstrate whether or not this variability is genetically controlled?

82 a Does the phenotype of one generation ever affect the genotype of the next? Explain.

   b "Individuals of identical phenotype may have different genotypes, and vice versa." State whether this statement is true or false and why.

83 a Define and explain Mendel's law of segregation.

   b State, giving evidence, whether or not segregation can occur at either of the meiotic divisions.

   c Does segregation occur in asexual reproduction? In sexually reproducing homozygotes? Explain.

84 a What evidence do we have that genes are self-replicating (produce exact copies of themselves) and arise only from preexisting genes?

   b What evidence is there that one allele does not alter the nature of its partner allele when both are together (in the diploid condition in somatic or meiotic cells)?

85 In the early part of this century Bateson and Punnett postulated the *presence-absence theory*, stating that a dominant trait is due to the presence of the gene and the recessive one is due to its absence.

   a Do you think this is adequate to explain all instances of monogenic inheritance with dominance?

   b How would you proceed to determine whether it is correct or not?

86 a What genetic phenomenon that is important in diploids cannot usually be investigated in haploids?

   b What are some of the advantages of a genetic analysis of haploid organisms over that of diploid ones?

87 A cross is made between a *black* and a *white* guinea pig. The F_1's are all *black*, and the F_2's show a 3:1 ratio of *black:white* (Wright, 1917).

   a List and discuss the conditions required to give these results.

   b Why do reciprocal crosses between true-breeding lines with different traits of the same character (e.g., one line *black*, the other *white*) give the same results?

88 Explain why phenotypically identical, or at least very similar, parents may produce very different kinds of offspring.

89 a Between 1875 and 1900 the cytological basis of Mendelian inheritance was established, and in the period 1902 to 1904 on the basis of parallelisms between gene and chromosome transmission the *chromosome hypothesis of heredity* was formulated by Sutton and Boveri. What were these parallelisms?

    b The acceptance of the chromosome hypothesis required that two conditions be fulfilled: (1) chromosomal continuity from cell to cell throughout the life cycle and (2) qualitative differences between the chromosomes of the male and female gametes insofar as their genetic effect is concerned.

        (1) Explain how Boveri showed that these conditions are cytologically fulfilled.

        (2) Outline an experiment to show that chromosomes are continuous from cell to cell and that they are qualitatively different. See Wilson (1925) for a discussion of these classic experiments.

90 Explain why undesirable traits due to dominant alleles are more easily eliminated from a population than those due to recessive ones.

91 Discuss the *gene concept* as derived only from a knowledge of meiosis and monohybrid inheritance.

92 In Shorthorn cattle the allele R for *red* coat is neither dominant nor recessive to the one for *white*, R^1 (Jones, 1947). Outline the results expected in the F_1 and F_2 generations of a cross between homozygous *red* and homozygous *white* individuals.

93 In guinea pigs *black* vs. *albino* is due to alleles of one gene, with the allele for *black* dominant to that for *albinism*. Castle and Phillips (1909) replaced the ovaries of an *albino* female with those from a homozygous *black* female. The female was then mated with an *albino* male and produced two *black* offspring.

    a Show whether these results would be expected on the basis of Mendelian inheritance.

    b What bearing have they on the validity of the Lamarckian doctrine of inheritance of acquired characters?

94 Explain why, in human families, many traits, e.g., *albinism, blue eyes*, and *phenylketonuria*, skip generations while traits such as *polydactyly, free earlobes*, and *A* and *B* blood groups do not.

95 One form of *peroneal atrophy*, a rare human trait, occurs in families of unrelated parents; another, more severe form occurs almost exclusively in families where the parents are first cousins. Explain.

96   **a** What are the major difficulties encountered in the study of the genetics of man?

    b If you had several human pedigrees for a simply inherited trait, what criteria would you use to determine whether the trait was due to a dominant or a recessive allele?

    c What criteria would you apply to decide whether or not a certain human anomaly was controlled by a rare recessive allele? How would these criteria have to be modified to apply to a rare dominant?

97 Human geneticists are continually trying to develop methods by which individuals heterozygous for harmful or lethal recessive alleles can be detected.

a What is the practical use of such efforts?

b Discuss the possible disadvantages that might accrue in future generations if these methods are used for human betterment.

98 **a** Does a study of pedigrees always permit a person to determine whether an allele is dominant or recessive? Explain.

b Why is it much easier to work out human pedigrees for autosomal dominant traits than for autosomal recessive ones?

c Why are parents of individuals homozygous for rare recessive alleles likely to be related?

d Briefly discuss the conditions under which a recessive trait may appear to be inherited as a dominant one and vice versa and the precautions necessary in drawing conclusions from pedigree analysis.

99 A population geneticist collects 400 deer mice (*Peromyscus maniculatus*) from an area near Jasper. All these mice have *normal* ears. He raises about 800 animals in each of the following generations (all matings occur at random) and finds that all except one male in the fourth generation have *normal* ears. The exceptional male is *earless* but otherwise *wild type* and fertile.

a Suggest three possible types of causes that could lead to the appearance of this *earless* male.

b Briefly discuss how you would proceed to distinguish between these alternatives and show the results expected with each cause in this experiment.

PROBLEMS

100 The dorsal pigmentary pattern of the common leopard frog (*Rana pipiens*) is gene-controlled (Volpe, 1960). When *kandiyohi* frogs (mottling between the dorsal spots) are mated with *wild-type* frogs (white between the dorsal spots), all the progeny are *kandiyohi*. The F_1 females mated with F_1 males produced 55 *kandiyohi* frogs and 17 *wild type*.

a Which of these alternative patterns is due to a dominant allele?

b How many of the F_2 of dominant phenotype are expected to be heterozygous?

c Approximately how many of the F_2 of recessive phenotype are expected to be homozygous?

d How can you determine which of the F_2 individuals with the dominant phenotype are homozygous and which are heterozygous?

101 **a** Mendel (1866) crossed a true-breeding strain of peas (*Pisum sativum*) bearing *yellow* seeds with a true-breeding strain bearing *green* seeds. The F_1's produced *yellow* seeds only. A large number of F_1's were self-fertilized, giving progeny consisting of 6,022 plants with *yellow* seeds and 2,001 with *green* ones. Of the former 519 were self-fertilized; 166 bred true for *yellow*, while 353 produced *yellow* and *green* seeds in a 3:1 ratio. Give a genetic explanation of these results and show how it explains the ratios obtained.

b The F_1 with *yellow* seed (from a cross between a *yellow* and a *green* strain) were crossed reciprocally by Mendel with the *green* parent with the following results:

$F_1 ♀ × green ♂ → 58\ yellow : 52\ green$
$F_1 ♂ × green ♀ → 46\ yellow : 52\ green$

What conclusions regarding the contributions of male and female gametes to the offspring can be drawn from these results?

102 An observant gardener finds that some of his bean plants have *pubescent* leaves and others have *glabrous* leaves. He crosses different plants and obtains the results shown.

Table 102

Cross	Parents	Progeny	
		Pubescent	*Glabrous*
1	*pubescent* × *glabrous*	56	61
2	*pubescent* × *pubescent*	63	0
3	*glabrous* × *glabrous*	0	44
4	*pubescent* × *glabrous*	59	0
5	*pubescent* × *pubescent*	122	41

a Explain these results.
b Using your own gene symbols, give the genotypes of the parents of each cross.
c How many of the *pubescent* progeny in crosses 2, 4, and 5 would you expect to produce *glabrous* progeny when self-fertilized?

103 *Peroneal muscular atrophy*, the onset of which occurs between the ages of ten and twenty, consists of a slow progressive wasting of the distal muscles of the limbs. A study of a series of family pedigrees shows that a person never has *peroneal muscular atrophy* unless at least one of the parents has also had it (Macklin and Bowman, 1926). How is this trait most probably inherited? Explain.

104 *Spastic paraplegia* (a rare trait) vs. *normal* is controlled by different alleles of an autosomal gene (Rechtman and Alpers, 1934). An *affected* woman marries a *normal* man, and they have 5 children, 3 *normal* and 2 *affected*. Is the disease controlled by a dominant or by a recessive allele? Explain.

105 In rabbits certain *short-haired* individuals when crossed with *long-haired* ones produce only *short-haired* progeny. Other *short-haired* individuals when crossed with *long-haired* ones produce approximately equal numbers of *short-haired* and *long-haired* offspring. When *long-haired* individuals are inter-crossed, they always produce progeny like themselves.

a Outline a hypothesis to explain these results and show the genotypes of all individuals.

b How would you proceed to test this hypothesis? Show the results you would expect in the crosses you describe.

106 *Woolly* hair is a rare trait among Caucasians (Mohr, 1932). Approximately half the progeny of *woolly* × *normal* matings have this trait. What proportion of the offspring of marriages between *woolly* individuals would you expect to have this trait?

107 Wild (*red*) foxes occasionally have *silver-black* pups appearing as sports. These sports have long been prized by trappers, after it was found that two such animals reared and mated in captivity at once established a true-breeding *silver-black* strain. What is the most likely explanation of the rather frequent appearance of the sports, and why was it so easy to establish true-breeding *silver-blacks* that did not throw *red* progeny?

108 In Holstein-Friesian cattle *black-and-white* vs. *red-and-white* coat is controlled by a pair of alleles (Cole and Jones, 1920). In a *black-and-white* herd the tabulated results of breeding were recorded. Explain these results and show how the recessive allele could be eliminated from the herd.

Table 108

Cross	Results
Bull 1 mated with 80 cows	80 *black-and-white* F_1 calves (49 heifers, 31 bulls)
Bull 2 mated with the 49 F_1 heifers	40 *black-and-white* calves, 7 *red-and-white* calves

109 When 20 purebred *Himalayan* female rabbits are mated with a *gray* male of unknown ancestry, 46 of the offspring are *Himalayan* and 52 are *gray*. A single pair of alleles is involved.

a Is *Himalayan* controlled by a recessive or dominant allele?

b According to your answer in (a), how many offspring would you expect in each phenotypic class?

c Test your explanation using the chi-square method and indicate whether you would accept or reject your hypothesis.

110 In sheep, fat color and ear size are each controlled by one gene (Rae, 1956). The allele for *white fat* is dominant to that for *yellow fat*; the allele for *normal-ear* size is incompletely dominant to that for *earless*, the heterozygous individuals being *short-eared*. Both *yellow fat* and *earless* are undesirable traits.

a In some Icelandic flocks the proportion of *yellow-fat* lambs reaches 25 percent in some years. Breeders attempt to eliminate *yellow-fat* animals from their flocks since the meat cannot be sold and the trait cannot be detected until the animals are slaughtered.

(1) If animals that have produced *yellow-fat* offspring are not used for further mating, will this eliminate the recessive allele from a flock?

(2) How could a breeder eliminate the recessive allele from his flock most efficiently?

(3) A *white-fat* ram is mated with two *white-fat* ewes. The first ewe has 3 *yellow-fat* and 2 *white-fat* offspring. The second has 9 offspring, all *white-fat*. What are the genotypes of the three parents?

b (1) Which of these undesirable traits would be easier to eliminate from the flock? Explain.

(2) In a flock in which all phenotypes occur for both characters how would you establish a true-breeding line for (a) *white fat* and (b) *normal ears*?

(3) Would it be possible to establish a true-breeding line of *short-eared* sheep? If not, what cross would you make to obtain the maximum number of these individuals?

111 Eriksson (1955) found that 65 of 143 foals from 124 mares mated with the pedigreed Belgian stallion Godvan were afflicted with *aniridia* (complete lack of the iris) and developed cataracts about 2 months after birth. Godvan had *aniridia*, but his parents were *normal*. Account for the appearance of the affliction in both the foals and the stallion.

112 In pigeons (*Columba domestica*) feather pattern is controlled by one gene; *checkered* pattern is dependent on allele *C* and *barless* on *c* (Hollander, 1938). A series of crosses are made between homozygous *checkered* and homozygous *barless* individuals.

a Briefly describe, using diagrams, the mechanism of transmission of the alleles as far as the F_2 generation. In your outline give the cytogenetic explanation for the results expected. Include:

(1) Genotypes of parents, F_1's, and F_2's.

(2) Genotypic ratio in the F_1 gametes and in the F_2 progenies.

(3) A reason for each of these ratios.

b Discuss the possible relationships that can exist between the two alleles, *C* and *c*. For each relationship state the F_1 phenotype and the F_2 phenotypic ratio.

c What genetic law do these F_2 results illustrate? Do they provide conclusive proof of the law? If not, how would you provide this?

113 The *polled* (hornless) condition in cattle is dominant over the *horned* (Lloyd-Jones and Evvard, 1916). A cattleman tries to keep *polled* cattle only, but occasionally *horned* ones are born. They are removed from the herd before they can reproduce. Assuming his herd is well isolated from all others, how can you explain the occasional appearance of a *horned* animal in the herd? Outline a breeding program that would eliminate *horned* cattle from his herd.

114 You have a single individual of dominant phenotype and wish to determine whether it is homozygous or heterozygous. Explain how you would do this if the organism is:
a A higher animal.
b A cross-fertilizing plant.
c A self-fertilizing plant.

115 In Guernsey cattle *normal* vs. *incomplete hairlessness* is due to a single pair of alleles, *H* vs. *h* (Hutt and Saunders, 1953). The following dispute is referred to you for solution.

   Mr. Burch, who never has had any *incompletely hairless* calves in his herd of Guernseys, finds that of 46 calves from the matings of the heifers from his own bull to one bought from Mr. Fraser, 7 are *incompletely hairless*. He insists that Mr. Fraser's bull is responsible, and Mr. Fraser states that the bull is only partially responsible. Burch does not agree. Fraser refers Burch to you for verification of his contention.
a What would you tell Burch?
b What evidence would you cite to absolve Fraser's bull from full responsibility? What matings would you suggest and what results would you require to confirm your stand?
c If Fraser's explanation is correct, what numbers of *normal* and *incompletely hairless* calves should be expected among the 46?
d How many of the 39 *normal* calves would you expect to be heterozygous?

116 A cross between two sweet pea plants produced 41 plants with *pink* flowers, 18 with *white* flowers, and 19 with *red* flowers.
a What is the phenotype of each parent and why?
b What phenotypes do you expect and in what proportions among the progeny of the following crosses?
   (1) *white* × *pink*   (2) *red* × *red*   (3) *pink* × *pink*

117 The *palomino* horse has a golden yellow coat with flaxen mane and tail, the *cremello* is almost white, and the *chestnut* is brown. These horses are identical with respect to the basic color genotype: all are homozygous for a recessive allele (*b*) for *brown*. The table shows the results obtained upon mating these types in various combinations (Castle and King, 1951).

Table 117

Cross	Parents	Offspring
1	*cremello* × *cremello*	all *cremello*
2	*chestnut* × *chestnut*	all *chestnut*
3	*chestnut* × *cremello*	all *palomino*
4	*palomino* × *palomino*	1 *chestnut* : 1 *cremello* : 2 *palomino*
5	*palomino* × *chestnut*	1 *palomino* : 1 *chestnut*
6	*cremello* × *palomino*	1 *cremello* : 1 *palomino*

a Describe the genetic control of coat color as revealed by these results, including the allelic relationship.

b Diagram the last three matings.

c If you raised *palominos*, how would you set up your breeding program and why?

d If a *chestnut* individual is mated with a *cremello* one, what is the chance of getting a *chestnut* animal? A *cremello* animal? A *palomino* animal? Explain.

118 In sheep, crosses between homozygous *earless* and homozygous *normal* always result in an intermediate *small-eared* condition (Lush, 1930). When a series of matings are made between F_1 rams and ewes, 79 *small-eared*, 42 *earless*, and 37 *normal* progeny are obtained.

a Propose a hypothesis to explain these results and diagram the cross to illustrate the hypothesis.

b If one of the F_2 *small-eared* individuals is mated with a *normal* one, what is the chance of getting a *normal* lamb? A *small-eared* lamb? An *earless* lamb? Explain.

119 A hen with *blue* feathers is mated with a rooster of the same phenotype. They produce 14 offspring with *black* feathers, 24 with *blue* feathers, and 11 with *white* feathers. How does feather color appear to be inherited? Illustrate the cross with genotypes.

120 In a certain species one strain is able to synthesize the amino acid arginine and another is not. These strains reproduce themselves true to type. The progeny of a cross between them consists of both types, in the following proportions:

Arginine synthesizers   85
Arginine nonsynthesizers 89

a Are the two strains diploid or haploid? Explain.

b What genetic law do these results support?

121 In *Neurospora crassa* a true-breeding strain unable to synthesize the vitamin thiamine is obtained by irradiating spores with x-rays. This strain is crossed with the *wild-type* (thiamine-synthesizing) one, and the vegetative mycelia from each of 200 ascospore pairs, isolated in order from 50 asci, are tested for their ability to synthesize thiamine; 100 are synthesizers, and an equal number are not.

a Diagram this cross as far as the meiotic products of the hybrids, illustrating your diagram with the pair of homologs and alleles concerned.

b Why is the phenotypic ratio exactly 1:1?

c Which of the alleles, if any, is dominant?

122 In *Neurospora*, tetrad analysis of a cross between a leucine, *leu*$^+$, and a leucineless, *leu*$^-$, strain showed a one-to-one segregation for *leu*$^+$:*leu*$^-$ (Regnery, 1944). It was also found that 21 of the 31 tetrads analyzed showed second-division segregation.

a How could these tetrads be recognized?

b What event during meiosis would lead to their production?

123   a *Normal* vs. *albino* in man is monogenically controlled, the allele for *normal* being dominant to that for *albinism* (Pearson, Nettleship, and Usher, 1911–1913).

    (1) A *normal* boy whose parents are *normal* has one *albino* grandparent. If none of the other grandparents carry the allele for *albinism*, what is the chance that this boy is heterozygous for the *albino* allele?

    (2) The boy marries a girl with the same kind of pedigree. What is the probability that their first child will be an *albino*?

b An *albino* woman whose father is *albino* and mother *normal* marries a *normal* man one of whose parents is *normal* and the other *albino*. He has an *albino* sister. The couple have a *normal* daughter. What are the genotypes of:

    (1) The parents of the mother and the father?

    (2) The mother and father?

    (3) Their child?

124   The pedigree is for *myopia* (solid symbols), a fairly common, simply inherited human trait occurring with equal frequency in both sexes.

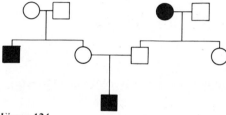

Figure 124

a Is the allele controlling expression of *myopia* dominant or recessive to the allele for *normal* sight? Explain.

b Give the genotypes of all individuals.

125   *Microphthalmia* (solid symbols) is a rare human trait which occurs as frequently in males as in females. It frequently exhibits the kind of inheritance pattern illustrated.

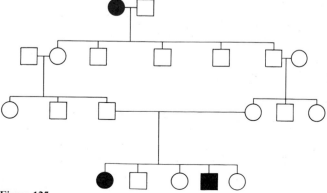

Figure 125

a Suggest the most likely mode of inheritance of this condition.

b Briefly outline the reasoning which leads to your conclusion.

126 a Figure 126*A* is part of a pedigree for *alkaptonuria* (from Khachadurian and Feisal, 1958), an inborn error of metabolism that is simply inherited.   State whether this pedigree shows that the allele for *alkaptonuria* can be (1) dominant only, (2) recessive only, or (3) either dominant or recessive.

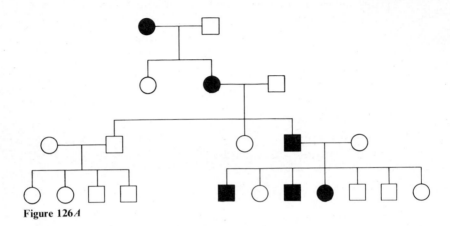

Figure 126*A*

b The trait is a very rare one.   State how this observation will change your answer to (a) and why.

c The complete pedigree of transmission in the family reported on by Khachadurian and Feisal is shown in Fig. 126*B*.   Does this pedigree confirm your answer to (b) and why?

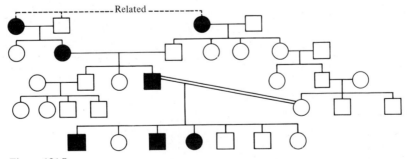

Figure 126*B*

127 In the human pedigrees illustrated the traits (represented by solid squares and circles) are controlled by alleles of single genes.   For each:
a State whether it is due to a dominant or recessive allele, giving your reasons.

b Determine the genotype, or alternative genotypes, of each individual.
These traits are either very rare or found in a small minority of the population.

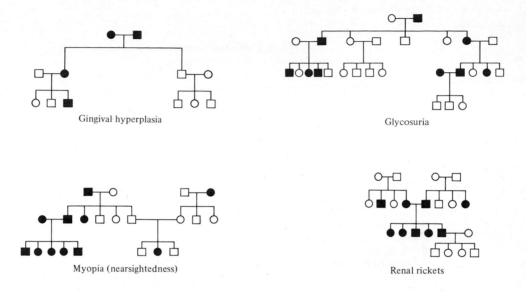

Gingival hyperplasia

Glycosuria

Myopia (nearsightedness)

Renal rickets

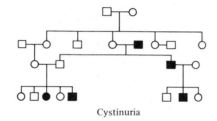

Cystinuria

Figure 127

REFERENCES

Bateson, W. (1909), "Mendel's Principles of Heredity," Cambridge University Press, London.
———, E. R. Saunders, and R. C. Punnett (1905), *Rep. Evol. Comm. R. Soc.*, **2**:1, 80.
Bell, J. (1943), in R. A. Fisher (ed.), "The Treasury of Human Inheritance," vol. IV, pp. 283–342, Cambridge University Press, London.
Bergsma, D. R., and K. S. Brown (1971), *J. Hered.*, **62**:171.
Bhatnagar, M. K. (1969), *Biochem. Genet.*, **3**:85.
Bouwkamp, J. C., and S. Honma (1970), *J. Hered.*, **61**:19.
Carter, C. O. (1969), "An ABC of Medical Genetics," Little, Brown, Boston.
Casady, A. J., and A. Q. Paulsen (1971), *J. Hered.*, **62**:193.
Castle, W. E., and F. L. King (1951), *J. Hered.*, **42**:61.
——— and J. C. Phillips (1909), *Science*, **30**:312.
——— and W. R. Singleton (1961), *Genetics*, **46**:1143.

Catcheside, D. G. (1951), "The Genetics of Micro-organisms," Pitman, New York.

Chiao, J. W., and S. Dray (1969), *Biochem. Genet.*, **3**:1.

Cleaver, J. E. (1969), *Proc. Natl. Acad. Sci.*, **63**:428.

Cole, L. J., and S. V. H. Jones (1920), *Wis. Agric. Exp. Stn. Bull.* 313.

Dean, G. (1969), *Br. Med. Bull.*, **25**(1):48.

De Lorenzo, R. J., and F. H. Ruddle (1969), *Biochem. Genet.*, **3**:151.

Eriksson, K. (1955), *Nord, Vet, Med.*, **7**:773.

Fuhrmann, W., and F. Vogel (1969), "Genetic Counseling," Springer-Verlag, New York.

Gates, R. R. (1946), "Human Genetics," 2 vols., Macmillan, New York.

Goodman, R. H. (ed.) (1970), "Genetic Disorders in Man," Little, Brown, Boston.

Grüneberg, H. (1952), *Bibliogr. Genet.*, **15**:1.

Hollander, W. F. (1938), *Genetics*, **23**:24.

Holzel, A., and G. M. Komrower (1955), *Arch. Dis. Child.*, **30**:155.

Hsia, D. Y. (1959), "Inborn Errors of Metabolism," The Year Book, Chicago.

Hutt, F. B. (1949), "Genetics of the Fowl," McGraw-Hill, New York.

———— and L. Z. Saunders (1953), *J. Hered.*, **44**:97.

Jones, I. C. (1947), *J. Genet.*, **48**:155.

Khachadurian, A., and K. A. Feisal (1958), *J. Chronic Dis.*, **7**:455.

Knudson, A. G. (1965), "Genetics and Disease," McGraw-Hill, New York.

———— (1969), *Annu. Rev. Genet.*, **3**:1.

Little, C. C. (1957), "The Inheritance of Coat Color in Dogs," Comstock, Ithaca, N.Y.

Lloyd-Jones, O., and J. M. Evvard (1916), *Iowa Agric. Exp. Stn. Res. Bull.* 30.

Lush, J. L. (1930), *J. Hered.*, **21**:107.

Macklin, M. T., and J. T. Bowman (1926), *J. Am. Med. Assoc.*, **86**:613.

McKusick, V. A. (1968), "Mendelian Inheritance in Man," 2d ed., Johns Hopkins, Baltimore.

———— (1969), "Human Genetics," 2d ed., Prentice-Hall, Englewood Cliffs, N.J.

Mendel, G. (1866), *Verh. Naturforsch. Ver. Bruenn*, **4**:3; trans. and reprinted in J. A. Peters (ed.), "Classic Papers in Genetics," pp. 1–20, Prentice-Hall, Englewood Cliffs, N.J., 1959.

Mohr, O. L. (1932), *J. Hered.*, **23**:345.

Najjar, S. S. (1964), *J. Pediatr.*, **64**:372.

Neel, J. V., and W. J. Schull (1954), "Human Heredity," pp. 83–86, 89–91, 240–241, University of Chicago Press, Chicago.

Pearson, K., E. Nettleship, and C. H. Usher (1911–1913), "A Monograph on Albinism in Man," Draper's Company Research Memoirs, Biometric Series 6, 8, 9, Dulau, London.

Rabbini, M. G., and R. H. Baker (1970), *J. Hered.*, **61**:135.

Race, R. R., and R. Sanger (1968), "Blood Groups in Man," 5th ed., Blackwell, Oxford.

Rae, A. L. (1956), *Adv. Genet.*, **8**:189.

Rao, D. C. (1970), *Hum. Hered.*, **20**:8.

Rechtman, A. M., and B. J. Alpers (1934), *Arch. Neurol. Psychiat.*, **32**:248.

Regnery, D. C. (1944), *J. Biol. Chem.*, **154**:151.

Rick, C. M., and L. Butler (1956), *Adv. Genet.*, **8**:267.

Robinson, R. (1959), *Bibliog. Genet.*, **18**:273.

Sager, R. (1955), *Genetics*, **40**:476.

———— and F. J. Ryan (1961), "Cell Heredity," Wiley, New York.

Schrode, R. R., and J. L. Lush (1947), *Adv. Genet.*, **1**:209.

Stanbury, J. B., J. B. Wyngaarden, and D. S. Fredrickson (eds.) (1966), "The Metabolic Basis of Inherited Diseases," 2d ed., McGraw-Hill, New York.

Stern, C. (1960), "Principles of Human Genetics," 2d ed., Freeman, San Francisco.

———— (ed.) (1950), "The Birth of Genetics," *Genetics Suppl.*, **35**:(5), pt. 2 (includes Mendel's letters to Carl Nageli, 1866–1873, and the papers by von Tschermak, De Vries, and Correns reporting the rediscovery of Mendel's laws in 1900).

Sutton, W. S. (1903), *Biol. Bull.*, **4**:231.

Tanaka, Y. (1953), *Adv. Genet.*, **5**:240.

Townes, P. L. (ed.) (1969), "The Medical Clinics of North America," vol. 53, no. 4, Saunders, Toronto.
Volpe, E. P. (1960), *J. Hered.*, **51**:151.
Wilson, E. B. (1925), "The Cell in Development and Heredity," 3d ed., Macmillan, New York.
Woolf, C. M. (1971), *Am. J. Hum. Genet.*, **23**:289.
Wright, S. (1917), *J. Hered.*, **8**:476.

4
Dihybrid
and
Multihybrid Inheritance

Second Law of Genetics:
The Law of Independent Assortment

QUESTIONS

128 What proportion of the progeny of the following crosses, in which allele pairs are segregating independently, will be homozygous?

  a *Aa Bb* × *Aa Bb*   **b** *Aa Bb CC* × *AA BB cc*

  c *aa bb cc* × *AA bb CC*

129 Show the types and expected frequencies of gametes produced by the following genotypes if all allelic pairs segregate independently:

  a *AA Bb Cc*   **b** *AA Bb Cc DD*   **c** *Aa BB CC*

  d *Aa Bb Cc DD*   **e** *Aa Bb Cc Dd*   **f** *Aa bb Cc dd*

130 A corn plant homozygous for dominant alleles at four loci on four different (nonhomologous) chromosomes (*AA BB CC DD*) is crossed with one homozygous for the corresponding recessive alleles (*aa bb cc dd*) to produce a tetrahybrid (*Aa Bb Cc Dd*) which is self-fertilized. The allele pairs control the expression of different pairs of traits.

  a Give:

    (1) The number of gametic genotypes produced by the tetrahybrid and their proportions.

    (2) The number of F_2 genotypes and their proportions.

    (3) The number of F_2 phenotypes and their proportions.

    (4) The proportions of the F_2 that would resemble each of the parents of the F_1 respectively.

  b What proportion of the tetrahybrid gametes will carry the dominant alleles at all four loci? What proportion will carry dominant and recessive alleles?

  c What is the probability that an F_2 individual will get all the dominant alleles? All the recessive alleles? All eight alleles?

  d What proportion of the F_2 genotypes are expected to show the recessive phenotype for all four loci? To be homozygous for all dominant alleles?

  e Would your answers be different if the cross was *AA bb cc DD* × *aa BB CC dd*?

131 Does independent segregation occur in homozygous individuals? In asexual reproduction? Give reasons for your answers.

132 What evolutionary advantages do segregation, independent assortment, and random union of gametes give a sexually reproducing species over an asexually reproducing one?

133 In 1865 Mendel postulated that the segregations of certain different allele pairs are independent of each other. The Sutton-Boveri hypothesis (1902–1904) states that different chromosome pairs align at metaphase I and segregate at anaphase I independently of each other. Proof of this hypothesis was not provided until 1913 by Carothers using males of *Brachystola* (Orthoptera).

  a Why was this proof difficult to obtain?

b See Carrothers (1913 or 1917) for confirmation or rejection of your answer to (a) and his proof of independent segregation of different chromosome pairs.

134 In the following cases the allele pairs segregate independently.

1   In the mouse *agouti* vs. *nonagouti*, *black* vs. *brown* hair color and *solid* vs. *white-spotted* are controlled by the allele pairs *Aa*, *Bb*, and *Ss*, respectively (Grüneberg, 1952). A series of crosses are made between *AA bb SS* and *aa BB ss* individuals. Complete dominance occurs in each allele pair.

2   In snapdragons (*Antirrhinum majus*) flower color and leaf shape in the following cross are each controlled by a single gene (Crane and Lawrence, 1952). A *red-flowered, broad-leaved* plant, *RR BB*, is crossed with a *white-flowered, narrow-leaved* one, *rr bb*. Incomplete dominance occurs in both pairs of alleles.

a Diagram each cross as far as the F_2 giving:
  (1) Gametes of parents and F_1 and the F_1 gamete ratio.
  (2) F_1 and F_2 genotypes and F_2 genotypic ratio.
  (3) F_2 phenotypes and their proportions.
b What would be the phenotypic proportion among the progeny derived from crossing F_1 individuals with the second parent in each case?

PROBLEMS

135 In silkworms (*Bombyx mori*) the hemolymph ("blood") may be *deep yellow* or *white* (light yellow or colorless). Fully developed larvae may be *plain* (white) or *moricaud* (heavily lined and dotted but not solid black) (Tanaka, 1953). A cross is made between the two true-breeding strains *deep-yellow plain* and *white, moricaud*. When 10 F_1 females are crossed with 10 F_1 males, the following progeny result:

Deep-yellow, moricaud	293
Deep-yellow, plain	96
White, moricaud	104
White, plain	38

a State how many allele pairs are involved and which alleles are dominant; give reasons for your answer.
b Would all F_1's have the same phenotype? Why?
c Using your own gene symbols, give the genotype of:
  (1) Each parent.
  (2) The F_1's.
  (3) The four F_2 phenotypes.
d Explain why the F_2 ratio is approximately 9:3:3:1. How many silkworms would you expect in each F_2 phenotypic class? Using the chi-square method, show that the actual results do not deviate significantly from the 9:3:3:1 ratio.

e What phenotypic ratio do you expect from crosses of F_1 males with females from the *white, moricaud* strain?   Of F_1 males with females of the true-breeding *deep-yellow, plain* strain?   Would you expect results different from these if F_1 females were crossed with males from the two strains? Why?

f Suppose you made a cross between a true-breeding *deep-yellow, moricaud* and a true-breeding *white, plain* strain.   Would you expect the F_2 results to closely approximate those given?   Explain.

136 In sesame both the number of seed pods per leaf axil and the shape of the leaf are monogenetically controlled.   The *one-pod* condition is dominant to *three-pod*, and *normal* leaf is dominant to *wrinkled* (Langham, 1945).   The two characters are inherited independently.   The results of five crosses, each between a single pair of plants, gave the results shown.   Determine the genotypes of the parents of each cross.

Table 136

		Number of progeny			
Cross	Parents	One-pod, normal	One-pod, wrinkled	Three-pod, normal	Three-pod, wrinkled
1	one-pod, normal × three-pod, normal	318	98	323	104
2	one-pod, normal × one-pod, wrinkled	110	113	33	38
3	one-pod, normal × three-pod, normal	362	118	0	0
4	one-pod, normal × three-pod, wrinkled	211	0	205	0
5	one-pod, wrinkled × three-pod, normal	78	90	84	88

137 In corn the endosperm ($3n$) may be *sugary* or *starchy* and *floury* or *flinty*. Reciprocal crosses are made between true-breeding, *sugary, flinty* and *starchy, floury* strains.   The phenotypes of the F_1's are shown.   Why do reciprocal crosses give the same results for the first but not the second pair of traits? Illustrate your answer diagrammatically.

Table 137

Parents		
♀	♂	F_1
sugary, flinty × starchy, floury		starchy, flinty
starchy, floury × sugary, flinty		starchy, floury

138 In corn each of the characters kernel color, height, and reaction to rust is controlled by a single gene.   The allele *P* for *purple* is dominant to *p* for *white*; the allele *T* for *tall* is dominant to *t* for *dwarf*; and the allele *R* for *resistance* is dominant to *r* for *susceptibility*.   The allele pairs are on different (nonhomologous) chromosome pairs.   A *PP TT RR* strain is crossed with a *pp tt rr* one, and the F_1 is testcrossed.

a With the aid of the branching method where needed, give:
   (1) Genotypes of F_1's and testcross progeny and testcross genotypic ratio.
   (2) The F_1 gametic ratio.
   (3) The phenotypic testcross ratio.
b Explain *cytogenetically* (viz. by illustrating the number and kinds of metaphase I orientations and anaphase I segregations expected in a large population of meiocytes and their consequent products) why the genotypic ratio in the F_1 gametes is as given in (2).
c How many genotypes and phenotypes do you expect in the F_2 and in what proportions?

139 A poultry geneticist made reciprocal crosses between two true-breeding strains of fowl, one with *rose combs* and *blue-splashed white* plumage, the other with *single combs* and *black* feathers.   After the cross was made, the investigator left for a year, instructing his technician to record F_1 phenotypes and to cross some of the F_1 birds with the *rose-comb, blue-splashed white* strain and other F_1's with the *single-comb, black* strain.   The technician omitted the recording of the F_1 phenotypes but did make the crosses.   The records of the offspring of the crosses are shown.   Analyze these results and state your conclusions concerning:

a The number of allele pairs segregating and the traits controlled by the alleles of each gene.
b The allelic relationship for the alleles of each gene.
c Whether inheritance is independent or interdependent.

Table 139

$F_1 \times$ *single-comb, black* strain		$F_1 \times$ *rose-comb, blue-splashed white* strain	
rose, blue	64	*rose, blue-splashed white*	108
rose, black	57	*rose, blue*	113
single, blue	61		
single, black	59		

140 The pedigree shows the pattern of transmission of two rare human traits, *cataract* and *pituitary dwarfism*.   Individuals with *cataract* are indicated by a solid upper half of the symbol; those with *pituitary dwarfism* are indicated by a solid lower half.

a What is the mode of inheritance of each of these traits?   Explain.
b IV-1 marries IV-6, and they have 5 children, 3 *dwarfs* with *no cataract* and 2 *dwarfs* with *cataract*.   Does this verify the hypothesis formulated in (a)?   Give the genotypes of the parents of this marriage.

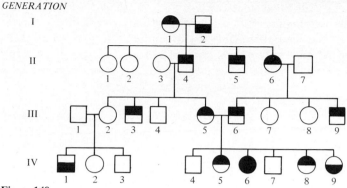

Figure 140

c What phenotypes and phenotypic ratios would you expect among progeny from the following marriages?
(1) III-5 × IV-1
(2) III-2 × IV-5
d What is the probability of:
(1) II-6 being heterozygous for the alleles for both traits?
(2) II-2 being homozygous for the allele for *dwarfism*?
(3) The first child of a mating of IV-2 and IV-4 being a *dwarf*?

141 Varieties of flax (*Linum usitatissimum*) differ in their reactions to specific races of the fungus *Melampsora lini*, which causes flax rust.   The phenotypes of two such varieties, their F_1 hybrids, and the F_2's obtained from self-fertilizing F_1's, are shown with respect to *resistance, R*, vs. *susceptibility, S*, to each of two fungal races.

Table 141

Parents and generations	No. of Plants	Reaction to rust race	
		22	24
Bombay	All	R	S
770B	All	S	R
F_1	All	R	R
F_2	128	R	R
	44	S	R
	14	S	S
	39	R	S

a Propose a hypothesis to account for *resistance* vs. *susceptibility* to the two races of rust.
b On the basis of your hypothesis how many plants would you expect in each of the four classes in the F_2?

142 In his stocks of Mexican swordtail fish (*Xiphophorus helleri*) a pet-shop owner has two *montezumas* (bright orange-red), one with a *crescent* spot at the base of the caudal fin and the other with *twin spots* at this position. When crossed, they gave 12 *montezumas* and 9 *wild-type* (olive-green) progeny all with *crescent* marking.   The *montezuma* F_1's when intercrossed produced

Montezuma, crescent	66
Montezuma, twin spot	24
Wild-type, crescent	31
Wild-type, twin spot	12

The *wild-type* F_1's when intercrossed produced 122 progeny, all of which were *wild type*, 88 of them *crescent* and 34 *twin spot*.
a Explain the manner of genetic control of the two pairs of traits.
b *Montezumas* with *twin spots* are in great demand.   Outline the breeding procedure a breeder should use to reap the maximum profit from these types.

143 In rabbits the allele for *Himalayan*, c^h, is dominant to that for *albinism* c (Sawin, 1932); that for *normal* hair, L, is dominant to that for *long* hair, l (Castle, 1940).   The two allele pairs are independently inherited.   A large number of *Himalayan, normal* F_1 males and females, from crosses between true-breeding *Himalayan, normal*, $c^hc^h LL$, and *albino, long*, cc ll, strains, are mated and produce 272 progeny.
a State the expected phenotypes and their proportions in this F_2 population.
b Would the proportions of these phenotypes be changed if the F_1 animals were derived from crosses between rabbits from the true-breeding *Himalayan, long* and *albino, normal* strains?   Explain.
c When F_1's of either cross are mated with rabbits from the *Himalayan, normal* strain, all animals are phenotypically the same.   Why is this so, and what is the phenotype of the progeny?

144 A testcross of an F_1 results in a 1:1:1:1 phenotypic ratio.
a How many genes are involved?
b Is it possible to determine the allelic relationship from these results?   If not, how would you obtain this information?
c What other information is given by these results?

145 Mendel (1866) crossed a pure-breeding strain with *round, yellow* seeds with a pure-breeding strain with *wrinkled, green* seeds.   The F_1 seeds were all *round, yellow*.   When he backcrossed the F_1 plants to the *wrinkled, green* parent, the progeny consisted of 55 *round, yellow*, 51 *round, green*, 49 *wrinkled, yellow*, and 53 *wrinkled, green* seeds.   The cross F_1 × *round, yellow* parent gave rise to *round, yellow* progeny only.   Explain his results and show the results you would expect in the F_2.

146 In the domestic fowl the allele pairs Rr and Bb, which are present on different chromosome pairs, carry the potential for the expression of *rose*

vs. *single* comb shape and *black* vs. *white* feather color, respectively.   What is the probability of obtaining:

a A *rose, black* bird from the cross *Rr bb* × *rr Bb*?

b A *R B* gamete from an *Rr BB* bird?

c A *rr bb* zygote from the cross *RR Bb* × *Rr BB*?

d A *rose, white* bird from the cross *Rr BB* × *Rr Bb*?

e A *r B* gamete from *Rr Bb*?

f A *Rr BB* zygote from the cross *Rr Bb* × *Rr Bb*?

147   *Paramecium aurelia* produces a variety of different antigens (Beale, 1952). Stock 90 at 25 and 29°C produces antigens G and D, respectively.   Stock 60 at corresponding temperatures produces the serologically related antigens 60G and 60D.   A cross between a stock-90 animal and a stock-60 one gives hybrid exconjugants that produce all four antigens.   Beale obtained the tabulated results by forcing the exconjugants to go through autogamy.

Table 147

Phenotype	Number
90D, 60G	18
90D, 90G	20
60D, 90G	19
60D, 60G	17

a Why does this F_2 progeny segregate in this manner?

b Diagram the cross, explaining why a 1:1:1:1 ratio is obtained.

148   Radishes (*Raphanus sativus*) may be *long, round,* or *oval* in shape.   The color may be *red, white,* or *purple* (Uphof, 1924).   A *long, white* variety crossed with a *round, red* one produced *oval* and *purple* F_1's.   The F_2 progeny segregated into nine phenotypic classes with characteristic proportions as follows: 9 *long, red* : 15 *long, purple* : 19 *oval, red* : 32 *oval, purple* : 8 *long, white* : 16 *round, purple* : 8 *round, white* : 16 *oval, white* : 9 *round, red*.

a How many allele pairs are involved?   Do they assort independently? What phenotypes would you expect in crosses between the F_1's and each of the parental strains?

b Give the genotypic and phenotypic ratios expected among the progeny of a cross between:

  (1) A *long, purple* and an *oval, purple* plant.

  (2) An *oval, purple* and a *round, white* plant.

c How many true-breeding varieties of radishes could be established?

d If *oval, purple* radishes were commercially preferred, state what lines should be maintained to produce them most profitably and why.

149   In *Drosophila melanogaster, gray* body is dominant to *black,* and *long* wings is dominant to *short* wings.   These pairs of traits are caused by two independently inherited pairs of autosomal alleles, *Bb* and *Ss*, respectively. Flies of known phenotypes but unknown genotypes are crossed and produce the results shown. What are the genotypes of the parents in each of the crosses and why?

Table 149

Cross	Parents	Gray, long	Gray, short	Black, long	Black, short
1	*gray, long* × *gray, short*	43	46	14	17
2	*gray, short* × *gray, short*	0	61	0	23
3	*black, long* × *gray, long*	40	0	43	0
4	*gray, long* × *gray, long*	48	18	0	0
5	*gray, short* × *gray, long*	31	33	0	0
6	*gray, long* × *gray, long*	93	34	30	12

(The column group header "Phenotypes of progeny" spans the four progeny columns.)

150 In man, *gingival hyperplasia* (overgrowth of the gums) is dominant to the *normal* condition (Garn and Hatch, 1950). *Cataract* is also dominant to the *normal* condition (Danforth, 1912; Halbertsma, 1934). A woman with *cataract* whose father suffered from it marries a man with *gingival hyperplasia* whose mother also had this latter condition.

    a What kinds of children could this couple have? State the probability for each kind.

    b A son with both *gingival hyperplasia* and *cataract* marries a *normal* woman, and their first child is *normal*. What is the genotype of the son? What is the chance of a child having *cataract* and *normal* gums?

151 Some dogs *bark* when trailing; others are *silent*. The *barker* trait is controlled by an allele dominant to that for *silent* (Whitney, 1929). *Normal* vs. *screw* tail is also monogenically controlled, the allele for *normal* being dominant to that for *screw* (Stockard, 1932).

    a How would you go about proving that the two characters are controlled by independently assorting genes?

    b You have a homozygous *barker, screw* male and a homozygous *silent, normal* female, and you wish to produce the following two true-breeding strains: (1) *barker, normal* and (2) *silent, screw*. Describe the breeding program required to obtain them. Which of the two would be the easier to establish and why?

    c Would it matter in any of your matings whether a particular parent chosen for a given genotype were the male or the female? Explain.

152 A man whose hobby is the rearing of Mexican swordtail fish (*Xiphophorus helleri*) crosses a strain true-breeding for *stippled* and *patternless* (lacking spots at the base of the caudal fin) with a strain true-breeding for *nonstippled* and caudal *twin spot*.   He finds that all the offspring are *stippled* and possess a caudal spotting pattern unlike that of either parent, called *crescent spot*. When the offspring are permitted to mate at random, he finds that the progeny fall into six categories in the following actual numbers:

Crescent, stippled	83
Crescent, nonstippled	27
Twin spot, stippled	29
Twin spot, nonstippled	9
Patternless, stippled	35
Patternless, nonstippled	12

He comes to you for a genetic explanation of these results, being particularly interested in raising true-breeding lines for the two *crescent* phenotypes. Using your own symbols, carry out the genetic analysis required, showing genotypes of parents and of F_1 and F_2 offspring and suggest how he might develop the true-breeding lines in question.

153   Determine the probable mode of inheritance of tongue color in dogs (*Canis familiaris*) from the pedigree chart.   State how it could be verified.

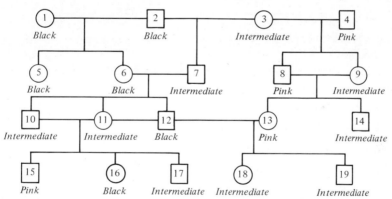

Figure 153

154   Two breeds of dogs, one with *straight*, *black* fur and the other with *curly*, *brown* fur, were crossbred.   The hybrids were intercrossed and produced the following progeny:

Brown, straight	35
Brown, curly	29
Brown, wavy	62
Black, wavy	179
Black, straight	95
Black, curly	89

a Propose a hypothesis to explain these results. What was the phenotype of the F_1 animals?

b Test your hypothesis using the chi-square method and indicate whether you accept or reject it.

c What fundamental principles or laws are illustrated by these data?

155 In mink (*Mustela vison*), as in most other mammals, many genes affect the color of the coat. These are of particular interest to breeders since new and different furs command premium prices. At least 13 genes are known to control coat color in mink. The results of crossing *wild type* (dark brown to nearly black) with *platinum* (blue-grey) and *aleutian* (blue-grey to steel-grey color) are shown (Shackelford, 1950).

Table 155

Cross	Parents	F_1	F_2
1	*wild type* × *platinum*	*wild type*	3 *wild type* : 1 *platinum*
2	*wild type* × *aleutian*	*wild type*	3 *wild type* : 1 *aleutian*
3	*platinum* × *aleutian*	*wild type*	133 *wild type*
			41 *platinum*
			46 *aleutian*
			17 *sapphire* (very light blue)

a Why is a dihybrid ratio obtained in the third cross and not in either of the first two?

b Explain how these results show that there is independent assortment but not independent gene action.

c What results would you expect if the *wild types* from the cross between *platinum* and *aleutian* were crossed with the *sapphires*? Why?

156 In *Neurospora crassa* the *thiamine-4* (thiamine requirement) gene is close to the centromere on chromosome 4 and the *adenine-1* (adenine requirement) gene is close to the centromere on chromosome 6. If only first-division segregation occurs at both loci, what kinds and proportions of tetrads are expected in the cross $thi^+ ad^+ \times thi^- ad^-$? Would the same results be obtained in the cross $thi^+ ad^- \times thi^- ad^+$?

157 **a** Illustrate by means of a diagram the origin of PD (parental ditype, viz. containing only parental types), NPD (nonparental ditype, viz. containing only recombinant types), and T (tetratype, viz. containing both parental and recombinant types) tetrads.

b In *Chlamydomonas reinhardi*, arg^+ (non-arginine-requiring) vs. arg^- (arginine-requiring) and thi^+ (non-thiamine-requiring) vs. thi^- (thiamine-requiring) are each controlled by a single pair of alleles (Levine and Ebersold, 1960). A cross between arg^+ thi^+ and arg^- thi^- gives the results shown from classifying 100 ordered (linear) tetrads. Are the two pairs of alleles segregating independently? Explain.

Table 157

1 PD	2 NPD	3 T	4 T	5 T	6 T
$arg^+ thi^+$	$arg^+ thi^-$	$arg^+ thi^+$	$arg^+ thi^-$	$arg^- thi^-$	$arg^- thi^+$
$arg^+ thi^+$	$arg^+ thi^-$	$arg^+ thi^-$	$arg^+ thi^+$	$arg^+ thi^-$	$arg^+ thi^+$
$arg^- thi^-$	$arg^- thi^+$	$arg^- thi^+$	$arg^- thi^-$	$arg^- thi^+$	$arg^- thi^-$
$arg^- thi^-$	$arg^- thi^+$	$arg^- thi^-$	$arg^- thi^+$	$arg^+ thi^+$	$arg^+ thi^-$

No. of tetrads	38	41	7	6	4	4

158 Tetrads of *Neurospora* are ordered within the ascus. Lindegren (1933) obtained the results shown from tetrad analysis of a cross between the strains *nonfluffy*, f^+, "*plus*" *mating-type*, mt^+, and *fluffy*, f^-, "*minus*" *mating-type*, mt^-.

Table 158

Spore order	Segregation class						
1	$f^+ mt^+$	$f^+ mt^-$	$f^+ mt^+$	$f^+ mt^+$	$f^+ mt^+$	$f^+ mt^-$	$f^+ mt^+$
2	$f^+ mt^+$	$f^+ mt^-$	$f^+ mt^-$	$f^- mt^+$	$f^- mt^-$	$f^- mt^+$	$f^- mt^-$
3	$f^- mt^-$	$f^- mt^+$	$f^- mt^+$	$f^+ mt^-$	$f^+ mt^+$	$f^+ mt^-$	$f^+ mt^-$
4	$f^- mt^-$	$f^- mt^+$	$f^- mt^-$	$f^- mt^-$	$f^- mt^-$	$f^- mt^+$	$f^- mt^+$

No. of tetrads	16	20	6	60	3	1	3

a Explain how the data reveal that the allele pairs segregate independently.
b Make a diagram to illustrate how each class originated and state whether segregation occurs at first or second division for each locus.
c Suppose that 200 ascospores from the cross, isolated at random, were analyzed. Show the results you would expect.

159 Working with *Chlamydomonas reinhardi*, Sager (1955) found that the allele pair *Ygr ygr* (*green* vs. *yellow-green*) segregated independently of the allele pair *Sr sr* (*streptomycin susceptibility* vs. *streptomycin resistance*). What types of tetrads (PD, NPD, or T) are possible from this cross? State the conditions under which each would occur. Show the various genotypes that are possible and the proportions in which they occur. (See Prob. 157 for definition of PD, NPD, and T.)

160 In yeast, *large* vs. *petite* colonies is controlled by the allele pair *Ll*; mating type (+ or −) by the allele pair $mt^+ mt^-$. A *large*, mt^+ strain crossed with a *petite*, mt^- strain produced the tetrads shown. Are the gene loci on the same chromosome or not? Explain.

Table 160

Kind of tetrads				Number
$L\ mt^-$	$L\ mt^-$	$l\ mt^+$	$l\ mt^+$	40
$L\ mt^+$	$L\ mt^+$	$l\ mt^-$	$l\ mt^-$	36
$L\ mt^+$	$l\ mt^+$	$L\ mt^-$	$l\ mt^-$	18

REFERENCES

Barratt, R. W., D. Newmeyer, D. D. Perkins, and L. Garnjobst (1954), *Adv. Genet.*, **6**:1.

Bateson, W. (1909), "Mendel's Principles of Heredity," Cambridge University Press, Cambridge.

———— and R. C. Punnett (1905–1908), *Rep. Evol. Comm. R. Soc.*, **2–4**; reprinted in Peters (1959), pp. 42–60.

Beale, G. H. (1952), *Genetics*, **37**:62.

Bonner, D. (1946), *Am. J. Bot.*, **33**:788.

Carrothers, E. E. (1913), *J. Morphol.*, **24**:487.

———— (1917), *J. Morphol.*, **28**:455.

Castle, W. E. (1940), "Mammalian Genetics," Harvard University Press, Cambridge, Mass.

———— (1954), *Genetics*, **39**:35.

———— and W. R. Singleton (1961), *Genetics*, **46**:1143.

———— and S. Wright (1916), *Carnegie Inst. Wash. D.C.*, Publ. 241.

Catcheside, D. G. (1951), "The Genetics of Micro-organisms," Sir Isaac Pitman & Sons, London.

Crane, M. B., and W. J. C. Lawrence (1952), "The Genetics of Garden Plants," Macmillan, London.

Danforth, C. H. (1912), *Am. J. Ophthalmol.*, **31**:161.

Dunn, L. C. (1965), "A Short History of Genetics," McGraw-Hill, New York.

Garn, S. M., and C. E. Hatch (1950), *J. Hered.*, **41**:41.

Grüneberg, H. (1952), *Genetica*, **15**:1.

Halbertsma, K. T. A. (1934), *Ned. Tijdschr. Geneeskd.*, **78**:4079.

Hall, K. W., K. R. Hawkins, and G. P. Child (1950), *J. Hered.*, **41**:23.

Hefner, R. A. (1941), *J. Hered.*, **32**:37.

Hogben, L., R. L. Worrall, and I. Zieve (1932), *Proc. R. Soc. Edinb.*, **52**:264.

Hutt, F. B. (1949), "Genetics of the Fowl," McGraw-Hill, New York.

———— and W. F. Lamoureux (1940), *J. Hered.*, **31**:231.

Iljin, N. A. (1932), "Genetics and Breeding of the Dog," Moscow and Leningrad (in Russian).

Langham, D. G. (1945), *J. Hered.*, **36**:245.

Levine, R. P., and W. T. Ebersold (1960), *Annu. Rev. Microbiol.*, **14**:197.

Lindegren, C. C. (1933), *Bull. Torrey Bot. Club*, **60**:133.

Little, C. C. (1957), "The Inheritance of Coat Color in Dogs," Comstock, Ithaca, N.Y.

———— (1958), *Q. Rev. Biol.*, **33**:103.

Mendel, G. (1866), *Verh. Naturforsch. Ver. Bruenn*, **4**:3; trans. and reprinted in Peters (1959), pp. 1–20.

Mitchell, H. K., and M. B. Houlahan (1946), *Fed. Proc.*, **5**:370.

Olby, R. C. (1966), "Origins of Mendelism," Constable, London.

Peters, J. A. (ed.) (1959), "Classic Papers in Genetics," Prentice-Hall, Englewood Cliffs, N.J.

Robinson, R. (1958), *Bibliog. Genet.*, **17**:229.

———— (1959), *Bibliog. Genet.*, **18**:273.

Rodriguez, V. A., and A. J. Norden (1970), *J. Hered.*, **61**:161.

Sager, R. (1955), *Genetics*, **40**:476.

Sawin, P. B. (1932), *Carnegie Inst. Wash., D.C., Publ.* 427, p. 15.

—— (1955), *Adv. Genet.*, **7**:183.

Shackelford, R. M. (1950), "Genetics of the Ranch Mink," Pilsbury, New York.

—— (1957), *J. Hered.*, **48**:129.

Shrode, R. R., and J. L. Lush (1947), *Adv. Genet.*, **1**:209.

Slome, D. (1933), *J. Genet.*, **27**:363.

Smith, L. (1951), *Bot. Rev.*, **17**:1, 133, 285.

Smith, P. G. (1950), *J. Hered.*, **41**:138.

Stockard, C. R. (1932), *Proc. 6th Int. Congr. Genet. Ithaca*, **2**:193.

Sturtevant, A. H. (1965), "A History of Genetics," Harper & Row, New York.

Sutton, W. S. (1902), *Biol. Bull.*, **4**:231.

Tanaka, Y. (1953), *Adv. Genet.*, **5**:240.

Uphof, J. C. (1924), *Genetics*, **9**:292.

Volpe, E. D. (1956), *J. Hered.*, **47**:79.

Whitney, L. F. (1929), *J. Hered.*, **20**:561.

Wright, S. (1917), *J. Hered.*, **8**:224.

5
Probability

QUESTION

161 Explain why actual genetic ratios in small populations frequently deviate considerably from those expected whereas ratios in large populations usually conform rather closely.

PROBLEMS

162 In man, the autosomal gene *A* for *absence of molars* (*affected*) is dominant to its allele *a* for *presence of molars* (*unaffected*). A certain couple, both heterozygous (*Aa*), have 5 children.
   a Expand the binomial $(p + q)^5$.
   b Derive the probabilities that:
   (1) None will be *unaffected*.
   (2) In any order 3 will be *affected* and 2 *unaffected*.
   (3) All will be *affected*.
   (4) 1 will be *unaffected* and 4 *affected* in any order.
   (5) The first 2 will be *unaffected*, the last 3 *affected*.
   (6) All will be *unaffected girls*.
   c What is the chance that the next child will be *unaffected*? The next 4 children will be *affected*?
   d In how many ways (sequences) can 3 *affected* and 2 *unaffected* children be produced?
   e What is the probability of the first 3 children being either *affected, unaffected, affected* or *unaffected, affected, unaffected*?

163 If 25 girls and 15 boys enter a 40-seat classroom and seat themselves at random:
   a What is the chance of a girl sitting in chair 18?
   b What is the probability of there being 2 boys and 2 girls in the first four chairs?

164 In rabbits the autosomal gene *S* for *short* hair is dominant to its allele *s* for *long* hair. Assuming that II-2 and II-5 are *SS*, what is the probability that any offspring of a mating between III-1 and III-2 will have (1) *long* hair, (2) *short* hair? Show how you arrive at your answer.

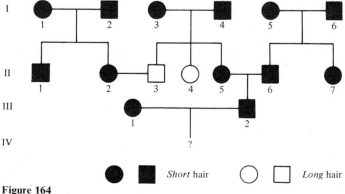

Figure 164

165   In sheep, the autosomal gene *B* for *black* wool is dominant to its allele *b* for *white* wool.   A heterozygous ram is mated with 44 *black* ewes each of which produce 3 lambs; 10 of the ewes produce 1 lamb with *white* and 2 with *black* wool, and 5 produce 2 *white* and 1 *black* lamb.   All the lambs from the other 29 ewes have *black* wool.
   a How many of the ewes are definitely heterozygous?
   b How many more heterozygous ewes probably exist in the flock?
   c If each of the 29 ewes had produced 6 lambs, what would be the chance of any heterozygous ewe producing only *black* offspring?

166   *Galactosemia* is a simply inherited autosomal recessive trait.   A *normal* couple have an *affected* child.   Find the probability that:
   a The next two children will be *galactosemic*.
   b Of the next 4 children, 1 will be *galactosemic*.
   c The father of the *galactosemic* child is heterozygous for the recessive allele.
   d The paternal grandmother is heterozygous for the recessive allele.
   e The next child will be a heterozygote.
   f A child of a *normal* sister of the *affected* child will be heterozygous.
   g Any sibling of the *galactosemic* will be *affected*.
   h There will be no *affected* individuals among the next 3 siblings of the *galactosemic*.
   i Of the next 3 children, 2 will be heterozygous and the other homozygous for the dominant allele.
   j Any child will be (1) a *galactosemic* girl, (2) a *normal* boy.
   k The first child will be a *normal* boy and the second a *galactosemic* girl.
   l In a family of 4 (1) at least 2 of the children will be *normal*, (2) 2 of the children will be *galactosemic* girls and 2 *galactosemic* boys.
   m Of the first 2 of these children either the 2 will be *galactosemic* or the 2 will be *normal*.

167   In sheep, the autosomal gene *Y* for *white* fat is dominant to *y* for *yellow* fat.
   a A *Yy* ram is mated with a *Yy* ewe.   What proportion of the *white* F_2's is expected to be heterozygous?
   b One of the F_2 *white* ewes is mated on three separate occasions with a *yellow*, *yy*, ram and produces 2 offspring in each of three litters.   All 6 develop *white* fat.
   (1) What are her possible genotypes?   Which is more probable?   Why?
   (2) With what degree of confidence can her genotype be specified?

168   A plant is heterozygous *Aa Bb*.   The genes are inherited independently.
   a Find the probability that a pollen-grain nucleus will contain:
   (1) An *A* allele.
   (2) An *A* allele and a *b* allele.
   (3) The *B* allele or the *b* allele.
   b If pollen grains of this plant are used to fertilize eggs of a plant of the same genotype, find the probability that a seed-embryo will contain:

(1) Two *A* alleles.
(2) One *A* allele and one *a* allele.
(3) Two *A* alleles and two *b* alleles.
(4) The genotype *Aa Bb*.

169 The synthesis of *normal* vs. *sickle-shaped* red blood cells in human beings is controlled by alleles Hb^A and Hb^S, respectively, of an autosomal gene.   A fairly large sample of Negro children in Africa of genotypes Hb^AHb^A and Hb^AHb^S were tested for degree of infection by the malarial parasite *Plasmodium falciparum*, with the results shown.

Table 169

	Hb^AHb^A	Hb^AHb^S
Heavily infected	206	340
Not infected or lightly infected	48	133

Are the heterozygotes more resistant to infection by the parasite than the normal homozygotes?

170 In Toronto in 1940, of 18,762 deaths, 424 were from *tuberculosis* and 622 were from *diabetes*.   From these data derive the probability that the next death will be:
a From *tuberculosis*
b From either *tuberculosis* or *diabetes*

171 Assume that the sex ratio at birth in man is 1:1.
   a Determine the probability that a family of 6 children will consist of:
   (1) 4 boys and 2 girls in any order.
   (2) All the same sex.
   (3) At least 3 girls.
   (4) No fewer than 2 girls and 2 boys.
   (5) 3 or more girls.
   (6) 4 boys and 2 girls or 2 boys and 4 girls.
   b What is the chance of the eldest child being a boy and the youngest a girl?
   c Determine
   (1) The most frequently expected number of males and females and
   (2) The percentage of all families of 6 expected to have 3 boys and 3 girls.

172 When a *wire-haired* rabbit is mated to a *normal-haired* one, all F_1 progeny are *wire-haired*.   When these F_1 rabbits are mated among themselves, they produce 22 rabbits, 14 *wire-haired* and 8 *normal-haired*.
   a How many of the F_2 *wire-haired* rabbits would you expect to be homozygous?
   b What results do you expect in the F_2?   Determine whether the actual results deviate significantly from the expected.   If they do, does this mean your hypothesis is wrong?   Explain.

173   In man, *hypotrichosis* (sparse body hair) is recessive to *normal*.
    a Two *normal* parents have 5 children, the first 2 with *hypotrichosis* and the others *normal*.
    (1) What are the genotypes of the parents?   Explain.
    (2) What is the chance of all the *normals* being heterozygous?
    b A *normal* man and *hypotrichotic* woman have 3 children, 1 *hypotrichotic* and 2 *normal*.
    (1) What is the man's genotype?
    (2) What is the probability of the *normal* children being heterozygous?
    (3) If one of these *normals* marries a *normal* from the marriage described in (a), what is the probability of their first child being *hypotrichotic*?   If they have 4 children, what is the chance of all being *normal*?

174   *Pituitary dwarfism*, due to deficiency of a growth hormone, is controlled by an autosomal recessive allele *p*.   Two *normal* parents have a son who is *affected* and a daughter who is *normal*.
    a What is the probability that the daughter is heterozygous?
    b If this daughter marries a *normal* male whose sister is a *pituitary dwarf*, what is the probability that the first offspring of this marriage will be *affected*?   If the first child is *affected*, what is the chance of the second child's also being a *dwarf*?

175   Two mated rabbits, a male and a female, have the genotype *Aa Bb Cc Dd*.
    a How many different kinds of gametes will each individual form?
    b What is the probability that their first offspring will be *Aa*?   *BB*?   *DD*?
    c What is the probability that the first two offspring will be *bb* and *Bb* respectively?   *Bb* and *Bb*?
    d If the first two offspring are *Bb* and *bb*, what is the chance that the next offspring will be *BB*?

176   In man, *free* vs. *attached* earlobes is controlled by a single gene, the allele for *free*, *F*, being dominant to that for *attached*, *f*.   A certain couple, both heterozygous and therefore *free-lobed*, have a family of 4 children.   Derive the following probabilities:
    a That the first 2 will have *attached* earlobes.
    b That 3 of the children will be girls and 1 a boy.
    c These 3 girls will have *attached* earlobes and the boy *free* earlobes.
    d That 1 or more will have *free* earlobes.
    e No fewer than 2 will have *free* and one will have *attached* earlobes.
    f 2 or more will have *attached* earlobes.

177   In a cross between two *Drosophila*, one from a true-breeding *yellow* body strain and the other with the same phenotype but of unknown origin, 80 offspring were formed, 42 of which had *grey* bodies.   Would you reject the hypothesis that the fly of unknown origin was heterozygous?

178   From a natural population of lady beetles a sample was classified for sex and elytra spots.   There were 81 *five-spot* males, 73 *five-spot* females, 150 *four-spot* females, and 55 *four-spot* males.

   a Does the ratio of *four-spot* : *five-spot* deviate significantly from 1:1?
   b Is the sex ratio in this sample significantly different from the expected 50:50?
   c Is color independent of sex in this population?

179 How many degrees of freedom would there be in:
   a An F_2 with a 3:6:3:1:2:1 ratio.
   b A testcross with a 1:1:1:1 ratio.
   c A testcross of a tetrahybrid in which there is no gene interaction.
   d An F_2 of a self-fertilized trihybrid in which there is no gene interaction.

180 The hypothesis that there is assortative mating with respect to eye color is tested by sampling a population. From a sample of 10,000 couples chosen at random, the data shown are gathered.

Table 180

Eye color		No. of couples
Wife	Husband	
Brown	*Brown*	6,190
Brown	*Blue*	1,033
Blue	*Brown*	1,667
Blue	*Blue*	1,110
Total		10,000

   a Use the observed ratio of *brown* : *blue* in men and women respectively to derive the theoretical ratio for $B/B : b/B : b/b$ matings.
   b Use the chi-square test to determine whether mating is random or assortative.
   c State what form the assortative mating may take.

181 A self-fertilized F_1 tomato plant from a cross between a plant with a *simple* inflorescence and a plant with a *compound* inflorescence produced 360 progeny, 75 of which developed a *compound* inflorescence. Test the hypothesis that a single autosomal pair of alleles is involved with dominance of *S* (*simple*) over *s* (*compound*).

182 In the silkworm, a true-breeding line in which females lay *white* eggs is mated with a true-breeding line in which females lay *pink* eggs. The F_1 females, all of which lay *white* eggs, were crossed with their brothers and produced 208 F_2 females; 141 laid *white*, 49 laid *black*, and 18 laid *pink* eggs.
   a Present a hypothesis to account for these results.
   b Test your explanation using the chi-square method and indicate whether your hypothesis is consistent with the data.
   c Crosses between F_1 males and females laying *pink* eggs produced 19 females that laid *pink*, 27 that laid *black*, and 54 that laid *white* eggs. Are these results consistent with the hypothesis in (a)? Explain.

183   Guinea pigs of a strain with *short, yellow* hair mated with those of a strain having *long, white* hair produced F_1's with *short, cream* hair, which, when crossed with *long, white* animals, produced 40 *short, cream*, 11 *long, cream*, 9 *short, white*, and 38 *long, white* progeny.

   a Propose a hypothesis to explain these results and indicate the results expected if:
   (1) F_1's were crossed among themselves.
   (2) F_1's were crossed with pigs from the *short, yellow* strain.

   b On the basis of your hypothesis indicate whether:
   (1) Alleles of each pair are transmitted by F_1's in equal (1:1) proportions.
   (2) The testcross results are consistent with the hypothesis that the different allele pairs are inherited independently.

184   When fowl from two true-breeding lines with *white* feathers were crossed, all the F_1's were phenotypically identical to the parents and the F_2 consisted of 124 *white* and 36 *colored* birds. These results may be explained by a single autosomal pair of alleles with dominance of the allele for *white* over that for *colored* and also by the interaction of two independently inherited pairs of alleles involving dominant and recessive epistasis.

   a Using the chi-square test, determine whether each hypothesis is consistent with the data.

   b Outline a genetic test that would clearly indicate which of the two hypotheses is the correct one.

185   The distribution of girls shown came from 250 families, each with 4 children.

Table 185

Family type	No. of boys	No. of girls	No. of families
1	0	4	12
2	1	3	69
3	2	2	84
4	3	1	57
5	4	0	18

Do the data indicate that the sexes occur in a 1:1 ratio?

186   In guinea pigs a series of matings are made between *black* heterozygotes, *Bb*. The first four matings produced 20 offspring: 13 *black* and 7 *white*. Twenty subsequent matings produced 200 animals: 130 *black* and 70 *white* (10 times as many as in the first four matings with the same relative magnitude of deviation from the expected 3:1 ratio as in the first population).

   a Do chi-square tests to show whether the results in the small and large samples are consistent with the expected 3:1 ratio.

   b Should one use large or small samples to test a hypothesis. Why?

REFERENCES

Bailey, N. T. J. (1951), *Ann. Eugen.*, **16**:223.

—————— (1959), "Statistical Methods in Biology," Wiley, New York.

Dixon, W. J., and F. J. Massey (1957), "Introduction to Statistical Analysis," 2d ed., McGraw-Hill, New York.

Falconer, D. S. (1961), "Introduction to Quantitative Genetics." Ronald, New York.

Goldstein, A. (1964), "Biostatistics: An Introductory Text," Macmillan, New York.

Kempthorne, O. (1957), "An Introduction to Genetic Statistics," Wiley, New York.

Levene, H. (1958), E. W. Sinnott, L. C. Dunn, and T. H. Dobzhansky (eds.) in "Principles of Genetics," 5th ed., pp. 388–418, McGraw-Hill, New York.

Mather, W. B. (1964), "Principles of Quantitative Genetics," Burgess, Minneapolis.

Morton, N. E. (1959), *Am. J. Hum. Genet.*, **11**:1.

Snedecor, G. W. (1956), "Statistical Methods," 5th ed., Iowa State College Press, Ames.

6
Gene
Interaction

QUESTIONS

187 Distinguish between epistasis and dominance. What does *gene interaction* mean?

188 Is the following statement true or false? Explain. "The segregation and independent assortment of genes are influenced by the manner in which genes express or fail to express their potentials."

189 **a** Does interaction of genes always produce modified F_2 and testcross phenotypic ratios? Explain.
b Is it possible to have a modified F_2 or testcross ratio without interaction? Explain.

190 Which of the following, if any, does the interaction of genes influence and how?
a Segregation of alleles.
b The kinds of gametes formed.
c The genotypes in F_2's.
d The phenotypes in F_2's.
e The phenotypic ratio in F_2's.

191 Is it possible by breeding procedures alone to detect the presence of a gene if it exists in only one allelic form? Explain.

192 State the expected phenotypic ratio from the testcross of F_1's giving the following dihybrid F_2 ratios and state the cause of modification of the 9:3:3:1 ratio for each:
a 13:3   **b** 3:6:3:1:2:1   **c** 9:3:4
d 12:3:1   **e** 15:1   **f** 1:2:2:4:1:2:1:2:1

PROBLEMS

193 Many wild mammals, including mice, have an *agouti* coat-color pattern, the individual hairs being black or dark brown with a yellow band just below the tip. Mutants in mice with coats entirely *black*, *yellow*, *cream*, *cinnamon* (brown agouti), *chocolate* (rich brown), or *albino* are known (Wright, 1917*a,b*; Little, 1958). The table shows results obtained in crosses between true-breeding strains.

Table 193

Cross	Parents	F_1	F_2
1	*agouti* × *black*	*agouti*	3 *agouti* : 1 *black*
2	*agouti* × *cinnamon*	*agouti*	3 *agouti* : 1 *cinnamon*
3	*cinnamon* × *black*	*agouti*	9 *agouti* : 3 *cinnamon* : 3 *black* : 1 *chocolate*

a What do the results of the first two crosses show about the inheritance of *agouti*, *black*, and *cinnamon*?

b From the F_2 of *cinnamon* × *black* determine the number of allele pairs controlling the differences between the parents, their genotypes and those of the F_1's, and the phenotypic classes in the F_2's.

c Describe the type of gene interaction illustrated.

d What phenotypes and phenotypic ratio would you obtain in the progeny of a backcross of an F_1 of *cinnamon* × *black* to the *cinnamon* strain? To the *black* strain? To the *chocolate* strain?

194 In poultry the shape of the comb may be *rose*, *pea*, *walnut*, or *single*. In crosses between two true-breeding lines, one for *rose* comb and the other for *pea* comb, Bateson and Punnett (Bateson and Punnett, 1906; Bateson, 1909) obtained F_1's all with *walnut* combs and F_2's with a total segregation ratio of 95 *walnut* : 26 *rose* : 38 *pea* : 14 *single*.

a Explain these results, giving the genotypes of the various phenotypes. Subsequently a *walnut*-combed male crossed with five *single*-combed females gave 22 *walnut* : 21 *rose* : 20 *pea* : 20 *single*.

b Show how your answer to (a) explains these results.

c What are the genotypes of the parents and offspring in each of the crosses in Table 194?

Table 194

Mating	Offspring
walnut × *walnut*	1 *single*, 1 *pea*, 1 *walnut*
walnut × *single*	42, all *walnut*
pea × *walnut*	7 *walnut*, 6 *pea*, 3 *rose*, and 2 *single*
walnut × *single*	3 *walnut*, 4 *rose*

195 In rats two independently inherited pairs of alleles *Aa* and *Rr* interact as follows:

A—R—	*grey*
A—rr	*yellow*
aa R—	*black*
aa rr	*cream*

The genotypes above are expressed only in the presence of the dominant allele of a third gene, *C*, since the recessive, *c*, causes *albinism*. Four different homozygous *albino* lines, each crossed with a true-breeding *grey* strain, produced *grey* F_1's, which in turn produced the F_2's shown.

a What type of interaction, if any, is involved in each cross?

b State the probable genotype of each *albino* line and give reasons for your answers.

Table 195

Albino line	Classification of F_2's				
	Grey	Yellow	Black	Cream	Albino
1	174	0	65	0	80
2	48	0	0	0	16
3	104	33	0	0	44
4	292	87	88	32	171

196 *Retinitis pigmentosa*, a form of blindness in man, may be caused either by a dominant autosomal gene, *R*, or a recessive autosomal gene, *a* (Gates, 1946). Thus only *A— rr* individuals are *normal*.   An *afflicted* man whose parents are both *normal* marries a woman of genotype *Aa Rr*.   What proportion of the children are expected to suffer from this affliction if *R* and *A* are inherited independently?

197 Before the development of fox farming two distinctly different mutations to *black* (*silver*) are known to have occurred in the common red fox (*Vulpes fulva* [= *vulpes*]).   The *Alaskan silver* is a mutant of Alaskan origin; the *standard silver* is of Canadian origin.   The two traits are very similar, the pelts being dark at the base and overlaid with white-banded guard hairs that give the silvery appearance.   The pelts of the *wild* red fox are primarily red.   A fourth type, the *cross* fox, has a pelt predominantly red over the back, shoulders, and head but black on the legs and belly.   The following breeding information has been assembled by Warwick and Hanson (1937):

1   *Alaskan silver* pups first appeared in *wild* × *wild* and *wild* × *cross* matings; whenever two *Alaskan silvers* were mated, they produced *Alaskan silver* offspring only.   This was also true of *standard silver*.

2   When *Alaskan silver* was crossed with *standard silver*, the F_1's showed a new phenotype, the *blended cross* (predominantly red, with a dark cross over the shoulders).

3   The numerous offspring from many matings of *blended cross* × *blended cross* segregated into nine distinct classes in approximately the following proportions:

1 *red* : 2 *smoky-red* : 2 *cross-red* : 4 *blended-cross* : 1 *standard silver* : 2 *substandard silver* : 1 *Alaskan silver* : 2 *sub-Alaskan silver* : 1 *double black*

a (1) Explain why it is so easy to establish true-breeding *silver* strains.

(2) Explain the obvious contradictions between the results of the cross between the *Alaskan* and *standard silver* and the observation that both these traits are recessive to *red*.

(3) (*a*) How many genes are segregating in the matings between *blended-cross* individuals, and what are the allelic relationships of each?

(*b*) Which of the progeny phenotypes should breed true?

b Why could you not establish a true-breeding line of *blended-cross* foxes? What true-breeding strains should be maintained to obtain the maximum proportions of these animals?

198 In the silkworm (*Bombyx mori*) true-breeding strains may be either *tri-*, *tetra-*, or *pentamolting* (larvae pass through three, four, or five molts before pupation). When certain *trimolting* strains are crossed with *penta* ones, the F_1 are *tri* and the F_2's consist of *tri* and *penta* types in a 3:1 ratio. Crosses between *tetra* and *penta* strains produce *tetra* F_1's and F_2's with a 3 *tetra* : 1 *penta* ratio. When certain *tri* strains are crossed with *tetra* ones, the F_1's are *tri* and the F_2 consists of *tri*, *tetra*, and *penta* types (Morohoshi, 1939; Nakamura, 1942).

a In what proportions would you expect the F_2 phenotypes of the third cross to appear?

b Why does this ratio differ from the classical 9:3:3:1 ratio?

c Give the genotypes of the parents and F_1's in the three crosses.

199 One homozygous strain of the silkworm (*Bombyx mori*) lays *white* eggs, another lays *pink*, and still another *black*. Suppose that *white* and *black* are not controlled by alleles of the same gene and that *white* is epistatic to *black*. In the F_2 of a *black* × *white* mating, one-sixteenth of the progeny lay *pink* eggs. Set up a hypothesis to explain the results and diagram the expected genotypes and phenotypes in the F_1 and F_2.

200 a In the Japanese morning glory (*Pharbitis nil* [= *Ipomoea hederacea* Jacq.]) crosses among true-breeding *purple-*, *blue-*, and *scarlet*-flowered strains gave the results shown in Table 200*A*.

Table 200*A*

Cross	Parents*	Progeny
1	*purple* 1 × *purple* 1	all *purple*
2	*purple* 2 × *purple* 2	all *purple*
3	*scarlet* × *purple* 1 or 2	all *purple*
4	*scarlet* × *scarlet*	all *scarlet*
5	*purple* 1 × *purple* 2	all *blue*
6	*blue* × *scarlet*	all *blue*

* Numbers 1 and 2 refer to different strains of *purple* which are phenotypically identical.

(1) How many genes appear to be involved, and what are their phenotypic effects?

(2) Can you determine from the above data whether they are inherited independently?

b Do the results of the crosses in Table 200*B* support your hypothesis? Give the genotypes of the parents of each cross.

Table 200*B*

Cross	Parents	Progeny
1	*blue* × *purple*	30 *blue* : 40 *purple* : 10 *scarlet*
2	*purple* × *purple*	26 *blue* : 54 *purple* : 23 *scarlet*
3	*blue* × *blue*	78 *blue* : 25 *purple*
4	*scarlet* × *blue*	52 *purple* : 28 *scarlet* : 25 *blue*
5	*purple* 1 × *scarlet*	43 *purple* : 46 *scarlet*
6	*purple* 2 × *scarlet*	49 *scarlet* : 44 *purple*

201 In the onion (*Allium cepa*) in the presence of *C* (dominant) an enzyme catalyzing the production of bulb-color pigment is formed; in *cc* onions the enzyme is not formed.   *K* (dominant) prevents this enzyme from functioning and hence prevents pigment production, but *k* does not (Rieman, 1931).   The two genes are independently inherited.   A true-breeding *colored* variety is crossed with a *white* one having the genotype *ccKK*.
a What are the expected phenotypes and phenotypic ratios in the F_1 and F_2?
b What type of interaction is involved?
c A pair of alleles, *Rr*, carried on a chromosome pair other than those carrying the *Cc* and *Kk* allele pairs, controls the specific color of the bulbs, *R* giving *red* and *r*, *yellow* bulbs.   If the true-breeding colored variety were homozygous for *R* and the white one homozygous for *r*, what would the phenotypic ratio in the F_2 of the above cross be?

202 In breeding experiments designed to study fruit color in summer squash (*Cucurbita pepo*) Sinnott and Durham (1922) found that strains with *white* (fruits) occasionally produced plants with *green* and plants with *yellow*.   Never did strains with *green* or *yellow* produce plants with *white*.   Moreover, strains with *green* never produced plants with *yellow*.   Crosses between true-breeding *green* and *yellow* strains produced plants with *yellow* only.   These *yellow* plants when intercrossed produced 81 *yellow* and 29 *green*.   Homozygous *white* × homozygous *yellow* produced plants with *white* only.   These when intercrossed produced *white*, *yellow*, and *green* in a 155:40:10 ratio.   Give a complete genetic explanation of these results.

203 **a** In Duroc Jersey swine the color of hair is *white*, *sandy*, or *red* (Wentworth and Lush, 1923).   Two *sandy* strains and a *white* one, crossed in all possible combinations, give the results shown.
    (1) Can these results be explained by dominant or recessive epistasis?   Give reasons for your decision.
    (2) Postulate a third hypothesis and show how it could explain the above results.

Table 203

Cross	Parents	F_1	$F_1 \times$ white
1	white × sandy strain 1	sandy	1 sandy : 1 white
2	white × sandy strain 2	sandy	1 sandy : 1 white
3	sandy strain 1 × sandy strain 2	red	1 red : 2 sandy : 1 white

(3) What is the expected ratio in the F_2 of *white* × true-breeding *red*?

b The three-generation pedigree shows the transmission of *red, sandy,* and *white* swine coat colors. Two independently inherited allele pairs, *Aa, Bb,* are involved. Determine the mode of inheritance of the three traits. *Sandy, red,* and *white* occur in a 4:3:1 ratio among the offspring of A × B. What are the genotypes of A and B and why?

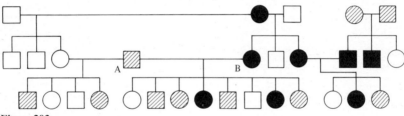

Figure 203

204 In the summer squash (*Cucurbita pepo*) *spherical* fruit is recessive to *disk*. *Spherical* races from different geographic regions were crossed. The F_1's were *disk*, and the F_2's segregated 35 *disk*, 25 *spherical*, and 4 *elongate* (Sinnott, 1927). Explain these results.

205 In the wild (*wild-type*) Mexican swordtail fish (*Xiphophorus helleri*) the color is olive-green (a mosaic skin pattern produced by two kinds of pigment-carrying cells, the "blacks" and the "yellows"). Many other color variants have been developed since 1909, including a *golden* and an *albino*. The *golden* owes its distinctive color to the fact that the skin has many yellow cells but few blacks. The eyes, in which pigment cells are more frequent, are black. The *albino* has a few yellow cells but no black ones, and the eyes are pink. Matings involving the three types give the tabulated result (Gordon, 1941).

a Explain the reversion, or atavism, to *wild type* in matings between *golden* and *albino*, giving the probable number of allele pairs and the form of gene interaction involved.

b Is it possible for the *wild types* to breed true? Explain.

c When a *golden* is mated with *montezuma*, the progeny consist of *montezuma* and *wild type* in the ratio of 1:1, and when the *montezuma* offspring of a *montezuma* × *golden* cross are mated back to the *golden*, about one-quarter

Table 205

Cross	Parents	Progeny
1	golden × golden	golden
2	albino × albino	albino
3	golden × albino	wild-type
4	wild-type × golden	68 wild-type : 52 golden
5	wild-type × albino	33 wild-type : 23 albino
6	wild-type* × wild-type*	202 wild-type : 65 golden : 72 albino

* Obtained from *golden* × *albino* matings.

of the young are *wild type*. Explain how this reversion can occur, giving genotypes of parents and offspring.

206 In the lady beetle *Propyrea japonica* the elytra may show *three, four,* or *five spots.* A homozygous *five-spot* strain is crossed with a homozygous *four-spot* one. The F_1 are *five-spot*, and in the F_2 an average of 1 in 16 individuals is *three-spot*. Set up a hypothesis to explain these results and show:
a The genotypes of the parents, F_1, and F_2.
b The F_2 phenotypes and their theoretical proportions.

207 When *red* cattle are crossed with *brindle* (irregular narrow stripes of black on a red background), the F_1 are *brindle* and the F_2 consists of *brindles, black-and-reds*, and *reds* in a 9 : 3 : 4 ratio. In certain crosses between the Angus (*black*) and Jersey (*black-and-red*) breeds, the F_1 are all *black*, and the F_2's segregate 12 *black* : 3 *black-and-red* : 1 *red*. In other crosses between these breeds the F_1's are also *black*, but in the F_2 four phenotypes appear: *black, brindle, black-and-red*, and *red* in a 48 : 9 : 3 : 4 ratio (Ibsen, 1933). Describe the genetic basis for coat color in these breeds, stating the number of allele pairs and the type of interaction and showing the genotypes of the Angus and Jersey parents in each of the three crosses described.

208 The larval body color in silkworms may be *chocolate* (reddish-brown), *normal* (black), or *dominant-chocolate* (like chocolate but with a black head) (Tanaka, 1953). When a *dominant-chocolate* strain was crossed with a *chocolate* one, the F_1 were all *dominant-chocolate*. In the F_2, the phenotypes and their proportions were 96 *dominant-chocolate* : 31 *normal* : 43 *chocolate*.
a How many allele pairs are involved in this cross, and how do they interact?
b What phenotypic ratio would be expected in a testcross of the F_1?

209 In the summer squash (*Cucurbita pepo*) *white* fruit is controlled by a dominant allele, *W,* and *colored* fruit by its recessive allele, *w.* In *ww* plants, the color alleles, *Y,* for *yellow* fruit, dominant to *y* for *green* fruit, are expressed (Sinnott and Durham, 1922).
a What is the phenotype of the F_1 of a *white* plant of genotype *WW yy* crossed with a homozygous *yellow* plant?
b What ratio is expected in the F_2? What type of interaction is involved?

c A cross between a plant with *yellow* fruit and a plant with *white* fruit produced 58 *white* : 39 *yellow* : 16 *green*.   What are the genotypes of the parents?

210   The self-fertilized F_1 from a cross of two true-breeding durum wheat varieties with *spring* growth habit produced 91 plants, 6 with *winter* growth habit and the remainder with *spring* growth habit.   Indicate:
a The number of allele pairs involved.
b The phenotype of the F_1's.
c The genotypes of the F_2 plants with *winter* growth habit.

211   The legs of Black Langshan fowl are *feathered*; those of Buff Rocks are *featherless*.   The F_1 of a cross between these two true-breeding strains are *feathered*, and in the F_2 a ratio of 15 *feathered* : 1 *featherless* is observed (Lambert and Knox, 1929).
a Explain these results and give the genotypes of the parents and the F_1 and the genotypes represented in each F_2 phenotype.
b What phenotypes would you expect in the offspring of crosses between the F_1's and *featherless* individuals?   In what proportions would they occur and why?
c Certain *feathered* F_2 birds when crossed with *featherless* ones produced only *feathered* progeny which when interbred gave progeny with a phenotypic ratio of 3 *feathered* : 1 *featherless*.   What phenotypic ratio would you expect in the progeny of a cross between two such *feathered* F_2 individuals? Explain.

212   A *green* barley plant when self-fertilized produces 182 *green* and 144 *albino* plants.   Explain genetically.

213   In garden stock (*Matthiola incana*), the results shown were obtained in crosses between *single*- and *double*-flowered varieties and among different *double*-flowered varieties.

Table 213

Cross	Parents	F_1	F_2
1	*single* × *double*	*single*	193 *single* : 152 *double*
2	*double* × *double*	*single*	281 *single* : 213 *double*
3	F_1 (*double* × *double*) × *double*		1 *double* : 1 *single*
4	F_1 (*double* × *double*) × *double*		3 *single* : 5 *double*

a From an examination of the F_1 results of the first cross, which of these traits appears to be recessive?   Why does it not breed true in crosses 2, 3, and 4?
b Describe the type of gene interaction involved and give the genotypes of the parents and F_1 and F_2 classes for the four crosses.

214 When certain true-breeding strains of *platinum* mink (*blue-grey* fur) are brought together and single-pair matings occur at random within the combined population, it is found that some of the matings produce only *blue-grey* offspring while others produce only *wild type* (Shackelford, 1950).

a On the basis of this information only, state the simplest hypothesis that will explain the inheritance of coat color in these animals.

b When *wild-type* individuals are intermated, they produce 192 *wild-type* : 148 *blue-grey*; when mated with *blue-greys*, they give a ratio of 125 *wild-type* : 131 *blue-grey*. Does this alter or confirm your previous explanation?

c A mink breeder finds that *wild types* occasionally appear in his *platinum* stock. Since these throwbacks reduce his profits, he wishes to eliminate them entirely. How should he proceed?

215 The seed capsule of shepherd's purse (*Capsella bursa-pastoris*) may be either *triangular* or *top-shaped* (Shull, 1914). A cross between *triangular* and *top-shaped* gave a *triangular* F_1. In the F_2, 146 plants were *triangular* and 12 *top-shaped*.

a How many allele pairs control *triangular* vs. *top-shaped* in this cross, and how do they interact?

b Assigning your own symbols to the genes, state the genotypes of the parents that would produce the following offspring:
(1) 3 *triangular* : 1 *top-shaped*.
(2) All *triangular*.
(3) 7 *triangular* : 1 *top-shaped*.
(4) 1 *triangular* : 1 *top-shaped*.

216 Two true-breeding *white*-flowered lines of sweet peas (*Lathyrus odoratus*) when crossed produced *purple*-flowered F_1's and 100 *purple*- and 72 *white*-flowered F_2's (Bateson et al., 1905).

a Show statistically that this could not be considered a random deviation from a 1:1 ratio.

b Describe the form of inheritance involved, indicating the most likely theoretical ratio. Diagram the cross, indicating the genotypes for each phenotype.

c Suppose that all the seed you had of the two true-breeding *white*-flowered lines is lost. How could you establish two lines from these F_2's that possess genotypes identical to these two parental lines?

217 Stevenson and Cheeseman (1956) have shown that *deaf-mutism* vs. *normal* has the following characteristics of inheritance:

1 Many *normal* couples have *normal* children only.
2 Occasionally a *normal* couple have *deaf-mute* and *normal* children in approximately a 1:1 ratio.
3 Still other *normal* couples also have *deaf-mute* and *normal* children but in approximately a 1:3 ratio.
4 Some *deaf-mute* couples have *deaf-mute* offspring only.

5   Other *deaf-mute* couples have *normal* offspring only.

6   Still other *deaf-mute* couples have both types of progeny.

Offer a genetic explanation for these observations.

218   Most pedigree studies reveal that *albinos* are homozygous for a recessive autosomal allele for *albinism.*   In some rare pedigrees, however, *albino* couples have been found to have both *albino* and *nonalbino* children or *nonalbinos* only. Explain genetically.

219   In the yellow daisy (*Rudbeckia hirta*) the *wild-type* strain has yellow flowers with *purple centers.*   Two true-breeding mutant strains of independent origin are discovered which have *yellow centers.*   Each of the mutants when crossed with the *wild-type* strain gives F_1's with *purple-centered* flowers.   The F_2 in the first cross segregates 88 *purple* : 28 *yellow*, and that in the second cross consists of 61 *purple* : 22 *yellow*.   However, when the two mutant strains are crossed, all the F_1's produce *purple-centered* flowers and 75 of 173 F_2 plants obtained by self-fertilizing F_1's have *yellow centers.*

a Account genetically for the origin of the two mutant strains and why they produce *purple-centered* F_1's when crossed.

b What phenotypic ratio would you expect in the cross F_1 (mutant 1 × mutant 2) × mutant 1?   × mutant 2?

220   Two *left-handed* parents may have a *right-handed* child, and two *right-handed* parents sometimes have a *left-handed* child (Rife, 1940).   Can *left-handedness* vs. *right-handedness* be explained on the basis of a single pair of alleles?   If not, suggest a satisfactory explanation for these statements based on gene interaction.

221   a In a true-breeding *purple-eye, pr,* strain of *Drosophila melanogaster* a *wild-type* individual appears.   What could be a possible cause of this reversion to *wild type*?

b A true-breeding strain of the reverted *wild type* is produced and crossed with an unrelated true-breeding *wild-type* strain.   Explain F_1 and F_2 results genetically.

Table 221

F_1	F_2	
wild type	*wild type*	80
	purple	6

222   If "fancy" breeds of animals or plants are allowed to mate at random, the population will in time contain many "wild" (or "ancestral") types.   This can be perceived in flocks of pigeons whose ancestors escaped from pigeon fanciers' stocks and whose mating is not controlled by man.   In New York City's Battery Park, among the thousands of pigeons that come fluttering

down to pick up crumbs along the paths, a few birds with *black-and-white* or *red-and-white* mottling and some that are predominantly *red* or *deep blue* are found.   The majority, however, resemble the ancestral, *wild* blue rock pigeon (*Columba livia*) (Gordon, 1941).

a Propose a possible explanation for this reversion to *wild type* in the New York pigeons.

b How would you test your hypothesis?

223   The inheritance of *scrotal hernia* in swine, which appears within 1 month after birth, was studied by Warwick (1931).   He tentatively concluded that the trait is due to homozygosity for recessive alleles at two loci.   Outline how you would proceed to test the validity of Warwick's hypothesis, delineating the results expected for:

a Control by a recessive allele of one gene.

b Control by recessive alleles of two genes.

Note: The trait appears in males only; *affected* males are sexually functional. Establishment of large F_2 populations to obtain specific ratios is not feasible with pigs.

224   A *prolineless Neurospora* (haploid) mutant (having a nutritional requirement for proline) reverts to a *wild type* which breeds true for the reversion. This reverted *wild type* is mated with an unrelated *wild type*, and the progeny segregate 1 *prolineless* : 3 *wild type*.   Explain.

225 a In *Neurospora crassa* the difference between *wild type* and each of the arginine-requiring (*arginineless*) mutants *arg-1* and *arg-3* is due to alleles of different genes (Perkins, 1959).   A cross was made between *arg-1* and *arg-3* strains, and 400 ascospores analyzed at random showed a ratio of 1 *wild type* : 3 *arginineless*.   Explain these results.

b Suppose the ratio had been 3 *wild type* : 1 *arginineless*.   How would you explain these results?

226   The genetics of host-parasite interactions has been extensively studied in flax by Flor (1956).   Genes for resistance to rust were identified by testing the

Table 226

Flax variety	No. of plants	Reaction to rust race*		
		16	52	7
Williston, Golden		*S*	*S*	*S*
Bison		*R*	*R*	*R*
F_1		*R*	*R*	*R*
F_2	92	*R*	*R*	*R*
	26	*S*	*R*	*R*
	28	*R*	*S*	*R*
	10	*S*	*S*	*S*

* *Key: S = susceptible; R = resistant.*

reaction of many varieties to individual races of rust.   The genes for parasitism were identified by testing many different host varieties.   The results of testing rust races 16, 52, and 7 on the varieties Williston, Golden, and Bison were as shown.   (See Table 226.)

Interpret these results genetically, stating:

a The number of allele pairs in the host governing *resistance* vs. *susceptibility* to races 16, 52, and 7.

b Their allelic relationships.

c The type of interaction.

d The genotypes of the three races of rust.

227 a What are suppressor genes?

b Suppressors may be dominant, *Su*, or recessive, *su*, and may suppress the phenotype given by a dominant or a recessive allele.   Thus we may have a recessive or dominant suppressor of a recessive or dominant gene.   Some suppressors in *Drosophila melanogaster* and the locus they affect are listed.

Table 227

Suppressor	Locus and allele suppressed	Reference
1 *Su-S* (dominant)	*star* (*S*) (dominant)	Morgan, Bridges, and Schultz, 1937
2. *Su-ss* (dominant)	*spineless* (*ss*) (recessive)	Morgan, Bridges, and Schultz, 1937
3. *su2-Hw* (recessive)	*hairy-wing* (*Hw*) (dominant)	Wagner and Mitchell, 1964
4. *su-v* (recessive)	*vermilion* (*vv*) (recessive)	Bridges and Brehme, 1950

What dihybrid phenotypic F_2 ratio would you expect with each of the four possible combinations shown if the suppressor has no directly observable phenotypic effect and the suppressor and suppressed genes assort independently?

228 a When dogs from two true-breeding *white* lines were crossed, all the F_1's were *white*.   However, when the F_1's were intermated, 142 of the progeny were *white* and 33 *colored*.   Explain these results showing the genotypes of the parental lines, the F_1, and the F_2 phenotypes.

b When dogs of the first true-breeding line were crossed with dogs from a true-breeding *brown* line, all the F_1's were *white*.   The F_2 from a number of F_1 intermatings consisted of 238 *white*, 63 *black*, and 20 *brown*.

(1) Explain these results giving the genotypes of the *brown* parent, the F_1, and each F_2 phenotype.

(2) What are the genotypes of the parents in the following matings?

Black × *brown* → 1 *black* : 1 *brown*

Black × *white* → 9 *white* : 7 *black* : 8 *brown*

c When dogs of the second true-breeding *white* line were crossed with true-breeding *brown*, the F_1's were unexpectedly *black* and the F_2's were mostly *black* with some *white* and fewer *brown*.   How would you explain this to a dog breeder?

229 In corn (*Zea mays*) a cross is made between a homozygous *red-kernel* and a homozygous *white-kernel* strain.   The F_1 individuals are *red-kernel* only, and the F_2 express both phenotypes in the ratio of 285 *red-kernel* : 378 *white-kernel*. Explain, stating the number of allele pairs involved and the type of gene interaction.

230 Different kinds of *white* phenotype are known in poultry.   White Silkies and Rose-comb Bantams are *white* because of a recessive allele, *c*, which causes the lack of a chromogen necessary for color.   White Wyandottes, Minorcas, Dorkings, and other breeds are *white* because of a different recessive plumage allele, *o*, which causes the lack of an enzyme converting the chromogen into pigment.   White Leghorns carry a dominant gene, *I*, inhibiting pigment formation, in the homozygous state (see Hutt, 1949).

a What phenotypes would you expect in the F_1 and F_2 of a cross between a Rose-comb Bantam and a Minorca?   Describe the type of interaction involved.

b What phenotypes would appear in the F_2 of a cross between a White Leghorn and a Minorca or Rose-comb Bantam?   Describe the type of interaction involved.

c What genotype must colored breeds possess?

231 A wheat plant with *red* kernels produces 320 seeds upon self-fertilization, 314 giving rise to plants with *red* kernels and 6 to plants with *white* kernels. The latter group upon self-fertilization breed true for *white* kernel.   Give a genetic explanation to account for these results.

232 The F_2 of a cross between a *dark-red* and a *virescent-yellow* strain of cotton segregated 38 *dark-red* : 92 *light-red* : 17 *dark-bronze* : 34 *light-bronze* : 45 *green* : 12 *virescent-yellow* (Killough and Horlacher, 1933).   Explain these results, giving the phenotype of the F_1 and the genotypes of parents, F_1, and each F_2 class.

233 Emerson (1921) worked out the genetic basis for stem and foliage color in corn.   The F_1 of a cross between a plant with *green* stems and one with *purple* stems was phenotypically like the second parent.   The F_2's segregated

Purple	265
Sun-red	85
Dilute-purple	93
Brown	91
Dilute-sun-red	32
Green	70

a How many allele pairs are probably involved, and what is the allelic relationship of each?

b Is a modification of an F_2 ratio involved?   If so, what is the method of interaction?

234 **a** Three *white*-kernel strains of corn were crossed in all possible combinations. The F_1's were always *red*, and the F_2 segregated approximately 9 *red* : 7 *white*. How many loci are involved in the control of kernel color in these strains? Answer by giving the genotypes of all three strains.

b A fourth *white* strain when crossed with a *purple* one produces *purple* F_1's, which upon self-fertilization produce approximately 81 *purple* : 27 *red* : 148 *white*. Moreover, when the *purple* strain is crossed with a fifth *white* strain, the F_1 are all *white* and one-fourth of the F_2 are *purple* (information based on work of Emerson, 1918). Determine the genetic basis of kernel-color control and give the complete genotype of each of the parental strains used in these crosses.

235 The self-fertilized *purple-pod* F_1 from a cross between true-breeding *green-pod* and true-breeding *purple-pod* strains of garden peas produce *purple-pod*, *green-pod*, and *yellow-pod* plants in close approximation to a 36:21:7 ratio.
a Why are 36 out of every 64 plants *purple-pod*?
b What genotypes have *green* and *yellow* pods respectively?

236 A *black*-coat strain of cattle is crossed with an *albino* strain. The F_1 are *brindle* (black stripes of hair on a background of red), and the F_2 consist of 29 *black* : 62 *brindle* : 31 *red* : 41 *albino*. What type of intergenic relationship can explain these results? Outline the crosses you would make to test your explanation and show the results expected in each case.

237 In the sweet pea, two *white*-flowered strains when crossed produced *purple*-flowered F_1's which upon self-fertilization gave 265 *purple*-, 92 *red*-, and 282 *white*-flowered plants (Bateson and Punnett, 1906). Is it possible to explain these results on the basis of two pairs of interacting alleles? If not, what is the basis for the F_2 segregation ratio?

238 This fairly extensive pedigree of *deaf-mutism* in man is from a family observed in Northern Ireland by Stevenson and Cheeseman (1956). State the most probable mode of inheritance of *deaf-mutism*. Give reasons for your conclusions.

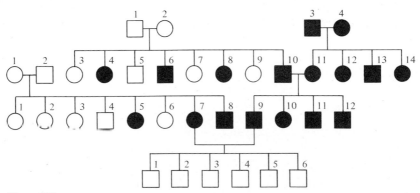

Figure 238

239 The pedigree illustrates the type of result obtained in the horse (*Equus caballus*) when *black, chestnut, bay,* and *liver* individuals are mated (based on the work of Castle, 1948, 1954, and Castle and Singleton, 1960).

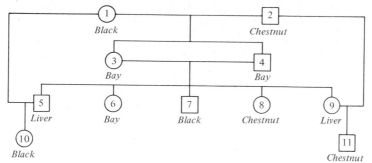

Figure 239

a Determine the genetic basis for coat color.

b When true-breeding *chestnut* horses are mated with *albinos* that appear within the *chestnut* strain, the F_1's are all *palomino*. The *palominos*, when mated, produce *palominos, chestnuts,* and *albinos* in a 2:1:1 ratio. These *albinos* when mated with *livers* produce *palomino* progeny only and when mated with *bays* produce *buckskin* F_1's, which, mated inter se in sufficient numbers, produce *albinos, buckskins, bays, palominos,* and *chestnuts* in a 4:6:3:2:1 ratio. No *blacks* or *livers* appear. A novice horse breeder, eager to establish a *palomino* herd, is somewhat baffled by all this. Outline the essential components of the *palomino* genotype and the matings he should make to ensure the greatest return for his efforts.

240 In the deer mouse (*Peromyscus maniculatus*) *grey* vs. *ivory* vs. *brown* vs. *albino*

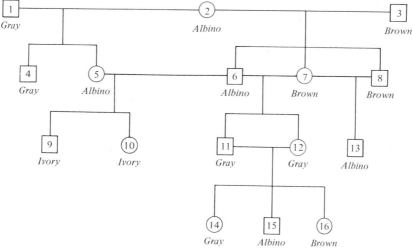

Figure 240

coat color is controlled by two pairs of alleles.   Which of the two types of interaction, recessive or dominant epistasis, is the more probable and why?

241   When a mouse fancier seeking strains with new combinations of traits crosses individuals from a true-breeding *white*-coated strain with *naked* mice obtained from another fancier, 41 of the offspring are *naked* and the remaining 23 have *black* hair.   The *naked* offspring when intercrossed produced 164 *naked* and 45 *black*-, 16 *brown*-, and 22 *white*-coated progeny.   The fancier, confused by these results, come to you for an explanation.   Assuming that the ratios do not depart appreciably from expectation, provide a satisfactory explanation giving the number of allele pairs, their allelic relationships, and gene interaction, if any.

242   A gladiolus breeder is eager to incorporate the night fragrance (*night-jasmine*) of *Gladiolus tristis* and the day fragrance (*violetlike*) of *G. recurvus* into his garden varieties and wants to know whether the inheritance of fragrance is simple or complex.   On the basis of the F_1 and F_2 results obtained by McLean (1938) from crosses between these very closely related species, what would you tell him with regard to the number of genes concerned and their interrelationships?

Table 242

Phenotype	F_1	F_2, %	$F_1 \times$ *G. tristis*, %
Violetlike	All	51.9	39.1
Violetlike + *night-jasmine*		2.2	18.7
Nonfragrant		37.3	6.0
Night-jasmine		8.6	36.2

Note: The percentages are based on large numbers of progeny.   Since the interaction is complex, you should analyze first for *presence* vs. *absence* of each type of fragrance, then consider them simultaneously.   Four pairs of alleles are involved.

REFERENCES

Ashri, A. (1964), *Genetics*, **50**:363.
Bateson, W. (1909), "Mendel's Principles of Heredity," Cambridge University Press, England.
——— and R. C. Punnett (1906), *Rep. Evol. Comm. R. Soc.*, **2**:11.
———, E. R. Saunders, and R. C. Punnett (1905), *Rep. Evol. Comm. R. Soc.*, **2**:80.
Black, G. (1970), *Euphytica*, **19**:22.
Blakeslee, A F. (1921), *Z. Indukt. Abstamm.-Vererbungsl.*, **25**:211.
Bridges, C. B., and K. S. Brehme (1950), The Mutants of *Drosophila melanogaster*, Carnegie Inst. Wash., D.C., Publ. 552.
Butler, L. (1948), *Can. J. Res.*, **d25**:190.
Castle, W. E. (1948), *Genetics*, **33**:22.
——— (1951), *J. Hered.*, **42**:48.

Castle, W. E. (1954), *Genetics*, **39**:35.

—— and W. R. Singleton (1960), *J. Hered.*, **51**:127.

Demerec, M. (1923), *Genetics*, **8**:561.

Emerson, R. A. (1918), *Cornell Univ. Agric. Exp. Stn. Mem.* 16, p. 231.

—— (1921), *Cornell Univ. Agric. Exp. Stn. Mem.* 39, p. 1.

Flor, H. H. (1956), *Adv. Genet.*, **8**:29.

Gates, R. R. (1946), "Human Genetics," vol. I, Macmillan, New York.

Goodman, R. H. (ed.) (1970), "Genetic Disorders in Man," Little, Brown, Boston.

Gordon, M. (1937), *J. Hered.*, **28**:221.

—— (1941), *J. Hered.*, **32**:385.

Grüneberg, H. (1952), *Bibliog. Genet.*, **15**:1.

Harris, H., and K. Hirschhorn (ed.) (1971), "Advances in Human Genetics," vol. 2, Plenum, New York.

Hutt, F. B. (1949), "Genetics of the Fowl," McGraw-Hill, New York.

Ibsen, H. L. (1933), *Genetics*, **18**:441.

Kidwell, J. F. (1962), *Can. J. Genet. Cytol.*, **4**:37.

Killough, D. T., and W. R. Horlacher (1933), *Genetics*, **18**:329.

Konigsmark, B. W. (1969), *N. Engl. J. Med.*, **281**:713.

Lambert, W. V., and C. W. Knox (1929), *Poult. Sci.*, **9**:51.

Little, C. C. (1958), *Q. Rev. Biol.*, **33**(2):103.

McLean, F. T. (1938), *J. Hered.*, **29**:115.

Mitchell, A. L. (1935), *J. Hered.*, **26**:425.

Mitchell, M. B., and H. K. Mitchell (1952), *Proc. Natl. Acad. Sci.*, **38**:205.

Morgan, T. H., C. B. Bridges, and J. Schultz (1937), *Carnegie Year Book*, **36**:301.

Morohoshi, S. (1939), *Bull. Sci. Fak. Terkult. Kyushu Univ.*, **8**:232.

Nakamura, T. (1942), *Rep. Seric. Exp. Stn. Chosen*, **3**:39.

Ouchi, S., and C. C. Lindegren (1963), *Can. J. Genet. Cytol.*, **5**:257.

Perkins, D. D. (1959), *Genetics*, **44**:1185.

Rieman, G. H. (1931), *J. Agric. Res.*, **42**:251.

Rife, D. C. (1940), *Genetics*, **25**:178.

Robinson, R. (1959), *Bibliog. Genet.*, **18**:273.

Rodriguez, V. A., and A. J. Norden (1970), *J. Hered.*, **61**:161.

Schultz, J., and C. B. Bridges (1932), *Am. Nat.*, **66**:323.

Shackelford, R. M. (1949), *Am. Nat.*, **83**:49.

—— (1950), "Genetics of the Ranch Mink," Pilsbury, New York.

Shrode, R. R., and J. L. Lush (1947), *Adv. Genet.*, **1**:209.

Shull, G. H. (1914), *Z. Indukt. Abstamm.-Vererbungsl.*, **12**:97.

Singleton, W. R. (1959), *J. Hered.*, **50**:261.

Sinnott, E. W. (1927), *Am. Nat.*, **61**:333.

—— and G. H. Durham (1922), *J. Hered.*, **13**:177.

Snyder, L. H., and P. R. David (1957), "The Principles of Heredity," 5th ed., Heath, Boston.

Stevenson, A. C., and E. A. Cheeseman (1956), *Ann. Hum. Genet.*, **20**:177.

Stroman, G. N., and C. H. Mahoney (1925), *Texas Agric. Exp. Stn. Bull.* 333.

Tanaka, Y. (1953), *Adv. Genet.*, **5**:239.

Wagner, R. P., and H. K. Mitchell (1964), "Genetics and Metabolism," 2d ed., Wiley, New York.

Warwick, B. L. (1931), *Ohio Agric. Exp. Stn. Bull.* 480.

—— and K. B. Hanson (1937), in "The Yearbook of Agriculture," pp. 1315–1349, USDA.

Wentworth, E. N., and J. L. Lush (1923), *J. Agric. Res.*, **23**:557.

Whaley, W. G. (1939), *J. Hered.*, **30**:335.

Wright, S. (1917*a*), *J. Hered.*, **8**:475.

—— (1917*b*), *J. Hered.*, **8**:224.

—— (1963), in W. J. Burdette (ed.), "Methodology in Mammalian Genetics," pp. 159–192, Holden-Day, San Francisco.

7
Lethal
Genes

QUESTIONS

243 a Distinguish between incompletely dominant lethal genes and recessive lethal genes. Which are easier to detect and why?

b Are incompletely dominant lethals easier or harder than recessive lethals to eliminate from a population? Give reasons to support your answer.

c Why are lethals much less common in haploid organisms than in diploid ones?

d Under what condition may dominant lethal genes be transmitted? Cite an example from man.

244 *Albinism* is lethal in plants, yet many species produce *albinos* among their progeny. If *albinos* always die before reproducing, why is the trait not eliminated?

245 What kind of experimental data (viz. types of crosses and expected ratios) would be required to demonstrate each of the following types of traits?

a A sex-linked recessive, lethal in birds, acting before birth.

b A sex-linked dominant in foxes, lethal shortly after birth.

c An environmentally induced congenital and lethal abnormality in cats.

d A previously unknown autosomal recessive trait in horses, lethal just before or shortly after birth.

246 Most lethals that have been discovered in man, regardless of whether they are dominant, incompletely dominant, or recessive, produce their effects shortly before or after birth. On the other hand, many lethals in organisms such as mice and *Drosophila* have been found that produce their effects in early embryonic stages. Why should it be difficult to detect early acting lethals in man but not in organisms like mice?

PROBLEMS

247 In mice the data shown represent the summed results of several investigators.

Table 247

	Offspring	
Parents	Yellow	Agouti
yellow × *yellow*	2,396	1,235
yellow × *agouti*	2,378	2,398

Propose a hypothesis to explain these data and outline how you would test your explanation.

248 In the swordtail fish (*Xiphophorus helleri*) a very colorful variety called the *montezuma* occurs, having a dazzling bright orange-red topcoat spotted with black like a leopard's skin.   When two *montezumas* are mated, about one-third of the progeny are throwbacks, or reversions, to *wild type* (an olive-green color finely mosaic for black and yellow) and the remainder are *montezumas*. It is impossible to develop a true-breeding strain of this variety.   Explain the reversion to *wild type* and why the *montezumas* fail to breed true.

249   In cattle and sheep, certain matings between normal parents produce only *normal* progeny, but when the daughters are mated with their fathers, the results are as shown.   For each character state:

Table 249

Animal	Classification of the progeny of father × daughter matings
cattle	98 *normal* : 12 *hairless* (lethal)
cattle	102 *normal* : 13 *amputated* (lethal)
sheep	29 *normal fetal muscle development* : 4 *fetal muscular degeneration* (lethal)

a (1) The genotypes of the father and mother, the daughters, and the progeny of the father × daughter matings.   Diagram each cross.
    (2) The type of lethal allele involved.
b If the allele produced its effects during early embryonic development, would it have been detected?   What conditions are necessary to detect this type of lethal?

250   In crosses between two *crested* ducks approximately three-quarters of the eggs hatch.   The embryos of the remaining quarter develop nearly to hatching and then die.   Of the ducks that do hatch about two-thirds are *crested*, and one-third are *crestless*.
a Explain these results genetically, stating the kind of lethal involved.
b What would you expect from the cross of a *crested* with a *crestless* duck?
c Is it possible to establish a true-breeding strain of *crested* ducks?   Explain.

251   *Blufrost* mink have light-blue underfur and silver guard hairs; *normal* mink (of wild origin) have dark fur.   In 1947 Moore and Keeler obtained the results shown.   Propose a hypothesis to explain these data and outline how you would test your explanation.

Table 251

Parents		Offspring	Average litter size
♀	♂		
blufrost × *normal*		78 *blufrost* : 80 *normal*	5.27
normal × *blufrost*		345 *blufrost* : 325 *normal*	5.11
blufrost × *blufrost*		19 *blufrost* : 10 *normal*	3.65

252 In bees, the wings of worker females may be *droopy* or *normal*; those of the males (drones) are never *droopy*. In hives where *droopy-winged* workers occur, a number of the eggs fail to hatch (within a hive all individuals are the offspring of a single mating between a queen and a male). Why do *droopy-winged* individuals occur among females only?

253 The first ancestor of the *platinum* fox was born on a silver fox ranch in Wisconsin. It had a silvery blue coat with symmetrical white spots. When *platinum* foxes are mated, the offspring usually appear in the ratio of 2 *platinum* : 1 *silver* : 1 *white* (white pups die during embryonic life or before weaning).
  a Could a fox breeder establish a true-breeding strain of *platinum* foxes?
  b The *platinum* coat is in great demand. What matings should the rancher make to obtain the maximum number of *platinum* foxes without getting any of the *white* pups?

254 The F_1's of a cross between two breeds of turkey were mated with each other. The females laid 642 eggs, of which 603 produced *normal* birds and 39 did not hatch. The embryos in these 39 eggs terminated development on about the seventh day of incubation. How would you explain these results?

255 Some of the seeds produced by a certain maple tree develop into *albino* plants. Explain genetically. Would it be possible to select offspring from this tree that would not produce any *albino* seedlings? If so, explain how it could be done.

256 In a certain breed of dogs, traceable back to a single bitch, about one-half the females of the line produced only half as many male as female offspring. What genetic explanation can you suggest for this?

257 *Multiple telangiectasia* (Snyder and Doan, 1944) and *hereditary sebaceous cysts* (Munro, 1937) are rare traits which never appear in a child unless also present in at least one of the parents. In extremely rare cases two *affected* individuals have married. In none of these marriages were the children all phenotypically like the parents. In some such marriages one or more of the offspring regularly died at about one year of age.
  a What is the most likely mode of inheritance of this kind of trait?
  b Describe the type of pedigree that would definitely substantiate your hypothesis.

258 In *Drosophila melanogaster* flies may have *red* (wild-type) or *plum* eye color and *normal* or *stubble* bristles (less than half normal length). The progeny of a mating between two *plum, stubble* flies gave the following results:

Plum, stubble   162
Plum, normal   80
Red, stubble   84
Red, normal   42

a Suggest a hypothesis to account for these results.

b Outline how you would test your explanation and indicate the results expected in your experiment.

259   *Achondroplasic*, or "bulldog," calves, dead at birth, have very short legs and vertebral columns, round short heads, protruding tongues, and numerous other anatomical abnormalities.   The Dexter breed is heterozygous for the incompletely dominant lethal allele responsible for the trait (Hutt, 1934). Wriedt and Mohr (Mohr, 1926) discovered a similar but less extreme condition, shown to be recessive, to be a characteristic of the Telemark cattle of Norway.   Punnett (1936) reported the results of crosses between Dexters and Telemarks made by Riches.   Only *Dexter types* and *normals* (long-legged) occurred in the 24 F_1 offspring.   The progeny from mating five *Dexter-like* F_1 females with a single *Dexter-like* F_1 male consisted of 5 *Dexter types*, 3 "*bulldog*," 2 *normal*, and 1 calf with the *achondroplasia* characteristic of the Telemark breed.   Using your own symbols, give a complete genetic explanation of these results.

260   Cole (1961) reported on genetic studies involving *paroxysm*, a new mutant in fowl, characterized by seizures, poor growth, and stilted gait.   Individuals with the condition hatch normally and do not show any of the symptoms until about the age of two weeks or more.   Death occurs in 14 weeks or less after birth.   A *normal* rooster mated with 10 *normal* unrelated hens produced 58 male and 52 female offspring; 3 males and 4 females died within a week of hatching, and by the fourteenth week a further 28 birds, all females, had died from *paroxysm*.

a Explain these results genetically.

b What proportion of the progeny would you expect to be heterozygous for *paroxysm*?   Explain.

c Outline a breeding program to eliminate the condition in one generation.

261   In Iowa in the 1930s Mr. Peck and his neighbor each bought a Poland China boar from a certain breeder to service their high-quality herds, in which no defects had occurred for many generations (Johnson, 1940).   Boar A, bought by Peck, was mated to all the females in both farmers' herds and produced *normal* pigs.   The next year the daughters of A, on both farms, were mated with boar B, and produced 207 *normal* : 25 *legless* pigs.

a What is the mode of inheritance of the *legless* trait?

b Diagram the pedigree to substantiate your explanation.

262   A poultryman buys a registered tom turkey and turns him out with the hens of his flock.   One-fourth of the eggs fail to hatch.   Later it is found that about two-thirds of the progeny are males.   Explain these results genetically; illustrate your explanation with a diagram of the cross.

263   Homozygous line A crossed with homozygous line B produced an F_1 all of which die as embryos.   Matings between these lines and homozygous line C produced fully viable F_1's.
a   What is the simplest genetic explanation for these results?
b   What results would you expect in the testcross when the F_1 resulted from A × C?   B × C?   Would this verify your hypothesis?

264   In common wheat (*Triticum aestivum*) the varieties Trumbull, Marquillo, and F.H. 27 are *normal* (in color and development).   The seedlings from crosses between Trumbull and Marquillo are *inviable* (the first leaves develop normally but soon wilt and die; no new leaves are formed, and death of the plant occurs about the sixth to eighth week after seeding).   However, when F.H. 27 is crossed with these two varieties, the F_1's of both crosses are *normal* (Caldwell and Compton, 1943).
a   Suggest an explanation for these results.
b   Would it be possible to test your hypothesis?   If so, explain how and outline the results expected within each cross.

265   In tomatoes (*Lycopersicon* [= *Lycopersicum*]) plants of the species *esculentum* and *hirsutum* are *normal*.   The F_1 of a cross between the two species is *inviable* when grown outdoors but *viable* when grown in the greenhouse (under special conditions) (Sawant, 1956).
a   Suggest an explanation for these results.
b   On the basis of your hypothesis what results would you expect in the F_2 grown:
(1) In the greenhouse?
(2) Outdoors?
c   How would you test your hypothesis?

266   In barley (*Hordeum vulgare*) two *normal* green plants when self-fertilized each produced some *albino* seedlings but when intercrossed produced only *normal* plants.
a   Explain these results genetically.
b   How would you proceed to test your explanation?

267   The pedigree shown is a modification of one presented by Eldridge et al. (1953) of *streaked hairlessness* in Holstein-Friesian cattle, a condition characterized by an abnormality of the hair coat in which narrow, irregular hairless streaks, running transversely around the trunk, appear.   What is the most likely mode of inheritance?   Give reasons for your decision.

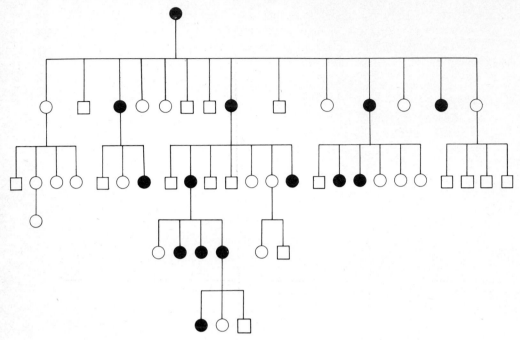

Figure 267

268   The pedigree shows the inheritance of *brachydactyly* (short fingers and toes).
All individuals examined were ten years of age or older.
a Is it due to an autosomal or sex-linked gene?   Is it dominant or recessive?

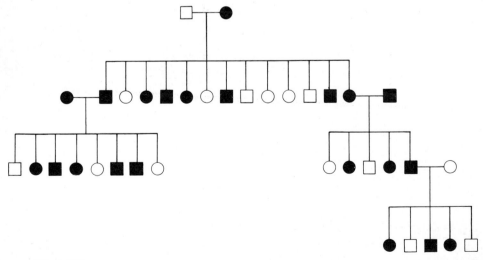

Figure 268

b Compare the progeny of marriages between *affected* individuals and those between *affected* and *unaffected* individuals.   What genetic ratios do you obtain?   Explain how this could possibly occur.

c What is the probable genotype of each *affected* individual?   Why?

269   *Multiple telangiectasia*, a rare disease in man, is characterized by an enlargement of fine blood vessels in the nose, tongue, lips, face, and fingers.   A typical pedigree for the transmission of the trait is shown in Fig. 269*A*.

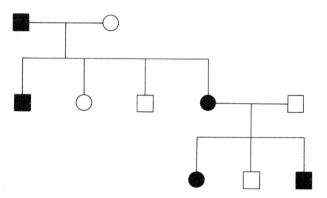

Figure 269*A*

In 1944 Snyder and Doan presented a pedigree showing the mating of two *affected* individuals (Fig. 269*B*).   The offspring of this mating had extremely enlarged fine blood vessels, which eventually ruptured and caused death within a few months after birth.   How is *multiple telangiectasia* inherited?

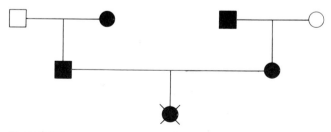

Figure 269*B*

REFERENCES

Abbott, U. K., L. W. Taylor, and H. Abplanalp (1960), *J. Hered.*, **51**:195.
Baur, E. (1908), *Z. Vererbungsl.*, **1**:124.
Blunn, C. T., and E. H. Hughes (1938), *J. Hered.*, **29**:203.
Caldwell, R. M., and L. E. Compton (1943), *J. Hered.*, **34**:65.
Cole, R. K. (1942), *J. Hered.*, **33**:83.
———— (1961), *J. Hered.*, **52**:47.
———— and R. M. Shackelford (1943), *Am. Nat.*, **77**:289.

Cuenot, L. (1905), *Arch. Zool. Exp. Gen.*, **3**:123.
Dunn, L. C., and D. Bennett (1971), *Evolution*, **25**:451.
Eldridge, F. E., and F. W. Atkeson (1953), *J. Hered.*, **44**:265.
Gluecksohn-Waelsch, S. (1953), *Q. Rev. Biol.*, **28**:115.
Gordon, M. (1941), *J. Hered.*, **32**:385.
Gruneberg, H. (1952), "The Genetics of the Mouse," 2d ed., pp. 1–650, M. Nyhoff, The Hague.
Hadorn, E. (1961), "Development Genetics and Lethal Factors," Wiley, New York.
Hunt, D. M. (1970), *Genet. Res.*, **15**:29.
Hutt, F. B. (1934), *Cornell Vet.*, **24**(1):1.
———— and G. P. Child (1934), *J. Hered.*, **25**:341.
Kidwell, J. T. (1962), *Can. J. Genet. Cytol.*, **4**:37.
Johnson, L. (1940), *J. Hered.*, **31**:239.
Landauer, W. (1956), *J. Genet.*, **54**:219.
Mohr, O. L. (1926), *Z. Indukt. Abstamm.-Vererbungsl.*, **41**:59.
———— and C. Wriedt (1919), *Carnegie Inst. Wash., D.C., Publ.* 295.
———— and ———— (1928), *J. Genet.*, **19**:315.
Moore, L., and C. E. Keeler (1947), *J. Hered.*, **38**:380.
Munro, T. A. (1937), *J. Genet.*, **35**:61.
Nachtsheim, H. (1950), *J. Hered.*, **41**:131.
Neel, J. V. (1949), *Science*, **110**:64.
Padgett, G. A., R. W. Leader, J. R. Gorham, and C. C. O'Mary (1964), *Genetics*, **49**:505.
Pontecorvo, G. (1942), *J. Genet.*, **43**:295.
Punnett, R. C. (1936), *J. Genet.*, **32**:65.
Sawant, A. (1956), *Evolution*, **10**:93.
Schröder, J. H. (1971), *Genetics*, **68**:35.
Snyder, L. H., and C. A. Doan (1944), *J. Lab. Clin. Med.*, **29**:1211.
Sturtevant, A. H. (1956), *Genetics*, **41**:118.
Welshons, W. J. (1971), *Genetics*, **68**:259.

8
Multiple Alleles

8
Multiple
Alleles

QUESTIONS

270 **a** Discuss the implications of multiple allelic series with regard to the Bateson and Punnett presence-absence theory.

b Discuss the significance of multiple allelism with regard to the genetic variability of a species.

271 State whether the following are true or false and why:

a It is possible to prove the existence of multiple allelism in an organism that reproduces by asexual means only.

b Blood groups can be used to exclude, but not to prove, paternity.

c Complementation as a test for allelism can be applied only to recessive mutations.

272 In the course of breeding experiments with *wild-type* (dark black to brown) mink you obtain two true-breeding lines; in one the coat color is *blue-gray* (platinum) and in the other it is *golden-palomino*.

a Outline the procedure you would follow in determining whether coat color is determined by multiple alleles or by genes at different loci.

b State briefly what results you would expect if coat color were controlled by:
(1) Multiple alleles (2) Genes at two loci

273 In what respects did the discovery of multiple alleles contribute to the concept of the gene?

PROBLEMS

274 The *striped* and *plain* types of larval color patterns in the silkworm were the first to be studied genetically. Many other patterns exist, one of which is *moricaud*, in which the full-grown larva is finely marked with dark lines and dots. The table shows the results of crosses between *striped*, *plain*, and *moricaud* in all possible combinations.

Table 274

Parents	F_1	F_2
striped × *plain*	*striped*	998 *striped* : 314 *plain*
striped × *moricaud*	*striped*	1,300 *striped* : 429 *moricaud*
moricaud × *plain*	*moricaud*	763 *moricaud* : 243 *plain*

a Determine the number of genes controlling these differences and the number of alleles involved.

b Describe the dominance relationships and state the different genotypes for each phenotype.

275 In rabbits, the blood agglutinogens H_1 and H_2 are genetically controlled: H_1 is produced by individuals in blood group H_1, and H_2 by those in group H_2.

Both agglutinogens are synthesized by H_1H_2 individuals, and neither is present in animals belonging to group O. Tests with anti-H_1 and anti-H_2 sera (containing antibodies agglutinating H_1 and H_2 antigens respectively) gave the results shown (Castle and Keeler, 1933).

Table 275

| | | Phenotypes | | | |
| | | Progeny | | | |
Mating	Parents	H_1H_2	H_1	H_2	O
1	$O \times O$				7
2	$O \times H$		9		12
3	$O \times H$			10	9
4	$H \times H$	17	13	15	20

a Present two hypotheses to explain these data, showing the genotypes for each of the phenotypes in each case.

b When H_1H_2 individuals were mated with those of group O, 140 H_1 and 162 H_2 individuals were obtained. Which hypothesis do these results support and why?

276   Johansson (1947) presented data on the inheritance of *silver, platinum,* and *white face* in the fox. The average litter size from each of the matings of *silver* × *silver, silver* × *platinum,* and *silver* × *white face* was 4.48; that from all other matings was reduced to 3.56. Give a complete genetic explanation of these results with reasons for your explanations.

Table 276

| | Number of progeny | | |
Parents	Silver	Platinum	White face
silver × platinum	4,157	3,842	
silver × white face	3,038		2,986
platinum × platinum	58	127	
white face × white face	267		483
platinum × white face	167	182	188

277   In the common red fox (*Vulpes vulpes* [= *fulva*]) two dominant autosomal mutations, *platinum* and *white-marked,* have been discovered. Both are phenotypically very similar, there being only a slight difference in the concentration of black pigment. Breeding data were collected by Cole and Shackelford (1943).

Table 277

Parents	Offspring	Average litter size*
platinum × silver	434 platinum : 418 silver	4.6
platinum × platinum	12 platinum : 8 silver	2.1
white-marked × silver	86 white-marked : 78 silver	4.6
white-marked × white-marked	183 white-marked : 79 silver	3.3
platinum × white-marked	7 platinum : 6 white-marked : 5 silver†	

* Average litter size for most crosses in foxes is 4.6.   † No average.

a Describe the genic basis of lethality in these crosses.

b Derive the theoretical outcome of the crosses described for the two hypotheses:

(1) Control by three alleles of one gene.

(2) Control by alleles of two genes.

Show whether both hypotheses suitably explain the data.

c Describe the dominance relationships involved in each hypothesis.

d What cross would you make to decide whether multiple alleles or two independently inherited genes are involved and why?   Show the types of progeny expected from this cross with each.

278 In mallard ducks the plumage pattern may be *restricted* (ventral wing surface white, dorsal spotted, neck ring white in males), *dusky* (ventral and dorsal wing surface colored, no neck ring), or *mallard* (*wild type*).   True-breeding lines exist for each type.   The table shows the results of crosses between the three lines (Jaap, 1934).

Table 278

Parents	F_1	Testcross
restricted × mallard	restricted	1 restricted : 1 mallard
restricted × dusky	restricted	1 restricted : 1 dusky
mallard × dusky	mallard	1 mallard : 1 dusky

a State the number of genes and alleles involved and the allelic relationships, giving reasons for your conclusions.

b These plumage-pattern alleles express themselves only in the presence of the dominant allele *C*, which permits the formation of colored pigments; in *cc* individuals the plumage is *white*, regardless of the genotype at the plumage-pattern locus.   A female with mallard pattern was mated to a *white* male.

She laid 16 eggs, 13 of which hatched and gave the following progeny: 6 *white* : 3 *restricted* : 2 *mallard* : 2 *dusky*. What were the probable genotypes of the parents?

279 The alleles Hb^S and Hb^A, controlling the synthesis of sickle-cell and normal hemoglobin, respectively, are codominant. Heterozygotes, $Hb^A Hb^S$, produce both types of hemoglobin, referred to as hemoglobins *A* and *S*. The pedigree shows the phenotypes of individuals for these two hemoglobins, and also a new one designated *G*.

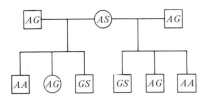

Figure 279

a Is it possible that the synthesis of the new hemoglobin is controlled by a new allele of the *Hb* gene?

b How would you prove that the gene for hemoglobin *G* is allelic to Hb^A and Hb^S?

280 In the nasturtium (*Tropaeolum majus*) the flowers may be *single*, *double*, or *superdouble*. *Superdoubles* are female-sterile; they originated in a *double-flowered* variety. In a series of crosses designed to determine the genetic relationships of the three traits, Eyster and Burpee (1936) obtained the following results:

True-breeding *single* × true-breeding *double* → F$_1$ *single* → F$_2$ 78 *single* : 27 *double*

True-breeding *double* × *superdouble* → 112 *superdouble* : 108 *double*

Homozygous *single* × *superdouble* → 8 *superdouble* : 7 *single*

True-breeding *doubles* × the 8 *superdouble* → 18 *superdouble* : 19 *single*

True-breeding *double* × the 7 *single* → 14 *double* : 16 *single*

a These results indicate that the genes for *single*, *double*, and *superdouble* flowers form a series of three alleles. Why?

b What is the order of dominance of the alleles?

281 Werret et al. (1959) obtained the tabulated results in the fowl starting with a *mutant* light-brown male, which appeared in the F$_1$ of a cross between a *brown* Leghorn male and a *silver* (white) Light Sussex female. Give a complete genetic explanation of these results, including the number of genes involved, their mode of interaction if any, allelic relationships, location of gene or genes, and genotypes and sex of the unsexed classes. Give reasons for your answers.

Table 281

Year	Parents ♂	Parents ♀	Sex	Silver	Brown	Semialbino
1956	mutant ×	silver (from true-breeding line)	♂	27	0	0
			♀	0	12	8
1957	mutant ×	semialbino (from previous mating)	♂	0	18	6
			♀	0	13	12
			Unsexed*	0	5	2
1958	semialbino ×	semialbino	♂	0	0	4
			♀	0	0	8
			Unsexed*	0	0	12
	semialbino ×	brown (from true-breeding line)	♂	0	18	0
			♀	0	0	23
			Unsexed*	0	0	3
1959	silver ×	semialbino	♂	1	0	0
			Unsexed*	0	1	0

* Sex not determined.

282 In mink (*Mustela vison*) animals in the wild may have *black, platinum* (blue-grey), or *sapphire* (very light blue) coats. One of many domesticated breeds has a *sapphire* coat. Three *black* animals (A, B, C), one *sapphire* one (D), and one *platinum* one (E) from the wild mated with animals from the domesticated breed produced the results shown. A × C matings produced progeny which were *apparently* true-breeding.

Table 282

Parents Wild	Domesticated	F_1
A *black* ×	*sapphire*	26, all *black*
B *black* ×	*sapphire*	14 *black* : 15 *sapphire*
C *black* ×	*sapphire*	21, all *platinum*
D *sapphire* ×	*sapphire*	21, all *sapphire*
E *platinum* ×	*sapphire*	16 *platinum* : 14 *sapphire*

a Outline the simplest hypothesis possible to account for these results. Using your own symbols, give the genotypes of the domesticated animals and the five wild ones.

b If your hypothesis is correct, what types of progeny and in what proportions would you expect them in the following matings?

A × E   B × E   C × D

283   The pedigree typifies the transmission characteristics of the *black, tortoise-shell* (black with red spots), and *red* coat traits in the guinea pig controlled by alleles of one gene (Wright, 1925, 1927). State whether two or three alleles are involved and why.

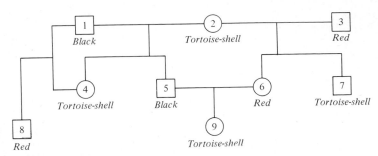

Figure 283

284   In *Drosophila melanogaster wild-type* (w.t.) flies have *dull-red* eyes because of the presence of both *red* and *brown* pigments, which are produced via specific but different pathways

$$\begin{array}{l} \to A \to B \to C \to red \\ \to D \to E \to F \to brown \end{array} \Big\rangle dull\text{-}red \; (wild\text{-}type)$$

Five recessive true-breeding *brown*-eyed mutant strains, designated *l, m, n, o, p,* arose independently after treating *wild-type* flies with the mutagen ethyl methanesulfonate (EMS). Phenotypes of F_1's, from crosses between the mutants in all possible combinations, were as shown.

Table 284

Male parent	Female parent				
	l	*m*	*n*	*o*	*p*
l	all	all	all	all	♀ w.t.
	brown	w.t.*	w.t.	brown	♂ brown
m	all	all	all	all	♀ w.t.
	w.t.	brown	brown	w.t.	♂ brown
n	all	all	all	all	♀ w.t.
	w.t.	brown	brown	w.t.	♂ brown
o	all	all	all	all	♀ w.t.
	brown	w.t.	w.t.	brown	♂ brown
p	all	all	all	all	♀ brown
	w.t.	w.t.	w.t.	w.t.	♂ brown

* w.t. = wild type.

a In how many different genes did the five mutations occur? Explain.

b What is the chromosome location (autosome or X) of each of the mutant genes? Explain.

c If F_1's from mutant l × mutant m were crossed, what results would you expect in the F_2?

285 In the house mouse (*Mus musculus*) coat color is genetically controlled. The results of matings involving *yellow*, *tan*, and *black* are shown. Give a complete genetic explanation of these results.

Table 285

Cross	Parents	Offspring
1	*black* × *yellow*	391 *black* : 402 *yellow*
2	*tan* × *yellow*	214 *tan* : 210 *yellow*
3	*yellow** × *tan*	184 *yellow* : 188 *blackish-tan*
4	*yellow†* × *black*	106 *yellow* : 110 *blackish-tan*
5	*yellow** × *yellow†*	218 *yellow* : 113 *blackish-tan*
6	*blackish-tan* × *tan*	132 *blackish-tan* : 138 *tan*
7	*black* × *blackish-tan*	141 *black* : 144 *blackish-tan*

* From cross 1.   † From cross 2.

286 In cats (*Felis domesticus*) *full*, *silver* (or *smoke*), and *Siamese* represent some of the known coat-color patterns. Progeny of matings between *full-colored* cats may be *silver* or *Siamese* as well as *full-colored*. Matings of *silver* cats produce progeny consisting of *silver* or *silver* and *Siamese* but never *full-colored*. Matings of *Siamese* cats give *Siamese* progenies only (Keeler and Cobb, 1933).

a Propose a hypothesis to explain these results and outline how you would test it.

b (1) A *full-colored* female mated with a *silver* male gives birth to a litter of two *full-colored* and three *silver* kittens. If the same two parents were mated again, could they produce some *Siamese* kittens? Explain.

(2) A litter of kittens contains all three types. What are the genotypes and phenotypes of the parents?

287 In the platyfish (*Platypoecilus* [= *Xiphophorus*] *maculatus*) the various tail patterns illustrated are genetically controlled. The result of matings involving *one-spot*, *crescent*, and *unspotted* are given in Table 287.

a Outline a hypothesis to explain these results and state which mating influences your decision regarding the mode of inheritance and how it does so.

b Describe the dominance relationships and show all possible genotypes for each of the four phenotypes.

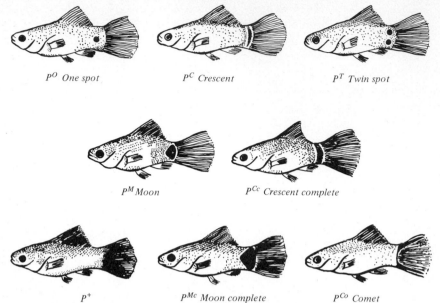

P^O *One spot* P^C *Crescent* P^T *Twin spot*

P^M *Moon* P^{Cc} *Crescent complete*

P^+ P^{Mc} *Moon complete* P^{Co} *Comet*

Figure 287   The eight tail patterns in wild *Platypoecilus maculatus*. [*From M. Gordon, Speciation in Fishes, Adv. Genet.,* **1** : 96 (1947).]

Whenever fish with either of the remaining five tail patterns are mated with *unspotted*, the results are similar to those in matings 2 and 4. When they are mated to *one-spot* or to *moon*, results similar to those in mating 6 between *one-spot* and *crescent* are obtained.

c Are the various tail patterns controlled by a series of multiple alleles or not?

d Outline the allelic relationships and give the basis for your outline.

Table 287

Cross	Parents	Offspring
1	*unspotted* × *unspotted*	all *unspotted*
2	*unspotted* × *one-spot*	all *one-spot* or 1 *unspotted* : 1 *one-spot*
3	*one-spot* × *one-spot*	all *one-spot* or 3 *one-spot* : 1 *unspotted*
4	*unspotted* × *crescent*	all *unspotted* : 1 *crescent*
5	*crescent* × *crescent*	all *crescent* or 3 *crescent* : 1 *unspotted*
6	*one-spot* × *crescent*	all *one-spot–crescent* or 1 *one-spot–crescent* : 1 *one-spot* or 1 *one-spot–crescent* : 1 *crescent* or 1 *one-spot–crescent* : 1 *one-spot* : 1 *crescent* : 1 *unspotted*

288 In the house mouse (*Mus musculus*) a series of multiple alleles affect the expression of coat color (Dunn, 1928; Grüneberg, 1952). Among these are A^Y (*yellow*), A^L (*agouti light belly*), A (*agouti*), a^+ (*black and tan*), and a (*black*). Aa^+ mice are *agouti with light belly*; otherwise dominance is complete in the order given above. Another gene controls pigment formation, the dominant allele C allowing pigment formation and the recessive c prohibiting it.

a What are the phenotypes of the offspring of the following matings?

(1) $a^+a \times A^LA$ (2) $Aa^+ \times A^Ya$

(3) $A^YA \times A^La$ (4) $AA \times A^LA^Y$

b Determine the genotypes of the parents in Table 288.

Table 288

Parents	Offspring
1. *agouti × agouti light belly*	14 *black* : 17 *agouti* : 31 *agouti light belly*
2. *yellow × agouti*	62 *agouti light belly* : 57 *yellow*
3. *albino × agouti*	F_1 all *agouti light belly*
	F_2 84 *albino* : 61 *agouti* : 122 *agouti light belly* : 64 *black and tan*
4. *agouti light belly ×* *agouti light belly*	28 *agouti* : 52 *agouti light belly* : 25 *black and tan*

289 In matings of *Drosophila melanogaster* with *wild-type* (dull-red) eyes to lines true-breeding for other eye colors the offspring are always *wild type*. When homozygous *cinnabar* flies are mated to homozygous *sepia*, the progeny are all *wild type*. Are *sepia* and *cinnabar* controlled by alleles of the same gene? Diagram the crosses described.

290 In 1910 Dungern and Hirszfeld suggested that the *ABO* blood groups were controlled by two independent pairs of alleles, viz. *Aa* and *Bb* with *aabb* individuals being of type *O*, *A—bb*'s of type *A*, *aaB—*'s of type *B*, and *A—B—*'s of type *AB*. Possible phenotypes of the offspring of specific types of matings according to the Dungern and Hirszfeld hypothesis are given in Table 290*A*.

Table 290*A*

Other parent	One parent			
	AB	A	B	O
	Offspring			
AB	AB, A, B, O			
A	AB, A, B, O	A, O		
B	AB, A, B, O	AB, A, B, O	B, O	
O	AB, A, B, O	A, O	B, O	O

a Show how the possibilities for the $AB \times AB$ mating in this table are derived.

b Table 290B, prepared from extensive data, shows the types of children found among the offspring of the various matings. Compare the two tables to show whether the hypothesis is valid.

Table 290B

Blood groups of parents	Blood groups which may occur in children	Blood groups which do not occur in children
$O \times O$	O	A, B, AB
$O \times A$	O, A	B, AB
$A \times A$	O, A	B, AB
$O \times B$	O, B	A, AB
$B \times B$	O, B	A, AB
$A \times B$	O, A, B, AB	
$O \times AB$	A, B	O, AB
$A \times AB$	A, B, AB	O
$B \times AB$	A, B, AB	O
$AB \times AB$	A, B, AB	O

291 Suppose you have extensive data from many human families in which you can study two generations of progeny from parents of known genotypes. List (1) the parental genotypes, (2) the expected first-generation phenotypes, and (3) the expected F_2 or testcross ratios for those matings necessary to prove that the A, B, AB, and O blood groups are controlled by a series of multiple alleles.

292 In each of two separate true-breeding *wild-type* (olive-green) populations of swordtail fish a mutant *golden* individual appears. When these mutants are mated with *wild type*, the F_1's are *wild type* and the F_2's segregate 3 *wild type* : 1 *golden*. Outline the procedure to determine whether the mutant genes are allelic or not. Show the phenotypes expected when the genes are allelic and when they are not.

293 a How many different diploid genotypes are possible when a gene exists in five allelic forms?

b Six alleles (B^1, B^2, B^3, B^4, B^5, and B^6) occur at a certain gene locus. How many genotypes are possible in a population in which all six alleles occur?

294 The dorsal pigmentary pattern of the common leopard frog (*Rana pipiens*) is genetically controlled. It may be *kandiyohi* (mottling between the dorsal spots), *burnsi* (complete absence of dorsal spots), *wild type* (spaces between dorsal spots are white), or *mottled burnsi*. Table 294A shows the results of matings among these types (Volpe, 1960).

Table 294A

Cross	Parents	F₁	F₂
1	*kandiyohi* × *wild type*	*kandiyohi*	3 *kandiyohi* : 1 *wild type*
2	*burnsi* × *wild type*	*burnsi*	3 *burnsi* : 1 *wild type*
3	F₁ *kandiyohi* × F₁ *burnsi*	1 *mottled burnsi* : 1 *kandiyohi* :	
		1 *burnsi* : 1 *wild type*	

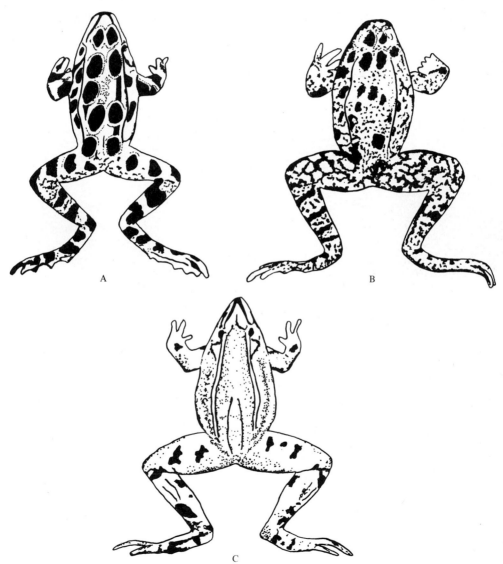

Figure 294   A. *Wild type;* B. *kandiyohi;* C. *burnsi.*   [*From E. P. Volpe, J. Hered.,* **51** : *150 (1960).*]

a What conclusions can you draw from the results of the first two crosses regarding the inheritance of pigmentation pattern and dominance relationships?

b The results of the third cross do not permit a distinction between multiple allelism and independent assortment of two allele pairs. Using your own symbols, show how these results can be explained by each of these mechanisms.

c What mating would you make to distinguish between these two alternatives? Show the results you would expect for the two mechanisms.

d Table 294B shows the results obtained by Volpe in matings of *mottled burnsis* with *wild types*. Which hypothesis do these results support? Perform a chi-square test to substantiate your answer.

Table 294B

Parents		Phenotype of offspring			
♀	♂	Mottled burnsi	Kandiyohi	Burnsi	Wild type
frog 3A × wild-type		14	18	24	20
wild-type × frog 3B		16	19	22	25

295 In mink the *wild type* is dark in color; two mutant color patterns called *black cross* and *royal silver* have originated in recent times. The patterns are similar but not identical. The following breeding data were recorded by Shackelford (1950):

1 *Wild type* (true-breeding) × *black cross* gives 1 *black cross* : 1 *wild type*.

2 *Wild type* (true-breeding) × *royal silver* gives 1 *royal silver* : 1 *wild type*.

3 When *black cross* mink from the first cross were crossed with *royal-silver* from the second cross, the kits appeared in the ratio 2 *black cross* : 1 *royal silver* : 1 *wild type*.

a These results do not permit a distinction between (1) multiple allelism and (2) independent assortment of two allele pairs. Using your own symbols, show how these results are explained by each of these two mechanisms.

Some of the *black-cross* progeny of the third cross, mated with *wild type*, produced *black-cross* and *wild-type* kits only, whereas two other *black-cross* from the same cross when intermated produced *royal-silver* and *black-cross* offspring only.

b Do these results support either of the hypotheses outlined above? Explain.

c The two *black-cross* offspring that produced *royal-silver* kits when intermated were also mated to *wild-type* mink. A small number of offspring were produced, consisting of *royal-silver* and *black-cross* kits only.

(1) Which hypothesis do these results support?

(2) What would constitute critical evidence that *royal silver* and *black cross* are controlled by different alleles of the same gene?

296 **a** How many crosses are necessary to prove that three alternative traits of a character, e.g., *full-color*, *himalayan*, and *albino* coat in rabbits, are controlled by multiple alleles?

b What must the results of any cross show to prove this?

c Suppose another form of the character appeared. How many crosses would be necessary to prove that it was controlled by another allele in this series?

297 In collies, coat pattern may be *tricolor* (black-and-tan plus a few white markings); *blue-merle* (a dilute color with black spots); *white* (an extremely dilute blue-merle, indistinguishable from the *white* phenotype caused by other genes, except that these dogs are usually blind and deaf); *sable* (a sooty yellow which may be dark or light); and *sable-merle* (light sable blotched with dark sable spots; one or both eyes may be blue or contain blue sectors). The data given are modified from Mitchell (1935).

Table 297

Cross	Parents	Offspring
1	*blue-merle* × *blue-merle*	20 *white* : 29 *blue-merle* : 20 *tricolor*
2	*dark-sable* × *dark-sable*	7 *light-sable* : 18 *dark-sable* : 9 *tricolor*
3	*dark-sable* × *blue-merle*	6 *sable-merle* : 7 *blue-merle* : 5 *dark-sable* : 6 *tricolor*
4	*light-sable* × *sable-merle*	7 *light-sable* : 9 *sable-merle*

a These data lend themselves to interpretations involving either multiple alleles or more than one gene. Illustrate, using your own symbols.

b A *white* crossed with a *light-sable* had four *sable-merle* offspring. Mated with *tricolor*, these offspring produced 3 *sable-merles*, 4 *sables*, 5 *blue-merles*, and 4 *tricolors*. Using these data, present arguments tending to eliminate one of the interpretations.

c A *white* offspring from cross 1 mated with a *white* from a cross of two *tricolors* produced only *blue-merles*. Mated with the second *white* parent, these *blue-merles* produced 8 *blue-merles*, 18 *whites*, and 7 *tricolors*. Are the *whites* controlled by alleles of the same or different genes? Explain. With this information give the complete genotype of all the coat-color types that occur.

298 In cattle, as in other mammals, the presence of antigen is dominant to its absence. Because of this, blood-group antigens have been very useful in solving cases of doubtful parentage. The table (Irwin, 1956) shows the blood types (antigens) of a cow, her calf, and two bulls, either of which, according to the owners, could have been the father of the calf. Which bull can be excluded and why?

Table 298

	Blood type																					
Individual	A	B	C_1	F	H	J	O	R	S	V	W	X_2	Y_1	Y_2	Z	A^1	E_3^1	H^1	I^1	J	K^1	L^1
Mother (S_3)	+	−	−	+	+	+	+	−	−	+	−	+	−	−	−	−	−	+	−	+	+	+
Calf (S_4)	+	−	−	−	+	−	+	+	−	+	+	+	−	+	+	+	−	−	+	−	+	+	+
Possible father 1	+	+	−	+	−	+	+	−	−	−	−	+	−	−	+	−	−	−	+	−	−	+
Possible father 2	+	+	+	+	−	+	+	+	+	+	+	+	+	+	+	+	+	+	+	−	+	−

299 Proxy motherhood is frequently used in cattle breeding. The table shows the blood types of a bull, a cow bred only to this bull, and the calf she gave birth to. State whether the calf resulted from the transplantation of a fertilized egg from another animal into this cow and why.

Table 299

	Blood type														
Individual	A	B	O_2	Y_1	A_1	P	Q	E_3^1	D	E_1^1	F	J	W_3	S	R
Bull	−	+	+	+	+	−	−	+	−	−	+	+	−	−	−
Cow	−	−	−	−	−	+	+	−	+	+	+	−	−	−	+
Offspring	+	+	+	+	+	−	−	+	−	+	+	+	+	+	−

300 In some studies of human twins a reliable means of classifying twin pairs as monozygotic or dizygotic is essential. In many studies, this reliability is accomplished by comparing each twin pair for many genetic characters, among which the blood groups play a relatively important role.

a A pair of twins are blood-typed as shown. Classify each of these twin pairs for monozygosity vs. dizygosity.

Table 300

	Anti $-A$	Anti $-B$	Anti $-N^S$	Anti $-M^s$	Anti $-M^S$	Anti $-N^s$	Anti $-Rh^+$
Twin 1	+	+	+	−	+	−	+
Twin 2	+	−	+	−	−	−	+
Twin 1	−	−	−	+	+	−	−
Twin 2	−	−	−	+	+	−	−

b Most geneticists would insist on blood-typing the parents as well.
(1) Why?
(2) Which of the twin pairs above would necessitate this? Would the parents have to be tested for the *MNS* blood types?
(3) Both sets of parents were tested and found to be of the following blood type:

Parents of pair 1   *A $M^S N^S$ Rh$^+$ B N^S Rh$^+$*
Parents of pair 2   *O M^s Rh$^-$ O M^S Rh$^-$*

Are the twins dizygotic or monozygotic? Explain why your answer for one twin pair in (a) might have been in error and list the ways in which the validity of assessment can be increased.

301 If you were on a jury before which the case outlined below was being tried, what would you conclude? Family X claims that baby C, given to them at the hospital, does not belong to them but to family Y, and baby D, in possession of family Y, is really theirs. It is alleged that the two babies, both boys, were accidentally exchanged soon after birth. Family Y denies such an exchange was made. Blood-group determinations show:

X mother   *AB*   Y mother   *A*   C baby   *A*
X father   *O*   Y father   *O*   D baby   *O*

302 a A person of blood group *A* is injured and must be given a blood transfusion. The attending physician is provided with blood from all four groups: *O, A, B,* and *AB*. Which should be accepted and which rejected, and why?

b In the choice of donors for blood transfusion, a patient's brother or sister is frequently preferred.
(1) Could a brother's or sister's blood be transfused without typing if the parents were as follows? Explain your answer in each case.
(*a*) Both of group *O*.
(*b*) One of group *A* the other of group *O*.
(*c*) Both of group *AB*.
(2) For those progenies in which typing is required state which genotypes among the sibs could be donors.

303 a To what blood group or groups can the father belong in the families.

Table 303

Family	Children	Mother
1	*A*	*O*
2	*AB*	*B*
3	*O, B*	*B*
4	*O, A, B, AB*	*A*

b To which blood groups can children from the following matings belong?

(1) $O \times A$ (2) $A \times AB$ (3) $A \times B$

(4) $B \times B$ (5) $AB \times O$

304 The father of a certain family belongs to blood group AB, and the mother to blood group O. They have four children, one belonging to group AB, one to A, one to B, and one to O. One of these children is adopted, and another is a child from an earlier marriage of the mother. State which is the adopted child and which is the child from an earlier marriage and why.

305 **a** Where one parent belongs to blood group AB and the other to group O, how many times in families of 3 children would you expect 1 child of group A and the other 2 children of group B?

b A father of group B and a mother of group O have a child of group O. What are the chances that their next child will be O? B? A? AB?

c In a family of 4 children the first is of blood group A, the second O, the third AB, and the fourth B.

(1) What are the genotypes and phenotypes of the parents?

(2) What is the probability of such a couple having 4 children, one in each of the four blood groups:

(*a*) In any order?

(*b*) In the order given?

306 An *Rh-positive* woman, one of whose parents was *Rh-negative*, marries an *Rh-negative* man. Is there any possibility of their having *Rh-negative* children?

307 **a** A woman learns that her fiance was the sixth child in a family in which the third and fifth children were born with *erythroblastosis fetalis*. She is told that it is impossible for her to have an *erythroblastotic* child and asks whether this is true or not. What should your answer be and why?

b A woman bears a child with *erythroblastosis fetalis* at her second delivery. She has never had a blood transfusion. On the basis of this information, what are the expected *Rh* blood types of the woman, her husband, and both children?

308 **a** A woman sues a man for the support of her child. She has type B blood, her child has type O blood, and the man has type A blood. Could the man be the father of this child? Explain.

b Further tests show that both the man and woman are *Rh-negative* while the child is *Rh-positive*. Show what bearing this information has on the case.

c Columns 1 and 2 show the blood type of a mother and child. Answer the questions in columns 3 and 4.

Table 308

Mother (1)	Child (2)	The father could be? (3)	Is it impossible for the father to be? (4)
A	AB	A, B, AB, O	A, B, AB, O
Rh⁻	Rh⁺	Rh⁺, Rh⁻	Rh⁺, Rh⁻
M	MN	M, MN, N	M, MN, N

309   A woman telephones the police that her baby has disappeared from its carriage outside a department store.   The police later ask her to go to a foundling home to identify a baby left there.   She claims the child is hers. Evaluate her claim on the basis of the following blood-typing results:

Table 309

Mother	A	MN	Rh-negative
Father	O	N	Rh-positive
Foundling	A	M	Rh-positive

310   The alleles C, c^k, c^d, c^r, c^a in the *albino* series of the guinea pig have the effects shown in the table on the relative amounts of melanin pigmentation in the coat (Wright, 1949, 1959).   Classify each allele as normal (fully

Table 310

Genotype	Percent melanin pigment	Phenotype
$C—$	100	Full color
$c^k c^k$	88	Intermediate color
$c^k c^d$	65	
$c^k c^r$	54	
$c^k c^a$	36	
$c^d c^d$	31	
$c^d c^r$	19	
$c^d c^a$	14	
$c^r c^r$	12	
$c^r c^a$	3	
$c^a c^a$	0	White

morphic), hypomorphic, or amorphic.   Explain your classification and arrange the alleles in order of decreasing activity with respect to phenotypic effect.

311 In *Drosophila* the phenotypic effects shown are noted when homozygous and heterozygous individuals with varying doses of the alleles *Hw* (*hairy wing*) and *Hw*⁺ (*wild type, hairless*) are studied (Muller, 1932).

Table 311

Genotype	Degree of hairiness
HwHw	Standard number of hairs
HwHw⁺	Half as hairy as *HwHw*
Hw/Y	Half as hairy as *HwHw*
Hw⁺*Hw*⁺	No hairs (*wild type*)
HwHw + segment of chromosome carrying *Hw*⁺	Standard number of hairs
HwHw⁺ + segment of chromosome carrying *Hw*⁺	Half as hairy as *HwHw*⁺
Hw⁺*Hw*⁺ + segment carrying *Hw*	Half as hairy as *HwHw*

a Classify the alleles as neomorphic, hypomorphic, hypermorphic, or amorphic with respect to the character studied and explain your classification.

b Is the amorph an allele or a deficiency of *Hw*? Explain.

312 A mutant allele may perform no function, or it may act in a direction opposite to that of the standard (normal) allele, or it may act in the same direction as the standard allele but less effectively. To determine which of these alternatives is correct, one studies individuals with the chromosome or chromosome segment bearing the allele singly, doubly, in triplicate, or quadruplicate. The effects shown were noted by Schultz (1935) in *Drosophila* with different doses of the fourth chromosome recessive allele *shaven, sv,* which reduces the number of bristles on the body in comparison with *wild type*, controlled by its dominant allele *Sv*. State whether *sv* is a hypomorph, an amorph, or an antimorph and why.

Table 312

Genotype	Phenotype
sv/—	*Very few bristles*
sv/sv	*Few bristles*
sv/sv/sv	*Almost wild-type number of bristles*
sv/Sv	*Wild type*
Sv/Sv	*Wild type*
sv/sv/sv/sv	*More bristles than wild type*

313   In *Drosophila* a series of multiple alleles produces a graded series of eye colors from *red* to *white*: *red*, *W*; *blood*, *w^{bl}*; *coral*, *w^{co}*; *apricot*, *w^a*; *buff*, *w^{bf}*; and *white, w*.   Suppose it is possible to obtain viable individuals with different doses (0, 1, 2, etc.) of any of the alleles.   Outline an experiment to determine whether the recessive alleles are functionless (amorphs), acting against the standard (antimorphs), or acting in the direction of the normal allele, *W*, but less effectively (hypomorphs; leaky).   See Crow (1964) for discussion of allele activity.

314   The human hemoglobin molecule consists of two identical halves, each containing an α and a β chain.   In addition to *normal* hemoglobin (*Hb-A*) abnormal ones occur: *Hb-S* (found in sickle-cell anemics), *Hb-C* (in sickle-cell C disease), *Hb* HO-2 (in Hopkins-2 hemoglobinemics), and others.   The first two abnormal hemoglobins are defective in the α chain, the latter in the β chain.   From the pedigrees, which are typical for these traits, determine whether the defects in the different chains, and therefore *Hb-S* and *Hb-A* on the one hand and HO-2 on the other, are controlled by alleles of the same or different genes.

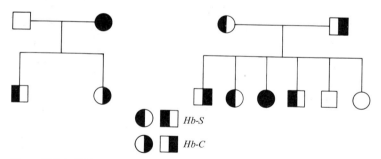

Figure 314 A   [*After Smith and Krevans, Bull. Johns Hopkins Hosp.,* **104**: 22 (1959).]

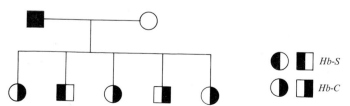

Figure 314B   [*After Hays and Engle, Ann. Intern. Med.,* **43**: 415 (1955).]

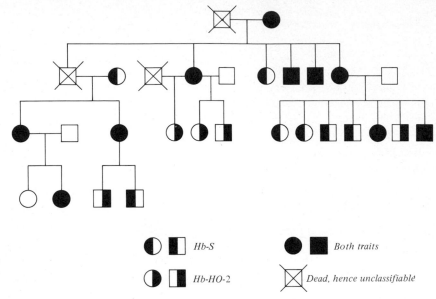

◑ ◧	*Hb-S*	● ■	*Both traits*
◐ ◨	*Hb-HO-2*	⊠	*Dead, hence unclassifiable*

Figure 314C

REFERENCES

Azen, E. A., O. Smithies, and O. Hiller (1969), *Biochem. Genet.*, **3**:215.
Berg, K., and A. G. Bearn (1968), *Annu. Rev. Genet.*, **2**:341.
Bowman, J. E., P. E. Carson, and H. Frischer (1969), *Hum. Hered.*, **19**:25.
Buettner-Janusch, J. (1970), *Annu. Rev. Genet.*, **4**:47.
Castle, W. E. (1954), *Genetics*, **39**:35.
—— and C. E. Keeler (1933), *Proc. Natl. Acad. Sci.*, **19**:92.
Cold Spring Harbor Symp. Quant. Biol. (1967), **32**.
Cole, L. J., and R. M. Shackelford (1943), *Am. Nat.*, **77**:289.
Cole, R. K., and T. K. Jeffers (1963), *Nature*, **200**:1238.
Crow, J. F. (1964), "Genetics Notes," 5th ed., Burgess, Minneapolis.
Cuenot, L. (1911), *Arch. Zool. Exp. Gen.*, **8**:40.
David, C. S., M. L. Kaeberle, and A. W. Nordskog (1969), *Biochem. Genet.*, **3**:197.
Davison, J. (1961), *J. Hered.*, **52**:301.
De Winton, D., and J. B. S. Haldane (1933), *J. Genet.*, **27**:1.
Dungern, E. V., and L. Hirszfeld (1910), *Z. Immunitaetsforsch.*, **6**:284.
Dunn, L. C. (1928), *Proc. Natl. Acad. Sci.*, **14**:816.
Edwards, R. H. (1971), *J. Hered.*, **62**:239.
Emerson, R. A. (1911), *Rep. Nebr. Agric. Exp. Stn.*, **24**:58.
Eyster, W. H., and D. Burpee (1936), *J. Hered.*, **27**:51.
Fisher, R. A. (1947), *Am. Sci.*, **35**:95.
Giblett, E. R. (1969), "Genetic Markers in Human Blood," Blackwell, Oxford.
Gordon, M. (1947), *Adv. Genet.*, **1**:95.
Grüneberg, H. (1952), *Bibliog. Genet.*, **15**:1.
Harris, H. (1969), *Br. Med. Bull.*, **25**(1):5.
—— (1971), "The Principles of Human Biochemical Genetics," American Elsevier, New York.
Hays, E. F., and R. L. Engle (1955), *Ann. Intern. Med.*, **43**:412.

Hrubant, H. E. (1955), *Am. Nat.*, **89**:223.

Hunt, J. A., and V. M. Ingram (1958), *Nature*, **181**:1062.

Hutt, F. B. (1949), "Genetics of the Fowl," McGraw-Hill, New York.

Irwin, M. R. (1956), *7th Int. Congr. Anim. Husb. Madrid*, pp. 7–32.

Itano, H. A., and E. A. Robinson (1960), *Proc. Natl. Acad. Sci.*, **46**:1492.

Jaap, R. G. (1934), *Genetics*, **19**:310.

Johansson, I. (1947), *Hereditas*, **33**:152.

Keeler, C. E., and V. Cobb (1933), *J. Hered.*, **24**:181.

Komai, T. (1956), *Adv. Genet.*, **8**:155.

Landsteiner, K., and P. Levine (1928), *J. Exp. Med.*, **48**:731.

—— and A. S. Wiener (1940), *Proc. Soc. Exp. Biol. (N.Y.)*, **43**:223.

—— and —— (1941), *J. Exp. Med.*, **74**:309.

Levine, P. (1954), *Adv. Genet.*, **6**:184.

Little, C. C. (1958), *Q. Rev. Biol.*, **33**:103.

Lush, I. E. (1964), *Genet. Res.*, **5**:39.

McKinnell, R. G. (1964), *Genetics*, **49**:895.

Mitchell, A. L. (1935), *J. Hered.*, **26**:425.

Mitchell, M. B., and H. K. Mitchell (1952), *Proc. Natl. Acad. Sci.*, **38**:205.

Møller, D. (1970), *J. Fish. Res. Can.*, **27**:1617.

Morgan, T. H., C. B. Bridges, and J. Schultz (1925), *Bibliog. Genet.*, **2**:1.

Morgan, W. T. J., and W. M. Watkins (1969), *Br. Med. Bull.*, **25**(1):30.

Muller, H. J. (1932), *Proc. 6th Int. Congr. Genet. Ithaca*, **1**:213.

Race, R. R., and R. Sanger (1968), "Blood Groups in Man," 5th ed., Blackwell, Oxford.

Sanger, R., and R. R. Race (1951), *Am. J. Hum. Genet.*, **3**:332.

Scandalios, J. G. (1969), *Biochem. Genet.*, **3**:37.

Searle, A. G. (1968), "Comparative Genetics of Coat Colour in Mammals," Logos, London.

Schnell, G. D., and J. H. Stipling (1966), in E. L. Green (ed.), "Biology of the Laboratory Mouse," 2d ed., pp. 457–491, McGraw-Hill, New York.

Schultz, J. (1935), *Am. Nat.*, **69**:30.

—— and C. B. Bridges (1932), *Am. Nat.*, **66**:323.

Shackelford, R. M. (1950), "Genetics of the Ranch Mink," Pilsbury, New York.

Smith, E. W., and J. R. Krevans (1959), *Bull. Johns Hopkins Hosp.*, **104**:17.

—— and J. V. Norbert (1958), *Bull. Johns Hopkins Hosp.*, **102**:38.

Stern, C. (1943), *Genetics*, **28**:441.

—— and E. W. Schaeffer (1943*a*), *Proc. Natl. Acad. Sci.*, **29**:351.

—— and —— (1943*b*), *Proc. Natl. Acad. Sci.*, **29**:361.

Stormont, C. (1955), *Am. Nat.*, **89**:105.

—— (1962), *Ann. N.Y. Acad. Sci.*, **97**:251.

Sturtevant, A. H. (1913), *Am. Nat.*, **47**:234.

Tan, C. C. (1946), *Genetics*, **31**:195.

Tanaka, Y. (1953), *Adv. Genet.*, **5**:240.

Volpe, E. P. (1960), *J. Hered.*, **51**:151.

Wagner, R. P., and H. K. Mitchell (1964), "Genetics and Metabolism," 2d ed., Wiley, New York.

Watkins, W. M. (1966), *Science*, **152**:172.

Werret, W. F., A. J. Candy, J. O. L. King, and P. M. Sheppard (1959), *Nature*, **184**:480.

Wiener, A. S., and I. B. Wexler (1958), "Heredity of the Blood Groups," Grune & Stratton, New York.

Wright, S. (1925), *Genetics*, **10**:223.

—— (1927), *Genetics*, **12**:530.

—— (1949), *Genetics*, **34**:245.

—— (1959), *Genetics*, **44**:1001.

9
Polygenic Inheritance

Note that arithmetic means additive and geometric means multiplicative; multiple genes are the same as polygenes. At polygene loci, alleles are of two kinds: contributors, which may act additively or multiplicatively, and neutrals, which have no effect.

QUESTIONS

315 Describe the kinds of observations that would lead you to suspect that a certain human character was controlled by polygenes.

316 **a** Are quantitative characters restricted to sexually reproducing organisms? Explain.
  b Define the multiple-gene hypothesis, evaluate its practical and theoretical significance, and describe some of its limitations.
  c Describe briefly the evidence supporting the multiple-gene hypothesis.
  d Discuss the statement: No new principles of genetics have originated from the study of quantitative characters.
  e Why is it more difficult to study the inheritance of quantitative characters such as size, weight, and intelligence than qualitative ones such as *ABO* and Rh blood antigens?

317 State the proportions of the F_2 that should resemble one parent when there is no dominance for the following numbers of genes: 2, 3, 5, *n*. What proportion resembles one parent or the other with *n* genes? Is this true of both polygenic and qualitative characters?

318 Occasionally a bimodal or a multimodal frequency distribution curve is obtained when an F_2 progeny is studied for a quantitative character. What genetic explanation can be given for this?

319 Two true-breeding lines of corn are $4\frac{1}{2}$ and $5\frac{1}{2}$ feet tall, respectively. In the F_2 of a cross between these lines, raised under the same kind of environmental conditions as the parents and the F_1, individuals appear which are 3, $3\frac{1}{2}$, $6\frac{1}{2}$, and 7 feet tall. Explain.

320 Mather (1949) championed the idea that euchromatin contains major genes controlling qualitative characters and heterochromatin contains polygenes controlling quantitative characters.
  a State whether or not you agree with this and why.
  b State whether or not you consider the hypothesis merits further investigation and why.

321 **a** Distinguish between the terms *polygene* and *modifying gene*. What have the two kinds of genes in common?
  b Describe an example of modifying-gene action showing the effects of these genes on the expression of the character.

322   In man *normal* vs. *brachydactyly* (short fingers) is genetically controlled (Farabee, 1905; Mohr and Wriedt, 1919). If individuals are classified for *normal* vs. *brachydactylous*, we find that the character is monogenically controlled, with the allele for *brachydactyly* incompletely dominant to the one for *normal*. *Brachydactylous* individuals, however, show a range in the length of the index finger from extremely short to only slightly short. What might explain this variation in the expression of *brachydactyly*?

PROBLEMS

323   You are given two mature plants of the bean (*Phaseolus vulgaris*), one of which is *1 foot* tall, the other *6 feet* tall. The species can be crossed as well as self-fertilized. You are asked to determine whether the height variation in the two plants is due to environmental factors or is genetically caused. For this experiment you are provided with a growth chamber with a single set of environmental conditions (a certain constant amount of light, temperature, etc.).

   a How would you proceed to determine whether the height differences between the two plants and among their progeny are genetically or environmentally caused?

   b If due to the genotype, how would you proceed to determine the number of allele pairs that may be involved in controlling the expression of the character?

324   In a cross between a *large* and a *small* strain of rabbits the F_1's are phenotypically uniform. On the average, their size is intermediate (midway) between that of the two parental strains. Among 2,025 F_2 individuals, 8 are the same as the *small*, and 7 are the same as the *large* strain.

   a How many polygenes are probably involved in the cross?

   b What does each contributing allele do?

325   In poultry, the average weight of F_1's from a cross between a *large* and a *small* breed closely approximates the arithmetic mean of the average weights for the two breeds. Among 2,000 F_2's from intercrosses of the F_1's, no individuals as *small* as the *small* breed or as *large* as the *large* breed were obtained. What conclusions can be drawn concerning the minimum number of polygenes involved?

326   The Flemish breed of rabbit has a mean weight of 3,600 g, while the Himalayan race averages 1,875 g. In a cross between these two races Castle (1931) found the standard deviation of the F_1 to be ± 162 g and the standard deviation of the F_2 to be ± 230 g. Determine the probable number of genes involved in the cross.

327   According to Davenport (1913), the difference in skin color between Negroes and whites is caused by two pairs of alleles with equal and additive effects.

On this basis five grades of pigment concentration are recognized as phenotypes in the progenies of marriages between Negroes and whites: *white, lightly pigmented, moderately pigmented, darkly pigmented*, and *black* (as in Negroes). Symbolizing the genotype of *black AA BB* and that of *white aa bb* write out the possible genotype and phenotype of the F_1 offspring of *black-white* matings. Show the genotypes and phenotypes of F_2's produced by the matings between these F_1's (assume independent assortment).

a Why should the offspring of some *moderately pigmented* parents differ in skin color while those of other such parents do not?

b (1) If those with various degrees of skin pigmentation married *white* individuals could they have *black* children? Explain.

(2) Could a *nonpigmented* couple with Negroid ancestry have a *black* child? Explain.

c Can matings among the first-generation progeny of *white-black* matings produce *black* offspring? *White* offspring?

d If the number of genes controlling skin pigmentation is actually four (or even five or six), as suggested by recent studies (see Stern, 1960):

(1) How would the expected results of crosses between *whites* and *blacks* differ from those considered above?

(2) Why has there been so much uncertainty concerning the number of genes controlling skin pigmentation?

e The first-generation progeny of *black-white* matings, called *mulattoes*, are not uniform in pigmentation, some being considerably lighter than others. Besides the effect of varying environment, what other cause would you suggest for this variation?

f Do you think all *whites* are identical in genotype with respect to pigment genes? Explain.

328 In common wheat (*Triticum aestivum*) kernel color varies from *red* to *white*, the genes for kernel color acting additively. Phenotypic classification of segregating progenies between *red*- and *white*-kernel varieties is possible (Nilsson-Ehle, 1909).

a A cross is made between a *red*-kernel, $R_1R_1 R_2R_2$, and a *white*-kernel, $r_1r_1 r_2r_2$, variety. Give the genotype and phenotype of the F_1 and the F_2 genotypes and ratios for the phenotypes *red, dark, medium, light*, and *white*.

b From a second cross between *red* and *white* varieties, $\frac{1}{64}$ of the F_2 progeny have *red* and $\frac{1}{64}$ *white* kernels. How many allele pairs control kernel color in this cross? What is the formula for determining the proportion of parental types in the F_2?

c Different F_2 plants from the second cross are crossed with the *white* parent. Give the F_2 genotypes that would give the following kinds of progenies:

(1) 1 *colored* : 1 *white*   (2) 3 *colored* : 1 *white*   (3) 7 *colored* : 1 *white*

d Different F_2 plants from the cross in (b) were self-fertilized. Give the genotypes of the F_2's that produced the following progenies:

(1) All *white* (2) All *colored*
(3) 15 *colored* : 1 *white* (4) 63 *colored* : 1 white

329 In 1913 Emerson and East recorded the data shown for ear length in corn.

Table 329

Parent and generation	Length of ear, cm																
	5	6	7	8	9	10	11	12	13	14	15	16	17	18	19	20	21
P60	4	21	24	8													
P54									3	11	12	15	26	15	10	7	2
F_1					1	12	12	14	17	9	4						
F_2			1	10	19	26	47	73	68	68	39	25	15	9	1		

a For each of the parents and for the F_1 and F_2 generations calculate the mean and the standard deviation for ear length.

b Compare the standard deviation (1) of the two parents and (2) of the F_1 and F_2. Offer an explanation for any large differences.

c Calculate the theoretical and geometric mean of the F_1. Compare these values with the observed F_1 mean. What type of gene action is probably involved?

d Note that neither of the parental extremes was recovered in the above F_2 population of 401 plants. What conclusion can be drawn from this about the minimum number of genes controlling ear length?

e Determine the approximate number of genes involved with the use of the formula

$$N = \frac{(\text{difference between parental means})^2}{F_2 \text{ variance} - F_1 \text{ variance}} = \frac{D^2}{8}$$

f Delineate the experimental conditions required for the valid use of this formula.

330 Heston (1942) studied the development of pulmonary tumors in mice induced by injecting the carcinogen 1,2,5,6-dibenzanthracene, by counting the tumorous nodules on the surface of the lungs of animals 16 weeks after injection. The

Table 330

Group	Mean	Standard deviation
Strain L	0	0
Strain A	75.4	15.7
F_1	12.5	5.3
F_2	10.0	14.1

results for a *resistant* strain L, a *susceptible* strain A, and the F_1 and F_2 from a cross between them are shown.   Approximately 100 mice were studied in each group.   Estimate the number of allele pairs that may be inferred to control *resistance* vs. *susceptibility* to pulmonary-tumor development in this cross.

331 An experiment carried out under uniform environmental conditions to determine whether two true-breeding strains of dogs were genotypically identical with respect to weight gave the data shown.

Table 331

Strain A, lb		Strain B, lb
97		97
	F_1 97 lb	

Testcross progeny	Weight, lb	No. of animals
F_1 × a 16-lb	97	32
true-breeding	43	124
strain	25	190
	19	134
	16	32

Note: (1) The segregating generation is a testcross, not an F_2.
(2) The 16-lb strain is the smallest strain known.

a Determine the number of polygene loci at which the parents carry different alleles.
b Using your own symbols, give the genotypes of the parents of the F_1.
c What type of cumulative gene action (additive or multiplicative) appears to be expressed in this cross?   Why?
d How much of the weight above that produced by the residual genotype (16 lb) was each contributing allele responsible for?

332 The data shown are from a series of crosses between two true-breeding strains, under uniform environmental conditions, to study fruit weight (in grams) in oranges.   Assuming that the contributing alleles of all genes had equal effects:
a Determine the number of polygene loci at which the parents carry different alleles.
b Using your own symbols, give the genotypes of the parents of the F_1.
c What type of cumulative gene action (additive or multiplicative) appears to be expressed in this cross?   Why?
d How much of the weight above that produced by the residual genotype (20 g) was each contributing allele responsible for?

Table 332

Strain A, g	Strain B, g
24	32
F$_1$ 28 g	

F$_2$	Weight, g	No. of plants
	36	2
	34	14
	32	60
	30	108
	28	140
	26	114
	24	52
	22	18
	20	2
		510

333 Nilsson-Ehle (1909) observed that only plants with *red* kernels were produced in both the F$_1$ and the 78 F$_2$ progeny of a cross between the *red*-kernel wheat variety Swedish Velvet and a *white* variety. The 78 F$_2$ plants were self-fertilized. Eight gave a 3:1 F$_3$ ratio (307 *red* : 97 *white*), 15 gave a 15:1 ratio (727 red : 53 *white*), 5 gave a 63 : 1 ratio (324 *red* : 6 *white*), and the remaining 50 F$_2$ plants bred true (2,317 *red*).

a Offer a genetic explanation to account for the F$_2$ results.

b On the basis of your hypothesis give the genotypes of the parents of the cross.

c Do the F$_3$ results agree with those expected on the basis of your hypothesis?

334 Assume that an ideal polygenic system occurs in which the contributing alleles *A*, *B*, *C*, and *D* of the independently inherited allele pairs *Aa*, *Bb*, *Cc*, and *Dd* each contribute 2 cm to height in a certain species of plant. In addition, an allele *E*, of a gene with a major effect on height, produces height of 20 cm when homozygous. A cross is made between plants of genotype *aa BB cc dd EE* and *AA bb CC dd EE* and carried to the F$_2$. All generations are raised under almost identical environmental conditions.

a Compare the mean of the F$_1$ with that of the F$_2$ for both arithmetic and geometric types of gene effect.

b What proportion of the F$_2$ populations will have the same height as the *AA bb CC dd EE* parent? As the *aa BB cc dd EE* parent? As both parents?

c What proportion of the F$_2$ population will breed true for the height shown by the *AA bb CC dd EE* parent?

d What proportion of the F$_2$ population will breed true for a height equivalent to that of the F$_1$?

335 Three watermelon plants bear 4-lb fruits. Plants A and B when self-fertilized breed true, but C when self-fertilized produces progeny with fruits ranging

in weight from 3 to 5 lb.   A cross of plant A with plant B produces F_1's with
4-lb fruits and an F_2 with fruits ranging from 3 to 5 lb.   Selection in this
and succeeding generations cannot increase fruit weight above 5 lb.   A cross
of plant A with plant C gives F_1's with fruits ranging from $3\frac{1}{2}$ to $4\frac{1}{2}$ lb.
Selection among the F_2 can raise the fruit weight to 6 lb.   A cross of plant B
with plant C produces F_1's with fruits from $3\frac{1}{2}$ to $4\frac{1}{2}$ lb.   By selection among
the F_2 it is possible to raise the weight of fruit to 5 lb.   Explain these
results, giving the genotypes of the three parents.

336 In tomatoes, fruit weight is a quantitative character.   One variety, of genotype
AA BB CC, produces tomatoes averaging 14 oz in weight, and another, of
genotype *aa bb cc*, produces tomatoes averaging 8 oz in weight.
  a If all contributing alleles (symbolized by capital letters) are equal in effect,
  what amount does each allele contribute to average fruit weight?
  b These varieties are crossed, and the F_1 plants are self-fertilized.   Show the
  genotypes and phenotypes expected in the F_1 and F_2 generations.   The
  genes are inherited independently.
  c What proportion of the F_2 should have the same fruit weight as the *small-
  fruit* parent?   The same as the *large-fruit* parent?   Both parents?   Give the
  formula for these derivations and explain how it is derived.
  d Graph the F_2 results in the form of a frequency curve.

337 **a** A corn breeder has 10 corn plants 70 in. tall.   He crosses them in pairs and
  self-fertilizes the offspring in each of his five crosses for several generations, selecting the taller plants in each generation.   His results for the five crosses are:

  Cross 1 produces only 70-in. offspring; selection fails to raise their height.
  Cross 2 produces offspring varying from 50 to 90 in.; selection among these fails to
    raise the height above 90 in.
  Cross 3 produces progeny varying from 60 to 80 in., and by selection it is possible
    to raise the height to 110 in.
  Cross 4 produces progeny varying from 50 to 90 in.; selection raises the height to
    110 in.
  Cross 5 produces offspring varying from 50 to 90 in.; selection raises the height to
    130 in.

Explain these results, giving the genotypes of the parents.
  b *Smooth* vs. *wrinkled* seed is monogenically controlled, with the allele *S* for
  smooth dominant to *s* for *wrinkled*.   A corn breeder crosses a true-breeding
  50-in. high, smooth variety with a true-breeding *70-in., wrinkled* one.   His
  objective is to produce a *90-in., wrinkled* variety.
  (1) The breeder wants this new variety in 2 years.   How many plants
  should he raise in the F_2 of the cross between the parental varieties to be
  reasonably certain of getting at least one plant of the new variety?
  (2) Assume the breeder has 4 years within which to obtain the new variety.
  What would you advise him to do to avoid having to raise such a big
  crop in the F_2 and in subsequent generations?

338 Burton (1951) presented extensive data on the inheritance of quantitative characters in pearl millet (*Pennisetum glaucum* [= *americanum*]).   The data for one of these characters, number of leaves per stem, from one of six crosses, are given.

Table 338

Parent and generation	Number of leaves per stem																N
	8	9	10	11	12	13	14	15	16	17	18	19	20	21	22	23	
P19	7	36	62	37	23	3											168
P782					2	4	4	11	24	37	26	21	26	13	9	2	179
F_1				2	1	3	17	52	73	52	17	3					220
F_2		4	16	49	163	282	379	375	161	72	19	3					1,523

a Calculate the mean and the standard deviation for each parent and for the F_1 and F_2.

b Why would you expect the standard deviation for the F_2 to be greater than each of the other standard deviations?

c Compare the mean of the F_1 and that of the F_2 with the two parental means. Do your findings agree with the results expected according to the polygene hypothesis?

d Is it possible to determine whether the action of the genes is arithmetic or geometric?   Explain.

e Suggest reasons why it is not possible to determine the number of genes involved with the data.

339 Data from a series of breeding experiments involved in the study of fruit weight in tomatoes are summarized in the table.

Table 339

Parent and gener-ation	Midclass values, oz																				
	3.6	3.8	4.0	4.2	4.4	4.6	4.8	5.0	5.2	5.4	5.6	5.8	6.0	6.2	6.4	6.6	6.8	7.0	7.2	7.4	7.6
P	3	6	17	19	8	5															
P														5	7	12	23	29	11	9	6
F_1							4	8	11	15	14	11									
F_2		1	3	8	12	18	19	24	27	28	29	25	23	18	16	13	11	8	4	2	1

a Why does the F_1 have less phenotypic variability than the F_2?

b Why is the range of phenotypes in the F_2 greater than in either parental strain?

c Do you expect some F_3's from crosses between F_2's producing fruits weighing 5.6 oz to produce fruits weighing more than 7.6 oz?

340 **a** A breeder in China has a strain of rice (*Oryza sativa*) that has been self-fertilized for 20 generations. He has repeatedly tried to increase the average seed yield per acre of this strain by selection, without success. Explain why.
b He crosses this strain with another that yields the same. The F_1's yield the same as their parents, but by selection in the F_2 and F_3 he is able in a few generations to increase the yield considerably. Explain how this is possible.

341 The table gives the mean fruit weight (in grams) for three tomato crosses (data from MacArthur and Butler, 1938).

Table 341

Cross	Larger parent	Weight, g	Smaller parent	Weight, g	Weight of F_1, g
1	Large Pear	54.1	Red Currant	1.1	7.4
2	Putnam's Forked	57.0	Red Currant	1.1	7.1
3	Honor Bright	150.0	Yellow Pear	12.4	47.5

a What type of cumulative gene action is expressed in these crosses?
b Outline how you would proceed to verify your answer to (a).
c Indicate the type of result you would expect in the cross or crosses outlined.

342 Breeds of cattle like the Holstein-Friesian have a *white-spotted* coat. There is, however, considerable variation in the amount of white from one individual to another. Some animals are *solid-colored* (black), others have varying amounts of *white spots*, and still others are almost completely white with few black spots. In certain crosses between *solid-colored* and *spotted* individuals all the F_1 are *solid-colored*, and their F_2 progeny consist of *solid-colored* and *spotted* individuals in a 3:1 ratio. Selection within the *spotted* group of animals can increase or decrease the amount of white. Selection has no effect in the parents of the F_1 and the F_1's that give a 3:1 ratio. Criticize the polygene hypothesis for these results, and suggest an alternative one that suits them better.

343 According to Dunn (1937), "*Variegated* mice are characterized by very finely dispersed spotting patterns consisting of small clumps of colored and white hairs." The amount of white spotting and the areas and the distribution of the patches vary considerably in *variegated* individuals and strains. Dunn reported the results shown from a study of spotting pattern in the house mouse. The genotypes of the parents of each cross at the W locus are given in the table. Describe the genetic control of variegation in these crosses and its effect on the expression of the W and w alleles.

Table 343

		Progeny	
Parents	Phenotype	Percentage of white area in dorsal region	Proportion among progeny
light-variegated (*Ww*)	*white*	90–100	$\frac{3}{4}$
	unspotted	0–5	$\frac{1}{4}$
unspotted (*Ww*)	*white*	100	$\frac{1}{4}$
	unspotted	0	$\frac{3}{4}$
medium-variegated (*Ww*)	*white*	100	$\frac{1}{4}$
	spotted	20–80	$\frac{1}{2}$
	unspotted	0	$\frac{1}{4}$
white (*Ww*)	*white*	100	$\frac{3}{4}$
	spotted	50–90	$\frac{1}{4}$
unspotted (*ww*)	*unspotted*	0	all

344 In the common pigeon (*Columba domestica*) the *wild* form has a smooth head with no crest and is said to be *crestless*. Many of the domestic breeds have a *crest*. When *crestless* × *crested* matings are made, the F_1's are all *crestless* and the F_2's and testcrosses show 3:1 and 1:1 ratios, respectively, of *crestless:crested*. This, however, is not the complete picture (Staples-Browne, 1905). Among individuals that are *crested* the size and appearance of the structure varies from one individual to another. The types that occur are illustrated. Outline what you consider to be the most plausible genetic explanation of the inheritance of *crestless* vs. *crested*. Briefly discuss how you would test your hypothesis and state the results expected.

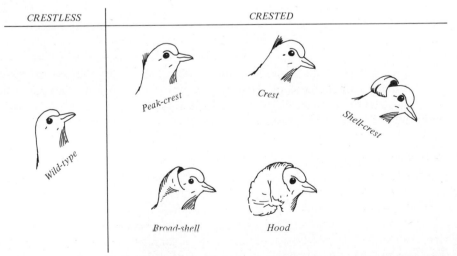

Figure 344 Pigeon heads illustrating the various types of crests. (*From W. M. Levi, "The Pigeon," R. L. Bryan Co., Columbia, S.C., 1945.*)

REFERENCES

Allard, R. W. (1960), "Principles of Plant Breeding," Wiley, New York.
Burton, E. W. (1951), *Agric. J.*, **43**:409.
Castle, W. E. (1931), *J. Exp. Zool.*, **60**:325.
Chai, C. K. (1956), *Genetics*, **41**:157.
Charles, D. R., and H. H. Smith (1939), *Genetics*, **24**:34.
Cockerham, C. C. (1956), *Brookhaven Symp. Biol.*, **9**:53.
Davenport, C. B. (1913), *Carnegie Inst. Wash., D.C., Publ.* 188.
Dunn, L. C. (1937), *Genetics*, **22**:43.
East, E. M. (1910), *Am. Nat.*, **44**:65.
—————— (1916), *Genetics*, **1**:164.
Emerson, R. A., and E. M. East (1913), *Nebr. Agric. Exp. Res. Bull.* 2.
Falconer, D. S. (1960), "Introduction to Quantitative Genetics," Ronald, New York.
Farabee, W. C. (1905), *Pap. Peabody Mus. Harv. Univ.*, **3**:65.
Harrison, G. A., and J. J. T. Owen (1964), *Ann. Hum. Genet.*, **28**:27.
Heston, W. E. (1942), *J. Natl. Cancer Inst.*, **3**:69.
Holt, S. B. (1961), *Br. Med. Bull.*, **17**(3):247.
Johannsen, W. (1903), "Über Erblichkeit in Populationen und in reinen Linien," G. Fischer, Jena.
Levi, W. M. (1945), "The Pigeon," R. L. Bryan, Columbia, S.C.
MacArthur, J. W. (1941), *J. Hered.*, **32**:291.
—————— and L. Butler (1938), *Genetics*, **23**:254.
Mather, K. (1943), *Biol. Rev.*, **18**:32.
—————— (1949), "Biometrical Genetics," Dover, New York.
Mohamed, A. H. (1959), *Genetics*, **44**:713.
Mohr, O. L., and C. Wriedt (1919), *Carnegie Inst. Wash., D.C., Publ.* 295.
Nilsson-Ehle, H. (1909), *Lunds Univ. Aarskr.*, N.F. Afd., (2) **3**(2):1.
Powers, L. (1941), *J. Agric. Res.*, **63**:149.
Roberts, J. A. F. (1961), *Br. Med. Bull.*, **17**(3):241.
Schertz, K. F., N. A. Sumpter, I. V. Sarkissiam, and G. E. Hart (1971), *J. Hered.*, **62**:235.
Scowcroft, W. R., and A. P. James (1971), *Genetics*, **68**:59.
Slatkin, M. (1970), *Proc. Natl. Acad. Sci.*, **66**:87.
Staples-Browne, R. (1905), *Zool. Soc. Lond. Proc.*, pp. 550–558.
Stern, C. (1953), *Acta Genet. Stat. Med.*, **4**:281.
—————— (1960), "Principles of Human Genetics," 2d ed., Freeman, San Francisco.
—————— (1970), *Hum. Hered.*, **20**:165.
Wright, S. (1934), *Genetics*, **19**:537.

10
Sex Determination and Sex Differentiation

Questions on sex determination and recombination in bacteria and viruses are in Chaps. 16 and 17.

Bisexual species are those in which each individual possesses both male and female reproductive organs. *Unisexual species* are those in which each individual possesses either male or female reproductive organs. Bisexual is synonymous with hermaphroditic and monoecious.

QUESTIONS

345 Distinguish between:
  a Primary and secondary sex characters.
  b Sex determination and sex differentiation.
  c Autonomous and nonautonomous sex differentiation.
  d Intersexes and gynandromorphs.
  e Hermaphrodites and monoecious individuals.
  f Sex mosaics and gynandromorphs.
  g Heterogametic and homogametic sex.
  h Autosome and sex chromosome.
  i Freemartin and pseudohermaphrodite.

346 The classical works of Bridges, Goldschmidt, Correns, and Hartmann led to the formulation of a basic theory of sex determination, expressed by Hartmann as "the law of the bisexual potentiality of both sexes." This states that all cells are bipotential with respect to sex.
  a Do you think there is a logical necessity for assuming this to be so, or do you not? Present arguments favoring your position.
  b Do you think that the acceptance of Hartmann's "law" carries a corollary acceptance of the requirement for a triggering mechanism to swing development toward one sex or the other?

347 Definite answers to many questions regarding sex depend upon whether a student uses specific definitions of terms such as *intersex*, *gynandromorph*, and *hermaphrodite*. Define the above terms and answer the following questions in accordance with your definitions, showing how your answer is supported by known human abnormalities.
  a Are human beings with Turner's or Klinefelter's syndrome intersexes?
  b Do you think human gynandromorphs might exist with a clear mosaicism for primary and secondary male and female characters?
  c Can human beings be hermaphrodites?
  d How might you explain the occurrence of intersexes in human beings?

348 Since 1959 many sex-chromosome mosaics in man have been recorded (Hirschhorn et al., 1960; Klinger and Schwarzacher, 1962; Mittwoch, 1967).
  a How do such mosaics originate?
  b Do you think that such individuals would show a sharp mosaicism for primary and secondary male and female traits, as is the case in insect gynandromorphs? Provide a reason for your answer.

349 In moths, butterflies, and *Drosophila*, gynanders (gynandromorphs) are not intersexes. Why? Is this also true in mammals? Explain.

350 Klinefelter's syndrome (XXY, XXXY, XXYY, etc.) occurs approximately once in every 400 to 600 male births, whereas Turner's syndrome (XO) occurs only about once in every 5,000 female births. Discuss with the aid of diagrams the mechanisms by which these conditions could arise and suggest what relation the great difference in frequency has to your explanation.

351 Suggest a hypothesis to account for the fact that the number of hetero-chromatic spots in mitotic nuclei in human beings is always less (by 1) than the number of X chromosomes.

352 **a** Cite evidence for the statement that the gonads of vertebrates are potentially dual in function.
   b Suggest an explanation for the fact that no YO human beings are known.
   c State whether it is possible for the members of a pair of monozygotic twins to differ in sex and why.
   d Mantids have either an XX–XO or an $X_1X_1X_1X_1$–X_1X_1Y sex-determining mechanism (White, 1941, 1962; Hughes-Schrader, 1950, 1953). Suggest how the latter system could evolve from the former.
   e Discuss the evolutionary advantages of the haplodiploid scheme of sex determination.

PROBLEMS

353 In *Drosophila melanogaster* flies with two sets of autosomes and two X chromosomes (AA XX) are female, while those with two sets of autosomes plus one X and one Y chromosome (AA XY) are males.
   a Why is this information alone insufficient to determine the location of the sex-determining genes?
   b Tetraploid *Drosophila* have been produced with the following karyotypes: AAAA XXYY; AAAA XXXY and AAAA XYYY. What should the sex of each of these individuals be and why?
   c AAA XX *Drosophila* are intersexes. The addition of fragments of X chromosomes to the AAA XX complement shifts sexuality toward femaleness (Dobzhansky and Schultz, 1934). What does this show regarding the location of sex-determining genes?

354 The following statements concern sex determination in a certain animal species. Females fertilized with sperm from a closely related hermaphroditic species produce progeny consisting of males and females in equal proportions. Individuals carrying the sex homolog of the homogametic sex in duplicate together with three sets of autosomes are intersexes.
   a Which sex is heterogametic?
   b What hypothesis of sex determination is supported by the data, and what are the probable locations of the male- and of the female-determining genes?

c Where the sex homolog of the homogametic sex is arbitrarily symbolized X and its heterologous mate, Y, what sex would you consider the following to be?

(1) AA XXY   (2) AA X

355 *Melandrium dioicum* (a member of the pink family) is a dioecious species in which the female is homogametic and the male heterogametic. Many polyploid and aneuploid plants have been obtained and used by Warmke (1946) and Westergaard (1958) to study the roles of the X chromosome, the Y, and the autosomes in sex determination. Some of these karyotypes, together with their corresponding sexes are shown in the table. Determine the location and relative strength of the male- and the female-determining genes.

Table 355

Chromosome constitution	Sex	Chromosome constitution	Sex	Chromosome constitution	Sex
AA XX	♀	AAA XY	♂	AA XXY	♂
AA XXX	♀	AAAA XY	♂	AAA XXY	♂
AAA XX	♀	AAAA XXYY	♂	AAA XXXY	♂
AAAA XXX	♀	AAAA XXXYY	♂	AAAA XXY	♂
AA XY	♂			AAAA XXXXYY	♂
AA XYY	♂			AAAA XXXY	♂
				AAAA XXXXY	♂ or ♀

356 Since 1959 a variety of sex mosaics have been found in man. The somatic-cell chromosome numbers and sex-chromosome constitutions for some of these are as shown (autosomal number in all individuals is 44). Suggest a common mechanism by which these individuals originate. Use any two of these mosaics to illustrate the mechanism.

Table 356

Somatic karyotype			
Chromosome number	Sex-chromosome constitution	Phenotype	Reference
45–47	XO–XXY	*Turner's syndrome*	Cooper et al. (1962)
47–48	XXX–XXXY	*Klinefelter's syndrome*	Crawfurd (1961)
46–46–45	XX–XY–XO	*Male pseudohermaphrodite*	Schuster and Motulsky (1962)
46–47	XX–XXY	*Klinefelter's syndrome*	Ford et al. (1959)
45–46	XO–XY	*True hermaphrodite*	Hirschhorn et al. (1960)

357 Schuster and Motulsky (1962) described triple mosaicism (XX/YY/XO) in an individual with male pseudohermaphroditism.   Outline a scheme to account for such an individual.

358 In certain species with male heterogamety, the heterogametic sex is XO, and the homogametic one is XX.   In the insect *Protenor belfragei* the female has 12 autosomes and two X chromosomes and the male 12 autosomes and one X chromosome (Wilson, 1906).

   a Diagrammatically outline meiosis in both the male and female and show how a 1:1 sex ratio is achieved.

   b From a comparison of AA XX and AA XO karyotypes and their associated sex, suggest a hypothesis regarding the location of genes for femaleness and genes for maleness.   List two further karyotypes with their associated sex that would corroborate your hypothesis.

   c It might be suggested that two different X chromosomes, one carrying genes for femaleness and one carrying alternative alleles of these genes controlling maleness, would explain the *Protenor* situation, the autosomes being neutral regarding sex.   Show why this explanation is unsatisfactory.

   d Does the *balance theory of sex* apply to these organisms?

359 **a** Man ($2n = 46$) has an XX-XY sex chromosome constitution (Tjio and Levan, 1956; Ford and Hamerton, 1956).   Individuals with Turner's syndrome (female external genitalia, short stature, webbed neck, small uterus, ovaries represented by fibrous streaks, etc.) are classed as being essentially female, and have the chromosome constitution AA XO (Ford et al., 1959).   Individuals with Klinefelter's syndrome (male external genitalia, testes small, body hair sparse, femalelike breast development, etc.) are classed as males, and usually have the chromosome constitution AA XXY (Jacobs and Strong, 1959).   From this information determine the roles, if any, of the X, the Y, and the autosomes in sex determination in man, showing how you arrive at your conclusions.

   b The table shows other types of sex-chromosome variants found by various workers (data from McKusick, 1964; Mittwoch, 1967).

Table 359

Chromosome number	Chromosome constitution	Sex phenotype	Fecundity
47	AA XXX	Female	Many fertile
48	AA XXXX	Female	Sterile
47	AA XYY	Some normal male, others Klinefelter's	Some fertile
48	AA XXYY	Klinefelter's	Sterile
48	AA XXXY	Klinefelter's	Sterile
49	AA XXXXY	Klinefelter's	Sterile

(1) Do these findings give additional information on the location of sex-determining genes? Explain.

(2) The extent to which the autosomes determine femaleness is not clear from these findings. What types of individuals with respect to chromosome number and constitution should provide information bearing on the problem?

360 In silkworms (*Bombyx mori*) the larval epidermis is *opaque* or *translucent*. In a series of reciprocal crosses involving these traits, Tanaka (1922) obtained the results shown.

Table 360

Cross	Parents ♀	♂	F_1	F_2
1	*translucent* × *opaque*		♀ 1,374 *opaque*	♀ 1,362 *opaque*, 1,301 *translucent*
			♂ 1,354 *opaque*	♂ 2,510 *opaque*, 0 *translucent*
2	*opaque*	× *translucent*	♀ 774 *translucent*	♀ 610 *opaque*, 456 *translucent*
			♂ 782 *opaque*, 1 *translucent*	♂ 524 *opaque*, 453 *translucent*

a Which sex is heterogametic? Give reasons for your answer.

b Account for the *translucent* male in the F_1 of the second cross.

361 a *Asparagus officinalis*, the garden asparagus, is normally dioecious with staminate (male) and pistillate (female) plants occurring in approximately equal numbers. Occasionally stamens develop in pistillate flowers and pistils in staminate flowers. Although usually nonfunctional, pistils of this nature sometimes produce viable seed by self-fertilization. From 61 such plants Rick and Hanna (1943) obtained 198 seeds in this way; when planted and grown to maturity, 155 proved to be male and 43 female.

(1) What do these results suggest regarding the genetic control of sex in this species?

(2) Which is the heterogametic sex?

b When 25 of the 155 male plants were crossed with pistillate plants, 8 gave male progeny only and 17 gave progeny with males and females in a 1:1 ratio.

(1) Do these results support your explanation to (a)?

(2) Is the allele for maleness dominant or recessive?

(3) Show the kinds of progeny and their expected proportions among seed (obtained by self-fertilization) of rare pistillate flowers with functional stamens.

(4) Designating the chromosome pair carrying the sex-determining locus XX in the homogametic sex and XY in the heterogametic sex, show how Rick and Hanna's results can be explained by an XY mechanism.

c Contrast the Y in *Asparagus* with that in *Drosophila* with respect to sex-determining genes and "inertness"(for the latter compare AA YY types in both organisms).

d According to certain investigators, staminate plants outyield pistillate ones by about 25 percent. Outline a method by which seed could be obtained that would give staminate plants only.

362 Hemp (*Cannibus sativa*) is normally a dioecious species. Occasionally a male plant develops flowers which produce female as well as male gametes. When such plants are self-fertilized, they produce males and females in a 3:1 ratio. Explain genetically.

363 a In papaya (*Carica papaya*) there are three main sexual forms: males and females (of dioecious strains) and hermaphrodites. The results of crosses between these three sex forms are summarized (from Hofmeyr, 1938, 1939;

Table 363

| | Progeny | | |
| | Dioecious | | |
Cross	♀	♂	☿
♀ × ♂	1	1	1
♀ × ♂	1	0	1
"♂" × "♂" or selfed*	1	2	0
♀ × ♂ or selfed	1	0	2
♀ × ♂ or reciprocal	1	1	1

* Male plants that occasionally produce pistils in staminate flowers.

Storey, 1953). Assuming a multiple allelic system C, c^+, c, controlling sex, give the genotypes of the three sexual forms.

b Since in the third and fourth crosses a 1:2 instead of a 1:3 ratio occurs, we may make one further assumption, that a pair of alleles, Vv, controls viability, the recessive v in the homozygous condition being lethal. Assuming that complete linkage exists between the C and V genes, show which alleles are present on the X and which on the Y chromosome.

364 According to Galan (1951), two varieties of *Ecballium elaterium* (a cucurbit) are found in Spain. Northern populations consist of monoecious types only (var. *monoicum*); those of southern Spain (var. *dioicum*) are dioecious. The results of extensive intra- and intervarietal crosses by this worker are given.

a Outline a genetic hypothesis to explain these results, stating which is the heterogametic sex of *E. elaterium* var. *dioicum*.

Table 364

Parents	Progeny		
	Dioecious		
♀ ♂	♀	♂	Monoecious
dioicum × *dioicum*	1	1	0
monoicum × *monoicum*	0	0	all
dioicum × *monoicum*	0	0	881
monoicum × *dioicum*	0	423	382
dm × md	16	0	34
md × md	0	105	109
md × d	35	48	30
m × md	0	483	550
dm × m	0	0	59

Key: d = dioecious; m = monoecious; dm = dioecious × monoecious cross; md = monoecious × dioecious cross.

b Designating the sex chromosomes of the homogametic sex XX and those of the heterogametic one XY, give the sex genotypes of male, female, and monoecious plants.

c Contrast the roles of the Y chromosomes in *Ecballium* and *Drosophila* with regard to sex and viability control.

365 In the fish *Lebistes reticulatus* there is a *black spot* on the dorsal fin. In crosses of *black-spot* males with *normal* females half the progeny have the *black spot* and the other half do not. *Black spot* occurs only in males. Moreover, *black spot* does not occur in males unless the father expressed the trait (Winge, 1922*a,b*). Present a diagrammatic explanation of these results showing which is the heterogametic and which the homogametic sex. Show what would be expected if the gene-controlling *black spot* were located on the alternative sex chromosome to that you have postulated.

366 The sex phenotypes of normal and aneuploid individuals in three animal species are shown. Discuss the roles of the X chromosome, the Y, and the autosomes in sex determination in each of these species.

Table 366

Chromosome constitution	Species 1	Species 2	Species 3
AAXX	♀	♀	♀
AAXO	♀	♀ (undersexed)	♂
AAXXY	♂	♂ with intersexual tendencies	♀
AAAXXY	♂	Intersex but more like male than female	Intersex

367   In *Drosophila* species, the male is heterogametic (XY) and female homogametic (XX).   In birds the mechanism is reversed.   In *Drosophila* species such as *parthenogenetica*, individuals that arise parthenogenetically are almost always females, but birds that arise by parthenogenesis are, with few exceptions, males.

    a Offer a genetic explanation to account for this difference.

    b Explain the occurrence of rare parthenogenetic males in *Drosophila*.

368   **a** Ford (1955) illustrates a bilateral gynandromorph of the waved umber moth (*Hemerophila abruptaria*) in which the right side is male and the left female. In addition, a color difference is involved, the left side being *normal* and the right a *deep-chocolate* shade.   This color difference is autosomally controlled.   How would you account for the occurrence of this moth?

    b Ford also showed a different coloring on the two sides of a *normal* female currant moth (*Abraxas grossulariata*).   Suggest an explanation for the occurrence of this bicolored female.

369   **a** A study of X-chromosome traits in pedigrees showing *afflicted* persons and both their parents provides some indication of the source of the X chromosome in XO, XXY, and other aneuploid types.   For each of the following pedigrees indicate the source of the X or X's and the reasons for your decision.   In each case assume aneuploidy is due to meiotic nondisjunction.

    (1) Glucose-6-phosphate dehydrogenase *deficiency* (G6PD *deficiency*) is controlled by a recessive allele (Kirkman and Hendrickson, 1963).

Figure 369A   (*After Gartler, Vuzzo, and Gandini, 1962.*)

    (2) A blood antigen, *Xg*, is controlled by Xg^a, an allele dominant to *Xg*, which apparently causes lack of the antigen (Mann et al., 1962).

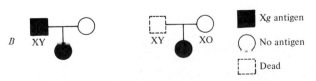

Figure 369B   (*After Lindsten et al., 1963.*)

(3) Red-green *color blindness.*

Figure 369C   *(After Stewart, 1957, from Polani, 1961.)*

(4) Red-green *color blindness.*

Figure 369D

(5) *Xg* blood antigen.

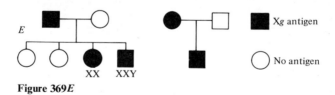

Figure 369E

(6) Red-green *color blindness.*

Figure 36 F

b Is it possible to explain the origin of each of these XO and XXY types by a mechanism other than meiotic nondisjunction?   If so, how?

370 The following data were obtained from studies on sex determination in a certain animal species:

1 *Normal females* crossed with *neomales* (genetically female) but converted into phenotypic and functional males by high temperature in early embryogenesis) produce progeny of both sexes.

2 As a consequence of nondisjunction in the homogametic sex, in addition to *normal* males and females individuals are obtained with the chromosome constitutions and sex phenotypes tabulated.

Table 370

No. of autosomal sets	No. of sex chromosomes of homogametic sex	Sex phenotype
2	1	Female (sterile)
3	2	Intersex
2	3	Supermale

3 All males have a *chocolate*-colored body; females are *chocolate* or *gray*. Body color is controlled by a single pair of autosomal alleles, the allele for *chocolate* being dominant to that for *gray*. An occasional individual appears that is *chocolate* on one side of the body and *gray* on the other side.

State, giving reasons for your statements:
a Which sex is heterogametic.
b The location of male- and female-determining genes.
c The mechanism of sex determination.
d The mechanism of sex differentiation.
e How *chocolate-gray* individuals could arise, whether the individual started out as male or female, and one possible genotype of the zygote.

371 **a** In the creeping vole (*Microtus oregoni*) both sexes are gonosomic mosaics. Eight pairs of autosomes are a constant feature of all somatic and germinal cells in both sexes. The males have an XY pair in somatic cells but only a Y in primary spermatocytes; they are described as OY/XY. The females have one X in somatic cells and two X's in the primary oocytes; they are described as XX/XO (Ohno et al., 1963). Account for this unusual sex-determining mechanism by diagrammatically illustrating meiosis and its consequences in both sexes, the sex-chromosome constitution of the zygotes for both sexes, and the probable mode of origin of the OY and XX meiocytes.

b A pair of X-linked alleles *Aa* is present in a population of voles. What genetic results would you expect in any three-generation cross that would help you support the cytological findings described above when:
(1) female was *a*, male was *A*   (2) female was *A*, male was *a*

If no cytological studies had been made on this organism, what other interpretation might you give such results?   Explain.

c Considering that mammalian females appear to require only one functional X chromosome, would you consider this form of sex determination more advanced or less advanced than that in other mammals?   Explain.

372   Corn (*Zea mays*) is normally monoecious.   A recessive allele, *sk* (*silkless*), eliminates functional female flowers, making *sksk* plants phenotypically and functionally male.   On another chromosome several recessive alleles of a multiple allelic locus, *ts* (*tassel seed*), ts_2, etc., cause pistillate flowers to replace staminate ones, making the plants effectively female (Jones, 1934). The ts_2 allele is epistatic to *sk* and *Sk*.   Plants ts_2ts_2 are female and fertile.

a What is the genotype of *normal* corn?   Of *silkless* plants?   Of female plants?

b A monoecious corn plant of genotype *Sksk* Ts_2 *ts* is self-fertilized.   Show the types of progeny that can occur and their frequency.   What type of gene interaction is involved?

c Using these two genes, show how a stable sex-determining system can be established in which the male is the heterogametic sex.

d What conclusions can be drawn from results such as these regarding the complexity of the differences between unisexual and bisexual species?

e What further mutational steps are needed for the conversion of such a dioecious system into one resembling the XY mechanism in *Melandrium* and man?

f Tassel seed, Ts_3, a dominant allele of ts_2, converts staminate flowers into functional female flowers.   Using Ts_3 and *sk*, illustrate whether it is possible to establish a stable sex-determining system with males and females in a $1:1$ ratio.   If so, which sex would be heterogametic?

373   In a certain insect with the X-Y method of sex determination it is not known which of the following sexual mechanisms exist:

1   Male-determining genes on autosomes; female-determining genes on the X

2   Male-determining genes on the Y chromosome; female-determining genes on the X

3   Male-determining genes on the Y chromosome; female-determining genes on the autosomes

4   Male-determining genes on the Y chromosome; female-determining genes on the X and autosomes

Using a pair of sex-linked alleles, *M* vs. *m* for *normal* vs. *vestigial* wings, outline an experiment to determine the actual mechanism.   (Include genetic tests and cytological studies in your answer.)   Show the results expected in your experimental progeny on the basis of each condition.

374   Evidence suggesting that all stages of the diploid sexual process may be under strict and specific gene control has been found in the plant *Parthenium*

argentatum, one of the Compositae, by Powers (1945).   His data support the hypothesis that three genes *A*, *B*, and *C* control normal meiotic *reduction* (in megasporocytes), *fertilization*, and *nondevelopment* of unfertilized eggs, respectively.   The recessive alleles *a*, *b*, and *c* cause female *nonreduction*, *nonfertilization* of eggs, and *parthenogenetic development* of unfertilized eggs.

a What type of breeding behavior (viz. sexual, producing diploid offspring; sexual, producing polyploids; sterility; or obligatorily apomictic, reproducing only by some asexual method substituted for the sexual process) would each of the following genotypes express?

(1) *AA bb CC*   (2) *AA BB CC*   (3) *aa BB CC*
(4) *AA BB cc*   (5) *aa bb cc*   (6) *aa BB cc*

b What phenotypic ratios are expected among the offspring resulting from a cross between two *Aa Bb Cc* plants?

375 In corn (*Zea mays*) the recessive allele *ba* (barren stalk) makes plants male (staminate) by eliminating the ears (pistillate inflorescence); the recessive allele *ts* (tassel seed) converts the tassel (staminate inflorescence) into an ear. Outline how you would manipulate these genes to convert corn from a monoecious to a dioecious species.

376 a How would you explain the origin of a sexual-mosaic butterfly that is approximately $\frac{3}{4}$ male and $\frac{1}{4}$ female?

b What do gynanders of butterflies and *Drosophila* tell us regarding the role of hormones in sex determination and the type of sex differentiation in these insects?

377 In the mouse, which has the X-Y mechanism of sex determination, fertile XO females are known to occur (Welshons and Russell, 1959; Welshons, 1963). What does this imply regarding the roles of the Y chromosome and the X in sex determination and why?

378 According to the complementary-allele hypothesis of sex determination in Hymenoptera (bees, wasps, ants, etc.) (Whiting, 1933) a multiple-allelic series at one locus (viz. X^a, X^b, ...) controls sex so that males are either haploids or diploids homozygous for a sex allele; females are always diploids and always heterozygous for two of the sex multiple alleles.

a To test the hypothesis, Whiting used the alleles *Fufu* (*fused* bristles vs. *normal*) which were closely linked to the sex locus and which would therefore segregate in much the same way as the sex locus itself.   To distinguish haploid from diploid males he used the autosomal allele pair *Vlvl* (*veined* vs. *veinless*).   The results modified after Whiting (1933, 1943), with idealized ratios substituted for the closely agreeing experimental ones, are shown in the table.

(1) Show whether the results of the first two crosses support Whiting's hypothesis by tracing the consequences of close linkage of a pair of sex alleles, $X^a X^b$, with the marker pair *Fufu*.

Table 378

Mother	Father	Haploid sons (vl)		Daughters (all $Vlvl$)		Diploid sons (all $Vlvl$)	
		Fu	fu	$Fufu$	$fufu$	$Fufu$	$fufu$
$Fufu$	fu	5	5	9	1	1	9
$Fufu$	fu	5	5	1	9	9	1
$Fufu$	fu	5	5	10	10	0	0

    (2) On the basis of your answers to (1) interpret the results of the third cross. Why is linkage complete in this cross?

b Use this hypothesis to explain the following:

    (1) A *Habrobracon* (wasp) female of genotype $X^d X^e$ is mated to a male of genotype X^a. Show the possible types of progeny she can have with respect to genotype and sex (*a*) if the eggs are not fertilized and (*b*) if they are fertilized.

    (2) In a small isolated colony of wasps would you expect diploid males to be present? Of how many different kinds?

    (3) *Black* vs. *orange* eyes is controlled by a single pair of alleles with *B* for *black* dominant over *b* for *orange*. Show how you could use this pair of traits to distinguish between diploid and haploid males.

379 In the Hymenoptera (bees, wasps, ants, etc.), some mites, some scale insects (Coccidae), and a few other organisms females are diploid and males haploid (Whiting, 1945).

    a Discuss sex determination in these organisms, explaining the diploid-haploid scheme and its coexistence with a mechanism involving female heterogamety.

    b Is it possible to obtain diploid males? If so, how? Would they be fertile?

    c Some people think that haploid individuals are male because they are not heterozygous rather than because they are haploid. What do you think? How would you determine which hypothesis is correct?

380 A certain species of frog possesses the XX-XY mechanism of sex determination. When the animals are raised at approximately 70°F, the progeny of a normal mating consists of males and females in equal proportions. When the tadpoles are raised in water at about 90°F, they produce male progeny only.

    a Suggest a reason why only male progeny are produced at 90°F.

    b These males are of two types. What are these types, and what will be the results of mating each type with XX females with regard to the karyotype and sex of the offspring?

381 The table shows the results obtained in crosses within and between unisexual (dioecious) and bisexual (monoecious) species of *Bryonia* (Heilbronn, 1953; Heilbronn and Basarman, 1942; Bilge, 1955).

Table 381

Parents		Progeny		
		Unisexual (dioecious)		
♀	♂	♀	♂	Bisexual (monoecious)
B. dioica × B. dioica		1	1	0
B. dioica × B. alba		All	0	0
B. alba × B. dioica		1	1	0
B. dioica × B. macrostylis		All	0	0
B. macrostylis × B. dioica		1	1	0
B. multiflora × B. dioica		1	1	0
B. macrostylis × B. macrostylis		0	0	All

a Show which species are dioecious, which are monoecious, and which are unclassifiable. For the dioecious species state which sex is heterogametic and why.

b This genus of plants lacks heteromorphic sex chromosomes. Outline the simplest gene control system that will explain the occurrence of unisexuality in this genus. (Note that this is a triggering system acting on an original sexual bipotentiality.)

c Since plants of the monoecious species develop both male and female organs, it may be assumed they are homozygous for the gene(s) for staminate development as well as the gene(s) for pistillate development. Assume that this genus carries a gene *M* (for *male*) and another *F* (for *female*), for which *effective* and *ineffective* alleles may exist. Show the genotypes and sex phenotypes for the first two crosses that will satisfactorily account for sex inheritance. Is it likely that these two genes lie on a single pair of (sex) chromosomes?

d How do you think such a dioecious system might evolve? Would you say that the dioecious species have evolved fairly recently? Present arguments to support your contention.

382 a That a given species has one sex heterogametic and one homogametic can be experimentally verified in several ways, e.g., the mating of two individuals of the same genetic sex; i.e., they have opposite functional sexes although their sexual genotypes are identical. To perform this kind of test involves the experimental sex reversal of one of the parents so that its sex phenotype is either opposite to that normally associated with the genotype or hermaphroditic. The results of such experiments, actually performed with amphibians, are described below.

1 Witschi (1923) intercrossed hermaphrodites of the frog *Rana temporaria*. He also crossed these with normal males and females. The sex distributions found in the progenies are tabulated.

Table 382

Parents		Progeny		
♀	♂	♀	♂	☿
normal ♀ × ☿		182	0	0
☿ × ☿		45	0	0
☿ × normal ♂		132	135	0
normal ♀ × normal ♂		128	127	0

2   Humphrey (1945) working with the Mexican axolotl (*Ambystoma* [= *siredon*] *mexicanum*) succeeded in masculinizing genetic females by heterologous embryonic grafts (graft of testis on a genetic female).   These neomales were crossed with normal females and gave a sex ratio of 3 female : 1 male.   Seventeen F_1 females, chosen at random and mated with normal males, produced the following offspring:

6 F_1 ♀ × normal ♂ → all (833) ♀ progeny

11 F_1 ♀ × normal ♂ → 370 ♂ : 378 ♀

Some F_1 females that gave only female progeny were similarly masculinized.   These neomales mated with normal females gave only female progeny (Humphrey, 1948).

For result 1 or result 2 or both, verify that one sex is homogametic and the other heterogametic.   State which is the heterogametic sex and whether the hermaphrodites (sex-reversed) individuals are heterogametic or homogametic.

b What do the results of these experiments tell us regarding the role of hormones in sex determination and sex differentiation in these organisms?

383 In *Drosophila melanogaster* a recessive gene, *tra*, on chromosome 3, when homozygous transforms diploid females into sterile males (Sturtevant, 1945).
a What sex ratio would you expect in the progeny of a cross between a *Tratra* female and a *tratra* male?
b Briefly discuss the significance of the *tra* effect.
c Suggest how *tra* may act on the sex phenotype.

384 In the silkworm (*Bombyx mori*) with the ZZ-ZW type of sex determination both sexes have a $2n = 56$ chromosome number.   Aneuploid and polyploid types have been obtained which have helped locate sex-determining genes.   Some of these and their sex expression are (Yokoyama, 1959):

AA ZO	♂	AAA ZZ	♂	AAAA ZZZW	♀
AA ZZW	♀	AAA ZZZ	♂	AAAA ZZWW	♀
AA ZZ	♂	AAA ZZW	♀		
		AAA ZWW	dies		
		AAA ZZZW	♀		

　　　a Determine the location of female-determining genes.　Is it possible to do the same for male determiners?　Explain.

　　　b What parallel do you see between the silkworms and man in terms of sex determination?

385 In the axolotl (*Ambystoma*) the female is heterogametic (ZW) and the male homogametic (ZZ).　Humphrey (1945) reported that in a cross between the sex-reversed female and normal female, 76 percent of the progeny were female and 24 percent male.

　　　a What do these results suggest regarding the viability and sex of WW animals in this genus?

　　　b How could you test your hypothesis further?

386 In an organism with the Z-W method of sex determination, genetic studies reveal that it is possible for fertile males to be hemizygous for sex-linked (Z) genes.　What are the possible chromosome constitutions of such males? How would you determine the correct chromosome constitution?　Would any of these karyotypes provide information regarding the role of the W in sex determination?　Explain.

387 Certain organisms like the moths, butterflies, birds, etc., have the *Abraxas* (ZZ-ZW or ZZ-ZO) mechanism of sex determination.

　　　a Which is the heterogametic sex in all these organisms?

　　　b Are the sex chromosomes in the heterogametic sex always cytologically distinguishable?　If not, how can the heterogametic sex be identified?

　　　c In the moth *Fumea casta* the male possesses 62 chromosomes and the female 61 (Gallien, 1959).　Diagram sex determination in this organism to show that basically it is the same as in the X-Y and X-O types.

　　　d Where are the sex-determining genes located in birds?　Moths (like *Fumea*)? How would you proceed to determine whether your statements are correct? See Gallien (1959), Crew (1965), or Ohno (1967) for an extensive treatment of the ZZ-ZW and ZZ-ZO mechanisms.

388 In *Blaps polychrestra*, a beetle in which the sex ratio is 1:1, the constitutions of the male and female karyotypes are as tabulated.　Show the constitutions of the two types of male gametes and the one type of female gamete.

Table 388

Sex	Autosomes	Sex chromosomes
♂	18	12X + 6Y
♀	18	24X

389 In man, *normal* vs. red-green *color blindness* is controlled by a pair of sex-linked alleles with the allele *r*, for the latter trait, recessive.　A woman with *normal* color vision marries a *color-blind* man.　Their first child has *normal* color vision in one eye but is *color blind* in the other.   With the aid of diagrams offer an explanation for this mosaicism.   What is the probable sexual phenotype of this child?

390   The early embryo of mammals and of amphibians such as the salamander is sexually neutral.   The rudimentary gonad consists of a *cortex* (an outer layer of tissue which in AA XX individuals dominates and develops into an ovary) and a *medulla* (an inner mass which in AA XY individuals develops into testes). Witschi (1934), Witschi and McCurdy (1929), and Witschi et al. (1931) carried out some very elegant grafting experiments with salamanders and found that when male and female embryos in the same early stage of development are grafted together, the male dominates in development, the testes are normal, and ovary development in the female graft partner is often entirely suppressed. If, however, the male graft is small in comparison with the female graft partner, the female influence predominates and the male gonads are often converted into ovaries.   Explain how these results illustrate that individuals are bipotential with respect to sex and that differentiation is under the control of hormones.

391   a How can sex reversal in chickens be explained?

   b Assume that a hen has undergone sex reversal and becomes a functional male, as reported by Crew (1925).   What sex ratio would be expected in the offspring from a mating between such a "male" and a normal female?

392   Administration of female hormone (estradiol) to male larvae of the African water frog (*Xenopus laevis*) causes them to develop as functional females (called *neofemales*).   Thirteen such females mated with normal males produced 1,624 males and no females (Gallien, 1956; Chang and Witschi, 1955).   State what conclusions can be drawn regarding:

   a The chromosomal mechanism of sex determination and the heterogametic sex.

   b The evolutionary status of the mechanism.

   c The kind of sex-differentiation process.

393   In 1922 Haldane proposed the rule that when individuals of a particular sex are either sterile, rare, or absent among the progeny of a cross between two strains, they are heterogametic.   Discuss the evidence that supports this rule.

394   In many organisms, e.g., the bedbug (*Cimex lectularius*), the beetle *Blaps lusitanica*, the praying mantis *Sphodromantis viridis*, and the hop *Humulus japonica*, sex is determined by "multiple sex chromosomes."   In the praying mantis the male sex-chromosome complement is X_1X_2Y, and the female complement is $X_1X_1X_2X_2$; while in hops the male and female complements are XY_1 and XX, respectively.

   a Show how you think the sex chromosome would probably behave during male and female meiosis in these two organisms to give a stable sex-determining system.

   b Suggest how the sex-chromosome complement in the praying mantis could have originated (see White, 1940, 1954).

395 Assuming that the bisexual state is the original one:

a Outline a hypothesis that would explain the evolution of a monoecious species into a dioecious one like the cucurbit *Ecballium elaterium* with the XX-XY method of sex determination in which the Y is morphologically the same as the X and in which YY individuals are viable. Does the hypothesis explain the results of the crosses outlined in Prob. 364?

b What are the possible stages in the evolution of the XX-XY mechanism described in (a) to one in which the X and Y differ morphologically and genetically to such a degree that YY individuals are inviable but the Y still plays a decisive role in sex determination (as in man and *Melandrium*)?

c How do you think that an XX-XY mechanism like that of man and mouse might change to one in which the Y chromosome is present but nonfunctional in sex determination and possibly completely inert, as in *Drosophila* and *Lygaeus*?

d Outline a possible origin of the XX-XO system, which occurs in many grasshoppers and in other organisms, from an XX-XY system.

e Certain mantid species with the X-O method of sex determination have one chromosome pair more than related species with an X_1X_2Y system (White, 1941). Suggest how the latter arose from the former.

396 a In 1940 Murray provided extensive data on sex inheritance from crosses between dioecious species of the genus *Acnida* and monoecious ones of *Amaranthus*. Some of the data are shown in the table. Which species of *Acnida* are dioecious, and which sex is heterogametic and why?

Table 396

Parents		Offspring		
♀	♂	♀	♂	⚥
Ac. cuspidata × *Am. retroflexus*		All		0
Ac. cuspidata × *Am. caudatus*		All		0
Am. retroflexus × *Ac. cuspidata*		$\frac{1}{2}$	$\frac{1}{2}$	0
Am. caudatus × *Ac. cuspidata*		$\frac{1}{2}$	$\frac{1}{2}$	0
Ac. turberculata × *Am. retroflexus*		All		0
Ac. tuberculata × *Am. caudatus*		All		0
Am. retroflexus × *Ac. tuberculata*		$\frac{1}{2}$	$\frac{1}{2}$	0
Am. caudatus × *Ac. tuberculata*		$\frac{1}{2}$	$\frac{1}{2}$	0

b Although the chromosomes cannot be cytologically identified, one of the chromosome pairs nevertheless carries sex-determining genes. Representing the pair in the homogametic sex as XX and in the heterogametic sex as XY (Y differing from X in the sex-determining alleles only), outline a hypothesis to account for sex control in *Acnida*. If possible, state the location of the male- and the female-determining genes.

c In 1940 Murray produced tetraploids in *Ac. tamariscina* by colchicine treatment.   Crosses between AAAA XXXX females and AAAA XXYY males gave 131 females : 1,633 males.   Assuming that female gametes of these tetraploids carry two X's and male gametes are almost always XY and only rarely XX or YY, describe the chromosomal constitutions of the progeny and state what information this may provide regarding the role of the Y in sex determination.

d When tetraploid females were crossed with males produced by the first mating, one half of the progeny were male, the other half female.   Do these results substantiate your previous conclusions?

QUESTIONS INVOLVING FURTHER READING

397 In grasshoppers the great majority of the species are XX-XO (the X being acrocentric), but a few have an XX-XY sex-determining system in which the X is metacentric and the Y completely or almost completely heterochromatic (Helwig, 1941).   Propose a method by which a stable XX-XY mechanism could arise from an XX-XO.   Illustrate the sex-chromosome pairing in the heterogametic sex produced by this method.   See Hughes-Schrader (1947), Helwig (1941), or Smith, (1949, 1952) for an explanation of the mechanism by which XX-XY arises from XX-XO.

398 a What is the Barr body (*sex chromatin*)?
b Discuss the relationship between this body and the X chromosome with supporting evidence.
c What is the functional significance of the Barr body?   Discuss cytogenetic evidence to support your answer.
For a full treatment of the Barr body, see Barr (1959 or 1963), Mittwoch (1967), or Ohno (1967).

REFERENCES

Allen, C. E. (1917), *Science*, **46**:466.
——— (1940), *Bot. Rev.*, **6**:277.
Allen, E. (ed.) (1939), "Sex and Internal Secretions," Williams & Wilkins, Baltimore.
Baker, R. J., and T. C. Hsu (1970), *Cytogenetics*, **9**:131.
Baltzer, F. (1935), *Collect. Net*, **10**:3.
Barr, M. L. (1959), *Science*, **130**:679.
——— (1963), in C. Overzier (ed.), "Intersexuality," pp. 48–71, Academic, New York.
Bartlett, D. J., W. P. Hurley, C. R. Brand, and E. W. Poole (1968), *Nature*, **219**:351.
Bellamy, A. W. (1936), *Proc. Natl. Acad. Sci.*, **22**:531.
Bilge, E. (1955), *Rev. Fac. Sci. Univ. Istanbul*, **B20**:121.
Bridges, C. B. (1916), *Genetics*, **1**:1.
——— (1921), *Science*, **54**:252.
——— (1925), *Am. Nat.*, **59**:127.
Burns, R. K. (1961), in Young (1961), pp. 76–158.
Carr, D. H., and M. L. Barr (1961), *Can. Med. Assoc. J.*, **84**:873.

Chang, C. Y., and E. Witschi (1955), *Proc. Soc. Exp. Biol. Med.*, **89**:150.
———— and ———— (1956), *Proc. Soc. Exp. Biol. Med.*, **93**:140.
Chemke, J., J. M. Carmichael, R. H. Geer, and A. Robinson (1970), *J. Med. Genet.*, **7**:105.
Cooper, H. L., H. S. Kupperman, O. R. Rendon, and K. Hirschhorn (1962), *N. Engl. J. Med.*, **266**:699.
Correns, C. (1928), in E. Baur and M. Hartmann (eds.), "Handbuch der Vererbungwissenschaften," Heft II, Band C, Borntraeger, Berlin.
Crawfurd, M. d'A. (1961), *Ann. Hum. Genet.*, **25**:153.
Crew, F. A. E. (1923), *Proc. R. Soc. (Lond.)*, **B95**:256.
———— (1965), "Sex-determination," 4th ed., Methuen, London.
Darlington, C. D. (1939), *J. Genet.*, **39**:101.
Dobzhansky, T., and J. Schultz (1934), *J. Genet.*, **28**:349.
Drescher, W., and W. C. Rothernbuhler (1964), *J. Hered.*, **55**:91.
Ferrier, P. E., and V. C. Kelley (1967), *J. Med. Genet.*, **4**:288.
Ford, C. E. (1969), *Br. Med. Bull.*, **25**(1):104.
———— et al. (1959), *Lancet*, **1**:711.
———— and J. L. Hamerton (1956), *Nature*, **178**:1020.
Ford, E. B. (1955), "Moths," Collins, London.
Fredga, K. (1964), *Hereditas*, **52**:411.
———— (1970), *Philos. Trans. R. Soc. Lond.*, **B259**:15.
Galan, F. (1951), *Acta Salmanticensia, Cliencias, Secc. Biol.*, **1**:7.
Gallien, L. (1956), *Bull. Biol. Fr. Belg.*, **90**:163.
———— (1959), in I. Brachet, J. Mirsky, and A. E. Mirsky (eds.), "The Cell," vol. I, pp. 399–436, Academic, New York.
Gardner, L. I. (ed.) (1961), "Molecular Genetics and Human Disease," Charles C. Thomas, Springfield, Ill.
Gartler, S. M., C. Vuzzo, and S. Gandini (1962), *Cytogenetics*, **1**:1.
————, S. H. Waxman, and E. Giblett (1962), *Proc. Natl. Acad. Sci.*, **48**:332.
George, W. L. (1970), *Genetics*, **64**:23.
Goldschmidt, R. (1934), *Bibliog. Genet.*, **11**:1.
Gordon, M. (1947), *Genetics*, **32**:8.
Gowen, J. W. (1961), in Young (1961), pp. 3–75.
Heilbronn, A. (1953), *Rev. Fac. Sci. Univ. Istanbul*, **B18**:205.
———— and M. Basarman (1942), *Rev. Fac. Sci. Univ. Istanbul*, **B4**:138.
Helwig, E. R. (1941), *J. Morphol.*, **69**:317.
Hirschhorn, K., W. H. Decker, and H. L. Cooper (1960), *N. Engl. J. Med.*, **263**:1044.
Hofmeyr, J. D. J. (1938), *Union S. Afr. Dept. Agric. For. Sci. 3 Bull.* 187.
———— (1939), *S. Afr. J. Sci.*, **36**:288.
Hughes-Schrader, S. (1947), *Chromosoma*, **3**:52.
———— (1948), *Adv. Genet.*, **2**:127.
———— (1950), *Chromosoma*, **4**:1.
———— (1953), *Chromosoma*, **6**:79.
Humphrey, R. R. (1945), *Am. J. Anat.*, **76**:33.
———— (1948), *J. Exp. Zool.*, **109**:171.
Jacobs, P. A. (1969), *Br. Med. Bull.*, **25**(1):94.
———— and J. A. Strong (1959), *Nature*, **183**:302.
————, W. M. C. Brown, et al. (1959), *Lancet*, **2**:423.
Jones, D. F. (1934), *Genetics*, **19**:552.
———— (1939), *Am. J. Bot.* **26**:412.
Hsu, T. C., R. J. Baker, and T. Utakoji (1968), *Cytogenetics*, **7**:27.
Kallman, K. D. (1968), *Genetics*, **60**:811.
Kirkman, H. N., and E. M. Hendrickson (1963), *Am. J. Hum. Genet.*, **15**:241.
Klinger, H. P., and H. G. Schwarzacher (1962), *Cytogenetics*, **1**:266.

Krishan, A., and R. N. Shoffner (1966), *Cytogenetics*, **5**:53.
Lewis, D. (1942), *Biol. Rev.*, **17**:46.
Lilienfeld, F. A. (1936), *Mem. Coll. Agric. Kyoto Univ.*, **38**:1.
Lillie, F. R. (1916), *Science*, **43**:611.
——— (1917), *J. Exp. Zool.*, **23**:371.
Lindsten, J., P. Bowen, et al. (1963), *Lancet*, **1**:558.
Lyon, M. F. (1969), *Cytogenetics*, **8**:326.
McClung, C. E. (1902), *Biol. Bull.*, **3**:43.
McKusick, V. A. (1964), "On the X-chromosome of Man," American Institute of Biological Sciences, Washington.
Mann, J. D., A. Cahan, A. G. Gelb, N. Fisher, J. Hamer, P. Tippett, R. Sanger, and R. R. Rao (1962), *Lancet*, **1**:8.
Matthey, R. (1951), *Adv. Genet.*, **4**:159.
Mittwoch, U. (1967), "Sex Chromosomes," Academic, New York.
——— (1971), *Nature*, **231**:432.
Montgomery, T. H. (1910), *Science*, **32**:120.
Morgan, T. H., and C. B. Bridges (1919), *Carnegie Inst. Wash., D.C., Publ. 278.*
Müller, H. J. (1925), *Am. Nat.*, **59**:346.
Murray, M. J. (1940), *Genetics*, **25**:409.
——— (1940), *J. Hered.*, **31**:477.
Ohno, S. (1967), "Sex Chromosomes and Sex-linked Genes," Springer-Verlag, New York.
——— (1969), *Annu. Rev. Genet.*, **3**:495.
———, J. Jainchill, and C. Stenius (1963), *Cytogenetics*, **2**:232.
——— and C. Weiler (1961), *Chromosoma*, **12**:362.
Overzier, G. (1963), "Intersexuality," Academic, New York.
Pennock, L. A., D. W. Tinkle, and M. W. Shaw (1969), *Cytogenetics*, **8**:9.
Polani, P. E. (1961), in L. I. Gardner (ed.)., "Molecular Genetics and Human Disease," Charles C. Thomas, Springfield, Ill., pp. 153–178.
——— (1969), *Nature*, **223**:680.
Powers, L. (1945), *Genetics*, **30**:323.
Pyle, R. L., D. F. Patterson, W. C. D. Hare, D. F. Kelly, and T. Digiulio (1971), *J. Hered.*, **62**:220.
Race, R. R., and R. Sanger (1969), *Br. Med. Bull.*, **25**(1):199.
Ray-Chaudhuri, S. P., L. Singh, and T. Sharma (1971), *Chromosoma*, **33**:239.
Rick, C. M., and G. C. Hanna (1943), *Am. J. Bot.*, **30**:711.
Russell, L. B. (1961), *Science*, **133**:1795.
Sarto, G. E., J. M. Opitz, and S. L. Inhorn (1969), in K. Benirschke (ed.), "Comparative Mammalian Cytogenetics," pp. 390–413, Springer-Verlag, New York.
Schrader, F. (1947), *Evolution*, **1**:134.
Schuster, J., and A. G. Motulsky (1962), *Lancet*, **1**:1074.
Short, R. V., J. Smith, T. Mann, E. P. Evans, J. Hallett, A. Fryer, and J. L. Hamerton (1969), *Cytogenetics*, **8**:369.
Smith, S. G. (1949), *Evolution*, **3**:344.
——— (1952), *J. Morphol.*, **91**:325.
Sonneborn, T. M. (1947), *Adv. Genet.*, **1**:262.
Stern, C. (1963), *Am. J. Med.*, **34**:715.
Stevens, N. M. (1905), *Carnegie Inst. Wash., D.C., Publ. 36.*
Stewart, J. S. S. (1957), personal communication to P. E. Polani.
Storey, W. B. (1953), *J. Hered.*, **44**:70.
Sturtevant, A. H. (1945), *Genetics*, **30**:297.
Tanaka, Y. (1922), *J. Genet.*, **12**:163.
Tjio, J. H., and A. Levan (1956), *Hereditas*, **42**:1.
Warmke, H. E. (1946), *Am. J. Bot.*, **33**:648.

Weiler, C., and S. Ohno (1962), *Cytogenetics*, **1**:217.

Welshons, W. J. (1963), *Am. Zool.*, **3**:15.

────── and L. B. Russell (1959), *Proc. Natl. Acad. Sci.*, **45**:560.

Westergaard, M. (1958), *Adv. Genet.*, **9**:217.

White, M. J. D. (1940), *J. Genet.*, **40**:303.

────── (1941), *J. Genet.*, **42**:143, 173.

────── (1954), " Animal Cytology and Evolution," 2d ed., Cambridge University Press, New York.

────── (1962), *Evolution*, **16**:78.

Whiting, P. W. (1933), *Science*, **78**:537.

────── (1940), *J. Morphol.*, **66**:323.

────── (1943), *Genetics*, **28**:365.

────── (1945), *Q. Rev. Biol.*, **20**:231.

Wilson, E. B. (1906), *J. Exp. Zool.*, **3**:1.

Winge, O. (1922*a*), *J. Genet.*, **12**:145.

────── (1922*b*), *J. Genet.*, **12**:137.

────── (1931), *Hereditas*, **15**:127.

────── and E. Ditlevsen (1947), *Heredity*, **1**:65.

Witschi, E. (1923), *Biol. Zentralbl.*, **43**:83.

────── (1934), *Biol. Rev.*, **9**:460.

────── (1960), *Am. Nat.*, **48**:399.

──────, W. Gilbert, and G. O. Andrew (1931), *Proc. Soc. Exp. Biol. Med.*, **29**:278.

────── and H. M. McCurdy (1929), *Proc. Soc. Exp. Biol. Med.*, **26**:655.

Yamamoto, T. (1953), *J. Exp. Zool.*, **123**:571.

────── (1961), *J. Exp. Zool.*, **146**:163.

────── (1963), *Gen. Comp. Endocrinol.*, **3**:101.

Yokoyama, T. (1959), "Silkworm Genetics Illustrated," Japanese Society for the Promotion of Science, Tokyo.

Young, W. C. (ed.) (1961), "Sex and Internal Secretions," 3d ed., Williams & Wilkins, Baltimore.

Sex
Linkage

11
Sex
Linkage

QUESTIONS

399 By means of a diagram show how sex-chromosome transmission in the homogametic and heterogametic sex causes crisscross inheritance of sex-linked characters. Briefly discuss the deviations from autosomal transmission that are responsible for the reciprocal-cross differences and explain why dominant traits are more frequent in the homogametic and recessive ones in the heterogametic sex.

400 In man the relative proportion of observed autosomal traits which are dominant is much greater than the proportion of those which are recessive. The reverse is true for X-linked traits. Experience suggests, however, that dominant mutations are less numerous than recessive ones. What are some possible reasons why these two observations regarding autosomal traits are at variance?

401 Would you expect recessive autosomal mutations to show up more quickly after their occurrence than recessive sex-linked mutations? Explain.

402 **a** Explain why most of the sex-linked traits thus far discovered in man are recessive whereas most of the autosomal ones discovered are dominant.
b Would you expect sex-linked recessive traits in organisms with an X-Y sex-mechanism to be more frequent in males than in females? Why?

403 Discuss the various lines of evidence that suggest that a Barr body is derived from one X chromosome.

404 State whether true or false and why:
a For a rare X-linked recessive trait, one-fourth of the sons of all daughters of carrier females are expected to be *affected*.
b The critical feature of the pedigree pattern of a rare sex-linked trait is absence of male-to-male transmission.
c Father-to-son transmission of X-linked traits can occur.

405 You are told that *microphthalmia* (small, nonfunctional eyes) in a certain family, for which records going back several generations are available, is controlled by a sex-linked recessive allele. What transmission characteristics would you look for in the pedigree to determine whether this information is correct?

406 In certain organisms males have grandfathers but no fathers. Explain, with the aid of diagrams, how this is achieved.

407 In the domestic fowl (*Gallus domesticus*) the gene for plumage color is sex-linked (Sturtevant, 1912). The dominant allele G determines *gold*-colored, and its recessive allele g determines *silver*-colored plumage. A cross is made between a homozygous *gold*-colored male and a *silver*-colored female. The F_1 males are mated with F_1 females. Give the genotypes and phenotypes for each sex in the F_1 and F_2.

408   In *Drosophila melanogaster* the recessive nonallelic genes *v* (*vermilion*) on the X chromosome and *cn* (*cinnabar*) on chromosome 2 have the same phenotypic effect: they cause a bright-red eye color.   Wild-type eye color is dull red. A *vermilion* male is crossed with a female from a true-breeding *cinnabar* strain. Show the genotypes, phenotypes, and phenotypic proportions in the F_1 and F_2.

PROBLEMS

409   The table shows (in idealized form) the results obtained by Morgan (1922) from reciprocal crosses between true-breeding *gray* and true-breeding *yellow* lines of *Drosophila melanogaster*.   Analyze these reciprocal crosses to show the manner of inheritance, allelic relationships, and the relationship of the traits with sex.

Table 409

Cross	Parents ♀ ♂	F_1	F_2
1	*gray* × *yellow*	♀ *gray* ♂ *gray*	♀ *gray* ♂ $\frac{1}{2}$ *gray* : $\frac{1}{2}$ *yellow*
			3 *gray* : 1 *yellow*
2	*yellow* × *gray*	♀ *gray* ♂ *yellow*	♀ $\frac{1}{2}$ *gray* : $\frac{1}{2}$ *yellow* ♂ $\frac{1}{2}$ *gray* : $\frac{1}{2}$ *yellow*
			1 *gray* : 1 *yellow*

410   In the moth *Abraxas grossulariata* the *wild type* has fairly large spots on the wings; in the *lacticolor* mutant these spots are greatly reduced in size.   The results of reciprocal crosses made by Doncaster and Raynor (1906) are summarized in idealized form.

Table 410

Cross	Parents ♀ ♂	F_1	F_2
1	*lacticolor* × *wild type*	♀ *wild type* ♂ *wild type*	♀ $\frac{1}{2}$ *lacticolor* : $\frac{1}{2}$ *wild type* ♂ — : *all wild type*
			1 *lacticolor* : 3 *wild type*
2	*wild type* × *lacticolor*	♀ *lacticolor* ♂ *wild type*	♀ $\frac{1}{2}$ *wild type* : $\frac{1}{2}$ *lacticolor* ♂ $\frac{1}{2}$ *wild type* : $\frac{1}{2}$ *lacticolor*
			1 *wild type* : 1 *lacticolor*

a Explain cytogenetically why the reciprocal crosses give different results.
b Which sex is heterogametic and why?

411 A virgin *Drosophila* female whose thorax bristles are very *short* is mated with a male having normal (*long*) bristles. The F_1 progeny are $\frac{1}{3}$ *short*-bristle females, $\frac{1}{3}$ *long*-bristle females, $\frac{1}{3}$ *long*-bristle males. A cross of the F_1 *long*-bristle females with their brothers gives only *long*-bristle F_2 progeny. A cross of *short*-bristle females with their brothers gives $\frac{1}{3}$ *long*-bristle females, $\frac{1}{3}$ *short*-bristle females, $\frac{1}{3}$ *long*-bristle males. Explain genetically.

412 The cinnamon variety of canary, which has brown-tinted feathers, due to the presence of chocolate-colored melanin and the absence of black melanin, has *pink* eyes. The green variety (a misnomer since it has both these pigments in its feathers and is black in color) has *black* eyes. Eye color is monogenically controlled. When two *pink*-eyed birds are mated, the progeny are always *pink*-eyed. When *pink*-eyed hens are mated with *black*-eyed cocks, all the offspring of both sexes are *black*-eyed; and when *black*-eyed hens and *pink*-eyed cocks are bred together, all male offspring are *black-eyed* and females, with a few exceptions, are *pink*-eyed (Durham, 1927).
a *Black* eyes is dominant and Z-linked. Why?
b What results would you expect in the F_1 and F_2 of the cross *black* male × *pink* female?
c Describe the breeding procedure you would use with the F_2 progeny from the cross in (b) to establish a true-breeding variety of *pink*-eyed canaries. Would it be easier to establish a true-breeding variety of *black*-eyed canaries? Explain.
d In matings of *black*-eyed hens and *pink*-eyed cocks, *black*-eyed hens occasionally appear in the progeny. This is thought to be the result of a meiotic abnormality. Suggest what this abnormality may be and how it would lead to this occurrence.

413 Mothers *afflicted* with the rare trait *ocular albinism* (nearly complete absence of eye pigment) always have *afflicted* sons, but the sons of *afflicted* fathers are hardly ever *afflicted* (Gillespie, 1961). Explain.

414 In the fish *Lebistes reticulatus* certain individuals possess a black spot on the dorsal fin as a result of the presence of the *maculatus, Ma*, gene. The trait has the following transmission characteristics. A male having it mated with a female lacking it transmits the trait to all the male offspring but none of the females. These females, moreover, do not transmit the trait to their offspring. The offspring of *unspotted* parents never show the trait. The trait does not appear in sons unless the father also expresses it (Winge, 1922a).
a Describe and illustrate diagrammatically the mode of inheritance of this trait.
b State which is the heterogametic sex and why.
c Is the *maculatus* gene dominant or recessive? Explain.

415 a In the domestic fowl (*Gallus domesticus*) a gene affecting the rate of feather
development is sex-linked; the allele *K* for *slow* feathering is dominant to
k for *rapid* feathering (Warren, 1925). *Rose* comb vs. *single* comb is controlled by an autosomal gene, the allele *R* for *rose* comb being dominant to
r for *single* comb (Bateson and Punnett, 1905, 1908). A *KK RR* rooster is
mated with a *rapid*-feathering, *single*-comb hen. Give the genotypes and
phenotypes of the F_1 and the F_2 for each sex.

b A *slow*-feathering, *rose*-comb male is crossed with a *slow*-feathering, *single*-
comb female. All the male progeny are *slow*-feathering, $\frac{1}{2}$ being *rose*-comb
and $\frac{1}{2}$ *single*-comb. Among the female progeny $\frac{1}{4}$ are *slow*-feathering, *rose*
comb; $\frac{1}{4}$ *rapid*-feathering, *rose* comb; $\frac{1}{4}$ *slow*-feathering, *single*-comb; and $\frac{1}{4}$
rapid-feathering, *single*-comb. Determine the genotypes of the parents.

416 What evidence would be required to prove that a character in man is due
to a gene on the Y chromosome that has no allele on the X?

417 Discuss the data given in an animal species with the X-Y mechanism of sex
determination, describing briefly the number of genes segregating, their
location, types of interallelic and intergenic relation, and any other features you
may notice.

Table 417

Parents		F_1
♀	♂	
albino × brown	♀	25 *black*, 50 *brown*, 24 *cream* (dead at birth)
	♂	96 *albino*

418 Four common varieties of platyfish are *ruber* (grayish-red with spots), *golden
ruber* (gold with spots), *gray* (grayish red without spots), and *gold* (gold
without spots). All wild platyfish are *ruber*. From these the other three

Table 418

Cross	Parents		Offspring
	♀	♂	
1	*ruber* × *ruber*		all *ruber*
2	*ruber** × *ruber* F_1		all *ruber*
	ruber† × *ruber* F_1		$\frac{3}{4}$ *ruber* : $\frac{1}{4}$ *golden ruber*
3	*golden ruber* × *golden ruber*		all *golden ruber*
4	*golden ruber* × *gray*		♀ *gray*, ♂ *ruber*
5	*gray*‡ × *ruber*‡		3 *ruber*, 1 *golden ruber*, 3 *gray*, 1 *gold*
6	*gold* × *gold*		*gold* only

* Certain F_1 females.   † Other F_1 females.   ‡ Offspring of previous cross.

types originated. The series of crosses tells the story of the origin of the *gold* variety (Gordon, 1935). Discuss the underlying genetic control of these phenotypes, diagraming the six-generation path leading to the *gold* variety.

419 In *Melandrium,* a flowering plant belonging to the pink family, the males are XY, the females XX. In 1931 Winge reported studies involving the heritable trait *aurea* (foliage is a blotchy yellowish-green). *Aurea* males crossed with *normal* females produced either all *normal* progeny or 1 *normal* female : 1 *normal* male : 1 *aurea* male. *Normal* males in crosses with *normal* females always produced *normal* female progeny, but in some such matings the males, instead of being all *normal*, consisted of *normals* and *aureas* in a 1:1 ratio. No *aurea* females were ever found. What is the genetic basis for this condition?

420 *Notch* is a dominant trait, and *vestigial* is recessive. A female *Drosophila* with *long, notch* wings crossed with a male with *vestigial* (very short, stubby) wings produced the following unusual progeny:

Long, notch females	10
Long, nonnotch females	11
Vestigial females	22
Long, nonnotch males	11
Vestigial males	10

Note: Classification for *notch* vs. *nonnotch* is not possible with *vestigial*-winged flies.
a What else is unusual about the ratio, and what does it suggest?
b Is more than one allele pair involved? If so, classify each as autosomal or X-linked. Explain your decision.
c Is gene interaction involved? What kind?
d Using your own symbols, diagram the cross, giving the genotypes of the parents and the progeny.

421 In *Drosophila,* two *wild-type* females were crossed to the same *wild-type* male. Although both females produced *wild-type* and *vermilion*-eye progeny, the ratios were different in the two populations. Why do the two females give different results?

Table 421

♀ parent	F₁			
	♂		♀	
	Wild type	*Vermilion*	*Wild type*	*Vermilion*
1	63	58	111	0
2	0	62	119	0

422   The enzyme glucose-6-phosphate dehydrogenase (G6PD) commonly occurs in at least three forms (*a*, *ab*, *b*) in most tissues of the deer mouse (*Peromyscus maniculatus*). This genetically controlled polymorphism was studied in four families by Shaw and Barto (1965), who obtained the results shown. Is the gene for G6PD autosomal or sex-linked. Why?

Table 422

	G6PD phenotypes						
	Parents		Offspring				
			a		*ab*		*b*	
Family	♀	♂	♀	♂	♀	♂	♀	♂
1	*b*	*b*					4	2
2	*a*	*ab*	1	3	1	1		
3	*ab*	*b*				4	2	1
4	*ab*	*ab*	1	2	2	5		2

423   Glucose-6-phosphate dehydrogenase (G6PD) specific to the erythrocytes of hares was studied by Ohno et al. (1965) in each of the two species of wild hares of Europe, *Lepus europaeus* and *L. timidus*, and in the reciprocal hybrids between them. Starch-gel electrophoresis revealed the following:

1   The single sharp band of the enzyme of *L. europaeus* was faster than that of *L. timidus* at both pH 7.0 and 8.6.
2   Each male hybrid had a single band of enzyme identical with that of its mother. Both parental types of G6PD coexisted in female hybrids.

Do these data indicate that G6PD is controlled by a sex-linked gene? Explain.

424   Recent studies of *resistant rickets* (the patient does not respond to normal doses of vitamin D) indicate that the trait is controlled by an X-linked dominant allele (Winters et al., 1958).
   a If this is true, what phenotypes would you expect among the male and among the female children of the following marriages:
   (1) An *affected* male and a *normal* female?
   (2) An *affected* female from marriage in (1) and a *normal* male?
   b Explain why more women than men show the trait.

425   In cats, a sex-linked pair of alleles, *Bb*, controls the color of the fur. The allele *B* for *yellow* is incompletely dominant over *b* for *black*, so that *Bb* individuals are *tortoise-shell*, a splotchy mixture of yellow and black hairs.
   a You and a friend are strolling down the street and see a *tortoise-shell* cat and you tell your friend the cat is a female. Your friend doubts your statement. How would you explain to him why you are correct?

b A *yellow* male is crossed with a *tortoise-shell* female. If the female has a litter of 6 males, what colors might they be?

c A *yellow* cat has a litter of 2 *tortoise-shell* and 1 *yellow*. What is the probable sex of the *yellow* kitten?

d A *tortoise-shell* female has a litter of 7: 2 *yellow* females, 2 *tortoise-shell* females, 1 *black* and 2 *yellow* males. What was the probable genotype and phenotype of the father?

e A *black* female has a litter of 6: 3 *black* males, 2 *tortoise-shell* females, and 1 *black* female. What were the genotypes and phenotypes of the parents?

f A *tortoise-shell* cat brings home her litter of *black*, *yellow*, and *tortoise-shell* kittens. By what criteria might it be possible to incriminate the *yellow* tom living next door as the probable father if there are no other *yellow* toms in the neighborhood? What criteria would exonerate him?

426 Sterile human males with the *testicular-feminization* syndrome have female external genitalia and breast development but a blind vagina and no uterus. They carry inconspicuous testes located in the abdomen or the groin and are of normal male karyotype *AAXY*. The pedigree, from Schreiner (1959), is typical for this trait.

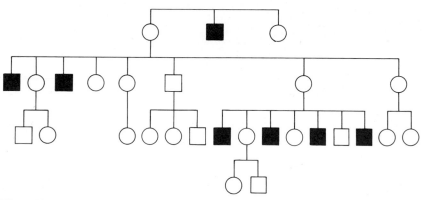

Figure 426

a What are the possible modes of inheritance of *testicular feminization*? Explain. What modes are definitely eliminated? Explain.

b By what means could you distinguish between the possible modes of inheritance?

427 The pedigree shown is typical for *incontinentia pigmenti*, a congenital human skin abnormality usually associated with skeletal and other malformations. In the fully developed disease, the skin shows swirling patterns of melanin pigmentation, especially in the trunk, giving a "marble cake" appearance.

The pigmentation fades gradually and usually disappears completely by age twenty (Haber, 1952; McKusick, 1964).

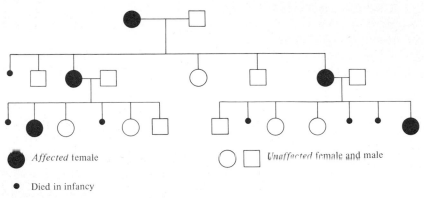

Affected female

Unaffected female and male

● Died in infancy

Figure 427

a What is the most likely mode of inheritance and why?

b What other modes are possible, and what criteria would be needed to eliminate them as possibilities?

428 A cross is made between a female *Drosophila* heterozygous for the recessive genes *cn* (*cinnabar* eye) and *y* (*yellow* body) and a *cinnabar* male. Among the female progeny the phenotypes are $\frac{1}{2}$ *wild type* and $\frac{1}{2}$ *cinnabar*. Among the male offspring the phenotypes were $\frac{1}{4}$ *wild type*, $\frac{1}{4}$ *yellow*, $\frac{1}{4}$ *cinnabar*, and $\frac{1}{4}$ *cinnabar* and *yellow*. Which, if either, of the genes is sex-linked? Show how the experimental results support your conclusion. *Note: wild type = red* eye, *gray* body; these normal traits are frequently omitted in descriptions (viz. *cinnabar = cinnabar, gray*).

429 In man *pseudohypertrophic muscular dystrophy* is characterized by a gradual wasting away of the muscles. It begins in childhood and usually ends in death in the early teens. The trait is controlled by a recessive sex-linked allele (Pearson, 1933; Morton and Chung, 1959).

a It occurs in boys but never in girls. Explain why.

b Since all boys with the trait die before attaining sexual maturity, why does the allele persist in the population?

c A couple have 2 *normal* daughters and 4 sons between the ages of nine and fourteen, the youngest being *afflicted*.

   (1) If this couple came to you for advice, what would you tell them regarding their chances of having another *muscular dystrophic* child?

   (2) If the *normal* children married *normal* individuals, could some of their children suffer from this condition? Explain.

430 *Hemophilia B* (*Christmas disease*) in man is controlled by a recessive sex-linked allele (Whittaker et al., 1962).

a Could a father and son both be *hemophilic?*

b Explain why *hemophilic* mothers always have *hemophilic* sons yet this is rarely true when fathers are *hemophilic.*

c What are the chances that a *normal* daughter of a *hemophilic* father and *normal* mother who marries a *normal* man will have a *hemophilic* child? Would it be male or female?

d A father and his son are both *hemophilic.* Could the son have inherited the allele from his father? Explain.

431 In canaries, *congenital loco* is a hereditary condition characterized by an inability to feed or drink, and death occurs within a few days of hatching. A canary breeder appeals to you for advice in eliminating this condition from his breeding flock. He has found that two of four males used in breeding carried the causative gene. When mated with unrelated normal females, they produced 152 offspring, 37 of these being *affected* (all females). He could test his males in single-pair matings but does not want to discard any more breeding stock than is necessary to eliminate the gene. Briefly state what he should do to achieve his ends and why.

432 The Rosy Grier Carrier Pigeon has *cream*-colored body plumage and a *gray* head. A male with a *cream*-colored head appears in the variety, and the owner wants to incorporate the trait into the variety to get an *all-cream* pigeon. The cross of a *cream-head* male with a *gray-head* female gives progeny in the ratio of 1 *gray-head* male : 1 *cream-head* male : 1 *gray-head* female. Approximately one-fourth of the eggs fail to hatch (Lienhart, 1937).

a Is the trait controlled by a sex-linked gene?

b Describe the phenotypic action of the *cream-head* allele.

433 In man, the presence of a fissure in the iris (*coloboma iridis*) is controlled by a sex-linked recessive allele. An *afflicted* daughter is born to a *normal* couple. The husband sues his wife for divorce on the grounds of infidelity. You are asked to explain the manner of inheritance of the affliction to the jury and to state the conditions under which the husband's lawyer could use the birth of the *afflicted* daughter as evidence of infidelity. Present your testimony.

434 In *Drosophila melanogaster* the allele pairs *Ww*, *Yy*, *Mm*, and *Bb* controlling *red* vs. *white* eyes, *gray* vs. *yellow* body, *normal* vs. *miniature* wing, and *bar* vs. *nonbar* eye, respectively, are sex-linked. A bilateral gynandromorph (left half male, right half female) is found with the male half showing the traits *white*, *yellow*, and *miniature* and the female half exhibiting all the dominant traits.

a How could such an individual arise?

b What is the genotype of the X chromosome in the male portion of the fly? Show diagrammatically how this fly may have been produced.

435 In certain coccids many independently inherited genes are known. All known genes show crisscross inheritance; i.e., when males from a true-breeding strain with the dominant traits are crossed with females from a true-breeding strain with the recessive traits, the female offspring express the dominant traits and the male offspring the recessive ones. Offer an explanation to account for these facts. How would you proceed to test your explanation? See Brown and Wiegmann (1969).

436 In the squash bug (*Anasa tristis*) the progeny of a certain mating consists of 402 females and 194 males. Suggest a possible explanation for these results.

437 In the moth *Papilio glaucus* the male always has a *black-and-yellow* pattern. There are two types of females, one like the male and a *black* form which mimics other insects.

  a What is the probable mode of inheritance of *black*?

  b From a mating between a *black* female and a *black-and-yellow* male a gynandromorph is produced, *black* on the female side and *black-and-yellow* on the male side. Give the genotypes of the zygote and the two parts of the gynandromorph and explain how this corroborates or fails to corroborate the explanation provided for the mode of inheritance. Is it possible to determine from these results which is the heterogametic sex?

438 **a** A person with Klinefelter's syndrome (XXY) whose parents are *normal* is red-green *color-blind* (a condition known to be caused by a sex-linked recessive). Explain this anomaly of sex-linked inheritance.

  b A second Klinefelter individual whose parents are also *normal* is red-green *color-blind* in one eye and *normal* in the other. How does the probable course of events leading to this condition differ from that in (a)?

439 In *Drosophila melanogaster* the eye colors *red*, *wine*, *coral*, and *white* are controlled by alleles of a single sex-linked gene. A *red* female is crossed with a *white* male, and the following progeny are produced: 32 *red* males : 37 *coral* males : 38 *red* females : 35 *coral* females. When a *red* female is crossed with a *coral* male, their progeny consists of 21 *red* males : 23 *wine* males : 26 *red* females : 22 *wine* females. Determine the dominance relationships of the alleles involved.

440 In cats the sex-linked pair of alleles *Bb* is responsible for coat color. *B/Y* and *BB* cats are *yellow*, *Bb* are *tortoise-shell*, and *b/Y* and *bb* cats are *black*. Occasionally sterile *tortoise-shell* males occur with 39 chromosomes instead of the normal number of 38 chromosomes (Thuline and Norby, 1961; Chu et al., 1964). Advance a hypothesis to explain these findings. Show the genotype of the parents.

441 Two true-breeding *white*-eyed strains of *Drosophila melanogaster* reciprocally crossed give the results shown. Give a complete explanation of these results, including:

a The number of allele pairs involved.
b The type of gene interaction.
c The location of allele pairs.
d The genotypes of the parents in both crosses.

Table 441

Parents				F_2			
♀	♂	F_1	Sex	Wild type	Scarlet	Brown	White
white strain 1 × white strain 2		all wild type	♀	182	62	60	20
			♂	90	35	35	164
				272	97	95	184
white strain 2 × white strain 1		♀ wild type ♂ white	♀	85	32	29	166
			♂	95	28	31	174
				180	60	60	340

442 A *white*-feathered hen is crossed with a *colored, nonbarred* cock. The F_1 birds are all *white*-feathered. In the F_2, 130 birds are *white*-feathered, 15 are *colored, barred,* and 15 are *colored, nonbarred*. *Note: Barred* vs. *nonbarred* feather pigmentation pattern can express itself only in *colored* birds. Offer a genetic explanation to account for these results.

443 In *Drosophila melanogaster, vermilion* eye is controlled by a recessive sex-linked gene. In exceptional cases, *vermilion* female × *red* (wild-type eye) male produce, in addition to the usual *vermilion* males and *red* females, a few *vermilion* females and *red* males.
  a What are the possible explanations of such results? Which is the most satisfactory?
  b Show all the classes of offspring that can be predicted according to the latter explanation when the *vermilion* F_1 females are crossed with *red* males.

444 The incompletely sex-linked pair of alleles *Bb* in *Drosophila* controls bristle size; the dominant allele *B* causes *wild-type* (long) bristle whereas the recessive allele *b* (bobbed) causes *short*, thin bristles.
  a Is it possible for this pair of traits to show crisscross inheritance? Illustrate.
  b How would you prove that both the X and the Y chromosomes carry the locus for this pair of alleles?

445 In *Drosophila* the eye colors *red, blood,* and *white* are controlled by alleles of the same gene. From the pedigree determine the order of dominance. Do the results indicate whether the locus is autosomal or sex-linked? Give reasons for your answers.

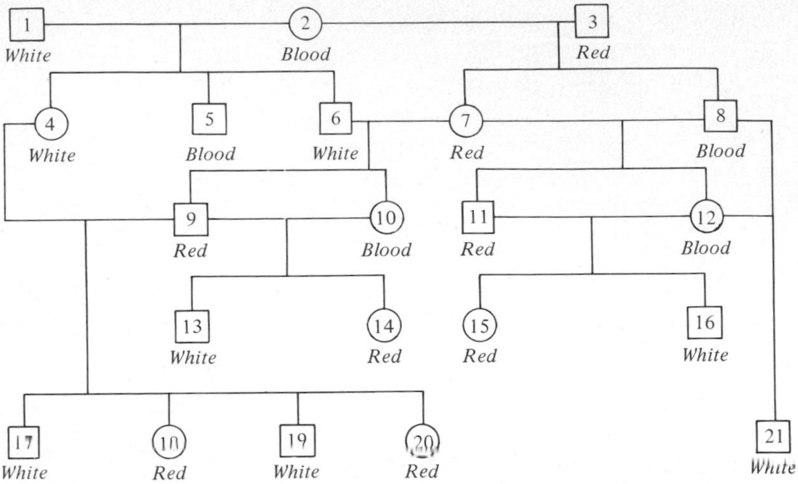

Figure 445

446 All the offspring of a *long*-winged, *gray*-bodied *Drosophila* female crossed with a *short*-winged, *yellow*-bodied male have *long* wings and *gray* bodies. The F_1's when intercrossed produce

Long, gray males	30
Long, gray females	62
Long, yellow males	30
Short, gray males	11
Short, gray females	24
Short, yellow males	10

Advance an explanation for these data and show the genotypes of the parents and F_1 and F_2 phenotypes.

447 In *Bryonia dioica*, a dioecious species with the X-Y method of sex determination, females from a clone with *yellow* flowers are crossed with males from a clone with *white* flowers. The *orange*-flowered F_1 males and females when crossed produced the types of progeny shown. Account completely for the F_1 and F_2 results.

Table 447

Flower color	Sex	Number
orange	♂	232
white	♂	303
orange	♀	450
yellow	♀	161
yellow	♂	76

448 In the mouse, which has an X-Y mechanism of sex determination, *Tata* is a sex-linked pair of alleles; *Ta* is incompletely dominant to *ta*. The genotypes may be distinguished phenotypically as follows:

TaTa and *Ta*/Y   *tabby* (patterned coat)

Tata   heterozygous tabby (mosaic phenotype; some patches of the coat are *tabby*, others *wild type*, and still others intermediate)

tata and *ta*/Y   *wild type*

In 1959 Welshons and Russell reported the occurrence of unexpected phenotypes among the female progeny of several crosses, two of which are presented in Table 448*A*.

Table 448*A*

| Cross | ♀ progeny | |
	Expected	Exceptional
1. *tata* × *Ta*/Y	$\frac{152}{154}$ *heterozygous tabby*	$\frac{2}{154}$ *wild type*
2. *Tata* × *ta*/Y	$\frac{501}{503}$ *wild type and tabby heterozygous*	$\frac{2}{503}$ *tabby*

a Outline two hypotheses (one implicating events in the male parent and the other the female parent as being responsible for the conditions) to explain the occurrence of the exceptional females and give the genotypes of these females according to each hypothesis. In this question it is assumed that the roles of the X and Y in sex determination are not known as they are in man.

b Outline genetic tests required to distinguish between these hypotheses and show the expectations on the basis of each.

c In a cross of exceptional *tabby* females derived from cross 2 with *wild-type* males the offspring shown in Table 448*B* were obtained. Which hypothesis do these results support? Are the results of this cross evidence that your hypothesis is correct? Explain.

Table 448*B*

| Cross | | Offspring | |
♀	♂	♀	♂
tabby × *wild*		7 *heterozygous*, 7 *wild*	12 *tabby*

d What chromosome counts are possible in the exceptional females if your hypothesis is correct? What is the most probable mechanism causing these exceptional females? On the basis of the data given, does it occur primarily in the male or female? Why?

e Is it possible on the basis of the above genetic data and cytological observations to determine the location of male-determining genes? Female-determining genes? Explain.

f Compare the mechanism of sex determination in the mouse with that in man and *Drosophila*.

449 A holandric gene is known in man which causes *hypertrichosis* (long hair growth) of the ears. If *hypertrichotic* men marry *normal* women:

a What proportion of the sons would you expect to be *hypertrichotic*?

b What proportion of the daughters would you expect to express the trait?

c What ratio of *hypertrichotic* : *normal* adults is expected?

450 In the pigeon *almond* vs. *blue* plumage is controlled by a Z-linked pair of alleles; the allele for *almond* is dominant. *Crested* vs. *crestless* is controlled by an autosomal pair of alleles; the allele for *crested* is dominant.

a A homozygous *almond, crested* male is crossed with a *blue, crestless* female. What are the expected phenotypes of the male and female offspring and the F_2?

b An *almond, crestless* male mated with a *blue, crested* female produces 1 *blue, crested* male, 1 *blue, crestless* male, and 2 *almond, crested* females. What are the genotypes of the parents?

c Two *almond, crested* birds mate to produce 1 *almond, crested* male and 1 *blue, crestless* female. Give the genotypes of the parents.

451 In *Drosophila*, the allele *g* for *garnet* eye is recessive to *G* for *wild type*. Many matings have been made between the true-breeding strains for these two traits. One very unusual result was obtained from a mating between a female of the *garnet* strain and a male of the *wild-type* strain. This mating yielded rather poorly: only 136 flies were produced, of which 72 were *garnet* females and 64 were *wild-type* males. Present a hypothesis to explain these unusual results and indicate how you might test it.

452 A useful technique in studying the ability of various mutagens to induce mutations producing visible effects is to mate treated males with X-X (attached-X) females. What are the advantages of this technique over the use of normal females?

453 In the domestic fowl the allele pair *B* (*barred* feathers) vs. *b* (*nonbarred* feathers) is sex-linked. The chicks that will develop *barred* feathers later in life have a whitish spot in the occipital region of the head. The *nonbarred* chicks lack this spot. Rhode Island Reds (*nonbarred*) are crossed in common hatcheries with *barred* Plymouth Rocks to produce F_1 females with high egg production. Outline how chick hatcheries can utilize the *Bb* pair of alleles for sexing the chicks as soon as they come out of the incubator so that egg producers, who require only the females, will get no males.

454 Normally marriages between *normal* males and women with *ocular albinism* (sex-linked recessive trait) (Gillespie, 1961) result in *normal* daughters and *affected* sons, and the reciprocal mating produces only *normal* offspring. Occasionally, however, in a marriage between a woman with *ocular albinism* and a *normal* man, a son is *normal* and a daughter has *ocular albinism*. Explain how such results may occur and how you might confirm your hypothesis cytologically (studying the somatic-cell karyotype).

455 Nowakowski et al. (1959) found 3 instances of red-green *color blindness* in 34 Klinefelters with XXY (similar results have been obtained by other workers). None of these boys had a *color-blind* father, but one had a *color-blind* mother.

   a Outline with diagrams what you consider to be the most probable manner in which each of these three examples originated. See Nowakowski and Lenz (1961) for their explanation of the origin of these types.

   b From these data, which parent appears to be responsible for the Klinefelter's syndrome?

456 In man *migraine* is dominant to the *normal* condition and autosomal. *Agammaglobulinemia* (a condition causing susceptibility to bacterial infection) is recessive and sex-linked (Garvie and Kendall, 1961; Janeway et al., 1953). A man suffering from *migraine* marries a *normal* woman whose father was afflicted with *agammaglobulinemia*.

   a Would any of their children be expected to have *migraine*? *Agammaglobulinemia*? Both? Explain diagrammatically.

   b You find that a couple have 7 children, 3 girls and 4 boys, all suffering from *migraine*. The girls have *normal* vision, but two of the boys have *agammaglobulinemia*. What are the genotypes of the parents?

   c One of the daughters is married to a *normal* man and wonders what the chances are that her children will suffer from the one or the other of these conditions. What would you tell her?

457 *Nystagmus*, a condition in which involuntary oscillation of the eyes causes impaired vision, is controlled by a sex-linked recessive allele (Billings, 1942; Waardenburg, 1953). A man and his wife with *normal* vision have 4 offspring, all married to *normal* individuals. The first son is *affected* with *nystagmus* but has a *normal* daughter; the second son, who is *normal*, has a *normal* daughter and a *normal* son; the first daughter, also *normal*, has 8 *normal* sons; the second daughter has 2 *normal* daughters and 2 sons, one *normal*, the other *affected*.

   a What are the genotypes of all individuals in this family?

   b If the original parents had had another child, what are the chances that this child would have been a *normal* boy? An *affected* boy? An *affected* girl?

   c If the original mother was *affected* and the father was *normal*, could any of their children show the recessive sex-linked traits? Explain.

458 In man, *normal* color vision vs. red-green *color blindness* is controlled by allelic genes *C* (dominant) and *c* (recessive) at an X-linked locus.

  a Can a *normal* daughter have a *color-blind* father? A *normal* father? A *color-blind* mother? A *normal* mother?

  b Answer the same questions for a *normal* son.

  c Can two *normal* parents have a *color-blind* son? A *color-blind* daughter?

  d Can two *color-blind* parents have a *normal* daughter? A *normal* son?

  e A brother and sister are both *color-blind*. Is it possible for them to have:

  (1) A *normal* brother?

  (2) A *normal* sister?

  (3) One parent *normal*, one *color-blind*?

459 The following pedigrees show typical transmission patterns of rare or fairly rare hereditary human traits. For each trait state, with reasons, whether the gene responsible is (1) dominant or recessive and (2) autosomal or X-linked. *Note:* In each case assume the gene is completely penetrant. If the mate of a parent is not shown, assume the mate is of *normal* phenotype and genotype.

  a *Ectodermal dysplasia* (absence of teeth, hypotrichosis, and absence of sweat glands).

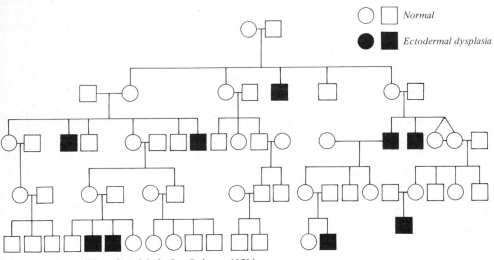

Figure 459*A*   (*Modified after Roberts, 1929.*)

  b *Agammaglobulinemia* (individuals lack important blood proteins and are extremely prone to infection).

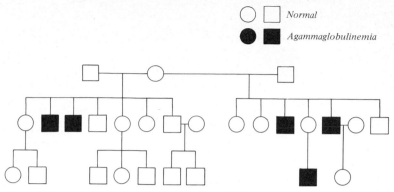

Figure 459B   (*Modified after Garvie and Kendall, 1961.*)

c *Hypophosphatemia* (low serum phosphorus), usually accompanied by vitamin D–resistant rickets (skeletal rickets not corrected by feeding vitamin D).

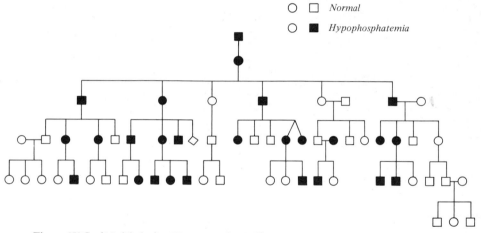

Figure 459C   (*Modified after Winters et al., 1958.*)

d Deficiency for the enzyme glucose-6-phosphate dehydrogenase.

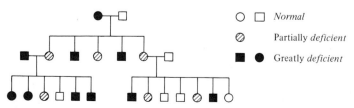

Figure 459D   (*Based on data of Kirkman and Hendrickson, 1963.*)

e *Xg* blood antigen.

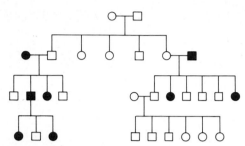

Figure 459E (*Modified after Sanger, 1965*)

f Reifenstein's syndrome (individuals with male karyotype that express hypo-
gonadism with gynecomastia and hypospadias).

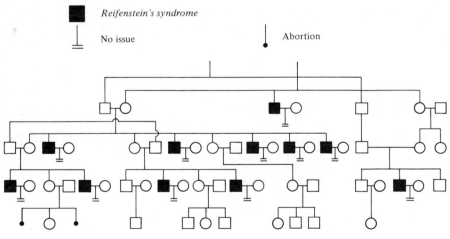

Figure 459F (*Modified after Reifenstein, 1947, and McKusick, 1964.*)

460 State with reasons for your decision whether the *rare* trait (solid symbols)
in each of the following pedigrees is more likely to be due to an autosomal
dominant gene whose expression is limited to males or to an X-linked
recessive gene:

a

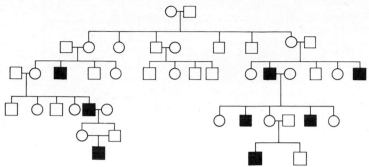

Figure 460*A*

b

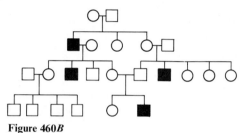

Figure 460*B*

REFERENCES

Abbott, U. K., R. M. Craig, and E. B. Bennett (1970), *J. Hered.*, **61**:95.
Aida, T. (1921), *Genetics*, **6**:554.
Asmundson, V. S., and U. K. Abbott (1961), *J. Hered.*, **52**:99.
Becker, P. E. (1955), *Acta Psychiatr. Neurol. Scand.*, **193**:427.
Bannerman, R. M., G. B. Ingall, and J. F. Mohn (1971), *J. Med. Genet.*, **8**:291.
Bateson, W., and R. C. Punnett (1905), *Proc. Camb. Philos. Soc.*, **13**:165.
———— and ———— (1908), *Poult. Rep. Evol. Comm. R. Soc.*, **4**:18.
Beutler, E., M. Yeh, and L. V. F. Fairbanks (1962), *Proc. Natl. Acad. Sci.*, **48**:9.
Billings, M. L. (1942), *J. Hered.*, **33**:457.
Bridges, C. B. (1916), *Genetics*, **1**:1, 107.
Brinkhous, K. M., and J. B. Graham (1950), *Science*, **111**:723.
Brown, S. W., and L. I. Wiegmann (1969), *Chromosoma*, **28**:255.
Chu, E. H. Y., H. C. Thuline, and D. E. Norby (1964), *Cytogenetics*, **3**:1.
Chung, C. S., N. E. Morton, and H. A. Peters (1960), *Am. J. Hum. Genet.*, **12**:52.
Cooper, H. L., H. S. Kupperman, O. R. Rendon, and K. Hirschhorn (1962), *N. Engl. J. Med.*, **266**:699.
Crawfurd, M. D'A. (1961), *Ann. Hum. Genet.*, **25**:153.
Csorsz, B. (1933), *Ung. Med.*, **2**:180, 1929; cited by E. A. Cockayne, in "Inherited Abnormalities of the Skin and Its Appendages," Oxford University Press, London.
Davenport, C. B. (1912), *J. Exp. Zool.*, **13**:1.
Doncaster, L., and G. H. Raynor (1906), *Proc. Zool. Soc. (Lond.)*, **1**:125.

Dronamraju, K. R., and J. B. S. Haldane (1962), *Am. J. Hum. Genet.*, **14**:102.

Durham, F. M. (1927), *J. Genet.*, **17**:19.

Ford, C. E., P. E. Polani, J. H. Briggs, and P. M. F. Bishop (1959), *Nature*, **193**:1030.

Franceschetti, A., and D. Klein (1957), *Acta Genet. Stat. Med.*, **7**:255.

Garvie, J. M., and A. C. Kendall (1961), *Br. Med. J.*, **1**:548.

Gillespie, F. D. (1961), *Arch. Ophthalmol.*, **66**:774.

Goldstein, J. L., J. F. Marks, and S. M. Gartler (1971), *Proc. Natl. Acad. Sci.*, **68**:1425.

Gordon, M. (1935), *J. Hered.*, **26**:97.

Graham, J. B., V. W. McFalls, and R. W. Winters (1959), *Am. J. Hum. Genet.*, **11**:311.

Haber, H. (1952), *Br. J. Dermatol.*, **64**:129.

Henderson, J. F., W. N. Kelley, F. M. Rosenbloom, and J. E. Seegmiller (1969), *Am. J. Hum. Genet.*, **21**:61.

Hirschhorn, K., W. H. Decker, and H. L. Cooper (1960), *N. Engl. J. Med.*, **263**:1044.

Hoefnagel, D., D. Andrew, N. G. Mireault, and W. O. Berndt (1965), *N. Engl. J. Med.*, **273**:130.

Holm, E. (1926), *Acta Ophthalmol.*, **4**:20.

Hutt, F. B. (1960), *Heredity*, **15**:97.

Janeway, C. A., L. Apt, and D. Gitlin (1953), *Trans. Assoc. Am. Physicians*, **66**:200.

Johnston, A. W., and V. A. McKusick (1962), *Am. J. Hum. Genet.*, **14**:83.

Kirkman, H. N., and E. M. Hendrickson (1963), *Am. J. Hum. Genet.*, **15**:241.

Lenz, W. (1961), "Medizinische Genetik: Eine Einfuhrung in ihre Gundlagen und Probleme," p. 89, Thieme, Stuttgart.

Lesch, M., and W. L. Nyhan (1964), *Am. J. Med.*, **36**:561.

Lienhart, R. (1937), *C. R. Soc. Biol.*, **126**:336.

Lyon, J. B. (1970), *Biochem. Genet.*, **4**:169.

Lyon, M. C., and S. G. Hawkes (1970), *Nature*, **227**:1217.

Lyon, M. F. (1961), *Nature*, **190**:372.

——— (1962), *Am. J. Hum. Genet.*, **14**:135.

McKusick, V. A. (1964), "On the X-Chromosome of Man," American Institute of Biological Sciences, Washington.

——— (1968), "Mendelian Inheritance in Man," 2d ed., Johns Hopkins, Baltimore.

Mann, J. D., et al. (1962), *Lancet*, **1**:8.

Marks, P. A., and R. T. Gross (1959), *J. Clin. Invest.*, **38**:2253.

Migeon, B. R. (1971), *Am. J. Hum. Genet.*, **23**:199.

———, V. M. Der Ka Loustian, W. L. Nyhan, W. J. Young, and B. Childs (1968), *Science*, **160**:425.

Miller, O. J., et al. (1971), *Proc. Natl. Acad. Sci.*, **68**:116.

Morgan, L. V. (1922), *Biol. Bull.*, **42**:267.

Morgan, T. H. (1910), *Science*, **32**:120.

——— (1914), *Am. Nat.*, **48**:577.

——— and C. B. Bridges (1916), *Carnegie Inst. Wash., D.C., Publ.* 237.

——— and H. D. Goodale (1912), *Ann. N.Y. Acad. Sci.*, **22**:113.

———, A. H. Sturtevant, H. J. Muller, and C. B. Bridges (1922), "The Mechanism of Mendelian Heredity," 2d ed., Holt, New York.

Morton, N. E., and C. A. Chung (1959), *Am. J. Hum. Genet.*, **11**:360.

Nowakowski, H., and W. Lenz (1961), *Recent Prog. Horm. Res.*, **17**:53.

———, ———, and J. Oarada (1959), *Acta Endocrinol.*, **30**:296.

Ohno, S. (1967), "Sex Chromosomes and Sex-linked Genes," Springer-Verlag, New York.

———, J. Poole, and T. Gustavsson (1965), *Science*, **150**:1737.

——— and C. Weiler (1962), *Chromosoma*, **13**:106.

Orel, H. (1929), *Z. Kinderheilkd.*, **37**:312.

Pearl, R., and F. M. Surface (1910), *Science*, **32**:870.

Pearson, K. (1933), *Ann. Eugen.*, **5**:179.

Pola, V., and J. Svojitka (1957), *Folia Haematol.*, **75**:43.

Race, R. R., and R. Sanger (1968), "Blood Groups in Man," 5th ed., Blackwell, Oxford.

Reifenstein, E. C. (1947), *Proc. Am. Fed. Clin. Res.*, **3**:86.

Rhodes, K., R. L. Markham, P. M. Maxwell, and M. E. Monk-Jones (1969), *Br. Med. J.*, **3**(Aug. 23):439.

Roberts, E. (1929), *J. Am. Med. Assoc.*, **93**:277.

Russell, L. B. (1961), *Science*, **133**:1795.

Sanger, R. (1965), *Can. J. Genet. Cytol.*, **7**:202.

Scholfield, R. (1921), *J. Hered.*, **12**:400.

Schreiner, W. E. (1959), *Geburtshilfe Frauenheikd.*, **19**:1110.

Schuster, J., and A. G. Motulsky (1962), *Lancet*, **1**:1074.

Shannon, M. W., and H. L. Nadler (1968), *J. Med. Genet.*, **5**:326.

Shaw, C. R., and E. Barto (1965), *Science*, **148**:1099.

Spillman, W. J. (1908), *Am. Nat.*, **42**:610.

Stern, C. (1926), *Biol. Zentralb.*, **46**:344.

——— (1960a), "Principles of Human Genetics," 2d ed., Freeman, San Francisco.

——— (1960b), *Nature*, **187**:905.

Sturtevant, A. H. (1912), *J. Exp. Zool.*, **12**:499.

Tanaka, Y. (1922), *J. Genet.*, **12**:163.

Tettenborn, U., R. Dofuku, and S. Ohno (1971), *Nat. New Biol.*, **234**:37.

Thuline, H. C., and D. E. Norby (1961), *Science*, **134**:554.

Townes, P. L. (ed.) (1969), "The Medical Clinics of North America," vol. 53, no. 3, Saunders, Toronto.

Vandenberg, S. G., V. A. McKusick, and A. B. McKusick (1962), *Nature*, **194**:505.

Voshell, A. F. (1933), *South. Med. J.*, **26**:156.

Waardenburg, P. J. (1953), *Acta Genet. Stat. Med.*, **4**:298.

——— and J. Van den Bosch (1956), *Ann. Hum. Genet.*, **21**:101.

Warren, D. C. (1925), *J. Hered.*, **16**:13.

Weinstein, E. D., and M. M. Cohen (1966), *J. Med. Genet.*, **3**:17.

Weiss, L., and R. C. Mellinger (1970), *J. Med. Genet.*, **7**:27.

Welshons, W. J. (1971), *Genetics*, **68**:259.

——— and L. B. Russell (1959), *Proc. Natl. Acad. Sci.*, **45**:560.

Whittaker, D. L., D. L. Copeland, and J. B. Graham (1962), *Am. J. Hum. Genet.*, **14**:149.

Winge, O. (1922a), *J. Genet.*, **12**:145.

——— (1922b), *J. Genet.*, **12**:137.

——— (1922c), *J. Genet.*, **13**:201.

——— (1931), *Hereditas*, **15**:127.

Winters, R. W., J. B. Graham, T. F. Williams, V. W. McFalls, and C. H. Burnett (1958), *Medicine*, **37**:97.

Yamamoto, T. (1963), *Genetics*, **48**:293.

12
Sex-influenced and Sex-limited Inheritance

QUESTIONS

461 **a** Distinguish between the following, giving the location of the genes, the mode of transmission, and relationship to sex:
(1) Sex-influenced and holandric characters.
(2) Sex-limited and sex-influenced characters.
(3) Sex-linked and sex-influenced characters.
b Maternal-influence genes and cytoplasmic genes also show an association with sex. Describe their location, transmission, and relationship to sex.

462 Sex-limited and holandric traits occur in one sex only. What features of transmission or phenotypic expression serve to distinguish the two types of traits?

463 In mammals and birds the sex hormones have been shown to be the main environmental factors governing the expression of characters that exhibit sexual dimorphism (a constant difference between the sexes, e.g., beards in men, not in women). Does this mean that these characters have no genetic basis? Explain.

464 The male buffalo differs from the female in having a well-developed mane. What genetic explanations might be advanced to explain this dimorphism?

465 A sex-limited trait cannot be expressed unless the individual carries the controlling allele and shows the appropriate sex-chromosome–autosome ratio. Discuss.

466 State whether the following are true or false and why:
a The mechanism of sex determination triggers the process of sex differentiation in all sexually reproducing organisms.
b In organisms with the Z-W method of sex determination, if a trait is found only in the males, the gene is not on the nonhomologous part of the W chromosome.
c The mode of transmission of genes controlling sex-influenced characters is identical to that of genes not associated with sex. It is only the expression that differs in the two sexes.
d Sex-influenced traits occur less frequently among females than males because the alleles causing them are less frequent among females.

467 In sex-influenced inheritance, one may note certain features of inheritance that are also characteristic of sex-linked recessive traits. For example, both *baldness* (sex-influenced) and red-green *color blindness* (sex-linked, recessive) are more common in men than in women. In what other features are these two types of inheritance similar? In what important respect do they differ?

468 An inherited trait occurs in approximately 20 percent of the males but only 4 percent of the females. Is it sex-linked or sex-limited?

PROBLEMS

469   Reciprocal crosses between true-breeding *bearded* and *beardless* breeds of goats give the results shown (Asdell and Smith, 1928).   State, giving reasons for your answers:

a The number of pairs of alleles involved and whether they are autosomal or sex-linked.

b The type of association with sex.

Table 469

Parents					
♂	♀	F_1		F_2	
bearded × *beardless*		♂	*bearded*	♂	3 *bearded* : 1 *beardless*
		♀	*beardless*	♀	3 *beardless* : 1 *bearded*
beardless × *bearded*		♂	*bearded*	♂	3 *bearded* : 1 *beardless*
		♀	*beardless*	♀	3 *beardless* : 1 *bearded*

470   **a** In a certain family both parents have *Heberdon's nodes*, a single-gene trait characterized by enlargement of the terminal joints of the fingers.   There are 8 children: 4 boys, all with *Heberdon's nodes*, and 4 girls, 2 of whom express the trait.   State which of the following types of genes cannot be responsible for the inheritance of *Heberdon's nodes* and why:

(1) Sex-linked dominant.

(2) Sex-limited recessive.

(3) Holandric.

(4) Autosomal dominant.

(5) Sex-influenced, dominant in males, recessive in females.

(6) Sex-linked recessive.

b One of the two *normal* daughters marries a *normal* man.   They have 5 children, 3 *normal* girls and 2 boys, 1 with *Heberdon's nodes*.   One of the *affected* daughters marries a *normal* man and has 6 children, 4 *affected* boys and 2 *normal* girls.   Using this information together with that from the earlier generations of this family, determine the mode of inheritance of the trait and explain your decision.

471   In Ayrshire cattle the allele M^m for *mahogany-and-white* is dominant in males, recessive in females.   The reverse is true for the allele M^f for *red-and-white*.

a A *red-and-white* male is crossed with a *mahogany-and-white* female.   Show the expected F_1 and F_2 genotypic and phenotypic proportions.

b A dairy farmer who raises this breed of cattle would like to be able to distinguish the male from the female calves in his herd by a difference in color.   How should he proceed?

c An Ayrshire breeder has used a *red-and-white* bull to sire his herd. A geneticist visiting the farm comments that all the *mahogany-and-white* calves are males. The farmer is impressed to find that this is true in all cases. Explain why the geneticist was able to predict the sex of the *mahogany-and-white* calves.

d A *mahogany-and-white* cow has a *red-and-white* calf. What is the calf's sex?

472 An inherited human abnormality shows the following features of inheritance:

1 The trait may occur in both sexes; i.e., both sexes may be *affected*.
2 *Affected* persons are usually of one sex.
3 A pair of *normal* parents may have an *affected* child.

a State which of the following modes of inheritance are *eliminated* by the above information, and specify one feature by which you eliminate each:

(1) Sex-limited   (2) Autosomal recessive
(3) Sex-linked recessive   (4) Sex-influenced, dominant in males
(5) Autosomal dominant   (6) Sex-influenced, dominant in females
(7) Holandric   (8) Sex-linked dominant

b The pedigree for this same trait should enable you to decide among the remaining alternatives. State what the mode of inheritance is and why.

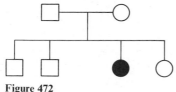

Figure 472

473 In the clover butterfly (*Colias philodice*) the wing color of males is always *yellow*, while that of females may be *yellow* or *white*. The data shown are summarized from studies by Gerould (1941), Hovanitz (1944), Komai and Ae (1953), and Remington (1954). Answer the following questions, giving reasons for your answers:

a Is *white* a sex-limited or a hologynic trait?
b How many genes control its expression?

Table 473

| | Parents | | F_1 | |
Cross	♀	♂	♀	♂
1	yellow × yellow		all yellow	yellow
2	yellow × yellow		$\frac{1}{2}$ yellow : $\frac{1}{2}$ white	yellow
3	yellow × yellow		all white	yellow
4	white × yellow		all white	yellow
5	white × yellow		$\frac{1}{4}$ yellow : $\frac{3}{4}$ white	yellow

c Is *white* dominant or recessive?

d Is the gene (or genes) on (1) the W chromosome, (2) the Z chromosome, or (3) the autosomes?

474   **a** State, with reasons for your decision, whether the rare trait in the pedigree (Fig. 474*A*) is probably due to a sex-limited dominant autosomal allele or to a sex-linked recessive.

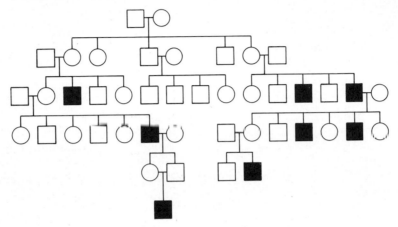

Figure 474*A*

b In the pedigree in Fig. 474*B* the blacked-in symbols refer to girls with transverse vaginal septum. The trait is due to an autosomal recessive gene rather than to an X-linked recessive gene.   Why?

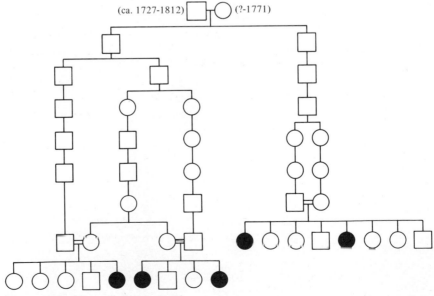

Figure 474*B*   (*From McKusick, 1968.*)

475   In sheep *horned* vs. *hornless* is controlled by a pair of autosomal alleles; *H*, *horned*, is dominant in males and recessive in females, the opposite being true of H^1 for *hornless* (Wood, 1909).   *White* vs. *red* fleece is also controlled by a single pair of autosomal alleles, *W* for *white* being dominant to *w* for *red*.

a A female from a cross between a homozygous *hornless, white* ram and a *horned, red* ewe is mated with a *hornless, red* male.   Show the phenotypes possible in the progeny and the probability of each.

b A *hornless, white* ram is crossed with a *hornless, white* ewe.   Their first lamb is a *horned, red* male.   What are the genotypes of the parents?   If a series of matings between individuals of these genotypes were made, what phenotypes would the progeny exhibit and in what proportions?

c Several matings are made between *hornless, red* ewes and *hornless, white* rams.   Among the male progeny *horned, white* and *hornless, red* occur in a 1:1 ratio.   The females are all *hornless*, but half of them are *white* and the other half *red*.   What are the genotypes of the parents?

d A *horned, red* ram is mated to a *hornless, white* ewe over a number of years; they have several offspring in which the following phenotypes are represented:

Males   *horned, white; horned, red; hornless, white; hornless, red*
Females   *hornless, red; hornless, white*

What are the genotypes of the parents?

476   In man, the absence of upper lateral incisors may be a sex-influenced trait dominant in males, recessive in females.   Red-green *color blindness* is a sex-linked recessive trait.   A woman of *normal* vision whose parents had *no incisors* has seven brothers and two sisters, all with *normal* vision.   She marries a red-green *color-blind* man *with incisors*.

a Give the possible genotypes and phenotypes of the offspring.

b Show why if she had married a *normal* man, any red-green *color-blind* or *no-incisor* children would be males.

477   In a certain species of butterfly all males have *rich-brown* wings, but the females have either *rich-brown*, *olive-green*, or *yellow* wings.   Each color trait is characteristic of a distinct, true-breeding race.   When the three races are

Table 477

Racial origin of parent*		F_1	Female offspring of testcross
♀	♂ (*rich-brown*)	♀	(F₁ × *yellow* race)
rich-brown × *olive-green*		*brownish-green*	1 *rich-brown* : 1 *olive-green*
rich-brown × *yellow*		*rich-brown*	1 *rich-brown* : 1 *yellow*
olive-green × *yellow*		*olive-green*	1 *olive-green* : 1 *yellow*

* The results from reciprocal crosses of parental races are identical.

intercrossed and the F_1's are crossed with the *yellow* strain, the results obtained are as shown.   Analyze these results, and include:

(1) The number of genes, the allelic relationship at each locus, and the type of gene interaction if any.

(2) The chromosome (autosome, Z, or W) involved in transmission.

(3) The relationship, if any, of wing color to sex.

478 In man, the allele B for *baldness* is dominant in males and recessive in females.   The reverse is true for B^1 for *nonbaldness*.

  a A *nonbald* woman whose mother was *bald* marries a *bald* man whose father was *nonbald*.   Give the genotypes of these two and show the types of children they can have with respect to this pair of sex-influenced traits.

  b A *nonbald* woman and a *nonbald* man marry and have 5 children.   The first one, a male, becomes *bald* at the age of 23.   Of the remaining children 2 are boys and 2 are girls.   What are the chances of either of the boys becoming *bald*?   Of both showing the trait?   The girls becoming *bald*?

  c In a certain population 1 percent of the women are *bald*.   How many men are *bald*?   How many women are heterozygous?

479 **a** In the four-spotted cowpea weevil (*Bruchus quadrimaculatus*) the body and elytra colors may be *red*, *black*, *white*, or *tan*.   In 1921 Breitenbecher, crossing homozygous strains, obtained the results shown in Table 479A.

Table 479A

Parents		F_1		F_2	
♀	♂	♀	♂	♀	♂
red × *tan*		53 *red*	57 *tan*	138 *red* : 57 *tan*	221 *tan*
black × *tan*		27 *black*	29 *tan*	153 *black* : 61 *tan*	201 *tan*
white × *tan*		4 *white*	7 *tan*	15 *white* : 8 *tan*	32 *tan*

Table 479B

Parents		F_1		F_2	
♀	♂	♀	♂	♀	♂
red	× *tan* (from a line in which all females are *black*)	850 *red*	923 *tan*	6,992 *red* : 2,335 *black*	9,224 *tan*
red	× *tan* (from a line in which all females are *white*)	420 *red*	418 *tan*	3,371 *red* : 1,253 *white*	4,817 *tan*
black	× *tan* (from a line in which all females are *white*	292 *black*	272 *tan*	1,413 *black* : 502 *white*	1,766 *tan*

Describe the two kinds of association with sex that may account for these results.

b Which of these explanations do the results in Table 479B support and why?

c If your explanation based on results in (b) is correct, would you expect the results shown in Table 479C? Explain giving genotypes.

Table 479C

Parents		Progeny	
♀	♂	♀	♂
F$_1$ *black* (from *black* × *tan*)	*tan* (from *white* strain)	122 *black* : 121 *white*	248 *tan*
F$_1$ *red* (from *red* × *white*)	*tan* (from *tan* strain)	78 *red* : 70 *white*	184 *tan*
F$_1$ *red* (from *red* × *tan*)	F$_1$ *tan* (from *black* × *tan*)	151 *red* : 49 *black* : 61 *tan*	249 *tan*
F$_1$ *red* (from *red* × *tan*)	F$_1$ *tan* (from *red* × *black*)	36 *red* : 16 *black*	54 *tan*

480 Stehr (1955) in studies of intraspecific variability in eastern North American populations of the spruce budworm (*Choristoneura fumigerana* (Clem.)) obtained the tabulated results in crosses between *brown* and *gray* moths from true-breeding strains. Discuss, with the aid of diagrams, the genetic basis for color dimorphism in this organism.

Table 480

	Parents			Progeny	
Generation	♀	♂	Generation	♀	♂
Cross 1:					
P$_1$	*brown* ×	*gray*	F$_1$	*brown* 0 : *gray* 76	*gray*
F$_1$	*gray* ×	*gray*	F$_2$	*brown* 43 : *gray* 44	*gray*
F$_2$	*brown* × *gray*⎫ *gray* × *gray*⎭		F$_3$	*brown* 0 : *gray* 128	*gray*
F$_2$	*brown* × *gray*⎫ *gray* × *gray*⎭		F$_3$	*brown* 50 : *gray* 50	*gray*
Cross 2:					
P$_1$	*gray* ×	*gray*	F$_1$	*brown* 39 : *gray* 0	*gray*
F$_1$	*brown* ×	*gray*	F$_2$	*brown* 118 : *gray* 123	*gray*
F$_2$	*brown* × *gray*⎫ *gray* × *gray*⎭		F$_3$	*brown* 81 : *gray* 70	*gray*
F$_2$	*brown* × *gray*⎫ *gray* × *gray*⎭		F$_3$	*brown* 126 : *gray* 0	*gray*

481 In 1951 Finne and Vike reported on breeding experiments with *atresia isthmi* in White Leghorns, a trait characterized by closure of the oviduct in the region of the isthmus, preventing unlaid eggs from passing through the oviduct. The eggs return to the body cavity and accumulate there, causing the birds to assume an upright penguinlike posture. Peritonitis sets in and causes death at five to six months of age.

One male (with an interruption of the right vas deferens) sired 50 *normal* and 43 *nonlaying* females (dissection of 36 of the latter showed *atresia isthmi* in every case). Among the 4 male offspring tested, 2 sired only *normal* females, a third produced 2 with *atresia isthmi* among 26 females, and the fourth produced 30 *normal* and 24 *nonlaying* females (21 of these were *atresic*). The fourth, mated with hens of a breed known to be free of the defect, sired 2 *atresics* among 5 female offspring. All sons were fertile.

What fairly definite conclusion does the transmission pattern suggest regarding the genetic basis of *atresia*? Regarding other aspects of the genetic basis no conclusions are possible. Why? How would you proceed to distinguish between these alternatives?

482 In 1937 Petterson and Bonnier reported genetic studies of a rare human heritable type of intersex known as *testicular feminization* (see Prob. 426 for a description of the syndrome). Progenies in which the trait appeared consisted of 8 *normal* males, 28 *normal* females, and 22 *affected*. Outwardly and physiologically the *affected* individuals are females and consider themselves *normal* females.

a What chromosome constitution might you expect in the *affected* individuals?
b What are the possible modes of inheritance of *testicular feminization*? Explain.

483 In the damselfly *Ischnura damula* the males all have the same (*andromorphic*) color pattern on the dorsal synthorax; the females have two patterns, one like the male and another termed *heteromorphic*. Johnson (1964) reported studies of the genetic basis of this female dimorphism. In all crosses a virgin female was mated with a single male. The results were as follows:

1 From 14 crosses the ratio of male to female progeny was approximately 45:55.
2 Certain females of either phenotype produced female progeny that included both *andromorphs* and *heteromorphs*. When both types of females occurred in the progeny, the ratios were 1:1 from *andromorphic* and 1:1 or 3:1 from *heteromorphic* females, respectively.
3 Other females of either phenotype produced only one type of female progeny. The female progeny of *andromorphic* females were either all *andromorphic* or all *heteromorphic*; the female progeny of *heteromorphic* females were all *heteromorphic*.
4 No *heteromorphic* males were encountered.

Pairs of traits having unequal frequencies between the sexes could result from a variety of mechanisms, including:

1   Differential lethality between the forms in one sex before classification

2   Higher frequencies of a recessive sex-linked trait in the heterogametic males than in females, which could be homozygous for the recessive

3   Expression of a sex-linked gene in only one sex

4   Expression of an autosomal gene in only one sex

Which of these mechanisms is the most likely explanation of the female dimorphism present in *Ischnura*?

484   In the platyfish, two lines occur in separate geographic regions.   In one line males have a *crescent spot* on the caudal fin, absent in females.   In the other neither sex has this *spot*.   Crosses between a female of the first line and a male of the second produce males with *spots* and females without.   Three-quarters of the F_2 males have *spots*, but none of the females have them. Describe the mode of inheritance of this pair of traits and diagram the crosses presented, using your own gene symbols.

485   In the bug *Euchistus variolarius*, which has the X-Y mechanism of sex determination, the male has a *black spot* on the abdomen; the female has not.   In the related species *E. servis* neither sex has this *spot*.   A female of the *variolarius* species was crossed with a male of the *servis* species.   The F_1 females did not have an abdominal *spot*, but all the males did.   In the F_2 the females did not have *spots*, but approximately three-fourths of the males did (Foot and Strobell, 1914; Morgan, 1914).

a   What kind of trait is this?

b   Explain why the Y chromosome cannot carry the gene.

c   Is it correct to state that the gene is not on the X chromosome?   Why?

d   An *unspotted* female was mated to a male of the same phenotype, and all male progeny were *spotted*.   What were the genotypes of the parents?

486   In the silver-washed fritillary butterfly (*Argynnis paphia*) all males have *rich-brown* wings; the females have either *rich-brown* wings like the males or *dark olive-green* ones (Ford, 1955).

a   The *dark olive-green* trait may appear in the progeny of matings between two *rich-brown* individuals.   Can it be a holandric trait?   Explain.

b   In certain crosses the female progeny show a 3:1 ratio of *dark olive-green*: *rich-brown*.

   (1)   Is the gene sex-linked or autosomal?

   (2)   What is the genotype and phenotype of the female parent?

c   A gynandromorph shows a *rich-brown* color on one side and a *dark olive-green* one on the other.

   (1)   How could such an individual arise?

   (2)   Did it start out as a male or as a female?

   (3)   What are the possible genotypes of each side?

   (4)   If the female parent of the gynandromorph was *dark olive-green*, what were the genotypes of the male parent and the zygote?

487   In the shepherd's purse (*Capsella bursa-pastoris*), a monoecious plant species, the leaf lobes of certain true-breeding strains are *sharp*; in others they are *rounded*. The F_1's of crosses between the two types of strains have *rounded* leaf lobes when young, but as the plants mature, the lobes elongate and become pointed, as in the *sharp* parents. In the young F_2's the ratio of *rounded*- to *sharp*-lobed plants is 3:1, but at maturity this ratio is reversed, 3 *sharp* : 1 *rounded* (Shull, 1929). Draw an analogy between youth vs. maturity and male vs. female to show how this trait resembles a sex-influenced character.

488   **a** In poultry, *hen feathering* (feathers short, broad, blunt, and straight) vs. *cock feathering* (hackle and saddle feathers are long, narrow, and pointed, the feathers of the cape and back are pointed, and the sickle feathers of the tail are long and curving) is controlled by a single autosomal pair of alleles, *Hh*. Sexually normal *HH* and *Hh* males are *hen-feathered*, while *hh* males are *cock-feathered*. Sexually normal females of any genotype are *hen-feathered* (Morgan, 1920*a*; Punnett and Bailey, 1921).

(1) In the Leghorns the males are *cock-feathered* and females *hen-feathered*, whereas in the Sebright Bantams both sexes are *hen-feathered*.

(*a*) What would be the phenotype of the F_1 and the F_2 from a cross between a Leghorn male and a Sebright Bantam female?

(*b*) Would the reciprocal cross give the same results? Explain.

(2) A *cock-feathered* male is mated with three females, each of which produces 18 chicks. Among the 54 offspring 27 are *hen-feathered* females, 5 are *cock-feathered* males and 22 are *hen-feathered* males. What are the most probable genotypes of the three parental females?

(3) In the Campine and Hamburgh breeds both *hen-feathered* and *cock-feathered* males occur. The females are *hen-feathered*.

(*a*) Starting with a flock of *hen-* and *cock-feathered* Campines, outline a breeding system to establish a strain in which the males would be always *cock-feathered*.

(*b*) A *hen-feathered* male Campine crossed with a female of the same breed produces 6 *hen-feathered* males, 3 *cock-feathered* males, and 10 *hen-feathered* females. What are the probable genotypes of the parents?

(4) A *hen-feathered* male is crossed with a female and becomes the father of 14 *hen-feathered* males, 4 *cock-feathered* males, and 14 females.

(*a*) What are the probable genotypes of the parents?

(*b*) Why is it unnecessary for the phenotypes of the females to be specified to determine this?

(*c*) If this male is crossed with another female and the progeny consist of 8 *hen-feathered* males, 7 *cock-feathered* males, and 12 females, what is the female parent's probable genotype? If this female underwent sex reversal, would she develop *cock feathering*? Explain.

b (1) The expression of the *Hh* alleles depends on the presence or absence, and the types, of hormones.   Removal of testes in *hen-feathered* males, *H*—, or of the ovary in females (regardless of breed) results in *cock feathering*.   Castrated *cock-feathered* males, *hh*, remain *cock-feathered* (Roxas, 1926; Eliot, 1928; Pezard, 1928).   Roxas (1926) showed that a caponized Sebright Bantam, *HH*, cock that had become *cock-feathered* again became *hen-feathered* after transplantation of a Leghorn testis, *hh*. In the reciprocal transplantation the castrated Leghorn male continued to express *cock feathering*.   Danforth and Foster (1929) and Danforth (1930) showed that transplantation of skin from a *cock-feathered* male, *hh*, to a female of any breed, *H*— or *hh*, results in *hen feathering*. Transplantation of skin from females, *hh*, results in *cock feathering*. However, transplantation of skin from *hen-feathered*, *H*—, males and females, *H*—, to *cock-feathered*, e.g., Leghorn, males does not alter the type of feathering of the transplant.   Finally, transplants of skin from females, *HH*, to males of breeds in which both sexes are *hen-feathered* do not alter feathering type, but when the host was castrated, all its feathers, graft included, were of the male type.   On the basis of this information discuss, with evidence:

(*a*) The effects of the two alleles.

(*b*) The conclusion(s) one may draw from these experiments.

(2) Danforth and Foster (1929) grafted skins from a *hen-feathered* Campine onto a White Leghorn male; the graft grew *hen feathers*.   Would the same kind of feathers develop in the reciprocal graft, made from male to male?   Explain.

(3) What types of feathers should grow on the following grafts?   Give a reason for your answer in each case.

(*a*) Sebright Bantam female to a *hen-feathered* Hamburgh male.   To a Leghorn male.

(*b*) Leghorn female to a *cock-feathered* Campine male.

489

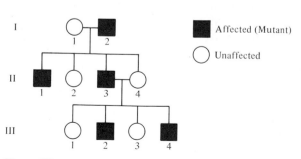

Figure 489

a Could the mutant trait be due to an X-linked recessive gene?   A holandric gene?

b If III-3 and III-4 mated and produced an *affected* female, would this allow an unequivocal determination of which explanation is correct?   Explain.

490   *Cryptorchidism* (the failure of one or both testes to descend into the scrotum) occurs in swine, horses, dogs, and some other mammals.   Bilateral *cryptorchids* in dogs are sterile; unilateral ones are fertile.   Some reports suggest that both conditions are due to the same autosomal recessive gene (British Veterinary Association, 1955).   Assume that this suggestion is correct.

a What are the possible genotypes of the parents in the matings shown, which involve unilateral *cryptorchids* in dogs?

Table 490

Parents		Offspring		
♂	♀	*Normal* ♂	*Cryptorchid* ♂	Normal ♀
normal × normal		52	0	56
normal × normal		33	18	47
cryptorchid × normal		20	0	20
cryptorchid × normal		29	31	59
cryptorchid × normal		0	28	30

b A female terrier produced in four litters 8 females and 8 *cryptorchid* males. The first two litters were from a mating of the female with her full brother, the latter two by a brother of a *cryptorchid* male.   What are the probable genotypes of the female parent?   The two male parents?

c If the males in (b) were *normal*, what ratio of *normal* to *cryptorchid* would you expect in the 8 sons:
(1) If the mother was homozygous for the causal gene?
(2) If she was heterozygous?
(3) Which of these is the more likely genotype of the female?

d How could such a female be of use to a breeder seeking to eliminate *cryptorchidism* from his kennels?

e What kind of dog would you use most easily to determine whether a female carries the gene for *cryptorchidism*?   How many *normal* sons would be required to prove that the tested female did not carry the gene for *cryptorchidism*?

f A dog breeder finds *cryptorchids* in his kennel.   What would you tell him to do to free his breeding stock of the defect without bringing in new dogs? (Assume the breeding stock is fairly large.)

491   Each of the 12 human pedigrees shown refers to a different human trait (represented by solid squares and circles).

a State which of the following kinds of genes could produce each trait and why:

(1) An autosomal dominant gene.
(2) An autosomal recessive gene.
(3) A sex-linked dominant gene.
(4) A sex-linked recessive gene.
(5) An incompletely dominant lethal gene.
(6) A holandric gene.
(7) A sex-influenced gene (dominant in males and recessive in females).
(8) A sex-influenced gene (dominant in females and recessive in males).

b In the pedigrees chosen, derive the genotypes of the original ancestors and of the individuals designated numerically.
c Calculate the probability that the trait in question will appear in the offspring of the matings indicated.

Table 491

Pedigree	Matings	Pedigree	Matings	Pedigree	Matings
B	4 × 6	E	4 × 5	F	4 × 6
	5 × 7		3 × 6		5 × 7
	7 × 8				
		I	4 × 8	J	4 × 7
			4 × 6		5 × 6

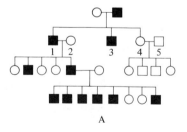

A

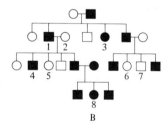

B

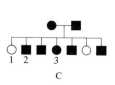

C

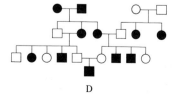

D

Figure 491 (A to D) *Continued overleaf.*

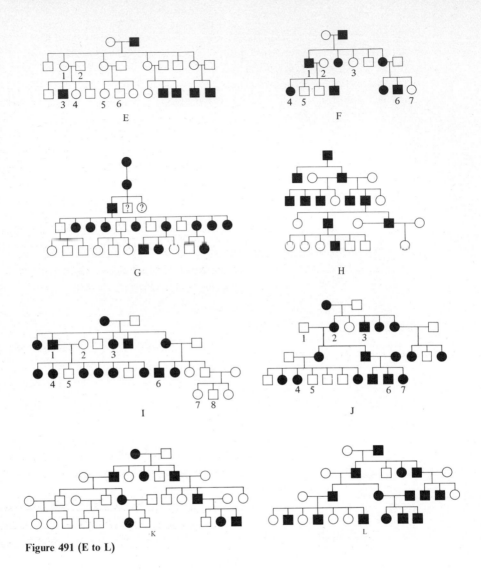

Figure 491 (E to L)

REFERENCES

Asdell, S. A., and A. D. B. Smith (1928), *J. Hered.*, **19**:425.

Bennett, J. H., F. A. Rhodes, and H. M. Robson (1959), *Am. J. Hum. Genet.*, **11**:169.

Breitenbecher, J. K. (1921), *Genetics*, **6**:65.

British Veterinary Association (1955), *Vet. Rec.*, **67**:472.

Cromwell, A. M. (1940), *J. Hered.*, **31**:94.

Danforth, C. H. (1930), *Biol. Gen.*, **6**:99.

———— and F. Foster (1929), *J. Exp. Zool.*, **52**:443.

Dobzhansky, T. (1960), *Science*, **132**:77.

Domm, L. V. (1939), in E. Allen (ed.), "Sex and Internal Secretions," pp. 227–326, Williams & Wilkins, Baltimore.

Dresher, W. (1964), *Am. Nat.*, **98**:167.

Eliot, T. S. (1928), *Physiol. Zool.*, **1**:286.

Finne, I., and N. Vike (1951), *Poult. Sci.*, **30**:455.

Foot, K., and E. C. Strobell (1914), *Arch. Zellforsch.*, **12**:485.

Ford, E. B. (1953), *Adv. Genet.*, **5**:43.

———— (1955), "Moths," Collins, London.

Gajdusek, D. C., C. J. Gibbs, Jr., and M. Alpers (1966), *Nature*, **209**:794.

Gerould, J. H. (1923), *Genetics*, **9**:495.

———— (1941), *Genetics*, **26**:152.

Goodlae, H. D. (1916), *Carnegie Inst. Wash., D.C., Publ.* 243.

Harris, H. (1946), *Ann. Eugen.*, **13**:172.

Holmes, S. J., and R. O. Schofield (1917), *J. Hered.*, **8**:359.

Hovanitz, W. (1944), *Genetics*, **29**:1.

Johnson, C. (1964), *Genetics*, **49**:513.

———— (1966), *Heredity*, **21**:453.

Kerr, R. W. (1960), *Nature*, **185**:868.

———— (1961), *Aust. J. Biol. Sci.*, **14**:605.

Komai, T., and A. S. Ae (1953), *Genetics*, **38**:65.

Lang, A. (1908), "Über die Bastarde von *Helix hortensis* Müller und *Helix nemoralis* L." (mit Beiträgen von Bosshard, Hesse, und Kleimer), Festschrift, Universität, Jena.

McArthur, N. (1964), *Ann. Hum. Genet.*, **27**:341.

McKusick, V. A. (1968), *J. Am. Med. Assoc.*, **204**:113.

Matthews, J. D., and F. M. Burnet (1965), *Lancet*, **1**:1138.

Morgan, T. H. (1914), *Am. Nat.*, **48**:577.

———— (1915), *Proc. Soc. Exp. Biol. Med.*, **13**:31.

———— (1920*a*), *Biol. Bull.*, **39**:257.

———— (1920*b*), *Biol. Bull.*, **39**:231.

Osborn, D. (1916), *J. Hered.*, **7**:347.

Petterson, G., and G. Bonnier (1937), *Hereditas*, **23**:49.

Pezard, A. (1928), Die Bestimmung der Geschlechtsfunktion bei den Hühnern, *Ergeb. Physiol.*, **27**:552.

Punnett, R. C., and P. G. Bailey (1921), *J. Genet.*, **11**:37.

Remington, C. L. (1954), *Adv. Genet.*, **6**:403.

Roxas, H. A. (1926), *J. Exp. Zool.*, **46**:63.

Schultz, A. H. (1932), *Hum. Biol.*, **4**:34.

———— (1934), *Hum. Biol.*, **6**:627.

Shull, G. H. (1929), *Proc. Int. Congr. Plant Sci. Ithaca*, **1**:837.

Snyder, L. H., and C. W. Cotterman (1936), *Genetics*, **21**:79.

———— and H. C. Yingling (1935), *Hum. Biol.*, **7**:608.

Stecher, R. M. (1941), *Am. J. Med. Sci.*, **201**:801.

———— and A. H. Hersh (1944), *J. Clin. Invest.*, **23**:699.

Stehr, G. (1955), *J. Hered.*, **46**:263.

Warwick, B. L., and P. B. Dunkle (1939), *J. Hered.*, **30**:325.

Wood, T. B. (1909), *J. Agric. Sci.*, **3**:145.

Zirkle, C. (1945), in F. A. Montague (ed.), "Studies and Essays in the History of Science and Learning in Honor of George Sarton," pp. 169–194, Henry Schuman, New York.

13
Linkage and Crossing-over[1]

Questions on linkage and recombination and the mechanism of crossing-over in bacteria and viruses are in Chaps. 16 and 17.

QUESTIONS

492 **a** Distinguish between linkage and sex linkage.

b Why is linkage an exception to Mendel's second law?

c In certain organisms the linkage relations between sex-linked genes are much better known than those between autosomal genes. Moreover, linkage maps of sex-linked genes can often be made long before the autosomal linkage groups are established. Why is this to be expected?

493 For short distances (approximately 15 map units or less) the percentage of meiocytes in which crossing-over occurs is twice the percentage of recombination gametes.

a Explain why.

b Why is this not true for genes fairly distant from each other?

494 Linkage studies provided the first proof that a gene occupies a fixed locus on a specific chromosome. Discuss.

495 State whether the following statements are true or false and why.

a Crossing-over always results in genetic recombination.

b Crossing-over occurs at the two-strand stage of meiosis.

c Segregation of alleles can occur only during the first meiotic division.

d Centromeres always separate reductionally at anaphase I.

e The percent recombination between genes is always half as great as the percent crossing-over.

f The percent recombination between any two genes is always a true estimate of the physical distance between them.

g Recombination involving linked genes is the consequence of exchange of homologous segments between nonsister chromatids.

h Linkage can be detected in homozygotes.

i It is possible to establish linkage relationships in organisms that reproduce exclusively by asexual means.

j Regardless of environmental conditions, the percent recombination between any two genes always remains the same.

496 Two silkworms both heterozygous for the same two pairs of alleles (viz. $Aa\ Bb$) are mated. In the offspring a 2:1:1:0 phenotypic ratio is obtained.

a Explain how genetic linkage can lead to this kind of ratio.

b Discuss the suitability of these kinds of data:

(1) For measuring the degree of linkage.

(2) For detecting linkage.

---

[1] There is no crossing-over in *Drosophila* males and silkworm females. An ascospore pair represents a single product of meiosis. Crossing-over refers to the cytological process by which recombinants between linked genes arise.

497 In haploid organisms like *Neurospora*, why is it not possible to determine the location of genes in relation to their centromeres by a random spore analysis?   What kind of analysis is necessary?

498 **a** What do the terms *equational division* and *reductional division* refer to? What chromosomal components will always separate reductionally at anaphase I?

   b How may segregation of a pair of alleles occur during the first meiotic division?   The second meiotic division?   Illustrate your answer.

499 **a** Why is linkage between genes on the X and Z chromosomes easier to detect than that between genes on the autosomes?

   b Under what conditions can an entire F_2 involving X- or Z-linked genes be treated as a testcross progeny?   Explain diagrammatically.

500 At what stage of meiosis does crossing-over occur?   Describe one of the classical experiments in which it was detected cytologically.

501 Explain why the frequency of second-division segregations of any gene, regardless of the organism, is a function of the distance between the gene and the centromere of the chromosome carrying it.

502 Discuss the effects that the assembly of genes into chromosomes confers upon:
   a Evolutionary change in a species.
   b Breeding for improvement.

PROBLEMS

503 In 1911 Morgan crossed a *white*-eyed, *yellow*-bodied *Drosophila* female with a *red*-eyed, *gray*-bodied male.   In the F_1 the females were phenotypically like the father and the males like the mother.   The F_1 flies were intercrossed, and the 2,205 F_2 progeny were classified as shown.

Table 503

Phenotype		Sex	
Eyes	Body	♀	♂
White	Yellow	543	474
Red	Gray	647	512
White	Gray	6	11
Red	Yellow	7	5

   a Outline a hypothesis to explain these results.
   b See Morgan's 1911 classic paper for his interpretation of these results.

504 In tomatoes (*Lycopersicon* [= *Lycopersicum*] *esculentum*) fruit pubescence and fruit length are each controlled by one gene; the allele *P* for *smooth* is dominant over *p* for *peach*, and the allele *R* for *round* is dominant over *r* for *long*.   The F_1 of a cross between homozygous *smooth, long* and homozygous *peach, round* was testcrossed by MacArthur (1928); the results obtained were as follows:

Smooth, round   12
Smooth, long   123
Peach, round   133
Peach, long   12

a Classify each of the phenotypes as either parental or recombinant and determine the percentage recombination.
b What evidence for linkage is shown in this cross?
c Give the percentage crossing-over and the map distance between the genes.

505 In *Drosophila melanogaster*, *ruby eye*, *ru*, is sex-linked and recessive to *red* (wild-type) eye.   *Miniature* wing, *m*, is also sex-linked and recessive to *normal* (wild-type) wing.   The genes for *ruby* and *miniature* show approximately 29 percent recombination.   The pair of alleles, *Bb*, controlling *gray* vs. *black* is autosomal (on chromosome 2).   A homozygous *ruby, normal, gray*, female is crossed with a homozygous *red, miniature, black* male, and the F_1 females are crossed with triple-recessive males.   Show the types and proportions of offspring expected from this mating and why these results are expected.

506 In mice, the allele pairs *Sh-2,sh-2* for *nonshaker* vs. *shaker* and *Wa-2,wa-2* for *straight* vs. *wavy* hair are on the same chromosome, 25 map units apart (Green and Dickie, 1959).   True-breeding *nonshaker, straight* mice are crossed with homozygous *shaker, wavy* ones, and the F_1's are testcrossed.   Diagram the parental, F_1, and testcross generations of this cross, showing the genes on the chromosome.   Show the expected frequencies of F_1 crossover and noncrossover meiocytes and demonstrate that the expected frequency of recombinant gametes is half that of crossover meiocytes.

507 In the house mouse (*Mus musculus*) *trembling* vs. *normal* and *rex* (short) vs. *long* hair are each controlled by an autosomal pair of alleles.   The alleles for *trembling* and *rex* are dominant.   In crosses of heterozygous *trembling, rex* females with *normal, long* males the following results were obtained by Falconer and Sobey (1953):

Trembling, rex   21
Trembling, long   52
Normal, rex   54
Normal, long   22

a State whether or not these two genes are located in the same chromosome and why.
b Were the *trembling, rex* females heterozygous in coupling or repulsion? Show how you determine this.
c Determine the percentage recombination.

d Would the results have differed if these workers had made the reciprocal cross? Explain.

e If heterozygous *trembling, rex* males and females were crossed, what phenotypic ratio would be expected in the progeny?

508 In the domestic fowl (*Gallus domesticus*) the three pairs of alleles *Kk* (*slow* vs. *rapid* feathering), *Dwdw* (*normal* vs. *dwarf*) and *Bb* (*barred* vs. *nonbarred* feathers) are Z-linked. The following data are from Hutt (1959, 1960):

1 F_1 males from the mating of a *slow, dwarf* female with a *rapid, normal* male were crossed with females homozygous for the recessive alleles *k* and *dw*. The following results were obtained:

Slow, normal	53
Slow, dwarf	98
Rapid, normal	102
Rapid, dwarf	52

2 Eight males heterozygous at the *K* and *Dw* loci were mated to *rapid, normal* hens. Only the daughters were classified. They fell into two clearly defined groups as shown.

Table 508

		Daughters			
Group	Parental males	*Normal, slow*	*Dwarf, rapid*	*Normal, rapid*	*Dwarf, slow*
1	J03, H33, M58, M01	186	223	20	17
2	M54, M51, M57, M10	17	10	252	245

3 Males heterozygous in the coupling phase for the *dwarfing* and *barring* genes produced 81 *barred, normal,* 79 *nonbarred, dwarf,* 95 *barred, dwarf,* and 104 *nonbarred, normal* female progeny.

a Determine the percent recombination from 1 and state whether this value is a true estimate of the map distance and why.

b Account for the significantly different progeny ratios in the two sets of results in 2 and determine the percent recombination from the total data of 2.

c Explain why only females were studied in determining percent recombination.

d There is a considerable discrepancy between the recombination values for 1 and 2. What feature of the crosses may account for this?

e Considering the finding regarding *Dw* and *K*, what do the results in 3 suggest regarding the linkage relationships of *Dw, K,* and *B*?

509 In corn, *starchy* vs. *sugary* endosperm and *susceptibility* vs. *resistance* to *Helminthosporium* are monogenically controlled.   The F_1 of a cross between *starchy, susceptible* and *sugary, resistant* are testcrossed, and the progeny phenotypes and their proportions are as follows:

Starchy, susceptible   92
Starchy, resistant   86
Sugary, susceptible   91
Sugary, resistant   88

a What are the possible explanations for these results?
b Many genes have been allocated to each of the ten linkage groups in corn. Knowing this, how would you determine whether these two genes are members of the same linkage group or not?

510 The allele pairs Ee (*gray* vs. *ebony* body) and Cc (*normal* vs. *curled* wings) in *Drosophila melanogaster* are located on chromosome 3 (an autosome).   A true-breeding *gray, normal* male was crossed with an *ebony, curled* female.   Two F_1 males mated with two F_1 females produced the following F_2 progeny:

Gray, normal   288
Gray, curled   14
Ebony, curled   88
Ebony, normal   10

Show how you would derive an estimate of the map distance between E and C. Explain your method and calculate the distance.

511 In the fowl, *silver* vs. *gold* feathering and *slow* vs. *rapid* feather development are each controlled by a single pair of alleles.   In crosses between males of a *silver, slow* strain and females of a *gold, rapid* strain the F_1's were phenotypically like the male parents.   The F_2 consisted of four phenotypes distributed as shown.

Table 511

Phenotype	Number ♂	♀
Silver, slow	300	120
Silver, rapid	0	35
Gold, rapid	0	123
Gold, slow	0	32

a State whether these two pairs of traits are controlled by independently segregating or by linked pairs of alleles.   Give reasons for your answer.
b Are they located on the autosomes or the sex chromosomes?

512 In the mouse, the mutant genes *Str* (*Striated*) and *Ta* (*Tabby*) each cause the development of dark transverse stripes. Individuals carrying *Str*, however, show patches of short fur, absent in individuals carrying *Ta*. Lyon (1966) crossed two sets of *striated, tabby* females with *wild-type* males (*nonstriated, nontabby*). The results of these crosses are shown. Analyze these data to provide as complete a description as possible of the mode of action and the linkage relationships of these genes.

Table 512

| Cross | Sex | Testcross phenotype | | | |
		Striated tabby	*Striated nontabby*	*Nonstriated tabby*	*Wild type*
Set 1	♀	6	117	106	16
	♂	0	0	113	10
Set 2	♀	27	2	1	34
	♂	6	0	4	41

513 *Drosophila* females with *gray* (wild-type) bodies and *crossveinless* wings were crossed with males with *yellow* bodies and *crossveined* (wild-type) wings. The tabulated results were obtained in crosses between F_1 females and *wild-type* males. Explain the male-female discrepancy and derive the percent recombination for the two genes.

Table 513

Phenotype	♀	♂
Gray, crossveined	3,743	254
Gray, crossveinless	0	1,621
Yellow, crossveinless	0	250
Yellow, crossveined	0	1,625

514 In 1915 a recessive X-linked lethal, *sa*, was discovered in Morgan's laboratory. Morgan and Bridges (1916) mated *red*-eye lethal-bearing females with *white*-eye males; when the lethal-bearing *red*-eye daughters were again mated to *white*-eye males, 894 sons were *white*-eye and 256 were *red*-eye. What is the percent recombination between the two loci in this cross?

515 A cross of a homozygous *wavy-haired, nonbelted, himalayan* mouse with a *straight-haired, belted, albino* one produces F_1's that are *wavy-haired, nonbelted, himalayan*. The following distributions are obtained among the testcross progeny:

Wavy, himalayan, nonbelted	22
Wavy, himalayan, belted	4
Wavy, albino, nonbelted	20
Wavy, albino, belted	4
Straight, himalayan, nonbelted	3
Straight, himalayan, belted	21
Straight, albino, nonbelted	4
Straight, albino, belted	20

Explain these data, giving the strengths of linkages if they exist.

516 In *Drosophila*, *arc* (wings somewhat curved) is controlled by a recessive allele, *a*; *black body* is controlled by a recessive allele *b*. The F_2 of a cross between *arc* females and *black* males consisted of 923 *wild type*, 401 *black*, 387 *arc*, and 0 *black, arc* (Bridges and Morgan, 1919).
a Show whether these genes are linked.
b Is it possible to determine the percent recombination? Why?

517 In *Drosophila melanogaster* the allele pairs *Cncn* (*dull red* vs. *cinnabar* eyes) and *Roro* (*smooth* vs. *rough* eyes) are on chromosomes 2 and 3 respectively.[1] A mutant fly with *bent* wings is discovered in the homozygous double-recessive *cinnabar, rough* strain. A true-breeding *cinnabar, rough, bent* strain is established, and females from this strain are crossed with males from a true-breeding *dull-red, smooth, straight* stock. A series of crosses between F_1 males and *cinnabar, rough, bent* females give the progeny shown.

Table 517

Phenotype	♀	♂
Dull-red, smooth, straight	38	36
Cinnabar, smooth, straight	34	38
Dull-red, rough, straight	35	36
Dull-red, smooth, bent	39	34
Cinnabar, rough, straight	40	35
Cinnabar, smooth, bent	34	33
Dull-red, rough, bent	37	34
Cinnabar, rough, bent	38	36

a On which chromosome is the mutant gene for *bent* wings located and why?
b Would the results of the reciprocal of this cross allow the same unequivocal answer? Explain.

518 In the housefly (*Musca domestica*; $2n = 12$; XY sex determination) the results shown were obtained with the second-chromosome pair of alleles *Bb* (*gray* vs. *brown*-body) in crosses involving a mutant, *bb*, and three *BB* strains: the

---

[1] *Drosophila melanogaster* has four chromosomes: number 1-sex chromosome, numbers 2, 3, and 4-autosomes.

Table 518

Cross		F$_1$	F$_2$ or testcross segregation			
			Gray		Brown	
♀	♂		♀	♂	♀	♂
mutant × normal		gray	2,590	2,290	773	738
normal × mutant		gray	1,999	1,967	587	600
mutant × furen		gray	921	4,079	748	0
furen × mutant		gray	2,120	1,990	699	689
mutant × F$_1$ of mutant ♀ × ND ♂		gray	0	1,224	889	0
mutant × F$_1$ of ND ♀ × mutant ♂		gray	788	640	464	360
mutant × F$_1$ of mutant ♀ × furen ♂		gray	0	2,856	839	0
mutant × F$_1$ of furen ♀ × mutant ♂		gray	530	495	448	459

normal, the furen, and the ND (Hiroyoshi, 1964).   Outline a plausible genetical and cytological hypothesis to account for these results.

519   In 1913 Carothers found a heteromorphic chromosome pair in the grasshopper *Brachystola magna*.   At anaphase I the chromosome with the two long chromatids always segregated from the chromosome with the two short chromatids.   Show, by means of a diagram, where the centromere is located.

520   In a unicellular haploid ($n = 1$) alga species, strain A was grown in a medium containing tritiated thymidine for many generations to ensure that all DNA in all chromosomes was completely labeled.   Cells of strain B were grown on a nonradioactive medium.   The cells of the two strains were then mixed on a nonlabeled medium under conditions that favor gamete union and zygotic meiosis.   The four meiotic products (spores) of each of many such meiocytes

Table 520

Chromosome	No. of meiocytes			
	13	5	1	1
1	CL	PL	PL	CL
2	PL	PL	PL	CL
3	PL	PL	PL	NL
4	NL	PL	NL	NL

Key: CL = completely labeled; PL = partly labeled; NL = not labeled.

were then separated and their chromosomes tested for radioactivity.   The results are given for 20 meiocytes.   Discuss the significance of these results with respect to:

a Mechanism of crossing-over.

b Time of crossing-over relative to chromosome replication.

See Peacock (1970), Hotta and Stern (1971), Howell and Stern (1971), Taylor (1965), and Henderson (1970).

521   **a** In *Drosophila melanogaster*, *Pr* (*red* eye) is dominant to *pr* (*purple* eye), and *B* (*gray* body) is dominant to *b* (*black* body).   Bridges and Morgan (1919) crossed a *red*, *black* female with a *purple*, *gray* male.   The F_1 males when testcrossed produced the following progeny:

Red, black	74
Purple, gray	71
Red, gray	0
Purple, black	0

When the F_1 females were testcrossed, the results were

Red-black	383
Purple-gray	382
Red, gray	22
Purple, black	16

(1) Do these reciprocal testcrosses give evidence of linkage?   Explain how you determine this.   What is the percent recombination for each of the crosses?

(2) Compare the recombination values from reciprocal crosses and give a possible reason for the discrepancies.

(3) Was the cross made in the repulsion or coupling phase?

b When the F_1's were intercrossed, the surprising F_2 phenotypic distribution shown below was observed:

Red, gray	684
Red, black	300
Purple, gray	371
Purple, black	0

Are these results consistent with your hypothesis?   Diagram the cross in answering this question.

522   In man, *deuteranopia* (green *color blindness*), *c*, and *hemophilia*, *h* (McKusick, 1964; Whittaker et al., 1962), are X-linked traits controlled by recessive alleles.   When 25 women whose fathers were *color-blind* have 5 sons each, the phenotypes and proportions of these 125 male progeny are as follows:

Deuteranopia, normal	61
Normal, hemophilia	54
Normal, normal	5
Deuteranopia, hemophilia	5

a What is the genotype of all mothers?

b Determine the percentage recombination.

c Why is it not possible to determine linkage intensities from similar data for daughters?

523   In the Chinese primrose (*Primula sinensis*) *Bb* (*blue* vs. *slate* flowers) and *Gg* (*green* vs. *red* stigma) are two autosomal pairs of alleles. *B* is dominant to *b*; *G* is dominant to *g*. Could you detect linkage in the cross *Bb gg* × *bb Gg*? Explain.

524   In *Drosophila melanogaster* alcohol dehydrogenase (ADH) is an enzyme composed of two polypeptide chains (a dimer) which exists in three electrophoretically different forms (*isozymes*) controlled by two alleles of one gene:

$Adh^F Adh^F$ *fast* isozyme; both chains identical
$Adh^S Adh^S$ *slow* isozyme; both chains identical
$Adh^F Adh^S$ *intermediate* isozyme; both chains different, as well as *fast* and *slow* isozymes

Grell et al. (1965) crossed true-breeding *cy bl/cy bl*, *ubx vno/ubx vno* (*wild-type*) males that were $Adh^S Adh^S$ with females of the balanced lethal stock *Cy bl/cy Bl*, *Ubx vno/ubx Vno* that were true-breeding for $Adh^F Adh^F$.

F$_1$ *Cy bl/cy bl*, *Ubx vno/ubx vno*, $Adh^S Adh^F$ males were crossed with *wild-type* females that were $Adh^S Adh^S$. All F$_2$ *Cy/cy* flies produced the *intermediate* as well as the *fast* and *slow* isozymes. The *cy/cy* flies produced the *slow* isozyme only. Half the *Ubx* flies produced the same isozymes as the *Cy* flies; the other half produced the *slow* isozyme only. Which chromosome carries the *Adh* gene and why? *Note: Cy* (*curly* wings) and *Bl* (*short* bristles) are incompletely dominant lethal genes on the second chromosome; *Ubx* (*enlarged* halteres) and *Vno* (*veins missing*) are incompletely dominant lethal genes on the third chromosome.

525   In corn, the aleurone layer of the kernel may be *colored* or *colorless*, and the endosperm may be *starchy* or *waxy*. From crosses between true-breeding lines A and B and C with D, Bregger (1918) obtained F$_1$ seed with *colored* aleurone and *starchy* endosperm. He testcrossed one F$_1$ plant from A × B and one from C × D with a true-breeding *colorless, waxy* line. The single ear produced by the F$_1$ plant from A × B carried 403 kernels, and that produced by the F$_1$ plant from C × D carried 292. The table shows the phenotypes and phenotypic ratios on each of these ears.

Table 525

F$_1$ plant	Colored starchy	Colorless waxy	Colored waxy	Colorless starchy
From A × B	147	133	65	58
From C × D	46	32	103	111

a Explain why the two ears give different dihybrid distributions of phenotypes.

b State the genotypes of the two F_1's and the true-breeding parental lines from which they were derived.

c Combine the two sets of data to derive the percentage recombination.

526 In the freshwater fish *Aplocheilus latipes*, frequently used as an ornamental, Aida (1921) obtained the diagramed results beginning with crosses between true-breeding *red* and *white* varieties.

Table 526

Parents		F_1 cross		F_2 segregation	
♀	♂	♀	♂	♀	♂
white × *red*		*red* × *red*		41 *red* + 43 *white*	76 *red* + 0 *white*
		red × *white**		2 *red* + 197 *white*	251 *red* + 1 *white*

* Homozygous.

a Only one of the modes of inheritance listed below is possible.   State which it is and give reasons for rejecting the others:
(1) Holandric   (2) Autosomal
(3) Incompletely sex-linked   (4) Sex-linked

b What cytological evidence would be required to distinguish between crossing-over and nondisjunction as the cause of *red* females and *white* males in the testcross offspring?

c Assuming that crossing-over is the cause, with what locus is the color gene recombining, and what is the map distance between the two loci?

527 In corn, *tunicate* vs. *nontunicate*, *glossy* vs. *nonglossy* seedling, and *liguled* vs. *liguleless* are each controlled by a single pair of alleles. A trihybrid *tunicate, glossy, liguled* plant crossed with a *nontunicate, nonglossy, liguleless* one produces the following offspring:

Tunicate, liguleless, glossy	53
Tunicate, liguleless, nonglossy	10
Tunicate, liguled, glossy	50
Tunicate, liguled, nonglossy	8
Nontunicate, liguled, glossy	11
Nontunicate, liguled, nonglossy	48
Nontunicate, liguleless, glossy	9
Nontunicate, liguleless, nonglossy	54

a Which pairs of alleles are linked, and which are segregating independently? Explain.

b Using your own symbols, give the genotypes of both parents.

c What is the map distance between the linked genes?

528 In *Drosophila melanogaster* the allele pairs *Bb* (*gray* vs. *black* body) and *Arcarc* (*straight* vs. *forward-curved* wings) are 27 map units apart on chromosome 2. You have two strains, one *bb ArcArc*, the other *BB arcarc* which is also true-breeding for *w* (*white* eye), a sex-linked recessive which can be used as a sex marker. Show how you would obtain a double-recessive line, using the principle that crossing-over occurs in females but not males. For each step show the genotypes of the parents used in the cross and the genotype of the offspring used for the next cross.

529 In rabbits, *C* (*colored* coat) vs. *c* (*albino*) and *B* (*black*) vs. *b* (*brown*) are autosomal pairs of alleles and show dominance as indicated (Robinson, 1958). Rabbits from true-breeding *brown* strain are crossed with *albinos* of genotype *cc BB*; the F_1's are crossed with *albinos* of genotype *cc bb*, with the following results:

Black   102
Brown   198
Albino   300

a What type of gene interaction is involved?
b Determine whether the genes are linked or not.
c Calculate the percentage recombination.
d If the *brown* rabbits were intercrossed, what phenotypes would appear in the progeny and in what proportions? Explain.

530 In 1916 Castle and Wright reported the results of studies of eye-color inheritance in the Norway rat (*Rattus norvegicus*). The F_1's from a cross between a true-breeding *red*-eye strain and a true-breeding *pink*-eye strain had *black* eyes. The F_2 consisted of 162 *black* : 90 *red* : 72 *pink*. When 45 of the 90 *red*-eye F_2's were crossed with the *pink*-eye strain, 32 produced only *black*-eye offspring and 13 produced *black*- and *red*-eye offspring in approximately equal numbers. When 40 of the 72 *pink* F_2's were crossed with the *red*-eye strain, 27 produced only *black* offspring, 10 produced *black* and *pink* offspring, and 3 produced only *red* offspring. Two pairs of alleles are interacting in a recessive epistatic manner. With reasons for your answers:
a State whether the allele pairs are linked or not.
b Give the genotypes of the individuals in the different phenotypic groups.
c If linked, calculate the map distance between the two loci on the basis of the 85 F_2 genotypes.

531 In Norway rats (*Rattus norvegicus*) the allele pairs *Rr* and *Pp* are on the same chromosome pair. *R—P—* individuals have *dark* eyes; *R—pp*, *rr P—*, and *rr pp* have *light* eyes. Castle (1919) obtained the following results in two separate experiments:

Experiment 1: $\dfrac{R P}{r p} \times \dfrac{r p}{r p} \rightarrow 1{,}255$ *dark*-eyed + 1,752 *light* eyed

Experiment 2: $\dfrac{R p}{r P} \times \dfrac{r p}{r p} \rightarrow 174$ *dark*-eyed + 1,540 *light*-eyed

a Do these results indicate that the loci of genes R and P are linked? Explain your answer.

b What is the percent recombination?

532 a In rabbits, three *rex* mutants numbered 1, 2, and 3, each characterized by short, soft, plushlike fur, have been found. The first was discovered in 1919 by Abbé Gillet in France, the second in 1926 by a breeder in Hamburg, Germany, and the third in 1927 by Mme Du Barry in France. *Normal* is dominant to *rex*. Castle and Nachtsheim (1933) crossed *rex-1* with *rex-2* and obtained F_1's with *normal* fur and 391 F_2's, 195 of which were *normal* and 196 *rex*.

(1) What type of gene interaction is involved and why?

(2) State whether the genes are linked or not and give reasons for your answer.

(3) If the genes are linked, is it possible, from these data, to determine the distance between them?

b The genotypic ratio for 51 F_2 *rex* individuals as determined by back-crossing to one or the other or both of the pure races *rex-1* and *rex-2* was

18 $r_1 r_1 R_2 R_2$
21 $R_1 R_1 r_2 r_2$
5 $r_1 r_1 R_2 r_2$
7 $R_1 r_1 r_2 r_2$

What is the percentage of recombination? (*Hint*: *rex-1* is $r_1 R_2 / r_1 R_2$, *rex-2* is $R_1 r_2 / R_1 r_2$, and the F_1's are $r_1 R_2 / R_1 r_2$.)

533 In poultry, *white* is due to either or both of the genes c and I; individuals are *colored* only when both the dominant allele C and the recessive allele i (in the homozygous condition) are present. True-breeding *colored* birds are crossed with *white* fowl of genotype *cc ii*. The F_1's are testcrossed, and 65 *colored* and 206 *white* are obtained. Explain these results and show that the wrong cross was made if the purpose was to determine whether linkage exists between these loci.

534 a The brine shrimp *Artemia salina* has *black* (wild-type) eyes. Bowen (1963) found one *white*-eyed male, which he mated with a *wild-type* female. Figure 534*A* shows the distribution of *white* eye for six generations, including the initial cross. Black symbols represent the *white* phenotype, and open symbols represent *black*. The number under a symbol is the number of progeny in that class. *Note*: Males are homogametic, females hetero-gametic. Outline a hypothesis to account for:

(1) This mode of transmission of *black* vs. *white* eye.

(2) The origin of the *white*-eye female.

(3) The reasons why the cross *white* female × *white* male produces only *white*-eye progeny.

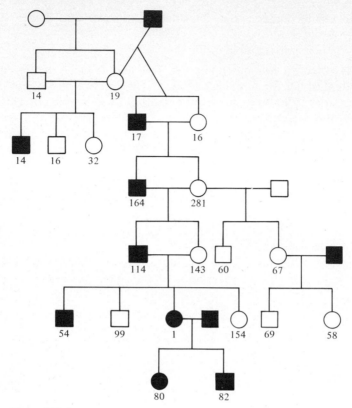

Figure 534 *A*

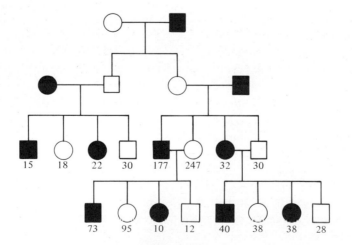

Quemado ♀ × white ♂

Figure 534 *B*

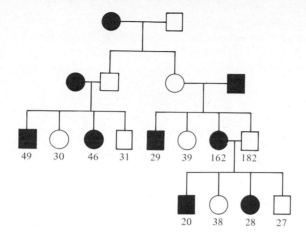

White ♀ x Quemado ♂

Figure 534 *C*

b Bowen (1965) further pursued the study of *black* vs. *white* in reciprocal crosses between selected *white*-eyed strains and *wild-type* strains from seven American races. Crosses between *wild type* from the salt lake near Quemado, New Mexico, and inbred stock number 11 gave the results illustrated in Figs. 534*B* and 534*C*. Are these results in agreement with the hypothesis proposed in (a)? Explain by giving the genotypes of all phenotypes in the Quemado female × *white* male cross. Also sketch the essential features of the chromosomes involved.

535 In the garden pea (*Pisum sativum*) *wrinkled* seed is recessive to *round* and *no tendrils* is recessive to *tendrils*. A true-breeding doubly recessive variety (*wrinkled, no tendrils*) crossed with a true-breeding doubly dominant variety (*smooth, tendrils*) by Vilmorin and Bateson in 1912 yielded the following results:

F$_1$	*Round, tendrils*	
F$_2$	*Round, tendrils*	319
	Round, no tendrils	4
	Wrinkled, tendrils	3
	Wrinkled, no tendrils	123

a Apply the chi-square test (see Appendix Table A-8) to each of the mono-hybrid ratios to determine whether they agree with the expected ratio for random segregation with dominance and state your conclusions.
b Test the observed four-class ratio obtained for linkage by applying the chi-square test for independent assortment and checking which classes are deficient in number. State your conclusions.
c Determine the percent recombination for the genes concerned, using the product-moment formulas in Appendix Table A-1.

536 In corn (*Zea mays*) allele pairs *Cc* (*colored* vs. *colorless* aleurone) and *Wxwx* (*starchy* vs. *waxy* endosperm) are on chromosome 9.   The following F_2 results were obtained by Bregger (1918) by crossing F_1's obtained from crosses between true-breeding *colored*, *starchy*, *CC WxWx*, and *colorless*, *waxy*, *cc wxwx*, strains:

Table 536

Phenotype	Number of plants
Colored, *waxy*	263
Colorless, *starchy*	279
Colorless, *waxy*	420
Colored, *starchy*	1,774

Calculate the percent recombination between these two allele pairs using the product-moment formulas in Appendix Table A-1.

537 In *Drosophila*, the mutant *purple*, *pr*, has *purple* eyes; wild-type flies have *red* eyes.   The mutant *vestigial*, *vg*, has wings that are minute in comparison with the *long* wings of the wild type.   In 1919 Bridges and Morgan reported two sets of reciprocal crosses involving these recessive mutants.   In each case the F_1 hybrids had *red* eyes and *long* wings.   These were testcrossed with the doubly recessive line, true-breeding for *purple* eyes and *vestigial* wings.

a The results when F_1 males were used for the testcross are shown in Table 537*A*.   Offer two explanations for these results.

Table 537*A*

Phenotype	F_1 ♂ (from *red*, *long* ♀ × *purple*, *vestigial* ♂) × *purple*, *vestigial* ♀	F_1 ♂ (from *red*, *vestigial* ♀ × *purple*, *long* ♂) × *purple*, *vestigial* ♀
Red, *long*	519	0
Purple, *vestigial*	552	0
Purple, *long*	0	346
Red, *vestigial*	0	358

b The results when F_1 females were used for the testcross are shown in Table 537*B*.

(1) Which of your two explanations do these results eliminate?

(2) What explanation is now necessary for the two sets of data and why?

(3) Account for the results of these two crosses by diagraming the crosses and designating parental and nonparental types among the testcross progeny.   What is the percent recombination?

Table 537*B*

Phenotype	F₁ ♀ (from *red, long* ♀ × *purple, vestigial* ♂) × *purple, vestigial* ♂	F₁ ♀ (from *red, vestigial* ♀ × *purple, long* ♂) × *purple, vestigial* ♂
Red, long	1,339	157
Purple, vestigial	1,195	146
Purple, long	154	1,067
Red, vestigial	151	965

538 In *Drosophila melanogaster* the allele pairs *Yy* (*black* vs. *yellow* body) and *Snsn* (*long* vs. *singed* bristles) are X-linked. Both loci are on the same side of the centromere; *sn* is closer to the centromere than *y*. Females of genotype *y Sn/Y sn* usually express the *wild-type* (*black, long*) phenotype throughout the entire abdomen and thorax. Stern (1936) found that a few such females exhibited patches (twin spots) of relatively equal size, of *yellow* and *singed* tissue adjacent to each other on the thorax or abdomen. Less frequently single *yellow* patches (spots) were found, and still less frequently single *singed* spots, without the adjacent patch of opposite phenotype.

a Show:

(1) Using appropriate diagrams, how the "twin spots" and the "single spots" could arise by crossing-over.

(2) Why the spots were infrequent.

(3) Why they occurred only in the females.

b In what ways is mutation ruled out as an explanation of these phenomena?

c Using diagrams, show whether mitotic crossing-over could be detected in females carrying these genes in coupling, e.g., *Y Sn/y sn*.

539 *Hemophilia A* and red-green *color blindness* in man are controlled by recessive alleles at different X-chromosome loci (see McKusick, 1964). Which of the

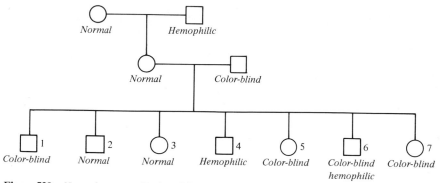

Figure 539 *Normal* = normal color vision, no hemophilia; *hemophilic* = normal color vision, has hemophilia A; *color-blind* = red-green color-blind, no hemophilia.

third-generation offspring in the pedigree are recombinants, which are non-recombinants, and which cannot be classified? Illustrate or explain your answers.

540 Porter, Schulze, and McKusick (1962) studied linkage between *color blindness* and *glucose-6-phosphate dehydrogenase* (G6PD) *deficiency* in eight families. Only one definite instance of recombination was discovered. Shown are three of the eight pedigrees, one of which has the recombinant. State which one this is and why.

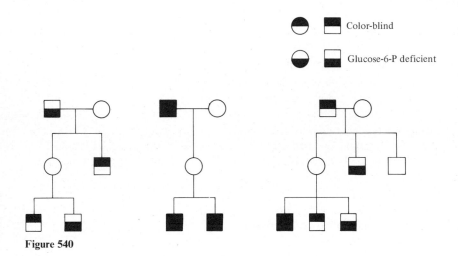

Figure 540

541 Graham et al. (1962) presented the pedigree shown, in which a phenotypically *normal* woman heterozygous for the three X-linked pairs of alleles Xg^aXg (*presence* vs. *absence* of the blood antigen Xg^a), Cc (*normal* color vision vs. red *color blindness*), and Hh (*normal* blood clotting vs. *hemophilia B*, prolonged blood clotting due to plasma-thromboplastin-component deficiency) gave birth to five phenotypically different males. The sequence of these genes is not known. Moreover, since the woman's parents were unavailable for examination, the genotypes (coupling relationships) of her two X chromo-

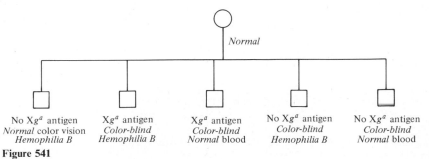

Figure 541

somes could not be determined. Nevertheless, the investigators concluded that at least one offspring is the result of a double crossover. Which one is it and why?

542 The *ABO* blood groups and the *nail-patella syndrome* (a rare anomaly involving abnormal fingernails, toenails, and kneecaps, together with other bone abnormalities) are controlled by different genes. A small part of one of the pedigrees studied by Renwick and Lawler is given. The blood-group genotypes appear below the symbols. Individuals with the syndrome are represented by solid symbols.

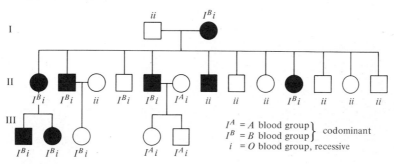

Figure 542   (*After Penrose, 1959.*)

a Is *nail-patella syndrome* controlled by a dominant or recessive gene? Is it autosomal or X-linked?

b Do the above data indicate there is linkage between the two genes? Explain.

c If linkage exists, give the genotypes of the homologous chromosomes of all individuals in the first two generations and indicate the second-generation individuals with recombinant phenotypes.

543 The blood of adult human beings contains two kinds of hemoglobin molecules: one (hemoglobin *A*) contains two α and two β polypeptide chains, the other (hemoglobin A_2) has two α and two δ polypeptide chains. The β and δ polypeptides are specified by different genes. β^A (in normal hemoglobin *A*) vs. β^S (in the *S* hemoglobin of sickle-cell anemics) are specified by the allele pair $\beta^A\beta^S$. δ^{A_2} (in normal hemoglobin A_2) vs. $\delta^{A_2^1}$ (a rare variant of hemoglobin A_2) are specified by the allele pair $\delta^{A_2}\delta^{A_2^1}$.

a Does this pedigree indicate linkage of the genes controlling the formation of the β and δ polypeptide chains? Justify your answer.

b Horton and Huisman also reported the inheritance patterns of the two genes in the children of five sets of parents in each of which one of the

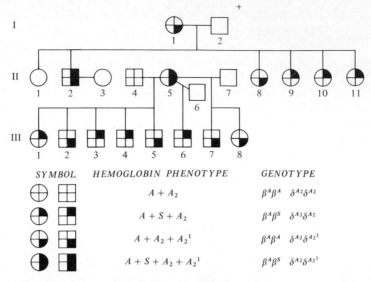

Figure 543   *(From Horton and Huisman, 1963.)*

parents was $\beta^A \beta^S \; \delta^{A_1} \delta^{A_2 1}$, and the other $\beta^A \beta^A \; \delta^{A_2} \delta^{A_2}$.   The phenotypes of the 21 offspring in the five families were:

$A + A_2$ 0
$A + S + A_2$ 13
$A + A_2 + A_2{}^1$ 8
$A + S + A_2 + A_2{}^1$ 0

Is the inheritance pattern beyond reasonable doubt of the type the pedigree in (a) suggests?   Explain.

544   In 1954 Marshall et al. studied the autosomal pairs of traits *elliptocytosis,* *E,* vs. *normal* blood cells, *e,* and presence, *R,* vs. absence, *r,* of the red-blood-cell antigen *D.*   *Elliptocytosis* is an apparently harmless condition in which some of the erythrocytes are elliptical instead of biconcave.   This trait is fully penetrant.   Part of a pedigree of family R is shown.

a What is the genotype of the deceased parent?

b Determine the number of children and grandchildren in each of the four phenotypic classes: *elliptocytosis, presence of D; elliptocytosis, absence of D;* *normal, presence of D; and normal, absence of D.*   Compare these values with those expected with independent assortment using the chi-square test (see Appendix Table A-8) and state your conclusions.

c How frequently did *E* segregate with *R*?   What is the probability of this on the basis of independent assortment?   Does this analysis corroborate your conclusions in (b)?

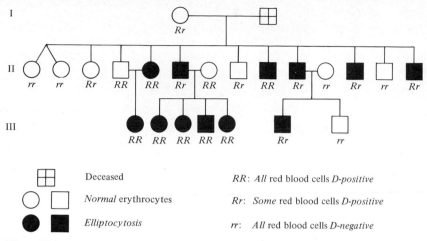

Figure 544

Deceased

Normal erythrocytes

Elliptocytosis

RR: *All* red blood cells *D-positive*

Rr: *Some* red blood cells *D-positive*

rr: *All* red blood cells *D-negative*

d Give the genotypes of all individuals and state whether they are definite or probable and also the genotype of the gamete received from the *elliptocytotic* parent by individuals whose genotype is definite.

e Explain why tests to distinguish heterozygotes from homozygous dominants are needed for complete studies of linkage in man.

545 a Penrose (1953) has shown that autosomal linkage can be demonstrated for any two loci by sib-pair data from the matings $AB/ab \times ab/ab$ and $Ab/aB \times ab/ab$. The results are shown in Table 545A. If the genes are

Table 545*A*

	Phenotype *A* or *a*	
Phenotype *B or b*	Sibs alike	Sibs unlike
Sibs alike	1	3
Sibs unlike	2	4

not linked, the four combinations should be equally frequent. If linked, some combinations will be more common than others regardless of the linkage phase in the heterozygous parent.

(1) Explain why this should be so.

(2) What other causes may produce evidence of linkage, by this test, when it does not in fact exist?

b Renwick and Lawler (1955) give the data in Table 545B for sib pairs involving the autosomal loci for *nail-patella syndrome* (described in Prob. 542) and for the *ABO* and *Rh* blood groups.

Table 545*B*

| | Nail patella and Rh loci | | | | Nail patella and ABO loci | | |
| | | Nail patella | | | | Nail patella | |
Rh	Like	Unlike	Total	ABO	Like	Unlike	Total
Like	53	57	110	Like	78	44	122
Unlike	59	58	117	Unlike	30	84	114
Total	112	115	227	Total	108	128	236

(1) Determine, using the chi-square method (see Appendix Table A-8), whether the genes are linked or not.

(2) Is it possible to determine the percent recombination?

546 a In *ordered*-spore (tetrad) analysis of fungi, e.g., *Neurospora*, it is necessary to halve the percentage of asci with second-division segregations to determine the map distance between a gene locus and the centromere of the chromosome carrying the gene.   Why is this division by 2 necessary?

b In 1949 Barratt and Garnjobst summarized data obtained by Hungate (1946), Regnery (1947), and themselves regarding the segregation of the leu^+leu^- (*normal* vs. *leucineless*) pair of alleles in *Neurospora*; 40 second-division segregations were observed among the 236 asci analyzed.

(1) What is the percentage of second-division segregations?

(2) Calculate the distance of the *leu* locus from the centromere and explain your calculations.

(3) How were these tetrads recognized, and what are the events in meiosis that lead to second-division segregation?

547 a In certain algae, bryophytes, and fungi all the products of a single meiosis can be analyzed genetically.   Such an analysis is called *tetrad analysis*.   In the alga *Chlamydomonas* the four products of a tetrad are *unordered*.   As a consequence three segregation patterns are possible if two pairs of alleles are segregating in a cross: parental ditype (PD), nonparental ditype (NPD), and tetratype (TT).   Assuming that two loci are linked and that no more than two crossovers ever occur between them, show with the aid of diagrams how each of the patterns can arise.

b In an organism with unordered tetrads the following types of tetrads were obtained from a cross between haploids of genotypes ABC and abc.

ABC	ABC	ABc	ABc
ABC	AbC	Abc	ABc
abc	aBc	aBC	abC
abc	abc	abC	abC
174	21	25	182

(1) Compute the relative frequencies of PD, NPD, and TT tetrads for each pair of alleles.

(2) Show which allele pairs are linked.

(3) For each gene show whether linkage exists with the centromere or not.

548 Lindegren (1932) studied the segregation patterns for the mt^+mt^- pair of alleles in 273 asci. Mt^+ and mt^- represent the two mating types. The classes of asci and the number in each class were as shown.

Table 548

	Classes of ascospore-pair arrangements					
	1	2	3	4	5	6
	A	a	A	a	A	a
	A	a	a	A	a	A
	a	A	A	a	a	A
	a	A	a	A	A	a
Number	105	129	9	5	11	14

a Which asci show second-division segregation? Diagram the origin of each.

b Calculate the distance between the mating-type locus and the centromere.

549 In 1933 Lindegren crossed a *Neurospora crassa* strain with *pale* conidia and *normal* growth with one showing *orange* conidia and *fluffy* growth habit. Of 109 asci the numbers showing first- and second-division segregation, respectively, for each of these two pairs of traits and for the mating-type

Table 549A

Pairs of traits	First division	Second division
Mt^+ vs. mt^- mating type	97	12
Pale vs. *orange*	73	36
Fluffy vs. *normal* growth form	42	67

traits were as tabulated. Lindegren's hypothesis to explain these results was as follows:

1 Crossing-over occurred at the four-strand stage.

2 It involved only two of the four chromatids.

3 The paternal and maternal centromeres of each chromosome pair segregated at the first division of meiosis.

a Do you agree with Lindegren's explanation?
b Calculate the gene-centromere distances.

In a cross between an *arginineless* and an *histidineless* strain of *Neurospora*, the spore arrangements shown in Table 549*B* are found.

Table 549*B*

| Spore arrangement | Ascospore pairs | | | | No. of asci |
	1	2	3	4	
1	$arg^+ hist^+$	$arg^+ hist^+$	$arg^- hist^-$	$arg^- hist^-$	78
2	$arg^- hist^-$	$arg^- hist^-$	$arg^+ hist^+$	$arg^+ hist^+$	83

c How are the allele pairs located in the chromosome with respect to their centromeres and with respect to each other?

550 In *Neurospora*, a cross between a double mutant, strains al^- (*albino*), $inos^-$ (unable to synthesize *inositol*), and *wild type* produced the following tetrads (Houlahan et al., 1949):

$al^+ inos^+$	$al^+ inos^-$	$al^+ inos^-$	$al^+ inos^-$	$al^+ inos^+$	$al^- inos^-$	$al^+ inos^-$
$al^+ inos^+$	$al^+ inos^-$	$al^+ inos^+$	$al^- inos^-$	$al^- inos^-$	$al^+ inos^+$	$al^- inos^+$
$al^- inos^-$	$al^- inos^+$	$al^- inos^-$	$al^+ inos^+$	$al^+ inos^+$	$al^+ inos^-$	$al^+ inos^-$
$al^- inos^-$	$al^- inos^+$	$al^- inos^+$	$al^- inos^+$	$al^- inos^-$	$al^- inos^+$	$al^- inos^+$
4	2	23	36	19	22	16

a Determine whether these two genes are linked. If linked, give the linkage distance between them.
b Which of the two genes is closer to its centromere?

551 The following ordered tetrads were produced in a cross between a thi^- (*thiamineless*) strain of *Neurospora crassa* which was otherwise *wild type* and a $pyro^-$ (*pyridoxineless*) strain which was otherwise wild type (Houlahan et al., 1949):

$thi^+ pyro^-$	$thi^+ pyro^-$	$thi^+ pyro^+$	$thi^+ pyro^+$
$thi^+ pyro^-$	$thi^+ pyro^-$	$thi^- pyro^-$	$thi^- pyro^+$
$thi^- pyro^+$	$thi^- pyro^+$	$thi^+ pyro^+$	$thi^+ pyro^-$
$thi^- pyro^+$	$thi^- pyro^+$	$thi^- pyro^-$	$thi^- pyro^-$
5	3	1	7

a Are the two genes linked?   Explain.
b Which ordered spore arrangements are missing and why?

552 A *Neurospora* strain with *yellow* conidia (ylo^-) and unable to synthesize *cystine* (cys^-) was crossed by Houlahan et al. (1949) with a *wild-type* strain. Analysis of 155 asci revealed the following ordered spore arrangements:

$ylo^+ cys^+$	$ylo^+ cys^-$	$ylo^- cys^-$	$ylo^- cys^-$	$ylo^+ cys^+$	$ylo^- cys^+$	$ylo^- cys^+$
$ylo^+ cys^+$	$ylo^+ cys^-$	$ylo^- cys^+$	$ylo^+ cys^-$	$ylo^- cys^-$	$ylo^+ cys^-$	$ylo^+ cys^-$
$ylo^- cys^-$	$ylo^- cys^+$	$ylo^+ cys^-$	$ylo^- cys^+$	$ylo^+ cys^+$	$ylo^- cys^+$	$ylo^+ cys^+$
$ylo^- cys^-$	$ylo^- cys^+$	$ylo^+ cys^+$	$ylo^+ cys^+$	$ylo^- cys^-$	$ylo^+ cys^-$	$ylo^- cys^-$
74	13	45	5	12	2	4

Determine:

a Whether these two genes are linked.
b If linked, the linkage distance between them.
c Whether they are in the same or different arms of the chromosome.

553 The results by Lindegren (1933) (presented by Catcheside, 1951) concern the four loci *pale, fluffy, crisp,* and *mating type* in *Neurospora crassa.*

Table 553

Cross	Classes of ascospore pair arrangements						
	1	2	3	4	5	6	7
	$a^+ b^+$	$a^+ b^-$	$a^+ b^+$	$a^+ b^+$	$a^+ b^+$	$a^+ b^-$	$a^+ b^+$
	$a^+ b^+$	$a^+ b^-$	$a^+ b^-$	$a^- b^+$	$a^- b^-$	$a^- b^+$	$a^- b^-$
	$a^- b^-$	$a^- b^+$	$a^- b^+$	$a^+ b^-$	$a^+ b^+$	$a^+ b^-$	$a^+ b^-$
	$a^- b^-$	$a^- b^+$	$a^- b^-$	$a^- b^-$	$a^- b^-$	$a^- b^+$	$a^- b^+$
$f^+ mt^+ \times f^- mt^-$	16	20	6	60	3	1	3
$p^+ mt^+ \times p^- mt^-$	62	1	10	34	0	1	1
$c^+ p^- \times c^- p^+$	2	52	12	3	0	9	0
$p^+ mt^- \times p^- mt^+$	3	46	8	17	0	3	1
$c^+ mt^+ \times c^- mt^-$	55	2	9	9	2	0	1

Note: Only one member of each pair of ascospores is indicated. *Key: a, b* = any pair of genes; + = *wild-type* allele; − = *mutant* allele.

a Determine the distance of each gene from the centromere.
b Determine the percent recombination between each pair of loci.
c Determine, for each cross, whether the genes are on the same or opposite arms of the chromosome and construct a linkage map. Explain your answer in each case.
d Do these results provide evidence for crossing-over at the four-chromatid stage? Explain, using the results of any one cross as an example.

554 Ebersold et al., (1962) studied allele pairs governing *prototrophism* (nutrilite independence) and *auxotrophism* (dependence) in the alga *Chlamydomonas reinhardi* for various metabolites, viz.:

Acac   acetate: *independence* vs. *dependence*
Pabpab   p-aminobenzoic acid: *independence* vs. *dependence*
Thithi   thiamine: *independence* vs. *dependence*
Nicnic   nicotinic acid: *independence* vs. *dependence*

and the allele pairs *Pfpf* governing *normal* vs. *paralyzed* flagella. In two-point crosses they observed the frequencies of PD, NPD, and TT tetrads shown.

Table 554

Cross	Observed number of tetrads		
	PD	NPD	TT
ac-14 × *ac-14b*	53	0	0
pf-15 × *pf-17*	22	23	46
ac-12 × *pf-12*	74	0	22
pab-1 × *thi-2*	36	2	47
ac-17 × *pf-17*	54	53	9
ac-51 × *pf-16*	65	0	41
pf-13 × *nic-11*	21	22	44

a Why are only PD types produced in the first cross?
b Show which pairs, if any, are linked.

555 According to Ebersold et al. (1962) in *Chlamydomonas reinhardi* the allele pair *Pfpf* (*normal* vs. *paralyzed* flagella) segregates independently of *Ac-1,ac-1* (*normal* vs. *acetate* requirement). Two mutant strains exist, the one true-breeding for *paralyzed*, the other for *acetate* requirement.
a What types of tetrads are expected in a cross between these two strains?
b How many genotypes will there be, and in what ratio will they appear?

556 Pascher (1918) crossed a strain of *Chlamydomonas* (a unicellular haploid alga) with *pear*-shaped cells and *lateral* chloroplasts with a strain with *spherical*-shaped cells and *basal* chloroplasts. The four cells (tetraspores) from each meiocyte (zygote) always segregated 1 : 1 for each of these pairs of traits. In some tetrads two spores were like one parent and two like the other, while in others the phenotypes of the four cells were *pear*, *lateral*; *pear*, *basal*; *spherical*, *lateral*; and *spherical*, *basal* (pairs of traits controlled by linked pairs of alleles). Show whether these results are in agreement with the hypothesis that crossing-over occurs at the four-strand stage.

557 **a** With the aid of diagrams explain how copy choice differs from chromatid exchange as an explanation of crossing-over.
b Moses and Taylor (1955), Taylor and McMaster (1954), Swift (1950), and Rossen and Westergaard (1966) have shown that the major period of meiotic replication of genes (viz. replication of DNA) is at a time prior to the onset of chromosome synapsis. Explain what bearing this has on the validity of the copy-choice hypothesis.

558 In *Aspergillus nidulans* the mutant genes *w* (*white*) and *acr* (*acriflavine*) are linked and on chromosome 2. *White* diploid mitotic segregants were selected from the diploid strains Y, *acr w/Acr W*, and Z, *Acr w/acr W*. All the *white* segregants from Y were homozygous *acr/acr* and all from Z were homozygous *Acr/Acr*. On the other hand, of the *acr/acr* segregants from diploid Y, 87 percent were *w/w*, and 13 percent were *W/w*. And of those from diploid Z, 82 percent were *W/W*, and 18 percent *W/w* (Pontecorvo and Kafer, 1958). What is the linkage relationship between *W*, *Acr*, and the centromere? Explain.

559 Analysis of diploid mitotic recombinants in *Aspergillus nidulans* indicated the linked sequence *ad-23* (*adenineless*)–*w* (*white*)–centromere. Other such studies revealed the linked order *pu* (*putrescine*)–*thi-4* (*thiamineless*)–centromere. Diploid mitotic recombinants did not reveal linkage between these two groups. A study of 85 *white* haploids from heterozygous diploids one of whose parents was *wild type* and the other true-breeding for all four mutant genes showed that all were genotypically identical, *ad-23 w pu thi-4*.

a Are the genes on the same or different chromosomes? Explain.

b If on the same chromosome, where is the centromere located relative to the four genes? Why?

560 The genes *ad-14* (*adenineless*), *pro-1* (*prolineless*), *pab-1* (*p-aminobenzoic acidless*), *y* (*yellow*) are in linkage group I in *Aspergillus nidulans*.

a Pontecorvo and Kafer (1958) selected 371 *yellow* mitotic recombinants (homozygotes) from a diploid strain heterozygous for the four markers and of genotype *ad-14 Pab-1 Y Pro-1/Ad-14 pab-1 y pro-1* (not necessarily in this order). Of these, 96 were prototrophs, i.e., *Pab-1— Ad-14— Pro-1—*; 245 required *p*-aminobenzoate, and 30 required proline as well as *p*-aminobenzoate. What is the sequence of these genes with respect to each other and the centromere? Show your reasoning.

b None of the 371 *yellow* homozygotes was homozygous for *ad-14*. Another diploid strain, *ad-14 y/Ad-14 Y*, produced *ad-14,ad-14* homozygous mitotic recombinants, all of which were *Yy*. Is *ad-14* on the same or different arm of chromosome 1 as the other three genes? Explain.

561 Analyses of recombinants due to mitotic crossing-over indicate the linkage sequence centromere–*pro-3*(*proline*)–*y*(*yellow*). A similar independent analysis indicated the sequence *ribo* (*riboflavin*)–*ad-14* (*adenine*)–centromere. Haploids of diploids heterozygous in the coupling phase were of two kinds only: *Pro-3 Y Ribo Ad-14* and *pro-3 y ribo ad-14*. Where is the centromere located relative to these genes? Show how you arrive at your answer.

562 Two diploid strains of *Aspergillus nidulans* which were heterozygous for the mutant genes *w* (*white*), *y* (*yellow*), *ad* (*adenineless*), *pab* (*p-aminobenzoic acidless*), and *bi* (*biotinless*) had the following genotypes. The parents of the first strain were *w ad pab y Bi* and *W Ad Pab Y bi*, and those of the second strain were *W ad pab y bi* and *w Ad Pab Y Bi*. Pontecorvo et al. (1954) selected *yellow* and *white* haploid segregants in each of the strains and tested them for presence of the other mutant genes, with the results shown. These data

Table 562

Phenotypes and genotypes of haploid segregants		In strain 1	In strain 2
yellow (y)	Ad pab Bi	8	0
	ad pab bi	0	7
white (w)	Ad pab bi	0	8
	Pab Bi	0	19
	pab Bi	2	0
	Pab bi	9	0

Note: w is epistatic to alleles at y locus; therefore *yellows* must all carry *W.*

indicate that the five genes belong to two linkage groups and that complete linkage is demonstrated by the 53 segregants in which only the parental combinations occur. Demonstrate that this is so by giving the genotypes of the homologs of each of the two chromosome pairs.

563 A diploid strain of *Aspergillus nidulans* heterozygous for the mutant genes *y* (*yellow*), *sm* (*small*), *pan*⁻ (*pantothenicless*), *nic-8* (*nicotinicless*), *phe-2* (*phenyl-alanineless*), and *cho*⁻ (*cholineless*) produced a large number of haploid segregants of which 113 *yellow* ones were selected and tested for the presence or absence of the other mutant genes. The genotypes of these *yellow* haploids and their numbers were as follows:

y nic-8 cho phe-2 pan sm   31
y Nic-8 Cho phe-2 pan sm   26
y nic-8 cho Phe-2 Pan Sm   24
y Nic-8 Cho Phe-2 Pan Sm   32

What are the linkage relationships of these genes? Illustrate.

564 According to Janssen's chiasmatype theory, crossing-over should occur after duplication and between only two of the four chromatids of a synapsed pair of homologs. Two other possibilities exist: (1) crossing-over might occur before duplication (at the two-strand stage). (2) It might occur after duplication and involve all four chromatids. To test the truth of Janssen's hypothesis, Bridges (1916) obtained *Drosophila* XXY females heterozygous (some in the coupling, others in the repulsion phase) for the X-linked allele pairs *Wwᵉ* (*red* vs. *eosin*) and *Vv* (*red* vs. *vermilion*) and crossed them with *bar*-eyed but otherwise *normal* males (*B/Y*).

The dominance of the *bar* trait permitted Bridges to select from the progeny those daughters which were both of an XXY constitution and had received both X chromosomes from their mothers (these would be *nonbar* females; XX females would be *bar*; and other XXY females derived from syngamy of an *XY* egg and an X sperm would also be *bar*). This distinction is necessary, since

Janssen's hypothesis can be tested with Bridges' method only in the XXY *nonbar* females.   A large number of these families were *red*-eyed and were left unclassified.   A few exceptional individuals showed recessive phenotypes for either or both genes, and they were subjected to progeny tests to determine the exact genotype with regard to both chromosomes and genes.   The results of Bridges' analysis are summarized in the table.   Analyze the data for the

Table 564   **Phenotypes and genotypes of the *nonbar* (XXY) females**

Phase of cross	Phenotype	Genotype	Number
Coupling	*Red*	Undetermined	414
	Vermilion-eosin	$\dfrac{w^e\text{-}v}{w^e\text{-}v}$	2
	Vermilion	$\dfrac{W\text{-}v}{w^e\text{-}v}$	3
	Eosin	$\dfrac{w^e\text{-}V}{w^e\text{-}v}$	3
Repulsion	*Red*	Undetermined	168
	Vermilion-eosin	$\dfrac{w^e\text{-}v}{w^e\text{-}v}$	0
	Vermilion	$\dfrac{W\text{-}v}{w^e\text{-}v}$	1
	Eosin	$\dfrac{w^e\text{-}V}{w^e\text{-}v}$	4

Note: In XXY females in 92 percent of meioses one X and the Y go to one pole and the remaining X to the other pole, while in the other 8 percent the two X's separate from the Y with the consequence that gametes are of four types: X, XY, XX, and Y.

recessive phenotypes on the assumption that rare crossing-over occurs between the two X's of an XXY female at meiosis, to show whether, and in what way, they support Janssen's theory.

565   The chiasma may be related to crossing-over in two ways.   On the partial chiasmatype, or one-plane, theory (Janssen, 1909; Belling, 1931; Darlington, 1937) each chiasma observed at diplotene has arisen as a result of crossing-over between nonsister chromatids of two homologs.   Before terminalization of chiasmata, the separation of homologs to produce the loops between successive chiasmata is always in the reductional plane.   Any reduction in the number of chiasmata from diplotene to metaphase I is due to terminalization.

According to the classic, or two-plane, hypothesis (McClung, 1927; Sax, 1932) chiasmata form as a result of the opening out of the four chromatids along two different planes in different regions. On one side of a chiasma there is a reductional separation, and on the other an equational (sister-from-sister chromatid) separation of the four chromatids. An exchange of homologous segments occurs when chiasmata break (disappear), giving rise to new unions producing crossover chromatids. Chiasma formation thus precedes crossing-over, and a reduction in chiasma number may be interpreted as the result of crossing-over.

Various critical configurations permit a decision to be reached between these rival theories.

a Brown and Zohary (1955) studied the cytological behavior of two plants in *Lilium formosanum* with a heteromorphic chromosome pair (a normal chromosome A and a chromosome A deficient for the terminal two-thirds of the short arm). The heteromorphic homologs always formed a bivalent with at least one chiasma in the long normal arm. They compared the frequencies of heteromorphic bivalents with and without chiasmata in the short arm at metaphase I with equational (1 short-arm chromatid + 1 long-arm chromatid to each pole) vs. reductional (short-arm chromatids to 1 pole, long-arm chromatids to the other) separation at anaphase I. The results are tabulated. (See Table 565.) Do these results support the classical or partial-chiasmatype theory of origin of chiasmata? Explain.

b Mather (1933) studied double-interlocking bivalents in *Lilium regale* microsporocytes produced as a result of the two homologs of one pair being caught between those of another pair during the zygotene pairing process. His evidence is presented in Fig. 565. Mather concluded that this provides support for the chiasmatype hypothesis. Do you agree? Why?

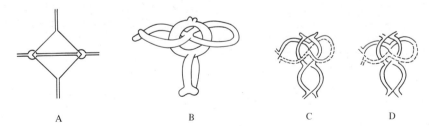

A   B   C   D

Figure 565   A. The probable arrangement of homologs at the end of zygotene (note that no twist is expected in the horizontal pair between the two homologs of the vertical pair) B. The observed configuration of a typical pair of interlocked bivalents at late prophase I (probably at diakinesis). C. The configuration required by the classic (two-plane) hypothesis (McClung, 1927; Sax, 1932). D. The configuration required by the chiasmatype (one-plane) hypothesis (Janssen, 1909; Belling, 1931; Darlington, 1937).

Table 565

Plant	Metaphase I				Anaphase I			
	Chiasma		No chiasma		Chiasma		No chiasma	
	No.	%	No.	%	No.	%	No.	%
1	452	70	183	30	172	71	69	29
2	120	51	114	49	81	55	67	45

QUESTIONS INVOLVING FURTHER READING

566 **a** All geneticists agree that chiasmata are somehow associated with crossing-over. Differences of opinion are held, however, about how the two are associated. Briefly discuss the hypotheses concerned with the origin of chiasmata.

b See Mather (1935), Swanson (1957), or Whitehouse (1969) for a discussion of these hypotheses and experimental data bearing on them.

567 Discuss the copy-choice and breakage-and-reunion hypotheses of crossing-over. Include in your discussion experimental data supporting each of these explanations and state which explanation you favor and why. See Meselson (1967); Taylor (1967); Peacock, Rhoades, and others in Peacock and Brock (1968); Whitehouse (1969); and papers by Tomizawa, Clark, Hurwitz et al. in (1967) for excellent discussions of the mechanisms of crossing-over.

568 Crossing-over is believed to involve the covalent union of DNA molecules derived from the two parents through a series of discrete steps. Describe the main steps that may be involved in this process and present the experimental evidence supporting this scheme. See Howard-Flanders and Boyce (1964, 1966), Anraku and Tomizawa (1965), Meselson (1967), Whitehouse (1970), and papers in Symposium (1967).

REFERENCES

References for this chapter are included in Chap. 14.

14
Chromosome Mapping

QUESTIONS

569 a How are chromosome maps[1] constructed?

    b What are the fundamental assumptions on which chromosome mapping is based?

    c Why are extremely short regions used in establishing genetic maps?

    d How many allele pairs are required to detect double crossovers? Explain.

    e Would you expect estimates of map distances to be increased or decreased by undetected double crossovers? Why?

    f Explain why, regardless of the number of double crossovers, the theoretical percentage recombination between any two genes can never be more than 50.

570 a In three-point testcrosses why are the parental types most frequent and the double-crossover types least frequent?

    b Explain how the knowledge that parental types are most frequent and double crossovers least frequent can be used to determine the relative positions of three genes on the linkage map.

    c What information regarding crossing-over can a single three-point testcross provide that is not available from all the two-point testcrosses involving the same genes?

571 You have two testcross populations, each numbering 150, one for two closely linked genes, for example, 15 map units apart, the other for two more distantly linked genes. Which of these two would you expect to give the more reliable estimate of map distance and why?

572 Barring chromosomal aberrations, the order of genes on a genetic map of a chromosome corresponds completely with the order on a cytological map. Why do the distances on the two maps not always correspond?

573 What is the difference between a linkage (genetic) map and a chromosome (cytological) map?

574 Does linkage mapping tell you which chromosome carries a particular linkage group? Explain. If not, how might you proceed to associate linkage groups with specific chromosomes?

575 In relation to the observations that follow discuss the statement: The number of linkage groups found in an organism is equal to the haploid number of chromosomes:

    a Some organisms are unisexual; others are bisexual.

    b In some organisms many genes have been studied; in others only a few.

576 In the mouse, the genes studied to date (approximately 300) fall into 20 linkage groups. How many chromosomes would you normally expect to find in a somatic cell of a mouse? Explain.

577 What are some of the difficulties involved in the construction of linkage maps

---

[1] Map distance = linkage distance.

in human beings?   Of what practical value would such a map be to medical practitioners?

578 Which of the following features must an organism possess to make the construction of genetic maps possible?

a Morphologically distinct sexes.
b Many monogenically inherited pairs of traits.
c Individuals heterozygous for at least three pairs of alleles.
d Standard environmental conditions.
e Homozygous recessive stocks.
f More than two alleles per locus.
g X and Y chromosomes.

PROBLEMS

579 Assume the following is a linkage (chromosome) map for genes y, sh, and c in corn.

(centromere)	y	sh	c
	18	28	48

Assume further that no double crossing-over occurs in either the Y-sh or the sh-c regions and that it occurs without interference in the y-c region.

a What proportion of the meiocytes in the trihybrid $Y\,Sh\,C/y\,sh\,c$ would you expect to show:
   (1) Double crossing-over in the y-c region?
   (2) Single crossing-over in the y-sh region?
   (3) Single crossing-over in the sh-c region?
   (4) No crossing-over in the y-c region?
b Answer the questions in (a) assuming the coefficient of coincidence is 0.5.

580 In *Drosophila simulans* the following three allele pairs are on chromosome 1 (the X chromosome):

Yy   *gray* vs. *yellow* body
Cmcm   *red* vs. *carmine* eyes
Ff   *normal* vs. *forked* bristles

F_1's heterozygous for these allele pairs were testcrossed by Sturtevant (1921). Parental and double-crossover-progeny classes were

cm y F   725
Cm Y f   719
Cm y F   34
cm Y f   32

a Show diagrammatically why the gene order cannot be *Cm Y F*.
b What is the correct gene order?   Explain how you determine it.
c What were the genotypes of the true-breeding parents and of the F_1?

581 In the mouse the following three allele pairs are found to be on chromosome 7:

Rere   *rex* (short) vs. *normal* hair
Sh-2 sh-2   *nonshaker* vs. *shaker*
Vtvt   *normal* vs. *vestigial* tail

In a testcross the two parental classes are *Vt Re sh-2* and *vt re Sh-2*, and the double-crossover classes are *Vt re sh-2* and *vt Re Sh-2*.

a Which of the double-crossover classes would you compare *vt re Sh-2* with to obtain the correct gene order? Would a comparison of the other parental and double-crossover classes give the same answer? Explain.

b Give the genotypes of the parents, using the correct gene order.

582 Genes at loci *K, L, M* are linked, but their order is unknown. F_1's from the cross *KK LL MM* × *kk ll mm* are testcrossed. The most frequent phenotypes in the testcross progeny will be *K L M* and *k l m* regardless of gene order. What phenotypic classes will be least frequent if locus *K* is in the middle? If locus *L* is in the middle?

583 In corn, the following allelic pairs of genes are on chromosome 9:

Cc   *aleurone* color vs. *no aleurone* color (*white*)
Shsh   *full* vs. *shrunken* endosperm
Wxwx   *starchy* vs. *waxy* endosperm

F_1 plants heterozygous for all three pairs of alleles were testcrossed by Hutchison, and the tabulated phenotypes were obtained (data from Emerson, Beadle, and Fraser, 1935).

Table 583

Phenotype	Number of progeny
White, shrunken, starchy	116
Colored, full, starchy	4
Colored, shrunken, starchy	2,538
Colored, shrunken, waxy	601
White, full, starchy	626
White, full, waxy	2,708
White, shrunken, waxy	2
Colored, full, waxy	113

a Determine the sequence of the genes on the chromosome, the map distances, and the genotypes and phenotypes of the homozygous parents of the F_1.

b Derive the coefficient of coincidence for interference.

c How many double-crossover phenotypes would you expect in this cross per 1,000 progeny? How many double-crossover chromosomes?

d Derive the amount of underestimation of the map distance between the end genes if the middle one had not been scored.

584 In the guinea pig *black* vs. *white* coat, *short* vs. *long* hair, and *wavy* vs. *straight* hair are each controlled by a single pair of alleles. Crosses are made between homozygous *white, short, wavy* and homozygous *black, long, straight* animals. The *black, short, wavy* F$_1$'s are crossed with animals from a true-breeding *white, long, straight* strain and produce the offspring shown. Are the allele pairs on the same chromosome pair or not? Explain, using your own symbols.

Table 584

Phenotype	Number of animals
White, short, wavy	46
White, short, straight	20
White, lonq, wavy	28
White, long, straight	6
Black, short, wavy	6
Black, short, straight	32
Black, long, wavy	20
Black, long, straight	42

585 In *Drosophila melanogaster* the *Adh* gene specifying the synthesis of alcohol dehydrogenase (ADH) is located on the second chromosome. The co-dominant alleles *Adh^F* and *Adh^S* specify the *fast* and *slow* electrophoretic variants (isozymes) of the enzyme. Grell et al. (1965) crossed males from a true-breeding *wild-type* stock homozygous for *Adh^S* with females homozygous for *Adh^F* and the second-chromosome recessive genes *b* (*black* body; map

Table 585

	Single-crossover phenotypes				Genotype of gamete received from F$_1$	Number of progeny in each class
Body color	Wing type	Bristle type	Eye color	ADH isozyme		
Black	*Normal*	*Normal*	*Red*	*Fast/slow*	*b El Rd Pr Adh^S*	10
Black	*Normal*	*Normal*	*Red*	*Fast*	*b El Rd Pr Adh^F*	0
Gray	*Elbow*	*Scraggly*	*Purple*	*Fast/slow*	*B el rd pr Adh^S*	0
Gray	*Elbow*	*Scraggly*	*Purple*	*Fast*	*B el rd pr Adh^F*	6
Black	*Elbow*	*Normal*	*Red*	*Fast/slow*	*b el Rd Pr Adh^S*	3
Black	*Elbow*	*Normal*	*Red*	*Fast*	*b el Rd Pr Adh^F*	25
Gray	*Normal*	*Scraggly*	*Purple*	*Fast/slow*	*B El rd pr Adh^S*	17
Gray	*Normal*	*Scraggly*	*Purple*	*Fast*	*B El rd pr Adh^F*	2
Black	*Elbow*	*Scraggly*	*Red*	*Fast/slow*	*b el rd Pr Adh^S*	0
Black	*Elbow*	*Scraggly*	*Red*	*Fast*	*b el rd Pr Adh^F*	5
Gray	*Normal*	*Normal*	*Purple*	*Fast/slow*	*B El Rd pr Adh^S*	5
Gray	*Normal*	*Normal*	*Purple*	*Fast*	*B El Rd pr Adh^F*	0

position 48.5), *el* (*elbow* wings; map position 50.0), *rd* (*scraggly* bristles; map position 51.0), and *pr* (*purple* eyes; map position 54.5). The F_1 females were backcrossed to males homozygous for Adh^F, *b*, *el*, *rd*, and *pr*. From among the testcross progeny individuals with single crossovers in one of the three regions *b-el*, *el-rd*, and *rd-pr* were selected and tested for their ADH isozymes. The results are tabulated.

a In which of the three regions is *Adh* located?

b What is its position on the genetic map?

586 In *Drosophila melanogaster* the dominant mutant gene *S* (*star*, for rough, small, narrow, and dark eyes) and the recessive mutant genes *a* (*aristaless* antennae) and *dp* (*dumpy* wings) are on the second chromosome. Stern and Bridges (1926) crossed a homozygous *star* strain with one homozygous for the two

Table 586

Phenotype	Number
Star	956
Aristaless, dumpy	918
Dumpy	5
Aristaless, star	7
Star, dumpy	100
Aristaless	132

recessive mutant genes. The *star* F_1 females produced the shown results when mated with *aristaless, dumpy* males. Which backcross phenotypes are missing and why?

587 In corn a true-breeding *liguled* (presence of leaf-base appendages), *starchy* (endosperm), *glossy* (leaf) plant was crossed with a true-breeding *liguleless*, *sugary, dull* one. The F_1's crossed with *liguleless, glossy, sugary* plants from a true-breeding line produced the tabulated results. A student in a genetics course concluded that these results are obtained because the three genes involved are on the same chromosome. State whether you agree and why. Give the genotype of the parents and of the F_1.

Table 587

Phenotype	Proportion
Liguled, starchy, dull	9
Liguled, starchy, glossy	38
Liguled, sugary, dull	9
Liguled, sugary, glossy	38
Liguleless, starchy, dull	38
Liguleless, starchy, glossy	9
Liguleless, sugary, dull	38
Liguleless, sugary, glossy	9

588   The data given are from a study by Bridges and Morgan (1919) of the percent recombination between the loci *dachs*, *black*, *vestigial*, *curved*, and *purple* on chromosome 2 of *Drosophila*. Map the chromosome for these five loci as accurately as possible. Note that determinations of percent recombination for short distances provide more accurate estimates of map length than those for long ones.

Table 588

Loci	Total number of *Drosophila*	Number showing recombinant phenotypes
Black, curved	62,679	14,237
Black, vestigial	20,153	3,578
Black, dachs	6,725	1,196
Black, purple	48,931	3,026
Curved, purple	51,136	10,205
Curved, dachs	462	145
Curved, vestigial	1,720	141
Dachs, vestigial	5,354	1,585
Dachs, purple	1,489	293
Purple, vestigial	13,601	1,609

589   In *Drosophila melanogaster* males from a true-breeding stock of flies homozygous for the recessive genes for *wrinkled* wings and *minute* eyes which is otherwise *wild type* were crossed with females from the balanced lethal stock $Cy/Pm\,D/Sb$ [the incompletely dominant lethal genes Cy (*curly*) = wings curled up and Pm (*plum*) = prune eye color are on the second chromosome; D (*dichaete*) = outstretched wings and Sb (*stubble*) = short thick bristles are on the third chromosome]. *Plum, stubble* F_1 males crossed with females from the true-breeding *wrinkled, minute* stock produced progeny of eight different phenotypes, as shown.

Table 589

	Phenotype			Sex	
Eye color	Bristle	Wing	Eye size	♀	♂
Plum	*Stubble*	*Straight*	*Normal*	201	0
Plum	*Stubble*	*Straight*	*Minute*	0	192
Red	*Long*	*Wrinkled*	*Normal*	209	0
Red	*Long*	*Wrinkled*	*Minute*	0	198
Red	*Stubble*	*Straight*	*Normal*	200	0
Red	*Stubble*	*Straight*	*Minute*	0	200
Plum	*Long*	*Wrinkled*	*Normal*	191	0
Plum	*Long*	*Wrinkled*	*Minute*	0	202

a Which chromosome carries the recessive gene for *wrinkled* wings?

b Which chromosome carries the recessive gene for *minute* eyes?   Explain or illustrate your answers.

590   In the mouse, the mutant genes *Str* (*striated*) and *Ta* (*tabby*) cause the development of dark transverse stripes.   Individuals carrying *Str*, however, show patches of short fur, absent in individuals carrying *Ta*.   The mutant gene *Blo* causes the development of patches of light fur.

Lyon (1966) crossed *Str Ta blo/str ta Blo* and *Str Ta Blo/str ta blo* females with normal *str ta blo/str ta blo* males.   The numbers of male and female offspring in each of the eight expected phenotypic classes are given in the table.

Table 590

Phenotype	Cross 1 ♂	Cross 1 ♀	Cross 2 ♂	Cross 2 ♀
Str Ta blo	0	43	0	3
str ta Blo	55	64	0	1
Str ta Blo	0	0	0	0
str Ta blo	3	4	0	0
Str ta blo	0	0	0	3
str Ta Blo	0	0	1	3
Str Ta Blo	0	5	0	48
str ta blo	2	5	39	42

a What is the sequence of the three loci?   Explain.

b Would you calculate recombination from pooled male and female data or not?   Why?

c Calculate the percent recombination for both regions.

591 a In tomatoes, the following three pairs of alleles are linked: *Mm* (*normal* vs. *mottled* leaf), *Pp* (*smooth* vs. *pubescent* epidermis), and *Awaw* (*purple* vs. *green* stem).   The percent recombination between *M* and *P* is 8, that between *M* and *Aw* is 25, and that between *P* and *Aw* is 17 (Butler, 1952).   In what order do these genes occur on the chromosome?

b The allele pair *Ss* (*simple* vs. *compound* inflorescence) shows 21 percent recombination with *Awaw*.   Where would you place *S* on the map?   What further data would you require to place it accurately?

c In a two-point testcross the percent recombination between *S* and *P* is 34.3.   Map all four genes.

d The sum of the recombination percentages between *S* and *Aw* and *Aw* and *P*, which gives the distance between *S* and *P*, is 21 + 17 = 38.   This value differs from that determined in the two-point testcross by 3.7 percent.   Explain the reason for this discrepancy.

592 In *Drosophila melanogaster* the allele pairs *Vv* (*red* vs. *vermilion* eyes), *Mm* (*normal* vs. *miniature* wings), and *Ff* (*normal* vs. *forked* bristles) are on the X chromosome at positions 33.0, 36.1, and 56.7, respectively, from the left end. A *vermilion, miniature, forked* female is crossed with a *red, normal, normal* male, and the F_1's are intercrossed.

a State the kinds of gametes produced by the F_1's and their frequencies, with no interference.

b What phenotypic ratio would be expected in the F_2?

c In the reciprocal of this cross would you expect results similar to those shown? Explain.

593 In rabbits, *white* fat, *Y*, is dominant to *yellow*, *y*, *himalayan* coat, c^h, is dominant to *chinchilla*, c^{ch}, and *black-tipped* fur, *B*, is dominant to *brown*, *b*. Castle in 1933 testcrossed $Yy c^h c^{ch} Bb$ individuals and obtained the results shown.

Table 593

Genotype	Phenotype			Number of progeny
	Coat	Fur	Fat	
$c^h B Y$	Himalayan	Black	White	151
$c^{ch} B Y$	Chinchilla	Black	White	33
$c^h b Y$	Himalayan	Brown	White	67
$c^{ch} b Y$	Chinchilla	Brown	White	11
$c^h B y$	Himalayan	Black	Yellow	2
$c^{ch} B y$	Chinchilla	Black	Yellow	48
$c^h b y$	Himalayan	Brown	Yellow	23
$c^{ch} b y$	Chinchilla	Brown	Yellow	142

a Which of these phenotypes are due to crossing-over between c^h and *B*? c^h and *Y*? *B* and *Y*?

b What is the coefficient of coincidence and degree of interference?

c What phenotypic ratio would you expect among the progeny of the cross $By/bY \times by/by$?

594 The distance between the *choline* locus and its centromere in *Neurospora* is approximately 24 map units (Horowitz et al., 1945). What is the expected frequency of second-division segregation for this locus?

595 In *Neurospora* the allele pairs *Lyslys* and *Adad* control the expression of *no lysine* vs. *lysine requirement* and *no adenine* vs. *adenine requirement*. From the cross *Lys Ad* × *lys ad* 150 asci had the spore sequences given. What are the positions of these allele pairs with respect to each other and their centromeres?

Table 595

Spore pair	Ascus type			
1 and 2	*lys Ad*	*Lys Ad*	*Lys ad*	*lys ad*
3 and 4	*lys Ad*	*Lys Ad*	*Lys ad*	*lys ad*
5 and 6	*Lys ad*	*lys ad*	*lys Ad*	*Lys Ad*
7 and 8	*Lys ad*	*lys ad*	*lys Ad*	*Lys Ad*
Number of asci	39	36	40	35

596 Asci from a single perithecium of a cross in *Neurospora* between *wild type* and a *lysine-requiring* mutant have the spore-pair arrangements shown.   What genetic conclusions can you draw from these distributions regarding types of segregation and gene-centromere distance?

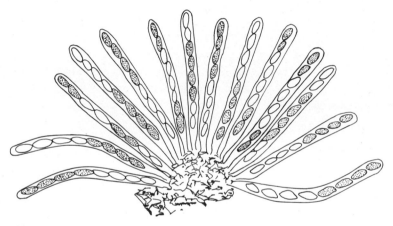

Figure 596   (*From Hayes, 1968.*)

597 The table gives a sample of the results obtained in *Neurospora* by Howe (1956) for three pairs of alleles: *Aa* for mating type *A* vs. mating type *a*, *Adad* for *adenine-independent* vs. *adenine-requiring*, and *Vv* for *slow* vs. *fast* growth, in a cross between parents that carried different alleles at all loci.   Answer the following questions regarding these data, giving reasons for your statements:
  a What are the genotypes of the parents?
  b Are the loci linked?
  c How are the three loci located with respect to each other and the centromere?
  d What types of exchanges (single, double, or triple; two-, three-, or four-strand) are responsible for each of the first four ascus types?   Explain the origin of each of the ascus types 5 to 15.

Table 597

No.		Ascus type			No. of asci
1	*A ad v*	*A ad v*	*a Ad V*	*a Ad V*	888
2	*A ad v*	*a ad v*	*A Ad V*	*a Ad V*	85
3	*A ad v*	*a Ad v*	*A ad V*	*a Ad V*	43
4	*A ad v*	*A ad V*	*a Ad v*	*a Ad V*	126
5	*A ad v*	*A Ad v*	*a ad V*	*a Ad V*	0
6	*A ad v*	*a Ad v*	*A Ad V*	*a ad V*	0
7	*A Ad v*	*a ad v*	*A ad V*	*a Ad V*	0
8	*A ad V*	*A Ad V*	*a ad v*	*a Ad v*	1
9	*A ad v*	*a Ad V*	*A ad v*	*a Ad V*	2
10	*A ad v*	*a Ad V*	*A ad V*	*a Ad v*	2
11	*A ad V*	*a Ad v*	*A ad V*	*a Ad v*	2
12	*A ad v*	*a ad V*	*A Ad v*	*a Ad V*	3
13	*A ad v*	*a ad V*	*A Ad V*	*a Ad v*	5
14	*A ad V*	*a ad v*	*A Ad v*	*a Ad V*	3
15	*A ad V*	*a ad v*	*A Ad V*	*a Ad v*	1

598 In the alga *Chlamydomonas reinhardi* Ebersold and Levine (1959) determined the number of two-, three-, and four-strand double exchanges for regions 1, 2, and 3 in linkage group I:

(centromere)	*arg*-1	*arg*-2	*pab*-2	*thi*-3
	5.2	15.4	29.9	
	1	2	3	

The results of three crosses are tabulated.   Is chromatid interference indicated (see Whitehouse, 1969)?

	Double exchanges		
Region	Two-strand	Three-strand	Four-strand
1–2	1	1	2
1–3	19	37	26
2–3	23	38	18

REFERENCES[1]

Adam, A., et al. (1969), *Ann. Hum. Genet.*, **32**:323.
———, C. Sheba, R. Sanger, R. R. Race, P. Tippett, J. Hampel, J. Gavin, and D. J. Finney (1963), *Ann. Hum. Genet.*, **26**:187.
Aida, T. (1921), *Genetics*, **6**:554.

[1] The following references apply to both Chaps. 13 and 14.

Allard, R. W. (1956), *Hilgardia*, **24**:235.

Anraku, N., and J. Tomizawa (1965*a*), *J. Mol. Biol.*, **11**:501.

—— and —— (1965*b*), *J. Mol. Biol.*, **12**:805.

Bannerman, R. M., G. B. Ingall, and J. F. Mohn (1971), *J. Med. Genet.*, **8**:291.

Barratt, R. W., and L. Garnjobst (1949), *Genetics*, **34**:351.

——, D. Newmeyer, D. R. Perkins, and L. Garnjobst (1954), *Adv. Genet.*, **6**:1.

Bateson, W., E. R. Saunders, and R. C. Punnett (1905), *Rep. Evol. Comm. R. Soc.*, **2**:1, 80.

Belling, J. (1928), *Univ. Calif. Publ. Bot.*, **14**:283.

—— (1929), *Univ. Calif. Publ. Bot.*, **14**:379.

—— (1931), *Univ. Calif. Publ. Bot.*, **16**:153, 311.

—— (1934), *Genetics*, **19**:388.

Bole-Gowda, B. N., D. D. Perkins, and W. N. Strickland (1962), *Genetics*, **47**:1243.

Bouwkamp, J. C., and S. Honma (1971), *J. Hered.*, **62**:37.

Bowen, S. T. (1963), *Biol. Bull.*, **124**:17.

—— (1965), *Genetics*, **52**:695.

Bregger, T. (1918), *Am. Nat.*, **52**:57.

Bridges, C. B. (1916), *Genetics*, **1**:1.

—— and K. S. Brehme (1944), *Carnegie Inst. Wash., D.C., Publ. 552.*

—— and T. H. Morgan (1919), *Carnegie Inst. Wash., D.C., Publ.* 278, p. 123.

—— and T. M. Olbrycht (1926), *Genetics*, **11**:41.

Brown, S. W., and D. Zohary (1955), *Genetics*, **40**:850.

Butler, L. (1952), *J. Hered.*, **43**:25.

Carothers, E. E. (1913), *J. Morphol.*, **24**:487.

Case, M. E., and N. H. Giles (1958), *Proc. Natl. Acad. Sci.*, **44**:378.

Cassuto, E. T. Lash, K. S. Sriprakash, and C. M. Radding (1971), *Proc. Natl. Acad. Sci.*, **68**:1639.

Castle, W. E. (1919), *Carnegie Inst. Wash., D.C., Publ. 288.*

—— (1933), *Proc. Natl. Acad. Sci.*, **19**:947.

—— and H. Nachtsheim (1933), *Proc. Natl. Acad. Sci.*, **19**:1006.

—— and S. Wright (1916), *Carnegie Inst. Wash., D.C., Publ. 241.*

Catcheside, D. E. A. (1968), *Genetics*, **59**:443.

Catcheside, D. G. (1951), "Genetics of Micro-organisms," Sir Isaac Pitman & Sons, London.

Clark, A. J., and A. D. Margulies (1965), *Proc. Natl. Acad. Sci.*, **53**:451.

Comings, D. E., and T. A. Okada (1970), *Nature*, **227**:45.

Craig-Cameron, T., and G. H. Jones (1970), *Heredity*, **25**:223.

Creighton, H. B., and B. M. McClintock (1931), *Proc. Natl. Acad. Sci.*, **17**:492.

Darlington, C. D. (1937), "Recent Advances in Cytology," 2d ed., McGraw-Hill, New York.

Davies, S. H., et al. (1963), *Am. J. Hum. Genet.*, **15**:481.

Donahue, R. P., W. B. Bias, J. H. Renick, and V. A. McKusick (1968), *Proc. Natl. Acad. Sci.*, **61**:949.

Dunn, L. C. (1920), *Genetics*, **5**:325.

Ebersold, W. T., and R. P. Levine (1959), *Z. Vererbungsl.*, **90**:74

——, ——, E. E. Levine, and M. A. Olmsted (1962), *Genetics*, **47**:531.

Emerson, R. A., G. W. Beadle, and A. C. Fraser (1935), *Cornell Univ. Agric. Exp. Stn. Mem.* 180.

Emerson, S. (1966), *Genetics*, **53**:475.

—— (1967), *Ann. Rev. Genet.*, **1**:201a.

Emery, A. E. H., C. A. B. Smith, and R. Sanger (1969), *Ann. Hum. Genet.*, **32**:261.

Emmerson, P. T. (1968), *Genetics*, **60**:19.

Falconer, D. S., and W. R. Sobey (1953), *J. Hered.*, **44**:159.

Fincham, J. R. S., and P. R. Day (1971), "Fungal Genetics," 3d ed., Blackwell, Oxford.

Fisher, R. A. (1935), *Ann. Eugen.*, **6**:187, 339.

Fortuin, J. J. H. (1971), *Mutat. Res.*, **66**:663.

German, J. (1964), *Science*, **144**:298.

Goldmark, P. J., and S. Linn (1971), *Proc. Natl. Acad. Sci.*, **67**:434.

Graham, J. B., H. L. Tarleton, R. R. Race, and R. Sanger (1962), *Nature*, **195**:834.

Green, M. C., and M. M. Dickie (1959), *J. Hered.*, **50**:3.

Grell, E. H., K. B. Jacobson, and J. B. Murphy (1965), *Science*, **149**:80.

Grell, R. F. (1962), *Proc. Natl. Acad. Sci.*, **48**:165.

Gurney, T., and M. S. Fox (1968), *J. Mol. Biol.*, **32**:83.

Hawthorne, D. C., and R. K. Mortimer (1960), *Genetics*, **45**:1085.

Hayes, W. (1968), "The Genetics of Bacteria and Their Viruses," 2d ed., Blackwell, Oxford.

Henderson, S. A. (1966), *Nature*, **211**:1043.

—— (1970), *Annu. Rev. Genet.*, **4**:295.

Hinton, C. W. (1970), *Genetics*, **66**:663.

Hiroyoshi, T. (1964), *Genetics*, **50**:373.

Holliday, R. (1964), *Genet. Res.*, **5**:282.

—— (1967), *Mutat. Res.*, **4**:275.

Horowitz, N. H., D. Bonner, and M. B. Houlahan (1945), *J. Biol. Chem.*, **159**:145.

Horton, B. F., and T. H. J. Huisman (1963), *Am. J. Hum. Genet.*, **15**:394.

Hotta, Y., M. Ito, and H. Stern (1966), *Proc. Natl. Acad. Sci.*, **56**:1184.

—— and H. Stern (1971), *J. Mol. Biol.*, **55**:337.

Houlahan, M. B., G. W. Beadle, and G. Calhoun (1949), *Genetics*, **34**:493.

Howard-Flanders, P., and R. P. Boyce (1964), *Genetics*, **50**:256.

—— and —— (1966), *Rad. Res. Suppl.* **6**:156.

Howe, H. B. (1956), *Genetics*, **41**:610.

Howell, S. H., and H. Stern (1971), *J. Mol. Biol.*, **55**:357.

Hungate, F. P. (1946), "The Biochemical Genetics of a Mutant of *Neurospora crassa* Requiring Serine or Leucine," thesis, Stanford University.

Hutt, F. B. (1949), "Genetics of the Fowl," McGraw-Hill, New York.

—— (1959), *J. Hered.*, **50**:209.

—— (1960), *Heredity*, **15**:97.

Hutton, J. L., and T. H. Roderick (1970), *Biochem. Genet.*, **4**:339.

Immer, F. R., and M. T. Henderson (1943), *Genetics*, **23**:419.

Ito, M., Y. Hotta, and H. Stern (1967), *Dev. Biol.*, **6**:54.

Janssens, F. A. (1909), *La Cellule*, **25**:387.

Jha, K. K. (1969), *Mol. Gen. Genet.*, **105**:30.

Kafer, E. (1958), *Adv. Genet.*, **9**:105.

Kayano, H. (1960), *Cytologia*, **25**:468.

Kitani, Y., L. S. Olive, and A. S. El-Ani (1962), *Am. J. Bot.*, **49**:697.

—— and —— (1969), *Genetics*, **62**:23.

Kozinski, A. W., P. B. Kozinski, and R. James (1967), *J. Virol.*, **1**:758.

Kushner, S. R., H. Nagaishi, A. Templin, and A. Clark (1971), *Proc. Natl. Acad. Sci.*, **68**:824.

La Cour, L. F., and S. R. Pelc (1959), *Nature*, **183**:1455.

Lawler, S. D., and J. H. Renwick (1959), *Br. Med. Bull.*, **15**:145.

Lawrence, C. W. (1967), *Mutat. Res.*, **4**:137.

Lederberg, J. (1955), *J. Cell Comp. Physiol.*, **45**(2):75.

Levine, R. P. (1962), "Genetics," Holt, New York.

Lindegren, C. C. (1932), *Bull. Torrey Bot. Club*, **59**:119.

—— (1933), *Bull. Torrey Bot. Club*, **60**:133.

—— (1936), *J. Genet.*, **32**:243.

Lissouba, P., J. Mousseau, G. Rizet, and J. L. Rossignol (1962), *Adv. Genet.*, **11**:343

—— and G. Rizet (1960), *C. R. Acad. Sci.*, **250**:3408.

Lyon, M. F. (1966), *Genet. Res.*, **7**:130.

MacArthur, J. W. (1928), *Genetics*, **13**:410.

McClung, C. E. (1927), *Qt. Rev. Biol.*, **2**:344.

McKusick, V. A. (1964), "On the X-Chromosome of Man," American Institute of Biological Sciences, Washington.
—— (1971), *Sci. Am.*, **224** (April):104.
Marshall, R. A., R. M. Bird, H. K. Bailey, and E. Beckner (1954), *J. Clin. Invest.*, **33**:790.
Mather, K. (1933), *Am. Nat.*, **67**:476.
—— (1935), *J. Genet.*, **30**:53.
—— (1951), "The Measurement of Linkage in Heredity," Wiley, New York.
Meselson, M. (1964), *J. Mol. Biol.*, **9**:734.
—— (1967), In R. A. Brink (ed.), "Heritage from Mendel," pp. 81–104, University of Wisconsin Press, Madison.
—— and J. J. Weigle (1961), *Proc. Natl. Acad. Sci.*, **47**:857.
Mitchell, M. B. (1955), *Proc. Natl. Acad. Sci.*, **41**:215, 935.
Moens, P. B. (1968), *Chromosoma*, **23**:418.
Morgan, T. H. (1911), *Science*, **34**:384.
—— (1912), *Science*, **36**:719.
—— (1914), *Biol. Bull.*, **26**:195.
—— and C. B. Bridges (1916), *Carnegie Inst. Wash., D.C., Publ.* 237.
——, ——, and A. H. Sturtevant (1925), *Bibliog. Genet.*, **2**:1.
—— and E. Cattell (1912), *J. Exp. Zool.*, **13**:79.
Morton, N. E. (1956), *Am. J. Hum. Genet.*, **8**:80.
Moses, M. J. (1958), *J. Biophys. Biochem. Cytol.*, **4**:633.
—— (1968), *Ann. Rev. Genet.*, **2**:363.
—— and J. H. Taylor (1955), *Exp. Cell Res.*, **9**:474.
Murray, N. E. (1968), *Genetics*, **58**:181.
Neel, J. V. (1941), *Genetics*, **26**:506.
Olive, L. S. (1959), *Proc. Natl. Acad. Sci.*, **45**:727.
Pascher, A. (1918), Über die Beziehung der Reduktionsteilung zur Mendelschen Spaltung, **Ber.** Dtsch. Bot. Ges., **36**:163.
Peacock, W. J. (1968), in Peacock and Brock (1968), pp. 242–252.
—— (1970), *Genetics*, **65**:593.
—— and R. D. Brock (eds.) (1968), "Replication and Recombination of Genetic **Material,"** Australian Academy of Science, Canberra.
Penrose, L. S. (1935), *Ann. Eugen.*, **6**:133.
—— (1953), *Ann. Eugen.*, **18**:120.
—— (1959), "Outline of Human Genetics," Wiley, New York.
Plough, H. H. (1921), *J. Exp. Zool.*, **32**:187.
Pontecorvo, G., E. T. Gloor, and E. Forbes (1954), *J. Genet.*, **52**:226.
—— and E. Kafer (1956), *Proc. R. Phys. Soc. (Edinb.)*, **25**:16.
—— and —— (1958), *Adv. Genet.*, **9**:71.
Porter, I. H., J. Schulze, and V. A. McKusick (1962), *Ann. Hum. Genet.*, **26**:107.
Pritchard, R. H. (1960), *10th Symp. Soc. Gen. Microbiol.*, London, pp. 155–180.
Rapley, S., E. B. Robson, H. Harris, and S. S. Maynard (1968), *Ann. Hum. Genet.*, **31**:237.
Regnergy, D. C. (1947), "A Study of the Leucineless Mutants of *Neurospora crassa*," thesis, California Institute of Technology.
Renwick, J. H. (1969), *Br. Med. Bull.*, **25**(1):65.
—— and S. D. Lawler (1955), *Ann. Hum. Genet.*, **19**:312.
Riley, R. (1969), *Sci. Prog.*, **54**:193.
—— and V. Chapman (1958), *Nature*, **182**:713.
Robinson, R. (1958), *Bibliog. Genet.*, **17**:229.
Rossen, J., and M. Westergaard (1966), *C. R. Trav. Lab. Carlsberg*, **35**:233.
Ruddle, F. H., V. M. Chapman, T. R. Chen, and R. J. Klebe (1970), *Nature*, **227**:251.
Sax, K. (1932), *J. Arnold Arbor.*, **13**:180.
Schultz, J., and H. Redfield (1951), *Cold Spring Harbor Symp. Quant. Biol.*, **16**:175.

Shaw, D. D. (1971), *Chromosoma*, **34**:281.

Slizynski, B. M. (1964), *Genet. Res.*, **5**:80.

Sobel, R. S., A. Tiger, and P. S. Gerald (1971), *Am. J. Hum. Genet.*, **23**:146.

Stadler, D. R., A. M. Towe, and J. L. Rossignol (1970), *Genetics*, **66**:429.

Stegeman, J. L., and E. Goldberg (1971), *Biochem. Genet.*, **5**:579.

Stern, C. (1931), *Biol. Zentralbl.*, **51**:547.

—— (1936), *Genetics*, **21**:625.

—— and C. B. Bridges (1926), *Genetics*, **11**:503.

Sturtevant, A. H. (1913), *J. Exp. Zool.*, **14**:43.

—— (1921), *Proc. Natl. Acad. Sci.*, **7**:235.

Sueoka, N., K. S. Chiang, and J. R. Kates (1967), *J. Mol. Biol.*, **25**:47.

Swanson, C. P. (1957), "Cytology and Cytogenetics," Prentice-Hall, Englewood Cliffs, N.J.

Swift, H. (1950), *Am. J. Bot.*, **37**:667.

Symposium on Chromosome Mechanics at the Molecular Level, *J. Cell. Physiol.*, **70**(1) (1967).

Taylor, J. H. (1958), *Genetics*, **43**:515.

—— (1965), *J. Cell. Biol.*, **25**:57.

—— (1967), in J. H. Taylor (ed.), "Molecular Genetics," pt. II, pp. 95–135, Academic, New York.

—— and R. McMaster (1954), *Chromosoma*, **6**:489.

Tomizawa, J., and N. Anraku (1964), *J. Mol. Biol.*, **8**:516.

—— and —— (1965), *J. Mol. Biol.*, **11**:509.

Vilmorin, P. D., and W. Bateson (1912), *Proc. R. Soc. (Lond.)*, **84**:9–11.

Walker, G. W. R., J. Dietrich, R. Miller, and K. Kasha (1963), *Can. J. Genet. Cytol.*, **5**:200.

Weiss, B., and C. C. Richardson (1967), *Proc. Natl. Acad. Sci.*, **57**:1021.

Westergaard, M. (1964), *C. R. Trav. Lab. Carlsberg*, **34**:359.

—— and D. von Wettstein (1970), *C. R. Lab. Carlsberg*, **37**:157.

Wettstein, D. von (1971), *Proc. Natl. Acad. Sci.*, **68**:851.

Wimber, D. E., and W. Prensky (1963), *Genetics*, **48**:1731.

Winge, O. (1923), *J. Genet.*, **13**:201.

Winkter, H. (1930), "Die Konversion der Gene," G. Fischer, Jena.

Whitehouse, H. L. K. (1963), *Nature*, **199**:1034.

—— (1969), "The Mechanism of Heredity," 2d ed., E. Arnold, London.

—— (1970), *Biol. Rev. Camb. Philos. Soc.*, **45**:265.

—— and P. J. Hastings (1965), *Genet. Res.*, **6**:27.

Whittaker, D. L., D. L. Copeland, and J. B. Graham (1962), *Am. J. Hum. Genet.*, **14**:149.

Wood, S. (1967), *Nature*, **216**:63.

15
Extranuclear
Inheritance and
Related Phenomena

15
Extranuclear
Inheritance
and
Related Phenomena

NOTATION[1]

mRNA = messenger RNA   A = adenine
rRNA = ribosomal RNA   C = cytosine
tRNA = transfer RNA   G = guanine
                               T = thymine

QUESTIONS

599   **a** What is the basic test by which cytoplasmic inheritance is distinguished from nuclear inheritance in almost all organisms? Suggest three forms of inheritance which behave in this test in a manner similar to cytoplasmic inheritance and show how they can be distinguished.

     b (1) What specific properties do chromosomal genes possess?

         (2) Explain how cytoplasmic genes show these properties, citing examples where possible.

600   In rabbits, strain A animals are *night-blind*, and strain B rabbits have *normal* night vision. All F_1's of the cross A female × B male are *night-blind*; those of B female × A male are *normal*. Among possible explanations of this reciprocal-cross difference are:

    (1) Sex linkage
    (2) Dauermodification
    (3) Maternal inheritance (influence)
    (4) Cytoplasmic inheritance
    (5) Maternal influence via placenta, milk, or egg
    (6) Rh-like situation

Outline the series of crosses you would make to determine which of these explanations is the correct one and show the results expected in each cross with each form of inheritance.

601   If an infectious disease were transmitted from the maternal parent to the offspring through the egg, the condition would closely mimic a cytoplasmically inherited trait. Encountering such a condition, how would you determine that it is an infectious disease?

602   You are given seed from a *male-sterile* plant in corn, a species in which it is possible to self-fertilize as well as cross-fertilize plants. Outline how you would proceed to determine whether the trait is controlled by cytoplasmic or nuclear genes. Show the results expected.

603   When a certain mutant strain of *Neurospora crassa* is the maternal parent in a cross with *wild type*, all the ascospores are *gray*. When the mutant is the paternal parent, all the ascospores are *white*. Offer two probable explanations

---

[1] Abbreviations occurring frequently in a chapter are listed for convenience. A more complete summary of the biochemical abbreviations used in this book will be found in Appendix Table A-7. Prefixes used in the metric system are explained in Appendix Table A-6.

for this pattern of inheritance and describe a simple genetic test for distinguishing them.

604 Van Wisselingh (1920), working with *Spirogyra*, found cells with two plastids, one with normal pyrenoids and the other lacking this structure. After many cell divisions both plastid types were still present in each cell. Why does this constitute evidence that plastids possess their own hereditary determiners?

605 Exposure of a culture of the alga *Chlamydomonas reinhardi* to the mutagen acriflavine produces some cells which are *resistant* to streptomycin. When the *resistant* mutants were propagated asexually, some *sensitives* reappeared. Explain these results.

606 Assume that extensive use of a certain drug leads to addiction and that children from matings between *addicted* mothers and *nonaddicted* fathers are *addicted* while those from reciprocal matings are *nonaddicted*. The reciprocal-cross differences are not due to cytoplasmic genes. Offer an explanation to account for these results.

607 In 1927 Hofmann (see Jollos, 1939) working with *Phaseolus* (beans), showed that leaf aberrations induced by chloral hydrate were transmitted through female gametes only. How would you demonstrate that the leaf aberrations were dauermodifications and were not controlled by cytoplasmic genes?

608 Occasionally one wants to determine whether a particular phenotype in the progeny (e.g., hemolytic disease in mammals) is due to the transmission of cytoplasmic substances through the female gamete or to the transfer of a substance or substances from mother to the fetus through the placenta. In rabbits and mice it is possible to remove ovaries and replace them with ovaries from other females. Many of the transplanted ovaries become established and produce and shed eggs normally. How might this technique be used to distinguish between the alternatives mentioned?

609 Answer each of the following as briefly and completely as possible:
a Which do you think would be easier to identify, the effects of plasmagenes or the effects of chromosomal genes? Explain.
b Why is it often difficult to distinguish between cytoplasmic genes (plasmagenes) and viruses?
c What conditions must be satisfied to prove that cytoplasmic genes are present in the chloroplast? Discuss fully.

610 A female meiocyte contains two alleles for each of two cytoplasmic genes. Would these alleles segregate from each other at meiosis in the manner that pairs of chromosomal alleles do? Why?

PROBLEMS

611 In the four-o'clock (*Mirabilis jalapa*) variegated plants occur in which certain branches have normal *green* leaves, other branches have leaves entirely *white*

or *pale*, and still others have leaves that are *green* with *pale* or *white* areas. Correns (1909) found that whatever the pollen parent, seed borne on flowers on *green* branches produce *green* plants, those on *white* branches produce *white* seedlings, and those on *variegated* branches gave *green, white,* and *variegated* seedlings in widely differing ratios.

a Explain these results.

b What is the critical observation for the acceptance of your explanation?

612 a Through inbreeding and selection, lines of mice have been developed that differ markedly in the incidence of *mammary cancer*.   The level of incidence in a given line is constant from generation to generation.   The inheritance of this character was studied by Bittner (1936, 1937, 1941, 1958; Bittner and Little, 1938) in reciprocal crosses involving a number of inbred lines. When lines with *high* incidence were crossed as females with lines possessing a *low* incidence of the disease, almost all the F_1 and F_2 females developed the cancer.   In the reciprocal crosses only a few F_1 and F_2 females were afflicted.   Suggest possible explanations for this reciprocal-cross difference.

b Newborn mice of the *high*-incidence strain were transferred, before they could be suckled by their mother, to foster mothers of the *low*-incidence strain; among them *mammary-cancer* development was no more frequent than in the *low*-incidence line.   When the reciprocal transfer was performed, many of the females developed *mammary cancer,* and the majority of their subsequent female progeny were also afflicted.   This information should permit a specific choice of one of the answers to (a).   State which it is, outlining in as much detail as possible the explanation for this mode of transmission of a disease from parent to offspring.

613 The life cycle of *Chlamydomonas reinhardi*, a unicellular haploid alga with two mating types mt^+ and mt^-, is illustrated.   The zygote contains, as far as can be detected, all the contents, both nuclear and cytoplasmic, of both gametes.

Wild-type cells are *streptomycin-sensitive, ss*.   Sager (1954, 1960) treated large populations of these with 500 μg/ml of this antibiotic and isolated a *resistant* strain, *sr*.   The cross *sr mt$^-$* × *ss mt$^-$* produced haploid progeny that segregated 1 : 1 for mating type.   The *sr* clones, after years of vegetative propagation under a variety of environmental conditions in the absence of streptomycin, showed the same degree of resistance as the *resistant* parental strain.   The *sr mt$^+$* F_1 clones backcrossed to *ss mt$^-$* produced *sr* offspring only (4:0), but mating type still segregated in a 1:1 ratio.   The same results were obtained in three subsequent backcross generations.   The reciprocal backcross (F_1 *sr mt$^-$* × *ss mt$^+$*), produced only *streptomycin-sensitive* (0:4) progeny.   In subsequent backcrosses resistance did not appear.

a What mode of transmission appears to be the most plausible explanation for the 4:0 segregation?   Explain.

b Does the same hypothesis account for the 0:4 segregation?   If not, outline alternative hypotheses to account for this segregation.

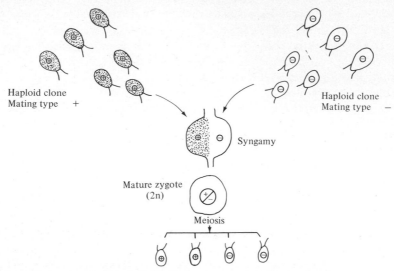

Four haploid offspring (sexual spores) from zygote

Figure 613

614 In the alga *Chlamydomonas* resistance to streptomycin may be due to cytoplasmic genes, a chromosomal gene, or both. A true-breeding *streptomycin-resistant* strain with both the cytoplasmic and chromosomal genes for *resistance* is crossed with a *streptomycin-sensitive* strain. What would be the proportions of *resistant* to *sensitive* strains (each strain from a single meiotic product) if:

a The *resistant* strain was of plus mating type and the *sensitive* strain of minus mating type?

b The *resistant* strain was minus and the *sensitive* one plus?

615 In the diploid protozoan *Paramecium aurelia*, two conjugants undergo a reciprocal exchange of gamete nuclei during short sexual fusion; during long fusion they exchange cytoplasm as well. Exconjugants that are heterozygous for a chromosomal gene can undergo an autogamous fission, producing individuals homozygous for each allele. The following data regarding the *killer* trait are from Sonneborn (1943a, c; see also Beale, 1954). *Killer* individuals contain 200 to 300 DNA-containing particles called kappa in their cytoplasm; *sensitive* animals lack these entities.

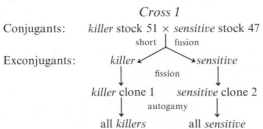

Cross 1

Conjugants: *killer* stock 51 × *sensitive* stock 47
                         short │ fusion

Exconjugants: *killer* ↙   ↘ *sensitive*
                            fission

killer clone 1   *sensitive* clone 2
                    autogamy

all *killers*   all *sensitive*

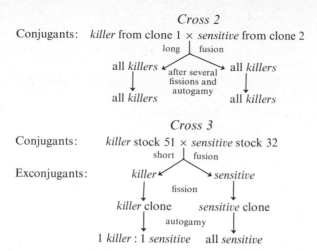

Cross 2

Conjugants: *killer* from clone 1 × *sensitive* from clone 2

long | fusion

all *killers* ← after several fissions and autogamy → all *killers*

↓ all *killers*

↓ all *killers*

Cross 3

Conjugants: *killer* stock 51 × *sensitive* stock 32

short | fusion

Exconjugants: *killer* ← → *sensitive*

↓ fission ↓

killer clone   *sensitive* clone

↓ autogamy ↓

1 *killer* : 1 *sensitive*   all *sensitive*

According to Sonneborn, kappa is responsible for the *killer* trait, but a nuclear gene *K* is necessary for the multiplication and maintenance of kappa. State which crosses support each of these conclusions, and why.

616 In corn, true-breeding mutants with *green* and *white striping* of the foliage are fairly common. One such mutant is called *iojap* (Jenkins, 1924). Reciprocal crosses between the true-breeding *iojap* line and true-breeding *green* lines gave the results shown.

Table 616

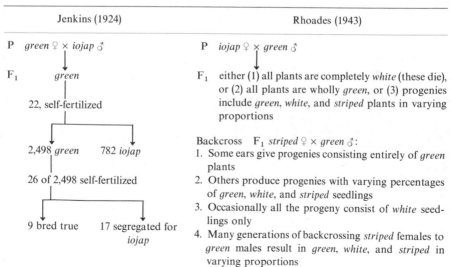

Jenkins (1924)	Rhoades (1943)
P *green* ♀ × *iojap* ♂	P *iojap* ♀ × *green* ♂
F$_1$ *green*	F$_1$ either (1) all plants are completely *white* (these die), or (2) all plants are wholly *green*, or (3) progenies include *green*, *white*, and *striped* plants in varying proportions
22, self-fertilized	
2,498 *green* 782 *iojap*	Backcross F$_1$ *striped* ♀ × *green* ♂:
26 of 2,498 self-fertilized	1. Some ears give progenies consisting entirely of *green* plants
	2. Others produce progenies with varying percentages of *green*, *white*, and *striped* seedlings
9 bred true 17 segregated for *iojap*	3. Occasionally all the progeny consist of *white* seedlings only
	4. Many generations of backcrossing *striped* females to *green* males result in *green*, *white*, and *striped* in varying proportions

Note: (1) *Green* plants contain normal chloroplasts, as do green tissues in *striped* plants. (2) The white tissue in *striped* and *white* plants contains minute colorless plastids. (3) Pale tissues contain cells with a mixture of green and colorless plastids.

a Show why it is believed that the *iojap* trait is controlled by both a nuclear gene and a cytoplasmic determinant.

b How might the chromosomal allele pair be related with a nonchromosomal determinant in the control of this form of striping?

c Do these results indicate whether or not the determinant is actually located in the chloroplast? Explain.

d Are the plasmagenes genetically autonomous in:

(1) Controlling the phenotype?

(2) Their duplication?

617 In 1938 L'Heritier and Teissier discovered an *ebony* strain of flies that was *sensitive* (killed by exposure) to carbon dioxide (L'Heritier, 1948). When females of the *sensitive* strain were crossed with males of *resistant* strains, all the offspring were *sensitive*. The progeny of F_1 females × *resistant*-strain males were all *sensitive.*

a Account for this transmission pattern of carbon dioxide sensitivity, showing that your explanation also accounts for the following:

(1) When the reciprocal of the first cross was made, some *sensitive* progeny were produced.

(2) Injection of extracts from *sensitive* flies into *resistant* ones can cause the latter to develop sensitivity to carbon dioxide. Suggest what kind of agent is likely to be concerned and how you might confirm your suggestion.

b What kinds of progeny would an injected female be likely to produce? Why?

618 a Corn (*Zea mays*) is a monoecious species which has male flowers at the top of the plant and female flowers along the side of the plant. A strain is developed in which almost all the plants are *male-sterile*; the few *male-fertile* ones produce very little pollen. The F_1's between the *male-sterile* line and different *male-fertile* ones are *male-sterile*. Male fertility is not restored by backcrossing the F_1 and subsequent backcross generations' offspring to the *male-fertile* lines even when all the chromosomes and their genes were from the *male-fertile* line. When female flowers on plants in the *male-fertile* line are fertilized by the rare pollen from plants in the *male-sterile* line, all the progeny are *male-fertile*. These plants when self-fertilized breed true. Offer a genetical explanation of these results.

b When the above *male-sterile* line is crossed with *male-fertile* strain A, all the progeny are *male-fertile*. These F_1's backcrossed as females to the *male-sterile* line produced 52 *male-sterile* and 55 *male-fertile* plants. The reciprocal of this cross also produces the same results. When the F_1's are self-fertilized, they produce 123 *male-fertile* and 42 *male-sterile* plants.

(1) Explain why these results are different from those in (a) despite the fact that the same *male-sterile* line is used.

(2) On the basis of your hypothesis would you expect *male-sterile* plants in the following crosses and, if so, in what proportions?

(a) F_1 (*male-sterile* × *male-fertile* strain A) × *male-fertile* strain A.

(b) F_1 (*male-sterile* × *male-fertile* strain A) × *male-fertile* strain discussed in (a).

619 Edwardson and Corbett (1961) showed that cytoplasmic male sterility in the petunia could be transferred from *male-sterile* to *male-fertile* plants by grafting. The cytoplasmic factors controlling sterility were transmitted subsequently through the seed. What do these findings suggest regarding the nature of the cytoplasmic factor?

620 A number of investigators have shown that treatment of plants and animals with various environmental agents (e.g., chemicals, high temperature) can induce the expression of abnormal phenotypes which are transmitted maternally for a few generations with slowly diminishing intensity of expression until they gradually disappear. For example, Hofmann in 1927 (see Jollos, 1939) found that treatment of bean (*Phaseolus vulgaris*) plants with chloral hydrate induced the formation of leaf aberrations which are transmitted by the female only and persist for six generations before disappearing. Offer an explanation for the origin of these modifications and the fact that they last for only a few generations.

621 In corn, leaves with green and white stripes can be produced by either of two recessive chromosomal genes, *j* (*japonica*) on chromosome 8 or *ij* (*iojap*) on chromosome 7. The striping caused by the two recessives is indistinguishable. *J— Ij—* plants have completely *green* leaves. All the possible true-breeding lines for these two genes are at your disposal, and you are given two *striped* plants, each homozygous for one recessive gene. What cross would you make to determine (1) whether both plants are genotypically identical, (2) which, if either, was homozygous for *j*? Show the expected F_1 genotypes and phenotypes in each case.

622 Among eight progeny of a self-fertilized *green* barley plant, one *albino* appears. (The *albino* was fed sugar solutions, matured, and formed seeds.) Outline experiments to determine whether the *albino* was the result of:
a Some environmental factor.
b Cytoplasmic inheritance.
c A gene mutation.

623 In barley, a self-fertilizing species that can be cross-fertilized, two true-breeding strains with *virescent* leaves occur. In strain A the trait is caused by a cytoplasmic gene; in strain B by a recessive chromosomal gene. What phenotypes would you expect among the progeny and in what proportions in each of the following?
a Reciprocal crosses between A and B.
b F_1 from each of the reciprocal crosses to each of the parental strains.
c Self-fertilization of F_1's of reciprocal crosses.
d Crossing between F_1's of reciprocal crosses.

624   In peanuts (*Arachis hypogaea* L.) commercial varieties possess either a *runner* or *bunch* growth habit.   Ashri (1964) obtained the tabulated results in reciprocal crosses between true-breeding lines with these two traits.   Propose a

Table 623

♀ ♂	F$_1$	F$_2$
V4 × NC2	runner	648 runner : 499 bunch
V4 × 123	runner	581 runner : 456 bunch
V4 × G2	runner	441 runner : 368 bunch
NC2 × V4	bunch	290 runner : 624 bunch
123 × V4	bunch	28 runner : 47 bunch
G2 × V4	bunch	89 runner : 180 bunch

model for intergenic and gene-cytoplasm interaction to account for these results.   See Ashri (1969).

625   In some perfect-flowered, cross-fertilizing species, e.g., onions and sugar beets, hybrid F$_1$ seed is commercially desirable but not economically feasible because of the high cost of emasculation (removal of anthers).   Male sterility obviates this necessity.   In onions, the inheritance of male sterility was studied by Jones and Clarke (1943) using the *male-sterile* line 13-53 and a large number of commercial varieties.   The results of some of the crosses are tabulated.

Table 625

Parents		Progeny
♀	♂	
1. *male-sterile* 13-53 × *male-fertile* variety 1		all *male-fertile*
2. *male-sterile* 13-53 × *male-fertile* variety 2		all *male-sterile*
3. *male-sterile* 13-53 × *male-fertile* variety 3		1 *male-fertile* : 1 *male-sterile*
4. F$_1$ *male-fertile* (self-fertilized)		3 *male-fertile* : 1 *male-sterile*
5. *male-sterile* 13-53 × F$_1$ *male-fertile*		1 *male-fertile* : 1 *male-sterile*
6. *male-fertile* F$_1$ from cross 3 × *male-fertile* parent		1 *male-fertile* : 1 *male-sterile*
7. reciprocal of cross in 6: *male-fertile* parent × *male-fertile* F$_1$		all *fertile*

a State with reasons whether male sterility is controlled by nuclear genes or not.

b Demonstrate that male sterility is controlled by cytoplasmic genes and show whether there is evidence of their nonautonomous duplication.

626   In barley two *yellow*-foliage true-breeding strains, A and B, were reciprocally crossed to a *green* strain C; the consequences are shown in Table 626*A*.

Table 626A

Cross	♀ ♂	F₁
1	A × C	*Green*
2	C × A	*Green*
3	B × C	*Yellow*
4	C × B	*Green*

Backcrossing of the F_1 of B female × C male as female parent to the true-breeding strain C (male parent) produced only *yellow* progeny.   The same results were obtained in successive backcrosses to the C parent.   The reciprocal backcrosses gave *green* offspring when the F_1 was used as male parent. Backcrosses of either of the *green* F_1's of the cross A × C used as either male or female to strain A produced progeny of which half were *yellow* and half *green*.   Another *green* strain D when crossed reciprocally to strain B gave the results in Table 626B.   What is the plasmatype and genotype of strain D? Explain.

Table 626B

♀ ♂	F₁	F₂
B × D	*Yellow*	*Yellow*
D × B	*Green*	All *green*

627 *Mammary cancer* in mice is maternally transmitted (Bittner, 1936; Bittner and Little, 1938).   How would you determine whether it was due to (1) a cytoplasmic gene, (2) a maternal effect, or (3) a virus or viruslike factor transmitted through the milk?

628 Mitochondrial DNA is approximately 15,000 base pairs (5 μm) long.
  a If the average length of proteins is 200 amino acids, how many proteins could this DNA code for?
  b Is this number sufficient to code for all mitochondrial structures and functions?   Explain.
  c How would you proceed to prove that mitochondrial DNA (1) can replicate, (2) specifies synthesis of its own *t*RNAs, *r*RNA, and some *m*RNAs?   See Tewari and Wildman (1970) and Goodenough and Levine (1970).

629 a Mitchell and Mitchell (1952) found an erratically slow-growing *Neurospora crassa* mutant called *poky*, *po*, which showed an unusual pattern of transmission (Table 629A).
  (1) How is the mutant inherited?
  (2) What further crosses would you make to confirm your conclusion and to determine whether the determiner is acting autonomously?

Table 629 A

Protoperithecial (♀) parent		Conidial (♂) parent	Progeny (ascospores)
wild type (normal)	×	poky	all wild type
poky	×	wild type	all poky

b In 1953 Mitchell et al. studied three other slow-growing mutant strains, mi-3, c115, and c117.   All were found to carry an abnormal system of respiratory enzymes.   The transmission pattern of these mutants is shown in Table 629B.

Table 629B

Protoperithecial (♀) parent		Conidial (♂) parent	Progeny (ascospores)	
			Wild type	Slow
mi-3	×	wild-type	0	2,071
wild-type	×	mi-3	1,113	3*
wild-type	×	c115†	596	590
wild-type	×	c117†	1,050	1,035

* Due to slow growth or other factors.
† Female-sterile.

(1) Give a genetic interpretation of these results.

(2) What results would you expect in crosses between c115 or c117 and mi-3 or *po*?

c The same workers in 1956 showed that certain strains of *poky* (here called *fast-poky* strains) had reverted to nearly *wild-type* growth, although they still retained the abnormal metabolism (and cytochromes).

(1) A cross with *fast-poky* as protoperithecial (female) parent and *wild type*, a conidial (male) parent, gives 1 *poky* : 1 *fast-poky* ascospores in all asci.

   (a) Interpret these results genetically, showing what they indicate about the autonomy of the cytoplasmic determiner in *poky*.

   (b) Show the results expected for the reciprocal of the cross described. Illustrate and explain your answer (assume no cytoplasm enters with the male nucleus).

(2) Reciprocal crosses were made between a *poky* and a *fast-poky* colony, both derived from ascospores of a single ascus of the cross described in (a).   The two crosses gave the same results, a 1:1 ratio of *poky* to *fast-poky*.   *Fast-poky* male × *wild-type* female produced only *wild-type* progeny.

   (a) Is your interpretation borne out?

   (b) What other conclusions can you draw from these results?

630 Ephrussi (1953), Yotsuyanagi (1962), and Nass and Nass (1963) in an elegant series of experiments on "petites" (*littles*) in diploid bakers' yeast (*Saccharomyces cerevisiae*) established the following:

a When *normal* cells are plated on agar medium, from 1 to 2 percent of the colonies formed are *littles*; when the medium contains acriflavine dye, all colonies are *littles* which vegetatively produce only *little* colonies. *Little* colonies are probably small because the cells must rely on anaerobic respiration for energy. The mitochondria (known to carry the enzymes of aerobic respiration) in these cells are devoid of many of these enzymes. Diploid F_1 hybrids from the crosses *little* × *normal* and *normal* × *normal* and their selfed progeny in subsequent generations are *normal*. Similarly, backcrosses of F_1's and subsequent hybrids to the *little* parent continue to express the *normal* trait. Describe the mode of inheritance of *littles*, showing how it explains these observations.

b Another diploid *little*, called *segregational little*, with a different inheritance pattern, was also found by these workers. Although it also showed similar deficiencies in respiratory enzymes and produced *normal* F_1's in crosses with *normal*, the ascospores of the F_1 hybrids gave rise to *normal* and *segregational-little* colonies in a 1:1 ratio. Describe the mode of inheritance of these *littles*.

c Mitochondria are known to contain DNA. The mitochondria of *littles* are structurally different, while those of *segregational littles* are indistinguishable from those in *normal* yeast. When cells from *segregational littles* are crossed with cells from *vegetative littles*, the diploid cells are *normal*. Asci formed by these diploids show a 1:1 ratio of *normal* to *little* ascospores. Interpret these results and construct a diagram that summarizes your interpretation.

631 *Killer* paramecia carry DNA-containing kappa particles in their cytoplasm. These particles, which are about the size of a small bacterium, produce paramecin, a poison that kills *sensitive* paramecia (those lacking kappa) except when *sensitives* conjugate with *killers*. The dominant chromosomal gene *K*, although necessary for the maintenance and replication of kappa, cannot initiate the production of kappa. Therefore *K*— *sensitives* do not produce kappa although they can acquire kappa during long fusion with a *killer* individual and thereafter maintain the particles. Paramecia of genotype *kk* that have acquired kappa during long fusion with *killers* are *unstable*. They lose kappa and eventually become *sensitives*. Cytoplasmic exchange (mixing) occurs if conjugation is long but not if it is short.

a A *killer* individual conjugates with a *sensitive* one without exchanging cytoplasm. Half the exconjugants are *killers*, half are *sensitives*. Autogamy of the *killer* exconjugants produces *killers* only.

  (1) What kinds of progeny would autogamy of *sensitives* produce?
  (2) Give the genotypes of the conjugants and the exconjugants.

b The conjugants of five different matings have the tabulated nuclear genotypes and cytoplasmic constitutions. For each cross answer the following questions for both short and long fusion:

Table 631

Mating	Nuclear genotype	Kappa	×	Nuclear genotype	Kappa
1	KK	Present		Kk	Absent
2	Kk	Present		kk	Absent
3	kk	Absent		kk	Present
4	KK	Absent		KK	Present
5	Kk	Present		Kk	Absent

(1) What ratio of *killer*:*sensitive*:*unstable* would you expect among the exconjugants?

(2) What proportion of the exconjugants will carry and maintain kappa through subsequent fissions?

c Two Kk paramecia conjugate. Can one of the exconjugants be kk and the other Kk? One KK, the other kk? Both KK? Both Kk? Why?

d A kk *sensitive* individual conjugates with a Kk *killer* without cytoplasmic mixing. The exconjugants go through autogamy. What proportion of the resulting progeny would be expected to carry and maintain kappa? Would the proportion be the same if cytoplasmic mixing occurred during conjugation?

632 In onions, *male sterility* is due to the interaction of a chromosomal allele pair *msms* and "sterile," *S*, cytoplasm. All other combinations (viz. *Ms*— and "sterile" cytoplasm, *Ms*— or *msms* and "fertile," *F*, cytoplasm) result in *male-fertile* plants. The *male-sterile* trait is incorporated into inbred lines to produce hybrid F_1 seed on a commercial scale.

a How would you perpetuate the *male-sterile* line?

b *Male-sterile* lines produce better or more commercial seed when used in combination with some inbred lines than with others. Outline the method of selecting the best inbred lines for commercial production.

c Sometimes it is desirable to transfer *ms* and the sterile cytoplasm to an inbred that produces desirable seed from crosses with other lines. How would you accomplish this transfer?

d Briefly outline the method of producing hybrid seed for the commercial crop. Does it matter whether the cytoplasm is fertile or sterile in the *male-fertile* inbred? Explain.

e How would you ascertain the nuclear genotype and the cytoplasmic type of a *male-fertile* plant?

633 Extensive studies have been made of reciprocal crosses between domestic cattle (*Bos taurus*) and the bison, or buffalo (*B. bison*). When domestic bulls are mated with bison cows, the fetuses develop normally and vigorous hybrids, called *cattalo*, are born, the females being fertile, the males sterile. In the reciprocal cross the hybrid calves are rarely born alive. One of the reasons for the high mortality associated with the male hybrids is the difficulty in parturition caused by the large hump of the male fetus. Another reason, associated with both sexes of the hydrid, is abortion caused by an early buildup of excess fluid within the fetal membrane. Offer two explanations to account for this *abortion* vs. *nonabortion* in the reciprocal crosses. Would it be possible to test your hypotheses? If so, explain how.

634 When certain stallions are repeatedly mated to the same mare, the occasional firstborn and many of the subsequent foals develop *hemolytic jaundice* (red-blood-cell destruction leading to death 3 to 4 days after birth). The condition can be avoided by preventing suckling by the mother (Bruner et al., 1948). It is thought that a single allele pair *Aa*, for *production* vs. *nonproduction* of antigen-6, is involved. The hemolysis occurs in second and subsequent *Aa* foals of *A—* male × *aa* female matings because the mother has developed antibodies that enter the milk. These react with the blood cells of the foal to cause agglutination, hemolysis, and subsequently the symptoms noted above.

a Is *hemolytic jaundice* in the foal an inherited trait? The result of maternal influence? Or neither? Elaborate.

b Suppose it is possible serologically to detect the A antigen and hence to determine whether a horse carries the dominant allele *A* but that it is impossible to tell whether the animal is homozygous or heterozygous. A mare that does not produce antigen-6, mated to a stallion that does, bears a foal that lacks the antigen. What is the genotype of the stallion?

c Bruner et al. (1948) found cases in which a mare produced two successive foals, the first one *normal* and the second one with *hemolytic jaundice*. The father of both foals was the same stallion. What does this information reveal about the genotypes of the stallion and the mare?

635 Bruner et al. (Bruner et al., 1948; Franks, 1962) found that sometimes a particular stallion and mare have a *normal* first foal but the occasional second and often the third and subsequent foals, although apparently normal at birth, would if nursed by the mother, develop *jaundice* within 12 to 48 hours after birth. Numerous red blood cells begin to be destroyed, and the foals die within 3 to 4 days. Bruner and colleagues discovered that if a foal born into such a jaundice-producing situation is prevented from nursing on its own mother and is given a foster mother for only 24 to 36 hours and thereafter returned to its own mother after her colostrum (first milk) is removed, it develops normally. If after having a foal or two that die the mare is bred to a different stallion, she frequently has no difficulty in rearing his foals. There is no indication that the factor responsible for the incompatibility is transmitted to its progeny by an *affected* individual that has recovered from this type of *jaundice*.

a Propose an explanation to account for this incompatibility.

b Compare this situation with the Rh one in man. How are they alike, and how are they different?

c Outline how you would test your explanation and the results expected if your hypothesis is correct.

636 The mutant *grandchildless* in *Drosophila subobscura* is caused by an autosomal recessive gene *gs* (Spurway, 1948).   The gene has no phenotypic effect on flies homozygous or heterozygous for it.   Although *gsgs* males are viable and fertile and have grandchildren, *gsgs* females, regardless of their fertile male mate, produce only sterile offspring: females with rudimentary ovaries and males 98 percent of which have no testes.   Offer an explanation to account for the delayed phenotypic effect of *gs* which occurs only in females.

637 The freshwater snail *Limnaea peregra* is a hermaphroditic species, capable of cross- or self-fertilization.   The species is polymorphic, some individuals having right-handed (*dextral*) coiling of the shell and others having left-handed (*sinistral*) coiling.   Boycott and Diver (1923) noted that all the progeny from single individuals by self-fertilization are of the same phenotype (viz. *dextral* or *sinistral*).   In 1930 Boycott et al. in crosses between true-breeding *dextral* and true-breeding *sinistral* lines found that the direction of coiling of the F_1 progeny was the same as that of their mother; the F_2 snails were all *dextral*, and in the F_3 1,192 *dextral* and 401 *sinistral* broods were obtained, each brood being the result of self-fertilization of an F_2 snail.   Offer an explanation to account for these results and see whether it agrees with the ingenious hypothesis of Sturtevant (1923).

638 In the meal moth *Ephestia kuhniella* the dominant allele *A* is necessary for the conversion of tryptophan to the diffusible hormonelike substance kynurenine, a precursor of brown pigment.   As a consequence, *A* larvae have a *pigmented* skin and *A* adults have *dark brown* eyes; *aa* larvae are *non-pigmented*, and *aa* adults have *red* eyes (no brown pigment).   Reciprocal crosses between true-breeding *pigmented*, *AA*, and *nonpigmented*, *aa*, moths both produce *pigmented* F_1's.   Reciprocal crosses between *Aa* and *aa* animals give the tabulated results.

Table 638

Cross	Result
$Aa\,♀ \times aa♂$	All larvae are *pigmented*, but half the adults have *dark-brown* eyes and half have *red* eyes
$aa\,♀ \times Aa♂$	Half of the larvae are *pigmented* and, as adults, have *dark-brown* eyes (Kuhn, 1936)

a Show diagrammatically that these results cannot be explained on the basis of sex-linked inheritance.
b For what reason do reciprocal crosses give different results?   What is this phenomenon called?

c A male and female, both *pigmented* as larvae, are crossed.   About half the adult progeny have *dark-brown* eyes, half *red* eyes.   What eye color did the male and female parents have as adults?

639 In the crustacean genus *Gammarus* (water flea or beach hopper), eye color is caused by a pair of alleles *Aa* (Sexton and Pantin, 1927).   The dominant allele, *A*, permits pigment formation; the recessive allele, *a*, does not.   The color of eyes of the young is controlled by the genotype of the mother but later by the genotype of the individual flea.

a Diagram reciprocal crosses between *AA* and *aa* fleas showing parental, F_1, and F_2 phenotypes and genotypes.   Distinguish phenotypes of young and adults in each generation.

b Explain why some of the young with *light* eyes develop *dark* eyes as they become older.

640 Hagberg (1950) reported on the relation between genotype and phenotype for *bitter* vs. *sweet* in the lupine, *Lupinus angustifolius*.   Reciprocal crosses were made between the *sweet* variety Borre and the *bitter* variety Blue.

a All F_1 plants from *sweet* female × *bitter* male were *sweet* when tested within 5 to 6 weeks after the seed was sown. F_1 plants from the reciprocal cross tested within the same period were *bitter*.   What forms of inheritance may be responsible for these F_1 results?

b The leaves after 5 to 6 weeks' growth and the ripe seeds of the F_1's were *bitter* regardless of which variety was used as the female parent.   Which of the forms of inheritance suggested in (a) are eliminated by this information?   Explain.

c Forty F_2 seeds, all *bitter*, from *sweet* female × *bitter* male gave rise to plants whose leaves tested *bitter* until approximately a month after seeding.   The mature F_2 plants segregated 28 *bitter* : 12 *sweet* for both their leaves and ripe seed (either *bitter* for both or *sweet* for both).

(1) At what stages in the ontogeny of lupines are the traits *bitter* and *sweet* affected by the genotype of the parental generation?   By their own genotype?

(2) Diagram the reciprocal crosses to illustrate your answer to (1).

641 The direction of coiling of the shells in snails may be *sinistral* (to the left) or *dextral* (to the right).   In the normally *sinistral* hermaphroditic snail *Laciniaria biplicata*, rare true-breeding *dextral* races exist.   Degner in 1952 found that reciprocal crosses between true-breeding *sinistrals* and true-breeding *dextrals* gave the following results:

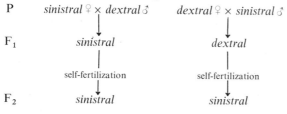

Three-quarters of the self-fertilized F_2's produced *sinistral*, and one-quarter produced *dextral* F_3's in both crosses.

a If hermaphrodites have no meiotic sex-determining mechanism (by genetic or chromosomal segregation), what must be the mode of inheritance of this trait difference and why? Give the genotypes of the parents, F_1's and F_2's.

b Account for the matroclinous nature of inheritance (shown by the difference in the F_1's from reciprocal crosses and in the F_2's).

642 In the hermaphroditic snail *Limnaea peregra*, capable of self- or cross-fertilization, direction of coiling of shells is influenced by the maternal genotype (Boycott et al., 1930). The allele S causing coiling to the right (*dextral*) is dominant over s for coiling to the left (*sinistral*).

a A homozygous SS female is crossed with a homozygous ss male. The F_2's, like the F_1's, are self-fertilized. Give the genotypes and phenotypes of the F_1's, F_2's, and F_3's and the proportions of these in each generation. Give a reason for each in each instance.

b What are the phenotypes of the parents of the F_1?

c A cross between two individuals produces a snail with *dextral* coiling. This F_1 snail upon self-fertilization produces only *sinistral* progeny. Determine the genotype of this snail and its parents.

643 In the Mexican axolotl (*Ambystoma* [$=$ *Siredon*] *mexicanum*) *fluid imbalance* is an embryonic condition in which localized deposits of jellylike fluid are found.

Table 643

Mother	Offspring	
	Genotype	Phenotype
Ff	FF, Ff, ff	Develop normally; exhibit *fluid imbalance* in its usual form (distention of the head, trunk, etc.); excess fluid disappears after circulation is established; most affected individuals hatch and survive to adult stage
ff	Ff	Fluid deposits accumulate during cleavage, eventually escapes; most individuals (about 85%) survive
	ff	Fluid deposits accumulating during cleavage do not escape later; circulation is not established; all individuals die before hatching

In 1948 Humphrey showed that *fluid imbalance* is caused by an autosomal recessive allele, *f*. In 1960 he reported detailed studies of the time and mode of action of this gene, as summarized in the table. The ovaries of *ff* donors transplanted to *FF* or *Ff* hosts behave autonomously. Propose an explanation of these results.

644 Riboflavin is essential for normal embryo development and hatching in the domestic fowl; its concentration in the blood of hens on adequate diets is multiplied some fiftyfold during the laying period.   In contrast, Maw (1954) found three hens (two full sisters and one half-sister) in a breeding stock that had been maintained on a very high riboflavin diet whose eggs failed to hatch; the embryos died between 10 and 14 days after incubation, even when the birds were fed riboflavin supplement.   Physiological studies revealed that the eggs were low in riboflavin and that if the vitamin was injected into the eggs at 200 mg per/egg after 3 days of incubation, they hatched normally.

1   The three hens were crossed with unrelated males.   After injection with riboflavin the eggs hatched, giving an F_1 of 7 males and 7 females.   The F_1 females mated with a single F_1 male laid normally *hatchable* eggs, leading to an F_2 of both sexes.   Of the 88 F_2 females, 67 laid *normal* and 21 laid *nonhatchable* eggs.

2   The three hens, crossed with the F_1 male that was used to produce the F_2 progeny, laid eight eggs, which, after injection with riboflavin, hatched into 3 females that laid *normal* and 5 that laid *nonhatchable* eggs.

a Describe the genic control of *nonhatchable* vs. *normal* eggs, as revealed by the breeding data alone, drawing analogies with other similar situations if possible.

b From an examination of the physiological data above and the table (from Buss et al., 1959), suggest the probable nature of the physiological action.

Table 644

Place where level of riboflavin measured	*Normal*-laying hens	F_1 hens	Hens laying *nonhatchable* eggs
In blood of nonlaying hens	0.008	0.008	0.008
In blood of laying hens	0.434	0.008	0.272
In egg yolk	4.3	0.41	2.50

645 a In *Drosophila melanogaster*, females from a true-breeding *normal*-winged strain crossed with *fused*-wing males produce F_1's with *normal* wings.   The F_1's when intercrossed produced 3 *normal* : 1 *fused*; all the *fused* offspring were males.   When the reciprocal cross was made (*fused* females × *normal* males), only females were produced in the F_1, the adult progeny being about half as numerous as in the first cross.   Analyze these results and state with reasons (1) the number of allele pairs involved and (2) whether located on an autosome or the X chromosome.

b Lynch (1919) made other crosses involving the allele *fused*.   The results of these are as follows:

Normal ♀ (from *normal* ♀ × *fused* ♂ or the reciprocal cross) × *fused* ♂ → 1 *fused* ♂ : 1 *fused* ♀ : 1 *normal* ♂ : 1 *normal* ♀

Fused ♀ × *fused* ♂ → no adult progeny

Fused ♀ × *normal* ♂ → *normal* ♀, 0 ♂

Propose an explanation to account for the action of the *fused* and *normal* alleles on viability.

646 Shires are a very large breed of horse, and Shetland ponies are comparatively small. Walton and Hammond (1938) used artificial insemination to produce reciprocal crosses between the two breeds. Weights of some of the animals concerned are tabulated. At three years of age the F_1 foals from the Shetland

Table 646

	Weight, kg		
1. Adult weight of parental mares	Shire	797	Shetland	214
2. Birth weight of purebred foals	Shire	71	Shetland	20
3. Birth weight of F_1's		Shire × Shetland	50
		Shetland × Shire	18

mares had not caught up to the F_1 foals from the Shire mares, and it appeared there was little chance of their doing so.
a Do these results suggest cytoplasmic inheritance?
b Can you think of an obvious alternative to such an explanation?
c Suggest additional information you would like to have and further experiments that might aid in arriving at a satisfactory explanation of the reciprocal-cross difference.

647 Spurway (1948) carried out extensive studies on an abnormal line of *Drosophila subobscura*. This line, otherwise very uniform, contained a few females which in matings with brothers, as well as unrelated males, produced *sterile* offspring of which all the females had rudimentary ovaries and 98 percent of the males had no testes. Only *fertile* offspring were produced by other females from the abnormal line in matings with brothers and related and unrelated males and by males from the abnormal line mated to sisters or unrelated females. Results of some of the initial matings that helped clarify this unusual pattern of inheritance are tabulated. Each cross involved one male and one female only. Determine the mode of inheritance of this sex-limited (grandchildless) trait and explain how and where the gene or genes may act to cause this effect.

Table 647

Mating		Offspring
♀	♂	
A × B		83 *fertile* ♂; when crossed with *fertile* sisters or unrelated females, these gave rise to *fertile* progeny only
		89 *fertile* ♀; when crossed with *fertile* brothers or unrelated males, these produced *fertile* progeny only
A × C		93 *fertile* ♂; breeding behavior the same as that of the 83 *fertile* ♂ from A × B
		97 *fertile* ♀, 24 of which when crossed to *fertile* brothers or unrelated males produced *sterile* offspring only; the other 73 ♀ bred like the 89 ♀ from A × B
D* × E†		58 *fertile* ♂; breeding behavior like that of *fertile* ♂ from A × B
		61 *fertile* ♀, 29 of which when crossed with *fertile* related or unrelated ♂ produced *sterile* offspring only ($\frac{1}{2}$ ♂ : $\frac{1}{2}$ ♀); the others produced *fertile* offspring

* Genotype identical to ♀ A in second cross.

† Brother of the 24 ♀ in second cross.

648 Sturtevant (1920, 1929) studied sex-linked inheritance in intercrosses between *Drosophila melanogaster* and *D. simulans*, both of which possess the X-Y method of sex determination. The hybrids were completely *sterile*, with rudimentary gonads. Two kinds of hybrids were identified: (1) in *regular* offspring X-linked traits inherit in normal manner, (2) in *exceptional* offspring the females inherit X-linked traits from the mother (matroclinously), and the males inherit them from the father (patroclinously). The table summarizes the analytical results. From these results what conclusions can you draw about the interaction between the nuclear genome and cytoplasm?

a Suggest a hypothesis to account for these results.

b If the hybrids were *fertile*, what experiments could be made to test the hypothesis. Show the results expected in each (show in outline form only).

Table 648

Parents		Type of offspring
♀	♂	
melanogaster (XX) × *simulans*		*regular* ♀
melanogaster (XXY) × *simulans*		*regular* ♀
		exceptional ♂
simulans (XX) × *melanogaster*		*regular* ♂
simulans (XXY) × *melanogaster*		*regular* ♂
		exceptional ♀

649 Two species of tRNA for each of the amino acids isoleucine and phenylalanine are detectable in light-grown but not in dark-grown *wild-type Euglena gracilis* cells or in the light- or dark-grown bleached mutant W_3BUL, which contains neither chloroplast DNA nor chloroplast structure.

In addition, light-grown *wild-type Euglena* contains two aminoacyl-tRNA synthetases for each of these amino acids. Isoleucyl-tRNA synthetase I (constitutive) is also formed by dark-grown *wild-type Euglena* cells and by W_3BUL cells under both conditions, whereas isoleucyl-tRNA synthetase II is present only in light-grown *wild-type* cells. The phenylalanyl-tRNA synthetases are present in dark-grown *wild-type* and W_3BUL cells as well as in light-grown ones. Only one of the two synthetases for each amino acid (light-inducible isoleucyl-tRNA synthetase II and phenylalanyl-tRNA synthetase I) is found in isolated chloroplasts. For each amino acid only the light-induced phenylalanine and isoleucine tRNAs can be acylated by the chloroplast synthetases (Reger et al., 1970).

a Which of these tRNAs and synthetases are coded by nuclear genes? Cytoplasmic genes?

b If it is impossible to answer the question for a particular tRNA or synthetase on the basis of the above data, design an experiment to obtain this information and show the results expected.

c Which chloroplast synthetase is definitely synthesized in the cytoplasm? Explain.

d Do chloroplasts possess a translational apparatus? Explain.

650 In certain strains of *Drosophila willistoni* the females produce progenies consisting, with few exceptions, of daughters only, regardless of the genetic nature of the males to which they are crossed; moreover, approximately half the eggs laid by these females fail to hatch (Malogolowkin and Poulson, 1957; Malogolowkin, 1958; and Malogolowkin, Poulson, and Wright, 1959). This capacity for the production of unisexual broods is inherited by their entire female progenies and is not transmitted by the few exceptional males that appear in some cultures. Many females of a normal strain when injected with cytoplasm from nonhatching eggs began producing all-female progenies after a latent period of about two weeks. How can you account for these results, and how would you proceed to test your hypothesis?

651 Among the Os, Og, and Ha strains of the mosquito *Culex pipiens* (from Osterberg, Oggelshausen, and Hamburg, Germany), reduced fertility occurs in some reciprocal interstrain crosses. The crosses between the Os and Og strains, for example, are both *fertile*; each gives male and female offspring in about a 1:1 ratio. On the other hand, crosses between strains such as those between Og and Ha are *fertile* only when performed in one direction (Laven, 1959). Suggest possible causes of these differences and describe how you would determine which explanation is correct.

652 In the fungus *Neurospora crassa* the heteroplasmon (fusion cytoplasm consisting of cytoplasm from both parents) between the poky and mi-4 mutants has the growth rate of *wild type*, unlike the heteroplasmon between poky and mi-3, which has a *mutant* phenotype (Pittenger, 1956).

a What do these observations tell us about the probable relationships between the *mutant* plasmagenes?

b Suggest experiments for clarifying this relationship.

653 Mitochondria and chloroplasts contain DNA (Edelman et al., 1965; Granick and Gibor, 1967; Luck and Reich, 1964; Sager and Ishida, 1963; Tewari and Wildman, 1968).   Is this proof that these DNAs carry genes responsible for the expression of cytoplasmically inherited traits?   Explain.

654 Densitometer measurements of whole cells and isolated chloroplasts may be used to identify different classes of DNA according to their molecular

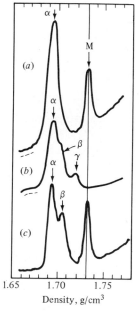

Figure 654 Densitometer tracings of DNA from beet leaves. [*Redrawn from E. H. L. Chun, M. H. Vaughan, Jr., and A. Rich, The Isolation and Characterization of DNA Associated with Chloroplast Preparations, J. Mol. Biol., 7:133 (Fig. 2) (1963).*]

weights.   Curve (*a*) represents DNA from whole leaves.   A major band (α) is present.   *M* represents marker DNA from *Micrococcus lysodeikticus*.   Curve (*b*) represents DNA isolated from presumably purified chloroplasts.   In addition to the α band, two heavier minor (satellite) bands (β and γ) are present.   Curve (c), the β band, has been further purified by repeated centrifugation in cesium chloride.

Similar studies with *Euglena* show only one band, that corresponding to α, to be present in mutant strains lacking chloroplasts (Leff et al., 1963) while both this and the satellite band β are present in strains carrying chloroplasts (Edelman et al., 1964).   The α- and β-band DNAs differ in base ratio as shown in the table (Edelman et al., 1964).   Do chloroplasts contain their own DNA?   What is the evidence?   (See Gibor and Granick, 1964.)

Table 654

	A	T	G	C
α	0.52	0.45	0.48	0.55
β	0.74	0.70	0.26	0.36

655   Chloroplasts are known to possess DNA (Granick and Gibor, 1967).   In 1962 Gibor and Granick irradiated the cytoplasm of *Euglena* with ultraviolet light at wavelengths close to 2600 Å while shielding the nucleus, and vice versa. Irradiation of the nucleus had no effect on the chloroplasts, but irradiation of the cytoplasm caused a hereditary bleached condition, in which bleached plastids remained tiny and colorless and reproduced as such from cell generation to cell generation.   Why do these results indicate that DNA of plastids is capable of replication?

656   Brawerman (1962) found a unique species of RNA in *Euglena* chloroplasts, which was associated with the ribosomes.   It was significantly higher in A + U and lower in G + C than cytoplasmic RNA (Brawerman et al., 1964; Ray and Hanawalt, 1964).   Kirk (1964) found that all four nucleotide triphosphates are required for chloroplast RNA synthesis.   Chloroplast treatment with actinomycin D or with deoxyribonuclease inhibited RNA synthesis.   The DNA of the satellite band of chloroplasts (see Prob. 654) is characterized by a higher ratio of A + T to G + C also.

a Do these data suggest that chloroplast DNA acts as template for RNA synthesis?

b How might you proceed to support your conclusion?

c What results would you expect?

657   What is the significance of the following finding?   Incorporation of radioactive amino acids by isolated chloroplasts of *Euglena* is inhibited by treating the chloroplasts with ribonuclease or with actinomycin D (Eisenstadt and Brawerman, 1963).

QUESTIONS INVOLVING FURTHER READING

658 "Never before in our experience had we encountered results suggesting the paradoxical conclusion that a hereditary difference was neither genotypic nor cytoplasmic in basis" (Sonneborn, 1963). For a challenging analysis of this paradox see Sonneborn (1963) and Tartar (1961) who discuss pioneering studies of the perpetuation of preformed structure.

659 Discuss the criteria that may be used in detecting extrachromosomal inheritance (see Jinks, 1964, and Srb, 1963, for a critical review of its detection and analysis in fungi).

REFERENCES

Aloni, Y., and G. Attardi (1971*a*), *J. Mol. Biol.*, **55**:251.
—— and —— (1971*b*), *J. Mol. Biol.*, **55**:271.
Ashri, A. (1964), *Genetics*, **50**:363.
—— (1969), *Genetics*, **60**:807.
Attardi, B., and G. Attardi (1969), *Nature*, **224**:1079.
—— and —— (1971), *J. Mol. Biol.*, **55**:215.
Attardi, G., and D. Djala (1971), *Nat. New Biol.*, **229**:133.
Barnett, W. E., D. H. Brown, and J. L. Epler (1967), *Proc. Natl. Acad. Sci.*, **57**:1775.
Bateson, W., and A. E. Gairdner (1921), *J. Genet.*, **11**:269.
Beale, G. H. (1954), "The Genetics of *Paramecium aurelia*," Cambridge University Press, Cambridge.
Bittner, J. J. (1936), *Proc. Soc. Exp. Biol. Med.*, **34**:42.
—— (1937), *J. Hered.*, **28**:363.
—— (1941), *Cancer Res.*, **1**:793.
—— (1958), *Ann. N.Y. Acad. Sci.*, **71**:943.
—— and C. C. Little (1938), *J. Hered.*, **28**:117.
Borst, P. (1970), *Symp. Soc. Exp. Biol.*, **24**:201.
—— and A. M. Kroon (1969), *Int. Rev. Cytol.*, **26**:107.
Boycott, A. E., and C. Diver (1923), *Proc. R. Soc. (Lond.)*, **B95**:207.
——, ——, S. L. Garstang, and F. M. Turner (1930), *Philos. Trans. R. Soc. Lond.*, **B219**:51.
Brawerman, G. (1962), *Biochem. Biophys. Acta*, **61**:313.
—— and J. M. Eisenstadt (1964), *Biochem. Biophys. Acta*, **91**:477.
Bruner, D. W., E. F. Hull, and E. R. Doll (1948), *Am. J. Vet. Res.*, **9**:237.
Buss, E. G., R. V. Boucher, and A. J. G. Maw (1959), *Poult. Sci.*, **38**:1192.
Chiang, K. S. (1968), *Proc. Natl. Acad. Sci.*, **60**:194.
Chun, E. H. L., M. H. Vaughan, and A. Rich (1963), *J. Mol. Biol.*, **7**:130.
Corneo, G., C. Moore, D. R. Sanadi, L. T. Grossman, and J. Marmur (1966), *Science*, **151**:687.
Correns, C. (1909), *Z. Indukt. Abstamm.-Vererbungsl.*, **1**:291.
Dawid, I. B. (1970), *Symp. Soc. Exp. Biol.*, **24**:227.
Degner, E. (1952), Der Erbgang der Inversion bei *Laciniaria biplicata* Mtg. nebst Bemerkungen zur Biologie der Art, *Mitt. Hamb. Zool. Mus. Inst.*, **51**:3.
Edelman, M., C. A. Cowan, H. T. Epstein, and J. A. Schiff (1964), *Proc. Natl. Acad. Sci.*, **52**:1214.
——, J. A. Schiff, and H. T. Epstein (1965), *J. Mol. Biol.*, **11**:769.
Edwardson, J. R., and M. K. Corbett (1961), *Proc. Natl. Acad. Sci.*, **47**:390.
Eisenstadt, J., and G. Brawerman (1963), *Biochem. Biophys. Acta*, **76**:319.
Ellis, R. J., and M. R. Hartley (1971), *Nat. New Biol.*, **233**:193.

Ephrussi, B. (1953), "Nucleo-cytoplasmic Relations in Microorganisms," Clarendon Press, Oxford.

Franks, D. (1962), *Ann. N.Y. Acad. Sci.*, **97**:235.

Fukasawa, H. (1953), *Cytologia*, **18**:167.

Gibor, A. (1965), *Am. Nat.*, **99**:229.

—— and S. Granick (1962), *J. Cell Biol.*, **15**:599.

—— and —— (1964), *Science*, **145**:890.

—— and T. Izawa (1963), *Proc. Natl. Acad. Sci.*, **50**:1164.

Gillham, N. W., J. E. Boynton, and B. Burkholder (1970), *Proc. Natl. Acad. Sci.*, **67**:1026.

Goodenough, U. W., and R. P. Levine (1970), *Sci. Am.*, **223**(November):22.

Granick, S., and A. Gibor (1967), *Prog. Nucleic Acid Res. Mol. Biol.*, **6**:143.

Hagberg, A. (1950), *Hereditas*, **36**:228.

Humm, D. G., and J. H. Humm (1966), *Proc. Natl. Acad. Sci.*, **55**:114.

Humphrey, R. R. (1948), *J. Hered.*, **39**:255.

Jenkins, M. T. (1924), *J. Hered.*, **15**:467.

Jinks, J. L. (1964), "Extrachromosomal Inheritance," Prentice-Hall, Englewood Cliffs, N.J.

Jollos, V. (1939), Grundbegriffe der Vererbungslehre, "Handbuch der Vererbungswissenschaft," Bornträger, Berlin.

Jones, H. A., and A. F. Clarke (1943), *Proc. Am. Soc. Hort.*, **43**:189.

Kirk, J. T. O. (1964), *Biochem. Biophys. Res. Commun.*, **14**:393.

Kuhn, A. (1936), *Naturwiss.*, **24**:1.

Laven, H. (1959), *Cold Spring Harbor Symp. Quant. Biol.*, **24**:166.

Leff, J., M. Mandel, H. T. Epstein, and J. A. Schiff (1963), *Biochem. Biophys. Res. Commun.*, **13**:126.

Leiter, E. H., D. A. LaBrie, A. Bergquist, and R. P. Wagner (1971), *Biochem. Genet.*, **5**:549.

L'Heritier, P. (1948), *Heredity*, **2**:325.

—— (1958), *Adv. Virus Res.*, **5**:195.

Luck, D. J. L., and E. Reich (1964), *Proc. Natl. Acad. Sci.*, **52**:931.

Lynch, C. J. (1919), *Genetics*, **4**:501.

Maan, S. S., and K. A. Lucken (1971), *J. Hered.*, **62**:149.

Malogolowkin, C. (1958), *Genetics*, **43**:274.

—— and D. F. Poulson (1957), *Science*, **126**:32.

——, ——, and E. Y. Wright (1959), *Genetics*, **44**:59.

Maw, A. J. G. (1954), *Poult. Sci.*, **33**:216.

Michaelis, P. (1954), *Adv. Genet.*, **6**:287.

Miller, P. L. (ed.) (1970), "Control of Organelle Development," Symposia of the Society for Experimental Biology, no. 24, Academic, New York.

Mitchell, M. B., and H. K. Mitchell (1952), *Proc. Natl. Acad. Sci.*, **38**:442.

—— and —— (1956), *J. Gen. Microbiol.*, **14**:184.

——, ——, and Tissieres (1953), *Proc. Natl. Acad. Sci.*, **39**:606.

Nass, M. M. K. (1966), *Proc. Natl. Acad. Sci.*, **56**:1215.

—— (1969), *J. Mol. Biol.*, **42**:529.

—— and C. A. Buck (1969), *Proc. Natl. Acad. Sci.*, **62**:506.

—— and —— (1970), *J. Mol. Biol.*, **54**:187.

—— and S. Nass (1963), *J. Cell Biol.*, **19**:593.

Owen, F. V. (1942), *Am. J. Bot.*, **29**:629.

—— (1948), *Proc. Am. Soc. Sugar Beet Tech.*, **5**:156.

Pittenger, T. H. (1956), *Proc. Natl. Acad. Sci.*, **42**:747.

Preer, L. B., and J. R. Preer, Jr. (1964), *Genet. Res.*, **5**:230.

Ray, D. S., and P. C. Hanawalt (1964), *J. Mol. Biol.*, **9**:812.

Reger, B. J., S. A. Fairfield, J. L. Epler, and W. E. Barnett (1970), *Proc. Natl. Acad. Sci.*, **67**:1207.

Reich, E., and D. J. L. Luck (1966), *Proc. Natl. Acad. Sci.*, **55**:1600.

Rhoades, M. M. (1943), *Proc. Natl. Acad. Sci.*, **29**:327.
────── (1946), *Cold Spring Harbor Symp. Quant. Biol.*, **11**:202.
────── (1955), in W. Ruhland (ed.), "Encyclopedia of Plant Physiology," vol. 1, pp. 19–57, Springer-Verlag, Berlin.
Roodyn, D. B., and D. Wilkie (1968), "The Biogenesis of Mitochondria," Methuen, London.
Sager, R. (1954), *Proc. Natl. Acad. Sci.*, **40**:356.
────── (1960), *Science*, **132**:1459.
────── (1965), *Sci. Am.*, **212**(January):70.
────── and M. R. Ishida (1963), *Proc. Natl. Acad. Sci.*, **50**:725.
────── and Z. Ramanis (1963), *Proc. Natl. Acad. Sci.*, **50**:260.
────── and ────── (1970), *Proc. Natl. Acad. Sci.*, **65**:593.
Sexton, E. W., and C. F. A. Pantin (1927), *Nature*, **119**:119.
Sherman, F. (1964), *Genetics*, **49**:39.
────── and P. P. Slonimski (1964), *Biochim. Biophys. Acta*, **91**:1.
Simpson, L., and A. da Silva (1971), *J. Mol. Biol.*, **56**:443.
Sonneborn, T. M. (1943*a*), *Proc. Natl. Acad. Sci.*, **29**:329.
────── (1943*b*), *Proc. Natl. Acad. Sci.*, **29**:338.
────── (1947), *Adv. Genet.*, **1**:263.
────── (1950), *Sci. Am.*, **183**(November):30.
────── (1963), in J. M. Allen (ed.), "The Nature of Biological Diversity," pp. 165–221, McGraw-Hill, New York.
Spurway, H. S. (1948), *J. Genet.*, **49**:126.
Srb, A. M. (1963), *Symp. Soc. Exp. Biol.*, **17**:175.
Sturtevant, A. H. (1920), *Genetics*, **5**:488.
────── (1923), *Science*, **58**:269.
────── (1929), *Carnegie Inst. Wash., D.C., Publ.*, **399**:1.
Surzycki, S. J., et al. (1970), *Symp. Soc. Exp. Biol.*, **24**:13.
Tartar, V. (1961), "The Biology of Stentor," Pergamon, New York.
Tewari, K. K., and S. G. Wildman (1968), *Proc. Natl. Acad. Sci.*, **59**:569.
────── and ────── (1970), *Symp. Soc. Exp. Biol.*, **24**:147.
Walton, A., and J. Hammond (1938), *Proc. R. Soc. (Lond.)*, **B125**(840):311.
Wilkie, D. (1964), "The Cytoplasm in Heredity," Methuen, London.
Wisselingh, C. van (1920), *Z. Indukt. Abstamm.-Vererbungsl.*, **22**:65.
Woodward, D. O., and K. D. Munkres (1966), *Proc. Natl. Acad. Sci.*, **55**:872.
Yotsuyanagi, Y. (1962), *J. Ultrastruct. Res.*, **7**:141.

16
Recombination in Bacteria

NOTATION[1]

1 Abbreviations of metabolites (viz. *met* for *methionine*) followed by the superscript + or − respectively distinguish phenotypically between the *ability* and *inability* to synthesize the substance.

2 A capital letter following a gene symbol (as in *met A*) signifies a group of allelic mutants. Numerical symbols, as in *met-53*, represent record numbers assigned to independently isolated mutants. It will be noted that where only the requirement for a certain essential metabolite has been tested, mutants with blocks at various points in the metabolic sequence ending in this metabolite are classified as being auxotrophic for it.

3 *Selected markers* In some recombination experiments, the *wild-type* (+) alleles at certain marker-gene loci are selected from among the progeny of crosses by growth on appropriate metabolite-deficient medium. These genes are called the *selected markers*.

4 *Unselected markers* After the above selection, the genotypes of the selected clones are determined for other genes segregating in the cross; these genes are termed the *unselected markers*.

QUESTIONS

660 Distinguish between:
    a Eukaryotes and prokaryotes.
    b Auxotrophs and prototrophs.
    c Generalized and restricted transduction.
    d Heterogenote and merogenote.

661 In a U-tube experiment involving two auxotrophic strains some of the offspring of the recipient strain showed certain donor traits.
    a Which mechanism of gene transfer can be eliminated as a cause of hereditary changes in the recipient strain?
    b How would you proceed to determine the mechanism causing the change?
    c What results would you expect with each mechanism?

662 λ (lambda) phage released from a gal^+ strain of *E. coli* may infect gal^- strains to produce heterogenotes. Morse et al. (1956) have shown that $gal^-/\lambda\text{-}gal^+$ heterogenotes may lose their prophage to yield not only nonlysogenic, haploid gal^- segregants but also a small number of gal^- segregants which are lysogenic and homozygous diploid for the gal^- allele ($gal^-/\lambda\text{-}gal^-$). How can the latter kind of bacteria arise. Can you think of a good use for them in bacterial genetics?

663 To detect bacterial recombinants in conjugation crosses two types of markers must be present in the donor strain, one to identify recombinants, the other to prevent cells of the donor strain from growing on selective media. This counterselection is necessary so that only recombinants are selected. For example, in the cross Hfr (met^+ thr^+ str^s) × F$^-$ (met^- thr^- str^r) the donor carries

---

[1] Abbreviations for the common amino acids are given, together with their structural formulas, in Appendix Table A-4.

the marker *str^s* (*streptomycin-sensitive*).   By growing the conjugants on a streptomycin medium all the donors are eliminated.   Should the gene controlling the counterselection trait, e.g., *streptomycin sensitivity*, be located near the *O* (origin) or the Hfr end of the chromosome?   Explain.

664   A *proline-requiring Salmonella* mutant is infected with phage from another *proline-requiring* mutant.   Minute colonies never arise, but large colonies are formed with very low frequency.   Account for these observations.

665   The chromosome of *E. coli* is circular, yet during conjugation it opens up and is transferred to the female as a linear structure with part of the sex factor at its trailing end.   Explain how this change from a circular to a linear structure apparently occurs.   See Jacob et al. (1963*b*), Cuzin et al. (1967), Gross and Caro (1966), Cohen et al. (1968), Ohki and Tomizawa (1968), and Ihler and Rupp (1969).

666   Genes controlling sequential steps in a number of biosynthetic pathways in bacteria are not only closely linked but also in the same sequence as the steps which they control.   Do you think that close linkage of genes with related effects is more advantageous in organisms like bacteria and viruses than in higher organisms like *Drosophila*?   Explain.

667   Crosses in *Escherichia coli* give the following results (Lederberg et al., 1952*a*; Cavalli et al., 1953; Hayes, 1953):

1   $F^+ \times F^+$: fertile (recombinants produced).
2   $F^- \times F^-$: sterile (no recombinants produced).
3   $F^+ \times F^-$: fertile (recombinants produced).
4   One F^+ cell × many F^- cells : all F^-'s rapidly converted to F^+'s that transmit F^+ (by fission) to all their progeny.
5   Hfr × F^- : fertile, produce 100 to 20,000 times as many recombinants as $F^+ \times F^-$ crosses.   The recombinants remain F^- with rare exceptions.
6   Hfr can arise from F^+ and revert to F^+.   Crosses between these reverted F^+ and F^- are fertile and rapidly convert F^- to F^+.

The ability of Hfr cells to produce high frequencies of recombinants is unaffected by pretreatment with streptomycin.   What properties of the F factor do these data reveal?

668   Two F^- auxotrophic strains of bacteria, *pro⁻ bio⁺* (*prolineless*) and *pro⁺ bio⁻* (*biotinless*) both produce prototrophic colonies (*pro⁺ bio⁺*) after growth in a common medium.

a Which mechanism (recombination, conjugation, transduction, or transformation) is ruled out as the means by which prototrophs are formed, and why?

b Describe the experimental procedure required to distinguish between the other two as the cause.

669   The process of genetic transformation in bacteria occurs in a series of discrete steps:

1   Competence
2   Binding of transforming DNA
3   Penetration of transforming DNA
4   Synapsis
5   Integration

Discuss the evidence that each is essential for transformation.   (See Fox, 1966; Hayes, 1968.)

670   **a** Outline the experimental evidence which indicates that F:
    (1) Can integrate at many different places on the *Escherichia coli* chromosome and can also deintegrate.
    (2) Is composed of genetic material.
    (3) Is carried extrachromosomally in F^+ strains.
  b The F factor is DNA and appears to be circular like the *E. coli* chromosome.   Illustrate how F can integrate into the chromosome and deintegrate. See Broda et al. (1964) and Hayes (1966, 1967).
  c What properties does F possess when it is integrated in a chromosome?

671 An *E. coli* cell is not lysed when placed in a medium containing λ particles. Is it lysogenic or not?

672   **a** The F factor can occasionally carry a chromosome segment from one *E. coli* strain to another, where it may be integrated into the recipient chromosome (sexduction or F-duction).   Beginning with an Hfr *gal*$^+$ strain and an F^- *gal*$^-$ one, diagram the process of sexduction.   (For details see Adelberg and Burns, 1960, and Jacob and Adelberg, 1959.)
  b How is sexduction similar to and different from transduction?
  c Why are *E. coli* harboring F' factors called *intermediate males*?
  d F$'$ carries a portion of the bacterial genome.   In what ways do you expect its breeding behavior to differ from that of F, which carries no bacterial chromosome DNA?

673 It appears that the replication of the *E. coli* genome is necessary for its transfer from Hfr to F^- strains (see Cuzin et al., 1967).   Explain with the aid of diagrams how the two may be related.   Include in your discussion other cellular structures and phenomena that may be involved in transfer and how they may be related to the process.

674 Answer the following briefly.
  a What features distinguish the bacterial chromosome from the chromosomes of eukaryotes?
  b Explain how sex is determined in bacteria.
  c Outline the events that occur from the time a portion of a donor chromosome enters an F^- conjugant to the appearance of a recombinant haploid segregant carrying it.

d What properties do proviruses confer on lysogenic bacteria?

e Both the sex factor F and the genetic material of temperate phages are episomes. Discuss.

f Why does simultaneous transduction of two loci indicate their close linkage?

PROBLEMS

675 The *wild-type* strain K12 of *E. coli* is a prototroph. By treating this strain with x-rays and ultraviolet light Lederberg and Tatum (1946*a,b*) obtained single-, double-, and triple-auxotrophic mutants which grow on minimal medium only if they are supplemented with appropriate growth factors. In pure cultures of the different mutant strains prototrophs were sometimes found. These proved to be back mutations for a single growth-factor allele only, e.g., *threonine-requiring* (thr^-) to *threonine-independent* (thr^+). Thus the triple-mutant strains required only two of the original three factors.

a Do you think that a mixing of two different true-breeding auxotrophs, e.g., $thr^-\ leu^+$ and $thr^+\ leu^-$, and the subsequent finding of prototrophs proves the occurrence of genetic recombination by conjugation?   Explain.

b The two triple-mutant strains, Y10, requiring *threonine, leucine,* and *thiamine,* $thr^-\ leu^-\ thi^-\ bio^+\ phe^+\ cys^+$, and Y24, requiring *biotin, phenylalanine,* and *cystine,* $thr^+\ leu^+\ thi^+\ bio^-\ phe^-\ cys^-$, were grown in pure and mixed cultures on appropriate selective media.   The only new types found in pure cultures of each of the triple mutants were those which reverted to prototrophy for a single factor.   In *mixed* cultures a variety of types were detected: many *wild types* (complete prototrophs) and some single- and double-mutant (requiring one and two nutrients, respectively) strains.   The prototrophs bred true on minimal medium.

(1) What are the possible explanations for the occurrence of the prototrophs in the mixture of the two triple strains?   Which is the most probable explanation and why?

(2) Outline tests that would rule out all but the most probable explanation.

676 Two triple auxotrophic strains of *E. coli,* $thr^-\ leu^-\ his^-\ pro^+\ pan^+\ bio^+$ and $thr^+\ leu^+\ his^+\ pro^-\ pan^-\ bio^-$, are allowed to conjugate in a liquid medium for 30 minutes.   After dilution of the broth the bacteria are plated onto a complete agar medium in a petri dish which serves as a master plate from which six replicas are made, each of which is plated onto a replica plate containing minimal medium and additional nutrient or nutrients as indicated in the figure.

a Determine the genotype of each of the first six clones.   Explain how you arrive at the genotype of each.

b What is the most likely genotype of clone 7?   Why?   How would you determine this?

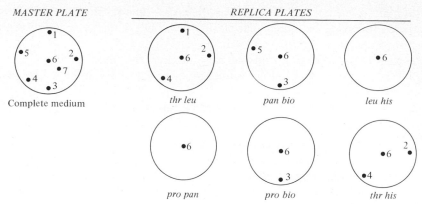

Figure 676

677 You are given two strains of *E. coli*. One is an F$^+$ strain, prototrophic for arginine, alanine, glutamine, proline, and leucine (symbolize the genes *arg$^+$*, *ala$^+$*, *gln$^+$*, *pro$^+$*, *leu$^+$*, respectively), whereas the other strain is F$^-$ and is auxotrophic for these amino acids (*arg$^-$*, *ala$^-$*, *gln$^-$*, *pro$^-$*, *leu$^-$*). The order of these genes on the bacterial chromosome is *arg-ala-gln-pro-leu* with the *arg* end entering first during conjugation. You find that the F$^-$ strain dies when exposed to penicillin (because it has an allele *pen^s* for penicillin *sensitivity*) whereas the F$^+$ strain does not (because it carries the alternative allele *pen^r*, for *resistance*). How would you locate the *pen* locus on the bacterial chromosome in relation to *arg, ala, gln, pro, leu*?

678 If the DNA in a bacterial nucleus is approximately 1,200 μm long and the entire chromosome is transferred from Hfr males in 90 minutes, how much of this molecule is transferred to a female, F$^-$, strain after 35 minutes of conjugation?

679 Jacob and Wollman (1957) crossed a large number of Hfr strains with appropriate F$^-$ ones and found that each transferred a different set of markers

Table 679

Hfr strain	Point of origin	Time sequence of injection (from left to right)
H	0	*thr leu azi T1 lac-T6 gal λ*
1	0	*leu thr bio met mtl xyl mal str^r*
2	0	*T1 azi leu thr bio met mtl xyl mal str^r*
3	0	*T6 lac-T1 azi leu thr bio met mtl xyl mal str^r*
4	0	*bio met mtl xyl mal str^r λgal*
5	0	*met bio thr leu azi T1 lac-T6 gal λ*

at high frequency and in different sequences.   The results in repeated experiments involving two particular strains were always the same, as shown (data from Jacob and Wollman, 1958b).

a How many linkage groups do these markers fall into?

b What is the form (straight, branched, circular) of the linkage group(s) and why?

c What do these results imply regarding the location of O?   Hfr?

680   In crosses between F^+ and F^- strains the majority of the unselected markers among recombinants are derived from the F^- parent.   To explain this Hayes (1953) suggested that the male usually transfers only a portion of its genetic material to the female to form a *merozygote*, whereas Lederberg (1949; Nelson and Lederberg, 1954) suggested that although the entire genome of the male was always transferred, only a randomly selected portion was integrated. To determine the true nature of gene transfer, Wollman and Jacob (1955) in one set of experiments mated an Hfr H strain carrying $thr^+ leu^+$ $azi^s T1^s lac^+ gal^+ \lambda$ with an F^- strain carrying the alternative markers. Conjugation was interrupted at different times by subjecting samples of the mixture to a shearing force in a blender, and the kinds of recombinants were then determined.   The results are tabulated for all experiments involving these two particular Hfr and F^- strains.

Table 680

Time after mixing, minutes	Recombinants with Hfr markers
0	0
8	thr^+
$8\frac{1}{2}$	$thr^+ leu^+$
9	$thr^+ leu^+ azi^s$
11	$thr^+ leu^+ azi^s T1^s$
18	$thr^+ leu^+ azi^s T1^s lac^+$
25	$thr^+ leu^+ azi^s T1^s lac^+ gal^+$
26	$thr^+ leu^+ azi^s T1^s lac^+ gal^+ \lambda$

a Which hypothesis do these results support and why?

b What other conclusions can you draw, e.g., with respect to probable number of linkage groups?

681   *E. coli* may contain from one to four nucleoids per cell, each with an entire (circular) genome.   Therefore, when mutations are induced, the purity of the resulting clones for mutant traits will be a function of the number of nucleoids in the bacterium at the time mutation occurred.   Witkin (1951) irradiated populations of *lactose-fermenting* (lac^+) *E. coli* with ultraviolet light. The bacteria were then plated on eosin–methylene blue medium containing

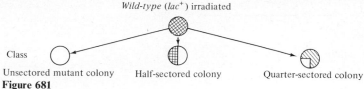

Wild-type (*lac*⁺) irradiated

Class   Unsectored mutant colony   Half-sectored colony   Quarter-sectored colony

Figure 681

lactose, on which *lac*⁺ and *lac*[−] (inability to ferment lactose) colonies were easily distinguished. She found the results sketched. Each colony arose from a single irradiated bacterium.

a How many nuclear bodies were present in the parent bacteria of each of the mutant types of colonies? Explain.

b The experiment appears to contain an anomaly if one assumes that all mutations are due to substitution or alteration of one base pair for another. What is this anomaly and why?

682 **a** Although by 1951 it had been shown that *E. coli* can reproduce sexually and that its genes are arranged on linear structures analogous to the chromosomes of higher organisms, there was no answer to the question: Are the members of a pair of conjugating bacteria sexually equivalent; i.e., can either conjugant act as donor or recipient of DNA? On the basis of the following results obtained by Hayes in 1952 answer this question and explain your answer. Mixed cultures of the two streptomycin-sensitive auxotrophs, 58-161, *met*[−], and W677, *thr*[−] *leu*[−] *thi*[−], produced prototrophs, *met*⁺ *thr*⁺ *leu*⁺ *thi*⁺, under normal conditions. When both strains were pretreated with streptomycin[1] before mixing, no recombinants were formed. When only W677 was pretreated with streptomycin, no recombinants were formed. When 58-161 was pretreated, prototrophs occurred.

b Pedigree studies of cells isolated following conjugation reveal that the vegetative progeny of one of the exconjugants never show recombination of genes that marked the conjugants. Cells derived by fission of the other exconjugant show recombinant types for marker genes of the original conjugants (Anderson, 1958). Do these results confirm the conclusions drawn in (a)? Explain.

683 The *E. coli* data shown are modified from Taylor and Adelberg (1960). Each of three different Hfr strains (A, B, and C) which arose independently and possess the same markers, *his*⁺, *gal*⁺, *pro*⁺, *met*⁺, *mtl*⁺, *xyl*⁺, *mal*⁺, *ser*⁺, *tyr*⁺, and *str*^s, were separately mixed with the same F[−] strain carrying the alternative markers. Conjugation was interrupted at various times after mixing over a period of 95 minutes. The bacteria were then plated on appropriate media and recipients scored for various Hfr markers. The times (in minutes after mixing) at which recombinants carrying particular donor markers first appear, and therefore the times at which these first enter the

[1] Streptomycin at first prevents fission and later kills the cells.

Table 683

Donor markers	Time of appearance, minutes		
	Strain A	Strain B	Strain C
his^+	65.0	68.5	29.5
gal^+	30.0	46.0	52.0
pro^+	34.5	41.5	56.5
met^+	58.0	20.0	76.0
mtl^+	67.0	11.0	85.0
xyl^+	68.5	9.5	86.5
mal^+	73.0	5.0	3.0
ser^+	81.5	86.5	11.5
tyr^+	88.0	78.5	19.5

F^- cell and then integrate into the recipient genome, are given.   What do these results suggest regarding:

a The type of linkage group (rodlike, branched, circular) in *E. coli*?

b The location of F in the different strains?

For confirmation of your answer to (a) see Cairns (1963, 1966).

684 Five Hfr strains (A to E), all carrying the same *wild-type* markers, are crossed to an F^- strain carrying the alternative set of alleles.   Using the conjugation-interruption technique, it is found that each Hfr strain transmits its genes in a unique sequence, which is different from that of the other four strains, as shown.

Table 684

	Hfr strain			
A	B	C	D	E
mal^+	ade^+	pro^+	pro^+	his^+
str^s	his^+	met^+	gal^+	gal^+
ser^+	gal^+	xyl^+	his^+	pro^+
ade^+	pro^+	mal^+	ade^+	met^+
his^+	met^+	str^s	ser^+	xyl^+

a What is the gene sequence in the Hfr strain from which these five strains are derived?   Explain.

b For each of the Hfr strains, state which donor trait should be selected in the recipients after conjugation to obtain the highest frequency of recombinants that will be Hfr?

685 In a conjugation-interruption experiment in which samples were removed after mixing at 2- to 5-minute intervals Taylor and Thoman (1964) obtained the results shown in six crosses between the Hfr strain AB259 carrying all *wild-type* markers and six different recipient strains carrying the alternate markers.

Table 685 **Time of entry markers transferred by strain AB259**

Recipient strain	thr$^+$	pyrA$^+$	leu$^+$	proA$^+$	proB$^+$	purE$^+$	lys$^+$ met$^+$	gal$^+$	aroA$^+$	purB$^+$
							Selected markers of the donor parent			
AT2213	7.25	7.5	7.75							
AT2217			8.0		16.0					
AT2270				15.0		19.5		24.0		
AT2036				14.5			21.5	23.5		
AB1321				15.0				23.5	29.0	
AB1325				15.0				24.0		32.0

a Map the chromosomal segment transferred by the Hfr strain AB259.

b Would you expect any of these markers to be cotransducible? Why?

c None of the recombinants were Hfr. What does this indicate regarding the location of the Hfr locus?

686 Certain mutants, *uv^r*, of *E. coli* are more *resistant* to ultraviolet irradiation than the parent strains, *uv^s*. To determine the genetic basis for *resistance* vs. *sensitivity* to ultraviolet light, Greenberg (1964) used the strains whose characteristics are shown in Table 686*A*. The frequency with which *radio-*

Table 686*A*

Strain	Sex	met	thi	lac	ara	mal	xyl	gal	T6	T1	val	str	uv
K12 W1895	Hfr$_1$*	−	+	+	+	+	+	+	s	s	s	s	r
K12 W4531	Hfr$_2$†	+	−	+	+	+	+	+	s	s	s	s	r
BPAM5	F$^-$	+	+	−	−	−	+	+	r	r	r	s	s
BPAM7	F$^-$	+	+	−	−	−	−	−	r	r	r	r	s

* Order of transmission of markers is *T6 lac T1 ara* ⋯.

† Order of transmission of markers is *ara T1 lac T6*.

resistance appeared as an unselected marker trait in crosses between W1895 and BPAM7 and in crosses between W4531 and various derivatives of strain B when progeny were selected for various other markers is summarized in Table 686*B*. The frequencies of other unselected markers are also shown.

Table 686*B* **Frequency of occurrences of donor markers among recombinants from the cross W1895 × BPAM7**

Recombinants selected	Number examined	Frequency of donor alleles among selected recombinants, %									
		uv^r	$T6^s$	lac^+	$T1^s$	ara^+	mal^+	met^-	val^s	xyl^+	str^s
$lac^+\ str^r$	100	86	87	100	72	63	22	8	4	3	—
$ara^+\ str^r$	100	50	50	57	65	100	15	12	6	1	—
$mal^+\ str^r$	100	57	51	62	61	76	100	62	46	29	—
$xyl^+\ str^r$	100	26	27	27	20	25	23	38	53	100	—

Table 686*C* **Frequency of occurrence of donor markers among recombinants between W4531 and BPAM5**

W4531 × female parent	Recombinants selected	Frequency of donor markers among recombinants, %							
		ara^+	$T1^s$	lac^+	$T6^s$	uv^r	gal^+	xyl^+	mal^+
BPAM5	$ara^+\ thi^+$	100	75	60	53	53	—	—	0
BPAM5	$lac^+\ thi^+$	69	62	100	78	73	—	—	—

a How many genes control *resistance* vs. *sensitivity* to ultraviolet irradiation, and where are they located? Indicate your reasoning.

b Is the point of origin the same in the two donor strains? Explain.

687 In one donor strain of *E. coli* the gene for regulation of RNA synthesis, *RC*, is transmitted into a recipient 8 minutes after initiation of conjugation. In another strain the same gene is not transmitted until 21 minutes after initiation of conjugation. How would these donor strains differ in order to have such a difference in time of transmission of the same gene?

688 The *E. coli* chromosome is 10^7 base pairs long. The DNA is continuous along the chromosome; if 1 time unit (1 minute) is equal to 20 recombination (map) units and the chromosome is 90 time units long, how many base pairs would there be per recombinational unit?

689 A dilute suspension of a *mutant* strain of *Salmonella typhimurium* is plated on complete medium. Replicas of the colonies are plated on various minimal media containing one, two, or three essential nutrients. The replicas grow only if the minimal medium contains *proline*, *threonine*, and *leucine*. One of the replicas of these colonies has a single colony on a minimal medium supplemented with *proline* and *threonine*; another has a single colony on minimal medium supplemented with *proline* and *leucine*.

a Determine the genotype of the original mutant strain.

b How did the latter two colonies arise, and what is their genotype? Explain.

690   An *E. coli* strain unable to synthesize leucine, *leu*⁻, is transduced by a strain unable to synthesize threonine, *thr*⁻. One half of the diluted culture medium is plated on minimal medium, the other half on minimal supplemented with threonine; 25 colonies grew on the minimal plates and 125 on the supplemented plates. Calculate the amount of recombination between *leu* and *thr*.

691   The sequence of entry of donor genes as determined by conjugation-interruption experiments for the Hfr strain H and the mutant strains HCR1 and HCR2, derived from it by treatment with nitrogen mustard, were as shown (Jacob and Wollman, 1961*b*). What were the mutational effects of the mutagenic treatment in each of the mutant strains?

Table 691

Strain	Sequence
H	*O thr leu azi T1 pro lac ad gal H S-G Sm mal xyl mtl ile met arg thi*
HCR1	*O thr leu azi T1 pro xyl mtl ile met arg ad gal H S-G Sm mal thi*
HCR2	*O lac ad gal H S-G thr leu azi T1 pro Sm mal xyl mtl ile met arg thi*

692   Each *E. coli* bacterium usually has two to four chromatin bodies and becomes uninucleate only when grown on a medium deficient in phosphorus. Driskell-Zamenhof and Adelberg (1963) found that when uninucleate F⁺ cells are mixed with F⁻, they lose F and become F⁻ at the same rate as the recipients are converted to F⁺. What conclusion can you draw from this finding?

693   Zygotes usually survive when Hfr bacteria conjugate with F⁻ cells lysogenic for an inducible phage such as λ. If the F⁻ cells are nonlysogenic, those zygotes which carry the prophage are induced to lyse (*zygotic induction*), forming plaques. Using this information, show how you would determine the locus of the prophage on the *E. coli* map.

694   When an F⁺ strain (*arg*⁺ *his*⁺ *xyl*⁺ *trp*⁺ *pro*⁺ *str*ˢ) and an F⁻ strain (carrying the alternative, −, markers) are mixed under appropriate conditions, a few F⁻ cells are found to express some of the male traits. How would you demonstrate that the few males that transferred their genes to the recipients were converted to Hfr males?

695   Exposure of *E. coli* to streptomycin initially prevents cell division but not transfer of the chromosome from Hfr to F⁻. Treated cells eventually die. Using a strain of *E. coli* susceptible to phage T5, design an experiment to determine whether this strain is F⁺, Hfr, or F⁻.

696   It has been shown (Hirota and Iijima, 1957; Hirota, 1960) that exposure of donor *E. coli* to acridine orange inhibits the replication of nonintegrated but not of integrated sex factor F. With these facts in mind design an experiment that will simultaneously:

a Identify colonies as F^-, F^+, or Hfr.

b Obtain F^- from F^+ cells.

697   When F is transferred by conjugation from *E. coli* to a different genus of bacteria, e.g., *Serratia* or *Proteus*, a DNA band appears in the cesium chloride density gradient that is not found in the host alone (Marmur et al., 1961; Falkow et al., 1964).

a What do these results indicate regarding the chemical characteristics of F?

b Why was it necessary to transfer F to a different genus to determine its chemical and physical properties by density-gradient centrifugation?

698   The F factor is DNA.   It contains approximately 3×10^5 base pairs, which is about 2 percent of the amount of DNA in the *E. coli* chromosome (Driskell-Zamenhof and Adelberg, 1963; Falkow and Citarella, 1965).   Moreover it is circular (Hickson et al., 1967) and consists of two regions; one region, making up 10 percent of F, contains DNA of 44 percent G + C; the other 90 percent of F contains 50 percent G + C.   Half of the latter region is hybridizable with *E. coli* chromosomal DNA (Falkow and Citarella, 1965). Show how these data can be used to explain F integration and chromosome transfer by Hfr cells.

699   Which of the following crosses would you make to determine the chromosome locus of prophage λ and why?

Hfr $\lambda^+ \times F^- \lambda^-$

Hfr $\lambda^+ \times F^- \lambda^+$

Hfr $\lambda^- \times F^- \lambda^-$

Hfr $\lambda^- \times F^- \lambda^+$

700   When F^- *E. coli* are lysogenic for λ and Hfr are nonlysogenic, the lysogenic trait segregates among the recombinants and can be located accurately on the chromosome.

a Assuming any location for λ you wish, show the results you would expect in a conjugation-interruption experiment between Hfr H (λ^-) $thr^+ leu^+ azi^+ lac^+ gal^+$ and F^- (λ^+) $thr^- leu^- azi^- lac^- gal^-$.

b What results would you expect if Hfr was λ^+ and F^-, λ^-?   Why?

701   The table shows the genetic constitution of thr^+, leu^+, str^s recombinants for various donor traits formed in various crosses between Hfr H str^s, which is *wild type* for all markers, and F^- P678 str^r, mutant at the loci studied.

a These results indicate that lysogeny is under genetic control.   Where is its determinant located in relation to *thr* and *leu*?

b Account for the disparity in the results of the second and third crosses.

c Is the genetic determinant causing lysogeny the prophage λ?   Which crosses support your hypothesis?

Table 701

Cross	azi^s	$T1^s$	lac^+	gal^+	$\lambda*$
	\multicolumn{5}{} $thr^+ leu^+ str^r$ recombinants				
Hfr λ^- × F λ^-	91	72	48	27	
Hfr λ^- × F λ^-	92	73	49	31	15
Hfr λ^+ × F λ^-	86	60	21	2.5	0.1
Hfr λ^+ × F λ^-	90	70	47	29	14 †

* The frequency of appearance among the recombinants of the λ trait of the Hfr parent, λ^- (nonlysogeny) in the first two crosses, and λ^+ (lysogeny) in the third cross.

† Hfr carried *wild-type* λ: plaques *small, uniformly turbid*; F⁻ carried a *mutant* λ: plaques *medium-sized, turbid center*.

702 A culture of *E. coli* is placed in a medium containing λ. Most of the bacteria are destroyed. Are these bacteria lysogenic or nonlysogenic, and how would you determine whether the surviving cells carry λ?

703 In a mixture of Hfr P4x1 ($pro^+ mal^+ met^+ lac^+$) and F⁻ ($pro^- mal^- met^- lac^-$) a few recombinants arise that carry only the pro^+ marker. All these recombinants are donors. Account for this result and state the location of F in P4x1.

704 Following treatment with nitrosoguanidine Clark and Margulies (1965) isolated two *E. coli* mutants unable to form recombinants after conjugation (and thus infertile in Hfr × F⁻ crosses) or to repair ultraviolet damage (dimer formation). True revertants from these mutants were able to repair ultraviolet damage (and were thus resistant to ultraviolet irradiation) and formed recombinants after conjugation like the *wild types*. Both crossing-over and ultraviolet repair (dimer removal) appear to involve a series of steps. What do these observations suggest regarding the nature of the genetic systems involved?

705 Jacob and Wollman (1961*b*) mixed F⁻ P678 ($thr^- leu^- azi^r T1^r lac^- gal^- mal^- xyl^- man^- str^r$) with HfrH carrying the alternative alleles of these genes. After 60 minutes $thr^+ leu^+ str^r$ recombinants were selected and tested for the presence of other donor traits, with the results shown.

Table 705

Type of recombinants selected	$thr^+ leu^+$	azi^s	$T1^s$	lac^+	gal^+	mal^+	xyl^+	man^+
		\multicolumn{7}{} Frequency of Hfr traits, %						
$thr^+ leu^+ str^r$	100	91	72	48	27	0	0	0
$gal^+ str^r$	83	78	79	81	100	0	0	0

   a With respect to *thr⁺ leu⁺* are the unselected markers distal or proximal to *O*? Explain.

   b Which selected recombinants, *thr⁺ leu⁺ str^r* or *gal⁺ str^r*, permit determination of gene sequence and why? What is the gene order?

706 Zinder and Lederberg (1952) performed a series of experiments with the mouse-typhoid bacterium (*Salmonella typhimurium*) to determine whether this organism undergoes conjugation. On the basis of their observations, presented below:

   a State which of the mechanisms of genetic transfer (conjugation, transduction, or transformation) can be eliminated and why.

   b Describe and illustrate the mechanism accounting for genetic transfer and explain all the facts stated.

Observations

1 Mixed cultures of *met⁻ thr⁺* and *met⁺ thr⁻* strains in appropriate medium produced numerous prototrophs (*met⁺ thr⁺*) that resembled the recipient parent (*met⁻ thr⁺*) except for the trait *met⁺* from the donor parent *met⁺ thr⁻*.

2 These prototrophs were much more frequent in mixed than in unmixed cultures.

3 In a U-tube experiment the strains were grown in opposite arms separated by a filter through which no bacteria could pass. After allowing the nutrient to pass from the donor end of the tube to the recipient end, some cells of the recipient strain on transfer to selective medium showed traits of the donor.

4 Treatment of the recipient culture with pure DNA from the donor strain produced no recombinants (prototrophs).

5 Adding deoxyribonuclease to the solution in the U-tube did not prevent the appearance of cells with donor traits in the recipient strain.

6 Broth from a culture of donor cells added to a culture of recipient cells resulted in no heritable changes.

7 Filtered broth from a culture containing both strains produced hereditary changes in fresh cultures of recipient cells.

8 After a culture of donor cells was exposed to fluids from the recipient strain and the fluid from this donor culture was mixed with other donor strains, it was found that all donors could cause hereditary changes in the recipient.

9 Viruses were found in donor cultures treated with fluids from the recipient strain.

707 **a** Distinguish between complete and abortive transduction.

   b In *Salmonella*, *flagellated* strains can swim and spread to produce a cloudy (spreading) swarm throughout the culture medium; *nonflagellated* strains cannot move and form new colonies. Stocker (1956) studied transduction of a certain strain from *nonmotility* to *motility*. He found that after transduction some of the transduced recipient cells produced spreading swarms whereas others produced trails, e.g., linear trail of colonies, each colony consisting entirely of *nonmotile* cells. Assume *motile* strains are *F b* and *nonmotile* ones *f b*.

(1) Which of these effects is complete and which abortive transduction? Explain with the aid of illustrations.

(2) What is the allelic relationship between F and f ?   Explain.

708   Ozeki (1956) found that when an *adenineless* recipient strain of *Salmonella typhimurium* was infected with phage from a *wild-type* strain or a strain auxotrophic for a different requirement, both slow-growing, *minute* colonies and *large* colonies appeared on plates containing minimal medium.   Subsequent tests gave the following results:

1   When the *large* colonies were restreaked on minimal agar, every cell of the colony was found to yield a *large* colony.

2   When the *minute* colonies were restreaked on minimal agar, only one cell of the colony was prototrophic, giving rise to another *minute* colony.   The other cells of the colony either did not reproduce or underwent only a few divisions.   The final progeny of such cells, like the original recipient, were not able to divide on minimal medium.

3   The *minute* colonies did not form when auxotrophic mutants were plated by themselves on minimal medium or when bacteria were infected with phage from the same strain.

4   Very infrequently a prototrophic colony appeared when an auxotroph was plated on minimal medium, each cell of which bred true.

5   When *minute* colonies were streaked, *large* colonies occurred with a frequency no higher than that expected for naturally occurring reversions.

a Why do *minute* colonies form in addition to *large* ones?

b Which of the facts support your hypothesis and why?

709   In *Salmonella*, *motility* vs. *nonmotility* (because of presence or absence of flagella, respectively) depends on two genes, F and P. The alleles Ff determine *presence* vs. *absence* of flagella, whereas Pp determine the kind of protein b vs. i, of which the tail is made up.   *Motile* (flagellated) strains swim and spread, producing a cloudy swarm throughout the culture medium. *Nonmotile* (nonflagellated) strains cannot move and hence form restricted colonies rather than swarming.   When Stocker, Zinder, and Lederberg (1953) treated the *nonmotile* strains SW543 and SL13, both of which carry protein b, with lysates of *motile* strain TM2 carrying protein i, (1) a number of swarms were observed; (2) most of these swarms had the protein b characteristic of the recipient, and only a few possessed the flagella protein i.

a State which of these types of swarms represents *linked transduction* and why.

b What kind of gene interaction is this an example of?   Explain.

710   **a** In *E. coli* the *ara* mutants (unable to utilize arabinose) are located between the *thr* and *leu* loci, close to *leu* and far from *thr* (Lennox, 1955).   To determine the sequence of the nonidentical mutant sites *ara-1*, *ara-2*, and *ara-3* with respect to each other and *thr* and *leu*, Gross and Englesberg (1959) made four-factor reciprocal crosses by transduction.   The recipient parent in each case was thr^- leu^- and the donor thr^+ leu^+.   Each mutant was used as donor in one cross and as a recipient in the reciprocal. *ara*$^+$ transductants were selected on medium containing mineral arabinose, threonine, and leucine. These transductants were then scored for both *thr*$^+$ and *leu*$^+$. The results are tabulated. What is the order of the *ara* mutant sites relative to each other and *thr* and *leu*? Explain with the aid of diagrams.

Table 710

Cross	Recipient	Donor	$\dfrac{ara^+\,leu^+}{\text{Total } ara^+} \times 100$	$\dfrac{ara^+\,thr^+}{\text{Total } ara^+} \times 100$
1	*ara-1*$^-$	*ara-2*$^-$	64.4	1.2
2	*ara-2*$^-$	*ara-1*$^-$	17.4	7.4
3	*ara-1*$^-$	*ara-3*$^-$	26.1	6.4
4	*ara-3*$^-$	*ara-1*$^-$	52.4	2.4
5	*ara-2*$^-$	*ara-3*$^-$	14.3	9.5
6	*ara-3*$^-$	*ara-2*$^-$	65.8	2.8

b When mutant 3 is infected with phage grown on mutants 1 and 2, in addition to *large* colonies many *minute* colonies are formed. No *minute* or *large* colonies are formed when phage are grown on mutant 1. *Minute* colonies are also formed when 1 is infected with phage grown on 2 and vice versa. What additional information do these data give you and why? Illustrate the arabinose region schematically as completely as possible.

711 It is claimed there is only one exogenote per microcolony in the *trail of microcolonies* formed as a result of abortive transduction of *nonmotile* cells of *Salmonella*. Outline an experiment to prove whether this is correct.

712 In 1959 Gross and Englesberg, working with the *E. coli* strain B/r, infected the recipient strains *thr*$^-$ *ara-3*$^-$ *leu*$^-$ and *thr*$^-$ *ara-2*$^-$ *leu*$^-$ with phage grown on the *wild-type* (*thr*$^+$ *ara*$^+$ *leu*$^+$) strain. Transduction by P1 phage of a given (selected) marker was detected on the appropriate medium, and the transduced colonies were subsequently typed for other loci by streaking on appropriate media. The results are summarized in Table 712.

a Are *ara-3* and *ara-2* linked with *thr* and *leu*? If so, is the linkage weak or strong? Explain.

b From conjugation studies it is known that the sequence of *thr* and *leu* is *thr-leu*. If the *ara* loci are linked to *thr-leu*, is the sequence *thr-ara-leu*, *ara-thr-leu*, or *thr-leu-ara*? Summarize the evidence supporting your answer.

OR

c Describe the linkage relationships of *ara-2*, *ara-3*, *leu*, and *thr*; viz. does linkage exist, and what loci, if any, show mutual linkage.

Table 712

Recipient	Selected marker	Number of transductants per P1 phage plated	Selected colonies containing unselected markers, %		
			leu^+	thr^+	ara^+
thr^- ara-3$^-$ leu^-	thr^+	2.5×10^{-5}	4.1		6.7
	leu^+	5.0×10^{-5}		1.9	55.4
	thr^+ leu^+	1.0×10^{-7}			80.0
	ara^+	3.5×10^{-5}	72.6	4.3	
thr^- ara-2$^-$ leu^-	ara^+	3.5×10^{-5}	59.6	5.2	

713 When each of three *ath* (*adenine-thymine-requiring*) mutants of *Salmonella typhimurium* were infected with phages from each of the other two mutants and the *wild-type* strain, the results shown were obtained (based on data of Ozeki, 1956).

Table 713

Recipient	Colonies	*Wild type*	Donor		
			ath-1	*ath-4*	*ath-6*
ath-1	large	many	none	very few	many
	minute	+	−	−	+
ath-4	large	many	very few	none	many
	minute	+	−	−	+
ath-6	large	many	many	many	none
	minute	+	+	+	−

Key: + = *minute* colonies formed; − = *no minute* colonies.

a Explain how the *large* colonies originate and why there are many in some transductions and very few or none in others.
b Why do *minute* colonies arise, and what is the significance of their occurrence?
c What is the minimum number of genes involved in adenine-thymine formation?

714 Hundreds of *histidine-requiring* mutants have been isolated in *Salmonella typhimurium* (Ames et al., 1967). Six of these, here numbered 1 to 6, have been found to belong to three cistrons (genes). Complementation tests, by abortive transduction, have given the results shown.
a Determine which mutants are in the same cistron.
b Explain how you group the mutants.

Table 714

	1	2	3	4	5	6
1	−	+	0	−	0	+
2		−	+	0	0	0
3			−	+	0	+
4				−	+	0
5					−	+
6						−

Key: + = complementation; − = no complementa-
ation; 0 = not tested.

715   In *Salmonella typhimurium* four genes control tryptophan synthesis. The sequence of biochemical steps in tryptophan synthesis and the steps controlled by each of the four genes are as follows:

Sequence: $\xrightarrow{A}$ anthranilic acid $\xrightarrow{B}$ indoleglycerol phosphate $\xrightarrow{C}$ indole $\xrightarrow{D}$ tryptophan

To determine whether these genes are linked to each other and the *cysB* locus and, if so, to determine the order of the four tryptophan genes and the *cysB* locus on the linkage map three-gene reciprocal crosses were made

Table 715

Reciprocal experiment	Genotype of recipient strain	Genotype of donor strain	Number of *wild-type* colonies on minimal medium
1	$trpD^- cysB^- trpB^+$	$trpD^+ cysB^+ trpB^-$	15
	$trpD^+ cysB^+ trpB^-$	$trpD^- cysB^- trpB^+$	292
2	$trpD^- cysB^- trpA^+$	$trpD^+ cysB^+ trpA^-$	16
	$trpD^+ cysB^+ trpA^-$	$trpD^- cysB^- trpA^+$	456
3	$trpA^- cysB^- trpB^+$	$trpA^+ cysB^+ trpB^-$	59
	$trpA^+ cysB^+ trpB^-$	$trpA^- cysB^- trpB^+$	43
4	$trpD^- cysB^- trpC^+$	$trpD^+ cysB^+ trpC^-$	26
	$trpD^+ cysB^+ trpC^-$	$trpD^- cysB^- trpC^+$	183
5	$trpA^- cysB^- trpC^+$	$trpA^+ cysB^+ trpC^-$	60
	$trpA^+ cysB^+ trpC^-$	$trpA^- cysB^- trpC^+$	286
6	$trpB^- cysB^- trpC^+$	$trpB^+ cysB^+ trpC^-$	24
	$trpB^+ cysB^+ trpC^-$	$trpB^- cysB^- trpC^+$	46

by Demerec and Hartman (1956) by transduction. Each parent (strain) was alternatively used as a donor and recipient, by growing the transducing phage on each mutant strain (donor) and then using it to transduce the other strain (recipient). The results of the crosses are tabulated. Assume (correctly) that *cysB* locus is to one side of the *trp* loci.

a Are the tryptophan loci linked to each other? To the *cysB* locus? What is the sequence of the *trp* loci on the genetic map with respect to each other and the *cysB* locus?

b Compare the sequence of the tryptophan genes with the steps they control in the tryptophan-biosynthesis pathway. What might your results suggest?

716 Transduction crosses among seven *proline-requiring* mutants of *Salmonella typhimurium* give the results shown (after Miyake and Demerec, 1960).

Table 716

Recipient strain	Donor strain						
	1	2	3	4	5	6	7
1	− 0	− 38	+ 521	+ 105	+ 1,920	+ 2,682	+ 12,342
2		− 0	+ 269	+ 194	+ 2,746	+ 2,898	+ 14,872
3			− 0	− 9	+ 1,396	+ 2,040	+ 8,236
4				− 0	+ 2,128	+ 2,440	+ 8,810
5					− 0	− 17	+ 5,388
6						− 0	+ 2,722
7							− 0

Key: Plus and minus signs indicate presence and absence of abortive transductants. Figures refer to the number of complete transductants.

a How many functional units (genes, cistrons) are represented by these seven mutants? Explain.

b Is it possible to deduce the sequence of these mutants and the approximate distance between them? Explain.

717   The data given are for pairwise crosses by transduction between seven *histidine-requiring* mutants in *Salmonella* (based on data of Hartman et al., 1960*b*; Ames et al., 1967).

Table 717

	41	55	101	102	129	134	152
41	−	−	+	+	+	+	−
55		−	−	+	+	+	−
101			−	−	+	−	−
102				−	−	+	−
129					−	+	−
134						−	−
152							−

Key: + = *wild-type* recombinants are formed; − = no recombinants are formed.

a What class of mutation event (viz. substitution, inversion, etc.) has given rise to these mutants?
b Draw a topological representation of these mutants.
Mutant 152 is unable to synthesize enzymes 1 (cyclose), 2 (isomerase), 3 (aminotransferase), 4 (dehydrase-phosphatase), 5 (transaminase), and 6 (dehydrogenase). Mutant 129 cannot synthesize enzymes 5 and 6, whereas mutants 101 and 41 cannot synthesize enzymes 1 to 4.
c Assuming each enzyme is controlled by one gene, what is the number of genes involved in these four mutations?
d What enzymes should mutants 55, 102, and 134 be unable to synthesize?

718 a The six *histidine-requiring, his,* mutants (41, 55, 134, 135, 150, 712) in *Salmonella typhimurium* have been crossed in all possible pairwise combinations. The progeny, when screened for the *presence* or *absence,* of *wild types,* gave the tabulated results (based on data of Hartman et al., 1960*b*).

Table 718*A*

	41	55	134	135	150	712
41	−	−	+	−	−	−
55		−	+	+	−	−
134			−	+	−	−
135				−	+	−
150					−	−
172						−

Key: + = *wild types* produced; − = no *wild types* produced.

Draw a genetic map of these mutations that will account for these results.

b Six *histidine-requiring* point mutations (1 to 6) were crossed in all possible pairwise combinations by transduction. Abortive transductants occurred in all crosses except 2 × 3 and 4 × 5. How many genes are involved? Why?

c Only some of the crosses between the six point mutations and the six previous mutations studied produced *wild-type* recombinants, as shown in Table 718B. Draw a map of the point mutations and draw lines below the map indicating the extent and end points of the six other mutations.

Table 718B

	41	55	134	135	150	712
1	+	+	+	−	+	−
2	−	+	+	−	+	−
3	−	+	+	+	+	−
4	+	+	+	+	−	−
5	+	+	−	+	−	−
6	+	+	+	+	−	−

Key: + = *wild-type* recombinants; − = no recombinants.

d Which of the following pairwise crosses by transduction would you expect to produce abortive transductants?
(1) 135 × 41   (2) 135 × 55   (3) 41 × 134   (4) 150 × 134

719 Taylor and Trotter (1967) in a five-factor transduction cross in *E. coli* obtained the following results:

P1 donor (AT2473): $thr^+ pyr^- pdx^+ ara^+ leu^+$
Recipient (AT2365): $thr^- pyr^+ pdx^- ara^- leu^-$

Table 719

Selected transductants	Transductants per 10^6 P1	No. scored	Percentage of transductants that scored as				
			thr^+	pyr^-	pdx^+	ara^+	leu^+
thr^+	13.2	176		25	17	12	7
pdx^+	13.2	176	24	32		68	56
$thr^+ pdx^+$	2.6	176		96		57	27
leu^+	16.9	264	3	10	21	66	
$leu^+ pdx^+$	5.6	264	8	28		91	
$leu^+ thr^+$	0.7	160		85	88	89	

Where are the *pyr* and *ara* loci on the linkage map relative to the other markers?

720 A few $ind^+ try^+ his^+$ colonies are formed when DNA from $ind^+ try^- his^+$ transforms $ind^- try^+ his^-$ cells. How many crossovers are involved?

721 The sequence of linked loci can be determined from reciprocal transduction experiments involving three genes. In *Salmonella typhimurium* the loci specifying tryptophan synthesis are linked to the *cysB* locus specifying one of the steps in cysteine synthesis (Demerec and Hartman, 1956). Consider only the *trpC* and *trpD* loci, which, like *trpA* and *trpB*, are to one side of the *cysB* region. Demerec and Demerec (1956) crossed the strains $cysB^+ trpC^+ trpD^-$ and $cysB^- trpC^- trpD^+$ reciprocally, using P22 transducing phage. They found that the number of *wild-type* recombinants, $cysB^+ trpC^+ trpD^+$, was about the same in reciprocal crosses. What is the sequence of the three loci? Explain.

722 In *E. coli*, z^- mutants are unable to synthesize β-galactosidase and therefore cannot ferment lactose. About $\frac{1}{2}$ hour after Hfr, $z2^- str^s pan^+$, bacteria[1] are added to a culture medium containing F^-, $z1^- str^r pan^+$, cells, the medium is diluted and plated on minimal medium containing streptomycin; 40 percent of the pan^+ colonies are able to ferment lactose, indicating they are z^+. Only 4 percent of pan^+ colonies from a reciprocal cross are z^+. What is the sequence of the mutant sites $z1$ and $z2$ relative to the pan locus? Show how you arrive at your answer.

723 Certain strains of *Salmonella* have flagella with which they can swim through liquids or semisolid medium such as soft gelatin agar. These motile cells migrate outward from the point of inoculation as they multiply and produce a *cloudy swarm* throughout the culture medium. Other strains are tailless and cannot move. As a consequence their growth into colonies remains circumscribed (they grow into colonies wherever they are placed). The swimming strains can be distinguished by the type of antigens of which

Table 723

Donor strain	Recipient strain	Antigenic types of motile swarms induced by lysate of donor strain*
TM$_2$	SW543	Many *b*, few *i* (A)
TM$_2$	SL13	Many *a*, few *i* (B)
B	SW543	Some *b*, some *i*
A	SL13	Some *a*, some *i*

* A and B refer to the induced swarms with antigen *i* in the first two transductions respectively. Cells isolated from swarms breed true.

[1] *pan*$^+$ vs. *pan*$^-$ indicates *ability* vs. *inability* to synthesize pantothenic acid; *str*^{s} vs. *str*^{r} indicates *sensitivity* vs. *resistance* to streptomycin.

their flagella are composed. The *nonmotile* strains SW543 and SL13 of *S. paratyphi* B and *S. paratyphi* A respectively possess antigens *b* and *a*. *S. typhimurium* TM$_2$ is a flagellar strain that produces antigen *i*.

Stocker et al. (1953) treated the *nonmotile* strains with phage lysates of TM$_2$, induced swarms A and B, and obtained *motile* progeny classified as shown. Offer a plausible explanation of these results. Using appropriate symbols, give the genotypes of all strains involved in the study.

724   The topological map of five deletion mutants (1 to 5) in the same proline gene of *Salmonella typhimurium* and of five point mutants (a to e) is shown below. Pairwise crosses by transduction are made in all possible combinations. Using + to indicate that *wild-type* recombinations are produced and − that no such recombinants are formed, complete the table.

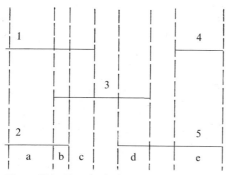

Figure 724 Topological map.

Table 724   Expected *wild-type* recombinants.

	a	b	c	d	e
1					
2					
3					
4					
5					

725   The process of transformation was discovered by Griffith in 1928 and clarified in the classic experiment by Avery, MacLeod, and McCarty in 1944.

   a Explain how transformation is accomplished in *Diplococcus* using the traits *streptomycin resistance* and *streptomycin susceptibility* or *encapsulated* and *unencapsulated*.

   b Explain how mutation is ruled out as the cause of the genetic change in the recipient strain.

   c How did Avery, MacLeod, and McCarty (1944) show that DNA is genetic material?

   d Speculate on the possible manner of incorporating the transformed genetic segments into the chromosome of the recipient bacterial cell.

726   **a** In *Diplococcus pneumoniae* Hotchkiss and Marmur (1954) exposed a recipient strain *unable to utilize mannitol, mtl$^-$*, and *susceptible to streptomycin, str^s*, to DNA from a strain *able to utilize mannitol, mtl$^+$*, and *resistant to streptomycin, str^r*. The transformed phenotypes and their proportions are shown in Table 726*A*. Discuss these results, including in

Table 726A

Single, %	Double, %
$mtl^+ str^s$ 2.0	$mtl^+ str^r$ 0.1
$mtl^- str^r$ 0.32	

your discussion the expected frequency of double transformation, the ratio of observed to expected double transformations, and a possible explanation to account for the double transformants.

b On the basis of your explanation state whether you should expect a smaller, an equal, or a greater proportion of double transformants ($mtl^+ str^r$) if a mixture of two DNAs, one extracted from an $mtl^+ str^s$ and one from an $mtl^- str^r$ strain, were introduced into a growing culture of recipient ($mtl^- str^s$) bacteria?   Explain.

c A mixture of DNAs, one from an $mtl^+ str^s$ and one from an $mtl^- str^r$ strain, when applied to an $mtl^- str^s$ recipient gave the results shown in Table 726B.

Table 726B

Transformants	
Single, %	Double, %
$mtl^- str^r$ 1.4	$mtl^+ str^r$ 0.003
$mtl^+ str^s$ 0.87	

Do these results confirm your expectations?   Explain.

727 When a bacterium is transformed, only a segment of one strand of donor DNA appears to be integrated into the recipient DNA by displacing the corresponding segment of the recipient duplex (Fox and Allen, 1964; Bodmer and Ganesan, 1964; Fox, 1966).   Working with *Diplococcus*, Guild and Robison (1963) found that one of the DNA strands was heavier than the other and that strands, as well as duplexes, can transform recipient cells.   Some *Diplococcus* of a *novobiocin-susceptible* strain, when exposed to the light strand from a *novobiocin-resistant* strain, show the *resistant* phenotype almost immediately.   These *resistant* segregants breed true.   In contrast, transformation with the heavy strand does not alter phenotypic expression until after one generation of replication, at which time some of the bacteria show *resistance*.   These cells also breed true.

a State whether either strand can be incorporated into the recipient or only one.

b Explain why the expression of donor phenotype is immediate with the light strand but requires one replication with the heavy strand.

728 **a** Bodmer and Ganesan (1964), Fox (1966), and Lacks (1962) found that only one strand of the donor transforming DNA segments replaces the corresponding segment of the recipient, by hybridizing (annealing) with the recipient's complementary strand. The product of integration, a physical "hybrid" duplex (one strand donor DNA and one strand host DNA for a certain segment of the genome) carries transforming acitivity. Two possibilities might be advanced regarding the nature of the hybrid duplex segment.

1 It remains genetically heterozygous, i.e., the two single strands remain unchanged, and either contains one particular donor strand or one or the other.

2 It becomes genetically homozygous; e.g., the recipient strand is lost in the following DNA synthesis or is subjected to a repair process rendering it complementary to the transforming strand. Such hybrids might occur as a consequence of integration of a unique strand or either strand of donor DNA fragments.

Guerrini (reported by Fox, 1966) treated d^+ (*sulfanilamide-sensitive, p-nitrobenzoic acid–utilizing*) *diplococci* with DNA from d^- (*p-nitrobenzoic acid–sensitive, sulfanilamide-utilizing*) donors. Each transformed bacterium (d^-) produced a mixed clone containing d^+ and d^- bacteria. Which of the models of hybrid DNA structure are excluded by these observations and why?

b Integration of a portion of a synapsed (donor) segment can theoretically occur in either of two ways:

1 *Breakage and reunion*, allowing exchange of genetic material between homologous segments of donor and recipient chromosomes

2 *Copy choice*, where a daughter chromosome is formed by the alternate use of the host chromosome and the donor DNA as template

Recent studies having a bearing on these alternatives in bacteria, using transforming DNA labeled with the isotope ^{32}P, have revealed that:

1 Incorporation is achieved in the absence of DNA synthesis (Fox, 1960, 1962).

2 When the DNA is extracted from recipient bacteria immediately after incorporation, transforming ability shows a gradual decline. The curve of decline is of the same shape as that generally found for DNA inactivation by ^{32}P decay (Fox, 1962).

Which hypothesis do these favor, and why?

729 *Highly-resistant, p^r*, strains of *Diplococcus* can grow in the presence of 0.3 unit/ml of penicillin; *sensitives, p^s*, can grow only if the concentration is reduced to 0.01 unit/ml or less. Hotchkiss (1951) showed that *high-level resistance* can be acquired by a p^s strain in a series of three successive transformation steps. DNA from a p^r strain can transform p^s recipients only to *low resistance* (grow in 0.05 unit/ml). These transformants could then be transformed with the p^r DNA to an *intermediate level of resistance* (tolerating 0.12 unit/ml). *High level of resistance* (tolerating 0.24 unit/ml) was achieved by treating the *intermediate-level* transformants with the p^r DNA.   Offer an explanation to account for these observations.

730   When genetic material carrying a particular marker is transferred from one bacterium to another which lacks it, the donor marker may replace its allele on the recipient chromosome so that the recombinant is haploid or it may be added to the recipient genome, forming a more or less stable diploid cell, with both the donor and recipient alleles at the locus or loci concerned.

Ravin (1959) showed that *noncapsulated Diplococcus* which have arisen from a *capsulated* type by mutation can reacquire the *capsulated* state by transformation with DNA from the *capsulated* strain.   Moreover the original *capsulated* as well as the *capsulated* transformants can in turn be transformed to *noncapsulated* types: all these types occur as true-breeding strains.

a Does the immigrant genetic material replace its homologous segment in the recipient, or is it added to the recipient genome?   Explain.

b What do these results tell you about the ploidy of *Diplococcus*?

731   Heating double-stranded DNA to about 100°C causes the strands to separate. If denatured DNA is cooled rapidly, the strands remain separate, but if cooled slowly, the complementary strands reunite to form the renatured normal DNA duplex (Marmur and Lane, 1960; Doty et al., 1960).

a In *Hemophilus influenzae* the genes specifying reaction to cathomycin and streptomycin are closely linked (capable of being carried by the same transforming fragment).   Herriott (1961) heated the DNA from *cathomycin-resistant, streptomycin-sensitive* and *cathomycin-sensitive, streptomycin-resistant* to 100°C and then cooled them slowly.   Some of the renatured double-stranded DNA molecules possessed the ability to transfer both *cathomycin resistance* and *streptomycin resistance* into a bacterium sensitive to both antibiotics.   The double-resistant transformed clones bred true. What is the inference regarding the constitution of the double-transforming DNA?   Should such DNA be capable of yielding true-breeding double transformants?   Explain.

b Can you think of a way in which the experimental procedure developed by Marmur, Lane, Doty, et al., can be used in clarifying taxonomic relationships in the bacteria?   See Marmur et al., 1963, Hoyer et al., 1964, for enlightening discussions bearing on the problem.

732   The pathway of tryptophan synthesis in *E. coli* is

Chorismic acid → anthranilic acid → PRA → CdRP → InGP → tryptophan

where PRA = *N*-(5′-phosphoribosyl) anthranilate
        CdRP = 1-(*O*-carboxyphenylamino)-1-deoxyribulose-5-phosphate
        InGP = indole-3-glycerol phosphate

a Anagnostopoulos and Crawford (1961) defined four types of *tryptophan-requiring* mutants in *Bacillus subtilis*.   One mutant, *trp⁻*, responded only to tryptophan; two, *ind⁻*, to either indole or tryptophan; and one, *ant⁻*, to anthranilic acid, indole, or tryptophan. One of the ind^- mutants, ind_a^-, accumulated CdRP and the other, ind_b^-, anthranilic acid. Is the pathway of tryptophan synthesis the same as in *E. coli*? Explain.

b By means of transformation, Anagnostopoulos and Crawford made a large number of crosses between these single-mutant strains and one unable to synthesize histidine, his^-. Each strain was *mutant* at one locus and *wild type* at the other. The results of some of these two-factor crosses are shown in the table. Determine whether these five genes are linked and, if linked, whether the sequence corresponds to the order of the biochemical steps. Show how your results might relate to the operon hypothesis.

Table 732

Cross	Donor	Recipient	Percentage* of *wild-type* (prototrophic) transformants
1	ind_a^-	trp^-	12.7
2	ind_b^-	trp^-	11.8
3	ant^-	trp^-	43.5
4	ant^-	ind_a^-	27.3
5	his^-	ind_a^-	44.9
6	trp^-	his^-	19.0
7	ind_a^-	his^-	26.3
8	ant^-	his^-	45.0

* Percentage *wild types* $= \dfrac{\text{no. of prototrophs}}{\text{total no. of transformants}} \times 100.$

733 In *Bacillus subtilis* the genes *ant*, *trp*, and *his* are linked. Anagnostopoulos and Crawford (1961) made reciprocal transformation crosses between single and double mutants in this species. The results of crosses between $ant^- trp^+ his^-$ and $ant^+ trp^- his^+$ were as shown. Show whether the gene sequence is *ant-his-trp*, *his-ant-trp*, or *ant-trp-his*, and why.

Table 733

Donor	Recipient	Progeny/class	Percent
$ant^- trp^+ his^-$	$ant^+ trp^- his^+$	$ant^+ trp^- his^+$	40.7
		$ant^- trp^+ his^+$	14.6
		$ant^+ trp^+ his^+$	3.2
		$ant^+ trp^+ his^-$	37.5
		$ant^- trp^- his^+$	4.0
$ant^+ trp^- his^+$	$ant^- trp^+ his^-$	$ant^- trp^+ his^-$	70.1
		$ant^+ trp^+ his^-$	10.6
		$ant^- trp^+ his^+$	14.7
		$ant^+ trp^+ his^+$	4.6

734 Reciprocal three-factor crosses in *Bacillus subtilis* were made by Wilson et al. (1966) to determine the sequence of the linked genes *met*, *thyB*, and *ile*. In each cross *ile⁺ thy⁺* transformants were selected and scored for the *met* marker. The results were as tabulated. What is the sequence of the three markers?

Table 734

DNA donor	Recipient	No. of *ile⁺ thy⁺* colonies picked	Recombinants			
			met⁺ ile⁺ thy⁺		*met⁻ ile⁺ thy⁺*	
			No.	%	No.	%
ile⁺ thyB⁻ met⁺	*ile⁻ thyB⁺ met⁻*	264	31	12	233	88
ile⁻ thyB⁺ met⁻	*ile⁺ thyB⁻ met⁺*	280	133	48	147	52

735 Nester et al. (1963) treated a *trp⁻ his⁻ tyr⁻* strain of *B. subtilis* unable to synthesize tryptophan, histidine, and tyrosine with DNA from the prototrophic strain *trp⁺ his⁺ tyr⁺* able to synthesize these amino acids. The number of colonies in each of the seven transformant classes is shown in the table.

Table 735

Transformant class	No. of colonies
trp⁺ his⁺ tyr⁺	11,940
trp⁻ his⁺ tyr⁺	3,660
trp⁻ his⁻ tyr⁺	685
trp⁺ his⁻ tyr⁻	2,600
trp⁺ his⁻ tyr⁺	107
trp⁺ his⁺ tyr⁻	1,180
trp⁻ his⁺ tyr⁻	418

What topographical information do these data permit you to deduce regarding the region of the genome carrying the genes *trp*, *his*, and *tyr*? Show your calculations and illustrate.

REFERENCES

Adelberg, E. A., and S. B. Burns (1960), *J. Bacteriol.*, **79**:321.
———— (ed.) (1966), "Papers on Bacterial Genetics," 2d ed., Little, Brown, Boston.
Ames, B. N., R. F. Goldberger, P. E. Hartman, R. G. Martin, and J. R. Roth (1967), in V. V. Koningsberger and L. Bosch (eds.), "Regulation of Nucleic Acid and Protein Biosynthesis," pp. 272–287, Elsevier, Amsterdam.
Anagnostopoulos, C., and I. P. Crawford (1961), *Proc. Natl. Acad. Sci.*, **47**:378.
Anderson, T. F. (1958), *Cold Spring Harbor Symp. Quant. Biol.*, **23**:47.

Avery, O. T., C. M. MacLeod, and M. McCarty (1944), *J. Exp. Med.*, **79**:137.
Beckwith, J. R., E. R. Signer, and W. Epstein (1966), *Cold Spring Harbor Symp. Quant. Biol.*, **31**:393.
Bodmer, W. F., and A. T. Ganesan (1964), *Genetics*, **50**:717.
Bonhoeffer, F., and W. Messer (1969), *Annu. Rev. Genet.*, **3**:233.
—— and W. Vielmetter (1968), *Cold Spring Harbor Symp. Quant. Biol.*, **33**:623.
Braun, W. (1965), "Bacterial Genetics," 2d ed., Saunders, Philadelphia.
Broda, P., J. R. Beckwith, and J. Scaife (1964), *Genet. Res.*, **5**:489.
Cairns, J. (1963), *J. Mol. Biol.*, **6**:208.
—— (1966), *Sci. Am.*, **214**(January):36.
Campbell, A. (1969), "Episomes," Harper & Row, New York.
Carlton, B. C., and D. D. Whitt (1969), *Genetics*, **62**:445.
Cavalli, L. L., J. Lederberg, and E. M. Lederberg (1953), *J. Gen. Microbiol.*, **8**:89.
Clark, A. J., and A. D. Margulies (1965), *Proc. Natl. Acad. Sci.*, **53**:451.
Cohen, A., W. Fisher, R. Curtiss, and H. Adler (1968), *Cold Spring Harbor Symp. Quant. Biol.*, **33**:365.
Cold Spring Harbor Symp. Quant. Biol., **33** (1968), "Replication of DNA in Micro-organisms."
Curtiss, R. (1969), *Annu. Rev. Microbiol.*, **23**:69.
Cuzin, F., G. Buttin, and F. Jacob (1967), *J. Cell. Physiol.*, **70**(suppl. 1):77.
De Haan, P. G., et al. (1969), *Mutat. Res.*, **8**:505.
Demerec, M., and L. E. Demerec (1956), *Brookhaven Symp. Biol.*, **8**:75.
—— and Z. Hartman (1956), *Carnegie Inst. Wash.*, D.C., Publ. 612, p. 5.
—— and H. Ozeki (1959), *Genetics*, **44**:269.
Doty, P., J. Marmur, J. Eigner, and C. Schildkraut (1960), *Proc. Natl. Acad. Sci.*, **46**:461.
Dove, W. F. (1968), *Annu. Rev. Genet.*, **2**:305.
Driskell-Zamenhof, P. J., and E. A. Adelberg (1963), *J. Mol. Biol.*, **6**:483.
Dubnau, D., and R. Davidoff-Abelson (1971), *J. Mol. Biol.*, **56**:209.
——, C. Goldthwaite, I. Smith, and J. Marmur (1967), *J. Mol. Biol.*, **27**:163.
Enomoto, M. (1966), *Genetics*, **54**:1069.
—— (1971), *Genetics*, **69**:145.
Ephrussi-Taylor, H. (1951), *Cold Spring Harbor Symp. Quant. Biol.*, **16**:445.
—— (1955), *Adv. Virus Res.*, **3**:275.
Falkow, S., and R. V. Citarella (1965), *J. Mol. Biol.*, **12**:138.
——, J. A. Wohlhieter, R. V. Citarella, and L. S. Baron (1964), *J. Bacteriol.*, **87**:209.
Fox, M. S. (1960), *Nature*, **187**:1004.
—— (1962), *Proc. Natl. Acad. Sci.*, **48**:1043.
—— (1966), *J. Gen. Physiol.*, **49**:183.
—— and M. K. Allen (1964), *Proc. Natl. Acad. Sci.*, **52**:412.
Freifelder, D. (1968), *Cold Spring Harbor Symp. Quant. Biol.*, **33**:425.
Fuerst, C. R., F. Jacob, and E. L. Wollman (1956), *C. R. Acad. Sci.*, **243**:2162.
Gingery, R., and H. Echols (1968), *Cold Spring Harbor Symp. Quant. Biol.*, **33**:371.
Goldthwaite, C., D. Dunbau, and I. Smith (1970), *Proc. Natl. Acad. Sci.*, **65**:96.
Goodgal, S. H., and R. M. Herriott (1961), *J. Gen. Physiol.*, **44**:1201.
—— and E. H. Postel (1967), *J. Mol. Biol.*, **28**:261.
Greenberg, J. (1964), *Genetics*, **49**:771.
Griffith, F. (1928), *J. Hyg. Camb.*, **27**:113.
Gross, J., and E. Englesberg (1959), *Virology*, **9**:314.
Gross, J. D., and L. G. Caro (1966), *J. Mol. Biol.*, **16**:269.
Guerrini, F., and M. S. Fox (1968), *Proc. Natl. Acad. Sci.*, **59**:1116.
Guild, W. R., and M. Robison (1963), *Proc. Natl. Acad. Sci.*, **50**:106.
Hartman, P. E., Z. Hartman, and D. Serman (1960*a*), *J. Gen. Microbiol.*, **22**:354.
——, J. C. Loper, and D. Serman (1960*b*), *J. Gen. Microbiol.*, **22**:323.
Havender, W. R., and T. A. Trautner (1970), *Molec. Gen. Genet.*, **108**:61.

Hayes, W. (1952), *Nature*, **169**:118.

―――― (1953), *Cold Spring Harbor Symp. Quant. Biol.*, **18**:75.

―――― (1966), *Proc. R. Soc. (Lond.)*, **B164**:230.

―――― (1967), *Endeavour*, **26**:33.

―――― (1968), "The Genetics of Bacteria and Their Viruses," 2d ed., Wiley, New York.

Herriott, R. M. (1961), *Proc. Natl. Acad. Sci.*, **47**:146.

Hickson, F. T., T. F. Roth, and D. R. Helinski (1967), *Proc. Natl. Acad. Sci.*, **58**:1731.

Hirota, Y. (1960), *Proc. Natl. Acad. Sci.*, **46**:57.

―――― and T. Iijima (1957), *Nature*, **180**:655.

―――― and P. H. A. Sneath (1961), *Jap. J. Genet.*, **36**:307.

Hotchkiss, R. D. (1951), *Cold Spring Harbor Symp. Quant. Biol.*, **16**:457.

―――― and M. Gabor (1970), *Annu. Rev. Genet.* **4**:193.

―――― and J. Marmur (1954), *Proc. Natl. Acad. Sci.*, **40**:55.

Howard-Flanders, P., and R. P. Boyce (1964), *Genetics*, **50**:256.

―――― and ―――― (1966), *Rad. Res. Suppl.* **6**:156.

――――, ――――, and L. Theriot (1966), *Genetics*, **53**:1119.

Hoyer, B. H., B. J. McCarthy, and E. T. Bolton (1964), *Science*, **144**:959.

Ihler, G., and W. D. Rupp (1969), *Proc. Natl. Acad. Sci.*, **63**:138.

Ikeda, H., and J. I. Tomizawa (1965), *J. Mol. Biol.*, **14**:85.

Jacob, F. (1955), *Virology*, **1**:207.

―――― and E. A. Adelberg (1959), *C. R. Acad. Sci.*, **249**:189.

―――― and ―――― (1966), trans. and reprinted in Adelberg (1966), pp. 437–439.

―――― and S. Brenner (1963*a*), *C. R. Acad. Sci.*, **248**:3219.

――――, ――――, and F. Cuzin (1963*b*), *Cold Spring Harbor Symp. Quant. Biol.*, **28**:329.

―――― and E. L. Wollman (1956), *Ann. Inst. Pasteur*, **91**:486.

―――― and ―――― (1957), *C. R. Acad. Sci.*, **245**:1840.

―――― and ―――― (1960), trans. and reprinted in Adelberg (1966), pp. 398–400.

―――― and ―――― (1958), *Symp. Soc. Exp. Biol.*, **12**:75.

―――― and ―――― (1961*a*), *Sci. Am.*, **204**(June):92.

―――― and ―――― (1961*b*), "Sexuality and the Genetics of Bacteria," Academic, New York.

――――, ――――, and W. Hayes (1956), *Cold Spring Harbor Symp. Quant. Biol.*, **24**:141.

Jones, D., and P. H. A. Sneath (1970), *Bacteriol. Rev.*, **34**:40.

Kaiser, A. D., and F. Jacob (1957), *Virology*, **4**:509.

Kloos, W. E., and N. E. Rose (1970), *Genetics*, **66**:595

Lacks, S. (1962), *J. Mol. Biol.*, **5**:119.

Lederberg, J. (1947), *Genetics*, **32**:505.

―――― (1949), *Proc. Natl. Acad. Sci.*, **35**:178.

―――― (1955), *J. Cell. Comp. Physiol.*, **45**(suppl. 2):75.

―――― (1956), *Genetics*, **41**:845.

――――, L. L. Cavalli, and E. Lederberg (1952*a*), *Genetics*, **37**:720.

―――― and E. Lederberg (1952*b*), *J. Bacteriol.*, **63**:399.

―――― and E. L. Tatum (1946*a*), *Nature*, **158**:558.

―――― and ―――― (1946*b*), *Cold Spring Harbor Symp. Quant. Biol.*, **11**:113.

Lennox, E. S. (1955), *Virology*, **1**:190.

Marmur, J., S. Falkow, and M. Mandel (1963), *Annu. Rev. Microbiol.*, **16**:329.

―――― and D. Lane (1960), *Proc. Natl. Acad. Sci.*, **46**:453.

――――, R. Rownd, S. Falkow, L. S. Baron, C. Schildkraut, and P. Doty (1961), *Proc. Natl. Acad. Sci.*, **47**:972.

Meselson, M. (1964), *J. Mol. Biol.*, **9**:734.

Miyake, T., and M. Demerec (1960), *Genetics*, **45**:755.

Momose, H., and L. Gorini (1971), *Genetics*, **67**:19.

Morse, M. L., E. Lederberg, and J. Lederberg (1956), *Genetics*, **41**:758.

――――, ――――, and ―――― (1956), *Genetics*, **41**:142.

Nelson, T. C., and J. Lederberg (1954), *Proc. Natl. Acad. Sci.*, **40**:415.

Nester, E. W., A. T. Ganesan, and J. Lederberg (1963), *Proc. Natl. Acad. Sci.*, **49**:61.

Novick, R. (1969), *Bacteriol. Rev.*, **33**(2):210.

Ohki, M., and J. I. Tomizawa (1968), *Cold Spring Harbor Symp. Quant. Biol.*, **33**:651.

Oppenheim, A. B., and M. Riley (1966), *J. Mol. Biol.*, **20**:331.

O'Sullivan, A., and N. Sueoka (1967), *J. Mol. Biol.*, **27**:349.

Ozeki, H. (1956), *Carnegie Inst. Wash., D. C., Publ.* 612, p. 97.

—— and H. Ikeda (1968), *Annu. Rev. Genet.*, **2**:245.

Pittard, J., and E. A. Adelberg (1964), *Genetics*, **49**:995.

Ravin, A. (1959), *J. Bacteriol.*, **77**:296.

—— (1960), *Genetics*, **45**:1387.

Rupp, W. D., and G. Ihler (1968), *Cold Spring Harbor Symp. Quant. Biol.*, **33**:647.

Sanderson, K. E. (1967), *Bacteriol. Rev.*, **31**:354.

Scaife, J. (1966), *Genet. Res.*, **8**:189.

—— (1967), *Annu. Rev. Microbiol.*, **21**:601.

Sicard, A. M., and H. Ephrussi-Taylor (1965), *Genetics*, **52**:1207.

Signer, E. R. (1968), *Annu. Rev. Microbiol.*, **22**:451.

—— and J. R. Beckwith (1966), *J. Mol. Biol.*, **22**:33.

Stent, G. S. (1963), "Molecular Biology of Bacterial Viruses," Freeman, San Francisco.

—— (1971), "Molecular Genetics," Freeman, San Francisco.

Stocker, B. A. D. (1956), *J. Gen. Microbiol.*, **15**:575.

——, N. D. Zinder, and J. Lederberg (1953), *J. Gen. Microbiol.*, **9**:410.

Stuy, J. H. (1965), *J. Mol. Biol.*, **13**:554.

Susman, M. (1970), *Annu. Rev. Genet.*, **4**:135.

Takano, T. (1971), *Proc. Natl. Acad. Sci.*, **68**:1469.

Taylor, A. L., and E. A. Adelberg (1960), *Genetics*, **45**:1233.

—— and —— (1961), *Biochem. Biophys. Res. Commun.*, **5**:400.

—— and M. S. Thoman (1964), *Genetics*, **50**:659.

—— and C. D. Trotter (1967), *Bacteriol. Rev.*, **31**:332.

Taylor, H. E. (1949), *C. R. Acad. Sci.*, **228**:1258.

Tatum, E. L., and J. Lederberg (1947), *J. Bacteriol.*, **53**:673.

Tomizawa, J., and N. Anraku (1964), *J. Mol. Biol.*, **8**:516.

Ward, C. B., and D. A. Glaser (1969), *Proc. Natl. Acad. Sci.*, **62**:881.

Wilson, M. C., J. L. Farmer, and F. Rothman (1966), *J. Bacteriol.*, **92**:186.

Witkin, E. (1951), *Cold Spring Harbor Symp. Quant. Biol.*, **16**:357.

Wollman, E. L., and F. Jacob (1955), *C. R. Acad. Sci.*, **240**:2449.

—— and —— (1956), *Sci. Am.*, **195**(July):109.

—— and —— (1957), *Ann. Inst. Pasteur*, **93**:323.

——, ——, and W. Hayes (1956), *Cold Spring Harbor Symp. Quant. Biol.*, **21**:141.

Yanofsky, C., and E. S. Lennox (1959), *Virology*, **8**:425.

Yoshikawa, H., and N. Sueoka (1963), *Proc. Natl. Acad. Sci.*, **49**:559.

—— and —— (1963), *Proc. Natl. Acad. Sci.*, **49**:806.

Zelle, M. R., and J. Lederberg (1951), *J. Bacteriol.*, **61**:351.

Zinder, N. D., and J. Lederberg (1952), *J. Bacteriol.*, **64**:679.

17
Recombination in Viruses

Questions on fine-structure analysis appear in Chap. 26.

QUESTIONS

736 When a nontransducing temperate phage infects a bacterium, it may take one of two distinct paths, depending on the conditions of infection. What are these paths, and what characteristics does each confer on the host cell? Illustrate. See Jacob and Wollman, 1961.

737 Schlesinger in 1933 showed that bacteriophages consist of approximately equal proportions of protein and DNA. On the basis of work in the early 1950s by Anderson and Herriott the detailed structure was revealed (Brenner et al., 1959). Illustrate diagrammatically the structure of T-even viruses and explain how they attach to and infect bacterial cells.

738 Using *host-range*, *h*, and *plaque-type*, *r*, mutants outline an experiment to demonstrate that recombination in phages does not occur until the phages are injected into the host.

739 *Amber* mutations in the same gene do not complement each other. Why? Explain in molecular terms.

740 What is meant when we say: The bacterial chromosome has temperate-phage memory, and temperate phage has bacterial-chromosome memory?

741 **a** Explain what is meant by a phage "cross."
b In what respects are the F particle and the λ phage similar, and in what respects are they different?
c Describe the lytic cycle of T-even phage.

742 In T2 and T4 phages the linkage maps are circular, but the chromosomes (DNA duplexes) are linear. Explain how this apparently paradoxical situation can exist. (See Thomas, 1967, and others referred to in his report for their explanation.)

743 In T2 (and T4) phage the chromosomes (DNA molecules) are circularly permuted and terminally repetitious (Thomas and MacHattie, 1964; Thomas and Rubenstein, 1964; MacHattie et al., 1967; Thomas, 1967). Explain what is meant by "circularly permuted and terminally repetitious."

744 Both "early" and "late" *temperature-sensitive* mutants have been isolated in phage T4 (Edgar and Lielausis, 1964*b*). What is meant by early and late when applied to these mutants, and what is their significance?

745 Certain phage, e.g., PL22, can transduce any gene in the donor genome. Others, e.g., λ, transduce only one or a few genes in a particular region of the genome. Account for these facts.

746 **a** Distinguish between virulent and temperate phage.
b Discuss the life cycle of λ phage and compare and contrast it with the life cycle of phage T4.

747   What is the evidence that a phage (in the prophage state) is attached to a bacterial chromosome?

PROBLEMS

748   Doermann (1953) infected *Escherichia coli* cells with two strains of T4 phage, one a triple-mutant strain of genotype *m* (*minute*), *r* (*rapid lysis*), *tu* (*turbid*) and the other *wild type* at each of the gene loci.   The numbers of plaques of each of the different kinds of lytic products were as shown.

Table 748

Genotype	No. of plaques	Frequency
$m\,r\,tu$	3,467	0.335
$m^+\,r^+\,tu^+$	3,729	0.361
$m\,r^+\,tu^+$	520	0.050
$m^+\,r\,tu$	474	0.046
$m\,r\,tu^+$	853	0.082
$m^+\,r^+\,tu$	965	0.093
$m\,r^+\,tu$	162	0.016
$m^+\,r\,tu^+$	172	0.017

a State whether these genes are inherited independently or linked and why.
b If linked, give:
   (1) The linkage distances between *m* and *r*, *r* and *tu*, and *m* and *tu*.
   (2) The sequence of the three genes.
c Calculate the coefficient of coincidence and state its significance.

749   In the bacteriophage T2, r^+ (*wild type*) produce *fuzzy-edged* plaques, whereas *r* mutants produce *sharp-edged* plaques; h^+ and *h* control the ability to infect certain *E. coli* strains (h^+ phages infect only strain B; *h* can infect both B and B/2).   In 1949 Hershey and Rotman carried out the first detailed study of genetic recombination in "crosses" between $h\,r^+$ and various $h^+\,r$ strains of T2.   The procedure in making the crosses and determining the progeny genotypes was as follows.   A growing culture of *E. coli* equally sensitive to both parental strains was infected with an input average of about 10 $h\,r^+$ and $h^+\,r$ phage per cell.   After 5 minutes the culture was diluted 10^4-fold to avoid reinfection (since some cells lyse earlier than others) and incubated for 60 minutes.   The progeny phage were plated on agar seeded with a mixture of strain B and strain B/2 as indicators.   The results of three of the crosses, involving the mutants *r1*, *r7*, and *r13* are presented in the table.   After a careful consideration of the data answer the following questions:

Table 749

Cross	Step	$h^+ r^+$	$h r^+$	$h^+ r$	$h r$
$h r^+ \times h^+ r1$	Input	0	53	47	0
	Yield	12	43	34	12
$h r^+ \times h^+ r7$	Input	0	49	51	0
	Yield	5.9	56	32	6.4
$h r^+ \times h^+ r13$	Input	0	49	51	0
	Yield	0.74	59	39	0.94

a Is T2 a haploid or a diploid phage?
b How many genes specify plaque type?
c Are these genes linked to the h locus?
d If linked, illustrate the conceivable arrangements of h, $r1$, $r7$, and $r13$.
e What type of gene interaction is involved among the r mutants?

750 Virulent T-phage strains differ in the range of bacterial hosts they can infect (h^+ many, h few) and the type of plaque (size and appearance of the area of lysed bacterial cells, r^+ causing *small* plaques with *rough* edges and r causing *large* plaques with *sharp* edges). If an agar plate contains a mixture of two or more suitable indicator bacteria, all of which can be attached and destroyed by h^+ strains but only one by the h strains, the h^+ strains produce *dark* plaques (because all bacteria are lysed) and h strains produce *light* plaques (because only one bacterial strain is destroyed). Hershey and Rotman (1948) exposed a sensitive bacterial strain to a highly concentrated mixture of $h^+ r$ and $h r^+$ phages. The progeny particles from the lysate plated on a mixture of *E. coli* strains B and B/2 (indicator bacteria) produced the following types of plaques:

Small, light
Large, dark
Small, dark
Large, light

The two nonparental types were present in approximately equal frequency but were less frequent than the parental types, which were also present in approximately equal numbers. Each progeny type bred true.
a Do the data indicate whether the phage is haploid or diploid? Explain.
b Would you expect similar results in the $h^+ r^+ \times h r$ cross? Why?

751 Hershey and Rotman (1949) concluded that T2 bacteriophage has three linkage groups, as shown in Fig. 751. In 1960 Streisinger and Bruce made crosses involving three genes. Each of their experiments involved two linked genes, e.g., $r2$ and $r7$, and a third gene, e.g., $r1$, apparently unlinked to the other two and therefore apparently belonging to a different linkage group. By using different

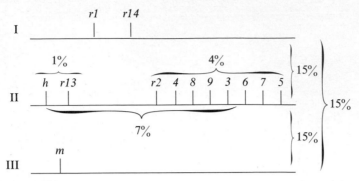

Figure 751

concentrations of the two infecting (parental) phages in mixed, two-phage infections and forcing premature lysis of the bacterial cell to limit the frequency of crossing-over they obtained the tabulated results, which are typical of all other such crosses.   Do the results of Streisinger and Bruce verify Hershey and Rotman's conclusions?   If not, how many linkage groups does T2 have?   Explain and give the sequence of the genes if possible.

Table 751

	Cross		Multiplicities of infection	Recombinants selected	Frequency of r allele of minority parent among selected recombinants
Phage	Minority parent	Majority parent			
T2	$r2\,r7^+\,r1 \times r2^+\,r7\,r1^+$		1: $\geqslant 10$	$r2^+\,r7^+$	Most $r1$; very few $r1^+$
	$r2^+\,r7\,r1^+ \times r2\,r7^+\,r1$		1: $\geqslant 10$	$r2^+\,r7^+$	Few $r1^+$; most $r1$

752   In 1958 Edgar carried out a genetic analysis of 13 rII mutants of bacteriophage T4D.

1   Strain K of *E. coli* (on which only r^+ phage grow) was infected with different rII mutants two at a time.   The results are given in Table 752*A*.

Table 752*A*

Cross	Progeny
$r6 \times r47, r64, r71, r70, r59, r69,$ or $r62$	Very few, if any, phage
$r51 \times r43, r60, r65,$ or $r73$	Very few, if any, phage
$r47 \times r73$	Normal number of phage
$r61 \times r51$	Normal number of phage

2   The 13 *r*II mutants were then studied in two-factor crosses in all possible combinations.   The results (frequency of recombinants) of the 68 *r* × *r* crosses are summarized in Table 752*B*.   With reasons for your answers:

(1) State the number of genes (cistrons) involved.

(2) Map the *r*II region as accurately as possible.

(3) State which mutants appear to possess mutations at identical sites.

Table 752*B*

	r61	r47	r64	r71	r70	r59	r69	r62	r51	r60	r43	r65
r73	8.2	0.0	0.0	0.0	0.0	0.0	0.0	0.0	0.0	1.4	1.4	0.5
r65	5.2	0.0	0.0	0.0	0.0	4.3	2.6	2.6	0.0	0.7	0.7	
r60	8.2	6.6	0.0	0.0	0.0	4.1	3.7	3.7	0.0			
r43	8.2	6.6	0.0	0.0	0.0	4.1	3.7	3.7	0.0			
r51	0.0	6.2	0.0	0.0	0.0	0.0	2.4	2.4				
r69	5.8	3.2	2.7	2.6	0.98	0.45						
r62	5.8	3.2	2.7	2.6	0.98	0.45						
r59	0.0	3.5	1.5	1.4	0.59							
r70	0.0	0.0	1.5	0.78								
r71	0.0	0.0	1.1									
r64	1.6	0.83										
r47	0.46											

Note: Frequency of recombinants $\dfrac{2 \times \text{number of } r^+ \text{ progeny}}{\text{total number of progeny}} \times 100.$

753   Edgar et al. (1962) carried out a series of mapping experiments with *r* (*rapid-lysis*) mutants of bacteriophage T4D.   Out of 194 such mutants, 64 failed to grow on *λ*-lysogenic strains of *E. coli* K12.   Two-factor crosses among these mutants showed a very low percent (0.16, 0.40, etc.) of *wild-type* recombinants.   In a few crosses no *wild types* were obtained.   When 74 other mutants were crossed to one or more "tester" strains, *r48*, *r ED b 50*, *r67*, the results were as tabulated.   Of the 64 mutants which failed to grow on K12, one, *r ED d f 41*, never reverted to r^+ and gave no recombinants with any of the other 63 in the group or with the 7 mutants which produced few *wild types* with *r ED b 50*.   All other mutants reverted to r^+ at a low frequency.

Table 753

Cross	Percentage *wild types*
65 of these *r* mutants × *r48*	Extremely low
× *r ED b 50* and *r67*	Extremely high
7 *r* mutants × *r ED b 50*	Extremely low
× *r48* and *r67*	High
2 *r* mutants × *r67*	Extremely low
× *r48* and *r ED b 50*	High

a At how many loci do these *r* mutants map?   Explain.

b What is the nature of (1) the *r ED df 41* mutant and (2) other *r* mutants? Explain.

754 By early 1964 the linkage map of T4 phage was shown to be

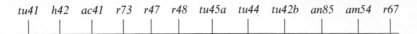

This information is compatible with either of two alternatives:

1   The linkage map is linear and rodlike (viz. two-ended).   In this case *r67* is closer to *ac41* than to *h42*.

2   The linkage map is circular, formed by joining the ends shown in the above map. In this case, *r67* is closer to *h42* than to *ac41*.

Streisinger et al. (1964) crossed a *r67 h42 ac41$^+$* phage with one of genotype *r67$^+$ h42$^+$ ac41*, and *h42 ac41* recombinants were selected.   About 65 percent of these were *r67* and 35 percent *r67$^+$*.   Which of the above hypotheses is correct and why?   See Streisinger et al.

755 Four *amber* mutants of phage T4 (*am-11*, *am-12*, *am-13*, and *am-14*) grow on strain CR63 of *E. coli* but not on strain B.   Three *temperature-sensitive* mutants (*ts-7*, *ts-8*, *ts-9*) grow on both strains at 25°C, but on neither at 42°C.   Strain B is infected at 42°C by the mutant strains in pairs, with the results shown (after Epstein et al., 1963; Edgar et al., 1964a):

Table 755

Strain	Burst size (average yield of phage per infected cell)
wild type	300
ts-7 × *ts-8*	250
am-11 × *am-12*	1
am-11 × *am-13*	250
am-13 × *ts-8*	<1
ts-7 × *ts-9*	30
am-11 × *am-14*	1
am-11 × *ts-9*	1
am-11 × *ts-8*	250

a In how many different genes did the seven mutations occur?

b Can any of these genes undergo an *amber* as well as a *temperature-sensitive* type of mutation?

c Is the protein controlled by the T7 gene more likely to be a monomer than a dimer, or is the reverse true?

756 Three *r* mutants of independent origin in the A gene of the *r*II region in T4 phage give the tabulated results.

Table 756

Cross	Recombinant (r^+) progeny
$r7 \times r8$	0
$r9 \times r8$	a few
$r9 \times r7$	0

a What is the linkage sequence of the three mutants? Explain.
b What is the molecular nature of the *r7* mutation?
c Would you expect a revertant of *r7* to possess a *wild-type* or *pseudo-wild-type* phenotype? Why?

757 A large number of *ts* (*temperature-sensitive*) and *sus* (*amber* or *suppressor-sensitive*) mutants of bacteriophage S13 have been isolated and classified into seven complementation groups, six of which (A to F) are considered here. To establish the linear order of the mutants three-factor crosses were carried out as follows. Each cross was between an *sus ts* double mutant and a *ts* or *sus*

Table 757

Cross No.	Cross*	Selected progeny	Fraction of selected progeny carrying wild-type allele of outside (unselected) marker
1	*sus8*(F) *ts2*(A) × *ts4*(A)	ts^+	0.15
2	*sus12*(C) *ts4*(A) × *ts2*(A)	ts^+	0.25
3	*sus13*(C) *ts4*(A) × *sus12*(C)	sus^+	0.3
4	*sus12*(C) *ts4*(A) × *sus13*(C)	sus^+	0.8
5	*sus12*(C) *ts3*(D) × *sus13*(C)	sus^+	0.23
6	*sus12*(C) *ts6*(D) × *ts3*(D)	ts^+	0.17
7	*sus12*(C) *ts3*(D) × *ts6*(D)	ts^+	0.76
8	*sus5*(E) *ts3*(D) × *ts6*(D)	ts^+	0.07
9	*sus10*(E) *ts3*(D) × *ts6*(D)	ts^+	0.3
10	*sus10*(E) *ts3*(D) × *sus5*(E)	sus^+	0.3
11	*sus5*(E) *ts9*(B) × *sus10*(E)	sus^+	0.13
12	*sus5*(E) *ts9*(B) × *ts11*(B)	ts^+	0.20
13	*sus10*(E) *ts11*(B) × *ts9*(B)	ts^+	0.90
14	*sus10*(E) *ts9*(B) × *ts11*(B)	ts^+	0.19
15	*sus8*(F) *ts11*(B) × *ts9*(B)	ts^+	0.14

* Letters in parentheses represent the complementation group in which the mutation

single mutant.   In each cross, two markers are in the same complementation group, and the third is an outside marker.   The progeny of each cross were plated, and those which were *wild type* (ts^+ or sus^+) for the two markers in the same complementation group were selected (see the third column).   The selected recombinants (which formed plaques on the assay plates) were picked and tested to determine the genotype for the unselected outside marker, e.g., *ts* or ts^+ if sus^+ progeny were selected (see fourth column).   The data from 15 such three-factor crosses are given (Tessman, 1965).

a Analyze and establish the order of the markers in each cross.

b Make a composite genetic map from all the data, and carefully note any peculiarities of this map.   (See Baker and Tessman, 1967.)

758 A series of two-factor crosses in λ phage are carried out by infecting λ-sensitive bacteria with pairs of λ mutants.   The mutants and their genotypes are shown in Table 758*A*.   The crosses and the recombination frequencies

$$\frac{\text{Yield of recombinant types} \times 100}{\text{Yield of parental plus recombinant types}}$$

Table 758*A*

Mutant	Phenotype	Genotype
λc	*Clear* plaque; c^+ (*wild type*) plaques are turbid	$ch^+ mi^+ m5$
λh	*Adsorbs* to, and grows in, strains of bacteria resistant to h^+ (*wild type*)	$c^+ h mi^+ m5^+$
λmi	*Small* plaque (considerably smaller than *wild type*, mi^+)	$c^+ h^+ mi m5^+$
$\lambda m5$	*Large* plaque (considerably larger than *wild type*, $m5^+$)	$c^+ h^+ mi^+ m5$

between the markers in each cross are shown in Table 758*B*.   Show how the markers can be arranged in a linear map.   Make a diagram of the map indicating distances between adjacent markers.

Table 758*B*

Cross	Recombination frequency
$\lambda c \times \lambda h$	7
$\lambda c \times \lambda mi$	7
$\lambda h \times \lambda mi$	14
$\lambda m5 \times \lambda h$	8
$\lambda m5 \times \lambda c$	15

759 a In λ phage, the *host-range* mutant *h* reproduces in bacteria resistant to *wild-type* (h^+) phage; *temperature-sensitive* (*ts*) phage are able to grow and develop to lysis at 30°C but not at 42°C; and the *plaque-type* mutant *c* makes *clear* plaques while c^+ (*wild type*) makes *turbid* plaques. A three-factor cross is made between $ts\,h\,c^+$ and $ts^+\,h^+\,c$ by infecting bacteria growing at 30°C. After the cells have been lysed, the lysate (of progeny phages) is diluted and then plaque-assayed for $ts^+\,h$ recombinants by incubation at 42°C on h^+-*resistant* bacteria (only these recombinants will form plaques under these conditions). The proportion of the plaques appearing under these plating conditions which are *turbid*, c^+, is found to be 0.95.
(1) What is the map order of the three markers?
(2) Is there evidence of linkage?

b The assay plates described above are incubated at 30°C instead of 42°C. *Clear* plaques are then spotted onto another assay plate seeded with λ-*resistant* indicator bacteria and the plates incubated at 42°C to determine the fraction of the $h\,c$ recombinants that are *temperature-sensitive*, i.e., are *ts* and will not lyse the indicator cells at 42°C.
(1) If the fraction is again 0.95, show that the same order of the three markers is obtained.
(2) What additional information can be obtained by this method which was not obtained by plating at 42°C?

c Can you devise another, more direct way of analyzing the proportion of $h\,c$ recombinants carrying the *ts* marker?

760 a The single-stranded DNA phage ϕX174 contains approximately 5,500 nucleotides (Sinsheimer, 1959). If all this DNA codes for proteins and the average polypeptide contains 200 amino acids, how many different polypeptides could this DNA specify?

b (1) The phage T4 chromosome is a single long continuous DNA duplex of molecular weight about 1.2×10^8. Since the average molecular weight of each nucleotide is 357 (assume 360) and base pairs are separated by 3.4 Å, what is the length of the T4 chromosome?
(2) The total genetic map of T4 is about 2,500 recombination units long. If two *h* mutants with mutations in adjacent bases are crossed, with what frequency do you expect h^+ recombinants?

761 a Table 761*A* shows results obtained by Kaiser (1955) in two-factor crosses

Table 761*A*

Parents	Progeny			
$s\,mi^+ \times s^+\,mi$	$647\ s\,mi^+$	$502\ s^+\,mi$	$65\ s^+\,mi^+$	$46\ s\,mi$
$s\,c^+ \times s^+\,c$	$808\ s\,c^+$	$566\ s^+\,c$	$19\ s^+\,c^+$	$20\ s\,c$
$co_1\,mi^+ \times co_1{}^+\,mi$	$459\ co_1\,mi^+$	$398\ co_1{}^+\,mi$	$17\ co_1{}^+\,mi^+$	$25\ co_1\,mi$
$c\,mi^+ \times c^+\,mi$	$1,213\ c\,mi^+$	$1,205\ c^+\,mi$	$84\ c^+\,mi^+$	$75\ c\,mi$
$s\,co_1 \times s^+\,co_1{}^+$	$46\ s\,co_1{}^+$	$53\ s^+\,co_1$	$1,615\ s^+\,co_1{}^+$	$1,774\ s\,co_1$

with λ phage involving the genes c, s, co_1, and mi.   Determine the percentage recombination in each cross and draw a linkage map showing the relative positions of the genes and the distances between them.

b A three-factor cross involving c, co_1, and mi was also performed by Kaiser (1955).   The progeny phenotypes and genotypes are given in Table 761B for the two least frequent recombinant types of plaques (classes).   What is the sequence of mi, c, and co_1?

Table 761B

Parents	Recombinant phenotypes	Total number of progeny plaques examined
$c\,co_1{}^+ mi^+ \times c^+ co_1\, mi$	$8\ c^+ co_1{}^+ mi^+,\ 1\ c^+ co_1{}^+ mi$	6,600

762 *Wild-type* recombinants in the T4 phage cross $am85^+ tu44\,am54 \times am85\,tu44^+ am54^+$ occur much less frequently than in either $am85\,tu44^+ am54 \times am85^+ tu44\,am54^+$ or $am85\,tu44\,am54^+ \times am85^+ tu44^+ am54$.   What is the sequence of the genes?   Show how you arrive at your answer.

763 In λ phage, c^+ (*wild type*) form *turbid* plaques and c form *clear* plaques. A *suppressed-sensitive* (*sus* or *amber*) mutant strain of λ c, $c\,sus_1\,sus_2{}^+$, and a second *sus* mutant, $c^+ sus_1{}^+ sus_2$, are crossed at $37°C$ by growing them together in a permissive host strain of bacteria (in which *sus* or *amber* mutants carrying the chain-termination codon UAG in the middle of the gene can grow; in this host, codon UAG is translated as sense, in the non-permissive host, as nonsense).   Cell lysates are diluted and plaque-assayed on the nonpermissive host.   The proportion of *clear* to *turbid* plaques is found to be 20:1.   What is the order of the three phage markers c, sus_1, and sus_2? *Note:* The three markers are closely linked.

764 You are given the following three strains of *E. coli* K12:

Wild-type (prototrophic) grows on minimal medium.

Mutants I and II (both auxotrophic) need substances A and B for growth on minimal medium.   The requirement for A in both strains is produced by mutation a, but the requirement for B is produced in *mutant* I by mutation b_1 and in *mutant* II by mutation b_2 (see below).

Genetic map

The genotypes of the strains may therefore be written

Wild-type $a^+ b_1{}^+ b_2{}^+$
Mutant I $a\,b_1$
Mutant II $a\,b_2$

Explain how to make (and positively identify) a double mutant of genotype $a^+ b_1 b_2$ using the generalized transducing phage P1, assuming that these markers (a, b_1, and b_2) are all cotransducible with high frequency.

765 You are given the following three phage strains:

1 λ tsA c^+ h^+, a *temperature-sensitive* mutant which grows at 30°C but not at 42°C
2 λ tsA^+ c h^+, a *clear* plaque mutant
3 λ tsA^+ c^+ h, a *host-range* mutant

and bacterial indicator strains C600 λ^S and CR63 λ^R. Describe how to isolate the triple mutant λ tsA c h.

766 Bacteriophages T2 and T4 possess different tail structures by which they adsorb to bacteria and which determine whether they can infect a particular strain of *E. coli*.

E. coli strain B is sensitive to both phages; i.e., both can infect and lyse it. Strain B/2 is sensitive to T4 and resistant to T2, and strain B/4 is sensitive to T2 and resistant to T4.

In 1951 Novick and Szilard mixedly infected *E. coli* strain B with T2 and T4 phages and found that about half the total first-generation progeny formed plaques on B/4 only (and showed T2 phenotype, including T2 tail structures) and the other half formed plaques on B/2 only (T4 phenotype including T4 tail structures). An unexpected result occurred when the B/2 bacteria were infected with only one first-generation phage, each from plaques on B/2. About half the second-generation progeny could no longer infect and form plaques on B/2 but could do so on B/4. Also, about half the first-generation phages which formed plaques on B/4 produced second-generation progeny that formed plaques on B/2 only. The second-generation progeny in both cases bred true.

Taking the physiology of phage reproduction into consideration, offer an explanation that will account for all these results.

767 When bacteria are mixedly infected with T2 r^+ and T2 r phages, which produce *fuzzy*-edged and *sharp*-edged plaques, respectively, and the lysate diluted (so that each phage produces its own separate plaque) and plaque-assayed, approximately 2 percent of the plaques are *mottled* (partly r^+ and partly r). The remaining 98 percent are either r^+ or r. When these *mottled* plaques are picked, the phage suspended in broth, and then diluted and plaque-assayed, not only are r- and r^+-type plaques obtained in about equal numbers but about 2 percent of the plaques are again *mottled*. The breeding behavior is repeated in subsequent generations. Noting that phage are haploid, offer an explanation to account for the origin of the *mottled* plaques.

768 The c mutant strains of T2 and T4 phage require the amino acid tryptophan to adsorb to cells of *E. coli*; other strains of T2 and T4 are c^+ (*tryptophan-independent*). When host cells are infected with a mixture of c^+ and c phage,

about half of the *tryptophan-independent* progeny are found upon subsequent testing to be *c*.   Explain how this comes about.

769 Streisinger (1956) found that the progeny viruses produced by mixed infection with T2 and T4 phage were capable of infecting both the B/2 and the B/4 strain of *E. coli*.   The viruses produced in the next generation were pure T2 or pure T4.   *Note:* T2 can infect only B/4; T4 can infect only B/2.

a Account for these events.

b What proportion of the progeny of a mixed infection of *E. coli* with equal amounts of T2 and T4 would you expect to be capable of infecting both B/2 and B/4?

770 Genetic recombination in bacteriophage corresponds to the production of a DNA molecule derived partly from one and partly from another parent.   This may occur by copy choice, breakage and reunion, or breakage and copy choice.   In 1961 experiments were conducted by Meselson and Weigle and by Kellenberger, Zichichi, and Weigle to determine whether there are any DNA genomes that are entirely derived from one parent in a recombinant phage.   Crosses were made between two strains of the temperate λ phage, carrying alternate alleles of two genes.   One strain was labeled with the isotopes ^{13}C and ^{15}C, and the other was not.   It was found that discrete amounts of the original parental DNA appeared in the recombinant phages.

a Which is the most plausible mechanism for recombination on the basis of these results?   Which is the least plausible?   Are any of the mechanisms excluded?   Explain.

b Are the copy-choice and breakage explanations of recombination in phage mutually exclusive?   Explain and illustrate.

771 Recent work at the molecular level with λ phage (Meselson and Weigle, 1961; Meselson, 1964) and T4 phage (Tomizawa, 1967; Shahn and Kozinski, 1966) demonstrates that recombinants due to crossing-over are the result of breakage and reunion of DNA segments.   The process involves several steps.   What may these be, and in what sequence may they occur?   [See papers pertinent to the question in *J. Cell Physiol.*, **70**, suppl. 1 (1967), and in *Cold Spring Harbor Symp. Quant. Biol.*, **33** (1968).]

772 Phages like S13 and ϕX174 are single-stranded.   Nevertheless it is highly unlikely that recombination takes place between single-stranded molecules.   Offer an explanation.   (See Pratt, 1969.)

773 λ phage has two linear linkage maps, the prophage and the vegetative maps:

Prophage map   *sus*j, *sus*B, *sus*A, *sus*r, *sus*Q, *i*   (Campbell, 1963)
Vegetative map   *sus*A, *sus*B, *sus*j, *i*, *sus*Q, *sus*r   (Campbell, 1961)

Explain how these two are interrelated.

774 In phage T2 you have a series of *r* mutants that form *sharp*-edged plaques in contrast to the *fuzzy*-edged plaques formed by *wild-type* (r^+) T2. How would you determine which of these are deletion mutants?

775 In phage T4, r^+ (*wild-type*) strains produce *small* plaques on both *E. coli* strains B and K. The *r* mutants in the *r*II region, which consists of two contiguous genes *A* and *B*, produce *large* plaques on strain B but no plaques (no progeny) on K when they infect these hosts individually. When strain K is infected with both *r75* and *r101*, no progeny are produced, but when infected with both *r75* and *r89*, many *small* plaques (r^+ phenotype) are produced. Explain why plaques are formed in the latter infection but not in the former. What phenotype would you expect in mixed infections by *r101* and *r89* and why?

776 *Wild-type* T2 h^+ phage forms plaques on *E. coli* strain B but not B/2, whereas the mutant phage strain, *h*, forms plaques on both hosts. If the *host-range* traits are to serve as a genetic marker in phage crosses, it is necessary to recognize both h^+ and *h* alleles among the progeny particles. In an $h r^+ \times h^+ r$ cross how would you proceed to determine whether a particular plaque was h^+ or *h*? Illustrate.

777 Edgar and Lielausis (1964*b*) isolated a large number of *temperature-sensitive* mutants of bacteriophage T4D. They form plaques at 25°C but, unlike the *wild-type* strain, cannot form plaques at 42°C. A total of 382 mutations were studied.
  a To map the mutations complementation tests were first performed. Why?
  b What procedure would you follow after complementation tests to locate the mutations precisely on the linkage map?

778 In T4 phage a nonsense mutant, which produces an incomplete polypeptide chain, is found in gene 23, which controls the synthesis of head protein. How would you detect and maintain such a mutant?

779 **a** Certain T2 phage mutants fail to produce coat protein when grown in strain B of *E. coli* and thus form no plaques on this host strain. These mutants are able to produce coat protein in a permissive strain C and thus will grow

Table 779*A*

	1	2	3	4	5	6	7	8	9	10
1	−	+ + + +	+ + + +	+	+	+	+	+	+	+ + + +
2		−	+	+ + + +	+ + + +	+ + + +	+ + + +	+ + + +	+	+
3			−	+ + + +	+ + + +	+ + + +	+ + + +	+ + + +	+	−
4				−	+	+	+	+	+	+ + + +
5					−	+	+	+	+	+ + + +
6						−	−	−	+	+ + + +
7							−	−	+	+ + + +
8								−	+	+ + + +
9									−	+
10										−

Key: − = no plaques; + = few plaques; + + + + = many plaques.

and form plaques.   Wishing to know whether one or more cistrons control the production of the coat protein, an investigator isolates 10 such mutants, designated 1 to 10.   A complementation analysis is then done by testing the mutants in pairs (trans) using the spot test, with the results shown in Table 779*A*.

(1) Which indicator bacterial strain did the investigator use in the spot test?   Explain.

(2) How many cistrons appear to be involved in the production of coat protein in T2?

(3) Group the mutations within the cistron or cistrons.

(4) Do there appear to be any hot spots?   If so, at what sites?

b The investigator measures the reversion frequency of each mutant and finds the results shown in Table 779*B*.

Table 779*B*

Mutant	1	2	3	4	5	6	7	8	9	10
Reversion frequency	10^{-6}	2×10^{-6}	10^{-5}	10^{-7}	5×10^{-7}	10^{-6}	2×10^{-6}	3×10^{-6}	*	*

* Do not revert.

(1) How were these results on reversion frequencies obtained?

(2) What do the reversion frequencies allow the investigator to conclude about each of the mutants?

(3) Draw a linear map summarizing the data in (a) and (b), indicating the extent to which the sequence is arbitrary.

c A single cistron, *H*, controls a certain *h* conditionally lethal mutant phenotype in this phage.   A number of *H*-cistron mutants are available, some of which are deletions.   From the latter, the investigator selects a set of deletion standards which divide the *H* cistron into three regions, *a-b*, *b-c*, and *c-d* as shown in Fig. 779*A*.   He now wishes to place nine suspected point mutants

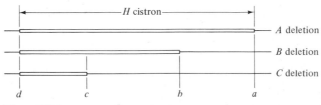

Figure 779*A*

(numbered 1 to 9) in the *H* cistron.   To do this, he spot-tests each of these mutants against the deletion standards with the results shown in Fig. 779*B*.

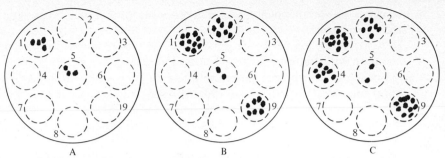

Figure 779B   Each circle (broken lines) circumscribes the area to which a particular mutant is localized.   A to C: deletion standard used.

(1) What is your interpretation of the results on plate A?   Quantitate your answer where possible.
(2) Localize the mutations within the regions *ab*, *bc*, and *cd* by filling in Table 779C with an X in the appropriate box.

Table 779C

Mutant	Location		
	a-b	*b-c*	*c-d*
1			
2			
3			
4			
5			
6			
7			
8			
9			

d How do you suppose the experimenter decided his mutants 1 to 9 were point mutations?

780   When a new phage is obtained, it is usually assumed that a single mutational event (e.g., one base substitution) has given rise to the mutant phenotype. Occasionally, however, a double mutant is obtained involving two separate mutational events at different sites on the chromosome, each capable of producing the mutant phenotype.   These double mutants can be recognized by certain properties.

a How would the frequency of reversion to *wild type* of a double-point mutant compare with that of a single-point mutant?

b Suppose that the double mutant in question is crossed with a set of single-point mutants (1 to 10) whose order on the genetic map is 1, 2, 3, ..., 10. Next, suppose that *wild-type* recombinants are obtained at a frequency of 10 percent in biparental crosses with 1 and 10; 5 percent with 2 and 9; 2 percent with 3 and 8; and <0.1 percent with 4, 5, 6, and 7. Analyze these data and describe the most probable location of the two mutations in the double mutant. (*Hint:* If a mutant is double, the above are all three-factor crosses.)

c Using the above example, describe how to separate the two mutations in the double mutant by means of a cross and describe how to differentiate the various genotypes produced in the cross given the additional information that a two-factor cross between 4 and 7 produces *wild-type* recombinants at a frequency of 10 percent.

781 a *Temperature-sensitive, ts,* alleles of at least 56 genes of T4 phage have been isolated. They constitute a class of conditionally lethal mutants that allow the phage to propagate at 25°C but not at 37 to 42°C (Edgar et al., 1964*a*; Edgar and Lielausis, 1964*b*). Provide an explanation in molecular terms to account for the conditional lethal nature of their action.

b Suppose you isolate three *temperature-sensitive* mutants (grown at 25°C, not at 35 to 42°C) in T4 phage. Outline the procedure you would follow to determine (1) whether the mutants are in the same or different genes and (2) the percent recombination between any two mutants. Indicate the results expected when the mutants are in the same and when in different genes.

c Do the same assuming that the three mutants are of the *amber* type.

782 A collection of T2 DNA molecules, each depicted by two parallel lines corresponding to the complementary chains of a double helix, is illustrated.

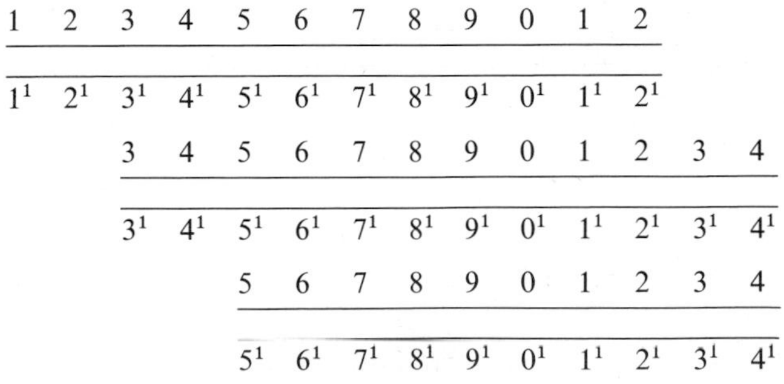

1	2	3	4	5	6	7	8	9	0	1	2		

1^1 2^1 3^1 4^1 5^1 6^1 7^1 8^1 9^1 0^1 1^1 2^1

| | | 3 | 4 | 5 | 6 | 7 | 8 | 9 | 0 | 1 | 2 | 3 | 4 |

3^1 4^1 5^1 6^1 7^1 8^1 9^1 0^1 1^1 2^1 3^1 4^1

| | | | | 5 | 6 | 7 | 8 | 9 | 0 | 1 | 2 | 3 | 4 |

5^1 6^1 7^1 8^1 9^1 0^1 1^1 2^1 3^1 4^1

Figure 782   (*After Thomas, 1967.*)

Show how circles can be formed after treatment with exonuclease III.

REFERENCES

Adams, M. H. (1959), "Bacteriophages," Interscience, New York.
Anderson, T. F. (1953), *Cold Spring Harbor Symp. Quant. Biol.*, **18**:197.
────── and A. H. Doermann (1952), *J. Gen. Physiol.*, **35**:657.
Anraku, N., and J. Tomizawa (1965a), *J. Mol. Biol.*, **11**:501.
────── and ────── (1965b), *J. Mol. Biol.*, **12**:805.
Baker, R., and I. Tessman (1967), *Proc. Natl. Acad. Sci.*, **58**:1440.
Baylor, M. B., A. Y. Hessler, and J. P. Baird (1965), *Genetics*, **51**:351.
────── D. D. Hurst, S. L. Allen, and E. T. Bertani (1957), *Genetics*, **42**:104.
Benjamin, T. L. (1970), *Proc. Natl. Acad. Sci.*, **67**:394.
Benzer, S. (1955), *Proc. Natl. Acad. Sci.*, **41**:344.
Bernstein, H. (1968), *Cold Spring Harbor Symp. Quant. Biol.*, **33**:325.
Brenner, S. (1957), *Virology*, **3**:560.
────── G. Streisinger, R. W. Horne, S. P. Champe, L. Barnett, S. Benzer, and M. W. Rees (1959), *J. Mol. Biol.*, **1**:281.
Broker, T. R., N. Anraku, and I. R. Lehman (1969), *Fed. Proc.*, **28**:348.
Burgi, E., and A. D. Hershey (1961), *J. Mol. Biol.*, **3**:458.
Cairns, J. (1961), *J. Mol. Biol.*, **3**:756.
Calef, E., and G. Licciardello (1960), *Virology*, **12**:81.
Campbell, A. M. (1961), *Virology*, **14**:22.
────── (1963), *Virology*, **20**:344.
────── (1969), "Episomes," Harper & Row, New York.
Chase, M., and A. H. Doermann (1958), *Genetics*, **43**:332.
Childs, J. D. (1971), *Genetics*, **67**:455.
Cohen, S. (1948), *J. Biol. Chem.*, **174**(1):295.
Cold Spring Harbor Symp. Quant. Biol., 33 (1968), "Replication of DNA in Micro-organisms."
Cox, J. H., and H. B. Strack (1971), *Genetics*, **67**:5.
Delbruck, M. (1940), *J. Gen. Physiol.*, **23**(5):643.
────── and W. T. Bailey (1946), *Cold Spring Harbor Symp. Quant. Biol.*, **11**:33.
Denhardt, D. T., D. H. Dressler, and A. Hathaway (1967), *Proc. Natl. Acad. Sci.*, **57**:813.
Dickson, R. C., S. L. Barnes, and F. A. Eiserling (1970), *J. Mol. Biol.*, **53**:461.
D'Herelle, F. (1922), *Br. Med. J.*, **2**(3216):289.
Doerfler, W., and D. Hogness (1968), *J. Mol. Biol.*, **33**:661.
Doermann, A. H. (1952), *J. Gen. Physiol.*, **35**(4):645.
────── (1953), *Cold Spring Harbor Symp. Quant. Biol.*, **18**:3.
────── and M. B. Hill (1953), *Genetics*, **38**:79.
Dressler, D. H., and D. T. Denhardt (1968), *Nature*, **219**:346.
Edgar, R. S. (1958), *Virology*, **6**:215.
────── H. G. Denhardt, and R. H. Epstein (1964a), *Genetics*, **49**:635.
────── and R. H. Epstein (1965), *Sci. Am.*, **212**(February):70.
────── R. P. Feynman, S. Klein, I. Lielausis, and C. M. Steinberg (1962), *Genetics*, **47**:179.
────── and I. Lielausis (1964b), *Genetics*, **49**:649.
────── and ────── (1965), *Genetics*, **52**:1187.
────── and W. B. Wood (1966), *Proc. Natl. Acad. Sci.*, **55**:498.
Ellis, E. L., and M. Delbruck (1939), *J. Gen. Physiol.*, **22**:365.
Epstein, R. H., A. Bolle, C. M. Steinberg, E. Kellenberger, E. Boy de la Tour, R. Chevalley, R. S. Edgar, M. Susman, H. G. Denhardt, and A. Lielausis (1963), *Cold Spring Harbor Symp. Quant. Biol.*, **28**:375.
Floor, E. (1970), *J. Mol. Biol.*, **47**:293.
Forsheit, A. B., D. S. Ray, and L. Lica (1971), *J. Mol. Biol.*, **57**:117.
Foss, H., and F. W. Stahl (1963), *Genetics*, **48**:1659.
Freifelder, D., and M. Meselson (1970), *Proc. Natl. Acad. Sci.*, **65**:200.

Gough, M., and M. Levine (1968), *Genetics*, **58**:161.

Hayes, W. (1968), "The Genetics of Bacteria and Their Viruses," 2d ed., Wiley, New York.

Hershey, A. D. (1947), *Cold Spring Harbor Symp. Quant. Biol.*, **11**:67.

―――― and E. Burgi (1956), *Cold Spring Harbor Symp. Quant. Biol.*, **21**:91.

――――, ――――, and L. Ingraham (1963), *Proc. Natl. Acad. Sci.*, **49** :748.

―――― and M. Chase (1951), *Cold Spring Harbor Symp. Quant. Biol.*, **16**:471.

―――― and ―――― (1952), *J. Gen. Physiol.*, **36**:39.

―――― and R. Rotman (1948), *Proc. Natl. Acad. Sci.*, **34**:89.

―――― and ―――― (1949), *Genetics*, **34**:44.

Hindley, J., D. H. Staples, M. A. Billeter, and C. Weissmann (1970), *Proc. Natl. Acad. Sci.*, **67**:1180.

Horiuchi, K., H. F. Lodish, and N. D. Zinder (1966), *Virology*, **28**:438.

Ikeda, H., and J. Tomizawa (1968), *Cold Spring Harbor Symp. Quant. Biol.*, **33**:791.

Jacob, F., and E. L. Wollman (1961), *Sci. Am.*, **204**(June):92.

―――― and ―――― (1965), trans. and reprinted in Stent (1965), pp. 340–341.

Jeng, Y., D. Gelfand, M. Hayashi, R. Shleser, and E. S. Tessman (1970), *J. Mol. Biol.*, **49**:521.

Jeppesen, G. N., J. Argetsinger Stertz, R. F. Gesteland, and P. F. Spahr (1970), *Nature*, **226**:230.

Kaiser, A. D. (1955), *Virology*, **1**:424.

―――― and R. B. Inman (1965), *J. Mol. Biol.*, **13**:78.

―――― and T. Masuda (1970), *J. Mol. Biol.*, **47**:557.

―――― and R. Wǔ (1968), *Cold Spring Harbor Symp. Quant. Biol.*, **33**:729.

Kellenberger, G., M. L. Zichichi, and J. J. Weigle (1961), *Proc. Natl. Acad. Sci.*, **47**:869.

Knippers, R., and R. L. Sinsheimer (1968), *J. Mol. Biol.*, **34** :17.

Kolstad, R. A., and H. H. Prell (1969), *Mol. Gen. Genet.*, **104**:339.

Komano, T., R. Knippers, and R. L. Sinsheimer (1968), *Proc. Natl. Acad. Sci.*, **59**:911.

Konings, R. N. H., R. Ward, B. Francke, and P. H. Hofschneider (1970), *Nature*, **226**:604.

Kozinski, A. (1968), *Cold Spring Harbor Symp. Quant. Biol.*, **33**:375.

Kumar, S., et al (1969), *Nature*, **221** :823.

Levine, M. (1969), *Annu. Rev. Genet.*, **3**:323.

Levinthal, C. (1954), *Genetics*, **39**:169.

―――― and N. Visconti (1953), *Genetics*, **38**:500.

Luria, S. E. (1951), *Cold Spring Harbor Symp. Quant. Biol.*, **16**:463.

Lwoff, A., and A. Gutmann (1950), *Ann. Inst. Pasteur*, **78**:711; trans. and reprinted in Stent (1965), pp. 316–335.

――――, L. Siminovitch, and N. Kjeldgaard (1950), *C. R. Acad. Sci.*, **231** :190; trans. and reprinted in Stent (1965), pp. 336–337.

McFall, E., and G. S. Stent (1958), *J. Gen. Microbiol.*, **18**:346.

MacHattie, L. A., D. A. Ritchie, C. A. Thomas, Jr., and C. C. Richardson (1967), *J. Mol. Biol.*, **23**:355.

Meselson, M. (1964), *J. Mol. Biol.*, **9**:734.

―――― (1967), in A. Brink (ed.), "Heritage from Mendel," pp. 81–104, University of Wisconsin Press, Madison.

―――― and J. J. Weigle (1961), *Proc. Natl. Acad. Sci.*, **47**:857.

Novick, A., and L. Szilard (1951), *Science*, **113**:34.

Parkinson, J. S. (1971), *J. Mol. Biol.*, **56**:385.

Pratt, D. (1969), *Annu. Rev. Genet.*, **3**:343.

Radding, C. M. (1969), *Annu. Rev. Genet.*, **3**:363.

Robinson, D. N. (ed.) (1970), "Heredity and Achievement," Oxford University Press, New York.

Rubenstein, I., C. A. Thomas, Jr., and A. D. Hershey (1961), *Proc. Natl. Acad. Sci.*, **47**:1113.

Sechaud, J., G. Streisinger, J. Emrich, J. Newton, H. Lanford, H. Reinhold, and M. M. Stahl (1965), *Proc. Natl. Acad. Sci.*, **54**:1333.

Shah, D. B., and H. Berger (1971), *J. Mol. Biol.*, **57**:17.

Shahn, E., and A. Kozinski (1966), *Virology*, **30**:455.

Shalitin, C., and F. W. Stahl (1965), *Proc. Natl. Acad. Sci.*, **54**:1333.

Signer, E., et al. (1968), *Cold Spring Harbor Symp. Quant. Biol.*, **33**:711.

―――― and J. Weil (1968), *Cold Spring Harbor Symp. Quant. Biol.*, **33**:715.

Sinsheimer, R. L. (1959), *Brookhaven Symp. Biol.*, **12**:27.

―――― B. Starman, C. Nagler, and S. Guthrie (1962), *J. Mol. Biol.*, **4**:142.

Spiegelman, S., et al. (1967), *J. Cell Physiol.*, **70**(suppl. 1):35.

Stahl, F. W., R. S. Edgar, and J. Steinberg (1964), *Genetics*, **50**:539.

―――― H. Modersohn, B. E. Terzaghi, and J. M. Crasemann (1965), *Proc. Natl. Acad. Sci.*, **54**:1341.

―――― and C. M. Steinberg (1964), *Genetics*, **50**:531.

Stent, C. S. (1963), "Molecular Biology of Bacterial Viruses," Freeman, San Francisco.

―――― (1971), "Molecular Genetics," Freeman, San Francisco.

―――― and C. R. Fuerst (1955), *J. Gen. Physiol.*, **38**:441.

―――― (ed.) (1965), "Papers on Bacterial Viruses," 2d ed., Little, Brown, Boston.

Stone, L. B., et al. (1971), *Nature*, **229**:257.

Streisinger, G. (1956), *Virology*, **2**:388.

―――― and V. Bruce (1960), *Genetics*, **45**:1289.

―――― R. S. Edgar, and G. H. Denhardt (1964), *Genetics*, **51**:775.

―――― J. Emrich, and M. M. Stahl (1967), *Proc. Natl. Acad. Sci.*, **57**:292.

―――― and N. C. Franklin (1956), *Cold Spring Harbor Symp. Quant. Biol.*, **21**:103.

Tessman, E. S. (1965), *Virology*, **25**:303.

―――― (1966), *J. Mol. Biol.*, **17**:218.

Thomas, C. A., Jr. (1966), *J. Gen. Physiol.*, **49**(2):143.

―――― (1967), *J. Cell. Physiol.*, **70**(suppl. 1):13.

―――― and L. A. MacHattie (1964), *Proc. Natl. Acad. Sci.*, **52**:1297.

―――― and I. Rubenstein (1964), *Biophys. J.*, **4**:93.

Tomizawa, J. (1967), *J. Cell. Physiol.*, **70**(suppl. 1):201.

―――― and N. Anraku (1964), *J. Mol. Biol.*, **8**:516.

――――, ――――, and Y. Iwama (1966), *J. Mol. Biol.*, **21**:247.

Visconti, N., and M. Delbrück (1953), *Genetics*, **38**:5.

Watson, J. D. (1970), "Molecular Biology of the Gene," 2d ed., Benjamin, New York.

Wollman, E. L., and F. Jacob (1954), *C. R. Acad. Sci.*, **239**:455; trans. and reprinted in Stent (1965), pp. 338–339.

Wood, W. B., and R. S. Edgar (1967), *Sci. Am.*, **217**(July):60.

18
Genotype, Environment, and Phenotype

QUESTIONS

783 In the latter part of the nineteenth century Weismann performed the following experiments with a purebred strain of mice with *normal* tail length. He cut the tails off a certain number of mice and mated them. This procedure was repeated each generation for a number of generations. The progeny in all cases developed tails of *normal* length. What is the significance of this experiment?

784 State whether you agree with the following statements and explain why:
a The genetic material is different in different species of organisms and in different lines or breeds of the same species.
b Variations in phenotype among individuals of a species may be due to genetic or environmental differences but not both.
c All genotypes possess the same norm of reaction.
d Two individuals rarely, if ever, have identical environments.

785 Assessments of relative importance of the following kind are sometimes made:

1 The relative contributions of heredity and environment to intelligence are 30 and 70 percent, respectively.
2 Environment is more important than heredity in determining a person's weight.

Discuss these statements, stating whether it is possible to determine the relative importance of the two components of any character. Cite evidence to support your contention.

786 Give some arguments for and against the view that schizophrenia is a hereditary disease.

787 The question is frequently asked whether a particular trait is the result of heredity or environment.
a How would you answer such a question? Cite an example.
b How would you state the question to make it meaningfully answerable?

788 **a** Mental characters are largely but not wholly determined by heredity. Discuss.
b A character is known to be inherited. Does this mean that the genotype alone is responsible for its expression? Cite an example to support your view.
c Distinguish between *congenital defects* and *hereditary defects*.
d Of what importance in medicine is a clear understanding of the concept of the *norm of reaction*?

PROBLEMS

789 Certain rats are *resistant* to rickets, i.e., they show no ill effects when fed a diet deficient in vitamin D. Other rats are *susceptible*; they suffer from the disease when fed the vitamin D–deficient diet:

a Suppose that 75 percent of the rats in a certain group fed a vitamin
D–deficient diet develop rickets. Would the disease be attributed to
hereditary susceptibility or to the environment? Would it be due to both?
Explain.

b Suppose that in a true-breeding rickets-*susceptible* line of rats 45 percent
of the individuals are *affected*. Should the disease be attributed to
heredity or to environment?

790 When rabbits are raised on a diet containing xanthophyll, some have *white*
and others *yellow* fat. However, rabbits raised on a xanthophyll-free diet
always have *white* fat.

a Explain these observations.

b How would you determine the mode of inheritance of the character?

c How would you eliminate the *yellow* trait from a breeding population if the
character is monogenically controlled and *yellow* is recessive?

791 In a population of rabbits some individuals are *very small*, others are
intermediate in size, and still others are *very large*. How would you demonstrate whether this variability is a result of genotypic differences?

792 An individual shows a condition caused by an external environmental agent,
e.g., diphtheria, caused by the bacterium *Corynebacterium diphtheriae*. Does
this mean that his genetic constitution plays no part in determining the
expression of the character? Explain.

793 a Distinguish between hereditary and nonhereditary diseases.

b Many people regard heritable diseases as inborn and incurable traits against
which there is no relief and nonheritable infectious diseases and even
accidental mutilations as much less horrible because they can be cured.
How would you explain to such people that their views are baseless, using
phenylketonuria and galactosemia as examples of heritable diseases and
malaria and measles as examples of nonheritable ones?

c Is it possible to determine whether heredity or environment is more important
in determining the final expression of any character? Explain. If it is not
possible, what is it possible to determine? Cite an example to support your
answer.

794 *Rumplessness* in the fowl can have the following causes (Landauer, 1948a,
1954):

1   A dominant allele at one autosomal locus and a recessive allele at another such locus
2   The injection of insulin into the yolks of eggs before or during incubation
3   A severe shaking of the eggs before incubation
4   An abrupt change in temperature during the first week of incubation

Suggest a possible way in which the dominant and recessive alleles might
induce *rumplessness* (see Landauer, 1954).

795 a Distinguish between monozygotic and dizygotic twins.

b How would you determine whether twins of like sex are monozygotic or dizygotic?

c Why are monozygotic twins especially favorable for comparing the relative roles of heredity and environment?

d Identical twins reared apart show a greater mean pair difference in IQ tests than in body weight when the values are compared with the population means. Suggest reasons for this.

e What genetic information may be furnished by human-twin data that could not be obtained from studies of other types of data, such as pedigree and population studies?

f In studies involving the comparisons of monozygotic and dizygotic twins, all the dizygotic twins investigated are of the same sex. Why are unlike-sexed twins not used in such comparisons?

g Discuss the limitations of the twin method for genetic analyses.

h Can twin studies by themselves tell us anything about genes? About genetic recombination?

796 Fraternal (nonidentical) pairs of twins may be of three types: both male, both female, or male and female. What are the expected frequencies of these types in a large population?

a Distinguish between concordance and discordance.

b If one is studying qualitative characters, one way of using twin data is to study the percentage of concordance between the members of twin pairs of like sex.

(1) What conclusions can be drawn when the percentage concordance is the same for dizygotic as for monozygotic twins?

(2) What conclusions can be drawn if concordance is 100 percent for monozygotic twins and 45 percent for dizygotic twins?

(3) Explain why concordance in identical twins may often be less than 100 percent.

797 A certain pair of identical twins reared apart are strikingly similar in one character and quite different in another. What conclusions can be drawn concerning the relative influence of environmental and genetic factors on the expression of these characters?

798 In studies involving many twin pairs it has been found that in 96 percent of the cases where one member of an identical twin pair had been infected with measles, the partner twin had also been *affected*. This held true for 91 percent of the dizygotic twins. What inferences regarding a hereditary basis for the disease can be drawn from the monozygotic twins? From the observations of both monozygotic and dizygotic twins?

799 The tabulated percent concordances were observed in twins.

a Interpret the data for each character using the terms *heritability* and *norm of reaction*.

Table 799

Character or trait	Concordance, %	
	Monozygotic (identical)	Dizygotic (fraternal)
ABO blood groups	100	65
Eye color (brown or blue)	100	53
Diabetes mellitus	85	36
Schizophrenia	87	16
Tuberculosis	87	28
Spina bifida	75	35
Clubfoot	32	3
Measles	95	92

b For which of the above characters or traits is a specific environment (a limited range of environments) required?

c For which of the characters is the genetic basis quite uniform throughout the population? Explain.

800 In the Chinese primrose (*Primula sinensis*) a certain true-breeding variety produces *white* flowers when the temperature is above 86°F during a critical period in early flower development and *red* ones if the temperature is below this level during the critical period. Another variety produces *red* flowers regardless of the temperature. You have two primrose plants, one with *red* and *white* flowers and another with *white* ones only.

a Outline an experiment to show that the genotypes of all flowers on a plant of the first variety are identical and that only their expression is altered.

b How would you determine the mode of inheritance of flower color using the two varieties?

801 Phenylketonuria is caused by a recessive allele that prevents phenylalanine from being converted to tyrosine. Recently it has been reported that it is possible to "cure" phenylketonurics by feeding them a phenylalanine-free diet.

a Would this "cure" modify the contention that phenylketonuria is caused by a recessive allele? Explain.

b Would the "cure" of phenylketonurics affect the frequency of these recessives in future generations?

802 When 100 progeny of a barley plant heterozygous for *Aa* [chlorophyll (*green*)

Table 802

Experiment 1 sunlight	Experiment 2 dark
72 *green* 28 *albino*	all *albino*

vs. nonchlorophyll (*albino*)] are grown in the sunlight and the same number are grown in the dark, the results are as shown.

a Interpret the results of each of these experiments separately and then together.

b Which, heredity or environment, if either, is more important in determining the expression of the character? Discuss.

803 In man, *galactosemics* are unable to metabolize galactose and consequently are mentally deficient; the condition is caused by a recessive allele, *g*.

a When *galactosemics* are fed galactose-free diets, they apparently become *normal* in most respects. What important principle does this illustrate?

b When two such *normals* mate, would you expect *galactosemic* children? Explain.

804 Designating individuals from poor families affording little opportunity for education as type A and individuals coming from well-to-do families as type B, answer the following:

a Would type A individuals as a rule rate as highly as those of type B in intelligence tests?

b Might type A nevertheless have equally as good genetic constitutions for intelligence as type B? Explain.

805 Some people believe the superstition that the emotional state of the parents at the time of conception has an important bearing on the nature of the resulting child. Comment on this belief. Could you prove that this viewpoint is incorrect? Explain.

806 A very talented woman musician, wishing to transmit some of this talent to her child, spends much time practicing, playing, and studying musical masterpieces during her pregnancy. As her child matures he shows great musical talent. Does this prove that the mother influenced her child through her activities before birth? Explain.

807 The proportion of defective calves from heifers (young cows) is somewhat greater than from mature cows. Does this mean that the age of the mother influences the heredity (genotype) of the calves? Explain.

808 Many people believe that drugs have an influence on male and female gametes and the genes they contain, which may lead to the production of defective offspring.

a What result may excessive use of alcohol have on a man's gametes? On his children? Is this effect on the offspring genetic or environmental?

b What result may excessive use of drugs by a mother have on the embryo? Is the effect genetic or environmental?

809 Because of recent advances in various branches of medicine, the effects of an increasing number of man's lethal and semilethal genes can be circumvented by providing the proper environment; e.g., phenylketonurics develop normally on a phenylalanine-free diet.   Discuss the moral and genetic issues arising from the point of view of the individual and society.

810   Some individuals affected with *porphyria variegata* may suffer mild porphyria only (skin abrasions and blistering) or the acute form, in which more variable features of the syndrome are expressed.   Acute porphyria apparently appears only after the use of drugs such as barbiturates and sulfonamides; the syndrome consists of abdominal and muscular pain, weakness, respiratory difficulty (patient talks with a whisper and may even die because respiration stops), and emotional disturbance involving crying very easily.   The basis of the disorder is a defect in porphyrin metabolism.   Porphyrin titers are high in feces in both the mild (quiescent) and acute phases (Dean, 1969).   The mode of inheritance of this disease is indicated in the pedigree, which is based on an extensive genealogical tree of G. R. van Rooyen in South Africa, presented by Dean (1963).   Assume you are a genetics counselor and the propositus (arrow) brings you this family history of the trait and asks for information and counsel.

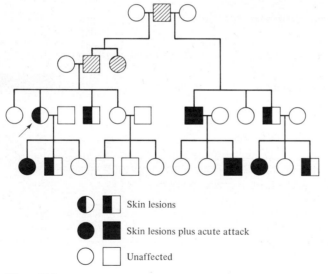

Skin lesions

Skin lesions plus acute attack

Unaffected

Figure 810

a What would you tell her regarding the mode of inheritance of this rare trait and the chance of her children expressing it?

b What counsel would you give the children of an affected parent?   What additional counsel would you suggest to those found to be carriers?

811   *Pyloric stenosis* is a condition in which there is a constriction of the opening between the stomach and the duodenum.   Since 1920 the Ramstedt operation has meant survival for most affected infants.

1 *Pyloric stenosis* occurs in 1 out of 300 births.

2 Of the *affected* infants, 80 percent are males.

3 Monozygotic twins show 22 percent concordance for the condition; dizygotic twins show 4 percent concordance.

4 A given sample of *affected* females have 3 to 4 times as many *affected* children compared to a similar sample of *affected* males. (Data adjusted to account for family size variation, etc.)

5 *Affected* males or *affected* females may have *affected* sons and daughters.

6 Consanguinity of parents is normal for the population.

7 *Affected* sibships do not fit either the 1 in 4 or the 1 in 2 ratios of *affected* to *normals*.

a Do you think this condition is genetically controlled?

b Advance a genetic hypothesis which could account for the above facts (a graph may be useful).

c Explain facts (2) and (4) in terms of your hypothesis.

812 *Drosophila melanogaster*, the fruit fly, normally has a *gray* body, but if a small quantity of silver nitrate is added to its larval food, its body will be *yellow*. Some strains of this fly, however, develop a *yellow* body regardless of the food on which they are raised. Would you say that body color depends on heredity or environment or both? Explain.

813 From a geranium plant we can take portions of roots, stems, or leaves, and from any of them we can grow an entire geranium plant which looks like the parent. Discuss this observation in relation to the theory of the isolation of the germ plasm.

814 In 1960 a sudden increase in the incidence of babies with anomalies of limb development (absence or markedly imperfect development of arms or legs but with hands and feet present) was noted in Germany and other European countries. These anomalies resemble the *phocomelia* phenotype in birds caused by a recessive autosomal allele; the inheritance of the condition is not definitely known in man. Many of the mothers were found to have

Table 814

	1960	1961	1962 Jan. –July	1962 Aug.–Oct.	Total
Total births	19,052	19,917	13,326	5,542	57,837
Phocomelialike malformations	28	60	40	2	130
History:					
Thalidomide taken	13	46	33	2	94
No evidence of intake	0	5	1	0	6
No history available	15	9	6	0	30

taken thalidomide, a drug used to ease the distress of pregnancy.   The table shows the total number of births and number of *phocomelialike* births at 18 obstetric hospitals in Hamburg, Germany, from January 1960 to October 1962 (data from Lenz, 1964).   What conclusions would you make from these data regarding the importance of the genotype in causing the effects?   The importance of the environment?

815   It is now occasionally possible to recognize or detect individuals that are heterozygous for certain recessive or incompletely dominant deleterious or lethal alleles fairly early in life.   Early detection is also possible with certain homozygous recessive phenotypes long before the phenotypic effect is manifested.   Discuss:

a The value of such detection to the individuals and to their progeny.

b The effect on the frequencies of such traits in future generations.

c Your opinion on the moral responsibility, if any, of individuals who know they are carriers of deleterious or lethal alleles.

REFERENCES

Baumgarten, A. (1965), *Nature*, **205**:109.
Boulton, A. A. (1971), *Nature*, **231**:22.
————— (1967), *Nature*, **215**:132.
Carter, C. O. (1969), "An ABC of Medical Genetics," Little, Brown, Boston.
Cattell, R. B., D. B. Blewett, and J. R. Reloff (1955), *Am. J. Hum. Genet.*, **7**:122.
Clausen, J., D. D. Keck, and W. M. Hiesey (1940), *Carnegie Inst. Wash., D.C., Publ.* 520, p. 1.
Dean, A. C. R. (1964), *Nature*, **202**:1046.
Dean, G. (1963), "The Porphyrias: A Story of Inheritance and Environment," Pitman, London.
————— (1969), *Br. Med. Bull.*, **25**(1):48.
Dobzhansky, T., A. M. Holz, and B. Spassky (1942), *Genetics*, **27**:464.
Dunn, L. C., and W. Landauer (1934), *J. Genet.*, **29**:217.
Edwards, J. H. (1964), *Proc. 2d Int. Conf. Congenital Malformations*, pp. 297–305.
Fleishman, E. A. (1969), *Annu. Rev. Psychol.*, **20**:349.
Fogel, B. J., H. M. Nitowsky, and P. Gruenwald (1965), *J. Pediatr.*, **66**:64.
Freeman, G. H., and J. M. Freeman (1971), *Heredity*, **27**:15.
Fuhrmann, W., and F. Vogel (1969), "Genetic Counseling," Springer-Verlag, New York.
Gedda, L. (ed.) (1963), in *Proc. Int. Congr. Hum. Genet.*, **2**:245.
Gloor, H. (1944), *Rev. Suisse Zool.*, **51**:394.
Gluecksohn-Waelsch, S. (1954), *Cold Spring Harbor Symp. Quant. Biol.*, **19**:41.
Goldschmidt, R. B. (1949), *Sci. Am.*, **181** (October):46.
————— and L. K. Piternick (1957), *J. Exp. Zool.*, **135**:127.
Goldstein, S., J. W. Littlefield, and J. S. Soeldner (1969), *Proc. Natl. Acad. Sci.*, **64**:155.
Gottesman, I. I., and J. Shields (1967), *Proc. Natl. Acad. Sci.*, **58**:199.
Hadorn, E. (1951), *Adv. Genet.*, **4**:53.
Haldane, J. B. S. (1938), "Heredity and Politics," Norton, New York.
Hersh, A. H., R. M. Stecher, W. M. Solomon, R. Wolpaw, and H. Hauser (1950), *Am. J. Hum. Genet.*, **2**:391.
Heston, L. L. (1970), *Science*, **167**:249.
Hirsch, J. (1967), "Behavior-Genetic Analysis," McGraw-Hill, New York.
Hoffer, A. (1965), *Nature*, **208**:306.

Huxley, J., E. Mayr, and H. Osmond (1964), *Nature*, **204**:220.
Jacobs, P. A., et al. (1968), *Annu. Hum. Genet.*, **31**:339.
Kallman, F. H. (1953), "Heredity in Health and Mental Disorder," Norton, New York.
Karlsson, J. L. (1964), *Hereditas*, **51**:74.
Kidwell, J. F. (1963), *Genetics*, **48**:1593.
Landauer, W. (1948a), *Growth*, **18**:171.
—— (1948b), *Genetics*, **33**:133.
—— (1954), *J. Cell. Comp. Physiol.*, **43**(suppl. 1):261.
—— and L. C. Dunn (1945), *J. Exp. Zool.*, **98**:65.
Lenz, W. (1964), *Proc. 2d Int. Conf. Congenital Malformations*, pp. 263–276.
Manosevitz, M., G. Lindzey, and D. D. Thiessen (eds.) (1969), "Behavioral Genetics: Method and Research," Appleton-Century-Crofts, New York.
Meade, J. E., and A. S. Parkes (eds.) (1966), "Genetic and Environmental Factors in Human Ability," Oliver & Boyd, Edinburgh.
Montagu, A. (1959), "Human Heredity," World, Cleveland.
Murphy, E. A., and G. S. Mutalik (1969), *Hum. Hered.*, **19**:126.
Nadler, H. L. (1969), *J. Pediatr.*, **74**:132.
Newman, H. H. (1932), *J. Hered.*, **23**:369.
—— (1933), *J. Hered.*, **24**:209.
——, F. N. Freeman, and K. J. Holzinger (1937), "Twins: A Study of Heredity and Environment," University of Chicago Press, Chicago.
Nylander, P. P. S. (1971), *Annu. Hum. Genet.*, **34**:409.
Polani, P. E. (1969), *Nature*, **223**:680.
Predescu, V., D. Florescu, and C. Radulescu (1968), *Nature*, **217**:1150.
Rapoport, J. A. (1939), *Bull. Biol. Med. Exp. URSS*, **7**:414.
Record, R. G., T. McKeown, and J. H. Edwards (1969), *Annu. Hum. Genet.*, **33**:61.
Reed, S. C. (1963), "Counseling in Medical Genetics," 2d ed., Saunders, Philadelphia.
Rife, D. C. (1933), *J. Hered.*, **24**:407, 443.
Rogers, S. (1970), *New Sci.*, **29**(January):194.
Sang, J. H. (1963), *J. Hered.*, **54**:143.
—— and B. Burnet (1967), *Genetics*, **56**:743.
—— and J. M. McDonald (1954), *J. Genet.*, **52**:392.
Saxen, L., and J. Rapola (1969), "Congenital Defects," Holt, Toronto.
Shields, J. (1958), *Eugen. Rev.*, **50**:115.
—— (1962), "Monozygotic Twins," Oxford Unversity Press, Toronto.
Stern, C. (1949), in Glenn L. Jepsen, Ernst Mayr, and George G. Simpson (eds.), "Genetics, Paleontology, and Evolution," pp. 13–22, Princeton University Press, Princeton, N.J.
—— (1960), "Principles of Human Genetics," Freeman, San Francisco.
Vandenberg, S. G. (1966), *Psychol. Bull.*, **66**:327.
Waddington, C. H. (1963), *New Sci.*, **18**(335):145.
Warkany, J. (1954), *J. Cell. Comp. Physiol.*, **43**(suppl. 1):207.
Winchester, R. S. (1957), "Heredity and Your Life," Vantage, New York.

19
Pleiotropism, Penetrance, Expressivity, and Phenocopies

QUESTIONS

816 **a** A gene present in a homozygous strain shows variable expressivity. State whether you think its penetrance would be likely to be complete or incomplete and why.

    b By what criteria can one decide whether a gene is fully penetrant or not?

    c Describe how a dominant trait may skip generations.

    d Suggest why there seem to be many more genes with incomplete penetrance and variable expressivity in man than in experimental organisms.

    e List and discuss the various environmental and genetic factors that can affect the penetrance and expressivity of genes, citing examples where possible.

817 Discuss each of the following statements using specific examples:

    a The degree of penetrance depends on the acuity of examination.

    b The pleiotropic effect of a gene is not a property of the gene and not a general property of gene action but an embryological consequence of the time, place, and type of the primary disturbance of development caused by the mutant gene.

818 If a gene has one primary phenotypic effect of a biochemical and perhaps enzymatic nature, how can it affect the expression of two or more characters?

819 Describe, using an example, how you would determine whether a specific phenotype is a phenocopy or the result of a particular genotype.

820 **a** In what way does the concept of phenocopy production influence the theory and practice of clinical medicine?

    b What advice, if any, should medical practitioners give patients with inherited anomalies who are known to be phenocopies of "normal"?

821 **a** Why are genes with complete penetrance and constant expressivity valuable in the study of gene properties?

    b How can phenocopies be used to study gene action? What value do they have for investigations of this kind?

822 State (with the aid of examples) whether or not:

    a An allele can affect (1) the penetrance, (2) the expressivity, and (3) the pleiotropy of its oppositional allele.

    b A gene can affect (1) the penetrance, (2) the expressivity, and (3) the pleiotropy of another.

823 You find that two traits in a cross are inherited together. What kind of evidence is required to determine whether the two traits are controlled by two closely linked genes or by the pleiotropic effects of one gene?

PROBLEMS

824 In human beings, *normal* vs. *blue* sclera (blue eyewhites) is monogenically controlled; the allele for *normal* is recessive to that for *blue sclera* (Bell, 1928). Matings between *normal* and *affected* give rise to an average of 50 percent *affected* children. Occasionally, however, *unaffected* offspring of such marriages married to unrelated *normals* have *affected* children. *Blue-sclerotics* may or may not have brittle bones and may or may not suffer from *otosclerosis* (deafness). *Affected* persons differ considerably in the intensity of blueness of sclera and degree of brittleness of the bones. What properties of expression of this dominant allele explain these observations?

825 *Himalayan* rabbits, $c^h c^h$, are all white at birth; as they mature, the extremities become black. These differences reflect the effects of temperature upon the action of an enzyme produced by this genotype that transforms a colorless substance into black pigment. The enzyme is active only at temperatures below 34°C. The temperature of the body proper is about 34°C; at the extremities it is about 27°C (Danneel, 1941).

a Does the c^h allele have more than one primary effect? Explain.

b Do the effects of the c^h allele differ in different parts of the body?

c If the answer to (b) is no, explain how the *Himalayan* phenotype is produced in the *Himalayan* breed.

d May the allele be termed pleiotropic? Explain.

826 *Camptodactyly* (stiff little finger) vs. *normal* in man is monogenically controlled; the allele for *camptodactyly* is dominant (Moore and Messina, 1936). However, *normal* couples sometimes have children who are *camptodactylous* in one hand. Explain these cases, giving the probable genotypes of the parents and the genotype of the *camptodactylous* children.

827 The pedigree is human for *polydactyly*, a condition involving more than five digits per limb.

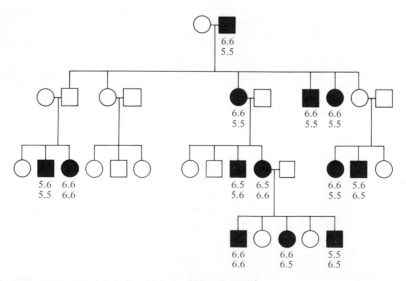

Figure 827 (*Modified after Mohr and Wriedt, 1919.*)

a State whether the trait is autosomal or sex-linked and whether dominant or recessive, giving reasons for your decision.

b Describe any unusual features of expression of the trait and suggest possible causes for them.

828 In a group of irradiated mice, Grüneberg isolated a line with a wide range of abnormalities of eyes, ears, and paws (clubfoot; too many or too few digits; or fused digits). He concluded from the breeding data shown that the traits were controlled by a single pair of allelic genes, *H* dominant, *normal*; *h*, recessive. (See Grüneberg, 1948, for details.) The abnormality was found

Table 828

Parents	Offspring
1. *Hh* × *Hh*	589 *normal* : 109 *affected*
2. *Hh* × *hh*	272 *normal* : 156 *affected*
3. The 156 *affected* intercrossed (*affected* × *affected*)	132 *normal* : 594 *affected*
4. The 132 *normals* intercrossed (*normal* × *normal*)	424 *normal* : 106 *affected*

(Bonnevie, 1934) to be caused by the extrusion in the eleven-day-old embryo of an abnormally large amount of cerebrospinal fluid, which migrated posteriorly in the form of liquid *vesicles*, or *blebs*. Their temporary resting on specific protrusions of the embryo, such as the eye or limb buds, caused the local abnormalities observed at birth. In some cases the blebs could miss these obstructions and be resorbed by the embryo.

a Explain how the embryological observations might be related to the genetic ratios through reduced penetrance and variable expressivity.

b Explain briefly how you would determine by breeding experiments whether reduced penetrance and variable expressivity was the result of the essentially random developmental effect described above or of modifying genes.

c If positive and negative modifying alleles were the cause of the reduced penetrance and variable expressivity, outline a breeding program whereby you might increase penetrance to 100 percent or reduce it to zero.

829 In man, *hemolytic jaundice* is caused by a dominant allele with about 10 percent penetrance (Gates, 1946). A woman heterozygous for the allele marries a *normal* man.

a If the first child is heterozygous, what is its chance of expressing the trait?

b If the couple had 10 children, what is the expected number of *affected* children?

830   In 1936 Danforth discovered a mutant in the mouse with a very short tail. When Dunn et al. (1940) performed crosses between *short-tail* descendants of this mutant and also between these and a *normal* strain, he obtained the following results:

1   *Short-tail* females × *short-tail* males give 159 *normal*, 365 *short-tail* (including *tailless*), and 153 *tailless*, which die within 24 hours after birth.

2   *Short-tail* males × *normal* females (reciprocal cross) give 899 *normal* and 857 *short-tail* (including some *tailless*).

The *short-tail* group of individuals, which varied in the expression of tail length from zero to one-half *normal* were mated with individuals from a true-breeding *normal* strain.   The *short-tail* progeny of the next five generations were crossed without selection for tail type with the *normal* strain.   As the number of the backcross generations increased, tail length among the *short-tail* individuals decreased.   After five backcrosses approximately 90 percent of the individuals in the *short-tail* class were *tailless*, and none of the tailed individuals had tails longer than one-eighth *normal*.

a How many allele pairs are involved?

b Are the alleles completely penetrant?   Constant in expressivity?   If not, what is the cause of incomplete penetrance and variable expressivity?   Give a reason or reasons for each answer.

c Embryological studies by Gluecksohn-Schoenheimer (1943) revealed that *tailless* individuals dying at birth lack kidneys, urethra, several of the lower vertebrae, and rectal and anal openings.   *Short-tailed* individuals and *tailless* ones living beyond the date of birth have, with few exceptions, kidneys and ureters of reduced size, and one or both kidneys may be absent.   Explain how the embryological observations might be related to the genetic ratios through reduced penetrance and variable expressivity.

831   In domestic fowl, breeding experiments beginning with the cross homozygous *normal*, *TT*, × homozygous *tremor*, *tt* (affected individuals shake almost continuously), gave the tabulated results (modified after Hutt and Child, 1934).

Table 831

	Offspring	
Parents	Generation	Phenotype
1. *Normal* × *normal*		All *normal*
2. *Normal* × *tremor*	F_1	All *normal*
3. F_1 × F_1	F_2	423 *normal* : 45 *tremor*
4. F_1, *normal* × *tremor* (from F_2)	Testcross	182 *normal* : 38 *tremor*
5. The 38 *tremor* (progeny of cross 4) intercrossed		190 *normal* : 110 *tremor*
6. The 190 *normal* (progeny of cross 5) intercrossed		370 *normal* : 210 *tremor*

Note: Some *tremor* fowls shake so violently that they have great difficulty in eating; in others the *tremor* is barely perceptible.

a State which allele shows incomplete penetrance and which shows variable expressivity. Give reasons for your answer.

b If modifying genes are responsible for the variations in expression, is it possible with suitable breeding procedures to establish strains with no penetrance or with 100 percent penetrance? Explain.

832 A certain variety of the Chinese primrose (*Primula sinensis*) is homozygous for an allele that causes *white* flowers when the temperature during a critical period during early floral development is above 86°F and *red* ones if the temperature is below this level. Since different flowers develop at different times, it is possible, by changing the temperature during inflorescence development, to obtain both *red* and *white* flowers on a single plant.

a Does the allele act differently in *red* and *white* flowers? If not formulate an explanation for the flower-color differences described.

b Is the allele pleiotropic? Fully penetrant? Constant in expressivity? Explain.

Another variety, homozygous for another allele at the same locus as the first, bears *red* flowers regardless of temperature during flower development.

c Would you expect the dominance relationship to be independent of temperature? Outline the genotypic and phenotypic results you would expect among the F_1 and F_2 of a cross between the two varieties grown at temperatures below and above 86°F.

833 In a *normal* (homozygous) breed of fowl, a *rumpless* bird appears. The abnormality is characterized by the absence of caudal vertebrae and of tail structures such as muscles, feathers, and preen gland (Landauer, 1945, 1955).

a Is one justified in stating from this information alone that *rumplessness* is a mutant trait? Support your answer.

b Outline an experiment to distinguish between gene control and environmental inducement (phenocopy) of the trait.

834 The pedigree shows individuals affected with *blue sclera* (bluish, usually thin outer wall of the eye) and those with *brittle* (fragile) *bones*.

a Is *blue sclera* controlled by the same gene as *brittle bones*?

b Is *blue sclera* controlled by an autosomal or sex-linked gene? Is it dominant or recessive?

c What is the percent penetrance of the allele? State the progenies from which you extracted your information and show your calculations.

d How would you explain the fact that individuals with *blue sclera* frequently do not have *brittle bones*?

e Give the probability of occurrence of a *blue-sclerotic* child for each of the following matings:

(1) IV-3 × IV-7   (2) IV-4 × IV-5
(3) IV-3 × III-9   (4) III-8 × IV-12

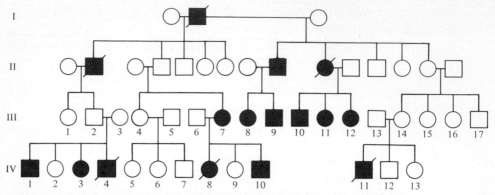

Figure 834   Solid symbol = *blue sclera*; diagonal bar = *brittle bones*.   (*After Bell, 1928.*)

835   In mice, *normal* vs. *pituitary dwarf* is controlled by a single pair of alleles *Dwdw* (Grüneberg, 1952).   Growth rate of *dwarfs* falls behind that of *normals* shortly after birth, and by the seventeenth day it stops.   Thereafter they lose weight, and some die.   Later the survivors begin to grow slowly again and eventually reach about a quarter of the normal adult weight.   At this stage their tails, snouts, and ears are unusually short and their thyroid, thymus, and pituitary glands are small (Boettiger and Osborn, 1938; De Beer and Grüneberg, 1940; and Dawson, 1935).

   Normal young surgically deprived of the anterior pituitary grow and develop exactly like *dwarfs*, and *dwarf* young surgically implanted with pituitaries from *normal* young approach *normals* in growth rate, adult size, glandular and morphological development, and fertility (Smith and MacDowell, 1930).
   a Using the terms pleiotropy and phenocopy, explain the above results.
   b How would you determine whether a *dwarf* arising in a laboratory stock was a phenocopy or homozygous for the recessive allele, *dw*?

836   In *Drosophila melanogaster*, the dominant allele *L* (*lobe eye*) exhibits incomplete penetrance and variable expressivity.   Most flies in the *LL* strain have eyes of greatly reduced size which vary in size, including some with no eyes and a few with *normal*-sized eyes (Bridges and Brehme, 1944).   While part of this variability in expression is developmental, some of the differences between individuals are due to the presence of modifying genes.
   a Outline an experiment to determine whether modifying genes affect penetrance and expressivity.   Remember you have flies homozygous for *L* with variations in eye size from completely *normal* to *no eyes*.
   b Show the results expected in the F_1 and F_2 if there is one pair of modifying alleles, *Mm*.

837   *Muscular dystrophy*, a fairly rare trait, occurs in many different forms and degrees of severity.   In some cases it is autosomal and in other cases X-linked. The two pedigrees from two different unrelated families in the same population show the typical transmission patterns for the two different types of severe

TYPE A

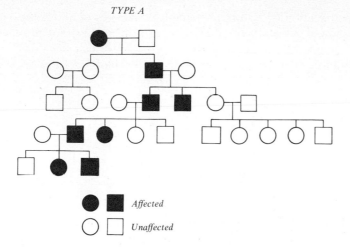

Figure 837*A*

TYPE B

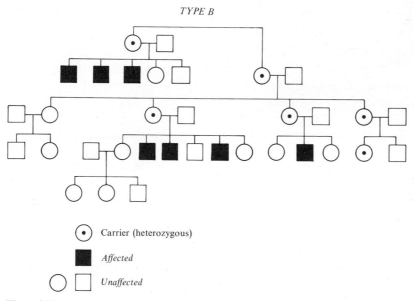

Figure 837*B*

muscular dystrophy.   With respect to *type B muscular dystrophy* it is possible to detect carriers (heterozygotes) of determining genes by serum creatine kinase levels, which are significantly higher in carrier than *normal* (noncarrier) individuals.   No such differences occur between heterozygotes and *normals* (homozygotes) with respect to *type A muscular dystrophy.*   For which of these two types of *muscular dystrophy* would counseling be more specific?   What could you tell individuals in such families and why?   See Carter (1969, pp. 61–68).

838   The pedigree is a modification of results presented by Winters et al. (1958). The family studied showed a history of *vitamin D–resistant rickets* (not preventable by normal vitamin D administration) and *hypophosphatemia* (unusually low concentration of inorganic phosphate in the blood serum), as

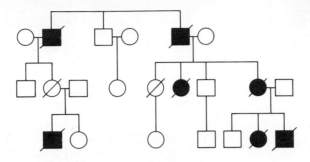

Solid symbols: Individuals *affected* with *rickets*
Diagonal bars: Individuals showing low concentrations of
                   phosphorus in serum tests

Figure 838

well.   Analyze this progeny with regard to the following pairs of alternatives, giving reasons for your decisions:

a Control of the two conditions by one gene or by two linked genes.
b Dominance vs. recessiveness of an allele for *hypophosphatemia*.
c Sex vs. autosomal linkage for *hypophosphatemia*.
d Recessiveness vs. reduced penetrance of the *rickets* trait.

REFERENCES

Bagg, H. J. (1929), *Am. J. Anat.*, **43**:167.
────── and C. C. Little (1924), *Am. J. Anat.*, **33**:119.
Bell, J. (1928), *Treas. Hum. Inher.*, **2**:269.
Boettiger, E. G., and C. M. Osborn (1938), *Endocrinology*, **22**:447.
Bonnevie, K. (1934), *J. Exp. Zool.*, **67**:443.
Bridges, C. B., and K. S. Brehme (1944), *Carnegie Inst. Wash., D.C., Publ.* 552.
Carter, C. O. (1969), "An ABC of Medical Genetics," Little, Brown, Boston.
Coulombre, J. L., and E. S. Russell (1954), *J. Exp. Zool.*, **126**:277.
Danneel, R. (1941), *Ergeb. Biol.*, **19**:55.
Dawson, A. B. (1935), *Anat. Rec.*, **61**:485.
De Beer, G. R., and H. Grüneberg (1940), *J. Genet.*, **39**:297.
Dunn, L. C., S. Gluecksohn-Schoenheimer, and V. Bryson (1940), *J. Hered.*, **31**:343.
────── and W. Landauer (1934), *J. Genet.*, **29**:217.
Gates, R. R. (1946), "Human Genetics," vol. 1, Macmillan, New York.
Gluecksohn-Schoenheimer, S. (1943), *Genetics*, **28**:341.
Goldschmidt, R. B. (1949), *Sci. Am.*, **181**(October):46.

Green, M. C. (1955), *J. Hered.*, **46**:91.

Grüneberg, H. (1948), *Symp. Soc. Exp. Biol.*, *Cambridge*, **2**:155.

—— (1952), "The Genetics of the Mouse," 2d ed., M. Nyhoff, The Hague.

Hadorn, E. (1956), *Cold Spring Harbor Symp. Quant. Biol.*, **21**:363.

—— (1961), "Developmental Genetics and Lethal Factors," Wiley, New York.

Hollingshead, L. (1930), *Genetics*, **15**:114.

Hutt, F. B., and G. P. Child (1934), *J. Hered.*, **25**:341.

Landauer, W. (1945), *J. Exp. Zool.*, **98**:65.

—— (1948), *Growth*, **18**:171.

—— (1954), *J. Cell. Comp. Physiol.*, **43**(Suppl. 1):261.

—— (1955), *Am. Nat.*, **89**:35.

—— (1958), *Am. Nat.*, **92**:201.

Mohr, O. L., and C. Wricdt (1919), *Carnegie Inst. Wash., D.C., Publ.* 295.

Moore, W. G., and P. Messina (1936), *J. Hered.*, **27**:27.

Osborn, R. H., and F. V. De George (1959), "Genetic Basis of Morphological Variation," Harvard University Press, Cambridge, Mass.

Pepkin, S. B., and A. C. Pepkin (1946), *J. Hered.*, **37**:93.

Russell, E. S., and E. C. McFarland (1966), *Genetics*, **53**:949.

Sang, J. H. (1963), *J. Heredity*, **54**:143.

Smith, L. T., and A. W. Nordskog (1963), *Genetics*, **43**:1141.

Smith, P. E., and E. C. MacDowell (1930), *Anat. Rec.*, **46**:249.

Snyder, L. H., and P. R. David (1953), Penetrance and Expression, chap. 2 in A. Sorsby (ed.), "Clinical Genetics," Butterworth, London.

Steinberg, A. G., and R. M. Wilder (1952), *Am. J. Hum. Genet.*, **4**:113.

Stern, C. (1960), "Principles of Human Genetics," 2d ed., Freeman, San Francisco.

Winchester, A. M. (1958), "Genetics," Houghton Mifflin, Boston.

Winters, R. W., J. B. Graham, T. F. Williams, V. W. McFalls, and C. H. Burnett (1958), *Medicine*, **37**:97.

20
Euploidy: Haploidy and Polyploidy

QUESTIONS

839 a Describe chromosome behavior in haploids.

b Why are they of interest to the cytogeneticist? Cytotaxonomist? Plant breeder?

c What formula is used to determine the frequency of balanced spores or gametes a haploid is expected to produce?

d What percentage of spores or gametes are expected to be balanced in haploid plants of a species with $2n = 10$ chromosomes?

840 a What are autotriploids and how may they arise?

b Describe diagrammatically behavior at meiosis in an autotriploid with $2n = 9$ showing:

(1) Pairing at zygotene.

(2) The configurations and orientations at metaphase I.

(3) Segregation at anaphase I.

(4) The chromosome constitution of the meiotic products.

c Label each stage carefully and state which of the meiotic products are balanced.

841 a Why are autotriploids highly sterile?

b If autotriploids are highly or completely sterile, how is it possible to maintain autotriploid strains such as the Keizerskroon tulip and the Baldwin apple?

842 a Although autotriploids frequently show greater vigor, yield, and vitamin content than the diploids from which they arise, they have been unsuccessful in establishing themselves in nature. Explain why.

b In which of the following types of organisms would you expect to find natural autotriploids?

(1) Cross-fertilizing.

(2) Self-fertilizing.

(3) Asexually reproducing.

843 Some of the most desirable apples are triploids. If a desirable mutation occurred in a branch of a triploid tree, how would you establish an orchard of trees with this mutation?

844 State which of the following would be the easiest and which the most difficult to obtain in a given species of plant:

1 Triploids if you have only diploids and tetraploids

2 Diploids if you have only haploids

3 Tetraploids if you have only diploids and tetraploids

Explain.

845 a In which group of plants would you expect polyploids to have a better opportunity of becoming established: those with or those without an asexual means of reproduction? Why?

b How widespread is the occurrence of polyploids in plants? What groups exhibit this phenomenon? See Stebbins (1950, 1966) for a treatment of this subject.

846 **a** Describe the morphological, cytological, genetic, and reproductive (e.g., fertility) criteria for distinguishing between autopolyploids and allopolyploids.

b Is any one of these criteria sufficiently reliable by itself for distinguishing the two types of polyploids? Cite examples to illustrate your answer.

847 In many groups of related plant species multiples of a basic chromosome number occur. The first of such series was discovered by Tahara (1915) in Japan in the genus *Chrysanthemum*, various species of which have the following multiples of $x = 9$ (the basic number): 18, 36, 54, 72, 90. Discuss the origin of series of this type. See Winge (1917) for the first correct interpretation of the manner of origin of such series and Clausen and Goodspeed (1925) for the first experimental verification of Winge's hypothesis.

848 **a** What arguments have been advanced to explain the relative lack of polyploids in animals as compared with plants? Name one animal or group of animals that appears to be an exception to this rule.

OR

b Polyploidy has been important in the evolution of the flowering plants. It is believed to have been unimportant in the evolution of the arthropods and the vertebrates (see Astaurov, 1969).

(1) What features in reproduction probably account for the lack of polyploids in these groups of animals?

(2) What sexual characteristics would you expect to find in animals where polyploids occur in nature?

849 Briefly discuss the significance and role of polyploidy in evolution. Be sure to state whether autopolyploidy or allopolyploidy has been the more important in speciation and why.

PROBLEMS

850 What percentages of gametes would be expected to be n in autotriploids that have 15 and 18 chromosomes, respectively? What is the expected percentage of the sum of n and $2n$ gametes? Show how you derive your answers.

851 In a certain diploid plant species unbalanced meiotic products lead to inviable gametes. A triploid plant of this species is 3 percent fertile (97 percent sterile) when pollinated by a diploid. What is the probable chromosome number of the triploid plant? (Show work.)

852 In 1922 Belling and Blakeslee reported on cytological studies of a triploid in *Datura stramonium* ($n = 12$). The tabulated chromosome numbers were found at metaphase II in the pairs of secondary meiocytes produced by 84 pollen-mother cells. Show whether the orientation of the trivalents at metaphase I was random or not.

Table 852

	Products of chromosome numbers						
Secondary meiocyte 1	12	13	14	15	16	17	18
Secondary meiocyte 2	24	23	22	21	20	19	18
Number of meiocytes	1	1	6	13	17	26	20

853 Rhoades (1936) found a triploid in the corn cross at $glgl\ Ws_3\ Ws_3 \times GlGl$ ws_3ws_3. The cross $3n \times glgl$ produced 89 Gl : 20 gl, and the cross $ws_3ws_3 \times 3n$ produced 42 Ws_3 : 90 ws_3.

a What was the genotype of the triploid?

b Which parent supplied the $2n$ gamete to form the triploid? Justify your answer.

854 In a self-fertilizing plant homozygous for gene A one of the alleles mutates to the recessive form, a. Would the recessive trait be most likely to appear if the plant was a diploid, an autotetraploid, or an allotetraploid? Explain.

855 The New World cotton species *Gossypium hirsutum* and *G. barbadense* have a $2n$ chromosome number of 52 (13 large and 13 small pairs). The American and Old World species *G. thurberi* (small chromosomes) and *G. herbaceum* (large chromosomes) each have a diploid chromosome number of 26. Cytological analysis of meiosis in various hybrids yields the data shown (Beasley, 1942).

Table 855

Cross	Hybrid: metaphase I pairing
G. hirsutum × *G. thurberi*	13 small bivalents + 13 large univalents
G. hirsutum × *G. herbaceum*	13 large bivalents + 13 small univalents
G. thurberi × *G. herbaceum*	13 large univalents + 13 small univalents

a Interpret these results with respect to the phylogenetic relationships of these species.

b How would you determine whether your interpretation is correct? Illustrate the results expected.

856 Polyploids are often said to be conservative in the evolutionary sense, i.e., they do not allow the phenotypic expression of mutant recessive alleles. (In allopolyploids, for example, this would be true for recessives of functionally identical genes, one in each of the different genomes.) What are the reasons for this? Illustrate your answer with an example from an autotetraploid or an allotetraploid and compare with the results expected in a diploid.

857 In the nightshade genus (*Solanum*) the basic n number appears to be 12, and different species of the genus are known to have 24, 36, 48, 60, 72, 96, 108, and 144 chromosomes (Jorgensen, 1928). Does this series of chromosome numbers suggest whether speciation has resulted from autopolyploidy, allopolyploidy, or both? Explain your decision and outline how you might test it.

858 **a** Cultivated tobacco, *Nicotiana tabacum*, has 24 pairs of chromosomes. *N. sylvestris* has 12 pairs of chromosomes, as does *N. tomentosiformis*. Meiotic data from studies of species relationships (Greenleaf, 1941; Goodspeed, 1945) are given. What do these observations reveal about the origins of these species and genomic relationships among them?

Table 858A

Crosses and haploids	Metaphase I
N. tabacum × *N. tomentosiformis*	12 bivalents + 12 univalents
N. tabacum × *N. sylvestris*	12 bivalents + 12 univalents
N. tomentosiformis × *N. sylvestris*	24 univalents
N. tabacum haploid	24 univalents

b Three diploid species in the genus *Brassica* are *campestris*, *nigra*, and *oleracea* with genomes AA, BB, and CC, respectively. Chromosome numbers and number of bivalents and univalents in other species and F_1 hybrids of various crosses are as follows:

Table 858B

Species and/or F_1 hybrid	Chromosome number	Number of bivalents	Number of univalents
B. juncea	36	18	0
B. carinata	34	17	0
B. napus	38	19	0
F_1 from *juncea* × *nigra*	26	8	10
F_1 from *napus* × *campestris*	29	10	9
F_1 from *carinata* × *oleracea*	26	9	8
F_1 from *juncea* × *oleracea*	27	0	27
F_1 from *carinata* × *campestris*	27	0	27
F_1 from *napus* × *nigra*	27	0	27

(1) What are the haploid chromosome numbers of the 3 diploid species? Explain.

(2) Are *juncea*, *carinata*, and *napus* autopolyploids or allopolyploids? If the latter, what kind and why?

(3) Give the genome constitutions of *juncea*, *carinata*, and *napus*. Explain your assignment.

859 Stadler (1929) measured the frequency of mutants in the progeny of self-fertilized diploid ($2n = 14$), tetraploid ($2n = 28$), and hexaploid ($2n = 42$) species of wheat and oats exposed to varying doses of x-rays.   For any particular dose of x-rays mutants appeared with the highest frequency in diploid species and with the lowest frequency in hexaploids.   How would you interpret these results?

860 Trujillo et al. (1965), using starch-gel electrophoresis, showed that the enzyme glucose-6-phosphate dehydrogenase (G6PD) of the horse was distinctly different from that of the donkey.   Examination of the reciprocal inter-specific hybrids, the mule (female horse × male donkey) and the hinny (female donkey × male horse) revealed the following:

1 All female mules and female hinnies contain both types of G6PD.
2 Male mules contain only horse G6PD; male hinnies contain only donkey G6PD.

Show whether the genes controlling G6PD in these species are located on an autosome or an X chromosome.

861 In the donkey (*Equus asinus*) the diploid chromosome number is 62, and in the horse (*E. caballus*) it is 64.   The chromosome complements of the two genera are morphologically different (Trujillo et al., 1962).   In crosses between male donkeys and mares, male offspring are always sterile, but in rare instances females are fertile.   When these fertile females are crossed with stallions, the offspring express only horse traits (Anderson, 1939) and have 63 chromosomes.
a Explain the complete sterility of the male and most female mules.
b Offer an explanation to account for the fertility of the rare female mules and the horselike nature of their offspring in backcrossing to stallions.

862 a The number of chromosome pairs at meiosis in haploids (monoploids) of five species are given, along with the somatic chromosome number and the number of genomes of each species (data from Brewbaker, 1964).

Table 862

Species	No. of pairs	No. of somatic chromosomes	No. of genomes in species
Medicago sativa	8	32	$4x$
Solanum tuberosum	12	48	$4x$
Digitalis mertonensis	7	112	$16x$
Triticum aestivum	1	42	$6x$
Nicotiana tabacum	1	48	$4x$

(1) Offer an explanation to account for the pairing in the haploids of each species.

(2) Which of these species would you classify as autotetraploid and which as allopolyploid? Why? Suggest a genomic formula for each of the haploids.

b Polyploids such as common wheat $(2n = 6x = 42)$ have three sets of chromosomes from three species which are somewhat closely related. In spite of this the chromosomes at meiosis form bivalents only. Outline plausible explanations to account for this cytological behavior even though some of the chromosomes in each genome may be partly homologous to those in the other genomes. (See Riley, 1960, Riley and Law, 1965, for excellent discussions of this topic.)

863 Stebbins (1947) recognizes four types of polyploids: autopolyploid, allopolyploid, segmented allopolyploid, and autoallopolyploid. Classify each of the four polyploid species below on the basis of the data given, with reasons for your classification.

a *Primula floribunda* $(2n = 2x = 18) \times$ *P. verticillata* $(2n = 2x = 18)$ produce a diploid hybrid with 9 chromosome pairs. The polyploid *P. kewensis* $(2n = 4x = 36)$ regularly forms 18 chromosome pairs, but occasionally zero to three quadrivalents, together with a reduced number of bivalents (Newton and Pellew, 1929).

b *Galeopsis pubescens* $(2n = 2x = 16) \times$ *G. speciosa* $(2n = 2x = 16)$ produce a hybrid with 16 univalents. The polyploid *G. tetrahit* $(2n = 4x = 32)$, produced by doubling the chromosome number of the hybrid, regularly forms 16 chromosome pairs (Muntzing, 1932).

c Hybrids of *Helianthus tuberosus* $(2n = 6x = 102) \times$ *H. annuus* $(2n = 2x = 34)$ represent a new polyploid species with 68 chromosomes, forming 34 bivalents (Kostoff, 1939).

864 Tetraploids of *Datura* are highly fertile (Blakeslee et al., 1923), whereas those in *Primula* are highly sterile (Somme, 1930). Assuming that events at meiosis are solely responsible for this difference, what are these events and in what ways can they affect fertility?

865 *Albinism* is recessive in all plant species in which it has been encountered. It occurs frequently in barley but very rarely in common wheat. Offer an explanation to account for this difference.

866 In autotetraploid *Primula sinenis*, Somme (1930), studying dominant-recessive allele pairs *Ss* (*short* vs. *long* style), *Gg* (*green* vs. *red* stigma), and *Bb* (*magenta* vs. *red* flowers) obtained the data shown.

a Determine:
(1) The probable theoretical phenotypic ratio to which each observed phenotypic ratio corresponds.
(2) The genotypes of the parents.

b Are these examples of chromosome or chromatid segregation? Explain, using one cross to illustrate.

Table 866

	Phenotypic ratio of progeny in F_2 or testcross generation	
Cross	Dominant	Recessive
short × *long*	287	45
short × *short*	1,000	342
green × *green*	1,659	52
green × *red*	240	212
green × *red*	1,175	223
magenta × *magenta*	205	5
magenta × *magenta*	264	79
magenta × *red*	272	48

867   In *Lotus corniculatus*, which with few exceptions forms bivalents only, some plants contain a large amount of hydrogen cyanide, HCN, and others have none.   Dawson (1941) carried out a series of crosses between *high*-HCN and *no*-HCN strains to determine the mode of inheritance of this pair of traits. The F_2 data are summarized.

Table 867

Cross no.	♀	Trait	♂	Trait	F_2 offspring *high* HCN	*no* HCN
1	A	*high* HCN	B	*no* HCN	512	89
2	C	*high* HCN	D	*no* HCN	696	689
3	E	*high* HCN	F	*high* HCN	1,039	354
4	F	*high* HCN	J	*high* HCN	788	78
5	G	*high* HCN	K	*high* HCN	384	12
6	H	*no* HCN	L	*no* HCN	0	485

a Outline a hypothesis to account for these results.   Show the parental genotypes and the expected genotype ratios of the offspring for any three of the matings listed.

b Discuss the cytotaxonomic implications of these data, viz. the euploid series to which the species probably belongs and whether interspecific hybridization is likely to be involved in its origin.

868   An autotetraploid forms bivalents only.   Will any of its genes show chromatid segregation?   Illustrate your answer using an autotetraploid of genotype *A Aaa*.

869   In autotetraploid Chinese primrose (*Primula sinensis*) the gene controlling stigma color is very near the centromere of the chromosome carrying it.   The allele *G* for *green* stigmas is dominant to *g* for *red* stigmas.   A homozygous *green* autotetraploid strain is crossed with a homozygous *red* autotetraploid strain.

   a What is the genotype of the F_1?

   b Show the types of gametes the F_1's may be expected to form and derive the expected proportion of each.

   c What phenotypic ratio of *green* to *red* is expected if:

     (1) The F_1's are intercrossed?

     (2) The F_1's are crossed with *red* plants?

   d If the *G* locus were 50 or more map units from the centromere, what types and proportions of gametes would the F_1 be expected to produce? Derive the expected F_2 phenotypic ratio.

870   Blakeslee et al. (1923) obtained the results shown for *purple* vs. *white* flowers, *Pp*, and *spiny* vs. *smooth* capsules, *Ss*, among the offspring of autotetraploid *Datura*.   There is complete dominance in each allele pair.

Table 870

Parents	Progeny	
	Dominant	Recessive
purple × *white*	160	1
purple × *purple*	1,280	0
purple × *white*	905	179
purple × *white*	696	682
purple × *purple*	7,547	2,619
spiny × *smooth*	257	6
spiny × *smooth*	518	137
spiny × *smooth*	3,383	118

   a Determine the probable phenotypic ratios.

   b On the basis of the phenotypic ratio determine the genotypes of the parents of each cross.

   c Which gene gives the better fit to chromosome segregation?   (Refer to specific crosses as evidence to support your answer.)

871   Self-fertilization of an autotetraploid, *AAaa*, gives a 35:1 ratio.   Under what cytological and genetical conditions are such results obtained?

872   Spinach (*Spinacia oleracea*) is a dioecious diploid; *staminate* and *pistillate* plants occur in a 1:1 ratio.   Janick (1955) induced male and female tetraploids with colchicine.   The progeny of the original tetraploid *pistillate* × tetraploid *staminate* cross consisted of 63 *staminate* and 10 *pistillate* plants.   One 4*n* F_1

staminate plant crossed with an F_1 $4n$ *pistillate* plant produced 60 *staminate* and 20 *pistillate* progeny; crossed with a $2n$ *pistillate* plant, it produced 36 *staminate* and 9 *pistillate* plants.   Four other $4n$ F_1 *staminate* plants were crossed with diploid *pistillate* offspring.   Similar results were obtained in the other three crosses.

Nineteen $4n$ *staminate* plants from the previous progenies were crossed with ten $4n$ *pistillate* sibs.   Of these 19 crosses, 2 gave ratios of 79:13 and 65:12 for *staminate* vs. *pistillate*, and 17 gave a 1:1 ratio, e.g., in 1 cross 56 *staminate* and 52 *pistillate* plants were produced.

Sex differences in spinach are probably due to a single determining gene; one sex being *Aa* the other *aa*.

a Using your own symbols, explain these results by indicating:
  (1) The sex genotypes of the diploid parents, colchicine-induced tetraploids, and *staminate* tetraploids in the first, second, and third generations.
  (2) Which sex is heterogametic and the dominance of the heterogametic determining allele.
  (3) Whether disjunction of the allele pair and therefore chromosomes carrying the sex-determining allele pair is random.
b Are these results typical of chromosome segregation?   If so, would the chromosomes involved necessarily pair to form quadrivalents?   If so, what conclusions re disjunction and gene-centromere distance would you make?

873   In the tomato (*Lycopersicon* [= *Lycopersicum*] *esculentum*) color of the fruit flesh is controlled by a gene close to the centromere; and allele *R* for *red* flesh is dominant to *r* for *yellow* flesh.   Both diploid and autotetraploid plants have been self-fertilized or testcrossed.   The table gives examples of the results obtained (modified data from Lindstrom, 1932).

Table 873

Parents	Progeny	
	Red	*Yellow*
red × *red*	362	11
red × *red*	98	34
red × *yellow*	158	32
red × *yellow*	134	141

a Determine the most probable theoretical phenotypic ratio for each cross and on that basis state whether the parents of a particular cross were both diploid or both tetraploid or could be either.   Give reasons for your answers.

b Where it is impossible to decide on the basis of phenotypic ratios alone whether the parents were diploid or tetraploid, what other criteria would help you decide the type of parents involved?

874 In *Datura stramonium* ($2n = 24$), a true-breeding strain, $p^1p^1 s^1s^1$, with *deep-purple* flowers and *spiny* capsules is crossed with a true-breeding strain, $p^2p^2 s^2s^2$, with *white* flowers and *smooth* capsules. Among 1,000 F_1 plants, 6 are found that are *larger* than the others and *highly sterile*. These *large, highly sterile* plants with *blue* flowers and *spiny* capsules are crossed to the diploid parent, $p^2p^2 s^2s^2$, with *white* flowers and *smooth* capsules. The progeny consists of

Spiny, deep purple	26
Spiny, blue	100
Spiny, white	24
Smooth, deep purple	5
Smooth, blue	19
Smooth, white	6

Explain these results, indicating:
a Why the 6 F_1 plants and their progeny are *large* and *highly sterile*.
b Why 6 testcross phenotypes are obtained in the proportions shown. Include in your explanation the F_1 gametic ratios and the cytological (chromosomal) behavior at meiosis that produces such types and proportions of gametes.
c Are the genes for flower color and capsule type on the same or different chromosomes? Explain.
d Are the alleles p^1 (*deep purple*) and p^2 (*white*) fully morphic, hypomorphs, or amorphs; why? Answer the same question for s^1 and s^2.

875 The mealybug *Planococcus citri* has chromosomes that are extremely small, so that it is impossible to distinguish the longitudinal from the broad axis of any chromosome.
a Outline an experiment to determine whether the sequence of meiotic divisions is *standard* (first reductional, second equational) or *inverse* (first equational, second reductional).
b After outlining your experiment to distinguish between the two alternatives, read the paper by Chandra (1962) which reports on an elegant experiment bearing on this question.

876 Heterozygous autotetraploid Chinese primrose plants, when self-fertilized or testcrossed, produced the results shown (Somme, 1930). If the chromosomes carrying these genes segregate at random, explain these results by giving the genotypes of the autotetraploids, the gametic types they produce and their frequencies, and the approximate distance of each gene locus from its centromere.

Table 876

Parents	Progeny
Palm vs. *fern leaf*	
palm (self-fertilized)	293 *palm* : 74 *extra lobes* : 10 *fern*
palm × *fern*	27 *palm* : 49 *extra lobes* : 18 *fern*
extra lobes (self-fertilized)	92 *palm* : 149 *extra lobes* : 76 *fern*
Normal vs. *primrose queen eye*	
normal × *primrose queen*	88 *normal* : 17 *primrose queen*
normal (self-fertilized)	191 *normal* : 8 *primrose queen*
normal × *primrose queen*	48 *normal* : 40 *primrose queen*
Green vs. *red stigma*	
green × *red*	1,175 *green* : 223 *red*
green (self-fertilized)	1,659 *green* : 52 *red*
green × *red*	240 *green* : 212 *red*

877   By 1931 preliminary data had suggested that *Dahlia variabilis* was a hybrid octaploid, also called a double autotetraploid ($2n = 8x = 64$), derived from a sterile tetraploid hybrid ($2n = 4x = 32$) between two tetraploid species by chromosome doubling. *Ivory* vs. *white* flower color in this species is controlled by the allele pair *Ii*, and *yellow* vs. *nonyellow* flower color is due to the allelic genes *Yy*. *Ivory* and *white* are completely masked by *yellow*. Lawrence (1931) obtained the tabulated results from a series of crosses.   Show

Table 877

Parents	Progeny		
	Yellow	*Ivory*	*White*
yellow × *ivory*	248	43	17
white × *yellow*	28	12	15
white × *yellow*	20	11	6
yellow × *ivory*	34	34	0
yellow × *yellow*	42	5	8
white × *yellow*	69	0	0
ivory × *white*	0	57	23

whether these results support the conclusion from the preliminary work.

878   In *Helianthus tuberosus*, a sunflower with bivalent chromosome pairing only, true-breeding *tall, purple*-flowered and *dwarf, white*-flowered lines are intercrossed.   The F_1's (*tall* and *blue*-flowered) crossed with the *dwarf, white* line produced the following progeny:

Tall, purple	32
Tall, blue	120
Tall, white	31
Dwarf, purple	11
Dwarf, blue	40
Dwarf, white	9

a Explain these results, indicating:

(1) The number of allele pairs controlling each pair of traits.

(2) Why the phenotypes were obtained in the proportions indicated (show the cytological basis for the F_1 gametic ratios).

(3) Whether you would classify the species as a diploid or polyploid; if the latter, specify the type.

b If some of the chromosomes formed quadrivalents, would the gene or genes for height or flower color be carried on such chromosomes?   Explain.

c (1) Indicate the direction of activity and types of alleles at each of the loci involved.

(2) Give a biochemical explanation of the action of the alleles at the locus (loci) specifying flower color.

879 Crane and Darlington (1932) obtained a fertile tetraploid RT4 ($2n = 4x = 28$) from a cross between the diploid species *Rubus rusticanus* var. *inermis* ($2n = 2x = 14$) and *R. thyrsiger*, a tetraploid ($2n = 4x = 28$).   The variety *inermis* is *spineless*, in contrast to other varieties of the species *rusticanus*, which are *furrow-spined* (the spines are confined to the furrows of the stem). *R. thyrsiger* has spines covering the entire stem.   RT4 was *furrow-spined*. When self-fertilized, it produced 835 *furrow-spined* and 37 *spineless* plants. When backcrossed to the variety *inermis*, it produced 33 *furrow-spined* and 10 *spineless* progeny.   With reasons for your answers, state whether:

a RT4 contained two genomes from *rusticanus*.

b RT4 was heterozygous for alleles at two loci.

c If two genes are involved:

(1) The chromosomes carrying both loci pair autosyndetically.

(2) At least one locus shows chromatid segregation.

880 In the Chinese primrose (*Primula sinensis*) the allele pair *Dd* specifies anthocyanin formation.   The allele *D* for *full* color is not completely dominant to *d* for *white*, giving a different phenotype for each tetraploid genotype (*DDDD*: *full color*; *DDDd, DDdd, Dddd*: intermediate and distinguishable; and *dddd*: *white*).   Determine the phenotypic ratio for each of the following matings if the gene is close to the centromere and homologs carrying it form quadrivalents only:

DDdd × *DDdd*

Dddd × *Dddd*

DDdd × *dddd*

881   The leaves of the common wheat variety H-44-24 are *pubescent*, whereas those of Marquis are *glabrous*. Neatby and Goulden (1930) found that of 270 F_3 lines obtained by self-fertilization of F_2's from crosses between these two varieties, 21 bred true for *glabrous*, 166 segregated for *pubescent* vs. *glabrous*, and 83 bred true for *pubescent*. How many allele pairs are involved, and how do they interact?

882   In the allopolyploid *Nicotiana digluta*, chromosome pairing is exclusively autosyndetic. Two pairs of alleles, Gg and $G_1 g_1$, specify the expression of *green* vs. *albino* plants. What phenotypic ratio would you expect in the F_2 of a dihybrid? *Note:* Only $gg\, g_1 g_1$ plants are *albino*.

883   In irises, the *plicata* pattern (as opposed to the *nonplicata* pattern) and *self color* (as opposed to *bicolor*) are controlled by recessive alleles at different loci (Sturtevant and Randolph, 1945). Assume that the two loci are on different chromosomes and that you have two tetraploid lines, one true-breeding for *plicata*, the other for *self color*. In the F_1 of the cross between the two lines 1 plant in 958 is *plicata, bicolor*; the others are *nonplicata, bicolor*. Outline how you would proceed to establish a line true-breeding for both recessive traits.

884   In alfalfa (*Medicago sativa*) certain varieties are *susceptible* to the root-knot nematode (*Meloidogyne haple*), a soil-infesting eelworm that causes severe damage to alfalfa in mild climates. Among a number of *resistant* plants from commercial fields in California Goplen found one plant (M-9) to be immune to all races of the organism. This plant was cloned, and a number of plants were self-fertilized. The S_1 (self-fertilized) generation consisted of 319 *resistant* and 119 *susceptible* plants. When M-9 was crossed with a *susceptible* clone, 46 *resistant* and 55 *susceptible* plants were obtained; 36 S_1 plants were self-fertilized. Among the 36 S_2 progenies the segregation patterns shown were obtained (data from Allard, 1960). Give a complete cytogenetic explanation of these results.

Table 884

Type of S_2 family	Number observed
Nonsegregating *resistant*	1
Segregating 21:1 to 35:1	3
Segregating approximately 3:1	20
Nonsegregating *susceptible*	12

885   In sweet peas two different true-breeding diploid *white*-flowered strains when crossed produced *purple*-flowered F_1's. Two of these F_1's when self-fertilized produced a total of 98 *purple*- and 72 *white*-flowered plants. Other F_1's of this cross had their chromosome number doubled. State the gametic genotypes

you expect to be produced by these tetraploids and their expected proportions if the gene loci are near the centromeres.

886   In 1928 Karpechenko crossed the cabbage (*Brassica oleracea*), a diploid ($2n = 18$), with the radish (*Raphanus sativus*), a diploid ($2n = 18$). The semi-sterile hybrid, with 18 univalents at meiosis, combined the worst economic features of both genera (it had the foliage characteristic of the radish and the tough root of the cabbage). It produced a few seeds from which vigorous, fertile plants looking like their parents and having 18 bivalents at metaphase I were obtained.

   a With the aid of appropriate illustrations, explain how these second-generation plants with 18 pairs of chromosomes were probably produced.

   b Explain (1) why the first-generation plants were sterile and (2) why the second-generation plants were fertile.

   c Can the second-generation plants be said to belong to either the species *B. oleracea* or the species *R. sativus*? Give reasons for your answer.

   d Would you expect the second-generation plants to breed true for all traits? Explain.

887   Chromosomes at metaphase I–anaphase I in a hybrid between two phenotypically different plants (one with a known somatic chromosome number of 4) behave as illustrated.

Figure 887

   a Are the parents of the hybrid likely to be members of the same species or not? Why?

   b What proportion of the meiotic products (e.g., microspores) of the hybrid would you expect to have the same chromosome number as its somatic cells?

   c What is the somatic chromosome number of the second plant?

888   The fertile allopolyploid *Nicotiana tabacum* probably originated from a highly sterile hybrid between *N. tomentosa* and *N. sylvestris*. If *N. tabacum* has 48 chromosomes in the nuclei of its somatic cells and *N. sylvestris* has a haploid number of 12 chromosomes:

   a What is the haploid chromosome number of *N. tomentosa*? Explain how you arrive at your answer.

   b Are bivalents expected at meiosis in *N. tabacum*? If so, how many and why?

889 Diploid species A and B are both true-breeding for *white* flowers. The F_1 between these two species is sterile and has *purple* flowers. Cytological examination shows that there is no chromosome pairing in the hybrids. Hybrids whose chromosome number has been doubled are fertile. State the types of offspring expected, with regard to color, if these doubled hybrids are self-fertilized, giving the expected ratio if possible. Assume that a dominant allele at each locus is required for *purple* flowers.

890 In white clover, either of the nonallelic dominant genes W_1 and W_2 on different (nonhomologous) chromosomes causes *white* flower color. Only the genotype $w_1w_1 w_2w_2$ causes *red* flowers. A *red*-flowered tetraploid is crossed with a *white*-flowered tetraploid of genotype $W_1 W_1 W_1 W_1 W_2 W_2 W_2 W_2$, and the F_1's are intercrossed. What are the expected proportions of *white* to *red* flowered plants in the F_2 if segregation of alleles of both genes is two by two and chromosomal (not chromatid)?

891 In the common (*vulgare*) wheat the F_2 segregation ratios of 63:1 and 15:1 occur frequently, but in barley 63:1 ratios have not been recorded and 15:1 are rare. Account for these differences.

892 Muntzing (1937) crossed four true-breeding *pubescent* lines of *Galeopsis tetrahit*, one of the hemp nettles, with one true-breeding for *glabrous* stems. All F_1 plants had *pubescent* stems. Of the F_2's, 1,404 were *pubescent* and 105 were *glabrous*.

  a Show whether, from these results only, the species should be classified as a diploid, an allopolyploid, or an autopolyploid.

  b The species *G. speciosa* and *G. pubescens* are also available. How would you proceed to determine whether the conclusion above is correct?

893 *Fragaria bracteata* and *F. helleri* are closely related diploid strawberry species with 14 chromosomes. *F. bracteata* has *white* and *F. helleri pink* flowers. An

Table 893

Mating		Offspring	
F_2 plant	Fertilized by	*Pink*-flowered	*White*-flowered
1	*F. bracteata*	17	17
2	*F. bracteata*	23	29
3	*F. bracteata*	15	4
4	*F. bracteata*	9	8
5	*F. bracteata*	16	15
6	*F. bracteata*	12	3
7	*F. bracteata*	42	0
2	Self	30	7
5	Self	13	3
6	Self	33	1
7	Self	8	0

F_1 plant with 14 bivalents was obtained by Yarnell (1931) from a cross between these two species. Upon self-fertilization it produced 7 F_2 plants, all with *pink* flowers. These were backcrossed to the *F. bracteata* and also self-fertilized. The segregations for flower color in the offspring are shown.

a When and how did the tetraploid F_1 plant arise?

b Analyze these data, to determine:

   (1) The tetraploid genotypes of the F_1 and the 7 F_2 plants.
   (2) The mode of pairing and segregation of the two pairs of chromosomes which carried the *Pp* (*pink* vs. *white*) allele pair.

c The F_1 plant showed 14 bivalents. Is this type of pairing expected from your genetic analysis? Explain what is expected and the segregation ratio that should follow.

894   A tetraploid variety of *Dahlia variabilis*, true-breeding for *yellow* flowers, was crossed with an F_1 plant obtained from crossing this variety with one true-breeding for *white* flowers. Lawrence (1931) crossed the resulting *yellow*-flowered hybrid with the true-breeding *white*-flowered variety; 378 of the progeny had *yellow* and 7 had *white* flowers. Are the *white*-flowered plants expected in such a cross? If so, outline the meiotic and breeding events that must occur for them to appear.

895   The Amazon molly fish (*Poecilia formosa*) is an all-female species which normally reproduces by gynogenesis after mating with males of related species such as *P. sphenops* ($2n = 46$). Rasch et al. (1965) estimated the amount of DNA per nucleus by cytophotometry in Feulgen-stained tissue sections in *P. formosa*, *P. sphenops*, *P. vittata*, and hybrids between *P. formosa* and the latter two species. The average amounts of DNA per nucleus in hepatocytes, closely comparable with those in other tissues, are shown. Account for the

Table 895

		DNA, Feulgen per nucleus	
Organism	Nuclear class	Mean	Standard error
P. formosa	$2n$	0.57	0.01
P. sphenops	$2n$	0.56	0.01
P. vittata	$2n$	0.69	0.01
P. formosa × *P. sphenops*	$2n$	0.83	0.01
P. formosa × *P. vittata*	$2n$	0.92	0.01

higher DNA content of the hybrids and explain how you would proceed to verify your explanation.

896   Muntzing (1936) states: "In short, the presence of multivalents indicates autopolyploidy, the absence of multivalents allopolyploidy." Discuss the following questions, illustrating your arguments with specific examples as far as possible:

a Is the chromosome pairing a sufficient basis on which to distinguish between different types of polyploids?

b If not, what other criteria should be used for establishing the nature of polyploids?

897 Autopolyploid plants are often larger and more vigorous and produce more seed and/or green matter than their diploid parents. If this is so, why do breeders not simply double the chromosome number of existing desirable varieties?

898 An F_1 hybrid between plant species A ($n = 5$) and B ($n = 7$) produced only a few pollen grains. When they were used to fertilize species B, a few plants were produced, all with 19 chromosomes. They were highly sterile but following self-fertilization produced a few plants with 24 chromosomes. These plants were different in phenotype from the original parents, and the progeny were fertile. What events were probably involved in the origin of these 24-chromosome plants?

899 With the purpose of more precisely defining interspecific relationships in the mantid genus *Liturgousa*, Hughes-Schrader (1953) carried out experiments that gave the tabulated results when she compared *L. maya* and *L. cursor*.

Table 899

Species	Male diploid chromosome complement	Relative total length of chromosomes	Relative amount of DNA per nucleus
L. maya	17 (XO)	1.0	1.00
L. cursor	33 (XO)	0.9	0.94

a Of the alternatives polyploidy, polyteny, and Robertsonian equivalence (centric fusion), which one is definitely ruled out as a possible cause of the relationship between the two species? Why?

b If you found genes that are in separate linkage groups in *L. cursor* to be in the same linkage group in *L. maya*, would this definitely settle the cause of the difference in chromosome number between the two species?

c If no structural rearrangements occurred after the initial changes, which of the two kinds of chromosome, acrocentric or metacentric, would you expect to find in each of these species? Explain.

900 Comparisons of chromosome number, chromosome size, and relative DNA content in 16 species representing 10 genera of the hemipteran tribe Pentatomini, which have diffuse centromeres, were made by Hughes-Schrader and Schrader (1956) to help clarify the taxonomic relationships of the organisms involved. The analysis revealed that these species could be placed in three groups, of which the species listed are representative. In view of the non-localized nature of the centromere, to what cause can we attribute (1) the

Table 900

Species	Diploid chromosome number and constitution of male	DNA value per nucleus
Loxa flavicollis	14 (12 + XY)	2.27
Thyanta perditor	14 (12 + XY)	1.10
T. calceata	27 (24 + XXY)	0.93

difference in DNA content between the two 14-chromosome species and (2) the difference in chromosome number between the two *Thyanta* species? Explain why you have eliminated alternative mechanisms and illustrate the one suggested.

REFERENCES

Allard, R. W. (1960), "Principles of Plant Breeding," Wiley, New York.
Anderson, W. S. (1939), *J. Hered.*, **30**:549.
Astaurov, B. L. (1969), *Annu. Rev. Genet.*, **3**:99.
Avery, A. G. (1959), in A. G. Avery, S. Satina, and J. Rietsma (eds.), "Blakeslee: The Genus *Datura*," pp. 71–85, Ronald, New York.
Babcock, E. B., and M. Navashin (1930), *Bibliog. Genet.*, **6**:1.
Bailey, G. S., G. T. Cock, and A. C. Wilson (1969), *Biophys. Res. Commun.*, **34**:605.
Beasley, J. O. (1940), *Am. Nat.*, **74**:285.
——— (1942), *Genetics*, **27**:25.
Beatty, R. A. (1957), "Parthenogenesis and Polyploidy in Mammalian Development," James Thin, Edinburgh.
——— and M. Fischberg (1952), *J. Genet.*, **50**(3):471.
Beçak, M. L., and W. Beçak (1970), *Chromosoma*, **31**:377.
———, ———, and M. N. Rabello (1966), *Chromosoma*, **19**:188.
———, ———, and ——— (1967), *Chromosoma*, **22**:192.
——— L. Denaro, and W. Beçak (1970), *Cytogenetics*, **9**:225.
Beçak, W., M. L. Beçak, D. Lavalle, and G. Schreiber (1967), *Chromosoma*, **23**:14.
Belling, J. (1921), *Proc. Natl. Acad. Sci.*, **7**:197.
——— and A. F. Blakeslee (1922), *Am. Nat.*, **56**:339.
——— and ——— (1923), *Proc. Natl. Acad. Sci.*, **9**:106.
——— and ——— (1924), *Am. Nat.*, **58**:60.
Bendich, A. J., and B. J. McCarthy (1970), *Genetics*, **65**:545.
Bernard, R., et al. (1967), *Ann. Genet.*, **10**:70.
Blakeslee, A. F., J. Belling, and M. E. Farnham (1923), *Bot. Gaz.*, **76**:329.
Brewbaker, J. L. (1964), "Agricultural Genetics," Prentice-Hall, Englewood Cliffs, N.J.
Bungenberg de John, C. M. (1957), *Bibliog. Genet.*, **27**:111.
Burnham, C. R. (1962), "Discussions in Cytogenetics," Burgess, Minneapolis.
Carr, D. H. (1967), *Am. J. Obstet. Gynecol.*, **97**:283.
——— (1971), *J. Med. Genet.*, **8**:164.
Catcheside, D. G. (1959), *Heredity*, **13**:403.
Chandra, H. S. (1962), *Genetics*, **47**:1441.
Clausen, R. E., and T. H. Goodspeed (1925), *Genetics*, **10**:278.
——— D. D. Keck, and W. M. Hiesey (1945), *Carnegie Inst. Wash.*, *D.C.*, *Publ.* 564.
Comings, D. E., and T. A. Okada (1971), *Nature*, **231**:119.

Crane, M. B. (1940), *J. Genet.*, **40**:129.
—— and C. D. Darlington (1932), *Nature*, **129**:869.
Darlington, C. D. (1931), *J. Genet.*, **24**:65.
—— (1937), "Recent Advances in Cytology," 2d ed., Blakiston, Philadelphia.
Dawson, C. D. R. (1941), *J. Genet.*, **42**:49.
Dawson, G. W. P. (1962), "An Introduction to the Cytogenetics of Polyploids," Blackwell, Oxford.
De Winton, D., and J. B. S. Haldane (1931), *J. Genet.*, **24**:121.
Digby. L. (1912), *Ann. Bot.*, **26**:357.
Douglas, C. R., and M. S. Brown (1971), *Am. J. Bot.*, **58**:65.
Edwards, J. H., et al. (1967), *Cytogenetics*, **6**:81.
Fankhauser, G. (1942), *Biol. Symp.*, **6**:21.
Fisher, R. G., and K. Mather (1943), *Ann. Eugen.*, **12**:1.
Ford, C. E., J. L. Hamerton, and G. B. Sharman (1957), *Nature*, **180**:392.
Frost, J. N. (1961), *Genetics*, **46**:373.
Golpen, E., and E. H. Stanford (1960), *Agron. J.*, **52**:337.
Goodspeed, T. H. (1945), *Bot. Rev.*, **11**:533.
Greenleaf, W. (1941), *Genetics*, **26**:301.
Gregory, R. P. (1914), *Proc. R. Soc. (Lond.)*, **B87**:484.
Haldane, J. B. S. (1930), *J. Genet.*, **22**:359.
Hamerton, J. L. (1971), "Human Cytogenetics," vols. I and II, Academic, New York.
—— et al. (1963), *Cytogenetics*, **2**:240.
Hughes-Schrader, S. (1953), *Chromosoma*, **5**:544.
—— (1958), *Proc. 10th Int. Congr. Entomol.*, **2**:935.
—— and F. Schrader (1956), *Chromosoma*, **8**:135.
Huskins, C. L. (1930), *Genetica*, **12**:531.
Irwin, M. R. (1971), *Genetics*, **68**:509.
Janick, J. (1955), *Am. Soc. Hortic. Sci.*, **66**:361.
Jones, K. (1964), *Chromosoma*, **15**:248.
Jorgensen, C. A. (1928), *J. Genet.*, **19**:133.
Karpechenko, G. D. (1928), *Z. Indukt. Abstamm.-Vererbungsl.*, **48**:1.
—— (1929), *Z. Indukt. Abstamm.-Vererbungsl.*, **52**:287.
Kawamura, T. (1951), *J. Sci. Hiroshima Univ.*, (B1)**12**:11.
Kihara, H. (1951), *Proc. Am. Soc. Hortic. Sci.*, **58**:217.
Kostoff, D. (1939), *Genetica*, **21**:285.
Lawrence, W. J. C. (1929), *J. Genet.*, **21**:125.
—— (1931), *J. Genet.*, **24**:257.
Lesley, M. M., and J. W. Lesley (1930), *J. Genet.*, **22**:419.
Leupold, U. (1956), *J. Genet.*, **54**:411.
Lewis, K. R., and B. John (1963), "Chromosome Marker," Churchill, London.
Lindstrom, E. W. (1932), *J. Hered.*, **23**:115.
—— (1936), *Bot. Rev.*, **2**:197.
Little, T. M. (1944), *Bot. Rev.*, **10**:60.
Lovis, J. D. (1968), *Nature*, **217**:1163.
Lutz, A. M. (1907), *Science*, **26**:151.
McClintock, B. (1929), *Genetics*, **14**:180.
McFadden, E. S., and E. R. Sears (1946), *J. Hered.*, **37**:81.
Maslin, T. P. (1962), *Science*, **135**:212.
Mather, K. (1935), *J. Genet.*, **30**:53.
—— (1936), *J. Genet.*, **32**:287.
Matthey, R. (1952), *Chromosoma*, **5**:113.
Moses, M. J., and G. Yerganian (1952), *Genetics*, **37**:607.
Mott, C. L., L. H. Lockhart, and R. H. Rigdon (1968), *Cytogenetics*, **7**:403.

Muller, H. J. (1914), *Am. Nat.*, **48**:508.
—— (1925), *Am. Nat.*, **59**:346.
Muntzing, A. (1932), *Hereditas*, **16**:105.
—— (1936), *Hereditas*, **21**:263.
—— (1937), *Hereditas*, **23**:371.
Neatby, K. W., and C. H. Goulden (1930), *Sci. Agric.*, **10**:389.
Newton, W. C. F., and C. Pellew (1929), *J. Genet.*, **20**:405.
Ohno, S. (1970), "Evolution by Gene Duplication," Springer-Verlag, New York.
—— W. A. Kittrell, L. C. Christian, C. Stenius, and G. A. Witt (1963), *Cytogenetics*, **2**:42.
—— U. Wolf, and N. B. Atkin, (1968), *Hereditas*, **59**:169.
Patterson, J. T., and W. S. Stone (1952), "Evolution in the Genus *Drosophila*," Macmillan, New York.
Pennock, L. A. (1965), *Science*, **149**:539.
Randolph, L. F. (1935), *J. Agric. Res.*, **50**:591.
Rasch, E. M., R. M. Darnell, K. D. Kallman, and P. Abranoff (1965), *J. Exp. Zool.*, **160**:155.
—— L. M. Prehn, and R. W. Rasch (1970), *Chromosoma*, **31**:18.
Rees, H. (1964), *Chromosoma*, **15**:275.
Rhoades, M. M. (1936), *J. Genet.*, **33**:355.
—— (1942), *Genetics*, **27**:395.
—— and E. Dempsey (1966), *Genetics*, **54**:505.
Riley, R. (1960), *Heredity*, **15**:407.
—— G. Kimber, and V. Chapman (1961), *J. Hered.*, **52**:22.
—— and C. N. Law (1965), *Adv. Genet.*, **13**:57.
Sansome, F. W. (1933), *J. Genet.*, **27**:105.
—— and J. Philip (1939), "Recent Advances in Plant Genetics," Churchill, London.
Satina, S., and A. F. Blakeslee (1937a), *Am. J. Bot.*, **24**:518.
—— and —— (1937b), *Am. J. Bot.*, **24**:621.
Schindler, A-M., and K. Mikamo (1970), *Cytogenetics*, **9**:116.
Schrader, F., and S. Hughes-Schrader (1956), *Chromosoma*, **7**:469.
Sears, E. R. (1944), *Genetics*, **29**:113.
Smith-White, S. (1948), *Heredity*, **2**:119.
Somme, A. S. (1930), *J. Genet.*, **23**:447.
Stadler, L. J. (1929), *Proc. Natl. Acad. Sci.*, **15**:876.
Stebbins, G. L. (1947), *Adv. Genet.*, **1**:403.
—— (1950), "Variation and Evolution in Plants," Columbia, New York.
—— (1951), *Sci. Am.*, **184**(April):54.
—— (1966), *Science*, **152**:1463.
Stephens, S. G. (1947), *Adv. Genet.*, **1**:431.
Sturtevant, A. H., and L. F. Randolph (1945), *Am. Iris Soc. Bull.* 99, p. 52.
Tahara, M. (1915), *Bot. Mag. Tokyo*, **29**:48.
Trujillo, J. M., C. Stenius, L. C. Christian, and S. Ohno (1962), *Chromosoma*, **13**:243.
—— B. Walden, P. O'Neil, and H. B. Anstall (1965), *Science*, **148**:1603.
Upcott, M. (1935), *Genetics*, **31**:1.
Uzzell, T. M. (1963), *Science*, **139**:113.
Volpe, E. P. (1970), *Cytogenetics*, **9**:161.
Wettstein, F. V. (1924), *Z. Indukt. Abstamm.-Vererbungsl.*, **33**:1.
White, M. J. D. (1954), "Animal Cytology and Evolution," 2d ed., Cambridge University Press, Cambridge.
White, M. J. F. (1957), *Aust. J. Zool.*, **5**:338.
Whitt, G. S., W. F. Childers, and T. E. Wheat (1971), *Biochem. Genet.*, **5**:257.
Winge, O. (1917), *C. R. Trav. Lab. Carlsberg*, **13**:131.
Wolf, U., H. Ritter, N. B. Atkin, and S. Ohno (1969), *Humangenetik*, **7**:240.
Yarnell, S. H. (1931), *Genetics*, **16**:455.
Yosida, T. H., et al. (1971), *Chromosoma*, **34**:40.

21
Aneuploidy

21
Aneuploidy

QUESTIONS

901 A single chromosome addition to the chromosome complement of a diploid organism usually produces a much more pronounced phenotypic effect than it does when added to the chromosome complement of an individual in a tetraploid species. Why?

902 Arrange the following in the order of the greatest to the least genic unbalance which they produce; assume the same chromosome is involved in each case:
a Monosomic   b Tetrasomic   c Nullisomic
d Primary trisomic   e Secondary trisomic

903 In man, since several inherited pathological syndromes such as *multiple congenital anomalies* and *Down's syndrome* have been shown to be caused by trisomy, it has been suggested that a further search for such conditions should be made among traits that are dominant and rare. Explain why this is a good suggestion, referring to the frequency of occurrence and transmission characteristics of trisomics.

904 The average age of marriage differs among different social groups. What effect might these differences have on the incidence of *mongolism* and of other trisomic types in these different groups?

905 Theoretically, sex-chromosome aneuploidy can result from nondisjunction at either or both stages of meiosis in either parent or from nondisjunction in the early mitotic divisions of the zygote. It may also result from chromosome loss between fertilization and the first cleavage division (Stewart, 1962; Russell, 1961; Stern, 1960). Illustrate these mechanisms in both sexes and male and female zygotes. Show the types of progeny expected in matings between *normals* and those in which nondisjunction occurs.

906 What can be concluded about the genetic and chromosomal constitution of the parents and the type of inheritance in each of the following.
a They produce three types of F_2 offspring in a 12:3:1 ratio.
b They produce two types of progeny in a 2:1 ratio.
c They produce one type of progeny only.
d They produce two types of progeny in a 5:1 ratio.
e They produce two types of progeny in a 35:1 ratio.

907 Illustrate the following meiotic events in a primary trisomic $(2n + 1)$ plant with five chromosomes (1) where the chromosome in triplicate is long and (2) where it is short:
a Pairing at zygotene and pachytene
b Configurations at diakinesis and metaphase I
c Segregation at anaphase I
d The chromosome constitution of secondary meiocytes
e Metaphase II and anaphase II
f Chromosome constitution of the meiotic products (megaspores and microspores). Show the proportions of the various types of spores.

908   How might you distinguish between a primary and secondary trisomic cytologically? See Belling and Blakeslee (1924, 1926).

909   What is the significance of aneuploidy (1) in studies of evolution and (2) in taxonomic studies?

910   Of what benefit would the discovery of further kinds of trisomics in man be to (1) human geneticists and (2) medical practitioners?

911   a In common wheat (*Triticum aestivum*; $2n = 6x = 42$) the transmission frequencies of n and $n - 1$ gametes in monosomics are, on the average, as follows: 25 percent n and 75 percent $n - 1$ through the female and 96 percent n and 4 percent $n - 1$ through the male (Sears, 1953).
   (1) Show the types of progeny expected from self-fertilization of a monosomic and their frequencies.
   (2) What are some possible reasons for the differences in transmission frequencies of n and $n - 1$ gametes in the two sexes? (See Sears, 1953.)
   b The transmission of $n + 1$ gametes in trisomic plants rarely reaches the expected 50 percent on the female side and is much lower (0 to 3 percent) on the male side. What are some of the possible reasons for the low transmission on the male side? For variations in transmission on the female side? Include in your answer events that occur at meiosis as well as postmeiotic factors. (See Einset, 1943; Buchholz and Blakeslee, 1932; Goodspeed and Avery, 1939.)

PROBLEMS

912   *Nystagmus* in man is an X-linked recessive trait causing involuntary eye movement associated with impaired vision. Two males with *Klinefelter's syndrome* (XXY), both of *normal* parents, have *nystagmus*. One has it in both eyes; the other in only one eye. Trace the probable course of events in each case, showing how they differ.

913   a In man, *normal* vs. *incontinentia pigmenti* (see Prob. 427) are due to an X-linked pair of alleles: *Nn* are *affected*, *nn* and *n/Y* are *normal*, and *N/Y*, abort. An *affected* woman and a *normal* man have an *affected* son. With the aid of diagrams show how such a male can arise and give the son's sex-chromosome karyotype.
   b *Normal* color vision vs. *color blindness* in man is controlled by the X-linked allelic genes *C* and *c*. A *color-blind* woman and a *normal* man have a *color-blind* daughter, contrary to the usual expectation. Explain how such a female can arise and show by chromosome constitutions of parents and daughter whether the abnormality occurred in the father or mother.

914   In a plant which has seven chromosomes in its somatic-cell nuclei, all meiocytes have three bivalents and one univalent at metaphase I. Is the plant monosomic or trisomic? Explain.

915   An individual with *Turner's syndrome* (AAXO) and *nystagmus* (X-linked recessive represented by solid symbols) is found in the third generation of a family segregating for the recessive trait.   Is it the paternal or maternal X chromosome that is missing in the female with *Turner's syndrome*?

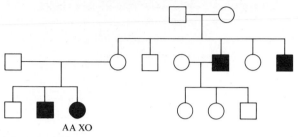

AA XO

Figure 915

916   In the mouse, *scurfy*, *sf*, is a recessive X-linked trait that causes death of both sexes before reproductive age.   Russell et al. (1959) found that on rare occasions heterozygous *normal*, *Sfsf*, females, when crossed with *normal*, *Sf/Y*, males produced some *scurfy* daughters.   Although *scurfy* females cannot reproduce, Russell and coworkers were able to obtain genetic progeny from such females by transplanting their ovaries to females of a *normal* line. The ovary-recipient females were subsequently mated with *normal* males.   The pooled results of six matings are shown.   Possible explanations for the origin of *scurfy* females which gave rise to the above results are:

1   High rate of mutation of *Sf* to *sf*.
2   Sex-reversal.   *Scurfy* females are of genotype X^{sf}/Y, converted to a female phenotype through the action of other factors.
3   *Scurfy* females are homozygous *sfsf* caused by nondisjunction in *Sfsf* mothers and are $X^{sf}X^{sf}Y$.
4   *Scurfy* females are genetically hemizygous $X^{sf}X^0$ (where 0 denotes a spontaneous deletion of a portion of the X^{sf} chromosome in heterozygous ($X^{sf}X^{sf}$) females).
5   Scurfy females are monosomic $X^{sf}/0$ due to nondisjunction in the male.

Table 916

Sons		Daughters	
Scurfy	*Normal*	*Normal,* transmit *scurfy*	*Normal,* do not transmit *scurfy*
19	0	16	10

a The data from ovarian-transplant offspring disprove three of these hypotheses.   Which are they and why?

b Which remaining explanation do you favor and why?

c What genetic and other tests would you make to confirm or reject your explanation?

917   In the mouse (X-Y sex-determining mechanism) the allele pair *Tata* for *tabby* vs. *wild-type* coat is sex-linked.   *Ta* is an incompletely dominant nonlethal gene; *TaTa* or *Ta*/Y show *tabby* (a patterned coat); *tata* or *ta*/Y show *wild type*; and *Tata* have a *heterozygous tabby* coat, a mosaic of *tabby*, *wild-type*, and *intermediate* areas.

a The data in Table 917*A* represent the results of experiments by Welshons and Russell (1959).

Table 917*A*

Parents		Female progeny			
♀	♂	Expected	Obtained	Exceptional	Obtained
1. $X^{ta}X^{ta} \times X^{Ta}/Y$		*heterozygous tabby*	152 in 154	*wild type*	2 in 154
2. $X^{Ta}X^{ta} \times X^{ta}/Y$		*heterozygous tabby* and *wild type*	501 in 503	*tabby*	2 in 503
3. $X^{ta}X^{Ta} \times X^{ta}/Y$		*heterozygous tabby* and *wild type*	123 in 125	*tabby*	2 in 125

Previous studies ruled out mutation and sex reversal as causes of exceptional progeny.   Outline two hypotheses to explain the occurrence of the exceptional females and give the genotypes of these females according to each hypothesis.   (The roles of the X and Y in sex determination were not known at the time this work was done.)

b This paper also presents the results of genetic tests (Table 917*B*) designed to distinguish between the alternative hypotheses in (a).

Table 917*B*

Cross		Offspring	
♀	♂	♀	♂
tabby (cross 2) × *wild-type, ta*/Y		2 *heterozygous tabby*	0
tabby (cross 3) × *wild-type, ta*/Y		1 *heterozygous tabby*; 1 *wild type*	1 *tabby*
tabby (cross 2) × *wild-type, ta*/Y		7 *heterozygous tabby*; 7 *wild type*	12 *tabby*
wild type (cross 1) × *tabby, Ta*/Y		7 *heterozygous tabby*; 3 *tabby*.	5 *wild type*

(1) Show which hypothesis these results support and why.

(2) Illustrate the meiotic chromosomal mechanism in the appropriate sex that most suitably explains these results and give the chromosome numbers of normal males and females and exceptional females.

c Is it possible on the basis of the above genetic data and the predicted chromosome constitutions of normal males and females and exceptional females to indicate the roles of the autosomes and X and Y chromosomes in sex determination?   If so, what conclusions can you draw?

918   *Nicotiana tabacum* is an allotetraploid (genome formula SSTT; $2n = 4x = 48$) which arose by chromosome doubling of a hybrid between the diploid species *N. sylvestris* (SS; $2n = 24$) and *N. tomentosa* (TT; $2n = 24$).   Clausen and Cameron (1944) crossed a monosomic *N. tabacum* plant with the *sylvestris* species and obtained 36-chromosome and 35-chromosome hybrids.   The latter consistently showed 12 bivalents and 11 univalents at metaphase I.

a Was this aneuploid monosomic for a chromosome in the S or the T genome?   Why?

b How many bivalents and univalents would you expect if the plant was monosomic for a chromosome in the other genome?   Illustrate.

919   The aneuploid individuals in each of the following pedigrees are probably due to meiotic nondisjunction.   For each aneuploid indicate (1) the source of the X or X's and (2) whether nondisjunction occurred during the first or second division, and why.

a The *Xgᵃ blood antigen* (solid symbol) is controlled by Xg^a, a dominant X-linked allele.

b *Glucose-6-phosphate dehydrogenase deficiency* (solid symbol) is a recessive X-linked trait.

c *Color blindness* (solid symbol) is a recessive X-linked trait.

d *Red-green color blindness* (solid symbol) is an X-linked recessive trait.

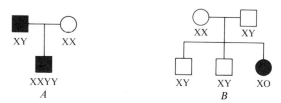

Figure 919A   (*Part A   after de la Chapelle et al., 1964; part B after Gartler et al., 1962.*)

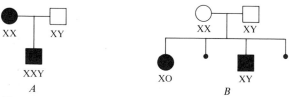

Figure 919B   (*After Ferguson-Smith et al., 1964.*)

920   The types of sex mosaics shown have recently been found in man (autosomal number in all individuals and in all tissues examined is 44).

Table 920

Chromosome number	Sex-chromosome mosaicism	Sex phenotype
45 46	XO XX	*Turner's syndrome*
46 47	XX XXY	*Klinefelter's syndrome*
45 47	XO XYY	*Amenorrhea*
46 47 48	XX XXY XXYY	*Hermaphrodite*

a Suggest a common mechanism by which these individuals may originate. Use any two of these mosaics to illustrate the mechanism.

b Explain how X-linked genes may be used to determine whether the event occurs in the male or female parent.

c Do you think that such individuals would show a sharp mosaicism for primary and secondary male and female characters, as is true in insect gynandromorphs? Explain your answer.

921   If both members of a pair of twins from normal parents suffer from multiple congenital anomalies, are they likely to be identical or fraternal? Explain.

922   In general, the consequences of trisomy and tetrasomy in animals are much more serious than in plants. For example, in man, only 4 of the possible 23 trisomics have been discovered, and of these only trisomics involving chromosome 21 (Lejeune et al., 1959a,b) and the X (and Y) chromosomes (Jacobs and Strong, 1959) are compatible with life. The others have numerous congenital abnormalities which result in death at an early age (Patau et al., 1960; Edwards et al., 1960). In contrast, complete series of trisomics have been established in diploids like the tomato (Rick and Barton, 1954; Rick et al., 1964) and *Datura* (Blakeslee, 1934), and tetrasomics are also viable. Why should plants and animals differ in the effects of extra chromosomes?

923   Trisomy for any chromosome may affect the expression of many different characters; e.g., in man, individuals trisomic for a chromosome in the 13 to 15 group produce, among others, the following congenital anomalies: cerebral defect, cleft palate, harelip, apparent anophthalmia, simian creases, "trigger thumbs," polydactyly, and heart defect (Patau et al., 1960). Why should trisomics have multiple effects on the phenotype?

924   In the medaka fish (*Oryzias latipes*) (XX-XY mechanism of sex determination) sex reversal is accomplished in either direction by feeding appropriate sex hormones to the larvae (Yamamoto, 1963). An incompletely sex-linked pair of alleles R vs. r controls *orange-red* vs. *white* skin, R being dominant to r. A heterozygous, sex-reversed female, X^rY^R, crossed with a heterozygous

X^rY^R male gave the results shown in Table 924A (modified after Yamamoto, 1963). *Note:* This fish has a primitive X-Y mechanism in which the X and Y are apparently completely homologous. This chromosome pair carries the *Rr* allele pair and the *Mm* (sex-determining) allele pair. Crossing-over may occur between the two allele pairs in both sexes.

Table 924A

Sex	Phenotypes
♂	72 *orange-red*, 2 *white*
♀	23 *white*, 3 *orange-red*

a Show which two of these four classes are expected and which are exceptional if the *Rr* locus behaves like a sex-linked gene.

b Since the locus is incompletely sex-linked, the following explanations for exceptional individuals may be given:

(1) Crossing-over in the region between *Rr* and the sex-determining allele pair (thus changing X^r and Y^R to X^R and Y^r).

(2) Mutation of *R* to *r* and *r* to *R* in either parent.

(3) Spontaneous sex reversal of some of the progeny, e.g., $X^rX^r \rightarrow$ *white* sons; $X^rY^R \rightarrow$ *orange-red* daughters.

(4) Nondisjunction.

Show diagrammatically how the exceptional individuals could arise by each of the four means listed above.

c Progeny tests of exceptional offspring gave the results shown in Table 924B.

Table 924B

Cross	Parents ♀	♂	White ♀	♂	Orange-red ♀	♂
1	*white* (X^rX^r) × exceptional *white*		20	14	0	0
2	exceptional *orange-red* × *red* (X^rY^R)		6	0	7	11

From this information, determine which of the above explanations can be eliminated and show which represents the most satisfactory explanation.

925 You have 12 trisomic strains of *Datura*, one for each of the 12 chromosomes ($2n = 24$). Outline your procedure for determining the chromosome carrying the locus of a new recessive gene mutation causing *dwarfism*.

926   In *Datura stramonium* ($2n = 24$) *purple* vs. *white* color is controlled by a single pair of alleles *Pp* located on the third smallest ("Poinsettia") chromosome (Blakeslee and Farnham, 1923). The allele *P* for *purple* is dominant to *p* for *white*. In plants trisomic for this chromosome, $n + 1$ pollen rarely ever functions, and only about 25 percent of the functional eggs are $n + 1$.
　　a What is the expected ratio of *purple* to *white* from the crosses shown if *P* is close to the centromere?

Table 926

♀	♂
PPp ×	*PPp*
Ppp ×	*Ppp*
PPp ×	*pp*
pp ×	*PPp*
pp ×	*Ppp*

　　b Would the ratio be the same if *P* is 50 or more map units from the centromere? Illustrate, using the cross *PPp* female × *pp* male.
　　c Why do trisomic ratios differ from those expected in disomic inheritance?

927   In barley (*Hordeum vulgare*) the number of rows of seed per spike is monogenically controlled. The allele for *six-rowed*, *v*, is recessive to that for *two-rowed*, *V*. A *vv* plant is crossed with one trisomic for chromosome 2 in a variety homozygous for *V*. Trisomic progeny are recovered and reciprocally crossed with *vv* (*six-rowed*) plants. What ratio of dominant to recessive phenotypes would you expect if the gene is not on chromosome 2? If it is near the centromere of chromosome 2?

928   In the cultivated tomato, crosses between a true-breeding *normal* and a true-breeding *chartreuse* strain (corolla yellowish green) produced *normal* F_1's and an F_2 ratio of 1,353 *normal* : 423 *chartreuse* (Rick et al., 1964). A plant trisomic for chromosome 8, found in a true-breeding *normal* line, was crossed with a diploid *chartreuse* plant. The F_1's consisted of *normal* diploid and *normal* trisomic plants. When self-fertilized, the latter produced

Diploid *chartreuse*	9
Diploid *normal*	82
Trisomic *normal*	26
Trisomic *chartreuse*	0

F_1's from *normal* plants trisomic for each of the other 11 chromosomes crossed with the *chartreuse* diploid strain produced F_2's in a ratio of 1 *chartreuse* : 3 *normal*. Determine:
　　a The frequency of transmission of $n + 1$ gametes on the female side.
　　b The location of the gene controlling *normal* vs. *chartreuse*:

(1) With respect to the chromosome.

(2) With respect to the centromere.

Note: Male $n + 1$ gametes do not function in the tomato.

929 In tomatoes (*Lycopersicon* [= *Lycopersicum*] *esculentum*; $2n = 24$), Lesley (1932) studied the following pairs of alleles in progenies obtained by self-fertilizing three different plants, each trisomic for a different chromosome (chromosomes A, B, and I) and each heterozygous for the following genes:

Dd	*standard* vs. *dwarf* growth habit
Ll	*normal green* vs. *virescent* plant color
Rr	*red* vs. *yellow* fruit flesh
Cc	*cut* vs. *potato* leaf form
Aa	*purple* vs. *nonpurple* stem
Yy	*yellow* vs. *nonyellow* fruit skin

His data are tabulated.

Table 929

Parent	Allele pair studied	Progeny			
		Diploid		Trisomic	
		Dominant	Recessive	Dominant	Recessive
Trisomic A	*Dd*	71	44		
Trisomic B	*Dd*	19	6	27	9
	Ll	23	2	35	0
	Rr	9	3	13	4
	Cc	19	6	27	9
	Aa	20	7	26	10
Trisomic I	*Yy*	68	24	6	1
	Rr	90	12	7	0

State, with reasons:

a Which of the allele pairs are on chromosome A, which are on B, and which are on I.

b The genotypes of the three parents.

c The percent transmission of n and $n + 1$ gametes on the female side (no $n + 1$ gametes function on the male side).

930 In spinach (*Spinacia oleracea*; $2n = 12$, $n = 6$) males are XY, and females XX. Sex is controlled by a single pair of alleles; *m* (female) is on the X, and *M* (male) is on the Y. The chromosomes carrying the sex-determining allele pair are morphologically alike (apparently completely homologous). In 1959 Janick et al. established all six primary trisomics and then crossed the male (staminate) plants in each of the trisomic lines with diploid female (pistillate) plants to determine which chromosome carried the sex-determining gene. *Note:* Rarely do $n + 1$ gametes function on the male side.   The results obtained were as shown.

Table 930

Trisomic line used as male parent	Number of progeny	
	♂	♀
1. *Savoy*	387	384
2. *Oxtongue*	169	173
3. *Star*	191	217
4. *Curled*	218	226
5. *Reflex*	74	143
6. *Wild*	1,116	1,148

a Which chromosome carries the sex-determining gene?   Explain.   Support your answer by performing a chi-square test.

b What was the chromosome constitution and genotype of the male parent trisomic for the sex-determining gene?

c If the staminate trisomics of the type mentioned in (b) arose from a disomic-trisomic cross, state whether the male or the female parent was trisomic.

931 The illustration shows the types of chromosome configurations possible in trisomic plants at diakinesis (chiasmata not shown in configurations).   One

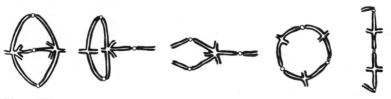

Figure 931

of these cannot occur in a primary trisomic.   Which is it and why?

932 In barley, *long* vs. *short* rachilla hairs are controlled by a single pair of alleles, *S* vs. *s*. Tsuchiya (1959) self-fertilized an F_1 plant trisomic for chromosome 7 and heterozygous at the *S* locus.   The $2n + 1$ F_2 progeny consisted of 45 plants with *long* rachillas and 2 with *short* ones.   The $2n$ F_2 plants segregated 100 *long* : 9 *short*.   Using your own symbols and diagrams, provide a satisfactory cytogenetic explanation for these results.

933 In mice, *male antigen* is produced by males, but none is produced by females. It could therefore be controlled either by a Y-linked or by an autosomal gene (hence present in both sexes but functionally suppressed by two X chromosomes of the female).   Celada and Welshons (1963) found that XX and XXY animals differ antigenically whereas XX and XO do not.   State whether the problem is resolved and why.

934   In corn, *purple* vs. *red* aleurone and *green* vs. *virescent* seedlings are each controlled by a single pair of alleles.   There were 246 *purple* and 61 *red* aleurone seeds on a trisomic plant that had been crossed with a *red virescent* plant.   When the seeds were grown, there were 181 *green* and 81 *virescent* seedlings.   Explain the two results.

935   a In organisms such as the mouse, XO individuals are healthy, fertile females, and OY individuals are inviable.   What is the expected ratio of XO, XX, XY, and OY offspring from XO female × XY male matings?
       b Cattanach (1962) found that in such matings the XO offspring were only one-third as frequent as the XX.   He further found that:

       1   The litter sizes of XO and XX females were the same.
       2   The sex ratios among the offspring of XO and XX females were the same.
       3   There is no postnatal inviability of the XO class, at least up to 3 weeks.

       What do Cattanach's observations suggest concerning the unexpected frequency of XO in such matings and its probable cause?

936   In 1934 Blakeslee and Avery reported their results in crosses between trisomics heterozygous for *sc* (*slender* capsules), *bd* ("*bad*" pollen), and *pl* (*pale* leaf) and diploids homozygous for the three recessives.   The number of dominant and recessive plants among both 2n and 2n + 1 progeny are shown.

Table 936

Parents		Phenotypic and karyotypic classification of progeny			
		2n		2n + 1	
Trisomic ♀	Diploid ♂	Dominant	Recessive	Dominant	Recessive
normal × *slender*		95	48	63	11
normal × *slender*		61	121	9	0
"*good*" × "*bad*"		988	111	310	9
"*good*" × "*bad*"		315	284	150	52
green × *pale*		190	119	45	3
green × *pale*		49	117	41	27

a For each of the genes studied show whether it is on the trisome.
b For genes on the trisome show whether the gene-centromere distance is great or small.

937 a What are the features by which B chromosomes can be distinguished from normal A chromosomes?
b How would you be able to distinguish between a B chromosome and an extra A chromosome in any $2n + 1$ organism?

938 In corn ($2n = 20$) two aneuploids occur, one a tetrasomic ($2n + 2$), the other a double trisomic ($2n + 1 + 1$).   How would you distinguish between the two cytologically?

939 A study of the breeding behavior of eight of the ten primary trisomics of corn by Einset (1943) revealed marked differences in the frequency with which the extra chromosome is transmitted to the progeny through the female gametes. Einset's data concerning the relation between chromosome length, frequency of trisomics in the progeny, and percentage of $n + 1$ microspores are given.

Table 939

Chromosome	Relative length	$2n + 1 \times 2n$ crosses		Microspores	
		Total	Percent $2n + 1$	Total	Percent $n + 1$
2	80	320	47	454	48
3	74	91	45	167	41
5	73	89	52	198	50
Average	76		48		46
6	60	155	38	109	34
7	56	80	41	193	50
8	57	191	31	245	36
Average	58		37		40
9	52	113	22	218	23
10	45	198	28	299	34
Average	49		25		29

Note: The ears of trisomic plants were as well filled as those of their disomic sibs, and their seeds were equally viable.
a For each chromosome, show whether segregation for $2n$ vs. $2n + 1$ progeny differs from that expected with two-by-one segregation of the three homologs.
b Discuss these segregations in the light of microspore-segregation and chromosome-size data.

940 In the medaka fish (*Oryzias latipes*) the allele pair Rr controls *red* vs. *white* body.   In 1921 Aida in Japan obtained the results given.

Table 940

Parents		F₁		F₂		Backcross (F₁ ♂ × white ♀)	
♀	♂	♀	♂	♀	♂	♀	♂
white × red		red	red	41 red, 43 white	76 red, 0 white	2 red,* 197 white	251 red, 1 white*
red × white		red	red	87 red, 0 white	42 red, 33 white	135 red, 2 white*	1 red,* 133 white

* Exceptional offspring.

a Is *red* vs. *white* controlled by a sex-linked or an autosomal pair of alleles?

b What is the probable method of sex determination in the medaka fish?

c The exceptional offspring in the backcross progenies may be due to:

1 Crossing-over between the sex chromosomes
2 Nondisjunction
3 Mutation
4 Spontaneous sex reversal

To determine which of these explanations best explains the appearance of the exceptional individuals Aida mated *red* females obtained from the cross *white* female × *red* male with *red* males obtained from the reciprocal mating *red* female × *white* male. The observed phenotypes of the two sexes in the progeny were:

Female   146 *red*, 2 *white*
Male   57 *red*, 80 *white*

Which of the above explanations is the most plausible? Explain. What further observations—genetical, cytological, or both—would you make to confirm your decision?

941 In man, about 0.5 percent of the population are heteromorphic for chromosome 1, the homolog carrying the *uncoiler* allele *Unl* being significantly longer

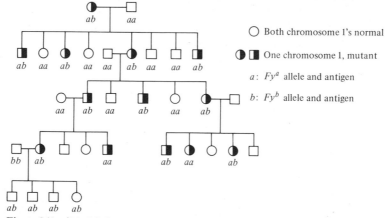

Figure 941 (*Modified after Donahue et al., 1968.*)

○ Both chromosome 1's normal
◑ ◨ One chromosome 1, mutant

a: *Fyᵃ* allele and antigen
b: *Fyᵇ* allele and antigen

than that carrying *unl* because of uncoiling in the long arm. The pedigree shows the segregation of mutant and normal chromosomes 1 and the co-dominant alleles Fy^a and Fy^b at the *Duffy* blood-group locus. Is the *Duffy* locus likely to be on chromosome 1? Explain.

942 In the allohexaploid wheat (*Triticum aestivum*; $2n = 6x = 42$) it is possible to establish 21 different nullisomic lines in any variety. *Red*-glumed varieties are homozygous for the dominant allele *R*, and *white*-glumed ones are homozygous for the recessive allele, *r*, at this locus.

  a Describe how the dominant allele for *red* glumes can be associated with a specific chromosome by the use of nullisomics.

  b Using a variety homozygous for the recessive allele for *white* glumes, would it be possible using the nullisomic method to determine which chromosome carries the gene? Explain.

943 In *Nicotiana tabacum* ($2n = 4x = 48$), *green* vs. *yellow-burley* is determined by two pairs of alleles on different (nonhomologous) chromosome pairs (Clausen and Cameron, 1944). *A—B—*, *A—bb*, and *aa B—* are *green*, and *aa bb* are *yellow-burley*.

  a A cross is made between all ($n = 24$) monosomics of an *aa bb* variety and the variety to be analyzed, whose genotype is *AA BB*. What ratios are expected in:

   (1) "Noncritical" F_2 families (univalent in monosomic does not carry either gene)?

   (2) "Critical" F_2 families (univalent in monosomic carries one of the genes)?

  b Clausen and Cameron crossed nine different *green* monosomic lines with a *yellow-burley* variety, selected F_1 monosomics, and backcrossed them to the *yellow-burley* variety. The results of the latter crosses are given.

Table 943

	Backcross-progeny phenotype	
Monosomic line	*Green*	*Yellow-burley*
M	36	9
N	28	8
O	19	17
P	33	9
Q	32	12
R	27	12
S	27	4
T	28	8
U	37	8

(1) Which chromosome carries one of the recessive genes? Why?

(2) Would you expect the other recessive gene to be in the same or in a different genome? Explain.

944 Clausen and Cameron (1950) found that crosses between plants in the varieties Purpurea and Chinchao in *Nicotiana tabacum* ($2n = 4x = 48$) produced *green* F_1's and F_2's segregating 626 *green* and 44 *white*. A cross between F_1's and a heterozygous Purpurea plant gave 87 *green* and 14 *white* offspring. Each of the 24 monosomics in Purpurea was crossed with Chinchao. The F_1 monosomics, all *green*, were self-fertilized. The segregation ratios in the F_2 for each of the 24 families as well as for one of the *normal* (euploid) crosses are shown. Give a cytogenetic explanation of these results, including in your

Table 944

Monosomic line	Phenotype		Monosomic line	Phenotype	
	Green seedlings	*White* seedlings		*Green* seedlings	*White* seedlings
A	87	7	M	88	3
B	85	8	N	92	5
C	94	5	O	88	8
D	83	4	P	196	13
E	92	8	Q	94	5
F	77	7	R	91	7
G	86	0	S	88	3
H	87	8	T	66	24
I	79	6	U	72	7
J	79	6	V	94	3
K	94	4	W	86	7
L			Z	86	7
Normal × *normal*	90	7			

explanation the mode of inheritance of chlorophyll production and the location of the gene or genes concerned.

945 In variety C.I. 12633 of common wheat ($2n = 6x = 42$), *resistance* to stem rust is governed by two linked duplicate dominant genes (alleles with identical effects at different loci). Nyquist (1957) crossed C.I. 12633 with all 21 monosomics of Chinese Spring, a variety *susceptible* to stem rust. The F_1 monosomics and one normal (euploid) F_1 were self-fertilized. The reactions to races 11, 17, and 56 in F_2 plants and F_3 families are given.

a Which chromosome pair carries the duplicate pairs of alleles? Diagram the cross to show why you reach the conclusion you do.

b What is the distance between the two linked genes? Indicate the results used in calculating map distances.

Table 945

	F$_2$ progenies			F$_3$ families	
Chromosome	Resistant	Susceptible	Adjusted chi square	Resistant and segregating	Susceptible
1	68	15	0.232	16	4
2	16	3	0.083		
3	43	8	0.028	19	1
4	75	14	0.010		
5	57	6	1.314	36	4
6	73	15	0.057	17	3
7	109	28	2.133	19	1
8	78	17	0.239		
9	83	14	0.027	19	1
10	86	12	0.584		
11	86	7	3.970	16	4
12	69	12	0.001	19	1
13*	131	3	17.084	24	
13†	58	20	5.301		
14	31	6	0.013		
15	80	16	0.026		
16	43	8	0.028	17	3
17	74	15	0.037		
18	70	9	0.748		
19	94	20	0.209	19	1
20	65	10	0.138	18	2
21	54	5	1.744		

* Plants from monosomic F$_1$ plants.   † Plants from normal F$_1$ plants.

946 Morgan (1922) found that when a *yellow*-body *Drosophila* female was crossed with a *gray*-body male, all the female offspring resembled the mother and all the sons (fertile) had *gray* bodies. The males and females appeared in a 1:1 ratio. The same results were obtained in subsequent generations. Propose a cytogenetic explanation to account for these results.

947 Prelude, a variety of common wheat ($2n = 6x = 42$) *susceptible* to race 56 of stem rust, was crossed by Campbell and McGinnis (1958) with the 21 monosomic lines in the *resistant* variety Redman. The phenotypes of the F$_1$ population are given in the table. How many genes control *resistance* vs. *susceptibility* to race 56? How do they interact?

Table 947

Redman monosomic line involved in cross with Prelude	Rust reaction			
	Disomics		Monosomics	
	Resistant	*Susceptible*	*Resistant*	*Susceptible*
1	4		11	1
2	2		4	
3	2		0	9
4	3		2	
5	6		7	
6	7		4	
7	2		11	
8	4		0	5
9	4		7	
10	5		8	
11	2		2	
12	6		6	
13	13		0	11
14	6		10	
15	6		6	
16	2		4	
17	2		3	
18	4		6	
19	2		20	
20	3		4	1
21	5		16	

948 *Triticum sphaerococcum* has short stems, dense heads, and small, spherical kernels (*sphaerococcum syndrome*), in contrast to *T. aestivum,* which has longer stems, laxer heads, and larger, nonspherical grains (the *nonsphaerococcum syndrome,* which is dominant to the *sphaerococcum syndrome*).   Sears (1947, 1953) noted the following:

1   Nullisomics, monosomics, and disomics in varieties with the *nonsphaerococcum* allele all show the dominant syndrome.
2   Nullisomics and monosomics in *T. sphaerococcum* also show the dominant syndrome.
3   F_1's from crosses between each of the 21 monosomic lines in Chinese Spring (*T. aestivum*) and *T. sphaerococcum* were self-fertilized, and the F_2 populations classified for *sphaerococcum* vs. *nonsphaerococcum.*   The F_2 ratios in all crosses were approximately 3 *nonsphaerococcum* : 1 *sphaerococcum.*   In the F_2 family involving chromosome 16, there were 6 *sphaerococcumlike* segregates (all disomic) and 14 *nonsphaerococcum* plants (13 monosomic, 1 nullisomic).   In all other F_2 families, the ratio of 3 *nonsphaerococcum* : 1 *sphaerococcum* occurred among the monosomics and nullisomics as well as among the disomics.

Note: In wheat monosomics, n and $n - 1$ gametes function in approximately a 1:3 ratio on the female side and 96:4 ratio on the male side, so that disomics, monosomics, and nullisomics are produced by self-fertilized monosomics in approximately a 24:73:3 ratio.

With reasons for your answers:

a For each of the alleles, *S* (*nonsphaerococcum*) and *s* (*sphaerococcum*), state whether it is:

(1) An amorph, hypomorph, or fully morphic (completely active).

(2) Effective when hemizygous, homozygous, or both.

b State whether *S* is on chromosome 16; if so, show how the results in the other F_2 populations would differ from these.   Use diagrams.

949 What proportion of abortions in man is due to chromosome anomalies? What proportion of these are aneuploids?   What proportion of liveborn children possess a clinically significant chromosome anomaly?   Which type of chromosome mutation is responsible for most of the clinically detectable diseases, malformations, and syndromes?   Discuss.   (See Carter, 1969.)

950 Common wheat[1] (*Triticum aestivum*; $2n = 6x = 42$) has three 7-chromosome genomes.   The A genome derives from *T. monococcum*, the B from *Aegilops speltoides*, and the D from *A. squarrosa*.   Of the 21 chromosomes of common wheat, those in the D genome were assigned the numbers 15 to 21 (Sears, 1954). Those in the A and B genomes were originally assigned numbers 1 to 14 arbitrarily.   To establish which of these belong to the A and which to the B genome, plants having 20 chromosome pairs plus a telocentric chromosome (one arm plus centromere) for each of the 14 chromosomes in the A and B genomes were crossed by Okamoto (1962) with the amphidiploid AADD (from *T. aegilopoides* × *A. squarrosa*).   The frequencies of heteromorphic bivalents (telocentric pairing with its normal homolog) in the F_1 35-chromosome plants are given.

Table 950

Chromosome telocentric	Percent with heteromorphic bivalents	Chromosome telocentric	Percent with heteromorphic bivalent
1	0.00	8*	0.00
2	1.70	9	9.72
3	0.00	10	0.00
4*	0.00	11	2.64
5	1.74	12	7.09
6	10.88	13	0.20
7	0.00	14	29.00

* Chromosomes 4 and 8 are homoeologous.

[1] In wheat each chromosome in the A genome is partly homologous (homoeologous) with one chromosome in each of the other two genomes.   When chromosome 5 is present, only homologs pair; when this chromosome is absent, homoeologous chromosomes may pair to form bivalents, trivalents, etc.

a Assign the first 14 chromosomes to their proper genomes, showing why you assign them as you do.

b Noting that a slight degree of similarity (homoeology) between chromosomes of different genomes may cause rare intergenomic pairing, which of your chromosome assignments would you consider to be subject to error and why?

c Since plants telocentric for chromosomes 2 and 13 formed heteromorphic bivalents with low frequency, to assign these chromosomes advantage was taken of the increased synapsis occurring in plants deficient for chromosome 5, which carries a diploidizing gene. Thus in F_1 plants selected cytologically for absence of chromosome 5, the percentages of heteromorphic bivalents in the crosses involving telocentrics 2 and 13 were 27.1 and 77.6 respectively.   Reassess your former classification in the light of these latter data.

REFERENCES

Aida, T. (1921), *Genetics*, **6**:554.

Barr, M. L., et al. (1962), *J. Ment. Defic. Res.*, **6**:65.

Bartalos, M., and T. A. Baramki (1967), "Medical Cytogenetics," Williams & Wilkins, Baltimore.

Bartlett, D. J., et al. (1968), *Nature*, **219**:351.

Baughan, M., R. Sparkes, D. Paglia, and M. G. Wilson (1969), *J. Med. Genet.*, **6**:42.

Bearn, A. G., and J. L. German (1961), *Sci. Am.*, **205**(November):66.

Beçak, W., M. L. Beçak, and B. J. Schmidt (1963), *Lancet*, **1**:664.

Belling, J. (1925), *J. Genet.*, **15**:245.

——— (1927), *J. Genet.*, **18** : 177.

——— and A. F. Blakeslee (1924), *Proc. Natl. Acad. Sci.*, **10**:116.

——— and ——— (1926), *Proc. Natl. Acad. Sci.*, **12**:7.

Berkeley, M. J. K., and M. J. W. Ford (1970), *J. Med. Genet.*, **7**:83.

Blakeslee, A. F. (1934), *J. Hered.*, **25**:81.

——— and A. G. Avery (1934), *J. Hered.*, **25**:393.

——— and M. E. Farnham (1923), *Am. Nat.*, **57**:481.

Blakeslee, J. (1927), *Proc. Natl. Acad. Sci.*, **10**:109.

Bloom, S. E. (1969), *Chromosoma*, **28**:357.

——— (1970), *Science*, **170**:457.

Brewer, G. J., C. F. Sing, and E. R. Sears (1969), *Proc. Natl. Acad. Sci.*, **64**:1224.

Bridges, C. B. (1914), *Science*, **40**:107.

——— (1916), *Genetics*, **1**:1.

——— (1921), *Proc. Natl. Acad. Sci.*, **7**:182.

Buchholz, J. T., and A. F. Blakeslee (1932), *Am. J. Bot.*, **19**:604.

Campbell, A. B., and R. C. McGinnis (1958), *Can. J. Plant Sci.*, **38**:184.

Carr, D. H. (1967), *Am. J. Obstet. Gynecol.*, **97**:283.

Carter, C. O. (1969), "An ABC of Medical Genetics," Little, Brown, Boston.

Caspersson, T., L. Zech, and C. Johansson (1970), *Exp. Cell Res.*, **62**:490.

———, ———, ———, and E. J. Modest (1970), *Chromosoma*, **30**:215.

Cattanach, B. M. (1962), *Genet. Res.*, **3**:487.

——— (1964), *Cytogenetics*, **3**:159.

——— and C. E. Pollard (1969), *Cytogenetics*, **8**:80.

Celada, F., and W. J. Welshons (1963), *Genetics*, **48**:131.

Chu, E. H. Y. (1963), *Am. Zool.*, **3**:3.

Clausen, R. E., and D. R. Cameron (1944), *Genetics*, **29**:447.

—— and —— (1950), *Genetics*, **35**:4.

Cooper, H. L., H. S. Kupperman, O. R. Rendon, and K. Hirschhorn (1962), *N. Engl. J. Med.*, **266**:699.

Court-Brown, W. M. (1967), "Human Population Cytogenetics," Wiley, New York.

——, P. Law, and P. G. Smith (1969), *Annu. Hum. Genet.*, **33**:6.

Darlington, C. D. (1940), *J. Genet.*, **39**:351.

de la Chappelle, A., H. Hortling, R. Sanger, and R. R. Race (1964), *Cytogenetics*, **3**:334.

Donahue, R. P., W. B. Bias, J. H. Renwick, and V. A. McKusick (1968), *Proc. Natl. Acad. Sci.*, **61**:949.

Driscoll, C. J. (1966), *Genetics*, **54**:131.

Edwards, J. H., D. G. Harnden, A. H. Cameron, V. M. C. Crosse, and O. H. Wolff (1960), *Lancet*, **1**:787.

Einset, J. (1943), *Genetics*, **28**:349.

Ferguson-Smith, M. A., W. S. Mack, P. M. Ellis, M. Dickson, R. Sanger, and R. R. Race (1964), *Lancet*, **1**:46.

Ford, C. E. (1969), *Brit. Med. Bull.*, **25**:104.

——, P. E. Polani, C. J. de Almeida, and J. H. Briggs (1959), *Lancet*, **1**:711.

Fraccaro, M., and J. Lindsten (1960), *Lancet*, **2**:1303.

Fraser, G. R. (1963), *Annu. Hum. Genet.*, **26**:297.

Fredge, K. (1968), *Chromosoma*, **25**:75.

Gartler, S. M., C. Vuzzo, and S. Gandini (1962), *Cytogenetics*, **1**:1.

German, J. (1970), *Am. Sci.*, **58**:182.

Giannelli, F., and R. M. Howlett (1966), *Cytogenetics*, **5**:186.

Goodspeed, J. H., and P. Avery (1939), *J. Genet.*, **38**:381.

Griffen, A. B., and M. C. Bunker (1967), *Proc. Natl. Acad. Sci.*, **58**:1446.

Hamerton, J. L. (1971), "Human Cytogenetics," vols. I and II, Academic, New York.

Hart, G. E. (1970), *Proc. Natl. Acad. Sci.*, **66**:1136.

Hauschka, T. S., et al. (1962), *Am. J. Hum. Genet.*, **14**:22.

Hecht, F., et al. (1969), *Am. J. Hum. Genet.*, **21**:352.

Jacobs, P. A., et al. (1959), *Lancet*, **2**:423.

—— et al. (1960), *Lancet*, **1**:1213.

—— and J. A. Strong (1959), *Nature*, **183**:302.

Janick, J., D. L. Mahoney, and P. L. Pfahler (1959), *J. Hered.*, **50**:47.

Klinger, H. P., and H. G. Schwarzacher (1962), *Cytogenetics*, **1**(5):266.

Kuspira, J., and J. Unrau (1959), *Can. J. Genet. Cytol.*, **1**:267.

—— and —— (1960), *Can. J. Genet. Cytol.*, **2**:301.

Lejeune, J. (1964), *Prog. Med. Genet.*, **3**:144.

——, M. Gautier, and R. Turpin (1959a), *C. R. Acad. Sci.*, **248**:1721.

——, R. Turpin, and M. Gautier (1959b), *Bull. Acad. Natl. Med.*, **143**:256.

Lesley, J. W. (1932), *Genetics*, **17**:545.

—— (1937), *Genetics*, **22**:297.

Lindsten, J., et al. (1963), *Lancet*, **1**:558.

Lisgar, F., et al. (1970), *Nature*, **225**:280.

Luig, N. H., and R. A. McIntosh (1968), *Can. J. Genet. Cytol.*, **10**:99.

McClintock, B., and H. E. Hill (1931), *Genetics*, **16**:175.

McIntosh, R. A., and E. P. Baker (1968), *Genet. Res.*, **12**:11.

Marshall, R. (1964), *Lancet*, **1**:556.

Mittwoch, U. (1963), *Sci. Am.*, **209**(July):54.

Morgan, L. V. (1922), *Biol. Bull.*, **42**:26 .

Muldal, S., and G. H. Ockey (1960), *Lancet*, **2**:492.

Neurath, P., et al. (1970), *Nature*, **225**:281.

Nielsen, G., and O. Frydenberg (1971), *Hereditas*, **67**:152.
Nyquist, W. E. (1957), *Agron. J.*, **49**:222.
Okamoto, M. (1962), *Can. J. Genet. Cytol.*, **4**:31.
O'Riordan, M. L., et al. (1971), *Nature*, **230**:167.
Pantelakis, S. N., et al. (1970), *Am. J. Hum. Genet.*, **22**:184.
Patau, K., et al. (1960), *Lancet*, **1**:790.
Pavan, C., C. Chagas, O. Frota-Pessoa, and L. R. Caldas (eds.) (1964), "Mammalian Cyto-genetics and Related Problems in Radiobiology," Pergamon, New York.
Payne, F. E., and R. D. Schmickel (1971), *Nat. New Biol.*, **230**:90.
Pearson, P. L., M. Bobrow, and C. G. Vosa (1970), *Nature*, **226**:78.
Polani, P. E. (1962), in J. L. Hamerton (ed.), "Chromosomes in Medicine," pp. 73–139, Little Club Clinic Developmental Medicine, 5, National Spastic Society, and Heinemann, London.
―――― (1969), *Br. Med. Bull.*, **25**:81.
Porter, I. H., W. Petersen, and C. D. Brown (1969), *J. Med. Genet.*, **6**:347.
Rhoades, M. M., and B. McClintock (1935), *Bot. Rev.*, **1**:292.
Rick, C. M., and D. W. Barton (1954), *Genetics*, **39**:640.
――――, W. H. Dempsey, and G. S. Khush (1964), *Can. J. Genet. Cytol.*, **6**:93.
Russell, L. B. (1961), *Science*, **133**:1795.
―――― (1962), *Prog. Med. Genet.*, **2**:230.
―――― and E. H. Y. Chu (1961), *Proc. Natl. Acad. Sci.*, **47**:571.
Russell, W. L., L. B. Russell, and J. S. Grower (1959), *Proc. Natl. Acad. Sci.*, **45**:554.
Schuster, J., and A. G. Motulsky (1962), *Lancet*, **1**:1074.
Sears, E. R. (1947), *Genetics*, **32**:102.
―――― (1953), *Am. Nat.*, **87**:245.
―――― (1954), *Mo. Agric. Exp. Stn. Res. Bull.* 572, p. 1.
―――― (1969), *Annu. Rev. Genet.*, **3**:451.
Slizynski, D. M. (1964), *Genet. Res.*, **5**:328.
Stene, J. (1970), *Hum. Hered.*, **20**:1.
Stern, C. (1960), *Nature*, **187**:905.
Stewart, J. S. S. (1962), *Nature*, **194**:258.
Sturtevant, A. H. (1936), *Genetics*, **21**:444.
Sumner, A. T., J. A. Robinson, and H. J. Evans (1971), *Nat. New Biol.*, **229**:231.
Suomalainen, E. (1962), *Annu. Rev. Entomol.*, **7**:349.
Taylor, A. I. (1968), *J. Med. Genet.*, **5**:227.
Thompson, M. W. (1961), *Can. J. Genet. Cytol.*, **3**:351.
Thuline, H. C., and D. E. Norby (1961), *Science*, **134**:554.
Tokunaga, C. (1970), *Genetics*, **65**:75.
Tsuchiya, T. (1959), *Jap. J. Bot.*, **17**:14.
―――― (1960), *Jap. J. Bot.*, **17**:177.
Warren, R. J., and J. I. Keith (1971), *J. Med. Genet.*, **8**:384.
Welshons, W. J. (1963), *Am. Zool.*, **3**:15.
―――― and L. B. Russell (1959), *Proc. Natl. Acad. Sci.*, **45**:560.
White, T. G., and J. E. Endrizzi (1965), *Genetics*, **51**:605.
Yamamoto, T. (1963), *Genetics*, **48**:293.
Zimmering, S., and C. K. Wu (1964), *Genetics*, **49**:499.

22
Chromosome
Aberrations

NOTATION

1   The symbol ⊙ indicates a metaphase ring of chromosomes. The number following it represents the number of chromosomes in the ring; the number preceding indicates the number of such rings.

2   *Interstitial segment* is the region of a chromosome between the centromere and the breakage point.

3   The term *semisterile* is not used in the strict sense, viz. 50 percent sterility, but refers to partial sterility, varying between 0 and 100 percent (semisterile = partially sterile).

4   Where chromosomes are designated with capital and small letters, e.g., A B ᐧ C D and a b ᐧ c d, unless otherwise stated, capital and small letters are not used to distinguish between dominant and recessive alleles but merely to distinguish a chromosome from its homolog. The dot represents the centromere unless otherwise stated.

5   A *standard* line is a euploid line free of any chromosomal aberration.

QUESTIONS

951 The frequencies of radiation-induced gene mutations and single-break chromosomal aberrations seem to be in direct proportion to the dosage of radiation over a very wide range, but this is not true of two-break aberrations. Explain.

952 Why are chromosomal aberrations considered to have less significance than gene mutations for subsequent generations?

953 a The chromosomes of salivary glands of *Drosophila* and of other dipterans, such as *Chironomus,* are much longer and wider than those of other somatic cells and meiocytes. What is the cause of this?

   b Although *D. melanogaster* has $2n = 8$ chromosomes, smears of gland chromosomes reveal only six arms extending out from the chromocenter. Explain.

   c Explain the banding in the salivary-gland chromosomes.

   d What features of salivary-gland chromosomes make them especially useful for cytogenetic studies?

954 How can salivary-gland chromosomes be used in localizing genes in *Drosophila* and in physical mapping of the chromosomes? See Painter (1934) and Mackensen (1935) for excellent discussions of the subject.

955 Many recessive mutations are minute deletions. Therefore the loci of the genes involved are lost rather than altered. What kinds of experiments and observations could be made to decide whether a particular mutation is the result of an alteration in the gene or the loss of a short segment of the chromosome carrying the gene?

956 a Describe and illustrate how (1) deletions, (2) inversions, and (3) reciprocal translocations arise in nature.

   b How can each be produced experimentally?

   c How can each be detected (1) genetically and (2) cytologically?

957 **a** Do inversions always suppress crossing-over? Give reasons for your answer.

b If your answer to (a) is no, why are inversions referred to as *crossover suppressors*? What term would be more appropriate to describe the effect?

c Show how (1) paracentric and (2) pericentric inversions can act as crossover suppressors.

958 A standard strain of a diploid plant, e.g., corn, is homozygous for the chromosomal gene sequence A B C D E F G H I J. Another strain is homozygous for the multiple recessive chromosome a b c d g f e h i j.

a Diagram the configuration of synapsing chromosomes at pachytene in the hybrid between these strains.

b Diagram the configurations you would observe at anaphase I and anaphase II of meiosis and show the types of spores that the hybrid would form if the inversion were a paracentric one and:

(1) There was one crossover within the inversion loop.

(2) There was one crossover between the inversion loop and the centromere.

(3) There were two crossovers within the inversion loop involving (*a*) all four chromatids, (*b*) the same two chromatids.

c Answer the questions in (*b*) if the inversion was pericentric.

d Genetically, what is the effect of crossing-over in the inversion loop? Explain?

For excellent discussions of cytological behavior in inversion heterozygotes see McClintock (1931, 1938).

959 Explain why gametophyte (e.g., pollen) lethals are common in plants whereas gametic lethals are unknown in animals.

960 **a** Under what conditions may pericentric inversions become established in nature?

b In wild populations of *Drosophila* almost all the inversions that have been found are paracentric. Explain why this should be so.

961 What may be an important role of chromosomal duplications in evolution?

962 A geneticist has a strain of diploid plants; the gene orders for two of the chromosomes in this strain are A B C D E F and K L M N O. He subjects the plants to x-ray treatments and induces a reciprocal translocation between these two chromosomes. The translocation chromosomes are A B C N O and K L M D E F.

a Diagram the pachytene configuration in translocation heterozygotes.

b Show the metaphase I orientations and anaphase I segregations when these are not directed and the types of spores resulting from each.

c Show diagrammatically the karyotypes and genotypes of offspring expected in a testcross of an F_1 heterozygous for the translocation and for the genes shown. Would you expect some of the progeny to be semisterile? Why?

963 Are viruses able to cause chromosome breakage? See Huang (1967) and Kato (1967).

964 In many human marriages numerous abortions occur. List one genic and two possible chromosomal causes of this condition. For one of the chromosomal causes, describe the chromosome mechanism underlying abortion.

965 Discuss the role of heterochromatin in karyotype evolution involving changes in chromosome number. Refer to specific examples if possible.

966 a Most persons with *Down's syndrome* are trisomic for chromosome 21 $(2n + 1 = 47)$; a few have been found with the normal number of 46 chromosomes. How can these latter cases be explained?

b A young couple's first child has *Down's syndrome*. They ask you whether it is advisable for them to have other children. What information would you need in the way of pedigrees and diagnostic tests before advising them, and what advice should be given? (See Carter, 1969; Valentine, 1969.)

PROBLEMS

967 All chromosomes in a certain diploid plant except one are normal. This abnormal chromosome has, as a result of chromosome breakage, genes a b g h i j in the order shown, whereas the gene content and gene order of the normal homolog of this chromosome is A B C D E F G H I J.

a Show how these two chromosomes would pair at pachytene.

b If such a plant was self-fertilized, show the types and proportions of zygotes expected.

c Deletion heterozygotes do not have the same probability of occurrence in plants and animals.

In which kingdom are they more likely to be found, and why?

968 In *Drosophila melanogaster*, at the X-linked *yellow* locus the allele *Y* (*gray* body) is dominant to *y* (*yellow* body). Deletion of the segment of the X chromosome carrying the *yellow* locus occurs naturally and can be induced by various mutagens. In both *YY* and *yy* strains, when a female (XX) is homozygous for such a deletion (does not carry the *yellow* locus on either X), she is, like individuals with the *y* allele, phenotypically *yellow* (Ephrussi, 1934). What does this tell you about *y*, knowing *y* is an allele, not a deficiency?

969 a In mice, *normal* vs. *waltzing* gait is controlled by a single gene; *v*, the allele for *waltzing*, is recessive to *V* for *normal*. In matings of true-breeding *normals* with *waltzers*, Gates (1927) found a single *waltzing* female among several hundred *normal* F_1's. When the F_1 *waltzer* was mated with two different males of the *waltzer* stock, 11 progeny were obtained, all *waltzers*. When mated with *normal* males, she produced 13 *normal* progeny and no *waltzers*. Three females of this *normal* progeny mated with two of their brothers produced 60 progeny, all *normal*. In a mating of one of these females with a third brother, 4 *normals* and 2 *waltzers* appeared in a litter of 6.   Possible explanations for the unexpected appearance of the original *waltzer* female are:

1   Mutation of a dominant to a recessive allele in the *normal* parent
2   Mutation of a suppressor of *v*
3   Nondisjunction of the chromosome carrying *V* in the *normal* parent
4   A deletion, in the *normal* parent, of a portion of the chromosome carrying *V*

Which two explanations are eliminated by the breeding data and why?

b Painter (1927) cytologically examined the chromosomes of two *waltzers* from the progeny of this female.   They had 40 chromosomes like *normal* mice (individuals homozygous for the *waltzer* allele also have 40 chromosomes), but one pair was heteromorphic, one chromosome of this pair being abnormally short.

(1) What was the cause of the original *waltzing* female?   Explain and illustrate.
(2) Show how the cross in which the unusual *waltzer* appeared exhibits pseudodominance.
(3) What was the chromosome constitution of the *normal* male and female sibs of the cross F_1 *waltzer* × *normal* male that gave only *normal* progeny?   Why?

970   A single *vestigial*-winged fly suddenly appears in a highly inbred strain of *normal*-winged *Drosophila*.

a Name three possible kinds of causes that could lead to the appearance of this fly.

b Briefly discuss your procedure in attempting to distinguish between any two of these alternatives and suggest any difficulties you might encounter in doing so.

c If the *Drosophila* had originated in a natural population (mating at random), what other explanation or explanations might be given for its appearance?

971   In corn, *Bmbm*, *Btbt*, V_3v_3, and *Bvbv* determine the expression of *green* vs. *brown* midrib, *soft* vs. *brittle* endosperm, *green* vs. *virescent* seedling, and *normal* vs. *half-normal* height, respectively.   In 1933 Stadler crossed a *bmbm btbt v_3v_3 bvbv* strain as female parent with a *BmBm BtBt V_3V_3 BvBv* strain.   A few plants in the progeny expressed all four recessive traits, and half their pollen was defective.   When crossed as males with the parental quadruple recessive strain, all progeny expressed the recessive traits but none had defective pollen.   Progeny from the reciprocal cross also showed the four recessive traits, and half of them produced 50 percent defective pollen. Plants with defective pollen showed a short buckled region in the long arm of chromosome 5 at pachytene.   Discuss these observations, giving reasons for your answers, with respect to:

a The condition causing the recessive F_1 phenotypes.

b The transmission frequency of this condition through male and female gametophytes.

972   In corn, a plant heterozygous for a deletion of a portion of the short arm of chromosome 9 is also heterozygous at the *sh* locus located in the short arm of this chromosome.   The deleted chromosome carries the recessive gene *sh* (*shrunken* endosperm), and its normal homolog carries the dominant allele *Sh* (*full* endosperm).   When a *shsh* plant was fertilized with pollen from this heterozygote, 14 percent of the kernels were *shrunken*.   Explain the appearance of these recessive types in view of the fact that the deleted chromosome is not transmitted through the pollen.

973   In corn *liguled*, *Lg*, is dominant to *liguleless*, *lg*.   In a cross between a true-breeding *liguled* and a true-breeding *liguleless* strain, among thousands of *liguled* progeny, a *liguleless* one appears.   In crosses with homozygous *liguled* it produces *liguled* F_1's which upon self-fertilization or intercrossing produce 3 *liguled* : 1 *liguleless*.   How would you proceed (1) genetically and (2) cytologically to determine whether the *liguleless* F_1 individual carried a deletion or a mutation of the dominant allele?

974   The linkage map of chromosome 2 of the tomato, together with standard chromosome 2 (heterochromatic short arm terminated by a heterochromatic satellite) and a heteromorphic variant, are shown below.

Linkage map of chromosome 2
s	*wv*	*me*	*aw*	*d*	*dv*
	7	4	13	14	5

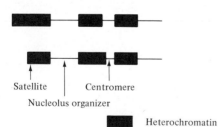

Satellite   Centromere

Nucleolus organizer

Heterochromatin

Figure 974   Structure of chromosome 2.

Moens and Butler (1963) crossed plants heterozygous for several chromosome-2 genes and heteromorphic for the variant chromosome with plants homozygous for all recessive alleles and isomorphic for the standard (long) chromosome. The results of the cross

$$\frac{d\,wv\;long}{D\;Wv\;short} \times \frac{d\,wv\,long}{d\,wv\,long}$$

are shown in Table 974.   Determine:
a The orientation of the linkage groups on the chromosome.
b The approximate distance of the genes from the centromere.

Table 974

Phenotype	Frequency
d wv long	13
D Wv short	14
d Wv short	5
D wv long	3
d wv short	11
D Wv long	9
d Wv long	1
D Wv short	1

975 The following data are from Rees and Jones (1967):

1   The chromosomal DNA content of the onion, *Allium cepa*, is about 27 percent greater than that of *A. fistulosum.*
2   The chromosomes of *A. cepa* are correspondingly larger than those of *A. fistulosum.*
3   In hybrids between the two species all bivalents at metaphase I are heteromorphic; the homologs of a chromosome pair are unequal in length.   Each of the bivalents shows at least one loop which is formed by one of the homologs only.

Both species have the same chromosome number.   If *A. fistulosum* is ancestral to *A. cepa*, what is the probable basis of the difference in chromosome size and DNA content between the two species?

976 Russell and Russell (1960) examined the F_1 progeny of a group of irradiated and unirradiated *wild-type* mice mated with a strain of mice homozygous for several recessive genes, including *d* (*dilute*) and *se* (*short ear*), which are on the same chromosome, 0.16 map unit apart; 15 individuals among several hundred F_1's were *dilute, short ear*.   From the following list of possible mechanisms that might produce these mutants show which should be accepted and which should not, and why:

1   Simultaneous mutation of *D* to *d* and *Se* to *se*
2   Inactivation of *D* and *Se* through position effect
3   Nondisjunction in the *wild-type* parent producing the monosomic *d se*/0
4   Deletion
5   Nondisjunction in both parents leading to an individual homozygous for *d se*

977 In mice ($2n = 40$) 20 linkage groups have been established, but only the sex-linked group has been associated with a specific chromosome.   Outline an experiment using x-rays or some other mutagen and appropriate genetic strains to establish these linkage groups by means of deletions.

978 In *Drosophila*, the X-linked genes *white, w; vermilion, v; rudimentary, r; forked, f; bar, B;* and *fused, fu,* have the following locations:

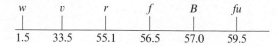

w	v	r	f	B	fu
1.5	33.5	55.1	56.5	57.0	59.5

a In 1917 Bridges crossed an XXY *nonbar* female heterozygous for *eosin, w^e,* and *vermilion, v,* in the repulsion phase, with a *white, bar* male. He obtained the F_1 progeny shown. Present two explanations for the exceptional female other than mutation or inactivation.

Table 978

Normal						Exceptional ♀
♀		♂ (all *nonbar*)				
White-eosin,* bar	Red, bar	Eosin	Vermilion	Eosin vermilion	Wild type	White-eosin,* nonbar
60	59	45	48	14	18	1

* *White-eosin = w/w^e.*

b The exceptional female crossed with a *nonbar, red* male produced 84 *nonbar, red* daughters and 51 *nonbar* sons (37 *eosin* and 14 *white*). Which of the explanations do these results support? Why?

c *Red, Ww, nonforked* females, chromosomally identical to the original exceptional female, were mated with *red, forked* males. The progeny consisted of 733 daughters (385 *red, forked,* 348 *red, nonforked*) and 353 *nonforked* males (207 *red* and 146 *white*). This cross definitely eliminates all explanations but one. Which is it and why? Illustrate.

979 By 1933 all the linkage groups except *j-v* in *Zea mays* had been associated with their specific chromosomes by the use of trisomics and reciprocal translocations. When x-rayed pollen from a homozygous *JJ* line (*normal green* leaves) was used to fertilize plants homozygous for the recessive *japonica* allele, *j,* which causes *white striping* of the leaves, a few *white-striped* plants were obtained. Meiocytes of the *white-striped* F_1's at pachytene showed a heteromorphic pair for chromosome 8, indicating that a portion of this chromosome had been deleted (Rhoades and McClintock, 1935). Explain these observations cytogenetically and state whether or not they indicate that the *j-v* linkage group is on chromosome 8, and why.

980 The pedigree, from Brøgger (1969), shows the phenotypes of the parents (I-1 and I-2) and of two of their sons, who are heterozygous for a deletion of a

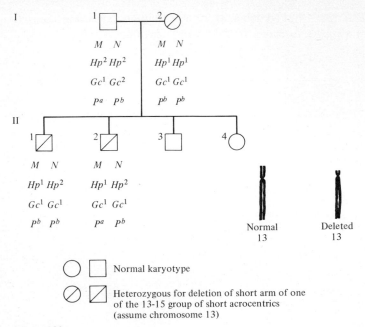

Figure 980

portion of the short arm of a chromosome in the 13 to 15 group. The alleles at each of the loci *MN* (MN blood groups), $Hp^1 Hp^2$ (haptoglobin type), $Gc^1 Gc^2$ (group specific component), and $P^a P^b$ (erythrocyte acid phosphatase), are codominant. *Note:* Since alleles are codominant, genotypes also represent phenotypes.

Assuming the deletion occurs in chromosome 13, is it possible to determine which, if any, of these genes is on this chromosome? Explain.

981   In man, the allele pair *Xgxg* (*presence* vs. *absence* of Xg^a blood antigen) is on the X chromosome. Lindsten et al. (1963) presented the results of a

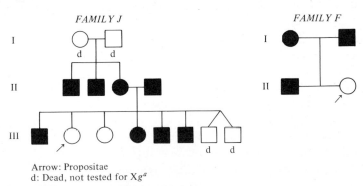

Arrow: Propositae
d: Dead, not tested for Xg^a

Figure 981

study of two families in which Xg^a antigen distribution was determined and in which the two propositae (showing ovarian dysgenesis) displayed, in the tissues sampled, chromosome mosaicism involving two kinds of cell lines: one, in the minority, had 45 chromosomes including a single X, and the other, in the majority, had 46 chromosomes, including a normal X and an isochromosome for the long arm of the X.

The pedigree of the J and F families are shown (solid symbols indicate presence of Xg^a antigen).

a Is the isochromosome of maternal or paternal origin?

b Is the Xg locus on the long or the short arm of the X chromosome?

982 **a** Using a hypothetical chromosome <u>A B C D . E F G H</u>, show how ring chromosomes can be produced which possess a centromere.

b Show how a ring chromosome (Fig. 982) with a centromere may increase,

Figure 982

decrease, or remain the same in size during mitotic divisions.   For confirmation of the usual behavior, see McClintock (1941).

c If the ring carries a dominant allele and is present in a plant along with normal chromosomes homozygous for the recessive allele, what phenotypic effects are expected?

983 **a** In *Drosophila melanogaster*, *bar eye* is caused by a tandem duplication of a segment of the X chromosome (*bar* females have the duplication in at least one X).   In *nonbar* flies this segment is present only once in each X; in *double bar* it is present in triplicate in at least one X.   In the offspring of homozygous *bar* females one out of 1,600 males is *nonbar*, and the offspring of *double bar* revert to *nonbar* at almost the same frequency (May, 1917; Zeleny, 1919, 1921).   Show diagrammatically how these reversions occur.

b In 1923 Sturtevant and Morgan testcrossed females of the constitution *F B fu/f B Fu* and found three *nonbar*, two *wild type* (*nonforked, nonfused*), one *forked, fused* among the male offspring.   In 1925 Sturtevant crossed *F B Fu/f B fu* females with *f B fu/Y* males, and classified the 18,999 progeny as shown.

(1) Show diagrammatically how this confirms your answer to (a) by using *forked* and *fused* as marker genes for the segments on either side of *bar*.

(2) Illustrate how both *nonbar* and *double bar* may originate from *homozygous bar* females by a single meiotic event.

Table 983

Bar								Nonbar				Double bar	
♀				♂				♀		♂		♀	♂
F Fu	f fu	F fu	f Fu	F Fu	f fu	F fu	f Fu	F fu	f Fu	F Fu	F fu	f Fu	f Fu	f Fu
5,413	3,749	94	140	5,218	4,160	93	124	0	2	1	2	1	1	1

984 In corn the allele pairs *Cc* (*colored* vs. *colorless* endosperm), *Shsh* (*full* vs. *shrunken* endosperm), and *Wxwx* (*starchy* vs. *waxy* endosperm) are at loci 26, 29, and 54, respectively, in the short arm of chromosome 9. McClintock in 1941 found a plant with a reversed duplication of the short arm of chromosome 9 in a strain homozygous for the dominant alleles *Wx, Sh, C*. Plants heterozygous for this chromosome and a normal 9 carrying the recessive alleles *c, sh, wx* were crossed with plants homozygous for the recessive alleles and normal chromosome 9 as female parents. Many of the kernels had a *variegated aleurone* (outer layer of endosperm) showing purple and white spots.

  a Illustrate with a drawing the chromosomal and genotypic constitutions of the meiotic products of a meiocyte after crossing-over between the reverted segment of duplicated chromosome 9 and its homolog as shown.

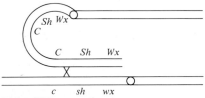

Figure 984

  b McClintock (1941) showed that the genotypic and phenotypic variegation in the endosperm of the testcross progeny was due to a *breakage-fusion-bridge-breakage* cycle. Explain with illustrative diagrams how this cycle, which occurs in the gametophytes and the endosperm tissue, gives rise to the genotypic and phenotypic variegation.

985 In *Drosophila melanogaster*, the paracentric inversion *scute-4* involves most of the X from just to the right of the *scute* locus to a point between *carnation* and *bobbed*. The progeny of female inversion heterozygotes contain no single-recombinant types in three-point tests for these genes with a standard chromosome line.

  a What are the two possible explanations for failure of the single recombinant to occur?

b Single crossovers do occur within the inverted segment, but no single recombinant is recovered and there are no zygotic abortions. Explain how this can happen.

986 In *Drosophila* the *Abab* and *Bb* pairs of alleles control the expression of *normal* vs. *abrupt* wing and *gray* vs. *black* body, respectively. In each of two different populations some homozygous *normal, gray* and *abrupt, black* individuals occur. The F_1's of crosses between the two homozygous phenotypes within each population were testcrossed with the results tabulated.

Table 986

Phenotype	No. in each class	
	Strain 1	Strain 2
Normal, gray	2,205	2,230
Normal, black	20	3
Abrupt, gray	24	2
Abrupt, black	2,245	2,265

Calculate the percent recombination in both strains and provide a cytogenetic explanation for the difference in the percentage recombination. Explain how you would verify your explanation cytologically.

987 a Do single crossovers in the inversion loop of a pericentric inversion produce both acentric and dicentric chromatids? Illustrate your answer.

b Why are *Drosophila* females heterozygous for a pericentric inversion less fertile than those heterozygous for a paracentric inversion?

c Carson and Stalker (1947) and Levitan (1950) have shown that flies carrying pericentric inversions are quite frequent in natural populations of *D. robusta*. Can you think of reasons for this?

988 In *Drosophila melanogaster* the allele pairs *Stst* (*red* vs. *scarlet* eye), *Srsr* (*normal* vs. *stripe*, thorax shape and pattern), *Ee^s* (*gray* vs. *sooty* body), *Roro* (*smooth* vs. *rough* eye) and *Caca* (*red* vs. *claret* eye) are on chromosome 3 as follows:

In 1926 Sturtevant crossed *wild-type* females heterozygous at all these loci with males homozygous for the corresponding recessives. The progeny fell into eight phenotypic classes, as shown.

Table 988

Phenotype	Number
Red, normal, gray, smooth, red	2,214
Scarlet, stripe, sooty, rough, claret	2,058
Red, stripe, sooty, rough, claret	219
Scarlet, normal, gray, smooth, red	238
Scarlet, stripe, gray, smooth, red	4
Red, normal, sooty, rough, claret	3
Scarlet, stripe, sooty, smooth, red	1
Red, normal, sooty, rough, red	1

a Is an inversion involved and if so, what is its effect on recombination? What is its position on the genetic map of chromosome 5?

OR

b Show how an inversion can account for these results and locate the appropriate position of the breakage point with respect to the gene loci.

989 Corn plants heterozygous for an inversion in chromosome 4 show no ovule abortion (Morgan, 1950).   How can you account for this?

990 The map distance between genes *A* and *B* in a standard strain is 25 map units. In another strain the distance is calculated as 10 map units.   The two strains when crossed produce an appreciable number of semisteriles in the progeny.
a What is the cause of the reduction in map distance in the second strain?
b How could you detect it cytologically?

991 **a** It is found that an individual heterozygous for a long inversion produces more double- than single-recombinant progeny.   Explain why this occurs.
b A certain paracentric-inversion heterozygote in barley shows approximately 25 percent pollen abortion but no ovule abortion.   Explain.

992 In *Drosophila melanogaster*, *Ss* (*spined* vs. *spineless*), *Ee* (*gray* vs. *ebony* body color), *Caca* (*red* vs. *claret* eyes), and *Roro* (*smooth* vs. *rough* eyes) are located as follows on the linkage map of chromosome 3:

The cross

$$\text{Female } \frac{S\,E\,Ro\,Ca}{s\,e\,ro\,ca} \times \text{male } \frac{s\,e\,ro\,ca}{s\,e\,ro\,ca}$$

gives the following testcross results (modified after Sturtevant, 1926):

S E Ro Ca	420			S e Ro Ca	0		
s e ro ca	410			s Ero ca	0		
S e ro ca	55			S e ro Ca	5		
s E Ro Ca	45			s E Ro ca	7		
S E ro ca	0			S Ero Ca	0		
s e Ro Ca	0			s e Ro ca	0		
S E Ro ca	38			S e Ro ca	0		
s e ro Ca	42			s Ero Ca	0		

Provide map distances for these data and a cytogenetic explanation for discrepancies between these and the distances listed above.

993 Strain 1 in barley, homozygous for genes *A*, *B*, *C*, *D*, *E*, *F*, *L*, *M*, and *N* on the same chromosome and in the sequence given, is crossed with strain 2, homozygous for the recessive alleles at these loci. The testcross progeny consisted of parental types as well as recombinants for the segments *A-B*, *L-M*, and *M-N* but no recombinants between *B-C*, *C-D*, *D-E*, *E-F*, or *F-L*.
a Suggest a hypothesis to acount for these facts.
b Show the gene order for the loci in strain 2.

994 A different sequence of chromosome-9 genes is found in each of four strains of corn, as shown below:

Strain 1 D^+ yg_2 *bz* sh_1 *c* *bp* *wx*
Strain 2 D^+ *bp* *c* yg_2 *bz* sh_1 *wx*
Strain 3 D^+ yg_2 *c*   *bp* *bz* sh_1 *wx*
Strain 4 D^+ yg_2 *c* sh_1 *bz* *bp* *wx*

The first is the ancestral one. In what sequence did the derived inversion strains arise? Illustrate the sequence of inversion events involved.

995 a In 1936 Sturtevant and Dobzhansky pointed out that overlapping inversions permit inferences regarding the phylogeny of different races with respect to the gene arrangements in a given chromosome. In *Drosophila pseudoobscura* the *Pike's Peak* and *Arrowhead* strains differ from the *standard* by one inversion each:

Standard	1	2	3	4	5	6	7	8	9	10	11	12	13
Arrowhead	1	2	3	4	9	8	7	6	5	10	11	12	13
Pike's Peak	1	2	7	6	5	4	3	8	9	10	11	12	13

What are the possible phylogenetic sequences?
b For a discussion of the use of overlapping inversions in determining the phylogeny of different races and of different species see Dobzhansky and Sturtevant (1938).

996 Snell (1933, 1935) x-rayed the testes of a number of male mice and mated each male with several females that in earlier matings gave litters of *normal* size (7 to 9 young). The matings of most of these males consistently resulted in *normal*-sized litters; a few males, however, whose matings consistently led to litters of *reduced* size (2 to 4 young) were found. About half the offspring of the *reduced* litters mated with individuals from *normal*-sized litters also produced *reduced* litters, whereas the other half, similarly mated, produced litters of *normal* size. This breeding behavior was consistently repeated over several generations. Suggest a hypothesis to explain these observations and describe genetic experiments and cytological observations that would test your explanation.

997 A plant heterozygous for a reciprocal translocation involving chromosomes 1.2 and 3.4 (chromosome arms are identified by numbers) has the two trans-located chromosomes 1.4 and 2.3. Only the following four kinds of gametes are formed:

(1.2, 3.4)   (1.2, 2.3)   (1.4, 2.3)   (1.4, 3.4)

Explain.

998 In the mouse, *Mus musculus*, whose chromosomes are acrocentric, the distances and sequence of linkage group 13 genes is as follows:

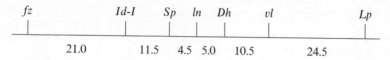

The alleles *Dip-I^a* and *Dip-I^b* (here symbolized D^a and D^b) of the dipeptidase-I gene are codominant, as are the alleles *Id-I^a* and *Id-I^b* (here symbolized I^a and I^b) of the isocitrate-dehydrogenase-I gene. *Sp* (*splotch*) is dominant to *sp* (*nonsplotch*).

Roderick (1971) crossed D^bD^b *spsp* I^bI^b males, after the lower half of their bodies were irradiated with 900 R of x-rays, with D^aD^a *SpSp* I^aI^a females. One "cytologically chosen" F$_1$ male crossed with two different females produced the results shown.

Table 998

| Parents | | Offspring | |
♀	♂	Number	Phenotype
$\dfrac{\text{a* } I^a\ sp\ D^a}{\text{a }\ I^a\ sp\ D^a} \times$	$\dfrac{\text{A* } I^b\ sp\ D^b}{\text{a }\ I^a\ Sp\ D^a}$	2	$I^a\ Sp\ D^a$
		7	$I^{ab}\ sp\ D^{ab}$
$\dfrac{\text{a } I^b\ sp\ D^a}{\text{a } I^b\ sp\ D^a} \times$	$\dfrac{\text{A } I^b\ sp\ D^b}{\text{a } I^a\ Sp\ D^a}$	7	$I^{ab}\ Sp\ D^a$
		12	$I^b\ sp\ D^{ab}$

* a = normal chromosome; A = aberrant homolog.

a What kind of chromosome mutation did the F$_1$ male possess in the hetero-zygous condition? Explain. Is it most likely paracentric or pericentric? Explain.

b Is the dipeptidase-I gene on chromosome 13?   Explain.   If the *Dip-I* locus is 24 map units from *Id-I*, where would you place *Dip-I* on the linkage map and why?

999   In barley, approximately half the progeny of a self-fertilized translocation heterozygote are *semisterile*, whereas all the progeny of a self-fertilized inversion heterozygote are *fully fertile*.   Explain why the progeny of the former but not the latter type of heterozygote are *semisterile*.

1000   In *Datura*, the allele pairs *Pp* (*purple* vs. *white* flowers) and *Ss* (*spiny* vs. *smooth* capsules) are usually independently inherited.   When the F_1's (all *fully fertile*) from a certain cross between *PP SS* and *pp ss* were testcrossed, however, two of the F_1 plants gave progeny deficient in recombinants.   When the two progenies were grouped, the total testcross offspring from these two plants was as follows:

PS   80
ps   90   all *fully fertile*
Ps   16
pS   14

Suggest a hypothesis to account for these results and explain why the F_1 parents of the above progeny and all the progeny are *fully fertile*.

1001   In *Drosophila melanogaster*, the allele pairs *Pp* (*purple* vs. *red* eyes) and *Mm* (*normal* vs. *miniature* wings) are located on chromosomes 2 and 3, respectively. Males heterozygous for both pairs of alleles were individually crossed with *pp mm* females.   The offspring of most testcrosses consisted of four classes of flies: (1) *purple normal*, (2) *purple, miniature*, (3) *red, normal*, and (4) *red, miniature*.   In a few progenies only *purple, normal* and *red, miniature* individuals were present in approximately equal numbers.
a Suggest a hypothesis to account for these results.
b Indicate what cytological observations would confirm your hypothesis.
c Would the same results be obtained in the reciprocal of this cross?   Explain.

1002   Anderson (1934) crossed a *semisterile* plant heterozygous for a reciprocal translocation and the alleles *Pl* (*purple* plant color) and *pl* (*green* plant color) with a *fully fertile* plant with the standard segmental arrangement and homozygous for *pl*.   The 402 offspring consisted of

Semisterile, green   55
Semisterile, purple   141
Fully fertile, green   137
Fully fertile, purple   69

How far from locus *pl* is the interchange (translocation) point?

1003   Working with a *semisterile* line of corn, Burnham (1948) found that from crosses with *fully fertile* lines it produced progeny in a ratio of 1 *fully fertile* : 1 *semisterile*.   Genetic tests with *Anan* (*normal* ear vs. *anther* ear) and *Rara*

Table 1003

Trait pair	Semisterile	Fully fertile
normal vs. anther ear	6 normal	196 normal
	188 anter	10 anther
normal vs. ramosa tassel	4 normal	204 normal
	190 ramosa	2 ramosa

(*normal* vs. *ramosa* tassel) gave results similar to those in the table.   In standard chromosome lines *an* is on chromosome 1, and *ra* is on chromosome 7.

a Formulate a plausible hypothesis to explain why these genes, which belong to two different linkage groups, fail to recombine at random with *semisterility* vs. *full fertility*.

b Give the genotypes and chromosome constitutions of the (1) *semisterile* and (2) *fully fertile* lines.

c What cytological configuration would you expect to see if chromosomes of *semisterile* plants were examined at pachytene?   At metaphase I?

d Diagram the meiotic synapsis of the chromosomes carrying these genes, showing one possible location of each gene.   Illustrate how aborted pollen originates.

1004   In the mouse, *Mus musculus*, the allele pairs *Bb* (*bent* vs. *straight* tail) and *Cc* (*normal* vs. *curly* whiskers) are usually inherited independently.   The F_1's from the cross *bent, normal* female × *straight, curly* male were all *fully fertile*. Two F_1 females produced unexpected results when mated with *bent, curly*

Table 1004

Phenotype	♀	♂
bent, normal	196	160
bent, curly	204	37
straight, normal	0	31
straight, curly	0	172

males.   As in the F_1, all the males and females were *fully fertile*.   Suggest a hypothesis to account for these results.   Be sure that the explanation accounts for the *full fertility* of all the F_1's and all the F_2's.

1005   Using Snell's technique (see Prob. 996), Koller and Auerbach (1941) and Koller (1944) produced three male mice giving *semisterile* offspring as evidenced by a consistent reduction in litter size from their matings with *normal* females.   In the *semisterile* lines A, B, and T established from each of these males, an association of four chromosomes was seen at spermatogenesis.   In lines B and T it was present as a chain of five and in A as a ring.   Suggest an interpretation of these observations accounting for the synaptic differences and indicate what genetic tests and cytological observations of somatic cells you would make to test your interpretation. *Note:* All 20 linkage groups have been established in the mouse, and each chromosome pair has a recognizable morphology in somatic cells.

1006 Burnham (1934) crossed a *semisterile* corn plant heterozygous for a reciprocal translocation and the recessive alleles for *shrunken, sh,* and *sugary, wx,* endosperm with a *fully fertile* plant with the normal segmental arrangement and homozygous for both recessive alleles.   The progeny of 574 segregated into eight phenotypic classes.

Table 1006

Phenotype	SS	FF
full, starchy	6	83
full, waxy	205	40
shrunken, starchy	17	171
shrunken, waxy	49	3

SS = semisterile; FF = fully fertile.

a Illustrate the cross cytogenetically for either *sh* or *wx* to account for the observed results.

b Is it possible from the above data alone to determine whether the genes in the standard line are on the same or different chromosomes?   Explain.   If not, how would you proceed to obtain this information?

1007 In corn, the allele pairs *Crcr* (*normal* vs. *crinkly* leaves) and *Dd* (*normal* vs. *dwarf* plant height) are on chromosome 3; *Ww* (*waxy* vs. *starchy* endosperm) and *Cc* (*purple* vs. *white* aleurone) are on chromosome 9.   Their positions relative to each other and the centromere, together with distances in map units, are as follows:

Chromosome 3
$$\frac{Cr \qquad D}{20 \qquad 1.5}$$

Chromosome 9
$$\frac{W \qquad C}{20 \qquad 20}$$

Line 1 is homozygous for all four dominant alleles

$$\frac{Cr \quad D}{Cr \quad D} \qquad \frac{W \quad C}{W \quad C}$$

and line 2 is homozygous for all four recessive alleles

$$\frac{cr \quad d}{cr \quad d} \qquad \frac{w \quad c}{w \quad c}$$

Pollen of line 1 was irradiated with x-rays and used to pollinate line 2.   The following results were obtained:

1   One F_1 had *crinkly* leaves.   When self-fertilized, 1/4 of the offspring showed this trait.
2   A second F_1 plant had *crinkly* leaves and was a *dwarf*.
When testcrossed with line 2, all the progeny were *dwarfed* and had *crinkled* leaves.
3   A third F_1 plant, dominant for all characters, was *semisterile*.   When testcrossed to line 2, it yielded offspring with the expected percent recombinants between *Cr* and *D* (1.5 percent) but no recombinants between *W* and *C*.
4   A fourth F_1 plant dominant for all characters was also *semisterile*; when testcrossed to line 2 and progeny classified for *Cr* vs. *cr* and *W* vs. *w*, it was found that these allele pairs were linked.

Describe the most probable nature of the effect of irradiation in each of the four F_1 plants and the results you would expect from a cytological examination of meiosis in each.

1008   **a**   In the screw-worm fly (*Cochliomyia hominivorax*) *pigmented* R cell (a wing trait) vs. *nonpigmented* R cell is autosomal.   *Nonpigmented* breeds true.   La Chance et al. (1964) found that in 18 successive generations of *pigmented* × *pigmented* matings, the progeny ratio consistently approximated 3 *pigmented* : 1 *nonpigmented*, the total ratio for all generations being 31,439 : 11,371.   The cross *pigmented* female × *nonpigmented* male gave 607 *pigmented* : 630 *nonpigmented*, and the reciprocal cross gave 627 *pigmented* : 598 *nonpigmented*. State which of the following is the more suitable explanation for these data and why:
(1) *Pigmented* controlled by an incompletely dominant lethal allele.
(2) *Pigmented* controlled by an allele conferring complete sterility on homozygotes.
b   Estimates of the percent fertility by means of egg hatchability of the crosses described above are tabulated.   Assuming these estimates are accurate:
(1) Show whether they are in agreement with the favored hypothesis in (a).
(2) Derive a conclusion regarding the nature of the sterility involved.

Table 1008

Cross		Egg hatchability, %
♀	♂	
nonpigmented × *nonpigmented*		94
pigmented × *nonpigmented*		49
nonpigmented × *pigmented*		67
pigmented × *pigmented*		39

c Subsequent studies showed that:

1 *Pigmented* was linked to the autosomal recessives *fused* (21.3 percent recombination) and *yellow* (17 percent recombination), which normally segregate independently.
2 Out of 99 males in early pupal stages 51 showed a cross configuration in prophase I of meiosis; the others showed bivalents only.
3 Normal emergence was 94 percent or higher, but in the progeny of *pigmented* × *pigmented* crosses emergence was 83 percent, and 77 percent of the adults were *pigmented*. Out of 74 adult males 70 showed a cross configuration; the remaining 4 flies showed six bivalents and were low in vigor and sterile.

Outline a hypothesis to explain why:
(1) *Pigmented* males are more fertile than *pigmented* females.
(2) *Pigmented* × *pigmented* crosses produce a 3 *pigmented* : 1 *nonpigmented* ratio among the progeny.
Note: (1) No crossing-over in males.   (2) In *pigmented* × *pigmented* crosses 61 percent lethality before hatching.   (3) Males 67 percent fertile in *nonpigmented* female × *pigmented* male cross; females 50 percent fertile in *pigmented* female × *nonpigmented* male cross.

1009 In a series of crosses between homozygous translocation lines, each of which differs from the standard by a reciprocal translocation, the following configurations are found at metaphase I in the F_1 hybrids.   Determine the number of chromosome pairs in common to the two translocations for each cross.   The species has a $2n = 22$ chromosome number.

Translocation 1 × translocation 2 → F_1 $\odot$ 6

Translocation 3 × translocation 4 → F_1 2 $\odot$ 4

Translocation 5 × translocation 6 → F_1, bivalents only

Translocation 7 × translocation 8 → F_1 $\odot$ 4

1010 a Heterozygosity for a reciprocal translocation is detected in a corn ($2n = 20$) plant that shows no *semisterility*.   Suggest an interpretation and indicate what cytological and genetic observations you would make to test your hypothesis.
b In a barley ($2n = 14$) plant with a reciprocal translocation about 10 percent *sterility* is present.   Explain how this is possible.

1011 The nonfamilial type of *Down's syndrome* (*mongolism*) (usually occurring among children of late conceptions) is due to trisomy of chromosome 21.   Individuals with the familial (transmissible) type, where more than one child may be *affected* and where the mother may be young, have 46 chromosomes (Polani, 1960; Carter et al., 1960).   The mother or, in rare instances, the father of such *mongoloids* is found to have 45 chromosomes, one of which represents a translocation between chromosome 21 and another autosome (in most cases 13, 14, or 15).   The origin of such a reciprocal translocation is illustrated.

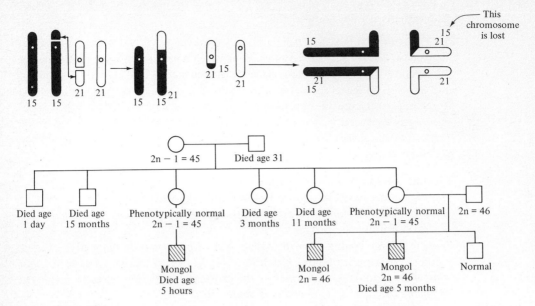

Figure 1011

Show the type of gametes the $2n - 1 = 45$ "carrier" mother can form and explain why she can produce a *normal* ($2n = 46$), a *mongoloid* ($2n = 46$), or a 45-chromosome child and why some of her offspring may die in early pregnancy or shortly after birth.

1012 Dobzhansky (1930) studied x-ray induction of chromosome aberrations using the allele pairs *Blbl* (*bristle* vs. *nonbristle*, on chromosome 2), *Dd* (*dichaete* vs. *nondichaete*, on chromosome 3), and *Eyey* (*normal* vs. *eyeless*, on chromosome 4) of *Drosophila melanogaster*.

a He carried out a large number of crosses (cultures) as follows: x-rayed males (*Blbl Dd EyEy*) × attached-X ($\hat{X}XY$) females (*blbl dd eyey*) and then crossed 153 male offspring selected for the phenotype *bristle, dichaete, normal* in single-pair matings to untreated *nonbristle, nondichaete, eyeless* females (*blbl dd eyey*). Of the 153 single-pair cultures 144 segregated for all but the body-color types and showed the expected eight phenotypic classes in approximately equal proportions in both males and females. Explain these distributions.

b The distributions in the remaining nine single-pair cultures were different and of two kinds. Two representative crosses of each type are presented in Table 1012A. Suggest:

(1) A hypothesis to account for each of the two sets of results.

(2) What cytological observations you would make to test your hypothesis.

Table 1012*A*

Exceptional class	Nonbristle, nondichaete, normal		Nonbristle, nondichaete, eyeless		Bristle, dichaete, normal		Bristle. dichaete, eyeless	
	♀	♂	♀	♂	♀	♂	♀	♂
1	28	28	24	26	22	24	15	21
2	22	16	20	20	28	21	14	19

	Nonbristle, dichaete, normal		Nonbristle, nondichaete, eyeless		Bristle, dichaete, normal		Bristle, nondichaete, eyeless	
	♀	♂	♀	♂	♀	♂	♀	♂
5	18	12	11	21	19	18	10	12
6	24	19	12	25	17	17	13	15

c *Bristle or nonbristle, dichaete, normal* males from each of cultures 5 and 6 were crossed separately with their *bristle or nonbristle, nondichaete, eyeless* sisters and produced the results in Table 1012*B*. The reciprocal of that cross gave the results in Table 1012*C*.

Table 1012*B*

From culture	Dichaete, normal	Nondichaete, eyeless
5	480	424
6	306	275

Table 1012*C*

From culture	Dichaete, normal	Nondichaete, normal	Dichaete, eyeless	Nondichaete, eyeless
5	1,080	2	2	915
6	1,240	52	58	842

(1) Are these results expected on the basis of your hypothesis?
(2) Account for the variations between the results of different crosses in the last mating.
For verification of your answer refer to Dobzhansky (1930).

1013   The chromosomes involved in a reciprocal translocation may be identified by crossing translocation homozygotes with a series of trisomics or monosomics and cytologically studying the F_1's.   This procedure has been used with trisomics in corn ($2n = 20$) by Burnham and McClintock (Burnham, 1930, 1948).   On the basis of the following metaphase I pairing associations, state, with your reasons, whether one of the chromosomes involved in reciprocal translocation is the same as the chromosome in triplicate in the trisomic.

Translocation 1 × trisomic A → $\odot 4 + 1^{III}$ (trivalent) + 17^{II} (bivalents)

Translocation 11 × trisomic B → chain of 5 chromosomes + 18^{II} (bivalents)

1014   In animals, e.g., the mouse, a large proportion of the zygotes and embryos formed by individuals heterozygous for a reciprocal translocation abort.   In higher plants, e.g., barley, however, very few, if any, zygotes or embryos formed by such individuals abort.   What is the reason for this difference?

1015   In *wild-type Haplopappus gracilis* ($2n = 4$), a reciprocal translocation occurs between the nonhomologous standard chromosomes 1.2 and 3.4 in two plants. It is found that 152 progeny of a cross between these two plants show a ring of four chromosomes at metaphase I and 48 show two bivalents.   What were the chromosome constitutions of these two plants?   Illustrate.

1016   The table (from Burnham et al., 1954) shows the synapsing configurations seen at metaphase I in meiocytes (microsporocytes) of F_1 plants from crosses between seven unknown translocation lines and a tester set of translocations in barley ($2n = 14$).   Determine the chromosomes in each unknown that are involved in the translocation.   The barley chromosomes are designated *a, b, c, d, e, f, g*.

Table 1016

Unknown	Translocation tester set				
	b-d	*c-e*	*e-f*	*c-d*	*a-b*
C1346	$\odot 6$	$\odot 6$	$\odot 6$		$\odot 6$
C1310	$\odot 6$	$2\odot 4$	$\odot 6$		$\odot 6$
C1462	$\odot 6$	$2\odot 4$	$2\odot 4$	$2\odot 4$	$\odot 6$
Erectoides-7	$\odot 6$	$2\odot 4$	$2\odot 4$		7^{II}

1017   A standard plant has the chromosomes 1.2, 3.4, 5.6, 7.8, 9.10, 11.12, 13.14, 15.16.   It is x-rayed, and a reciprocal translocation occurs between chromosomes 7.8 and 13.14 so that the new chromosomes are 7.14 and 13.8.   The homozygous translocation line 7.14, 13.8 is x-rayed, and further reciprocal translocations are produced.

1 One produces a ring of four with the 7.14, 13.8 translocaton line and only pairs (bivalents) with the standard line.

2 Another produces a ring of four with the 7.14, 13.8 lines and two rings of four with the standard line.

3 A third produces a ring of four with the standard line and a ring of six with the 7.14, 13.8 line.

What reciprocal translocations are present in these new lines?

1018 Starting with a standard diploid line (e.g., *Crepis capillaris*) with three pairs of chromosomes, how would you develop a multiple-translocation line which when crossed with the original species will give an F_1 hybrid showing a ring of six chromosomes at meiosis?

1019 Centric fusion and dissociation have both been shown to be important in karyotype reorganization, particularly in animals such as the rodent, *Gerbillus* (Wahrman and Zahavi, 1955), the mouse, *Mus* (Matthey, 1966), the shrew, *Sorex araneus* (Ford, Hamerton, and Sharman, 1957), several species of grasshopper (White, 1957, 1961), and *Drosophila* (Patterson and Stone, 1952), where acrocentrics are more common than in plants.

a Illustrate the two processes, paying particular attention to the kinds of chromosomes involved (acrocentric or metacentric), the importance of heterochromatin, and the source of the centromere involved in the dissociation process.

b In karyotype reorganization that involves a change in chromosome number, it is difficult to determine the direction of evolution, i.e., whether the chromosome numbers are increasing or decreasing. How might you determine whether a particular change in chromosome number between two species was due to dissociation or fusion?

1020 The table shows some of the results of a study by Hughes-Schrader and Schrader (1956) carried out to help clarify the taxonomic relationships among

Table 1020

Species	Diploid chromosome number	Karyotype of ♂	DNA value per nucleus
Banasa panamensis	14	12 + XY	1.4 ± 0.2
B. calva	27	24 + XXY	1.6 ± 0.02

10 genera of the hemipteran tribe Pentatomini, which have a diffuse centromere. One possible explanation of these data is that *B. calva* arose from *B. panamensis* by autotetraploidy.

a State whether or not you consider this a suitable explanation and why.

b Offer two other possible explanations and suggest which is the more satisfactory.

c How might you determine which explanation is the correct one?

1021 **a** In the genus *Carex*, chromosomes possess diffuse centromeres. The various species constitute an aneuploid series with the haploid chromosome numbers 30, 31, 32, 33, 34, 35, 36, etc. (Davies, 1956).

(1) Outline two mechanisms that could have led to this series.

(2) Suggest how information from extensive linkage-group studies might be used to decide which mechanism is the prevalent one with regard to the differences between any two species.

(3) How might studies of DNA content help to determine the mechanism involved?

b In the African pigmy mouse (*Mus minutoides*), in which all chromosomes possess localized centromeres, the basic karyotype has 36 acrocentrics. Various subspecies show a variety of karyotypes down to $2n = 18$, with all chromosomes metacentric (Matthey, 1966). There appears to be only one mechanism that can account for these results. Which is it and why?

1022 A marked degree of interspecific chromosomal polymorphism exists in the mantid genus *Ameles*. Extensive comparative cytotaxonomic and cytophotometric studies have been carried out by Wahrman and O'Brien (1956) to explain the evolutionary relationships of the species involved. The actual results of these studies for five species are shown.

Table 1022

| Species | ♂ diploid chromosome number | Autosomes | | X meta-centric | Basic arm no. | Mean DNA per spermatid |
		Meta-centric	Acro-centric			
Ameles sp. n.	19	10	8	1	30	2.07
	20	9	10	1	30	2.11
	21	8	12	1	30	1.96
A. heldreichi	27	2	24	1	30	1.84
A. andrei	28	1	26	1	30	1.71

Note: None of the chromosomes are telocentric.

a There appears to be but one satisfactory explanation for the differences in chromosome number and type between the different species. Discuss and illustrate this mechanism and state why alternative explanations will not suffice.

b If these species have been subjected to extensive genetic studies, i.e., many genes known in each, how might you utilize this information to provide support for your hypothesis?

1023 Frequently when two related species that have the same number of chromosomes are crossed, they fail to form zygotes or they produce sterile progeny. Suggest an explanation for this.

1024 In many groups of related plants, multiples of a basic chromosome number occur; e.g., in the genus *Chrysanthemum* different species have 18, 36, 54, 72, or 90 chromosomes.   Other groups of closely related species show a continuous range of chromosome numbers, e.g., the genus *Drosophila* with species numbers of 14, 12, 10, 8, and 6.   How might each of these conditions have come about?

1025 **a** In certain groups of related species or subspecies, a continuous range of diploid chromosome numbers occur; e.g., in *Mus minutoides*, different subspecies have 36, 34, 32, 30, etc., chromosomes.   The changes could be due to centric fusion, leading to a decrease in number, or to dissociation, leading to an increase in number.   Which is the more likely explanation of the two for such changes and why?

   b In the genus *Carex* (Davies, 1956) and also in *Liturgousa* (Hughes-Schrader, 1953) one species has a small number of large chromosomes, while a closely related species has a large number of small chromosomes.   Suggest explanations for these differences and ways of determining which is the most probable one.

1026 In *Drosophila melanogaster* the fourth chromosome is very small.   How might it be used for the development of a large metacentric chromosome from two acrocentric ones?

1027 Primates have 48 chromosomes; man has 46.   The tabulated karyotypes for four primate species, with man for comparison, are reported by Hamerton and Klinger (1963).

Table 1027

Species	Metacentric chromosome		Acrocentric chromosome		
			Autosomal		
	Autosomal	Sex	Large	Small	Sex
Homo sapiens (man)	34	X	6	4	Y
Pan troglodytes troglodytes (northern chimpanzee)	34	X	8	4	Y
P. t. paniscus (pigmy chimpanzee)	36	X	8	2	Y
Gorilla gorilla gorilla (lowland gorilla)	30	X, Y	12	4	
Pongo pygmaeus (orangutan)	26	X, Y	16	4	

a On the basis of numbers and kinds of chromosomes involved in the evolutionary changes, show how you would relate the species in paired comparisons (as distant, close, very close).

b Suggest how the human karyotype might have arisen from a primitive man-ape population having basically 48 chromosomes.

1028 A cross between two plants in *Crepis fuliginosa* ($2n = 6$), each with a ring of six chromosomes at metaphase I, produces 98 progeny with a ring of six, 52 with three bivalents and 50 with a ring of four and a bivalent. If the chromosome arms in the three nonhomologous chromosomes in the standard line are designated 1.2, 3.4, and 5.6, give the arm constitution of each of the six chromosomes in the parents of the cross.

1029 **a** The chromosome complements of six *Drosophila* species differing in chromosome number are shown, of which *trispina* is the putative ancestor.

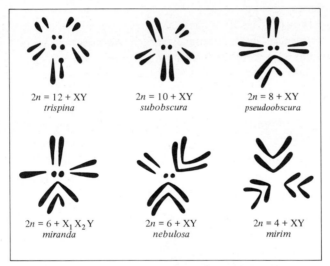

Figure 1029*A* Chromosome complements of six *Drosophila* species. That of *D. subobscura* is regarded as the basic type. [*Reproduced from K. R. Lewis and B. John, "Chromosome Marker," fig. 38, p. 82, J. and A. Churchill, Ltd., London, 1963; redrawn by them from the following figures in J. T. Patterson and W. S. Stone, "Evolution in the Genus Drosophila," The Macmillan Company, New York, 1952; fig. 42, p. 139 (trispina), fig. 40, p. 131 (Obscura, pseudoobscura, miranda), fig. 38, p. 122 (mirim), and fig. 39, p. 127 (nebulosa).*]

(1) Explain how each of these types of changes could come about and the conditions (e.g., chromosome and centromere type, importance of heterochromatin) under which the changes in chromosome number could occur.

(2) Indicate genetic experiments to test your explanation.

b The house mouse (*Mus musculus*; $2n = 40$) and the tobacco mouse (*M. poschiavinus*; $2n = 26$) are easily hybridized and produce healthy but semi-sterile offspring (Gropp et al., 1970; Tettenborn and Gropp, 1970). Somatic chromosome complements of both species and their interspecific hybrids and the typical chromosome behavior at meiosis (diakinesis) in the latter are shown in Fig. 1029*B*.

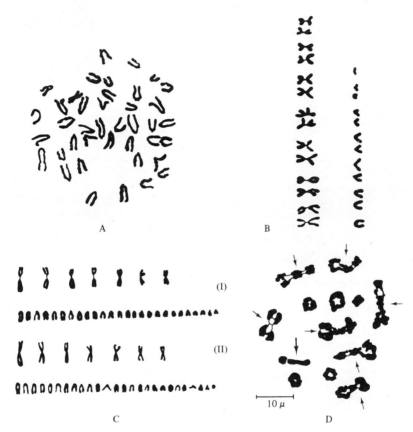

A   B

(I)

(II)

C   D

10 μ

Figure 1029*B* A. Normal house mouse (*Mus musculus*; $2n = 40$) complement. [*From M. C. Bunker, Chromosome Preparations from Solid Tumors of the Mouse: A Direct Method, Can. J. Genet. Cytol.,* **7** : 80 (*fig. 3*) (*1965*).] B. Normal tobacco mouse (*M. poschiavinus*; $2n = 26$) complement. C. Chromosome complement in (i) *M. poschiavinus* ♂ × *M. musculus* ♀ F_1 hybrids and (ii) *M. musculus* ♂ × *M. poschiavinus* ♀ F_1 hybrids. [*Reproduced from A. Gropp, U. Tettenborn, and E. von Lehmann, Chromosomenvariation vom Robertson'schen Typus bei der Tabakmaus, M. poschiavinus, und ihren Hybriden mit der Laboratoriumsmaus, Cytogenetics,* **9**:14 (*fig. 2*) (*1970*).] D. Meiosis (late diakinesis) showing seven chain trivalents (small arrows), five bivalents, and XY bivalent (large arrow). [*From U. Tettenborn and A. Gropp, Meiotic Nondisjunction in Mice and Mouse Hybrids, Cytogenetics,* **9**:276 (*fig. 2*) (*1970*).]

(1) Determine the number of acrocentric and metacentric chromosomes in each species from A and B of Fig. 1029*B*. By identifying the parental chromosomes in the hybrid, show whether the count obtained confirms the parental karyotypes.

(2) *M. poschiavinus* appears to be a species of more recent origin than *M. musculus* and therefore may have arisen from the latter. What appears to be the most plausible mechanism of origin of the *M. poschiavinus* karyotype from that of *M. musculus*? What is the basis for your explanation? Discuss fully. Note that the tobacco mouse probably has less DNA than the house mouse.

(3) How would you proceed to verify your hypothesis genetically? Indicate the results expected.

REFERENCES

Alexander, M. L. (1952), *Univ. Tex. Publ.* 5204, p. 219.
Anderson, E. G. (1934), *Am. Nat.*, **68**:345.
Atkin, N. B., G. Mattinson, W. Beçak, and S. Ohno (1965), *Chromosoma*, **17**:1.
Bahn, E. (1971), *Hereditas*, **67**:75.
Baker, R. H., R. K. Sakai, A. Mian (1971), *Science*, **171**:585.
Bartalos, M., and T. A. Baramki (1967), "Medical Cytogenetics," Williams & Wilkins, Baltimore.
Beadle, G. W., and A. G. Sturtevant (1935), *Proc. Natl. Acad. Sci.*, **21**:384.
Bearn, A. G., and J. L. German (1961), *Sci. Am.*, **205**(November) : 66.
Belling, J. (1914), *Z. Indukt. Abstammung.-Vererbungsl.*, **12**:303.
——— (1915), *Am. Nat.*, **49**:582.
——— (1925), *J. Genet.*, **15**:245.
——— and A. F. Blakeslee (1924), *Proc. Natl. Acad. Sci.*, **10**:116.
——— and ——— (1926), *Proc. Natl. Acad. Sci.*, **12**:7.
Bender, K., and S. Ohno (1968), *Biochem. Genet.*, **2**:101.
Bender, M. A., and E. H. Y. Chu (1963), in J. Buettner-Janusch (ed.), "Evolutionary and Genetic Biology of Primates," pp. 261–310, Academic, New York.
——— and M. A. Barcinski (1969), *Cytogenetics*, **8**:241.
Benedict, W. F., and M. Karon (1971), *Science*, **171**:68.
Benirschke, K. (ed.) (1969), "Comparative Mammalian Cytogenetics," Springer-Verlag, New York.
Bock, I. R. (1971), *Chromosoma*, **34**:206.
Boyer, S. H., et al. (1971), *Biochem. Genet.*, **5**:405.
Bridges, C. B. (1917), *Genetics*, **2**:445.
——— (1935), *J. Hered.*, **26**:60.
——— (1936), *Science*, **83**:210.
Brink, R. A., and D. C. Cooper (1931), *Genetics*, **16**:595.
Britten, R. J., and D. E. Kohne (1968), *Science*, **161**:529.
Brøgger, A. (1969), *Hereditas*, **62**:116.
Budnik, M., S. Koref-Santibanez, and D. Brncic (1971), *Genetics*, **69**:227.
Burnham, C. R. (1930), *Proc. Natl. Acad. Sci.*, **16**:269.
——— (1934), *Genetics*, **19**:430.
——— (1948), *Genetics*, **33**:5.
——— (1950), *Genetics*, **35**:446.
——— (1954), *Maize Genet. Coop News Lett.*, **28**:59.
——— (1956), *Bot. Rev.*, **22**:419.

Burnham, C. R. (1962), "Discussions in Cytogenetics," Burgess, Minneapolis.
———, F. H. White, and R. Livers (1954), *Cytologia*, **19**:191.
Carr, D. H. (1967), *Am. J. Obstet. Gynecol.*, **97**:283.
Carson, H. L. (1946), *Genetics*, **31**:95.
——— (1953), *Genetics*, **38**:168.
——— and H. D. Stalker (1947), *Evolution*, **1**:113.
Carter, C. O. (1969), "An ABC of Medical Genetics," Little, Brown, Boston.
———, J. L. Hamerton, P. E. Polani, A. Gunalp, and S. D. V. Waller (1960), *Lancet*, 2:678.
Carter, T. C., M. F. Lyon, and R. J. S. Phillips (1956), *J. Genet.*, **54**:462.
Catti, A., and W. Schmid (1971), *Cytogenetics*, **10**:50.
Chen, T. R. (1971), *Chromosoma*, **32**:436.
Cohen, M. M., V. J. Capraro, and N. Takagi (1967), *Ann. Hum. Genet.*, **30**:313.
Conger, A. D. (1967), *Mutat. Res.*, **4**:449.
Cooper, J., F. E. Arrighi, and T. C. Hsu (1970), *Cytogenetics*, **9**:468.
Court-Brown, W. M. (1967), "Human Population Cytogenetics," Wiley, New York.
——— and P. G. Smith (1969), *Br. Med. Bull.*, **25**:74.
Crandall, B. F., and R. S. Sparkes (1970), *Cytogenetics*, **9**:307.
Creighton, H. B. (1937), *Genetics*, **22**:189.
Darlington, C. D. (1937), "Recent Advances in Cytology," 2d ed., Blakiston, Philadelphia.
Davies, E. W. (1956), *Hereditas*, **42**:349.
De Grouchy, J., et al. (1964), *Ann. Genet.*, **7**:13.
Demerec, M., and M. Hoover (1936), *J. Hered.*, **27**:206.
Dishotsky, N. I., et al. (1971), *Science*, **172**:431.
Dobzhansky, T. (1930), *Genetics*, **15**:347.
——— (1931), *Genetics*, **16**:629.
——— (1944), *Carnegie Inst. Wash., D.C., Publ.* 554, p. 47.
——— and C. Epling (1944), *Carnegie Inst. Wash., D.C., Publ.* 554, p. 1.
——— and A. H. Sturtevant (1938), *Genetics*, **23**:28.
Eicher, E. M. (1970), *Genetics*, **64**:495.
Ephrussi, B. (1934), *Proc. Natl. Acad. Sci.*, **20**:420.
Ford, C. E. (1969), *Br. Med. Bull.*, **25**:81.
——— and H. M. Clegg (1969), *Br. Med. Bull.*, **25**:110.
———, J. L. Hamerton, and G. B. Sharman (1957), *Nature*, **180**:392.
Gates, W. H. (1927), *Genetics*, **12**:295.
German, J. (1969), *Am. J. Hum. Genet.*, **21**:196.
——— (1970), *Am. Sci.*, **58**:182.
——— and L. P. Crippa (1966), *Ann. Genet.*, **9**:143.
Giannelli, F., and R. M. Howlett (1967), *Cytogenetics*, **6**:420.
Grace, E., D. Drennan, D. Colver, and R. R. Gordon (1971), *J. Med. Genet.*, **8**:351.
Gripenberg, U. (1967), *Chromosoma*, **20**:284.
Gropp, A., U. Tettenborn, and E. von Lehmann (1970), *Cytogenetics*, **9**:9.
Gustavsson, I., and C. O. Sundt (1969), *Chromosoma*, **28**:245.
Hamerton, J. L. (1971), "Human Cytogenetics," vols. I and II, Academic, New York.
——— et al. (1963), *Cytogenetics*, **2**:240.
——— and H. P. Klinger (1963), *New Sci.*, **18**(341):483.
——— (ed.) (1962), "Chromosomes in Medicine," pp. 140–183, Little Club Clinic Developmental Medicine, no. 5, National Spastic Society, and Heinemann, London.
Hsu, T. C., and F. E. Arrighi (1966), *Cytogenetics*, **5**:355.
——— and ——— (1968), *Cytogenetics*, **7**:417.
Huang, C. C. (1967), *Chromosoma*, **23**:162.
Hughes-Schrader, S. (1953), *Chromosoma*, **5**:544.
——— and F. Schrader (1956), *Chromosoma*, **8**:135.
Ingram, V. M. (1961), *Nature*, **189**:704.

Insley, J. (1967), *Arch. Dis. Child.*, **42**:140.

Jackson, L., and M. Barr (1970), *J. Med. Genet.*, **7**:161.

Kato, R. (1967), *Hereditas*, **58**:221.

Khush, G. S., and C. M. Rick (1968), *Chromosoma*, **23**:452.

——, ——, and R. W. Richardson (1964), *Science*, **145**:1432.

Kistenmacher, M. L., and H. H. Punnett (1970), *Am. J. Hum. Genet.*, **22**:304.

Klinger, H. P. (1963), *Cytogenetics*, **2**:141.

Koehn, R. K., and D. I. Rasmussen (1967), *Biochem. Genet.*, **1**:131.

Koller, P. C. (1944), *Genetics*, **29**:247.

—— and C. A. Auerbach (1941), *Nature*, **148**:501.

La Chance, L. E., J. G. Riemann, and D. E. Hopkins (1964), *Genetics*, **49**:959.

Law, E. M., and J. G. Masterson (1966), *Lancet*, **2**:1137.

Lejeune, J. (1964), *Ann. Genet.*, **7**:7.

—— et al. (1963), *C. R. Acad. Sci.*, **257**:3098.

Lele, K. P., L. S. Penrose, and H. B. Stollard (1963), *Ann. Hum. Genet.*, **27**:171.

Léonard, T. H. (1971), *Mutat. Res.*, **11**:71.

Levitan, M. (1950), *Genetics*, **35**:674.

Lewis, K. R., and B. John (1963), "Chromosome Marker," Churchill, London.

Lindsten, J., M. Fraccaro, and H. P. Klinger (1965), *Cytogenetics*, **4**:45.

——, ——, P. E. Polani, J. L. Hamerton, R. Sanger, and R. R. Race (1963), *Nature*, **197**:648.

Mackensen, O. (1935), *J. Hered.*, **26**:163.

McClintock, B. (1931), *Genetics*, **25**:542.

—— (1931), *Mo. Agric. Exp. Stn. Res. Bull.* 163, p. 1.

—— (1938), *Mo. Agric. Exp. Stn. Res. Bull.* 290, p. 1.

—— (1941), *Genetics*, **26**:234.

—— (1944), *Genetics*, **29**:478.

McCracken, J. S., and R. R. Gordon (1965), *Lancet*, **1**:23.

Maguire, Marjorie P. (1966), *Genetics*, **53**:1071.

Mark, J. (1970), *Acta Ophthalmol.*, **48**:124.

Matthey, R. (1966), *Rev. Suisse Zool.*, **73**:585.

May, H. G. (1917), *Biol. Bull.*, **33**:361.

Miller, D. A., R. E. Kouri, V. G. Dev, M. S. Grewal, J. J. Hutton, and O. J. Miller (1971), *Proc. Natl. Acad. Sci.*, **68**:2699.

Moens, P., and L. Butler (1963), *Can. J. Genet. Cytol.*, **5**:364.

Morgan, D. T. (1950), *Genetics*, **35**:153.

Mukerjee, D., and W. J. Burdette (1966), *Am. J. Hum. Genet.*, **18**:62.

Muller, H. J. (1929), *Am. Nat.*, **63**:481.

Nadler, C. F. (1968), *Cytogenetics*, **7**:144.

Nance, W. E., and E. Engel (1967), *Science*, **155**:692.

Natarajan, A. T., and W. Schmid (1971), *Chromosoma*, **33**:48.

Nesbitt, M., and U. Francke (1971), *Science*, **174**:60.

Nowell, P. C., and D. A. Hungerford (1960), *Science*, **132**:1497.

Ohno, S. (1970), "Evolution by Gene Duplication," Springer-Verlag, New York.

—— and M. Morrison (1966), *Science*, **154**:1034.

—— and J. M. Trujillo (1963), *Acta Haematol.*, **29**:311.

—— U. Wolf, and N. B. Atkin (1968), *Hereditas*, **59**:169.

Painter, T. S. (1927), *Genetics*, **12**:379.

—— (1933), *Science*, **78**:585.

—— (1934), *J. Hered.*, **25**:464.

Palmer, C. G., N. Fareed, and A. D. Merritt (1967), *J. Med. Genet.*, **4**:117.

Patterson, J. T., and W. S. Stone (1952), "Evolution in the Genus *Drosophila*," Macmillan, New York.

Polani, P. E., J. H. Briggs, C. E. Ford, C. M. Clarke, and J. M. Berg (1960), *Lancet*, **1**:721.
Polito, L., F. Graziani, E. Boncinelli, C. Malva, and F. Ritossa (1971), *Nat. New Biol.*, **229**:84.
Prakash, S., and R. C. Lewontin (1971), *Genetics*, **69**:504.
Redfield, H. (1957), *Genetics*, **42**:712.
Rees, H., and R. N. Jones (1967), *Nature*, **216**:825.
Rhoades, M. M. (1951), *Am. Nat.*, **85**:105.
────── and B. McClintock (1935), *Bot. Rev.*, **1**:292.
Richards, G. K. (1964), *Chromosoma*, **15**:100.
Roberts, Paul A. (1967), *Genetics*, **56**:179.
Robson, E. B., P. E. Polani, S. J. Dart, P. A. Jacobs, and J. H. Renwick (1969), *Nature*, **223**:1163.
Roderick, T. H. (1971), *Mutat. Res.*, **11**:59.
Russell, L. B. (1971), *Mutat. Res.*, **11**:107.
────── and W. L. Russell (1960), *J. Cell Comp. Physiol.*, **56**(suppl. 1):169.
────── and C. S. Montgomery (1970), *Genetics*, **64**:281.
Rutishauser, A. (1960), *Heredity*, **15**:241.
────── and L. F. LeCour (1956), *Chromosoma*, **8**:317.
Scheel, J. J. (1971), *Hereditas*, **67**:287.
Schrader, F., and S. Hughes-Schrader (1958), *Chromosoma*, **9**:193.
Sjodin, J. (1971), *Hereditas*, **68**:1.
Smithies, O., G. E. Connell, and G. H. Dixon (1962), *Nature*, **196**:232.
Snell, G. D. (1933), *Am. Nat.*, **67**:24.
────── (1935), *Genetics*, **20**:545.
────── (1946), *Genetics*, **31**:157.
Somers, C. E., and T. C. Hsu (1962), *Proc. Natl. Acad. Sci.*, **88**:937.
Soudek, D., R. Laxova, and R. Adamek (1968), *Cytogenetics*, **7**:108.
Stadler, L. J. (1933), *Univ. Mo. Res. Bull.* 204, p. 3.
Stephens, S. G. (1951), *Cold Spring Harbor Symp. Quant. Biol.*, **46**:131.
Stern, C. (1943), *Genetics*, **28**:441.
Stewart, J. M., S. Go, E. Ellis, and A. Robinson (1970), *J. Med. Genet.*, **7**:11.
Sturtevant, A. H. (1921), *Proc. Natl. Acad. Sci.*, **7**:235.
────── (1925), *Genetics*, **10**:117.
────── (1926), *Biol. Zentralbl.*, **46**:697.
────── and G. W. Beadle (1936), *Genetics*, **21**:554.
────── and T. Dobzhansky (1936), *Proc. Natl. Acad. Sci.*, **22**:448.
────── and T. H. Morgan (1923), *Science*, **57**:746.
────── and C. R. Plunkett (1926), *Biol. Bull.*, **50**:56.
Tettenborn, U., and A. Gropp (1970), *Cytogenetics*, **9**:272.
Thompson, W. P., and I. Hutcheson (1942), *Can. J. Res.*, **C20**:267.
Valentine, G. H. (1969), "The Chromosome Disorders," 2d ed., Heinemann, London.
Van Dyke, H. E., A. Valdmanis, and J. D. Mann (1964), *Am. J. Hum. Genet.*, **16**:364.
Wahrman, J., and R. O'Brien (1956), *J. Morphol.*, **99**:259.
────── and A. Zahavi (1955), *Nature*, **175**:600.
Watts, R. L., and D. C. Watts (1968), *J. Theor. Biol.*, **20**:227.
Weiss, L. (1969), *J. Med. Genet.*, **6**:216.
White, M. J. D. (1957), *Aust. J. Zool.*, **5**:285.
────── (1961), "The Chromosome," 5th ed., Methuen, London.
────── and F. H. W. Morley (1955), *Genetics*, **40**:604.
Wurster, D. H., K. Benirschke, and H. Noelke (1968), *Chromosoma*, **23**:317.
Yamashita, K. (1951), *Cytologia*, **16**:164.
Yosida, T. H., K. Tsuchiya, H. Imai, and K. Moriwaki (1969), *Jap. J. Genet.*, **44**:89.
Zeleny, C. (1919), *J. Genet. Physiol.*, **2**:69.
────── (1921), *J. Exp. Zool.*, **34**:203.

23
Balanced Lethal Systems and Oenothera Cytogenetics

QUESTIONS

1030 Chromosomal aberrations appear to be a necessary adjunct to the development of balanced lethal systems. Discuss, with special reference to *Oenothera* and *Drosophila*.

1031 Balanced lethal stocks are maintained in certain *Drosophila* laboratories for genetic studies for a very good reason. What is it?

1032 Both zygotic and gametophytic lethals occur in *Oenothera*. How would you determine whether the one type or the other or both occur in a species?

1033 *Oenothera* gains the immediate advantage that comes from heterosis but sacrifices the chief advantage of sexual reproduction in that it does not allow beneficial mutations to concentrate in a given strain or race through recombination. Discuss.

PROBLEMS

1034 *Truncate Drosophila* (lacking wing tips) when mated together always produce some *normal* offspring. *Plum*-eyed flies mated together always produce some *nonplum* offspring. However, when *truncate, plum* flies are intermated, the offspring are always *truncate, plum* only. Explain all three results.

1035 Darlington and Gairdner (1937) found in the bellflower, *Campanula persicifolia* ($n = 8$), that translocation heterozygotes discovered in *wild* stocks when self-fertilized bred true. This was not true for translocation lines produced in *cultivated* varieties, which upon self-fertilization produce some progeny homozygous for the interchange and some homozygous for the standard segmental arrangement. Why do the *wild* and not the *cultivated* translocation stocks breed true for the ring configuration?

1036 *Beaded, Bd*, is a chromosome-3 mutant trait involving indented wing margins with beadlike remnants of the margin between indentations. Muller (1918) obtained the following results with this trait:

1 *Beaded* flies when intermated did not breed true but produced *beaded* and *normal* progeny in a 2:1 ratio.

2 After many generations a stock breeding true for *beaded* was obtained.

3 By appropriate crosses the alleles *ss* (*spineless*) and *se* (*sepia*) were introduced into the chromosome with *Bd*, in flies in which the homolog carried the dominant alleles at these loci. These flies

$$\frac{ss\ se\ Bd}{Ss\ Se\ bd}$$

when intermated bred true for *beaded* and the traits controlled by the dominant alleles. When this stock was crossed to one homozygous for *ss* and *se*, half of the F_1 flies were *spineless, sepia,* and *beaded*; the other half were *nonbeaded* but showed the dominant traits for the remaining characters. The F_1's of the first type when intercrossed segregated for *beaded* but bred true for the recessive traits of the remaining characters.   Those of the second type segregated approximately 2 *normal* : 1 *spineless sepia*.

4   In stocks in which the *Bd* chromosome carried *ss* and its homolog *Ss*, about 1 in 1,000 progeny showed the *spineless* trait when the flies were intercrossed.

Analyze and explain these data.   Be sure to state the genetic and cytological conditions that must be satisfied in each case.

1037   In the annual garden stock (*Matthiola incana*) *single* vs. *double* flowers is controlled by a single pair of alleles *S* vs. *s*.   While some *single*-flowered varieties breed true, others are "ever-sporting," i.e. when self-fertilized, they always produce both *single*- and *double*-flowered progeny.   The *double*-flowered plants are completely sterile.   The ever-sporting varieties also tend to degenerate, producing fewer *double* flowers in each succeeding year (Kappert, 1937).

a Show how this can be accounted for by a balanced lethal system.

b For alternative explanations accounting for the breeding behavior, see Kappert, 1937, 1951; Johnson, 1953.

1038   The results shown for leaf color are chosen from the results of many crosses in the snapdragon (*Antirrhinum majus*) (Gairdner and Haldane, 1929).   Reciprocal crosses gave approximately the same results.

Table 1038

Parents		Offspring		
♀ ♂		Yellow	Green	Albino
green × *yellow*		2,259	2,362	0
green × *green*		0	4,801	0
yellow × *yellow*		144	73	0
yellow × *yellow*		10,982	1,177	6,198
yellow × *green*		4,613	4,575	0
green × *green*		0	873	0
green × *green*		0	13,786	3,911*

* Some plants in this class may not have germinated.

State the number of allele pairs involved, the type of allele relationship for each, and the type of gene interaction, if any; show why in the fourth cross the yellows nearly bred true, producing about 9.5 *yellow* : 1 *green*.

1039   Two *Drosophila* with *curly* wings produce 102 *curly*-wing and 48 *straight*-wing offspring, the latter consisting of 23 fertile males and 25 sterile females.   The *curly* females mated with their *straight* male sibs produced *curly* and *straight* progeny in approximately equal proportions.   Half the *straight* progeny were fertile males; the other half were sterile females.   These fertile males when crossed to females from a true-breeding female-fertile *straight* stock produced only fertile *straight* progeny.   When males from this latter progeny were crossed with their sisters, they produced only *straight* progeny, one-fourth of the females being sterile.   Account for:

a The 2:1 ratio of *curly*:*straight* progeny when *curly* flies are intercrossed.

b The 1:1 ratio for these traits when *curly* females are mated with *straight* males.

c One-fourth of the females being sterile in the latter cross.

1040   In *Drosophila melanogaster*, *lg* is a recessive allele on the second chromosome that causes larvae to develop to a large size and then die.   In a stock of flies true-breeding for *curly* wings the adult progeny size is one-half that produced from *wild-type* matings.   Of the embryos that do not develop into adults, half die as giant larvae.   Why does the stock breed true for *curly* wings?

1041   In the mouse, *Mus musculus*, individuals may have *normal* or *short* (brachyuric) tails, or they may be *tailless*.   Extensive investigation by Dunn and coworkers (see Dunn, 1960; Dunn and Glueksohn-Waelsch, 1952) have shown that the *normal*, *short*, and *tailless* traits are controlled by the T locus (region) on chromosome 9.   The results of matings involving the different traits approximate theoretical ratios as shown (after data by Dunn and Glueksohn-Waelsch,

Table 1041

Parents	Offspring
normal × *normal*	all *normal*
normal × *short*	1 *normal* : 1 *short*
short × *short*	1 *normal* : 2 *short*
normal × *normal*	all *normal* (25% sterility)
normal × *short*	1 *normal* : 1 *tailless*[0]* : 1 *short* (25% sterility)
tailless[0]* × *tailless*[0]*	all *tailless* (50% sterility)
normal × *normal*	3 *normal* : 1 *tailless*[3]*
tailless[0]* × *tailless*[3]*	1 *tailless* : 1 *normal*
normal × *tailless*[0]*†	2 *tailless* : 1 *normal*

* *Tailless*[0] and *tailless*[3] are phenotypically similar but genotypically different.

† From previous progeny.

1952).   Offer an explanation of these results, showing the chromosomal and gene constitutions of the various phenotypes.   See Dunn (1939–1940) and Glueksohn-Schoenheimer (1938) for their explanations.   For a discussion of the embryological effects associated with these traits see Bennett and Dunn (1964), Bennett et al. (1959), and Silagi (1962).   The cytological basis underlying the genic control of these traits is discussed in an elegant paper by Geyer-Duszynska (1964).

1042 Reciprocal crosses between *Oenothera chicaginensis* and *O. cockerelli* produce a single, unique class of progeny in each direction; the progeny of reciprocal crosses are different. Reciprocal crosses between *O. lamarckiana* and *O. grandiflora* produce several hybrid phenotypes called *twin* or *multiple hybrids* in each direction; the results of reciprocal crosses are identical in this respect. Progeny of all crosses mentioned above breed true when self-fertilized (Cleland, 1936).

   a For each of the two sets of reciprocal crosses, show whether the lethals are gametic, zygotic, or both, and give reasons for your statements.

   b In which of the two sets of reciprocal crosses would you expect seed sterility? How much? Why?

1043 *Oenothera muricata* when self-fertilized produces only one type of zygote and shows no seed sterility. *O. lamarckiana* when self-fertilized produces three types of zygotes, but two of these abort to give 50 percent seed sterility. Why is there but one type of zygote and no seed sterility in *O. muricata* and three types of zygotes but 50 percent sterility in *O. lamarckiana*?

1044 In an organism with the chromosome number $2n = 8$, homozygous strains with the following segmental arrangements occur:

Standard	1.2	3.4	5.6	7.8
Race 1	1.2	3.8	5.4	7.4
Race 2	1.4	3.2	5.6	7.8
Race 3	1.2	3.6	5.4	7.8
Race 4	undetermined			

The following ring configurations are produced in crosses between race 4 and each of the other races:

Standard × race 4	Ring of 4
Race 1 × race 4	Ring of 6
Race 2 × race 4	Ring of 6
Race 3 × race 4	Two rings of 4

What is the segmental arrangement of the chromosomes in race 4? *Hint:* Two chromosomes must pair as bivalents with standard, only one with race 1, etc.

1045 The segmental arrangements of five chromosome complexes in *Oenothera* are as follows:

hookeri	1.2	3.4	5.6	7.8	9.10	11.12	13.14
flavens	1.4	3.2	5.6	7.8	9.10	11.12	13.14
velans	1.2	3.4	5.8	7.6	9.10	11.12	13.14
Johansen	1.2	3.4	5.6	7.10	9.8	11.12	13.14
acuens	1.4	3.2	5.6	7.10	9.8	11.12	13.14

a For each of the following combinations of complexes give the ring configurations and the chromosomes involved and the number of chromosome pairs forming bivalents at metaphase I of meiosis:

(1) *hookeri* × *velans* (2) *flavens* × *velans*
(3) *acuens* × *Johansen* (4) *hookeri* × *acuens*
(5) *flavens* × *Johansen* (6) *velans* × *acuens*

b *Rigens*, the egg complex of *muricata*, gives the following ring configurations and chromosome pairs:

With *hookeri*   Ring of 6 and four pairs
With *flavens*   Ring of 4, ring of 6, and two pairs
With *velans*   Ring of 8 and three pairs
With *acuens*   Ring of 4, ring of 8, and one pair

Show with illustrations:
(1) The chromosome configurations in a *rigens-Johansen* hybrid.
(2) The segmental arrangements in the *rigens* complex.
See Cleland (1936) for the type of problem presented here.

REFERENCES

Altenburg, E., and H. J. Muller (1920), *Genetics*, **5**:1.
Bennett, C., L. C. Dunn, and S. Badenhausen (1959), *J. Morphol.*, **105**:105.
Bennett, D., and L. C. Dunn (1964), *Genetics*, **49**:949.
Catcheside, D. G. (1931), *Proc. R. Soc. (Lond.)*, **B109**:165.
Cleland, R. E. (1935), *Proc. Am. Philos. Soc.*, **75**:339.
—— (1936), *Bot. Rev.*, **2**:316.
—— (1962), *Bot. Rev.*, **11**:147.
—— and A. F. Blakeslee (1930), *Proc. Natl. Acad. Sci.*, **16**:182.
Darlington, C. D., and A. E. Gairdner (1937), *Genetics*, **35**:97.
Dawson, P. S. (1967), *Heredity*, **22**:435.
de Vries, H. (1907), *Bot. Gaz.*, **44**:401.
Dunn, L. C. (1939–1940), *Harvey Lect.*, **35**:115.
—— (1960), *Genetics*, **42**:1531.
—— and S. Gluecksohn-Waelsch (1952), *Genetics*, **37**:577.
Emerson, S. H., and A. H. Sturtevant (1932), *Genetics*, **17**:393.
Gairdner, A. E., and J. B. S. Haldane (1929), *J. Genet.*, **21**:315.
Geyer-Duszynska, I. (1964), *Chromosoma*, **15**:478.
Gluecksohn-Schoenheimer, S. (1938), *Genetics*, **23**:573.
Johnson, B. L. (1953), *Genetics*, **38**:229.
Kappert, H. (1937), *Z. Indukt. Abstammung.-Vererbungsl.*, **73**:233.
—— (1948), "Die vererbungswissenschaftlichen Grundlagen der Pflanzen-Zuchtung," pp. 1–244, Berlin.
—— (1951), *Züchter*, **21**:205.
Muller, H. J. (1917), *Proc. Natl. Acad. Sci.*, **3**:619.
—— (1918), *Genetics*, **3**:422.
Renner, O. (1917a), *Ber. Dtsch. Bot. Ges.*, **34**:858.
—— (1917b), *Z. Vererbungsl.*, **18**:121.
Silagi, S. (1962), *Dev. Biol.*, **5**:35.
Sturtevant, A. H. (1926), *Q. Rev. Biol.*, **1**:282.

24
Gene
Mutation

NOTATION[1]

A	= adenine		HA	= hydroxylamine
AP	= 2-aminopurine		HMC	= 5-hydroxymethylcytosine
BD	= 5-bromodeoxyuridine		HX	= hypoxanthine
BU	= 5-bromouracil		NA	= nitrous acid
C	= cytosine		P	= proflavine
EES	= ethyl ethanesulfonate		T	= thymine
EMS	= ethyl methanesulfonate		U	= uracil
G	= guanine		X	= xanthine

QUESTIONS

1046 a What critical evidence is needed to distinguish a gene mutation from a minute deletion, i.e., one too small to be cytologically detectable?

b Show how you would determine the rate of mutation from a recessive to a dominant allele in man. (Use, for illustration, mutation of the recessive allele for *normal stature* to the dominant allele for *chondrodystrophy*.)

1047 A *hairless* mouse appears in a true-breeding strain of *normal* (*gray*-haired) mice in which *hairless* mice have not previously been found.

a Briefly describe how you would determine whether the *hairless* phenotype was a mutant one or not.

b Name two possible types of mutation that could lead to the appearance of this *hairless* mouse.

c Briefly discuss your procedure for distinguishing between these alternatives.

d If the *hairless* mouse had originated in a natural population (with mating more or less at random), what other explanation or explanations might be given for its appearance?

1048 It is generally found in microorganisms that the proportion of induced mutations that can revert is higher with ultraviolet as mutagen than with ionizing radiation or radiomimetic compounds. What conclusions can be drawn from this?

1049 a In a true-breeding *wild-type* (*red*-eyed) strain of *Drosophila melanogaster*, males are occasionally found that have one *red* and one *white* eye. (Mosaic females of this type are never found in this strain.) If the *white* eye is due to a mutation, why is only one eye *white*, and why did the mutated allele express itself in spite of being recessive?

b A man has one-*brown* and one-*blue* eye. What are the possible causes of this mosaicism?

1050 An *r*II mutant in phage T4 reverts (back-mutates) to produce a *pseudo-wild-type* strain. What tests would you employ to determine whether the

---

[1] Abbreviations and structural formulas of the common amino acids are given in Appendix Table A-4.

partial back mutant is due to a mutation: (1) in another gene, (2) at a second site in the same gene, or (3) in the same nucleotide-pair site as the first mutation?

1051 When 1,000 master plates of chloramphenicol agar are each replica-plated with about 1,000 colonies of a *chloramphenicol-sensitive* strain of bacteria, one plate is found with a single colony *resistant* to the drug. Explain how you would determine whether this mutation was post- or preadaptive.

1052 Explain:
  a Why geneticists find most mutations to be deleterious.
  b Why, nevertheless, the mutation process is considered to be the basis of evolutionary progress.
  c Why most mutations are recessive.
  d Why one mutation in a gene may cause a drastic phenotypic effect and another mutation in the same gene may have only a slight effect.

1053 In an experiment using the ClB method, a recessive lethal allele is induced in the X chromosome of a sperm that fertilizes an egg carrying a ClB chromosome. Why does this F_1 zygote not die? What should be the results in the F_2 of crossing such an F_1 female with a *wild-type* male?

1054 The coat of the Dalmatian breed of dogs is *spotted*.
  a What explanations can be given for this, other than somatic mutation?
  b Outline an experiment that would distinguish between the alternatives postulated. Show the results expected in each case.

1055 The homopteran insects, the plant genus *Luzula*, and some other organisms which have chromosomes with diffuse centromeres (Brown and Nelson-Rees, 1961; Hughes-Schrader, 1948) are less sensitive to radiation and chemical mutagens than organisms with localized centromeres. Explain why this should be so.

1056 One member of each of the following pairs of alternatives would probably suffer more from genetic damage from a large and constant increase in radiation exposure through fallout. State which one and why.
  1 A diploid or a polyploid
  2 A species with a long life cycle or one with a short life cycle
  3 A rapidly dividing tissue or a slowly dividing tissue of the same organism
  4 A species reproducing sexually or one reproducing asexually
  5 A germinal or a somatic tissue
  6 An adult insect or an adult mammal

1057 Is multiple allelism evidence for more than one mutational site per gene? Discuss.

1058 Answer each of the following questions as briefly as possible.
  a Which type of mutation, one induced by a base analog or one induced by proflavine, would you expect to be more deleterious to an organism and why?

b What evidence is there that ionization caused by x-rays need not occur in the gene itself to cause mutation?

c What are the possible mechanisms by which a gene may change to many different allelic forms?

d Is there a known finite dosage level of ionizing radiation below which no radiation-induced gene mutations are induced? Explain.

e Why are sex-linked lethal mutations easier to detect than autosomal lethals?

f Explain how the magnitude of phenotypic change caused by a gene mutation may be related to its significance in evolution.

g Why is the X chromosome more useful than the autosomes in studying the induction of recessive mutations?

h Which would cause greatest genetic damage, a nuclear explosion that exposed the population of a city of 500,000 to an average of 100 R or a radioactive fallout which exposed a population of 100 million to an average of 0.5 R? Explain.

i It has been recommended that in the event of an atom bomb attack adults over forty years of age should be responsible for rescuing exposed victims. Why?

1059 It has been observed that induced mutations occasionally appear in *Neurospora* and bacterial cultures several cell generations after treatment with a mutagen. Offer three explanations for this observation.

1060 Certain reports state that oxygen decreases radiation damage. Others state the opposite. Suggest a reason for these conflicting reports.

1061 There are two kinds of base replacements that can result from pairing mistakes during DNA replication (Freese, 1959*a*):

1 Transitions: a purine may replace a purine, or a pyrimidine may replace a pyrimidine.
2 Transversions: a purine may be replaced by a pyrimidine and vice versa.

Illustrate these two types of pairing mistakes, showing appropriate base-analog mutagens as replacement bases.

1062 **a** Describe what is meant by a tautomeric shift.

b Are the mutational consequences of tautomeric shifts base-pair transitions or base-pair transversions? Explain, using a specific example.

1063 There are two ways that base-analog incorporation during DNA synthesis can lead to mutation: (1) the analog may initially be incorporated opposite its correct partner but err in the choice of a partner during one of the subsequent replications (error during replication), or (2) the analog may pair with the wrong base during its incorporation (error during incorporation) (Freese, 1959*b*).

a Illustrate these two processes to show how transitions can be induced in either direction; e.g., A-T→G-C, and G-C→A-T.

b Explain how BU and other base analogs are able to cause two-way changes.

1064 Acridine dyes such as proflavine are mutagenic (De Mars, 1953).   According to Lerman (1961), the acridine molecules are inserted between adjacent base pairs in the DNA, forcing them apart by a distance of 6.8 Å rather than the normal distance of 3.4 Å.   On the basis of this and other observations, Brenner et al. (1961) suggested that acridine mutations may be caused by deletions or additions of base pairs rather than by base-pair substitutions of the transversion type.

a Show the types of mutation expected when an acridine molecule inserts itself between adjacent bases of (1) the template strand and (2) the new strand.

b With such a mechanism what types of change would you expect in proteins controlled by the gene, e.g., substitution of one amino acid by another, alteration in amino acid sequence, or other changes?   Discuss.

c See Freese (1959a) and Brenner et al. (1961) for alternative hypotheses of acridine action and expectations from them.   Which hypothesis do you think is the correct explanation and why?

1065 Three repair mechanisms are known in *Escherichia coli* for the repair of DNA damage (pyrimidine dimer formation) after exposure to ultraviolet light: (1) photoreactivation (Kelner, 1949; Wulff and Rupert, 1962); (2) excision (dark) repair (Howard-Flanders and Boyce, 1964, 1966); and (3) postreplication repair (Rupp and Howard-Flanders, 1968).   Compare and contrast these mechanisms, indicating how each achieves repair and how the events occurring in each may lead to gene mutations.

1066 Having the alkylating agent EES, the base analogs BU and AP, the acridine dye P, and HA to work with, outline how you would proceed to determine the nature of mutation in a spontaneously occurring *r*II mutant of T4 phage.   Show the results expected in each case.

PROBLEMS

1067 A *normal*-winged *ClB Drosophila* female crossed with an irradiated *normal*-winged male produces a large progeny.   One *short*-winged F_1 *ClB* female is found.   She is crossed with an F_1 *normal*-winged male and produces the F_2 progeny shown.

a Discuss the suggestion that the *short*-winged F_2's may have been pheno-copies.

Table 1067

	♀				♂	
ClB		Non-ClB				
Normal	Short	Normal	Short	Normal	Short
66	58	64	60	64	58

b If the change is genetic, is it autosomal or sex-linked? Recessive or dominant?

c What kinds of changes in the genetic material could produce this inherited change? How would you distinguish between these alternatives?

d A true-breeding strain of *short*-winged flies is established and treated with x-rays. Among 50,000 progeny, 1 *normal* revertant appears and gives rise to a true-breeding *revertant* strain. Crosses with the original *normal* strain give the following results.

F_1 *normal*

F_2 134 *normal* : 31 *short*

Did the reverse mutation occur at the original or at a suppressor locus? Explain.

1068 A study of vital statistics in a certain country reveals that 50 children among 735,000 born of *normal* parents were *abnormal*: 11 were *brachydactylic* (a dominant trait, involving short fingers and toes), 36 were *albino* (a recessive trait, involving complete lack of epithelial pigment), and 3 had *aniridia* (a dominant trait, absence of iris). Assume that none of these traits affect reproductive fitness (ability to marry and have children).

a For each of these traits, state whether or not a mutation rate can be determined and why.

b Estimate the mutation rates by the direct method where this is possible.

c Suggest two possible sources of error in your estimates.

1069 **a** A *normal* x-ray technician in an industrial plant whose wife is also *normal* has a son with *Duchenne muscular dystrophy* (X-linked). There are no cases of this disease in his ancestry or in that of his wife for the past four generations. He sues the industrial plant for failing to provide proper protection from radiation, claiming that his son's abnormality is the result of an induced mutation. Show whether his claim is justified.

b A man and a woman work in an atomic-energy plant, where they are exposed to small amounts of radiation daily. They marry and produce a child who is an *amaurotic idiot*. They find no cases of this disease in their ancestry and believe that this abnormality in their child resulted from the effects of radiation on their gametes. Evaluate this belief.

1070 The *purple adenineless* strain of *Neurospora* originated by a gene mutation at the *adenine*-3 locus. From a very large number of clones of this strain, Giles (1956) and Giles et al. (1955) obtained several that could synthesize adenine. Most of these were *colorless*, like *wild type*, but differed from *wild type* in the rate of adenine synthesis or in the effect of temperature on synthesis. One was *purple*, like the original mutant.

a Give one explanation for these exceptional individuals other than back mutation at the *adenine-3* locus. How could you distinguish between these alternatives for any mutant?

b Explain the differences among them.

1071 In the housefly, a diploid species, two true-breeding strains exist: one has *black* eyes, the other *white*. The difference in eye color between the two strains is due to a single pair of autosomal alleles; *B* controlling *black* is dominant to *b* for *white*. Of 250,000 F_1 progeny from a cross between the two strains, 4 are *white* and the remainder are *black*.

1 One of the F_1 *whites* when crossed with a fly from the true-breeding *white* strain produced 68 *black* and 71 *white* offspring.

2 Another *white* F_1 when crossed with an individual from the true-breeding *white* strain produced 33 *black* and 98 *white* flies.

3 The remaining two F_1 *whites* in crosses with true-breeding *white*-strain individuals produced *white* offspring only, all of which produced true-breeding strains, in each of which reversion to *black* occurred.

a State, with reasons, the cause of each of the three types of F_1 *white*.
b Calculate the mutation rate (1) per gamete and (2) from *B* to *b*.

1072 In the standard *C l B* test the "Florida inbred" strain of *Drosophila melanogaster* shows a natural X-linked lethal mutation rate of 23 per 2,108 X chromosomes, compared with 2 per 3,049 for the Oregon-R strain. The Florida X chromosome carries the mutant genes *g* and *pl*. When males from the cross, Florida inbred *C l B/g pl* females × *wild-type* Florida males were tested in the same way, the natural rate was much lower; only a few males carried lethals although their X chromosomes were Florida-inbred in origin. When autosomes (numbered 2 and 3; X is numbered 1) of a low-mutability-rate stock (Swedish-b; S) were substituted for Florida-inbred homologs (F) and the resultant lines tested for spontaneous X-linked lethals, the results were as shown (Demerec, 1937).

Table 1072

Constitution	Number of chromosomes tested	Lethals	
		No.	%
1S 2F 3S	1,707	17	1.00
1F 2F 3S	1,215	11	0.97
1S 2F 3F	790	9	1.14
1F 2S 3S	1,162	—	—
1S 2S 3F	804	—	—
1F 2S 3F	800	—	—

a Does the significantly higher rate of naturally occurring X-chromosome lethals in the first three stocks compared to the others have a genetic basis, and if so, what does it appear to be? How would you determine whether your explanation is correct?

b Explain two ways in which the mutator gene may cause a higher mutation rate (see Yanofsky et al., 1966).

1073   In an appropriate genetic background multiple alleles at the A locus of maize show serial dominance in the control of aleurone (kernel) color as follows: A and A^b give *deep* color, a^p gives *pale*, and a gives *colorless*. These aleurone phenotypes are sometimes associated with a pattern effect, *dotted* (*deep* spots on a *pale* or *colorless* background).

  a Rhoades (1936) found an exceptional self-fertilized ear of Black Mexican sweet corn, a true-breeding *deep* variety, on which a phenotypic ratio of 12 *deep* : 3 *dotted* : 1 *colorless* kernels occurred (the *dotted* had *colorless* backgrounds). Formulate a hypothesis that will account for this observation and for *dotted* being the result of somatic mutation.

  b The plants from the *colorless* kernels when self-fertilized or crossed with plants of any other *colorless*-kernel variety produced *colorless*-bearing plants, only. When crossed with true-breeding *dotted*-bearing plants, all progeny were *dotted*-bearing. When $a^p a^p$ plants (*pale*) were crossed with true-breeding *dotted* ones from the progeny shown in (a), the F_1's were *dotted* on a pale background and the progeny from $F_1 \times aa$ parent segregated 3 *pale* : 1 *dotted* on a pale background. Two-thirds of the *pale* plants bred true. On self-fertilization the other third segregated 3 *pale* : 1 *colorless*. The plants with *dotted kernels* on a pale background segregated 6 *pale* : 6 *dotted* on pale background : 3 *dotted* on colorless background : 1 *colorless*. The same results were obtained with A and A^b as with a^p. What can you conclude from these data regarding the effect of alleles at the *dotted* locus on the mutability of a, a^p, A^b, and A alleles?

  c The *dotted* kernels on *aa* plants have small spots of aleurone color which are fairly uniform in size and distributed at random over the aleurone layer. What does this suggest regarding the timing and physiological conditions of mutation induction? Suggest a mechanism through which colored cells can be produced in an *aa* plant.

1074   In corn, plants homozygous for the recessive allele a at the A locus on chromosome 3 have *colorless* aleurones (kernels). Rhoades (1938) showed that the "dotted" allele Dt of the $Dtdt$ pair is a *mutator* gene which effects the mutability of a in the aleurone. The frequency of somatic mutations was indicated by the number of *colored* (purple) dots per kernel, each dot representing a group of cells arising from a cell which carried A as the result of a

Table 1074

Parents		No. of dots per F_1 kernel
♀	♂	
dtdt aa ×	*DtDt aa*	7.2
DtDt aa ×	*dtdt aa*	22.2

mutation from *a* to *A*.   Reciprocal crosses between *DtDt aa* and *dtdt aa* lines gave the results shown.   Why should reciprocal crosses give different mutation rates for *a* to *A*?

1075   In the guinea pig, melanin-pigment content in the coat is affected by alleles of the *albino* gene as shown in Table 1075*A* (Wright, 1949).   From each of four

Table 1075*A*

Genotype	Percent melanin in relation to *CC*
Cc^a	100
$c^k c^a$	44
$c^d c^a$	15
$c^a c^a$	0

different populations, a true-breeding mutant strain is established, and in each strain rare revertants to *wild type* or to *pseudo wild type* are observed, as shown in the hypothetical data in Table 1075*B*.

Table 1075*B*

Mutant strain	No. of revertants found		Progeny of *revertant* × original *wild type* (*CC*)
	Wild type	*Pseudo wild type*	
$c^k c^k$	3	0	all *wild type*
$c^d c^d$	4	0	all *wild type*
	0	1	all *wild type* except 1 in 10,000 $c^d c^d$
$c^a c^a$	4	0	all *wild type*
$c^a c^a$	0	3	all *wild type* except 1 in 10,000 $c^a c^a$

a What kinds of point mutations (substitutions, deletions, insertions, or inversions) occurred to produce each of these mutant strains and why?

b Which reversions involved the same kind of point mutation as that producing the original forward mutant?

c How could you test your hypothesis if the amino acid sequence of the enzyme controlled by this gene were known?

1076   Several *methionineless Neurospora crassa* strains have been shown to be controlled by mutant alleles of the same gene.   In each an occasional colony is found that grows in the absence of methionine and breeds true for this reversion.   Possible causes of this phenotypic reversion to *wild type* are

1   Mutation at a suppressor locus
2   True reverse mutation of the mutant allele to *wild type*
3   Contamination of the culture with *wild-type Neurospora*
4   Physiological "adaptation"

Outline an experimental procedure to determine which of these explanations is the correct one.

1077   Stadler (1929) measured the frequencies of mutant types in self-fertilized progeny of diploid ($2n = 14$), tetraploid ($2n = 28$), and hexaploid ($2n = 42$) species of wheat and oats exposed to varying doses of x-rays.   Typical results (for a single dosage level) are shown.   Show how the frequency of mutants is

Table 1077

Species	Chromosome number n	Rate of induced visible mutations by x-rays
Avena brevis	7	4.1 ± 1.2
A. strigosa	14	2.6 ± 0.6
A. sativa	21	0
Triticum monococcum	7	10.4 ± 3.4
T. dicoccum	14	2.0 ± 1.3
T. durum	14	1.9 ± 0.5
T. aestivum	21	0

related to chromosome number and suggest a reason for the relationship.

1078   Conger and Johnston (1956) irradiated the flower buds of *Tradescantia paludosa* ($2n = 12$) with 500 R of x-rays and counted the number of deletions in two populations of microspores.   Which of the two cells is haploid ($n = 6$)

Table 1078

Cell type	No. of cells counted	Deletions per 100 cells		Total $\pm$ standard error
		Intercalary	Terminal	
1	50	292	60	352 ± 26
2	25	624	157	776 ± 56

and which diploid ($2n = 12$) and why?

1079   In all diploid organisms so far studied except man, most of the gene mutants that occur, whether of spontaneous or induced origin, are recessive to *wild type* regardless of the locus involved.   In man, although the same has been found to be true for most of the X-linked mutants, most of the autosomal mutants are dominant. Offer an explanation for this apparent contradiction between experimental and nonexperimental populations.

1080 In 1946 Mitchell and Houlahan obtained a single-gene *adenineless* mutant in *Neurospora crassa* by ultraviolet induction. Kolmark and Westergaard (1949) found that the rate of reversion to *wild type* could be increased by x-rays, ultraviolet light, and nitrogen mustard. A sample of the results obtained in crosses between *reverted* and *wild-type* colonies is shown.

Table 1080

Culture number	Treatment	Asci with 8 *wild-type* ascospores	Asci with 4 *wild-type* and 4 *adenineless* spores
1	ultraviolet light	4	0
2		0	3
8		6	2
13	x-rays	5	0
18		0	6
23		1	4
26	nitrogen mustard	3	1
27		3	0
34	spontaneous	5	0
35		´2	1

a Show why these results prove that gene mutation is the cause of the reversions.

b Noting that the haploid mycelium and the asexual spores (macroconidia) by which this organism reproduces are multinucleate, explain the results for cultures 1, 13, 27, and 34.

c Describe two ways in which gene mutation may lead to reversion.

d What bearing do these results have on Stadler's (1944) conclusion that x-ray-induced mutation consists of the loss of the gene?

1081 In the pigeon the feather colors *ash-red*, *blue*, and *brown* are controlled by the alleles B^A, B, and b, respectively, at a Z-linked locus (National Pigeon Association Information Booklet 1, 1950–1951).

a B^Ab males are usually a uniform *ash-red*, but some individuals have some *brown*-flecked (and in rare instances entirely *brown*) feathers. Suggest both a genic and a chromosomal explanation to account for the *brown* flecks.

b In B^A females flecks are usually *brown* but may occasionally be *blue*. *Blue* flecks have also been observed in B^Ab and B^AB males. Which explanation do these latter observations favor and why?

1082 Neel and Falls (1951) reported that in a Michigan population, 49 children from *normal* parents out of a total of 1,054,984 births were affected with *retinoblastoma*, a dominant trait.

a What is the mutation rate per gamete for this affliction? Explain.

b What is the mutation rate per gamete if the dominant allele of only one gene is involved? If dominant alleles at either of two loci can cause the abnormality?

1083 In the mouse the alleles C and c^h at the *albino* locus cause *black* and *chinchilla* respectively (Grüneberg, 1952). In a cross between the true-breeding strains *black* and *chinchilla*, a *black* male with *tan* spots was found. Crossed with females of a true-breeding *albino*, $c^a c^a$, strain, this male produced 20 *albino*, 80 *tan*, and 60 *black* F_1 offspring. When the F_1's were mated with the *albino* strain, their progenies segregated as shown. Suggest an explanation to account for these data.

Table 1083

Cross	Progeny
albino F_1's × *albino*	43 *albino*
tan F_1's × *albino*	153 *tan* : 159 *albino*
black F_1's × *albino*	121 *black* : 118 *albino*

1084 a Altenburg and Browning (1961) irradiated *wild-type* (*gray*) *Drosophila melanogaster* sperm with 3,000 R of x-rays before mating them with females homozygous for *yellow body* (*y*, X-linked). The majority of visible mutations were whole-body (*yellow* body); only a few were mosaics, and most of these were of the left-right type (one side *gray*, the other *yellow*). Explain how the mosaics may have arisen, stating whether the changes occurred in the already existing gene or during replication and how. (See Muller et al., 1961, for their explanation of how such types can arise.)

b In rare instances, mosaics showed mutant areas occupying approximately one-quarter of the body surface. Explain how these mosaics differ in origin from those described in (a).

1085 A *wild-type* (*gray-bodied*) $C l B$ *Drosophila* female was crossed with a *wild-type* male treated with mustard gas. All F_1's were *wild type*. One of the F_1 $C l B$ females mated to a *wild-type* F_1 male produced the progeny shown.

a Show why the *black* phenotype is not a phenocopy and state whether the change is autosomal or X-linked, dominant or recessive.

b A true-breeding strain of *black* flies was established. Among 40,000 flies in the true-breeding *black* strain two *wild-type revertant* males appeared. By appropriate crosses, true-breeding *wild-type* strains were established for these *revertants*. Females of both *revertant* lines mated with males of the

Table 1085

X chromo-some	Body color	Number
non-ClB♀	*wild type*	205
ClB ♀	*wild type*	198
♂	*black*	202

original *wild-type* strain produced *wild-type* F_1's.   When intermated, the F_1's of one line produced both *black* and *wild-type* F_2 progeny, but those of the other line produced only *wild types*.

(1) Did the reverse mutation occur at the original locus or at a suppressor locus in the first *revertant* strain?

(2) What can you conclude from the cross of the second *revertant* with *wild type* about the type of genetic change originally induced?

1086 a Using a *wild-type* strain of *Neurospora crassa*, three suspensions of 10^6 ascospores each are irradiated with doses of 100, 500, and 1,200 R respectively.   The table shows the numbers of *arginineless* mutants found.

Table 1086

Irradiation dose, R	Number of mutants
100	7
500	31
1200	80

Graph the mutation-response curve for dose vs. induced mutation.   Does mutation frequency appear to be linearly related to dosage?

b In a later study, rare *wild types* are found in every one of the mutant cultures from the first two treatments, but 10 percent of the mutant strains from the 1,200-R treatment failed to produce any *wild types*.

(1) What is the most likely cause of the difference in behavior between mutants of the 10 percent fraction and the other mutants and why?

(2) Show whether your conclusions about linearity in (a) should be altered.

1087 Nitrogen mustard induces both gene mutations and chromosome aberrations in *Drosophila*.   The frequency of reciprocal translocations due to this mutagen is approximately proportional to the square of the frequency of lethals (due to gene mutations and terminal deletions).   Why does this difference in frequency occur?

1088 Spencer and Stern (1948) observed that the x-ray dosage-response curve (Fig. 1088) for sex-linked lethals (corrected for spontaneous mutation) in *Drosophila melanogaster* was linear from 1,000 to 3,000 R.   The extrapolated curve passed

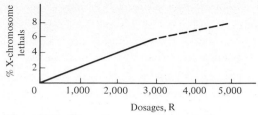

Figure 1088   (*From Spencer and Stern, 1948.*)

through the origin.   Above 3,000 R the curve (dotted portion) fell off from the straight line.

a How do you interpret the results for dosages 1,000, 2,000, and 3,000 R?

b Would you conclude there was a *threshold* dose (radiation intensity) below which no mutation is induced?

c Keeping in mind the fact that the sex-linked-lethal test detects only one lethal per chromosome tested, explain the falloff in the response curve for dosages above 3,000 R.

d X-irradiation causes breakage in the chromosome strand(s).   What kind of chromosomal aberration would simulate lethal gene mutation?   Would such aberrations tend to increase or decrease with increasing dosage?   What is their probable importance (frequency) relative to gene mutations?

1089   The number of chromosome breaks produced by x-rays increases linearly with radiation dosage, as shown in Fig. 1089*A*.   Chromosomal aberrations such as reciprocal translocations, however, do not increase in direct proportion to x-ray dosage but exponentially, as shown in Fig. 1089*B*.   How can you account for these two different relationships with dosage?

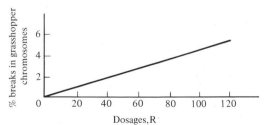

Figure 1089*A*   (*Modified after Carlson, 1941.*)

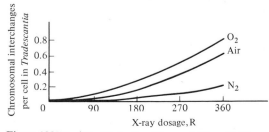

Figure 1089*B*   (*Modified after Giles and Riley, 1949.*)

1090 The percentage of X-linked lethals in *Drosophila melanogaster* increases in direct proportion to the amount of radiation at dosage levels up to approximately 4,000 R. At higher dosage levels the percentage of detectable lethal mutations falls below linear expectations. Account for these results.

1091 Flies from three different natural *Drosophila* populations are subjected to high-energy radiation and then released back into their respective populations.

Population 1   2,000 flies exposed to a dose of 25 R
Population 2   1,000 flies exposed to a dose of 50 R
Population 3   100 flies exposed to a dose of 500 R

Which population should carry the greatest genetic load of defective traits in future generations and why? The smallest?

1092 Benzer and Freese (1958) compared spontaneous and 5-bromouracil-induced mutants in the *r*II region of bacteriophage T_4. The mutagen not only increased the overall mutation rate several hundred times but acted at specific sites ("hot spots") where changes in the DNA were more frequent. Moreover, the map distribution of the induced mutations was different from that of naturally occurring *r* mutants, as shown in the illustration. Almost all (98 percent)

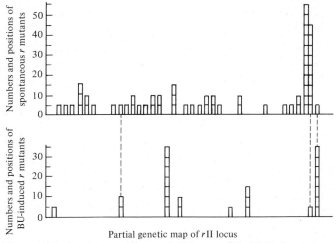

Figure 1092 Partial genetic map of *r*II locus. (*Modified after Benzer and Freese, 1958.*)

of the BU-induced mutants could be induced to revert (r to r^+) by BU treatment, but only 14 percent of the spontaneous mutants reverted by this treatment (Freese, 1959*b,c*). Spontaneous reversion also occurred for both base-analog-induced and spontaneous mutations, but at a very low rate. Discuss the implications of these findings.

1093 Nitrous acid is highly mutagenic for phage, bacteria, and tobacco mosaic virus (Tessman, 1959*b*; Mundry and Gierer, 1958), acting directly on the bases of

DNA that are not replicating, changing adenine to hypoxanthine, cytosine to uracil, and guanine to xanthine (see Appendix Table A-2).

a Illustrate the chemical reaction for each of these changes and suggest how they are related to NA mutagenicity.

b Why does nitrous acid have no effect on thymine?

c Would you expect xanthine to be mutagenic? Explain.

d NA acts directly on the bases and does not cause pairing mistakes during incorporation or subsequent replications, as the base analogs do. Nevertheless, it has been shown that mutations induced by base analogs such as BU and AP can be reverted by NA and vice versa. Thus NA appears capable of inducing transitions in both directions (A-T $\leftrightarrow$ G-C). Show how it may do so.

1094 Each of the four bases in DNA can exist in several alternative (tautomeric) chemical states. The most common state of these bases and the rare form suspected of being important in the origin of mutations are shown in Appendix Table A-2. Watson and Crick (1953*a*) suggested that the occurrence of the rare tautomer for each of the bases provides a mechanism for mutation during DNA replication. Outline the mechanism by which each of these tautomers might give rise to a mutation during DNA replication, stating the number of replications required to effect such a mutation. Are the changes transversions or transitions?

1095 Hydroxylamine acts mainly on cytosine, apparently causing unidirectional transitions from G-C to A-T. It causes reversion of mutations induced by AP but cannot cause true reversion of mutations induced by itself or most of those induced by BU. What kind of base-pair substitutions are primarily induced by BU and AP and why?

1096 5-Bromouracil, an analog of thymine that may occasionally pair with guanine, is mutagenic (Litman and Pardee, 1956; Freese, 1959*b*, 1963). For its alternative states and its chemical relation to thymine see Appendix Table A-2. Dunn and Smith (1954) grew phage on bacteria in a medium containing BU, and found that the analog replaced thymine in the phage DNA. Phage with BU incorporated in their DNA were allowed to infect bacteria growing on medium containing thymine. Among the progeny a few (many more than would occur spontaneously) were mutant in phenotype. These and other studies imply that the incorporation of BU itself does not constitute a mutation. On the basis of these findings and taking all the above facts (as well as the formulas of thymine and BU) into consideration:

a Illustrate why replacement of thymine by BU does not constitute a mutation.

b Suggest and illustrate possible mechanisms by which BU causes mutations. Illustrate.

c See Terzaghi et al. (1962), who discuss possible mechanisms by which BU causes mutations.

1097 When DNA is exposed to moderately low pH, only purine bases are removed. Removal of these bases has been shown to be mutagenic in phage T4 (Freese, 1959c). Outline (1) how this depurination might lead to mutation and (2) how you would test whether transitions or transversions or both are produced by this process by comparing the expectations with those for other forms of chemical mutagenesis (see Table 1102) knowing that base analogs, nitrous acid, and other chemicals have the effect shown in the table.

1098 **a** A DNA duplex, which contains the base pair A-T in an alternating sequence
A T A T, etc., is treated with EES, AP, BU, NA, and P.   Which mutagens
T A T A
can cause mutations in this duplex and why?

b A specific tryptophan synthetase A-gene mutant in *Escherichia coli*:

1   Can be induced by NA
2   Can be reverted to *wild type* by AP but not by NA
3   Cannot be induced to revert by P

A single base-pair change is involved.   What kind is it and why?

1099 Hydroxylamine is a chemical mutagen that reacts almost exclusively with C (or 5-hydroxymethylcytosine) replacing C in phage and causing the HA-reacted base to pair with A.   Thus HA predominantly induces the base-pair transition G-C → A-T or G-HMC → A-T (Freese et al., 1961a).

a In 1963 Drake found that 53 of 99 ultraviolet-induced rII mutations of phage T4 were reverted with base analogs.   Of these, 47 were reverted by AP and BU but not by HA.   The remaining 6 reverted with both mutagens. Which of the following changes does ultraviolet light primarily induce? Explain.

(1) A-T → G-C   (2) G-C → A-T
(3) A-T → C-G   (4) C-G → A-T

b Most of the remaining mutations were induced to revert by proflavine but did not respond to base analogs.   What was the probable mutagenic effect of ultraviolet light in these cases?   Illustrate.

c See Howard and Tessman's (1964) conclusions regarding the mutagenic specificity of ultraviolet light.

1100 2-Aminopurine, an analog of adenine, differs from adenine by having the amino group attached to the 2 position instead of the 6 position.   AP exists also in the rare tautomeric form (see Appendix Table A-2).

a Illustrate the base-pairing attributes of AP in (1) the normal state and (2) the rare imino state, due to a tautomeric shift of a hydrogen atom to the N-1 position.

b Should AP induce transitions or transversions?   In one or either direction? Elaborate.

1101 A study of forward and reverse host-range mutations induced in vitro by HA, EMS, and NA, in the single-stranded DNA phage S13 (Tessman et al.,

Table 1101

Mutational change	HA		EMS		NA	
	F	R	F	R	F	R
$h^+ \to h_i 1$	0	+ +	+	+ +	+	+ +
$h^+ \to h_i 2$	0	0	0	+ +	+ +	+ +
$h^+ \to h_i 1$	0	0	+ +	0	+ +	+ +
$h^+ \to h_i 2$	0	0	+ +	+	+ +	+ +
$h^+ \to h_i 65$	0	0	+ +	0	+ +	+ +
$h_i UR48 \to h_i UR48s$	+ +	—	+ +	—	+ +	—

Key: + + = high, + = low, 0 = zero induction frequency; F = forward; R = reverse.

1964) gave the data shown.   None of the h^+ revertants show recombination with any of the others.   Assume (1) that HA acts primarily on cytosine, and EMS on guanine, to produce 7-alkylguanine and (2) that only transitions are involved in these mutations.   Show:

a The base change for forward mutation for each mutant.

b Which transitions are induced by NA.

c Which bases (other than guanine) are acted on by EMS.

1102   The revertibility of phage T4 *r*II mutations induced by various chemical mutagens is shown in the table.   For each of the mutagens, state with reasons:

Table 1102

Mutations induced by	Relative frequency of reversions induced by				
	NA	HA	BU	AP	EES
NA	+ + +	—	+ +	+ +	—
HA	—	0	+ +	+ + +	—
BU	+ + +	0	+	+ + +	0
AP	+ + +	+ +	+ + +	+	+ +
EES	—	—	+ +	+ +	+

Key: + + + = complete reversion; + + = high reversion rate; + = low reversion rate; 0 = no or virtually no reversion.

Note: (1) Only seven BU and four HA mutants were tested; five out of nine mutants induced by AP were reverted by HA.   (2) The data are modified after Hayes, 1968, and based on extensive experimentation by Bautz-Freese (1961). Bautz-Freese and Freese (1960, 1961), Freese (1959*a*,*b*,*c*), and Freese et al. (1961*a*,*b*).   (3) BU induced mutations in both directions; HA acts on cytosine, altering its pairing properties; EES removes guanine preferentially; all reversions were true back mutations to *wild type*.

a Whether transitions, transversions, or both are induced.

b Whether the transitions or transversions occur in both directions or only one.

c If in both directions, whether they occur predominantly in one direction.

1103 AP causes transition, EES may cause transversion, and spontaneous mutations may be deletions or additions.   Describe all the possible types of effects of each of the mutagens, i.e., two-way transition, one-way transition A-T → G-C, one-way transition G-C → A-T, transversion, deletion or addition.

Table 1103

Inducing mutagen	Reversion mutagen						
	AP	BU	HA	EES	NA	P	Spontaneous
AP	+	+	$\frac{1}{2}$	$\frac{1}{2}$	+	—	
BU	+	+	$\frac{1}{2}$	$\frac{1}{2}$	+	—	
HA	+	+	—	—	+	—	
EES	$\frac{1}{3}$	$\frac{1}{3}$	—	—	$\frac{1}{3}$	—	
NA	+	+	+	$\frac{1}{3}$	+	—	
P	—	—	—	—	—	+	+
Spontaneous	$\frac{1}{2}$	$\frac{1}{2}$	$\frac{1}{2}$	$\frac{1}{2}$	$\frac{1}{2}$	$\frac{1}{2}$	+

Key: + = reversion induced in all; — = no mutants reverted; $\frac{1}{2}$ = approximately one-half of the mutants reverted; $\frac{1}{3}$ = approximately one-third of the mutants reverted.

1104 Exposure of DNA to low pH causes depurination, i.e., removal of A and G (Tamm et al., 1952), and occasionally backbone breakage (Tamm et al., 1953). Most of the mutations induced are of the point type, at least in phage. If the mutagenic effect is brought about by depurination, we are still unable to predict the nature of the changes in base sequence that could result.

a Using T4 phage and the pair of traits *wild-type* vs. *mutant* plaques (r^+ vs. r) outline an experiment to determine whether deletions, insertions, and/or substitutions are produced.   Show the consequences expected with each of these kinds of change in the DNA and protein controlled by this gene.

b (1) *Wild-type* phage T4 lysozyme has the amino acid sequence

Lys-Ser-Pro-Ser-Leu-Asn-Ala

in a portion of the chain in the *wild-type* protein.   Mutants induced by low pH show the following amino acid sequences:

Mutant 1   Lys-Gly-Pro-Ser-Leu-Asn-Ala
Mutant 2   Lys-Ser-Pro-Ser-Leu-Tyr-Ala
Mutant 3   Lys-Ser-Pro-Ser-Leu-His-Ala
Mutant 4   Lys-Val-His-His-Leu-Met

Using the code-word assignments in Appendix Table A-5, show the possible kinds of changes caused by depurination.

(2) Treatment of these mutants with certain mutagens produced revertants with the amino acid sequences shown. Do these confirm or reject your explanation in (b) (1)?

Table 1104

Mutant	Treated with mutagen	Amino acid sequence in revertant
1	HA	Lys-Ser-Pro-Ser-Leu-Asn-Ala
1	P	Lys-Ser-Pro-Ser-Leu-Lys-Cys
2	BU	Lys-Ser-Pro-Ser-Leu-Asn-Ala
2	P	Lys-Ser-Pro-Ser-Leu-Met-Leu
3	HA	Lys-Ser-Ser-Ser-Leu-Asn-Ala
4	P	Lys-Val-His-His-Ile-Asn-Ala

1105 a Alkylating agents, such as EES, EMS, and the nitrogen mustards, appear to react predominantly with the N-7 atom of guanine, adding an organic (R) group at this position (Brookes and Lawley, 1960, 1961; Krieg, 1963a). Alkylation in this position (see formula) has several possible consequences.

1 The alkylated guanine can lose a hydrogen atom from the N-1 atom (by ionization).

2 The bond joining this base to the sugar molecules is weakened so that the base tends to split away.

What kinds of changes (base-pair substitutions, deletions, etc.) might alkylating agents cause? Illustrate.

b Of the EES-induced mutations, about 70 percent are reverted by base analogs. The majority of AP-induced mutations (mainly A-T → G-C) are revertible by EES, but almost no BU-induced mutations (mostly G-C → A-T) are. EES-induced mutations which are revertible by EES cannot be reverted by NA (causes A-T → G-C or G-C → A-T). Some of EES-induced mutations nonrevertible by BU, AP, and NA were revertible by EES (Bautz-Freese, 1960, 1961; Krieg, 1963a,b; and other workers). Relate these observations to the types of changes you have postulated in (a). Explain by referring to the facts presented.

1106   Krieg (1963*b*) used EMS to induce reversions of *r*II mutants induced in T4 bacteriophage.   The table shows a selection from his results.   Outline a

Table 1106

Mutagen used to produce original mutant	Designation of original mutant	Frequency of revertants per 10^6 progeny phage	
		With EMS	Control (no EMS)
AP	114	290.0*s*	2.8*s*
		30.0*t*	0.9*t*
	275	50.0	1.0
	70	0.5	0.0
BU	90	23.3*s*	0.8*s*
		5.6*t*	0.4*t*
	7	0.6	0.01
	19	0.18	0.07
EMS	126	147.0	0.35
	34	22.0*s*	0.4*s*
		36.0*t*	15.0*t*
	30	0.3	0.0
P	85	60.8*s*	1.7*s*
		76.0*t*	27.0*t*
	28	0.7	0.3
	83	0.07	0.10

Key: s = standard plaque; *t* = tiny plaque.

hypothetical mechanism of mutagenic action for this chemical, showing which types of base-pair alterations are favored or possible.   *Note:* AP and BU cause transitions A-T $\leftrightarrow$ G-C; proflavine causes base-pair deletions and additions.

1107   Zamenhof and Greer (1958) showed that heating *Escherichia coli* strain W6 to 60°C is mutagenic.   Assume that you find that transitions, transversions, and deletions can occur by heating.   What is the most likely effect of this treatment?   See Greer and Zamenhof (1962).

1108   Tessman (1959*b*) found that in vitro treatment of the *wild-type* strains of two different kinds of DNA phages (which we call A and B), with nitrous acid produces *mutants*.   The mutation tested in the A phage was that for rapid lysis, $r^+ \rightarrow r$, and the test in the B phage was for host range $h^+ \rightarrow h$.   Viable phages were seeded very thinly on each of many plates to ensure that plaques with *mutants* would be completely isolated from *wild-type* plaques.   The majority of plaques were *wild type*.   The *mutants* of phage A always occurred in mottled plaques containing a mixture of *mutant* and *wild-type* phage, whereas all *mutants* of phage B arose in plaques that contain *mutants* only.

Offer an explanation in terms of DNA structure to account for the difference in response of the two phages to the mutagen.

1109 *Escherichia coli* possesses an allele at a certain locus which confers *resistance* on T1 phage irradiated with ultraviolet light. When a thymine analog (bromodeoxyuridine) is substituted for thymine in a phage, this protection is removed. What is the product and mechanism of action of the gene?

1110 The kind of change in the base pair at a DNA site induced by each of the major classes of chemical mutagens is shown. It can be seen that some

Table 1110

Mutagen	Substitution	Type of substitution
Base analog	A-T → G-C	Two-way transition
NA	A-T → G-C	Two-way transition
HA	G-C → A-T	One-way transition
EES	G-C → A-T	One-way transition
	G-C → T-A	Transversion
	G-C → C-G	Transversion

transitions are reversible; e.g., a second treatment by a base analog can restore the original base-sequence, by this group of chemicals.
a Show why:
  (1) Some EES-induced changes cannot be reversed by NA.
  (2) Some base-analog-induced changes can be reversed by HA.
  (3) Reversion by HA is not possible with EES-induced changes.
b Spontaneous mutants fall into the following classes when tested by various mutagens for reversibility (Freese, 1959*c*):

Reversible by AP and BU   14 percent
Reversible by proflavine   86 percent

Discuss the possibility that spontaneous mutations are a heterogeneous group consisting of the effects of a number of different kinds of mutagens present in the environment.

1111 Freese studied the effects various mutagens have on *r*II mutants in phage T4 induced by a variety of different chemicals. His results are summarized in the table. *Note:*(1) Most naturally occurring and proflavine-induced mutants are revertible by proflavine. (2) BU is known to cause transitions A-T → G-C and vice versa.
  a What kind of change is induced by each of the other chemical mutagens? Give a reason or reasons for your decision.
  b What kinds of change appear to constitute most spontaneous mutations in this organism? Justify your decision.

Table 1111

rII mutants induced by	Number of mutants tested	Mutants revertible by AP and/or BU, %	Mutants not revertible by AP and/or BU, %	Mutants of possible spontaneous origin, %
AP	98	98	2	2
BU	64	95	5	2
HA	36	94	6	4
NA	47	87	13	15
EES	47	70	30	10
P	55	2	98	
spontaneous	110	14	86	

From E. Freese, Molecular Mechanisms of Mutation, table III, p. 239, in J. H. Taylor (ed.), "Molecular Genetics," pt. I, Academic Press Inc., New York, 1963.

1112 A *mutator* gene in *Escherichia coli* increases the mutation rate of other genes (Treffers et al., 1954) including the *tryptophan synthetase A* gene. At some amino acid sites in the *A* protein different mutations may involve the substitution of any of several different amino acids for the original, making it possible to determine whether the mutator gene favors certain amino acid substitutions and hence to designate the type of base-pair change it favors. Yanofsky et al. (1966) studied five tryptophan *mutants*, each showing a single amino acid difference from *wild-type* A protein. Each *mutant A* gene was placed in a mutator background, and reversion to *wild type* was studied. *Revertants* were either full or partial. Reversion of all *mutants* was at a much

Table 1112

Mutant	Characteristics of revertants*	Amino acid change and RNA code words				
		Peptide	From	Code word	To	Code word
A223	F	TP4	Ile	AUU AUC	Ser	AGU AGC
A78	F	TP6	Cys	UGU UGC	Gly	GGU GGC
A58	P	TP6	Asp	GAU GAC	Ala	GCU GCC
A23	F	CP2	Arg†	AGA	Ser	AGC AGU
A46	F	CP2	Glu	GAA	Ala	GCA

* F = full, and P = partial.

† No reversions of Arg (AGA) to Ile (AUA) were observed.

greater rate in the mutator background. The results of the study, together with the data on amino acid substitutions and the corresponding RNA codons, are shown in the table.

    a Using the RNA codons assigned to the amino acids, determine whether:

      (1) The effect of the *mutator* gene is specific.

      (2) It is unidirectional in its action.

      (3) It induces transversions or transitions.

    b Suggest two mechanisms by which such mutator genes could produce their effects (see Pierce, 1966; Drake, 1970).

1113 *Wild-type* (r^+) T4 phage can grow on *Escherichia coli* strains B and K; *r*II mutants are able to form plaques on B only. Levisohn (1967) found that of 16 T4 *r*II mutants known to have G-C at the mutated site, 11 showed a high reversion rate when plated directly on strain K; 5 reverted only after exposure to hydroxylamine, which reacts with 5-hydroxymethylcytosine of T4. In which of these two classes of mutants did C occur at the mutated site of the transcribed strand?

1114 In T4 phage the enzyme lysozyme, a product of e^+, can occur in a variety of different mutant forms. The amino acid sequence in a specific peptide of the *wild-type* enzyme and the corresponding sequences in different naturally occurring *mutants* and their *revertants* are shown.

Table 1114

	Amino acid sequences of	
	Wild type and *mutants*	*Revertants*
Wild type	Lys-Ser-Pro-Ser-Leu-Asn-Ala	
Mutant A	Lys-Val-His-His-Leu-Met-Arg	Lys-Val-His-His-Leu-Asn-Ala
Mutant B	Lys-Ser-Pro-Ile-Lys-Cys-Ala	
Mutant C	Lys-Ser-Pro-Cys-Leu-Asn-Ala	1. Lys-Ser-Pro-Ser-Leu-Asn-Ala
		2. Lys-Ser-Pro-Cys-Leu-Asn-Thr
Mutant D	Lys-Gln-Ser-Ile-Thr $\cdots$	Lys-Gln-Ser-Ile-Thr-Asn-Ala

    a Classify the forward mutants as nonsense or missense.

    b What kind of mutation is responsible for each forward *mutant*? Explain.

    c Two true-breeding *mutant* strains, *e* and *e'*, arise naturally. In *e*, the protein formed is incomplete, in *e'* it is complete but has one arginine residue of the *wild type* substituted by lysine. Both *mutants* revert to produce *wild-type* strains, which produce no recombinants when crossed with the original *wild type*. Explain.

1115 The base-pair sequence and the amino acids it codes for in a small segment of a specific *wild-type* gene are shown. A single mutation in the *wild-type*

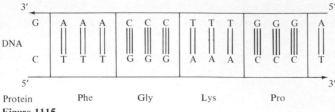

Figure 1115

gene produces a mutant protein which is identical to the *wild-type* one except that the sequence shown is replaced by Gly-Phe-Pro-Lys.

a What kind of mutation occurred to produce the mutant allele? Explain and illustrate.

b Suggest two alternative explanations and how they can be ruled out.

1116 Numerous *r*II mutants in the *A* gene of T4 phage have been crossed by Benzer (1959) by spot tests on *Escherichia coli* strain K plus *E. coli* strain B, in all possible combinations. The results for seven such mutants are tabulated.

Table 1116

	H23	250	C4	184	C33	221	882
H23	0	0	0	0	0	0	0
250	0	0	+	0	0	+	+
C4	0	+	0	+	+	0	0
184	0	0	+	0	0	+	+
C33	0	0	+	0	0	0	+
221	0	+	0	+	0	0	0
882	0	+	0	+	+	0	0

a What class of mutations do these mutants represent? What observations would verify your conclusions?

b Determine the sequence of the mutations.

1117 Westmoreland et al. (1969) separated the strands of the DNA duplexes of *wild-type* λ (λ^+) and the double mutant λ b2b5 by heating to 100°C and then hybridized the complementary strands of the two strains. An electron micrograph of a heteroduplex formed between strand *l* of λ^+ and strand *r* of λ b2b5 is shown, together with an interpretative drawing.

a What is the nature of the mutations in λ b2b5? Explain.

b Explain how they can be used in genetic studies.

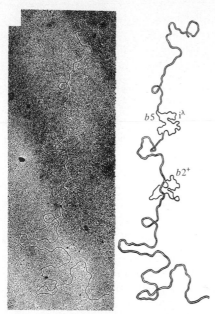

b5   i^λ

b2^+

Figure 1117 [*Electron micrograph from B. C. Westmoreland, W. Szybalski, and H. Ris, Mapping of Deletions and Substitutions in Heteroduplex DNA Molecules of Bacteriophage Lambda by Electron Microscopy, Science,* **163**:*1345 (1969), by permission.*]

1118 Four major kinds of ultraviolet-induced changes in DNA (see the excellent illustration in Deering, 1962) are

1   Splitting of a phosphate-sugar bond (single chain break)
2   Dimerization (bonding) between two adjacent thymines in the same strand
3   Dimerization between thymines in opposite strands
4   Disruption of hydrogen bonds between bases

a Which of these alterations probably cause gene mutations, and how?
b Can any one of these effects result in chromosome aberrations? If so, what kind and how?

1119 Thymine dimers block DNA synthesis in vitro and in vivo (Bollum and Setlow, 1963; Setlow et al., 1963). Certain strains of *Escherichia coli* are ultraviolet-*radiation-resistant* and even in the dark can recover to resume DNA synthesis. *Sensitive* cells cannot recover. Setlow et al. (1963) showed that thymine dimers are conserved in whole cells after ultraviolet irradiation. Boyce and Howard-Flanders (1964a) studied the fate of thymine dimers in DNA during incubation after ultraviolet light irradiation in a *resistant* strain (AB1157) and an ultraviolet-*sensitive* strain (AB1886) of *E. coli* K12. The two strains differ with respect to alleles at one locus.   They split the DNA into an acid-insoluble fraction (corresponding to macromolecular polynucleotides) and an acid-soluble fraction (corresponding to thymine and thymine dimers), as shown by autoradiography and chromatography.

1   Thymine dimers were identified in the acid-insoluble fractions of both strains before dark incubation.

2   After dark incubation, the acid-insoluble fraction (hydrolyzed) of the irradiated ultraviolet-*resistant* strain showed less thymine dimer than before incubation.   Moreover, as the thymine dimers disappeared from the acid-insoluble fraction, they appeared in the acid-soluble fractions.

3   In the irradiated ultraviolet-*sensitive* strain, which cannot synthesize DNA in the dark, the dimers remain in the acid-insoluble fractions; none are detectable in the acid-soluble fraction.

Photoreactivation involves the splitting of dimers.   A different mechanism occurs here, which permits *resistant* strains to recover and resume DNA synthesis.   What is this gene-controlled mechanism that occurs and what events follow in the *resistant* strains but not the *sensitive* ones to produce the original kind of DNA?   Illustrate.

1120   Breakdown of DNA is extensive in *ultraviolet-resistant* mutants of *E. coli* after treatment with x-rays (cause single-strand breaks) as in *wild-type* strains (Emmerson and Howard-Flanders, 1965).   The same mutants irradiated with ultraviolet light do not repair DNA.   What is the probable cause of lack of DNA in *ultraviolet-resistant* mutants after exposure to ultraviolet light?

1121   *Xeroderma pigmentosum* is an autosomal completely recessive disease in man in which the skin is extremely sensitive to ultraviolet light.   In *affected* individuals frecklelike lesions in exposed areas tend to become malignant, leading to a high incidence of skin cancer.   In vitro studies (Cleaver, 1969) show that skin fibroblasts from *xeroderma pigmentosum* patients do not perform DNA repair (of pyrimidine dimers) after exposure to ultraviolet light but do so after irradiation with x-rays (which cause polynucleotide chain breaks).   Fibroblasts from *unaffected* individuals (homozygous or heterozygous for the normal allele) undergo normal repair after ultraviolet exposure.   The process of postultraviolet DNA repair (dark reactivation) appears to proceed as follows (Howard-Flanders and Boyce, 1964, 1966):

1   An endonuclease recognizes the damaged region and breaks the relevant polynucleotide chain on one side of the dimer.

2   An endonuclease then digests away nucleotides adjacent to the break.

3   After digestion, a DNA polymerase resynthesizes the missing segment, using the portion of the chain opposite the gap as a template.   Specifically, 5'-nucleoside triphosphates hydrogen-bond to the single-stranded region and become linked by DNA polymerase to the 3' end of the broken chain.

4   The final gap is closed by a joining enzyme, probably polynucleotide ligase.

What appears to be the primary biochemical function of the *xeroderma pigmentosum* allele, and at which stage in the repair process does it appear to operate? Explain. (See Setlow et al., 1969, and Parrington et al., 1971.)

1122 Ultraviolet irradiation completely blocks colony formation in a *sensitive* strain (B1) of *Escherichia coli* but not in a *resistant* strain (B/r^s). It also permanently inhibits DNA synthesis in B1 but only temporarily inhibits that of strain B/r^s. The following additional data are from Setlow et al. (1963):

1 Inhibition of DNA synthesis was reduced by photoreactivation with blue light in the *sensitive* but not in the *resistant* strain.

2 Inhibition immediately after irradiation is complete in both strains, but the *resistant* strain rapidly recovers its ability to synthesize DNA.

3 Immediately after ultraviolet irradiation considerable amounts of thymine dimers are formed in the two strains (cross-linked thymines usually within one of the two DNA helices).

a Show how these observations are consistent with the concept that ultraviolet light is an excitatory rather than an ionizing radiation.

b Assuming that DNA synthesis is under the control of an enzyme produced by a single gene, suggest how the gene-controlled differences in recovery and the effects of photoreactivation may be related.

In the *resistant* strain, thymine dimers are not split in appreciable numbers during the time between inhibition and recovery of DNA synthesis. Photoreactivation of the *sensitive* strain, on the other hand, results in a considerable decrease in the number of thymine dimers.

c Explain how this information modifies (or changes) your answer to (b) and suggest a satisfactory mechanism of gene-controlled recovery.

1123 Ultraviolet irradiation at 2600 to 2800 Å of DNA causes the formation of thymine dimers (Beukers and Berends, 1960, 1961; Beukers et al., 1960). Short-wavelength (approximately 2400 Å) ultraviolet irradiation of thymine dimers causes reconversion to thymine. In 1949 Kelner showed that the killing and mutagen action of ultraviolet light may be reversed by exposure of irradiated organisms to visible light. In 1960 Rupert discovered an enzyme in bakers' yeast which when added to ultraviolet-treated *Escherichia coli* or other bacteria in the presence of visible light restored about 10 percent of the activity of DNA. Wulff and Rupert (1962) showed that while samples of irradiated DNA incubated with photoreactivating enzyme in the dark and samples incubated with heat-inactivated enzyme in the light both show the same amount of thymine dimers as present in the untreated irradiated DNA, incubation of irradiated DNA with this enzyme in the presence of visible light destroys over 90 percent of the dimers present and restores the normal DNA structure. The DNA damage (dimers) is repaired in one step. What appears to be the effect of the enzyme?

1124 Briefly describe how radiation can be used to regulate the populations of deleterious insects. (See Knipling, 1960*a,b*.)

REFERENCES

Alikhanian, S. T., et al. (1969), *Mutat. Res.*, **8**:451.
Altenburg, E. (1934), *Am. Nat.*, **68**:491.
—— and L. S. Browning (1961), *Genetics*, **46**:203.
Auerbach, C. (1949), *Biol. Rev.*, **24**:355.
—— and J. M. Robson (1946), *Nature*, **157**:302.
Baker, R., and I. Tessman (1968), *J. Mol. Biol.*, **35**:439.
Bautz-Freese, E. (1961), *Proc. Natl. Acad. Sci.*, **47**:540.
—— and E. Freese (1960), *Proc. Natl. Acad. Sci.*, **46**:1585.
—— and —— (1961), *Virology*, **13**:19.
Bender, M. A., and J. G. Brewen (1969), *Mutat. Res.*, **8**:383.
Benzer, S. (1959), *Proc. Natl. Acad. Sci.*, **45**:1607.
—— and E. Freese (1958), *Proc. Natl. Acad. Sci.*, **44**:112.
Beukers, R., and W. Berends (1960), *Biochem. Biophys. Acta*, **41**:550.
—— and —— (1961), *Biochem. Biophys. Acta*, **49**:181.
—— J. Ijlstra, and W. Berends (1960), *Rev. Trav. Chim.*, **79**:101.
Blatherwick, C., and C. Wills (1971), *Genetics*, **68**:547.
Bollum, F. J., and R. B. Setlow (1963), *Biochem. Biophys. Acta*, **68**:599.
Boyce, R. P., and P. Howard-Flanders (1964a), *Proc. Natl. Acad. Sci.*, **51**:293.
—— and —— (1964b), *Z. Vererbungsl.* **95**:345.
Brenner, S., L. Barnett, F. H. C. Crick, and A. Orgel (1961), *J. Mol. Biol.*, **3**:121.
Bridges, B. A., and R. J. Munson (1968), *Proc. R. Soc. (Lond.)*, **B171**:213.
Brookes, P., and P. D. Lawley (1960), *Biochem. J.*, **77**:478.
—— and —— (1961), *Biochem. J.*, **80**:496.
Brown, S. W., and W. A. Nelson-Rees (1961), *Genetics*, **46**:983.
Carlson, J. G. (1941), *Proc. Natl. Acad. Sci.*, **27**:46.
Cleaver, J. E. (1969), *Proc. Natl. Acad. Sci.*, **63**:428.
Conger, A. D., and A. H. Johnston (1956), *Nature*, **178**:271.
Cox, E. C. (1970), *J. Mol. Biol.*, **50**:129.
Crowther, J. A. (1924), *Proc. R. Soc. (Lond.)*, **B96** : 207.
Deering, R. A. (1962), *Sci. Am.*, **207**(December):135.
De Mars, R. I. (1953), *Nature*, **172**:964.
Demerec, M. (1937), *Genetics*, **22**:469.
—— (1960), *Proc. Natl. Acad. Sci.*, **46**:1075.
Drake, J. W. (1963), *J. Mol. Biol.*, **6**:268.
—— (1969), *Annu. Rev. Genet.*, **3**:247.
—— (1970), "The Molecular Basis of Mutation," Holden-Day, San Francisco.
—— and E. O. Greening (1970), *Proc. Natl. Acad. Sci.*, **66**:823.
Dunn, D. B., and J. D. Smith (1954), *Nature*, **174**:304.
Emmerson, P. T., and P. Howard-Flanders (1965), *Biochem. Biophys. Res. Commun.*, **18**:24.
Epstein, J. H., et al. (1970), *Science*, **168**:1477.
Freese, E. (1959a), *Proc. Natl. Acad. Sci.*, **45**:622.
—— (1959b), *J. Mol. Biol.*, **1**:87.
—— (1959c), *Brookhaven Symp. Biol.*, **12**:63.
—— (1963), in J. H. Taylor (ed.), "Molecular Genetics," pt. I, pp. 207–269, Academic, New York.
—— E. Bautz, and E. Bautz-Freese (1961a), *Proc. Natl. Acad. Sci.*, **47**:845.
——, ——, and —— (1961b), *J. Mol. Biol.*, **3**:133.
Freifelder, D. (1968), *J. Mol. Biol.*, **35**:303.
Genetics Conference (1947), Committee on Atomic Casualties, National Research Council, Genetic Effects of the Atomic Bombs in Hiroshima and Nagasaki, *Science*, **106**:331.
Giles, N. H. (1951), *Cold Spring Harbor Symp. Quant. Biol.*, **16**:283.

Giles, N. H. (1956), *Brookhaven Symp. Biol.*, **8**:103.

――― and H. P. Riley (1949), *Proc. Natl. Acad. Sci.*, **35**:640.

――― F. S. de Serres, and C. W. H. Patridge (1955), *Ann. N.Y. Acad. Sci.*, **59**:536.

Green, D. M., and D. R. Krieg (1961), *Proc. Natl. Acad. Sci.*, **47**:64.

Green, M. M. (1970), *Mutat. Res.*, **10**:353.

Greer, S. B., and S. Zamenhof (1962), *J. Mol. Biol.*, **4**:123.

Grüneberg, H. (1952), *Bibliog. Genet.*, **15**:1.

Hanawalt, P. C., and R. H. Haynes (1967), *Sci. Am.*, **216**(February):36.

Hayes, H. (1968), "The Genetics of Bacteria and Their Viruses," 2d ed., Blackwell, Oxford.

Hill, R. F. (1970), *Mutat. Res.*, **9**:341.

Hollaender, A. (ed.) (1954–1955), "Radiation Biology," 3 vols., McGraw-Hill, New York.

――― and C. W. Emmons (1941), *Cold Spring Harbor Symp. Quant. Biol.*, **60**:170.

Howard, B. D., and I. Tessman (1964), *J. Mol. Biol.*, **9**:364.

――― and ――― (1964), *J. Mol. Biol.*, **9**:372.

Howard-Flanders, P. (1968), *Annu. Rev. Biochem.*, **37**:175.

――― and R. P. Boyce (1964), *Genetics*, **50**:256.

―――, and ――― (1966), *Radiat. Res. Suppl.*, **6** : 156.

―――, ―――, and L. Theriot (1966), *Genetics*, **53**:1119.

――― and L. Theriot (1966), *Genetics*, **53**:1137.

Hughes-Schrader, S. (1948), *Adv. Genet.*, **2**:127.

Kapp, D. S., and K. C. Smith (1968), *Radiat. Res.*, **35**:515.

Kelner, A. (1949), *J. Bacteriol.*, **58**:511.

――― (1953), *J. Bacteriol.*, **65**:252.

――― (1949), *Proc. Natl. Acad. Sci.*, **35**:73.

Kihlman, B. A. (1961), *Adv. Genet.*, **10**:1.

Kirchner, C. E. J. (1960), *J. Mol. Biol.*, **2**:331.

――― and M. J. Rudden (1966), *J. Bacteriol.*, **92**:1453.

Knipling, E. T. (1960*a*), *Sci. Am.*, **203**(October):54.

――― (1960*b*), *Science*, **130**:902.

Knudson, A. G. (1971), *Proc. Natl. Acad. Sci.*, **68**:820.

Koch, R. E. (1971), *Proc. Natl. Acad. Sci.*, **68**:773.

Kolmark, G., and M. Westergaard (1949), *Hereditas*, **35**:490.

Krieg, D. R. (1963*a*), *Prog. Nucleic Acid Res.*, **2**:125.

――― (1963*b*), *Genetics*, **48**:561.

Kubitschek, H. E. (1966), *Proc. Natl. Acad. Sci.*, **55**:269.

Lawley, P. D. (1966), *Prog. Nucleic Acid Res.*, **5**:89.

――― (1967), *J. Mol. Biol.*, **24**:75.

――― and P. Brookes (1961), *Nature*, **192**:1081.

――― and ――― (1967), *J. Mol. Biol.*, **25**:143.

Lederberg, J., and E. Lederberg (1952), *J. Bacteriol.*, **63**:399.

Lemontt, J. F. (1971), *Genetics*, **68**:21.

Lerman, L. S. (1961), *J. Mol. Biol.*, **3**:18.

――― (1963), *Proc. Natl. Acad. Sci.*, **49**:94.

Levisohn, R. (1967), *Genetics*, **55**:345.

Litman, R. M., and A. B. Pardee (1956), *Nature*, **178**:529.

――― and ――― (1960), *Biochem. Biophys. Acta*, **42**:117.

Magni, G. E. (1963), *Proc. Natl. Acad. Sci.*, **50**:980.

Marmur, J., and L. Grossman (1961), *Proc. Natl. Acad. Sci.*, **47**:778.

――― et al. (1961), *J. Cell. Comp. Physiol.*, **58**(suppl.1) : 33.

Meselson, M. (1964), *J. Mol. Biol.*, **9**:734.

Mitchell, H. K., and M. B. Houlahan (1946), *Am. J. Bot.*, **33**:31.

Moorhead, P. S., and E. Saksela (1965), *Hereditas*, **52**:271.

Muller, H. J. (1927), *Science*, **66**:84.

Muller, H. J. (1928), *Science,* **67** : 82.

───── (1930), *J, Genet.,* **22**:299.

───── (1954), in Hollaender (1954–1955), vol. 1, pp. 351–626.

───── (1956), *Brookhaven Symp. Biol.,* **8**:126.

───── E. Carlson, and A. Schalet (1961), *Genetics,* **46**:213.

───── I. H. Herskowitz, S. Abrahamson, and I. I. Oster (1954), *Genetics,* **39**:741.

Mundry, K. W., and A. Gierer (1958), *Z. Vererbungsl.,* **89**:614.

National Pigeon Association Information Booklet 1 (Project on Genetics) (1950–1951), p. 1, Waterton, Wis.

Neel, J. V., (1963), "Changing Perspectives on the Genetic Effects of Radiation," Charles C Thomas, Springfield, Ill.

───── and H. F. Falls (1951), *Science,* **114**:419.

───── and W. J. Shull (1958), "Human Heredity," University of Chicago Press.

Newcombe, H. B., and A. P. James (1959), *Can. J. Pub. Health,* **50**:140.

Novick, A., and L. Szilard (1950), *Proc. Natl. Acad. Sci.,* **36**:708.

───── and ───── (1951), *Cold Spring Harbor Symp. Quant. Biol.,* **16**:337.

Orgel, A., and S. Brenner (1961), *J. Mol. Biol.,* **3**:762.

Parrington, J. M., J. D. A. Delhanty, and H. P. Baden (1971), *Ann. Hum. Genet.,* **35**:149.

Patterson, J. T., and H. J. Muller (1930), *Genetics,* **15**:495.

Pettijohn, D., and P. Hanawalt (1964), *J. Mol. Biol.,* **9**:395.

Pierce, B. L. S. (1966), *Genetics,* **54**:657.

Purdom, C. E. (1963), "Genetic Effects of Radiation," Academic, New York.

Resnick, M. A. (1969), *Genetics,* **62**:519.

Rhoades, M. M. (1936), *J. Genet.,* **33**:347.

───── (1938), *Genetics,* **23**:377.

───── (1941), *Cold Spring Harbor Symp. Quant. Biol.,* **9**:138.

Ronen, A., and J. Atidia (1971), *Mutat. Res.,* **11**:175.

Rupert, C. S. (1960), *J. Gen. Physiol.,* **43**:573.

───── (1961), *J. Cell. Comp. Physiol.,* **58**:57.

Rupp, W. D., and P. Howard-Flanders (1968), *J. Mol. Biol.,* **31**:291.

Russell, W. L. (1956), *Am. Nat.,* **90**:67.

───── and L. B. Russell (1959), in J. C. Bugher (ed.), "Progress in Nuclear Energy," ser. 6, vol. 2, pp. 179–188, Pergamon, Elmsford, N.Y.

Setlow, J. (1968), *Radiat. Res. Suppl.,* **6**:141.

Setlow, R. B., and W. L. Carrier (1964), *Proc. Natl. Acad. Sci.,* **51**:226.

───── J. R. Regan, J. German, and W. L. Carrier (1969), *Prog. Natl. Acad. Sci.,* **64**:1035.

───── and J. K. Setlow (1962), *Proc. Natl. Acad. Sci.,* **48**:1250.

───── P. A. Swenson, and W. L. Carrier (1963), *Science,* **142**:1464.

Shapiro, J. A. (1969), *J. Mol. Biol.,* **40**:93.

Siegel, E. C., and V. Bryson (1967), *J. Bacteriol.,* **94**:38.

Sideropoulos, A. S., and D. M. Shankel (1968), *J. Bacteriol.,* **96**:198.

Smith, D. W., and P. C. Hanawalt (1969), *J. Mol. Biol.,* **46**:57.

Spencer, W. P., and C. Stern (1948), *Genetics,* **33**:43.

Speyer, J. F. (1965), *Biochem. Biophys. Res. Commun.,* **21**:6.

Stadler, L. J. (1928), *Science,* **68**:186.

───── (1929), *Proc. Natl. Acad. Sci.,* **15**:876.

───── (1944), *Proc. Natl. Acad. Sci.,* **30**:123.

Stahl, F. (1969), "The Mechanics of Inheritance," 2d ed., Prentice-Hall, Englewood Cliffs, N.J.

Stern, C. (1960), "Principles of Human Genetics," Freeman, San Francisco.

Stone, W. S. (1956), *Brookhaven Symp. Biol.,* **8**:171.

───── O. Wyss, and F. Haas (1948), *Proc. Natl. Acad. Sci.,* **33**:59.

Strauss, B. (1968), *Curr. Top. Microbiol. Immunol.,* **44**:1.

Strelzoff, E. (1961), *Biochem. Biophys. Res. Commun.,* **5**:384.

Suzuki, D. T., T. G. Grigliatti, and R. Williamson (1971), *Proc. Natl. Acad. Sci.*, **68**:890.

Swanson, C. P. (1957), "Cytology and Cytogenetics," Prentice-Hall, Englewood Cliffs, N.J.

Tamm, C. M., E. Hodes, and E. Chargaff (1952), *J. Biol. Chem.*, **195**:49.

─── H. S. Shapiro, R. Lipschitz, and E. Chargaff (1953), *J. Biol. Chem.*, **203**:673.

Terzaghi, B. E., G. Streisinger, and F. W. Stahl (1962), *Proc. Natl. Acad. Sci.*, **49**:1519.

Tessman, I. (1959*a*) *Virology*, **7**:263.

─── (1959*b*), *Virology*, **9**:375.

─── H. Ishiwa, and S. Kumar (1965), *Science*, **148**:507.

─── R. K. Poddar, and S. Kumar (1964), *J. Mol. Biol.*, **9**:352.

Thoday, J. M., and J. M. Read (1947), *Nature*, **160**:608.

Treffers, H. P., V. Spinelli, and N. O. Belser (1954), *Proc. Natl. Acad. Sci.*, **40**:1069.

Vielmetter, W., and H. Schuster (1960), *Biochem. Biophys. Res. Commun.*, **2**:324.

von Borstel, R. C., K. T. Cain, and C. M. Steinberg (1971), *Genetics*, **69**:17.

Wacker, A. (1963), *Prog. Nucleic Acid Res.*, **1**:369.

Wagner, R. P., C. H. Haddox, R. Fuerst, and W. S. Stone (1950), *Genetics*, **35**:237.

Watson, J. D., and F. H. Crick (1953*a*), *Cold Spring Harbor Symp. Quant. Biol.*, **18**:123.

─── and ─── (1953*b*), *Nature*, **171**:964.

Weiss, B., and C. C. Richardson (1967), *Proc. Natl. Acad. Sci.*, **57**:1021.

Westmoreland, B. C., W. Szybalski, and H. Ris (1969), *Science*, **163**:1343.

Witkin, E. M. (1953), *Proc. Natl. Acad. Sci.*, **39**:427.

─── (1966), *Science*, **152**:1345.

─── (1969), *Mutat. Res.*, **8**:9.

─── (1969), *Annu. Rev. Genetics*, **3**:525.

─── (1971), *Nat. New Biol.*, **229**:81.

Wolff, S. (ed.) (1963), "Radiation-induced Chromosome Aberrations," Columbia, New York.

Wright, S. (1949), *Genetics*, **34**:245.

Wulff, D. L., and C. S. Rupert (1962), *Biochem. Biophys. Res. Commun.*, **7**:237.

Yanofsky, C., E. C. Cox, and V. Horn (1966), *Proc. Natl. Acad. Sci.*, **55**:274.

Yasuda, S., and M. Sekiguchi (1970), *J. Mol. Biol.*, **47**:243.

Zamenhof, P. J. (1966), *Proc. Natl. Acad. Sci.*, **56**:845.

Zamenhof, S., and S. B. Greer (1958), *Nature*, **182**:611.

25
Chemical Nature and Structure of Genes and Chromosomes

NOTATION

1   Heavy DNA = labeled DNA = ^{15}N; light DNA = normal DNA = ^{14}N.
2   Formulas for the building blocks of DNA are given in Appendix Table A-3.
3   A list of biochemical abbreviations used in this and subsequent chapters is given in Appendix Table A-7.
4   Prefixes used with the metric system are explained in Appendix Table A-6.

QUESTIONS

1125   **a** To carry genetic information any molecule must be structurally unique. What additional genetic properties must be accounted for in such a molecule?

        b What kinds of proof are there that DNA is genetic material in bacteria and most viruses?

        c What lines of evidence in eukaryotes indicate that DNA is the carrier of genetic information in these organisms?

1126   Mirsky (1947, 1953), Mirsky and Pollister (1946), and Mirsky and Ris (1951) have shown that the main constituents of chromosomes are DNA, RNA, and proteins (mostly basic) in the approximate percentages (by weight) of 14:14:72 (Maio and Schildkraut, 1967). Boivin et al. (1948) and Mirsky and Ris (1949) used the Feulgen test (specific for DNA) for measuring the quantity of DNA per nucleus in different cells of several animal species and obtained the kind of data shown. The quantity of other chromosome components varies

Table 1126

	DNA, picograms		
Animal	Nuclei ($2n$) of erythrocyte	Nuclei ($2n$) of hepatic cells	Nuclei (n) of sperm
Domestic fowl	2.34	2.39	1.26
Carp	3.49	3.33	1.64
Brown trout	5.79		2.67
Toad	7.33		3.70

with cell type, depending on cell function and the metabolic state of the cell. Discuss these findings with regard to the chemical nature of genes.

1127   Show how you would demonstrate that DNA polymerase recognizes only the regular sugar-phosphate portion of the nucleotide precursors and cannot determine base-sequence specificity.

1128   If genes in eukaryotes are composed of DNA, the distribution of DNA should parallel that of the genes, both in location and in quantity. With the aid of illustrations show the location, quantity, and distribution of DNA during mitosis, meiosis, and syngamy in a diploid animal with $2n = 2$. Would the distribution be different in a diploid plant? In a haploid plant? Explain.

1129 Although the nucleic acid in some DNA-containing viruses is single-stranded, the DNA of eukaryotes must be double-stranded. Why?

1130 In certain viruses the nucleic acid is single-stranded. For example, in ϕX174 the single-stranded DNA is circular. After infection of host *Escherichia coli*, progeny viruses are formed with circular DNA whose base content (ratio and sequence) is identical to that in the parent phage. Explain, with the aid of illustrations, how such DNA replicates. See Kornberg (1968), Sinsheimer (1962), Watson (1970).

1131 Circular duplex DNA replication begins at a fixed point and progresses sequentially from this location in the same general direction (Cairns, 1963, 1966). At this initiation-of-replication site the strands separate, and both act as templates for the synthesis of complementary fragments which grow in the 5'-to-3' direction. One or both strands are synthesized in vivo discontinuously (Okazaki et al., 1968; Sugimoto et al., 1969; Okazaki et al., 1970). Explain, with illustrations, how circular DNA molecules replicate. In your explanation indicate which of the enzymes endonuclease, exonuclease, DNA polymerase, and polynucleotide ligase are involved in the process and how and in what sequence they function. See Watson (1970), Okazaki et al. (1970), Richardson (1969).

1132 The building blocks of DNA, a polymer built up of a long chain of nucleotides, are shown in Appendix Table A-3.
  a Draw the four nucleosides and their corresponding nucleotides.
  b Diagram a polynucleotide chain four nucleotides (A, T, G, C) long. Be sure to show how the nucleotides are joined together.
  c The DNA molecule is 20 Å wide throughout its length. Illustrate a double helix consisting of four nucleotide pairs (A-T, G-C, T-A, C-G) showing how the bases pair to maintain this uniform width. Note its other most important features.

1133 A single nucleic acid–pentanucleotide chain contains the base sequence G-A-U-C-G. State whether it is from DNA or RNA and why.

1134 The DNA base ratios from cattle and rat are the same: 28 A : 22 G : 22 C : 28 T. Would you expect hybridization between single strands of the two species? Explain.

1135 The Watson-Crick model of DNA was a brilliant synthesis based on both chemical and x-ray-diffraction data, the former provided by Boivin, Vendrely, Chargaff, etc. (see Chargaff, 1955) and the latter by Wilkins and his colleagues (Wilkins and Randall, 1953; Franklin and Gosling, 1953). What fundamental features of DNA were revealed by each of these methods? Why were the chemical data insufficient to show how the DNA is built up?

1136 **a** Demonstrate why cytosine in its most common tautomeric form cannot pair with adenine in its most common form. Do the same for guanine and thymine.

b Show why a shift of a hydrogen atom in an adenine molecule from the 6-amino to the N-1 position permits hydrogen bonding with cytosine.

1137 The diagram shows the transfer of a single strand of DNA from a donor (Hfr) *Escherichia coli* into the recipient (F⁻) via a rolling circle model. Indicate which events have already occurred and which must still take place to replicate the chromosome in the donor and the single strand inside the recipient.

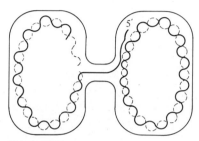

Figure 1137

1138 *Escherichia coli* possesses one circular chromosome consisting of one continuous DNA double helix. Cells are grown for many generations on a medium in which the only nitrogen source is ^{15}N. Thereafter the cells are transferred to a medium containing only ^{14}N and allowed to go through three cell divisions, i.e., DNA replications. The DNA density is measured by density-gradient centrifugation just before transfer and after each cell division on ^{14}N medium. Disregarding the circularity of the chromosome and using a solid line for labeled chains and a broken line for unlabeled ones, if replication were (1) semiconservative and (2) conservative, show the kinds of chromosomes expected after:

a Continued replication in ^{15}N.

b One generation of replication in ^{14}N.

c Two generations of replication in ^{14}N.

d Three generations of replication in ^{14}N.

1139 The DNA of a somatic cell in *Haplopappus gracilis* ($2n = 4$), an eukaryote, is labeled with tritiated thymidine during interphase.

a The labeled cell divides mitotically. Would you expect all or some of the chromosomes in one or both daughter cells to be labeled and why?

b If each of the daughter cells divides mitotically in a medium with normal thymine, what proportion of (1) the chromosomes and (2) the cells would you expect to be labeled? Illustrate.

1140 **a** What is heterochromatin? What are its chemical and cytological properties?

b Discuss the various kinds of supporting evidence that indicate that heterochromatin, regardless of whether it is a permanent feature of a genome or a transient one (as in B chromosomes), is genetically inert or almost so.

c Does heterochromatin replicate at the same time as euchromatin? Cite evidence to support your answer.

See Brown (1966), Brown and Nelson-Rees (1961), Cooper (1959), Lima-de-Faria and Jaworska (1968), and Schultz (1947).

1141 Mather (1944) championed the idea that euchromatin contains major genes (those controlling qualitative characters) and heterochromatin contains polygenes (controlling quantitative characters). Cytologists in general agree that heterochromatin is genetically inactive. State whether or not you agree with Mather's hypothesis and why.

1142 Barring somatic mutations, all cells of a multicellular individual contain the same genes: not only do nuclei of different organs have the same amount of DNA, but they also have the same $A + T : G + C$ ratios. On the other hand, base ratios of DNA from different species vary, and a specific base ratio is a species characteristic (Chargaff, 1955). Is this expected if DNA is genetic material? Discuss.

1143 What are the chemical units of DNA that correspond to the genetic units of replication, recombination, mutation, and function? Outline one form of evidence for each statement. Are the chemical units the same in both eukaryotes and prokaryotes? Discuss data that support your answer.

1144 **a** Discuss the various lines of evidence indicating that a chromosome is multistranded (in sense of classical chromatids).

b What are some of the possible disadvantages a multistranded chromosome may have in comparison with a unistranded one in recombination (crossing-over), segregation, and mutation?

c Do you think a multistranded chromosome may have advantages over a single-stranded one? If so, what are they?

1145 *Planococcus citri* ($2n = 10$) males have one heterochromatic (H) set and one euchromatic (E) set of chromosomes from the blastula stage on in most tissues. In 1966 Nur observed that polyploid nuclei of the cellular sheath surrounding each testis contain a polyploid number of E chromosomes (5, 10, 20, 40, or 80) but only a haploid number (5) of H chromosomes. Offer an explanation to account for this difference in the number of H and E chromosomes. How would you test your hypothesis? See Lorick (1970).

1146 Satellite DNA constitutes approximately 10 percent of the DNA of many eukaryotes. How does it differ from nonsatellite DNA, where is it located, and what functions may it perform? (See Eckhardt and Gall, 1971; Hsu and Arrighi, 1971; Jones and Robertson, 1970; Yasmineh and Yunis, 1970; Walker 1971; and Pardue et. al., 1970.)

1147 Species of the same genus with the same chromosome number frequently have significantly different DNA contents. For example, the toad *Bufo bufo* has 40 percent more DNA per somatic cell than *B. viridis* although both have the same chromosome number, $2n = 22$ (Ullerich, 1966). Related species in which one chromosome number is a multiple or near multiple of the other have the same DNA content. For example, *Thyanta perditor* (12 + XY), an insect belonging to the tribe Pentatomini, has a DNA value of 1.10, and *T. calceata* (24 + XXY) a DNA content of 0.93 (Hughes-Schrader and Schrader, 1956).

a Offer an explanation to account for the difference in *Bufo* and the similarity in *Thyanta*.

b How might you proceed to test your hypothesis?

1148 The fluorescent compounds quinacrine mustard and quinacrine hydrochloride bind preferentially to certain regions of chromosomes, producing a characteristic and reproducible pattern (Caspersson et al.,1968, 1969). O'Riordan et al. (1971) claim that all 22 pairs of autosomes in man can be identified visually without recourse to fluorometry after chromosomes are stained with quinacrine hydrochloride. What are the implications of this technique for human cytogenetics?

1149 Sinsheimer and coworkers (1959, 1962) showed that the DNA of mature phage ϕX174 is single-stranded and that immediately after entry into a susceptible host it displays the following features:

1 It has undergone several replications.

2 Limited replication can occur in the presence of the protein-synthesis inhibitor chloramphenicol.

3 It is of a lower density than the DNA of mature ϕX174 and about the same as that of its host, *Escherichia coli*.

4 Upon heating, it is denatured in a manner similar to double-stranded DNA.

5 It is more resistant to inactivation by ultraviolet irradiation than the DNA of mature viruses.

In chloramphenicol no mature progeny viruses are formed, whereas in normal infection these are present as early as 8 minutes after infection. These, like the parents, are single-stranded and possess the same base composition. These results suggest a sequence of events after infection which lead to the formation of single-stranded progeny phage. Illustrate this sequence.

1150 **a** Show whether it is possible to obtain an inversion in a single-stranded DNA molecule.

b The two strands of a DNA (or RNA) double helix can be separated and subsequently reannealed (Marmur and Lane, 1960; Schildkraut et al., 1961). Of what value is this technique in testing affinities between species?

c If the base sequence of a polynucleotide chain is G-C-A-G-T-C-T-A-A, what is the sequence of the bases on the complementary chain and why?

d Do the two chains of a DNA duplex carry the same genetic information? Explain.

e Since organisms contain thousands of different kinds of genes, do you think DNA is sufficiently complex to account for the diversity of genes within as well as between species?   Explain.

f What kind of evidence indicates (1) that DNA can reproduce itself and (2) that nucleotides occur in matched pairs in DNA molecules?

g The nucleic acid of a certain virus contains uracil.   What conclusions can you draw from this?

PROBLEMS

1151 The relationships between the bases adenine, cytosine, guanine, and thymine in the DNA of different species have been studied by Chargaff, Wyatt, and others.   The data are from Chargaff (1955).

Table 1151

Organism	A	T	G	C
Salmon	29.7	29.1	20.8	20.4
Locust	29.3	29.3	20.5	20.7
Sea urchin	32.8	32.1	17.7	17.7
Yeast	31.7	32.6	18.3	17.4
Serratia marcescens	20.7	20.1	27.2	31.9
Tuberculosis bacteria (avian)	15.1	14.6	34.9	35.4
Vaccinia virus	29.5	29.9	20.6	20.0
Man (thymus)	30.9	29.4	19.9	19.8

a Derive the quantitative relationships between:
  (1) Guanine and cytosine.
  (2) Adenine and thymine.
  (3) The two pyrimidines.
  (4) The two purines.
  (5) Total purines and total pyrimidines.
  (6) A + T and G + C.

b What conclusions would you draw regarding the probable pairing relationships between bases in DNA in all organisms?

c Derive the general rules for complementary pairing and for taxonomic differences with regard to DNA content.

1152 Ogur et al. (1952) determined the amount of DNA present in the nuclei of cells from four different strains of a certain species of yeast (*Saccharomyces*). Offer an explanation to account for the differences in DNA content between the four strains and suggest a way of testing your explanation.

Table 1152

Strain	DNA per cell, picograms
1	2.26 ± 0.23
2	4.57 ± 0.60
3	6.18 ± 0.54
4	9.42 ± 1.77

1153 Alfert and Swift (1953) give the following relative amounts of DNA for nuclei of various cells of the annelid worm *Sabellaria*:

First polar body	$127 \pm \;\; 3$
Sperm	$61 \pm \;\; 1$
Male pronucleus	$133 \pm \;\; 7$
Prophase of first cleavage	263 ± 10
Telophase of first cleavage	$124 \pm \;\; 3$

Show how these values relate to expected levels of ploidy and account for any major anomalies (viz. double or half the expected value).

1154 The table shows the average amount of DNA per nucleus in testis and malpighian tubules of the grasshopper *Dissosteira carolina* (Swift, 1950a). Explain these values.

Table 1154

Cell type	Class	Amount of DNA, arbitrary units
Spermatids		1.78
Malpighian tubule nuclei	I	3.25
	II	6.36
	III	14.10
	IV	26.40

1155 The base ratios of nucleic acids from eight different species in both the eukaryotes and prokaryotes are presented in the table.   For each species state whether its nucleic acid is:
a DNA or RNA.
b Single- or double-stranded and why.

Table 1155

Species	T	C	U	A	G	$\dfrac{A + T \text{ or } U}{G + C}$	$\dfrac{A + G}{C + T \text{ or } U}$
1	31	19		31	19		
2		19	31	31	19		
3	19	31		19	31		
4						1.00	1.26
5	32	18		25	25		
6		25	32	23	20		
7						1.26	1.00
8						1.00	1.00

1156 If all four kinds of ribonucleotides are present in the form of diphosphates, the enzyme RNA phosphorylase directs the synthesis of an RNA with the bases A, G, U, C incorporated at random even in the presence of a specific DNA, e.g., from *E. coli*, man, rat, etc.   How does the function of RNA phosphorylase differ from RNA polymerase?

1157 a The $(A + T)/(G + C)$ ratio in the DNA of *E. coli* is 1.00.   Is this information sufficient to decide whether the DNA is most likely to be double- or single-stranded?   If not, state the kind of information required to determine the strandedness of DNA.

b RNA with a 1 A : 1 G : 1 C : 1 U base ratio is enzymatically synthesized from the DNA duplex

G T C A
C A G T

Is it possible to determine whether the RNA is synthesized using one or both strands of the DNA as a template?   Explain.

1158 Autoradiography of T2 DNA labeled with ^{3}H-thymidine and extracted in the presence of a thousandfold excess of cold T2 shows an unbranched rod about 52 μm long (Cairns, 1961).   The distance between base pairs is 3.4 Å; 660 is the average molecular weight of a nucleotide pair.   If the molecule is a double helix throughout its length, what is its molecular weight?

1159 How many permutations are possible in a DNA molecule containing n nucleotide pairs?

1160 Virus f2 (RNA) is 7 times as resistant to ultraviolet as ϕX174 (DNA).   Moreover it does not undergo photoreactivation, as the DNA phages do (Zinder, 1965).   Offer an explanation to account for these observations.

1161 Tessman (1959) found that when phage S13 was labeled with ^{32}P, it was inactivated by the decay of its incorporated radioactive atoms with approximately 100 percent efficiency.   The inactivation of phages T1 and T2 occurred at a much slower rate.   Account for the difference in inactivation rate between the two groups of viruses, noting that all these viruses possess DNA.

1162   Bollum (1959) found that DNA polymerase from calf thymus glands would not catalyze the in vitro synthesis of DNA unless the primer DNA (of high molecular weight) was first heated to 99°C.   What is the significance of this finding?

1163   At any base-pair site in a DNA duplex any one of four possible base-pair combinations can occur:

A-T   T-A   G-C   C-G

Calculate the number of possible sequences for a DNA duplex 10^3 base pairs long (the average length of a structural gene) if no restriction is placed on the relative frequencies of the four kinds of base pairs.

1164   The molecular weight of the DNA (double helix) of a bacterial chromosome is about 3×10^9; that of a single nucleotide pair is 660, and the distance between adjacent bases is 3.4 Å.   How long is this DNA molecule?

1165   In an RNA duplex 500 out of 5,000 bases are uracil.   Calculate the approximate molecular weight of the duplex and determine the proportions of the bases present.

1166   The proportions of bases in DNA of yeast, which is double-stranded, are 31 A : 18 G : 18 C : 31 T.   Would you expect the $(A + G)/(C + T)$ and $(A + T)/(G + C)$ ratios to be the same and why?

1167   The vaccinia and M13 viruses have the base compositions shown.

Table 1167

	A	T	G	C
Vaccinia	29.5	29.9	20.6	20.0
M13	23.3	32.8	21.1	19.8

a Offer an explanation to account for these differences.
b How would you test your explanation?

1168   The pneumonia bacterium, *Diplococcus pneumoniae*, occurs in the *encapsulated* (*smooth*, *S*) or the *nonencapsulated* (*rough*, *R*) form.   *S* types are virulent; *R* types are not.   The capsule (a complex polysaccharide) exists in several different types (*SI*, *SII*, etc.).   Each type is true-breeding, but about 1 *S* cell in 10 million mutates to *R*, which, by fission, forms an *R* colony.   On rare occasions the *R* types can back-mutate to *S*, giving an *S*-type identical to the original *S* type (viz. an *R* type arising from a type *SII* culture mutates to *SII* only).

a Griffith (1928) heat-killed *S*III bacteria and injected them with living *R*II types into living mice.   Many of the mice died of pneumonia; their blood contained not only *R*II but also *S*III diplococci, each of which bred true-to-type.   *R* or *S* of any type, when heat-killed and injected alone, did not multiply.   In 1931 Dawson and Sia obtained the same results in vitro.
(1) Discuss mutation as a probable cause of the genetic change (RII → SIII).
(2) What alternative explanation can be given to account for the heritable change?

b Alloway (1932) produced in vitro hereditary changes of the kind described in (a) by exposing *R* bacteria to extracts of *S* cells from which all traces of structural components (mitochondria, ribosomes, etc.) had been removed. In 1944 Avery, MacLeod, and McCarty separated the extract from disrupted *S* cells into various chemical fractions and determined the ability of each fraction to cause specific hereditary changes in the *R* cells.   It was found that the fraction obtained after removing all proteins and all capsular polysaccharides continued to show this capability.   When treated with deoxyribonuclease, a DNA-destroying enzyme, however, it lost its capacity to change *R* into *S* cells.   What is the chemical component responsible for the hereditary changes and why?   Illustrate the probable mechanism by which the living *S*III bacteria are produced.

1169   Nucleic acids extracted from the phage MS2 can be used to infect bacterial protoplasts (bacteria with capsule removed) and to obtain mature phage particles from this infection.   This infectivity is destroyed by ribonuclease but not by deoxyribonuclease (Davis et al., 1961).   When the infecting nucleic acid or phage is labeled, no label is ever found in the progeny phage, which are identical in all respects to parental particles (Davis and Sinsheimer, 1963). What inferences are possible from this information with respect to:
a Kind of nucleic acid in MS2 chromosomes?
b How the chromosome replicates?

1170   To determine the nature of T2 phage genes Hershey and Chase (1952) designed an ingenious but simple experiment based on the fact that only DNA, and not protein, contains phosphorus and only protein, and not DNA, contains sulfur.   Two batches of T2 phages were used.   One infected bacteria whose protoplasmic constituents contained radioactive phosphorus, ^{32}P; the DNA in the progeny in this batch was labeled with ^{32}P.   The other batch infected bacteria labeled with ^{35}S, which was incorporated into the proteins of the labeled progeny.   In separate experiments, the two batches of labeled phage were then added to unlabeled bacteria.   Shortly after infection the portions of the phage remaining outside the bacteria were removed (separated) from the host cells by the shearing action of a blender, and two solutions, one containing washed bacteria and the other (the supernatant) containing the separated portions, were obtained.   When the infection was brought about by ^{32}P-labeled phage, no label was found in the supernatant (protein coats), the radioactivity being all in the bacteria.   Phage multiplication was *normal* in these bacteria.   On the average 30 to 50 percent of the parental ^{32}P was transferred to the progeny.   When ^{35}S-labeled phage were used, it was found that most of the radioactivity was in the supernatant (with the phage protein); the host bacteria were relatively unlabeled (actually about 3 percent of the phage protein is present).   The phage multiplied normally, but no ^{35}S appeared in the progeny.   Are the phage genes composed of both DNA and proteins or one or the other?   Why?

1171   Tobacco mosaic virus is composed of a central core of RNA surrounded by a protein coat (Schramm, 1958).

   a Fraenkel-Conrat, Singer, and Williams (1957) and Gierer and Schramm (1956) dissociated the protein from the RNA by shaking the particles with water and phenol.   When the separated fractions were tested for infectivity on tobacco leaves, only the RNA fraction retained this ability.   Cells of the tobacco plant infected with this naked RNA release complete virus particles. The infectivity is rapidly destroyed by ribonuclease.   What do these experiments indicate about the chemical repository of hereditary information in this virus?

   b In 1957 Fraenkel-Conrat and Singer separated the protein and RNA of the *normal* and the *Holmes-Ribgrass* (*HR*) strains.   Artificial reciprocal hybrids were then made by combining *HR* protein with *normal* RNA, and *normal* protein with *HR* RNA.   These and the separated fractions were applied to leaves of *Nicotiana tabacum* and *N. sylvestris* plants susceptible to both types of viruses.   The leaves were checked for infectivity, and when progeny viruses were released, their protein constitution was determined.   The results of these experiments are illustrated in simplified form in the table.   The

Table 1171

Infection		
Protein	RNA	Infection capacity
HR	*Normal*	+
Normal	*HR*	+
HR	None	−
Normal	None	−
None	*HR*	+
None	*Normal*	+

protein constitution in progeny viruses was always identical to the strain from which the nucleic acid was obtained.   What conclusions can you draw from this experiment regarding the hypotheses:

(1) That protein may be a carrier of genetic information?

(2) That RNA may be a carrier of genetic information?

1172   Three mechanisms of replication of DNA have been suggested (Delbrück and Stent, 1957):

1   *Conservative:*   The parental DNA double helix does not unwind (the integrity of the parental DNA is conserved).   The duplex serves as a template for the synthesis of a completely new DNA double helix.

2   *Dispersive:*   The DNA duplex fragments crosswise every half turn or so.   New molecules are synthesized along the fragments.   The new and old fragments reunite crosswise to form DNA double helices.   Thus both chains of each daughter DNA molecule in all generations are composed of alternating fragments of the old and the new chain.

3   *Semiconservative:*   The parental DNA double helix unwinds, and both chains serve as templates for the synthesis of new complementary chains.   New DNA molecules are thus composed of one chain from the parental DNA and one complementary to this one, newly synthesized.   The chains are conserved, but the parental complex disappears.   One of the two parental strands is passed on to each daughter molecule.

Meselson and Stahl (1958) grew *Escherichia coli* for several generations on a medium containing ^{15}N (nitrogen) as the sole source of nitrogen.   These bacteria, whose DNA was fully labeled with ^{15}N, were then transferred to a medium containing only ^{14}N, in which the cells multiplied synchronously for a number of generations (one DNA replication = one division).   Just before transfer and after each of the first three divisions, samples of the culture were removed, DNA extracted, and its density determined by density-gradient centrifugation (see various texts for technique).   In this method, molecules

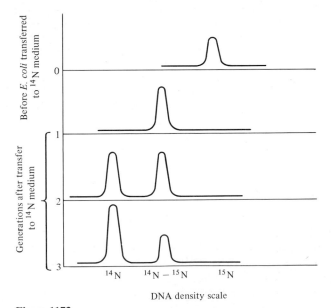

DNA density scale

Figure 1172

containing (1) ^{15}N only, (2) ^{14}N only, and (3) ^{14}N and ^{15}N sediment in three different bands as a result of differences in density. A schematic presentation of Meselson and Stahl's results is illustrated.

a Which mechanism do these data favor, and which do they definitely rule out? Explain fully.

b DNA extracted from the ^{14}N-^{15}N (hybrid) band was heated to 100°C (a procedure that completely separates the chains of DNA from each other) and centrifuged and its density determined. Two bands, one heavy and one light, appeared. What mechanism of replication is supported by these results?

c What information is necessary to prove that the most likely method is the correct one?

1173 Taylor et al. (1957) grew broad bean (*Vicia faba*; $2n = 12$) seedlings for 8 hours in a medium containing radioactive (^3H) thymidine (incorporated exclusively into DNA) during which time root-tip cells in interphase would go through one complete cycle of DNA synthesis and end in prophase of division. The seedlings were then washed and placed in a nonradioactive solution of thymine, containing colchicine, which delays division of the centromere so that at colchicine(c)-metaphase each chromosome has a crosslike appearance. The drug also inhibits spindle formation. Because of the latter effect, daughter chromatids, after separating, are incorporated in the same anaphase nucleus, with the result that the chromosome number is doubled during each mitosis. Hence, the number of mitoses which have occurred since the roots were placed in colchicine can be determined from the chromosome number. Some of the roots were kept in colchicine for 10 hours before fixing, staining, squashing, and autoradiographing. Others were kept in colchicine for 34 hours before treatment. Root cells at c-metaphase after 10 hours in colchicine had the normal diploid number of 12 chromosomes; all were labeled, equally and uniformly, in both chromatids. Many cells at c-metaphase after 34 hours in colchicine contained 24 chromosomes. All chromosomes were labeled, but only one of the two sister chromatids of each was radioactive. The distribution of the label after one and two replications is given in Fig. 1173.

a Interpret these results in terms of DNA replication if the chromosome is uninemic (one DNA double helix per chromosome in interphase before duplication).

b Would the interpretation of the method of DNA replication necessarily be the same if the chromosomes were polynemic? Illustrate your explanation with a binemic chromosome (two double helices per chromosome in interphase before duplication)?

c Some of the cells after 34 hours in colchicine had 12 chromosomes. In some of these, all chromosomes were unlabeled; in others, when labeling occurred, sister chromatids were uniformly labeled. When did these two types of cells replicate? Explain.

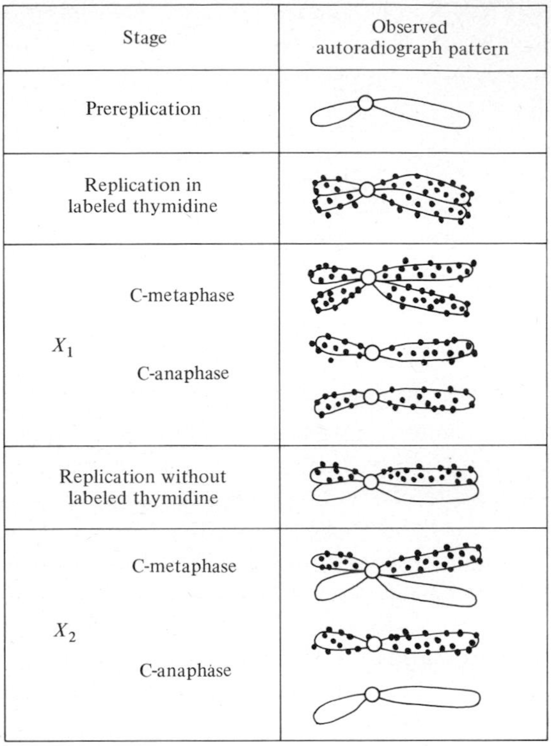

Stage	Observed autoradiograph pattern
Prereplication	
Replication in labeled thymidine	
X_1 C-metaphase C-anaphase	
Replication without labeled thymidine	
X_2 C-metaphase C-anaphase	

Figure 1173 Distribution of label in one chromosome after one round of replication in labeled medium followed by one round of replication in an unlabeled medium. (*Redrawn from K. R. Lewis and B. John, "Chromosome Marker," fig. 9, p. 30, J. and A. Churchill Ltd., London, 1963.*)

d A few of the metaphases after 34 hours in colchicine had 48 chromosomes. How many chromosomes in the complement would be completely unlabeled if the chromosome contains a single DNA duplex? How many of the other chromosomes would have a labeled and an unlabeled chromatid? Illustrate.

1174 a When heavy DNA (containing ^{15}N) and light DNA (containing ^{14}N) from different cultures of *Diplococcus pneumoniae* are density-gradient-centrifuged, two bands of DNA are formed in the tube. However, if the same mixture is heated to 100°C and cooled slowly before centrifuging, the results are somewhat different. The two bands are formed as before and in the same position in the tube as before (indicating that they represent the same substances), but in addition, an intermediate band, representing the mean of the densities of the other two bands, is formed. Account for (1) the two bands in the mixtures and (2) the band of intermediate density.

b When labeled DNA from *D. pneumoniae* and unlabeled DNA from *Escherichia coli*, another bacterial species, are mixed, heated to 100°C, slowly cooled, and then centrifuged, only two bands are formed in the cesium chloride density gradient.   Account for the absence of an intermediate band.

c What conclusions can be drawn from (a) and (b) regarding the nature and specificity of DNA base pairing and base sequence?

1175   Base analogs can replace their normal counterparts during DNA synthesis. Kornberg (1962) replaced each of the four DNA bases with various base analogs and studied the effects on in vitro DNA synthesis using *E. coli* DNA polymerase.   The amounts of DNA synthesized are expressed as percentages of that synthesized from normal bases only.   Which bases are analogs of adenine? Thymine?   Cytosine?   Guanine?   Explain.

Table 1175

Analog	Normal bases substituted by the analog			
	A	T	C	G
Uracil	0	54	0	0
5-Bromouracil	0	97	0	0
Hypoxanthine	0	0	0	25
5-Bromocytosine	0	0	118	0

1176   S13 and MS2 are single-stranded DNA- and RNA-containing viruses respectively.   The replicative behavior of both viruses, after infection of *Escherichia coli*, is the same.   The mature phage nucleic acid *plus* strand acts as a template for the formation of double-stranded replicative forms (*plus-minus*) from which *plus* strands are synthesized.   The isolated *minus* strands of the two viruses differ in behavior: those of S13 are infective, whereas those of MS2 show no such activity.   Outline a plausible explanation to account for the difference in behavior of the *minus* strands of the two viruses.

1177   When bacteriophage infect *E. coli*, the synthesis of host RNA is terminated and a new type of RNA is produced.   Volkin et al. (1958) found that the base ratios of T7 DNA and of the RNA formed shortly after this phage infects *E. coli* were as shown.   Is the RNA most probably synthesized from a DNA template?   Explain.

Table 1177

	A	T or U	C	G
T7 DNA	26.0	26.6	23.6	23.8
Labeled RNA	26.7	27.8	23.7	21.8

1178 Meselson and Weigle (1961) used λ phage, whose DNA is double-stranded, to show that crossing-over occurs by breakage and reunion of DNA molecules. In subsequent experiments, *E. coli* growing in a light medium (^{12}C, ^{14}N) were infected with heavy (^{13}C, ^{15}N) λ of genotype *AB* and heavy λ of genotype *ab*. Most λ DNA molecules replicated in the normal semiconservative manner. A few, however, did not do so but became enclosed in new protein shells. The DNA of these rare progeny was completely heavy. Using a cesium chloride solution, progeny of different density were separated and tested to determine whether they possess parental or recombinant genotypes. Some of the rare completely heavy DNA progeny possessed the genotype *Ab*, others the genotype *aB*. Are crossing-over and DNA replication independent phenomena? Explain.

1179 When ribonucleotides in the form of triphosphates are incubated in vitro in the presence of single- and double-stranded φX174 along with RNA polymerase and magnesium ions, RNA molecules are produced with the base ratios shown (Hurwitz and August, 1963; Hurwitz et al., 1962). What are the functions of RNA polymerase and the DNAs of this phage?

Table 1179

DNA	Bases incorporated							
	A		U		C		G	
	Obs.	Calc.	Obs.	Calc.	Obs.	Calc.	Obs.	Calc.
Single-stranded φX174	1.02	1.02	0.79	0.77	0.72	0.77	0.62	0.58
Double-stranded φX174	1.16	1.16	1.10	1.16	0.87	0.84	0.86	0.84

1180 The requirements for in vitro DNA synthesis, according to Kornberg (1960), are:

1   The enzyme DNA polymerase, extracted from *E. coli* (or other source).

2   All four bases present as substrates in the form of their deoxyribonucleoside 5'-(not 3'-) triphosphates. Neither nucleoside 5'-diphosphates nor nucleoside 5'-monophosphates were active as substrates.

3   The absence of any one of the nucleoside 5'-triphosphates caused the synthesis to drop almost to zero.

4   A primer DNA of high molecular weight, from any source.

5   Magnesium ions.

DNA synthesis proceeded until one of the above substrates was exhausted. The final DNA in some cases amounted to more than 20 times the primer added. It had previously been shown (Schackman et al., 1958) that synthetic and native DNA were alike in physical and chemical properties (same base ratios and sequences), and Lehman (1959) showed that when highly purified polymerase was used, no in vitro synthesis occurred unless the primer had been heated or treated gently with deoxyribonuclease or was initially single-stranded (from viruses ϕX174, S13).   From these data what conclusions can you draw regarding:

a The manner in which the polynucleotide chain grows; e.g., from one end by stepwise addition of nucleotides or in some other way.

b The specific function of polymerase.

c The function of the added DNA.

1181 **a** In *Drosophila melanogaster* all repeated (redundant) genes for 18S and 28S ribosomal RNA (*r*RNA) are located in the DNA (*r*DNA) of the nucleolus-organizer (NO) region, which occupies corresponding (homologous) positions on the X and Y chromosomes (Ritossa and Spiegelman, 1965).   This NO DNA, about 0.27 percent of the total *wild-type* genotype, carries at least 130 genes for each kind of *r*RNA.   This NO region is at the *bobbed, bb,* locus (Cooper, 1959), where many hypomorphic alleles occur which cause small bristles and slow development (Stern, 1929), the severity of the mutant phenotype being dependent on the mutant allele.   After denaturing DNA and using the technique of DNA-RNA hybridization along with RNAse to degrade any RNA that did not hybridize, Ritossa et al. (1966) determined the amount of NO DNA that would hybridize with *r*RNA in *wild-type* and four *mutant*-strain flies.   The saturation plateaus (all NO DNA completely paired with *r*RNA) are given in Table 1181*A*.

Table 1181*A* Summary of RNA/DNA ratios obtained with DNA from *wild-type* and *mutant* flies

Sex	Source of DNA*	Number of NO regions	(rRNA/DNA) $\times$ 100	
			At saturation	Per NO region
♀(XX)	*wild-type* (bb^+bb^+)	2	0.270	0.137
♀(XX)	$bb^{car}bb^{car}$	2	0.136	0.068
♀(XX)	$bb^{uco-3}bb^{uco\ 3}$	2	0.149	0.075
♀(XX)	$bb^{ds}bb^{ds}$	2	0.172	0.086
♀(XX)	bb^{ds}/bb^l†	2	0.098	0.033‡
♂(XY)	bb^{ds}/df§	1	0.065	0.065

* The superscripts *car, uco-3, ds* refer to different hypomorphic alleles affecting growth rate and bristle size.

† bb^l, lethal when homozygous.   § Deficiency of NO region.   ‡ Amount of NO DNA in bb^l.

(1) What kinds of mutations (single-site substitutions, deletions, etc.) appear to be responsible for the hypomorphic alleles?   Explain.

(2) Which of these mutant alleles would you expect to show the most and the least severe phenotype respectively, and why?

b Using the technique described in (a) to determine the ratio and sequence of genes for 18S and 28S *r*RNA, Quagliarotti and Ritossa (1968) studied flies homozygous for different hypomorphic alleles and *wild types*. The ratio of DNA complementary to 18S and 28S RNA in *wild-type* and different *mutant* flies is shown in Table 1181*B*. *Note:* (1) Genes for 28S are longer than

Table 1181*B*

	*r*RNA / DNA % in hybrid (1)	28S *r*RNA / DNA % in hybrid (2)	18S *r*RNA / DNA % in hybrid (3)	(1)/(3)	(1)/(2)	(2)/(3)
Wild type	0.373 ± 0.013	0.247 ± 0.012	0.122 ± 0.002	3.05 ± 0.15	1.51 ± 0.12	2.02 ± 0.13
yw bb S1	0.165 ± 0.009	0.111 ± 0.004	0.053 ± 0.003	3.11 ± 0.28	1.48 ± 0.13	2.09 ± 0.23
yw bb S2	0.205 ± 0.006	0.138 ± 0.004	0.066 ± 0.003	3.10 ± 0.24	1.48 ± 0.09	2.09 ± 0.18
car bb S1	0.254 ± 0.006	0.170 ± 0.003	0.082 ± 0.003	3.09 ± 0.18	1.49 ± 0.06	2.07 ± 0.11
car bb S2	0.168 ± 0.006	0.109 ± 0.005	0.053 ± 0.002	3.17 ± 0.23	1.54 ± 0.13	2.05 ± 0.17
UCO3 bb S1	0.185 ± 0.006	0.123 ± 0.004	0.060 ± 0.002	3.08 ± 0.20	1.50 ± 0.10	2.05 ± 0.13

those for 18S.   (2) True saturation plateau for 28S is about 2.5 to 5 percent higher.

(1) Suggest and explain the ratio between the number of genes for 18S *r*RNA and that for 28S *r*RNA.

(2) What is the ratio between the amount of DNA complementary to 28S *r*RNA and that complementary to 18S *r*RNA?

(3) The genes for the two *r*RNAs may be arranged as follows:

(*a*) Genes for 28S *r*RNA clustered in one block and those for 18S *r*RNA clustered in another block, the two blocks being adjacent to each other.

(*b*) The two kinds of genes are interspersed in the *bb* locus, e.g., in alternating sequence 18S-28S-18S-28S-18S-28S···.

Which arrangement do these data support and why?

(4) Would you expect complementation between mutant alleles?   Why?

1182 Following the original in vitro synthesis of DNA, Lehman et al. (1958) set out to answer the following questions:

a What is the role of primer DNA?   Is it simply extended in length during the synthesis by terminal addition of nucleotides, or does it serve as a template for the building up of molecules identical to itself?

b Does synthetic DNA show the equivalence of adenine to thymine and guanine to cytosine that characterizes natural DNA?

c Does the base composition of primer DNA influence the composition of the synthetic product?

Using DNA from different organisms as primer, they determined and compared the base composition of the newly synthesized DNA with that of the DNA primer with the results shown.   Varying the rate of synthesis or the deoxynucleotide concentration of the reaction mixture did not change these values.

Table 1182

$\dfrac{A + T}{G + C}$		$\dfrac{A + G}{T + C}$		Source of primer DNA
Primer DNA	Synthetic DNA	Primer DNA	Synthetic DNA	
0.49	0.48	1.01	0.99	*Mycobacterium phlei*
0.97	1.02	0.98	1.01	*Escherichia coli*
1.25	1.29	1.05	1.02	Calf thymus
1.92	1.90	0.98	1.02	Bacteriophage T2
—	>40	1.00	1.03	A-T copolymer*

* A synthetic DNA containing A and T in alternating sequence.

The data are sufficient to answer all the questions posed.   What are the answers and why?

1183   Simon (1961) grew human HeLa cervical carcinoma cells in a medium containing 5-bromodeoxyuridine, an analog of thymine, for two generations. After the first division all the DNA was found to be half-labeled; after the second division both half- and fully labeled material was found.
   a What do these results suggest regarding the method of DNA replication?
   b Do they explain the division of chromosomes?   Explain.   Illustrate if necessary.

1184   Subspecies *thummi* of the fly *Chironomus thummi* has 27 percent more DNA per 2*n* complement than subspecies *piger* in both salivary glands and meiocytes.   Comparisons of the banding patterns and dimensions of salivary-gland chromosomes of the two subspecies (same chromosome number) with those of 20 *Chironomus* species indicated that *piger* is phylogenetically older than *thummi* (Keyl, 1965).   Analysis of salivary-gland chromosomes of hybrids between the two subspecies showed that the banding patterns of their homologs are the same but that several bands of *thummi* are larger and take longer to replicate than their counterparts in *piger* (Keyl and Pelling, 1963). Bands of *thummi*, regardless of absolute amount, either had the same amount of DNA as the corresponding ones in *piger* or differed by a power of 2; viz. 2, 4, 8, or 16 times as much DNA (Keyl, 1965).   What appears to be the main, if not the only, cause of evolution of *thummi* from *piger*?

thummi piger

thummi thummi

Figure 1184   A demonstrable local and geometric increase in the chromosomal DNA of *Chironomus*. [*From H. G. Keyl, Experientia,* **21**:*192 (1965).*]

1185   The term *lecanoid system* describes a form of chromosome behavior peculiar to certain related groups of insects.   Both sexes develop from fertilized eggs and are diploid (in a typical species, the mealy bug *Pseudococcus nipae*, $2n = 10$ for both sexes).   Except that centromeres are diffuse, mitotic chromosomal behavior is normal; however, unusual phenomena are observed in these insects:

1   In females, meiosis is *inverse*.   The paired homologs, at metaphase I, autoorient in tandem so that one chromatid of one homolog passes with one chromatid of the other homolog to the same pole; thus disjunction is equational.   This is followed by a second synapsis of homologous daughters at prophase II and hence by a reductional division to form the gametic nuclei.

2   In males, one entire haploid set becomes heterochromatic before the blastula stage and remains so in all subsequent cells.   The first meiotic division is lecanoid (viz. the euchromatic set does not pair with its homologs in the heterochromatic set).   The resulting univalents divide mitotically, producing a $2n$ set at each pole.   In meiosis II the euchromatic set segregates as a unit from the heterochromatic one (by means of a monopolar spindle) so that of the four gamete nuclei, two contain heterochromatic sets, and two contain the euchromatic sets.   Only the euchromatic derivatives form sperm; those containing the heterochromatic set disintegrate.   The process may be illustrated as shown in Fig. 1185.

a (1)   Outline the essential differences between this and the normal mechanism of meiosis in males and females in species with true haplodiploidy.
(2)   How would you prove genetically that the heterochromatic chromosome set in the lecanoid system comes from the father and is largely or completely inert?   [Assume that in the mealy bug *P. nipae* ($2n = 10$) two

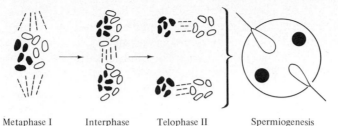

Metaphase I   Interphase   Telophase II   Spermiogenesis

Figure 1185 Heterochromatic chromosomes are shown in black. [*From S. W. Brown and W. A. Nelson-Rees, Radiation Analysis of a Lecanoid Genetic System, Genetics,* **46** : *984 (1961).*]

true-breeding lines are available, one homozygous for the dominant alleles at the loci under study, the other homozygous for the recessive alleles at these loci.]

(3) How might such a system transform into a true haplodiploid one?

b In 1961 Brown and Nelson-Rees carried out radiation studies with *Pseudococcus citri*; typical results are given in the table.

Table 1185

Experiment 1: Progeny of x-irradiated ♀ × untreated ♂

X-ray dose, R	Av. no. of survivors per mother after maternal treatment	
	♂*	♀
Control	172.4	295.1
1,000	59.9	45.7
2,000	26.4	40.6
4,000	1.4	4.1
8,000	0.2	0

Experiment 2: Progeny of x-irradiated ♂ × untreated ♀

X-ray dose, R	Av. no. of survivors per mother after paternal treatment	
	♂†	♀
Control	163.8	212.8
2,000	129.0	114.0
8,000	188.0	41.0
16,000	275.0	3.0

* In male progeny all chromosome breaks occurred in the euchromatic set.

† In male progeny all chromosome breaks occurred in the heterochromatic set.

(1) Discuss the bearing of these data on the hypotheses that:
    (*a*) The heterochromatic set is of paternal origin.
    (*b*) The heterochromatic set is genetically inert.
(2) Is heterochromatin a permanent characteristic of a chromosome or chromosome segment in the lecanoid system? Explain.

1186 Swift (1950*b*) estimated the amount of DNA in nuclei of various somatic and meiotic cells of corn and *Tradescantia* by microphotometric determinations on individual Feulgen-stained nuclei. The relative amounts of DNA are shown.

Table 1186

Mitosis (*Zea mays*)				Meiosis (*Tradescantia paludosa*)			
Tissue	Stage	DNA class	DNA, arbitrary units	Tissue	Stage	DNA class	DNA, arbitrary units
Leaf and root meristem	Interphase	2C	3.4	Sporog- enous	Interphase	2C → 4C	13.1
	Interphase	4C	6.6		Prophase	4C	16.0
	Prophase	4C	6.9		Telophase	2C	8.1
	Telophase	2C	3.2				
				Micro- spore mother cells	Preleptotene	2C	8.7
Root- elongation zone	Interphase	4C	6.6		Leptotene	2C → 4C	13.0
		4C → 8C	8.7		Zygotene	4C	16.1
		8C	12.5		Pachytene	4C	16.8
		8C → 16C	20.2		Diplotene	4C	16.9
		16C	26.1		Diakinesis	4C	16.3
		16C → 32C	33.8				
		32C	49.0	Micro- spores	Tetrad	C	4.4
					Early interphase	C	4.0
Embryo	Interphase	2C	3.6		Late interphase	C → 2C	7.4
		4C	7.1		Prophase	2C	9.2
Aleurone	Interphase	3C	4.8	Pollen	Tube nuclei at interphase	C	4.1
		6C	10.1				
		12C	20.5		Generative nuclei at interphase:		
					Early	C	4.3
					Late	2C	8.5

a Does the distribution of the DNA parallel that of the genes?   Illustrate.
b When does DNA replication occur in somatic tissues? Meiocytes? What is the basis for this conclusion?

1187 Swift (1950*a*) measured photometrically the amount of DNA during mitosis in Feulgen-stained pronephros cells and erythrocyte nuclei of a recently hatched

Ambystoma [=*siredon*] *opacum* larva.   Interphase nuclei of dividing tissues showed a large number of DNA values intermediate between two classes, I and II (amount of DNA in II twice that in I), as well as classes I and II.   All early prophase nuclei measured fell into class II, and all telophases fell into class I.

a When is DNA synthesized in these cells?

b Does its quantity and distribution in the various stages conform to expectations on the theory that genes are linear segments of DNA?

1188 In the pigeon (*Columba domestica*) several large (macro-) chromosomes are a constant feature of all cells; many microchromosomes, also present, are very difficult to identify and count.   Galton and Bredbury (1966), studying tritiated-thymidine uptake by autoradiography, showed that one of the macrochromosomes of the female pigeon completes replication after the others.   In the male all the macrochromosomes terminated synthesis simultaneously. Macrochromosomes replicate later than the microchromosomes in both sexes.

a Which macrochromosome (autosome Z or W) in the female is late-replicating?

b If late replication signifies genetic inactivation, how should homozygotes and heterozygotes for mutant genes on this chromosome compare in phenotype?

1189 In 1969 De Lucia and Cairns isolated a mutant strain (*pol-A1*) of *Escherichia coli* that has less than 1 percent of normal level of DNA polymerase I.   The mutant survives and replicates DNA normally, but unlike *wild-type E. coli* it is sensitive to ultraviolet light (possesses a defective repair mechanism).   Subsequently, Gross and Gross (1969) showed that the recessive mutant gene *pol-A1* which affects DNA polymerase activity is of the *amber* nonsense type and located between *met E* and *rha*, at approximately 75 minutes on the *E. coli* chromosome.   De Lucia and Cairns then isolated five additional *pol* mutants, all of which are allelic to *pol-A1*.   Kelley and Whitfield (1971) studied one of these, *pol-6*.   This temperature-sensitive mutant contains approximately 20 percent of *wild-type* DNA polymerase I activity, which is partially destroyed by preincubation at temperatures which have little effect on the *wild-type* form of the enzyme.   Whereas purified DNA polymerase I from *wild-type* strains has the same activity at 37 and 52°C and maximum activity at 55°C, the pure form of the enzyme from *pol-6* cells is less active at 52°C than at 37°C, and its optimum activity occurs at 45°C.   The *wild-type* and *pol-6* forms of this enzyme also differ in several other ways.   All *pol* mutants excise dimers and undergo genetic recombination normally.   Kornberg and Gefter (1971) isolated a second DNA polymerase (II) from *pol-A1* cells which is the same as that which Knippers and Strätling (1970), Knippers (1970), and Okazaki et al. (1970) found in a cell-free fraction from cells from the same mutant strain which can synthesize DNA rapidly.

a Are the *pol* alleles in the structural gene for DNA polymerase?   Explain.

b If so, is the enzyme involved in DNA replication?   Explain.

c Is it involved in DNA repair?   If so, at what stage?   See Klein and Niebch (1971) and Witkin (1971).

1190 **a** Mirsky and Ris (1951) found that somatic cells of the amphibian *Amphiuma* contain about 20 times as much DNA as those of a certain toad and 70 times as much as those of domestic fowl (ducks, geese, chickens).   Do you think it likely that *Amphiuma* has 20 times as many genes as this toad and 70 times as many as these birds?   If not, what other explanations can be offered for these differences?

b

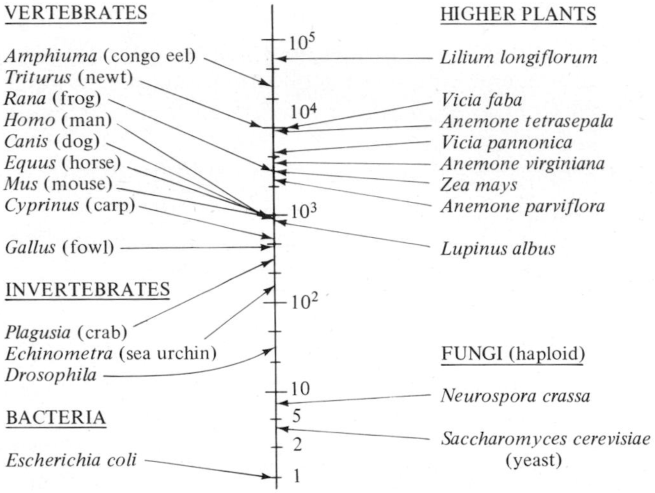

Figure 1190   The amounts of DNA per genome (chromosome set) in some fungi, vertebrates, and higher plants expressed as multiples of the amount in *E. coli* (mol. wt. = $2.4 \times 10^9 = 4 \times 10^{12}$ mg). [*Redrawn from R. Holliday, The Organization of DNA in Eukaryote Chromosomes, Symp. Soc. Gen. Microbiol.,* **20**: *362 (fig. 1) (1970).*]

(1) Determine the approximate amounts of DNA in the different organisms in milligrams.

(2) On the average, genes and proteins are of the same size in all organisms. If each kind of gene is present only once per genome, calculate the number of genes the chromosome set would contain in *E. coli*, *Neurospora*, *Drosophila*, man, *Amphiuma*, *Lilium*, *Vicia faba*, and *Lupinus*.

(3) Do you think these estimates of gene number per genome in these organisms are realistic?   Explain.

(4) Can DNA content be related to an organism's position on the evolutionary scale?   Explain.

(5) Significant differences in DNA amounts are not restricted to different genera, families, orders, etc., but often occur within plant and animal

genera.   Chooi (1971), Rees and Jones (1967), Keyl (1965), and many others have shown that in some cases closely related species with the same chromosome number differ significantly in their DNA content, sometimes by a factor of 5 to 10 or more.   What intrachromosomal mechanisms can cause such differences in DNA content?   Indicate some possible function(s) of redundant DNA, giving examples supporting your suggestions where possible.

1191   Interphase chromosomes usually respond to x-rays as though they were undivided (single chromatids) and prophase chromosomes as though they were divided into two chromatids.   What does this indicate regarding the relationship of strandedness to the cell cycle?

1192   A delay in the appearance of mutations following the use of some chemical mutagens has been noted in eukaryotes.
   a Can this be taken as evidence for a multistranded condition of the eukaryote chromosome?   Explain.
   b If not, what alternative explanations can you offer to account for the delay?

1193   Bivalents at diplotene of meiosis in the oocytes of amphibians and birds are called *lampbrush chromosomes*.   In the newts they are remarkably long (up to 800 μm); each homolog consists of two chromatids with hundreds of loops projecting laterally for 10 to 15 μm.   The loops are the result of a localized looping-out of the chromomeric regions; those arising from a particular chromomere pair are the same length and form.   At later stages of meiosis the bivalents shorten and the loops retract.   Like other chromosomes, these contain DNA, RNA, and protein (Gall, 1963; Callan and Macgregor, 1958).

Diplotene bivalent

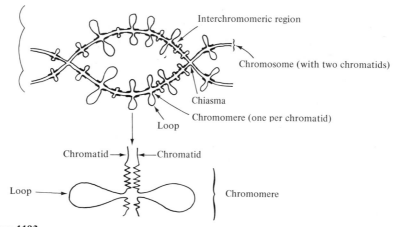

Figure 1193

a Callan and Macgregor (1958) treated isolated unfixed lampbrush chromosomes from the newt *Triturus cristatus* with trypsin, pepsin, and ribonuclease, first individually and then in various combinations. Although matrix material surrounding the axis of the chromosome was dissolved, the chromosome thread in the loop remained continuous. After treament with deoxyribonuclease these submicroscopic fibrils broke, and the lateral loops fragmented into small pieces. Which of the three macromolecules appears to be stable and indispensable to the maintenance of structural integrity and therefore to form the backbone of the chromosome, at least in amphibian oocytes? Explain.

b Gall (1963) showed that DNase causes breakage in the interchromomeric regions as well. Breakage kinetic studies in *T. viridescens* to determine the number of subunits per chromosome showed that there are two subunits in the loops and four in the interchromomeric regions. Studies by Miller (see Miller, 1965) revealed a fine fibril approximately 40 Å in diameter. What can you infer from these data regarding the number of DNA duplexes (continuous or discontinuous) per chromatid (or chromosome at anaphase, telophase, and interphase before replication)? See DuPraw (1965, 1966), Trosko and Wolff (1965), and Ris (1957) for their interpretation of similar data in eukaryotes without lampbrush chromosomes.[1]

1194 In an elegant experiment by Bibring (reported by Mazia, 1960) sea urchin eggs were treated with mercaptoethanol during prophase of the first zygotic cleavage division (zygotes receive one centriole from each parent). The treatment, which did not affect the separation of chromosome strands or centriole replication, completely suppressed the anaphase movement of the chromosomes to the poles and the subsequent DNA synthesis. During blockage a tetrapolar aster developed, as illustrated. Upon removal to seawater lacking

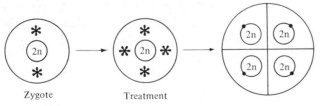

Zygote   Treatment

Figure 1194

mercaptoethanol, the cells immediately underwent division into four daughter cells, each of which received the full diploid complement of chromosomes but only half the diploid amount of DNA; viz. DNA determination of their subsequent prophases showed only half as much DNA as is found in (third-division) prophases in untreated eggs.

---

[1] From studies of DNA in solution DNase attacks the two DNA chains independently and at random. Scission of the molecule occurs only when breaks in the two strands fall within a few nucleotide pairs of each other.

a What conclusion can you draw from these results regarding the strandedness of mitotic chromosomes in this organism? Explain.

b What kind of evidence would indicate the same level of polyteny, i.e., number of strands per chromosome, in chromosomes in the meiocytes?

Be definite about strandedness (whether in the sense of classical chromatids or in terms of DNA strandedness).

1195 The details of eukaryote chromosome structure are not known. Three of the models postulated are illustrated. Which of these is the easiest to reconcile with the existing genetic, cytological, and cytogenetic facts, and why? Why are the others less probable? See Crick (1971) and Comings and Riggs (1971).

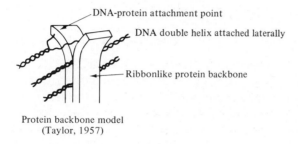

Protein backbone model
(Taylor, 1957)

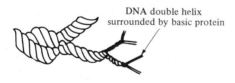

Multistranded model
(Steffensen, 1959; Kaufmann et al., 1960)

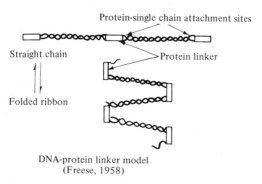

DNA-protein linker model
(Freese, 1958)

Figure 1195

QUESTIONS INVOLVING FURTHER READING

1196 The amount of DNA is not always proportional to genome number. Thus different diploid cells in the same individual may have significantly different amounts of DNA. On the other hand, cells of entirely different (though related) species having the same DNA content per genome may differ greatly in chromosome number; e.g., one may have twice as many chromosomes as the other.

    a Propose a hypothesis of chromosomal organization to account for these facts.

    b After outlining your explanation read critically one of the following papers, which discuss possible ways of accounting for these observations:

    1   Mirsky and Ris (1951).
    2   Schrader and Hughes-Schrader (1956, 1958).
    3   Hughes-Schrader and Schrader (1956).
    4   Hughes-Schrader (1957).

1197 Centromeres are of four types: localized, neocentric, diffuse (holokinetic), and polycentric (Swanson, 1957). Discuss the evolution of the centromere. Include in your discussion the criteria you use to decide which is the most primitive and which are more advanced. The following papers may aid you in discussing the subject: Hughes-Schrader and Schrader (1961); Lima-de-Faria (1949, 1958).

1198 Compare and contrast the patterns of DNA replication in:

    a Eukaryote and prokaryote chromosomes.

    b Euchromatin and heterochromatin.

    c Autosomes and sex chromosomes.

See Cairns (1966), Huberman and Riggs (1968), Hsu et al. (1964), Lima-de-Faria (1959, 1961), Morishima et al. (1962), Schmid (1963), Ris and Kubai (1970) for penetrating studies on replication in eukaryotes. Cairns (1963, 1966), *Cold Spring Harbor Symp. Quant. Biol.* (1968), Richardson (1969), and Watson (1970) discuss replication in *E. coli* and other prokaryotes.

REFERENCES

Alfert, M., and H. Swift (1953), *Exp. Cell Res.*, **5**:455.
Alloway, J. L. (1932), *J. Exp. Med.*, **55**:91.
Arrighi, F. E., T. C. Hsu, P. Saunders, and G. F. Saunders (1971), *Chromosoma*, **32**:224.
Avery, O. T., C. M. MacLeod, and M. McCarty (1944), *J. Exp. Med.*, **79**:137.
Baer, D. (1965), *Genetics*, **52**:275.
Baltimore, D. (1970), *Nature*, **226**:1209.
Berlowitz, L., D. Pallota, and C. H. Sibley (1969), *Science*, **164**:157.
Bessman, M. J., I. R. Lehman, E. S. Simms, and A. Kornberg (1958), *J. Biol. Chem.*, **233**:171.
Boivin, A., R. Vendrely, and C. Vendrely (1948), *C. R. Acad. Sci.*, **226**:1061.
Bollum, F. J. (1959), *J. Biol. Chem.*, **234**:2733.
Bostock, C. J., and D. M. Prescott (1971), *J. Mol. Biol.*, **60**:151.

Bram, S., and H. Ris (1971), *J. Mol. Biol.*, **55**:325.
Brewen, J. G., and W. J. Peacock (1969), *Proc. Natl. Acad. Sci.*, **62**:389.
Brinkley, B. R., and E. Stubblefield (1966), *Chromosoma*, **19**:28.
Britten, R. J., and D. E. Kohne (1968), *Science*, **161**:529.
Brown, S. W. (1966), *Science*, **151**:147.
—— and W. A. Nelson-Rees (1961), *Genetics*, **46**:983.
Cairns, J. (1961), *J. Mol. Biol.*, **3**:756.
—— (1963), *J. Mol. Biol.*, **6**:208.
—— (1966), *J. Mol. Biol.*, **15**:372.
Callan, H. G. (1963), *Int. Rev. Cytol.*, **15**:1.
—— (1967), *J. Cell Sci.*, **2**:1.
—— and L. Lloyd (1956), *Nature*, **178**:355.
—— and H. C. Macgregor (1958), *Nature*, **181**:1479.
Caspersson, T., et al. (1968), *Exp. Cell Res.*, **49**:219.
—— et al. (1970), *Chromosoma*, **30**:215.
—— et al. (1971*a*), *Hereditas*, **67**:213.
——, G. Lomakka, and L. Zech (1971*b*), *Hereditas*, **67**:89.
——, E. J. Modest, G. E. Foley, U. Wagh, and E. Simonsson (1969), *Exp. Cell Res.*, **58**:128.
Chargaff, E. (1955), in "The Nucleic Acids," vol. 1, chap. 10, edited by E. Chargaff and J. N. Davidson, Academic, New York.
—— B. Magasanik, E. Vischer, C. Green, R. Doniger, and D. Elson (1950), *J. Biol. Chem.*, **186**:51.
—— and J. N. Davidson (eds.) (1955 and 1960), "The Nucleic Acids," vols. 1 and 3, Academic, New York.
Chooi, W. Y. (1971), *Genetics*, **68**:195.
Chun, E. H. L., and J. W. Littlefield (1961), *J. Mol. Biol.*, **3**:668.
Clark, R. J., and G. Felsenfeld (1971), *Nat. New Biol.* **229**:101.
Cold Spring Harbor Symp. Quant. Biol., **3** (1968), "Replication of DNA in Microorganisms."
Coleman, J. R., and M. J. Moses (1964), *J. Cell. Biol.*, **23**:63.
Comings, D. E., and T. A. Okada (1970), *Chromosoma*, **30**:269.
—— and A. D. Riggs (1971), *Nature*, **223**:48.
Cooper, K. W. (1959), *Chromosoma*, **10**:535.
Corneo, E., and K. W. Jones (1971), *Nat. New Biol.*, **233**:268.
Crick, F. H. C. (1957), *Sci. Am.*, **197**(September):188.
—— (1971), *Nature*, **234**:25.
Davidson, J. N. (1965), "The Biochemistry of the Nucleic Acids," 5th ed., Wiley, New York.
Davis, J. E., and R. L. Sinsheimer (1963), *J. Mol. Biol.*, **6**:203.
—— J. H. Strauss, and R. L. Sinsheimer (1961), *Science*, **134**:1427.
Dawson, M. H., and R. H. P. Sia (1931), *J. Exptl. Med.*, **54**:681.
Delbrück, M., and G. S. Stent (1957), in W. D. McElroy and B. Glass (eds.), "The Chemical Basis of Heredity," pp. 699–736, Johns Hopkins, Baltimore.
De Lucia, P., and J. Cairns (1969), *Nature*, **224**:1164.
Denhardt, D. T., and R. L. Sinsheimer (1965), *J. Mol. Biol.*, **12**:647.
Derksen, J., and H. D. Berendes (1970), *Chromosoma*, **31**:468.
DuPraw, E. J. (1965), *Nature*, **206**:338.
—— (1966), *Nature*, **209**:577.
—— and P. M. M. Rae (1966), *Nature*, **212**:598.
Eckhardt, R. A., and J. G. Gall (1971), *Chromosoma*, **32**:407.
Fareed, G. C., and C. C. Richardson (1967), *Proc. Natl. Acad. Sci.*, **58**:665.
Feulgen, R., and H. Rossenbeck (1924), *Z. Physiol. Chem.*, **135**:203.
Forsheit, A. B., D. S. Ray, and L. Lica (1971), *J. Mol. Biol.*, **57**:117.
Fraenkel, F. R. (1968), *Proc. Natl. Acad. Sci.*, **59**:131.
Fraenkel-Conrat, H., and B. Singer (1957), *Biochem. Biophys. Acta*, **24**:540.

Fraenkel-Conrat, H., B. Singer, and R. C. Williams (1957), *Biochem, Biophys. Acta*, **25**:87.

Franklin, R. E., and R. Gosling (1953), *Nature*, **171**:740.

Freese, E. (1958), *Cold Spring Harbor Symp. Quant. Biol.*, **23**:13.

Gall, J. G. (1963), *Nature*, **198**:36.

—— (1968), *Proc. Natl. Acad. Sci.*, **60**:553.

—— (1969), *Genetics* **61**(suppl. 1):121.

——, E. H. Cohen, and M. L. Polan (1971), *Chromosoma*, **33**:319.

Galton, M., and R. P. Bredbury (1966), *Cytogenetics*, **5**:295.

—— and S. F. Holt (1965), *Exp. Cell Res.*, **37**:111.

Giannelli, F., and R. M. Howlett (1966), *Cytogenetics*, **5**:186.

Gierer, A., and G. Schramm (1956), *Nature*, **177**:702.

Gomatos, P. J., R. M. Krug, and I. Tamm (1965), *J. Mol. Biol.*, **13**:802.

Goulian, M., and A. Kornberg (1967), *Proc. Natl. Acad. Sci.*, **58**:1723.

——, ——, and R. L. Sinsheimer (1967), *Proc. Natl. Acad. Sci.*, **58**:2321.

Griffith, F. (1928), *J. Hyg.*, **27**:113.

Griffith, J., J. A. Huberman, and A. Kornberg (1971), *J. Mol. Biol.*, **55**:209.

Gross, J. D., and M. Gross (1969), *Nature*, **224**:1166.

Guthrie, G. D., and R. L. Sinsheimer (1960), *J. Mol. Biol.*, **2**:297.

Hahn, W. E., and C. D. Laird (1971), *Science*, **173**:158.

Hearst, J. E., and M. Botchan (1970), *Annu. Rev. Biochem.*, **39**:151.

Henderson, S. A. (1964), *Chromosoma*, **15**:345.

Hershey, A. D., and M. Chase (1952), *J. Gen. Physiol.*, **36**:39.

Holland, I. B. (1970), *Sci. Prog.* **58**:71.

Holliday, R. (1970), *Symp. Soc. Gen. Microbiol.*, **20**:359.

Hosoda, J., and E. Mathews (1971), *J. Mol. Biol.*, **55**:155.

Hotta, Y., M. Ito, and H. Stern (1966), *Proc. Natl. Acad. Sci.*, **56**:1184.

—— and H. Stern (1971), *J. Mol. Biol.*, **55**:337.

Howard-Flanders, P. (1968), *Annu. Rev. Biochem.*, **37**:175.

Howell, S. H., and H. Stern (1971), *J. Mol. Biol.*, **55**:357.

Hsu, T. C., and F. E. Arrighi (1971), *Chromosoma*, **34**:243.

——, W. Schmid, and E. Stubblefield (1964), in M. Locke (ed.), "The Role of Chromosomes in Development," pp. 83-112, Academic, New York.

Huang, C-C. (1971), *Chromosoma*, **34**:230.

Huberman, J. A., and A. D. Riggs (1968), *J. Mol. Biol.*, **32**:327.

Hughes-Schrader, S. (1948), *Adv. Genet.*, **2**:127.

—— (1957), *Chromosoma*, **8**:709.

—— and F. Schrader (1956), *Chromosoma*, **8**:135.

—— and —— (1961), *Chromosoma*, **12**:327.

Hurwitz, J., and J. August (1963), in J. N. Davidson and W. E. Cohn (eds.), "Progress in Nucleic Acid Research," vol. 1, pp. 59–92, Academic, New York.

—— J. J. Furth, M. Anders, and A, Evans (1962), *J. Biol. Chem.*, **237**:3752.

Hutchison, C. A., and R. L. Sinsheimer (1966), *J. Mol. Biol.*, **18**:429.

John, B., and G. M. Hewitt (1966), *Chromosoma*, **18**:254.

Jones, K. W. (1970), *Nature*, **225**:912.

—— and F. W. Robertson (1970), *Chromosoma*, **31**:331.

Kaufmann, B. P., H. Gay, and M. R. McDonald (1960), *Int. Rev. Cytol.*, **11**:76.

Kelley, W. S., and H. J. Whitfield (1971), *Nature*, **230**:33.

Keyl, H. G. (1965), *Experientia*, **21**:191.

—— and C. Pelling (1963), *Chromosoma*, **14**:347.

Kiger, J. A., and R. L. Sinsheimer (1971), *Proc. Natl. Acad. Sci.*, **68**:112.

Klein, A., and U. Niebch (1971), *Nat. New Biol.*, **229**:82.

Knippers, R. (1970), *Nature*, **228**:1050.

—— and W. Strätling (1970), *Nature*, **226**:713.

Kornberg, A. (1962), "Enzymatic Synthesis of DNA," Wiley, New York.
———— (1960), *Science*, **131**:1503.
———— (1968), *Sci. Am.*, **219**(October):64.
———— (1969), *Science*, **163**:1410.
Kornberg, T., and M. L. Gefter (1971), *Proc. Natl. Acad. Sci.*, **68**:761.
Kozinski, A. W. (1968), *Cold Spring Harbor Symp. Quant. Biol.*, **33**:375.
Laird, C. D. (1971), *Chromosoma*, **32**:378.
Lark, K. G. (1969), *Annu. Rev. Biochem.*, **38**:569.
Lee, J. C., and J. J. Yunis (1971), *Chromosoma*, **32**:237.
Lehman, I. R. (1959), *Ann. N.Y. Acad. Sci.*, **81**:745.
————, S. B. Zimmerman, J. Adler, M. J. Bessman, E. S. Simms, and A. Kornberg (1958), *Proc. Natl. Acad. Sci.*, **44**:1191.
Lima-de-Faria, A. (1949), *Hereditas*, **35**:422.
———— (1958), *Int. Rev. Cytol.*, **7**:123.
———— (1959), *J. Biophys. Biochem. Cytol.*, **6**:457.
———— (1961), *Hereditas*, **47**:674.
———— and H. Jaworska (1968), *Nature*, **217**:138.
Lodish, H. F., and N. D. Zinder (1966), *J. Mol. Biol.*, **21**:207.
Lorick, G. (1970), *Chromosoma*, **32**:11.
Maio, J. J., and C. L. Schildkraut (1967), *J. Mol. Biol.*, **24**:29.
———— and ———— (1969), *J. Mol. Biol.*, **40**:203.
Marmur, J., and D. Lane (1960), *Proc. Natl. Acad. Sci.*, **46**:453.
Mather, K. (1944), *Proc. R. Soc. (Lond.)*, **B132**:308.
Mazia, D. (1960), *Ann. N.Y. Acad. Sci.*, **90**:455.
Meselson, M., and F. W. Stahl (1958), *Proc. Natl. Acad. Sci.*, **44**:671.
———— and J. J. Weigle (1961), *Proc. Natl. Acad. Sci.*, **47**:857.
Miescher, F. (1955), trans. in M. L. Gabriel and S. Fogel (eds.), "Great Experiments in Biology," pp. 233–239, Prentice-Hall, Englewood Cliffs, N.J.
Miller, D. A., P. W. Allderdice, O. J. Miller, and W. R. Breg (1971), *Nature*, **232**:24.
Miller, G., L. Berlowitz, and W. Ragelson (1971), *Chromosoma*, **32**:251.
Miller, O. L. (1965), *Natl. Cancer Inst. Monogr.* 18, p. 79.
Miller, O. L., and B. R. Beatty (1969), *J. Cell Physiol.*, **74**(suppl. 1):225.
Mirsky, A. E. (1947), *Cold Spring Harbor Symp. Quant. Biol.*, **12**:143.
———— (1953), *Sci. Am.*, **188** (February) 47.
———— and A. W. Pollister (1946), *J. Gen. Physiol.*, **30**:117.
———— and H. Ris (1949), *Nature*, **163**:666.
———— and ———— (1951), *J. Gen. Physiol.*, **34**:451.
Moens, P. B. (1968), *Chromosoma*, **23**:418.
———— (1969), *J. Cell Biol.*, **40**:273.
Morishima, A., M. M. Grumbach, and J. H. Taylor (1962), *Proc. Natl. Acad. Sci.*, **48**:756.
Moses, M. J. (1956), *J. Biophys. Biochem. Cytol.*, **2**:215.
———— (1968), *Annu. Rev. Genetics*, **2**:363.
———— and J. H. Taylor (1955), *Exp. Cell Res.*, **9**:474.
Nur, U. (1966), *Chromosoma*, **19**:439.
Ogur, M., S. Minckler, G. Lindegren, and C. C. Lindegren (1952), *Arch. Biochem.*, **40**:175.
Ohno, S., and T. S. Hauschka (1960), *Cancer Res.*, **20**:541.
————, U. Wolf, and N. B. Atkin (1968), *Hereditas*, **59**:169.
Okazaki, R., T. Okazaki, K. Sakabe, and K. Sugimoto (1967), *Jap. J. Med. Sci. Biol.*, **20**:255.
————, ————, ————, ————, and A. Sugino (1968), *Proc. Natl. Acad. Sci.*, **59**:598.
————, K. Sugimoto, T. Okazaki, Y. Imae, and A. Sugimoto (1970), *Nature*, **228**:223.
O'Riordan, M. L., J. A. Robinson, K.   . Buckton, and H. J. Evans (1971), *Nature*, **230**:167.
Pardue, M. L., and H. G. Gall (1970), *Science*, **168**:1356.
————, S. A. Gerbi, R. A. Eckhardt, and J. G. Gall (1970), *Chromosoma*, **29**:268.

Pearson, P. L., M. Bobrow, and C. G. Vosa (1970), *Nature*, **226**:78.

Pelling, C. (1966), *Proc. R. Soc. (Lond.)*, **B164**:279.

Philips, D. M. P. (ed.) (1971), "Histones and Nucleohistones," Plenum, New York.

Quagliarotti, G., and F. M. Ritossa (1968), *J. Mol. Biol.*, **36**:57.

Ray, D. S., A. Preuss, and P. H. Hofschneider (1966), *J. Mol. Biol.*, **21**:485.

Rees, H. F., and R. N. Jones (1967), *Nature*, **216**:825.

Richardson, C. C. (1969), *Annu. Rev. Biochem.*, **38**:795.

Ris, H. (1957), in W. D. McElroy and B. Glass (eds.), "The Chemical Basis of Heredity,"
pp. 23–69, Johns Hopkins, Baltimore.

—— and D. F. Kubai (1970), *Annu. Rev. Genet.*, **4**:263.

Ritossa, F. M., and S. Spiegelman (1965), *Proc. Natl. Acad. Sci.*, **53**:737.

—— K. C. Atwood, and S. Spiegelman (1966), *Genetics*, **54**:819.

Rolfe, R. (1962), *J. Mol. Biol.*, **4**:22.

Schackman, H. K., et al. (1958), *Fed. Proc.*, **17**:304.

Schildkraut, C. L., J. Marmur, and P. Doty (1961), *J. Mol. Biol.*, **3**:595.

Schmid, W. (1963), *Cytogenetics*, **2**:86.

—— and M. F. Leppert (1969), *Cytogenetics*, **8**:125.

Schrader, F., and S. Hughes-Schrader (1956), *Chromosoma*, **7**:469.

—— and —— (1958), *Chromosoma*, **9**:193.

Schramm, G. (1958), *Annu. Rev. Biochem.*, **27**:101.

Schultz, J. (1947), *Cold Spring Harbor Symp. Quant. Biol.*, **12**:179.

Shaw, D. D. (1971), *Chromosoma*, **34**:19.

Simon, E. H. (1961), *J. Mol. Biol.*, **3**:101.

Sinsheimer, R. L. (1959), *J. Mol. Biol.*, **1**:43.

—— (1962), *Sci. Am.*, **207**(July):109.

Southern, E. M. (1970), *Nature*, **227**:794.

Stahl, F. W. (1962), *J. Chim. Phys.*, **58**:1072.

Steffensen, D. M. (1959), *Brookhaven Symp. Biol.*, **12**:103.

Stern, C. (1929), *Biol. Zentralbl.*, **49**:261.

Sueoka, N. (1961), *J. Mol. Biol.*, **3**:31.

Sugimato, K., T. Okazaki, and R. Okazaki (1968), *Proc. Natl. Acad. Sci.*, **60**:1356.

——, ——, Y. Imae, and R. Okazaki (1969), *Proc. Natl. Acad. Sci.*, **63**:1343.

Sumner, A. T., J. A. Robinson, and H. J. Evans (1971), *Natl. New Biol.*, **229**:231.

Swanson, C. P. (1957), "Cytology and Cytogenetics," Prentice-Hall, Englewood Cliffs, N.J.

Swift, H. H. (1950a), *Physiol Zool.*, **23**:169.

—— (1950b), *Proc. Natl. Acad. Sci.*, **36**:643.

Taylor, J. H. (1953), *Exp. Cell Res.*, **4**:164.

—— (1957), *Am. Nat.*, **41**:209.

—— (1958), *Sci. Am.*, **198**(June):36.

—— (1960), *J. Biophys. Biochem. Cytol.*, **7**:455.

—— (1965), *J. Cell. Biol.*, **25**:57.

—— (1968), *J. Mol. Biol.*, **31**:579.

—— and R. McMaster (1954), *Chromosoma*, **6**:489.

——, P. S. Woods, and W. L. Hughes (1957), *Proc. Natl. Acad. Sci.*, **43**:122.

Tessman, I. (1959), *Virology*, **9**:375.

Thomas, C. A., Jr., and L. D. MacHattie (1967), *Annu. Rev. Biochem.*, **36**(2):485.

Trosko, J. E., and S. Wolff (1965), *J. Cell Biol.*, **26**:125.

Ullerich, F. H. (1966), *Chromosoma*, **18**:316.

Volkin, E., L. Astrachan, and J. L. Countryman (1958), *Virology*, **6**:545.

Walker, P. M. B. (1971), *Nature*, **229**:306.

Warburton, D., et al. (1967), *Am. J. Hum. Genet.*, **19**:339.

Watson, J. D. (1970), "Molecular Biology of the Gene," Benjamin, New York.

—— and F. H. C. Crick (1953a), *Nature*, **171**:737.

Watson, J. D. (1970), and F. H. C. Crick (1953*b*), *Nature,* **171**:964.
────── and ────── (1953*c*), *Cold Spring Harbor Symp. Quant. Biol.,* **18**:123.
Werner, R. (1971), *Nature,* **230**:570.
Wettstein, R., and J. R. Sotelo (1965), *J. Ultrastruct. Res.,* **13**:367.
Whitehouse, H. L. K. (1969), "The Mechanism of Heredity," 2d ed., Macmillan, Toronto.
────── (1970), *Biol, Rev. Cam. Philos. Soc.,* **45**:265.
Wilkins, M. H. F. (1956), *Cold Spring Harbor Symp. Quant. Biol.,* **21**:25.
──────, R. G. Gosling, and W. E. Seeds (1951), *Nature,* **167**:759.
────── and J. T. Randall (1953), *Biochem. Biophys. Acta,* **10**:192.
Wimber, D. E., and W. Prensky (1963), *Genetics,* **48**:1731.
Witkin, E. (1971), *Nat. New Biol.,* **229**:81.
Wolfe, S. L. (1965), *J. Ultrastruct. Res.,* **12**:104.
Wolff, S. (1969), *Int. Rev. Cytol.,* **25**:279.
Wolfgang, H., I. Hennig, and H. Stein (1970), *Chromosoma,* **32**:31.
Wyatt, G. R. (1952), *J. Gen. Physiol.,* **36**:201.
Yasmineh, W. G., and J. J. Yunis (1970), *Exp. Cell Res.,* **59**:69.
Zamenhof, S., G. Brawerman, and E. Chargaff (1952), *Biochem. Biophys. Acta,* **9**:402.
Zinder, N. D. (1965), *Annu. Rev. Microbiol.,* **19**:455.

26
The Gene: Genetics of Gross and Fine Structure and Interallelic Complementation

QUESTIONS

1199 From about 1915 to the early 1950s it was commonly held that allelic genes were incapable of recombining or complementing each other (in heterozygotes or heterokaryons). Now we know that in some cases they can do both. What basis remains for calling such genes allelic?

1200 Is there generally a correlation between location of genes in a chromosome and their phenotypic effect? Give examples. Answer by first considering eukaryotes and then prokaryotes.

1201 Discuss the current concept of the gene and the evolution of this concept beginning with Mendel's *factors of inheritance*.

1202 **a** What tests should be performed to determine whether two independently isolated mutations are in the same or different genes?
b Discuss evidence indicating that functional and recombinational units are not materially equivalent.
c Is complementation of two recessive mutant phenotypes evidence that the two mutations are nonallelic? Is absence of complementation proof of allelism?

1203 *Hemophilia A* (classic hemophilia) and *hemophilia B* (Christmas disease) are both X-linked conditions in which the blood exhibits clotting failure. Explain why a mixture of blood from an A and a B *hemophiliac* has normal clotting capacity.

1204 **a** In which of the following would you expect intragenic (interallelic) complementation between mutant alleles?
(1) A gene controlling synthesis of a dimeric enzyme, e.g., *Escherichia coli* alkaline phosphatase.
(2) A gene controlling a formation of a monomeric enzyme, e.g., phage T4 lysozyme.
(3) A gene specifying a *t*RNA molecule.
(4) A gene controlling phage-head-protein synthesis.
b Would it matter whether the mutant alleles were due to missense or nonsense mutations? Explain.

1205 Which of the following would you not equate with the term (1) gene and (2) allele, and why?
a Recon   **b** Cistron   **c** Muton

1206 What is the cis-trans (complementation) test and what does it test for? Illustrate.

1207 Two mutations occurring in a certain gene which specifies a dimeric protein do not complement, whereas two others do. Explain these observations, showing what you would expect in the F_1 of a cross between one of the first pair of mutants and one of the second pair.

1208 a Why is it that abortive transduction is useful in complementation studies and complete transduction is not?
   b Why is transduction useful in fine-structure analysis?

1209 Heterozygotes in some cases are more vigorous than the homozygotes. With reasons for your arguments, state what role complementation within genes may play in this phenomenon?

1210 Do you think that the nature of genetic organization in eukaryotes, e.g., existence of mechanisms for intragenic recombination, linear order of mutational sites, etc., is the same as in prokaryotes? An excellent example to verify or refute your opinion is the *rosy* cistron in *Drosophila melanogaster* at position 52.00 in the right arm of chromosome 3. Read the papers by Chovnick (1966), Chovnick et al. (1964, 1969), Hubby (1961), Schalet et al. (1964), and Yen and Glassman (1965).

1211 R. B. Goldschmidt, one of the pioneers in physiological genetics, championed the idea that it may be the chromosome, rather than the individual gene, that is the basic unit of heredity. He viewed the chromosome as an integrated product of evolution and a mutation simply as a "disruption" in this integrated unit. Discuss the current information supporting or disputing this concept.

1212 "Our present knowledge of the physical basis of heredity is based on considerations and observations that have been made on many levels, from physics and chemistry through various biological disciplines to the comparison of evolutionary series. Enumerate methods on various levels that have proved fruitful in revealing the composition and structural organization of genetic determinants, citing a representative experiment or observation illustrating each mode of attack, and pointing out what each has contributed to our concept of the gene." (D. D. Perkins, Stanford University.)

PROBLEMS

Table 1213

Recipient	Donor	Number of *wild-type* recombinants
trp-10	wild type	1,822
	trp-1	4
	trp-3	270
	trp-4	602
	trp-7	7
	trp-8	208
	trp-9	12
	trp-10	0
	trp-11	0
trp-11	*trp-1*	22
	trp-2	240
	trp-3	280

1213 Brenner (1955) showed that ten independent, *tryptophan-requiring* auxotrophic mutants of *Salmonella typhimurium* fell into four clear-cut groups according to the steps blocked in the pathway of tryptophan synthesis. In 1956 Demerec and Hartman made transduction crosses between some of these mutants. The results of several are given in Table 1213. For each, state whether the two mutants involved are controlled by allelic genes or not, and why, and for allelic crosses whether the mutant alleles were due to mutations at identical or different sites.

1214 *Wild types*, r^+, of the virulent phage T4 can infect and lyse both B and K strains of *E. coli*. *r*II mutants form large plaques on B only. *Wild types* form small plaques on both strains. Benzer (1957) infected strains K with pairs of independently isolated *r*II mutants. The results with three such mutants are shown. For each infection explain the distribution of the phenotypes (plaques and no plaques) and indicate the phage genotypes of the different plaque types (disregard mutation).

Table 1214

Infecting phage	Plaque production on specific hosts	
	B	K
r^+	+	+
*r*IIa	+	−
*r*IIb	+	−
*r*IIc	+	−
*r*IIa + *r*IIb		−
*r*IIa + *r*IIc		+
*r*IIb + *r*IIc		+

Key: + = plaques; − = no plaques.

1215 Five different strains of *Salmonella typhimurium*, some of which are *cysteine-requiring* auxotrophs, grown on minimal medium and separately transduced with phage from each of the other strains give the results shown.

Table 1215

Recipient strain	Donor strain				
	A	B	C	D	E
A	L,	L	L	L	L
B	L, S	0,	L, S	L, S	L, S
C	L, S	L, S	0	L	0
D	L, S	L, S	L	0	L
E	L, S	L, S	0	L	0

Key: L = large colonies only; L, S = some large colonies and some minute ones; 0 = no colonies.

a Classify each of the strains as auxotrophic or prototrophic for *cysteine requirement*.

b State whether or not the auxotrophic strains are functionally allelic and why.

c What further tests would you conduct to test your classification of the strains as allelic or not?

1216 A *methionine-requiring* mutant in *Salmonella typhimurium* is infected with phage from another *methionine-requiring* mutant. Minute colonies never arise; large colonies are formed, with very low frequency. Account for these results.

1217 Phage from *wild-type*, *ath* (adenine-thymine requiring) and *ad* (adenine-requiring) mutants of *Salmonella typhimurium* were used to infect two *ath* mutants, seeded on adenine pantothenate–enriched minimal medium at concentrations of 1×10^7 for *ath A-4* and 5×10^7 for *ath C-5* bacteria per plate. Progeny phenotypes were determined after 4 days' incubation. Figures represent average numbers of colonies per plate (Ozeki, 1956).

Table 1217

				ath A								
Recipient	Colony size	Wild type	2	4	8	9	10	ath B-6	ath C-5	ath D-12	ad C-2	ad D-10
ath A-5	large	116	63	0	15	12	0	119	160	130	123	116
	minute	313	0	0	0	0	0	622	341	268	148	125
ath C-5	large	72	74	69	70	72	92	84	0	67	58	56
	minute	955	731	557	878	822	658	1,371	0	109	347	285

Note: All mutants can revert to *wild type*.

a Explain how the large colonies originate and why there are so many in some transductions and very few or none in others.

b Why do minute colonies arise, and what is the significance of their occurrence?

c Which of these mutants are functionally allelic? Why?

d Which of the mutants appear to be due to mutations at identical sites (homoalleles)?

e Which, the *wild-type* or mutant condition, is dominant?

f What is the minimum number of genes involved in adenine-thymine formation?

1218 The mutants in the *r*II region of the phage T4 form plaques with sharp edges on *Escherichia coli* strain B and no plaques on *E. coli* K. *Wild-type* phage form plaques with rough edges on both strains. Benzer (1955) infected K with various *r*II mutants, two at a time, and plaque-assayed the lysates obtained on *E. coli* B. Typical results are shown. These data are representative of all mutants in the *r*II region.

Table 1218

	47	51	101	102	104	106
47	−	+	−	+	−	−
51		−	+	−	+	+
101			−	+	−	−
102				−	+	+
104					−	−
106						−

Key: + = normal number of phagues formed; − = few, if any, phagues.

a Does the *r*II region consist of one, two, or more, cistrons? Explain.

b What results would you expect if *wild-type, r*⁺, and any one of the mutant phage infected *E. coli* K?

1219 *Neurospora crassa* has two mating types, *A* and *a*. Sexual reproduction occurs only in crosses between clonal strains of different mating type. The mutant strains *lysineless, nicotinicless-1, nicotinicless-2, nicotinicless-3* and *pantothenicless* do not grow on minimal medium. Beadle and Coonradt (1944) showed that if conidia of the same mating type from two strains are planted very close together on minimal medium, some of the hyphae of the two clones fuse. Such hyphae may form mycelia (heterokaryons) on minimal medium. The phenotypes of three such heterokaryons are shown. From each hetero-

Table 1219

Heterokaryon	Phenotype on minimal medium
pantothenicless + lysineless	*wild type*
nicotinicless-1 + nicotinicless-2	*wild type*
nicotinicless-1 + nicotinicless-3	*mutant* (little growth)

karyon hyphal tips were separately cultured and allowed to undergo sexual fusion with *wild-type* conidia to produce perithecia. The ascospores from individual asci were tested for phenotype in appropriate media. In each of the crosses, asci produced *wild types* vs. *mutants* in a 1 : 1 ratio. Moreover, in each cross, the mutant-spore phenotype was sometimes identical to one mutant parent, sometimes to the other mutant parent.   What do these results indicate regarding (1) nuclear type(s) in heterokaryons and (2) functional allelism?

1220   Fincham (1959) found that when two *am* (*alanineless*) mutants of *Neurospora crassa* of the same mating type were plated on the same minimal medium close together, they grew poorly or not at all.   Both mutants, however, responded to alanine, and when a minimal medium containing alanine was inoculated with the two double mutants *am²* *arg-1* and *am³* *arg-10*, an arginine-independent culture resulted.   Samples of mycelia of this culture transferred to minimal medium grew no better than *am²* or *am³* mutant strains.   Show whether or not a heterokaryon was formed on minimal plus alanine and state the conclusions to be derived regarding allelism (1) for the *am* mutants and (2) for the *arg* mutants.

1221   **a** Three *temperature-sensitive* mutants are induced with 5-bromouracil in *wild-type* phage T4.   These reproduce at 25°C, producing much larger plaques than *wild-type* T4, but do not reproduce at 42°C.
   (1) Indicate how you would determine whether any two mutants are functionally allelic and show expectations.
   (2) Indicate how you would determine the frequency of *wild-type* recombinants.
   (3) Show how you would calculate the percent recombination.

   b Two mutants are discovered in phage T2 which, although they can infect *E. coli* B, cannot reproduce because they produce an incomplete head protein.   They do, however, grow on *E. coli* CR63 and produce some head protein.   Outline the hosts you would use and the methods you would employ to make the determinations outlined in (a).

1222   The *A* gene in the histidine pathway[1] of *Salmonella typhimurium* specifies the synthesis of the enzyme isomerase, which is composed of a single polypeptide chain with a molecular weight of 29,000 (Ames et al., 1967).   Ten *A*-gene mutants (unable to grow on minimal medium) are isolated which produce *wild types* in all pairwise combinations with each other.
   a What is the most likely nature of all these mutations?
   b What genetical and biochemical tests would you conduct to refute or verify your answer to (a)?   Show the results expected with each test.
   c If some of the mutants were "leaky," would this give you a clue to the nature of the mutation involved?
   d Would any of the mutants when crossed in pairs be expected to exhibit abortive transduction?   Why?

[1] For a deep and detailed insight into the combined biochemical and genetic analysis of the histidine pathway in *Salmonella*, the most thoroughly and extensively investigated pathway so far undertaken in any organism, see Ames et al. (1960, 1967), Hartman et al. (1960*a,b*), Loper et al. (1964), Smith and Ames (1964, 1965).   These studies include the isolation of more than 900 independent *his* mutants, the mapping of 540 of these by means of transduction, the identification of nine genes in the pathway, the isolation and analysis of the nature of the enzymes involved, and analysis of numerous intragenic complementation tests.

1223 In *Aspergillus nidulans* the *adenineless* mutations *ad16* and *ad8* in cis and trans give the following phenotypes (Roper and Pritchard, 1955):

Mutant $\quad \dfrac{ad16^-}{ad16^+} \quad \dfrac{ad8^+}{ad8^-}$

Wild type $\quad \dfrac{ad16^-}{ad16^+} \quad \dfrac{ad8^-}{ad8^+}$

Are *ad16* and *ad8* functionally allelic or not and why?

1224 In a standard *magenta*-flowered variety of snapdragon (*Antirrhinum majus*) *white*-flowered mutants 1, 2, 3, and 4 are obtained by treating the pollen with ultraviolet light. The cis and trans phenotypes of the hybrids from crosses between these mutants are shown. The F_2's in all crosses except 3×4 segregated 9 *magenta* : 7 *white*.

Table 1224

| | Hybrid | | |
Combination	1×2	$(1 \text{ or } 2) \times (3 \text{ or } 4)$	3×4
cis	*magenta*	*magenta*	*magenta*
trans	*magenta*	*magenta*	*white*

a What is the minimum number of functional genes (cistrons) that determine flower color in *Antirrhinum*? Explain.
b Are they linked or not? Explain.
c Are they acting in the same or different biochemical pathways?

1225 In *Drosophila melanogaster* the eye colors *carmine*, *coral*, and *white* are determined by sex-linked mutant genes recessive to *wild type*. The results of two crosses between true-breeding strains with these phenotypes are shown.

Table 1225

| Cross | | Offspring | |
♀	♂	♀	♂
white × *carmine*		*wild type* (*red*)	*white*
white × *coral*		*light coral*	*white*

a Which mutant genes are allelic and why?
b What results would you expect in the cross *coral* female × *carmine* male? *Carmine* female × *coral* male?

1226 Two of the alleles at the sex-linked *white* locus in *Drosophila melanogaster* (position 1.5) are *w* and *w*^{co}, which cause *white* and *coral* eyes respectively. Females heterozygous for these alleles and *sc* (*scute*, position 0.0), *ec* (*echinus*, position 5.5), and *cv* (*crossveinless*, position 13.7) in coupling phase, viz.,

$$\frac{sc\ w\ \ ec\ cv}{Sc\ w^{co}\ Ec\ Cv}$$

were crossed with $\underline{sc\ w\ ec\ cv}$/Y males. Since males were *w*, their progeny were all expected to have *white* or *coral* eyes. The phenotypes of all but three of 21,067 progeny examined conformed to expectation. The three (females) had *wild-type* eyes and were of genotype

$$\frac{sc\ W\ Ec\ Cv}{sc\ w\ \ ec\ \ cv}$$

a Why can mutation be ruled out as a cause of these three females?

b What process is probably responsible for the formation of the *W* (*wild-type*) alleles in these three females?   Explain.

c Where would you place *w* and *w*^{co} on the linkage map relative to *sc* and *ec*, and why?

1227 In *Drosophila melanogaster* two mutant alleles for *star* (small, rough eyes), one dominant, *S*, and the other recessive, *s*, to *wild type*, are located at position 1.3 on chromosome 2 (Lewis, 1942, 1945). Females were obtained with the chromosome-2 constitution

$$\frac{al\ S\ ho}{Al\ s\ Ho}$$

where

al = *aristaless*, 0.0
ho = *held-out*, 4.0

These were crossed with *aristaless*, *star*, *held-out* males.   Among the 31,106 progeny classified all but 4 possessed the expected *star* phenotype.   The 4 unexpected flies had *normal* eyes and *held-out* wings.

a Are these results evidence for or against the bead hypothesis (the chromosome is a string of beads, each bead a gene representing a particle or region of the chromosome wherein the units of function, mutation, and crossing-over are coextensive)?   Explain by giving the genotype of the female gametes producing the 4 *normal* progeny, the process (viz. mutation or crossing-over) responsible for the *normal* progeny, and the position of the *S* and *s* mutant sites relative to the outside markers.

OR

b (1) Discuss mutation as an explanation for the *normal*-eyed flies, giving reasons why it is not a satisfactory one.

(2) Discuss crossing-over as a satisfactory explanation and show what this hypothesis reveals concerning the nature of the gene and the occurrence and position of mutant sites within it.

1228 Roper (1950, 1953) in studies of three *biotin-requiring* mutants (bi_1, bi_2, bi_3) of *Aspergillus nidulans*, found that:

1 The mutant strains did not respond to the immediate precursor of biotin, pimelic acid.

2 Prototrophic recombinants occurred in all possible crosses; their frequencies, for the $bi_1 \times bi_3$ and $bi_1 \times bi_2$ crosses, were approximately 1 per 5,000 progeny and 1 per 2,000 progeny, respectively.

3 Prototrophs always showed recombination for markers on either side of the biotin locus.

4 The sites at which the mutations responsible for bi_1, bi_2, and bi_3 occurred were all within a chromosomal segment approximately 0.2 map unit long.

5 Mycelia heterokaryotic for bi_1-bi_2, bi_1-bi_3, or bi_2-bi_3 were *biotin-requiring*, as were the resulting diploid mycelia.

a Diagram the chromosomal basis of these findings, showing the sequence of the bi_1, bi_2, and bi_3 mutant sites.

b According to Pontecorvo (1952) the unit of function has many sites arranged in a linear array. These sites can mutate independently, and crossing-over can occur between them. Discuss the relationship of Roper's findings to Pontecorvo's hypothesis and define the terms gene and allele.

1229 Chovnick (1961) assayed segregating X chromosomes of females heterozygous for various combinations of four *garnet* mutants and for adjacent marker genes. The locations of the genes involved are shown below.

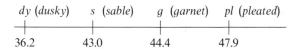

dy (dusky)	*s* (sable)	*g* (garnet)	*pl* (pleated)
36.2	43.0	44.4	47.9

The segregating X chromosomes were obtained by crossing these females with males whose X chromosomes carried a *garnet* mutant and a closely linked

Table 1229

Female genotype	*Nongarnet* chromosomes recovered	Total chromosomes assayed
$s\,g^1\,Pl/S\,g^2\,pl$	2 $s\,G\,pl$	68,054
$dy\,g^1\,Pl/Dy\,g^2\,pl$	1 $dy\,G\,pl$	65,251
$s\,g^{50e}\,Pl/S\,g^2\,pl$	2 $s\,G\,pl$	73,489
$dy\,g^1\,pl/Dy\,g^{50e}\,Pl$	2 $dy\,G\,Pl$	165,182
$s\,g^3\,Pl/S\,g^2\,pl$	4 $s\,G\,pl$	181,102
$S\,g^3\,pl/s\,g^{50e}\,Pl$	1 $S\,G\,Pl$	22,857
$Dy\,g^3\,Pl/dy\,g^1\,pl$	0 G	186,487

marker. The assay revealed the results given in the table. Since 1913 (Sturtevant) it has been known that genes are linearly arranged along the chromosome. Do these data indicate that this linear construction also extends to the intragenic organization?

1230 Mutant genes at the *vermilion, v,* locus in *Drosophila melanogaster* cause a bright-red eye color. Barish and Fox (1956) crossed trans heterozygous females with the genotype and chromosome constitution

ras^2	v^{36}	m
Ras^2	v^{48}	M

with $\mid ras^2 \mid v^{36} \mid m \mid$/Y males. Two *wild-type* (V/Y) males were obtained from 78,934 offspring which possessed the genotype $\mid ras^2 \mid V \mid M \mid$/Y.

a Is v^{36} to the left or to the right of v^{48} and why?
b Were the parent females likely *wild type* or *mutant* in phenotype? Explain.
c What is the frequency of crossover chromosomes among the 78,934 offspring? *Note: Vermilion* is X-linked.

1231 In 1944 McClintock described the allelic relationships of four seedling traits in corn: *Yg* (*green*), *yg* (*yellow-green*), *pyd* (*pale-yellow*), and *wd* (*white*) due to mutations in the short arm of chromosome 9. Plants homozygous for the chromosomal and genic types a, b, c, and d illustrated bred true for *green, yellow-green, pale yellow,* and *white* respectively. Crosses showed that *green* was dominant to all other phenotypes and *yellow-green* and *pale yellow* were dominant to *white.*

a Diagram the karyotypes and phenotypes expected in the F_1 and F_2 of the following crosses (*pale yellow* and *white* homozygotes can survive to maturity with appropriate nutrition).

(1) Homozygous b × homozygous d.
(2) Homozygous c × homozygous d.
(3) Homozygous b × homozygous c.

b

1 When Yg^2/pyd were crossed with Yg^2/wd, the progeny consisted of *green* and *pale yellow* in a 3:1 ratio.

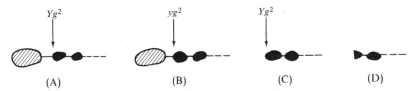

Figure 1231 The large oval represents the terminal heterochromatic knob. A thin chromatic segment joins the first chromomere with the knob. The small, solid ovals represent the two distal chromomeres. Chromosome type c is deficient for the knob and segment which joins the knob with the distal chromomere. Chromosome type D possesses a longer deletion than C which includes the terminal knob and a portion of the first chromomere. The arrow in A, B, and C points to the locus of Yg^2, yg^2, and Yg^2, respectively.

2 When Yg^2/pyd were crossed with yg^2/yg^2, the progeny were all *green*.

3 When Yg^2/wd were crossed with Yg^2/yg^2, the progeny consisted of *green* and *yellow-green* in a 3:1 ratio.

From these observations determine and illustrate the allelic relationships of Yg^2, yg^2, pyd, and wd.   Show your reasoning.

OR

(1) Map the functional sites of this color gene or genes.

(2) State whether you consider one gene to be involved or two, and why.

1232 *Adenine-requiring, ad^-,* auxotrophic mutants of *Saccharomyces cerevisiae* accumulate red pigment intracellularly and are therefore *pink*.   Diploid clones from crosses involving four different haploid mutants expressed the phenotypes shown (Bevan and Woods, 1962).   In how many different genes

Table 1232

Cross	Phenotypes of diploid clones
1 × 2	*auxotrophic, pink*
1 × 3 or 1 × 4	*prototrophic, white*
3 × 4	*auxotrophic, pink*

did the mutations producing these four strains occur?   Explain.   Show how you would verify your answer.

1233 The F_1 of a cross between two different true-breeding *golden* strains of swordfish is backcrossed to each of the parental strains.   About 1 in every 5,000 progeny in each backcross is *wild type* (*olive-green*).   Was the F_1 *wild type* in phenotype or not?   Explain.

1234 In the guinea pig the alleles C, c^k, c^d, and c^a in the *albino* series have the tabulated effects on the relative amounts of melanin pigmentation in the coat (Wright, 1959).

Table 1234

Genotype	Percent melanin	Phenotype
C—	100	*Full* color
$c^k c^k$	88	*Intermediate* color
$c^k c^a$	44	
$c^d c^d$	31	
$c^d c^a$	14	
$c^a c^a$	0	*White*

a Assuming that c^k is the *normal*, or *wild-type*, allele, classify each of the alleles (with reasons for your choice) as:

(1) *Normal* (*wild type*), *hypomorphic* (leaky), or *amorphic* (nonleaky).

(2) *Antimorphic* (active in direction opposite to that of the *normal* allele).

(3) *Hypermorphic* (more active than the *normal* allele) or *neomorphic* (activity qualitatively different from that of the *normal* allele).

b State the allelic relationships of C, c^k, and c^d with respect to c^a.

c Assuming that the gene controls the synthesis of the enzyme tyrosinase converting 3,4-dihydroxyphenylalanine (DOPA) to melanin, suggest differences in structure and function among these alleles to account for the tabular data.

d State, with reasons, whether any *full*-colored progeny are expected in the cross $(c^k c^k \times c^d c^d) \times c^a c^a$:

(1) According to the classical gene concept.

(2) According to the current concept.

1235 In *Neurospora crassa*, the *pan-2* (*pantothenic*) gene, which specifies conversion of ketovaline to ketopantoic acid, is closely linked to the *ylo* (*yellow*) and *trp-2* (*tryptophan*) loci.   To determine the linkage relationships of different specific-site mutant *pan-2* alleles, Case and Giles (1958a) crossed *yellow* strains carrying one *pan* allele with *trp* strains carrying a different *pan* allele.   Ascospores from each cross were plated on a medium deficient in pantothenic acid to detect *pan*⁺ recombinants.   The percent prototrophs (*pan*⁺) in each of the two noncrossover and crossover classes for *ylo* and *trp* in two sets of reciprocal crosses and one other cross, based on assays of over 200,000 ascospores per cross, are shown.   Derive the map for the linkage relationships among the four *pan* mutants and the two outside markers.

Table 1235

| Cross | Noncrossover | | Crossover | |
	ylo⁻ trp⁺	*ylo⁺ trp⁻*	*ylo⁺ trp⁺*	*ylo⁻ trp⁻*
pan 18⁻ ylo × pan 22⁻ trp	12.0	10.0	74.2	3.0
pan 22⁻ ylo × pan 18⁻ trp	10.0	13.0	6.0	71.0
pan 30⁻ ylo × pan 22⁻ trp	7.7	6.4	29.8	55.8
pan 22⁻ ylo × pan 30⁻ trp	5.0	14.0	51.5	29.0
pan 22 ylo × pan 36⁻ trp	0	0	0	0

Percent *pan* prototrophs in the four classes

1236 a In *Escherichia coli* the locus controlling *ara* mutants (unable to use arabinose) lies between *thr* and *leu*, close to *leu* (Lennox, 1955).   To determine the sequence of many independently ultraviolet-induced mutant sites in the *ara* region with respect to each other and *thr* and *leu*, Gross and Englesberg (1959) made reciprocal three-factor crosses by phage-mediated transductions, in which $thr^- leu^-$ and $thr^+ leu^+$ were used reciprocally as donors and recipients.   After selection for arabinose-positive, ara^+, transductants on medium lacking arabinose but containing threonine and leucine, these ara^+ transductants were scored for leu^+.   The results of some of these crosses are shown.   Derive the order of the *ara* mutants 2, 4, 7, 8, 12, 13, 14, and 16 with respect to each other and *leu*.

Table 1236

Recipient		Donor	$\dfrac{ara^+ leu^+}{\text{Total } ara^+}$	Percent
2	×	13	$\frac{15}{62}$	24.2
13	×	2	$\frac{104}{218}$	47.7
2	×	7	$\frac{13}{60}$	21.7
7	×	2	$\frac{48}{92}$	52.2
4	×	7	$\frac{74}{160}$	46.3
7	×	4	$\frac{34}{141}$	24.1
13	×	7	$\frac{69}{216}$	31.9
7	×	13	$\frac{113}{199}$	56.8
16	×	14	$\frac{4}{37}$	10.8
14	×	16	$\frac{52}{90}$	57.8
4	×	16	$\frac{30}{104}$	28.9
16	×	4	$\frac{67}{128}$	52.3
14	×	8	$\frac{26}{127}$	20.5
8	×	14	$\frac{45}{80}$	56.2
8	×	12	$\frac{29}{111}$	26.1
12	×	8	$\frac{46}{91}$	50.5

b When mutants 2, 4, 7, and 13 were transduced with phage from each of the other mutants (or vice versa), both large and small colonies were formed.   No small colonies appeared when 2, 4, 7, or 13 was infected with 2, 4, 7, or 13.   Moreover, when mutant 12 was transduced with phage grown on 8, 14, and 16, or vice versa, large and small colonies were formed.
(1) What additional information do these data give you and why?
(2) Illustrate the arabinose region schematically as completely as possible.

1237 The genome of bacteriophage R17 is a single-stranded RNA molecule of molecular weight 1.1×10^6.   The proteins coded by this genome are all likely to be essential for phage multiplication.   Gussin (1966) isolated nitrous acid–induced *amber* (conditionally lethal) mutants able to multiply on the permissive *Escherichia coli* strains S36RIE and C600 (containing suppressors *Su*I and *Su*II, respectively, which enable the mutants to multiply), but not on the nonpermissive strain S26 (no suppressors).   For complementation tests Su^- strain S26 was infected with mutants two at a time.   The results shown were obtained in single-burst experiments with the initial 12 mutants.

Table 1237

	1	3	8	4	2	5	11	6	7	13	9	12
1	−	−	+	−	+	−	+	−	−	+	−	−
3		−	+	−	+	−	+	−	−	+	−	−
8			−	+	+	+	+	+	+	−	+	+
4				−	+	−	+	−	−	+	−	−
2					−	+	−	+	+	+	−	−
5						−	+	−	−	+	−	−
11							−	+	+	+	+	+
6								−	−	+	−	−
7									−	+	−	−
13										−	+	+
9											−	−
12												−

Key: + = complementation; − = no complementation.

Mutants in the group to which mutant 1 belongs produce a normal yield of noninfective phage.   Their RNA, however, is infective to spheroplasts.   Mutants in the group to which mutant 8 belongs synthesize no RNA in S26 and approximately 50 percent of *wild-type* RNA in *Su*II; those in other groups produce normal levels of RNA during infection of both permissive and nonpermissive strains.

a If the average structural gene (cistron) is approximately 1,000 base pairs long, calculate the approximate number of genes in the R17 genome.

b Do the complementation data verify your conclusions?

c What do the different genes apparently code for?   Explain.

1238 *Wild-type*, *mot*⁺, *Salmonella typhimurium* possess flagella and are motile; i.e., they migrate outward as they multiply from their point of placement on agar medium to give an expanding "flare" of colonial growth.   *Paralyzed* mutants, *mot*⁻, also possess flagella but are nonmotile (growth of such colonies remains circumscribed).   Representative data for 15 of 97 *paralyzed* mutants from different genetic tests follow (Enomoto, 1966):

1   Results of complementation studies of eight *paralyzed* mutants using abortive transduction are shown in Table 1238*A*.   Complementation of two mutants is indicated

Table 1238A

	210	222	225	244	246	261	279	300
210	−	+	+	+	−	+	+	−
222		−	−	+	+	−	+	+
225			−	+	+	−	+	+
244				−	+	+	−	+
246					−	+	+	−
261						−	+	+
279							−	+
300								−

Key: + = complementation; − = no complementation.

by a linear trail of isolated colonies leading from a region of circumscribed growth on agar medium. Lack of complementation is indicated by failure of trails to appear when one mutant is transduced by another.

2 Seven nonreverting *mot* mutants crossed with each other in all pairwise combinations by transduction gave the results shown in Table 1238*B*. All these mutants produced *wild-type* recombinants when crossed with 244 and 279.

Table 1238B

	238	292	282	277	290	253	293
238	−	+	+	+	+	−	+
292		−	+	+	−	+	+
282			−	+	+	+	−
277				−	+	−	−
290					−	+	+
253						−	+
293							−

Key: + = *wild-type* (*mot*⁺) recombinants produced; − = no *wild-type* recombinants.

3 The results in Table 1238*C* were obtained when mutants 238, 253, 290, and 292 were crossed by transduction with the reverting mutants 210, 222, 225, 246, 261, and 300.

Table 1238C

	210	246	300	222	225	261
238	−	+	+	+	+	+
292	+	−	−	−	−	−
290	+	+	+	−	−	+
253	+	+	+	+	+	+

Key: + = *wild-type* (*mot*⁺) recombinants produced; − = no *wild-type* recombinants.

Note: (1) *Mot* 292 failed to produce motile *wild-type* recombinants with all 27 mutants that failed to complement it.   (2) Reverting mutants produced *wild-type* recombinants in all pairwise crosses with each other.   (3) Percent recombination between 246 and 261 is 2.75 and 3.82 between 300 and 261.

Referring to the data that support your contention:
a How many complementation groups (genes) do these mutants represent?
b Which genes, if any, are contiguous?
c Which mutants are multisite (deletions) and which are point mutants?
d Map the *Salmonella* genome ($n = 1$) as completely as possible.

1239 Fine-structure mapping of any short region can be carried out using a variety of techniques.   In *Bacillus subtilis*, Carlton (1966) used transformation to map mutant sites in the *trp B* locus, closely linked to the *anth* locus. Double- and single-mutant strains for the *trp B* mutants *B4*, *B6*, *B7*, and *B14* were crossed reciprocally and the percentage of $anth^+$ and trp^+ recombinants determined, with the results shown.
a Order the four *trp B* mutants relative to each other and *anth*.
b How would you determine whether a mutant is point or multisite in nature?

Table 1239

Recipient parent	Donor parent	Percent $anth^+$ among trp^+ recombinants
$anth^-$ $B4^-$ $B6^+$	$anth^+$ $B4^+$ $B6^-$	12.0
$anth^+$ $B4^+$ $B6^-$	$anth^-$ $B4^-$ $B6^+$	90.4
$anth^-$ $B4^-$ $B7^+$	$anth^+$ $B4^+$ $B7^-$	17.4
$anth^+$ $B4^+$ $B7^-$	$anth^-$ $B4^-$ $B7^+$	53.0
$anth^-$ $B6^-$ $B7^+$	$anth^+$ $B6^+$ $B7^-$	50.9
$anth^+$ $B6^+$ $B7^-$	$anth^-$ $B6^-$ $B7^+$	71.9
$anth^-$ $B6^-$ $B14^+$	$anth^+$ $B6^+$ $B14^-$	57.8
$anth^+$ $B6^+$ $B14^-$	$anth^-$ $B6^-$ $B14^+$	98.4
$anth^-$ $B7^-$ $B14^+$	$anth^+$ $B7^+$ $B14^-$	29.0
$anth^+$ $B7^+$ $B14^-$	$anth^-$ $B7^-$ $B14^+$	73.6

1240 In *Escherichia coli z* is closely linked with *i*.   In the cross $z_2^-\ i^+ \times z_1^-\ i^-$ most of the *wild-type*, z^+, recombinants are also i^+.   What is the order of the mutations in the two strains relative to *i*?

1241 In *Salmonella typhimurium* the *tryptophan* loci are closely linked and to one side of the *cys B* locus.   Demerec and Demerec (1956) crossed the strains *trp A*$^-$ *cys B*$^-$ *trp C*$^+$ and *trp A*$^+$ *cys B*$^+$ *trp C*$^-$ reciprocally, using transducing phage, and obtained the results shown.   What is the sequence of the three loci and why?

Table 1241

Recipient strain	Donor strain	Number of *wild-type* colonies on minimal medium
trp A⁻ cys B⁻ trp C⁺	*trp A⁺ cys B⁺ trp C⁻*	60
trp A⁺ cys B⁺ trp C⁻	*trp A⁻ cys B⁻ trp C⁺*	286

1242   In *Neurospora crassa* the *his-2* (histidine) and *nic-2* (nicotinamide) loci are approximately 2.0 and 3.0 map units to the left and right, respectively, of the *ad-3* (adenine-synthesis) locus.   Three *ad-3⁻* mutants found after ultraviolet light treatment were intercrossed as follows:

Cross 1   *his-2⁺ ad-3ᵃ nic-2⁺ × his-2⁻ ad-3ᵇ nic-2⁻*
Cross 2   *his-2⁺ ad-3ᵃ nic-2⁻ × his-2⁻ ad-3ᶜ nic-2⁺*
Cross 3   *his-2⁺ ad-3ᵇ nic-2⁻ × his-2⁻ ad-3ᶜ nic-2⁺*

Ascospores were plated on medium enriched by histidine and nicotinamide but lacking adenine; *ad-3⁺* recombinants were detected by growth on this medium.   Classification of these with respect to the other two loci gave the results shown.   What is the order of the *ad-3* mutants?

Table 1242

Ascospore genotypes	Cross 1	Cross 2	Cross 3
his-2⁻ ad-3⁺ nic-2⁻	0	6	0
his-2⁺ ad-3⁺ nic-2⁺	0	0	0
his-2⁻ ad-3⁺ nic-2⁺	15	0	5
his-2⁺ ad-3⁺ nic-2⁻	0	0	0
No. of ascospores from cross	41,236	38,421	43,600

1243   Ishikawa (1962) studied 308 independently obtained *adenine-requiring* mutants in *Neurospora crassa*.   All were unable to convert inosine monophosphate (IMP) to adenosine monophosphate succinate (AMPS) and either lacked the enzyme AMPS synthetase or had a defective form of it.   Pairwise crosses in all possible combinations with rare exceptions yielded only auxotrophic progeny.   Among the progeny of crosses between *wild type* and any *mutant*, one-half were *wild type*, the other half *mutant*.

a Is AMPS synthetase composed of one or more kinds of polypeptide chains? Explain.

b Two of the *mutants*, numbers 32 (induced by ultraviolet light) and 161 (induced by nitrous acid), never produce *wild-type* recombinants when crossed.   Explain in molecular terms.   Would you expect the polypeptide chains produced by these *mutants* to contain the same amino acid substitution if the mutation was missense?

1244 In *Neurospora crassa* the pathway leading to leucine synthesis consists of many steps, each catalyzed by a specific enzyme.   Absence or deficiency of any one of these enzymes results in a leucine requirement for growth.   The isomerization of β-carboxy-β-hydroxyisocaproate to α-hydroxy-β-carboxyisocaproate is catalyzed by a single enzyme, isomerase.   The following hypothetical results with two mutant strains unable to synthesize leucine are based on data of Gross (1962) and Gross and Webster (1963).

Mutant 1   ×   Mutant 2

Defective isomerase   ↓   Defective isomerase

$\frac{1}{4}$ of the ascospores produce mycelium on minimal medium with *wild-type* growth rate and normal enzyme activity

:

$\frac{3}{4}$ of the ascospores do not grow on minimal medium or grow slightly and slowly; isomerase is defective

a Show whether isomerase is composed of one kind of polypeptide chain or more than one.

b Describe the genetic experiments required to determine whether an enzyme that has more than one kind of polypeptide has each chain present once or more than once.

1245 The source of the α-amino groups of many of the amino acids is an amination reaction, catalyzed by glutamic dehydrogenase, in which nutritional ammonia is converted to the α-amino radical.   Fincham and Pateman (1957) discovered the following with respect to two amination-deficient mutants, am^{32} and am^{47}:

1 Both require exogenous α-amino radicals for normal growth.

2 Neither produces glutamic dehydrogenase, or it does so in very minute quantities.

3 *Wild-type* ascospores occur about 1 in 100,000 in crosses between am^{32} and am^{47}; the other ascospores are am^{32} or am^{47}.

4 Heterokaryotic, mycelium containing nuclei from am^{32} and am^{47} produces this enzyme, but its activity is only a fraction (25 percent) of that of *wild-type* enzymes.

a Explain these results in molecular terms, indicating the number of genes, the sites and nature of mutations, and the meric form of the enzyme.

b Indicate how Benzer's cistron fails to satisfy these data and suggest a redefinition of the cistron that will satisfy them.

1246 The *ad-4* gene in *Neurospora crassa* specifies the synthesis of adenylosuccinase, which catalyzes the conversion of AMPS to AMP and fumarate. Woodward et al. (1958) showed that each of the 123 allelic mutants of independent origin produce defective forms of this enzyme with activities less than 1 percent that of the *wild-type* enzyme.   Certain pairs of mutants complement each other, with a partial restoration of enzyme activity, when combined in heterokaryons. These also show complementation in vitro (Woodward, 1959). The enzyme activity in heterokaryons is always low, never exceeding 25 percent of *wild-type* activity. Assuming that adenylo-succinase is a dimer and that the allelic mutants *F*4 and *F*39 specify defective polypeptide chains altered as illustrated, show:

a The types of enzymes and their proportions in heterokaryons between *F*4 and *F*39.

b Which of these would express activity.

Table 1246

Allele	Location of protein defect (arrow)
*F*4	4 ↓ ————————————————
*F*39	————————————— 39 ↓ ————

1247   In *Escherichia coli* the enzyme alkaline phosphatase is composed of two identical subunits (polypeptides), each with a molecular weight of 40,000, whose structure is determined by a single structural gene *p* (Garen and Garen, 1963). These workers found that several p^- (phosphatase-negative) point mutants mapped at different sites in this gene. Each produced a functionally defective form of the enzyme with little or no activity. The mutants complemented each other in certain pairwise heterozygous combinations in vivo, but the resulting enzyme activity is much lower than in *wild-type* strains (see *U9* and *S33*). By acidification Schlesinger and Levinthal

Table 1247*A*

F′ complementing strain		Enzyme activity
Episome	Chromosome	
p^+ (CRM$^+$)*	p^+ (CRM$^+$)	240
U9 (CRM$^+$)	*S33* (CRM$^+$)	50

* CRM = cross-reacting material (see Prob. 1328 for a discussion).

(1963) prepared monomers from the defective enzymes of the p^- mutants *U9* and *S33*. The enzyme activity of the native mutant enzymes, a combination of these enzymes, enzyme monomers alone, and enzyme monomers together is shown in Table 1247*B*. The active enzyme of *U9-S33* reaction mixture,

Table 1247*B*

Protein	Concentration, μg/ml	Enzyme activity, units/ml in 50 minutes
U9 monomer	320	0.03
S33 monomer	320	0.30
U9 monomer + *S33* monomer	320	5.00*
U9 native	200	0.20
S33 native	200	0.02
U9 native + *S33* native	400	0.20

* Enzymatically active protein.

separated by electrophoresis, forms bands in a position intermediate between those of the two parental CRMs. *S33* CRM, which is able to carry out the complementation reaction with *U9*, sediments at the same rate as the *wild-type* monomer (mol wt = 40,000). Discuss the bearing of these data on:

a The hybrid-protein hypothesis of interallelic complementation.

b Whether active enzyme is a hybrid.

c Whether the material participating in complementation is a monomer.

1248 Interallelic complementation can occur if the enzyme specified by the gene is a multimer. In *Escherichia coli* alkaline phosphatase is a dimer, and certain p^- point mutants like *U9* and *S33* complement in heterozygotes both in vivo and in vitro because some enzyme molecules are formed from the association between polypeptide chains produced by different alleles. The enzyme activity of these enzymes is of the *wild type*, unlike the enzymes formed by association of chains specified by either mutant allele alone. Schlesinger and Levinthal (1963) mixed monomers from *U9* and *S33* in various proportions and determined the amount of enzyme activity in each mixture.

Table 1248

Ratio *S33 : U9*	Enzyme activity, units/ml
1:9	2.0
1:3	4.1
2:3	5.5
1:1	5.4
3:2	5.7
3:1	3.5
9:1	1.6

a Determine the relative amount of hybrid enzyme formed (in micrograms), assuming dimers are due to random associations of polypeptide chains, by using the binomial $(p + q)^n$, where p is the frequency of *S33* monomer and q that of the *U9* monomer, expressed as fractions of total monomer present in the mixture. Assigning a value of 100 mol/ml to the theoretical (relative) hybrid-dimer fraction formed in the 1:1 mixture, determine the theoretical concentration of hybrid dimer in the other mixtures accordingly.

b Calculate the specific activity of each mixture in units per milligram as follows:

$$\frac{\text{Enzyme activity}}{\text{Theoretical concentration of hybrid dimer}}$$

c Is the extent of complementation (enzyme activity) a function of the relative proportions of the monomers? Explain.

d Do the monomers produce dimers by random collisions? What is the evidence for your answer?

Note: For complementation enzyme of *U9-S33* the specific activity is 5 percent that of *wild-type* enzyme.

1249 Estimates by Chovnick et al. (1962) place the molecular weight of *Drosophila melanogaster* xanthine dehydrogenase near 400,000, which is close to estimates for this enzyme in the chicken (liver) and other vertebrates. Assuming that the average molecular weight of amino acid residues is 100 and that the enzyme consists of one polypeptide chain specified by one gene, *Ry*, derive the length of this gene in terms of nucleotides. (The distance between any two nucleotide pairs is 3.4 Å $= 3.4 \times 10^{-4}$ μm along the DNA double helix.)

1250 Forrest et al. (1956) have shown that at least two genes (*ma-1* and *ry*) in *Drosophila melanogaster* specify the synthesis of the enzyme xanthine dehydrogenase (XDH). Glassman (1962) demonstrated that enzyme extracts from either of these mutants show no detectable XDH activity; i.e., neither catalyzed the reaction 2-amino-4-hydroxypteridine $\rightarrow$ isoxanthopterin. However, the combined extracts from the two mutants showed considerable enzyme activity. Offer two explanations to account for these data. (See Glassman, 1965.)

1251 In *Drosophila melanogaster* two true-breeding *red-brown* eye mutant strains are partly deficient in red pigment and deficient in the enzyme xanthine dehydrogenase (XDH). As a consequence they cannot convert 2-amino-4-hydroxypteridine into isoxanthopterin. *Wild-type* (*red*-eyed) flies produce normal amounts of XDH. Reciprocal crosses between these strains give the results shown when flies eight days or older are classified (Glassman and Mitchell, 1959).

a On the basis of your knowledge of enzymes, together with the study of F_2 data, derive the number and location of genes controlling XDH.

b Explain the complementation phenomena for each of the four F_1 phenotypes (for sex and eye type) shown; e.g., is complementation present; if so, is it intergenic or intragenic; and does it occur at the level of

Table 1251

Parents		F₁		F₂	
♀	♂	♀	♂	♀	♂
Strain A × Strain B	*wild type*	*Mutant*	1 *wild type* : 2 *mutant*	1 *wild type* : 2 *mutant*
Strain B × Strain A	*wild type*	*wild type*	2 *wild type* : 1 *mutant*	1 *wild type* : 2 *mutant*

sequential metabolic blocks, like those of tyrosine metabolism in man, or at the single-enzyme level?

c Do these results support or refute:
  (1) the one-gene (one-cistron)–one-polypeptide-chain hypothesis?
  (2) the one-gene–one-enzyme hypothesis?

1252 Catcheside and Overton (1958) classified 40 *arginineless* (*arg-1*) mutants of *Neurospora crassa* into six groups, whose complementation relationships, determined by growth rate of heterokaryons on minimal medium, are shown.

Table 1252

	A	B	C	D	E	F
A	○	○	○	○	○	○
B		○	○	●	●	●
C			○	○	●	○
D				○	○	●
E					○	●
F						○

Key: ● = complementing;
       ○ = noncomplementing.

a Draw a complementation map of the gene represented by these mutants.
b Mutant A does not complement any of the other mutants. Does this indicate that it is a large deletion? Explain.
c Show whether you expect the complementation and genetic map of this gene to be collinear or not and why.

1253 A complementation matrix and map, typical of structural genes, each of which controls the synthesis of a specific multimeric enzyme, are shown for five *td* (= *trp-3*) mutants in *Neurospora crassa* (based on data from Lacy and Bonner, 1961). *td* controls tryptophan synthetase formation. All five mutants can back-mutate to *wild type*. Crosses between *wild type* and each of the revertants breed true. Since complementation occurs between different mutants within the *td* locus, should we regard the *td* region as consisting of several (i.e., 4) units, or blocks, with distinct physiological functions? Explain, using the data provided (which are sufficient to answer the question).

Table 1253A   Complementation matrix for *td* mutants

	1	2	6	7	24
1	○	○	○	○	○
2		○	○	●	●
6			○	○	●
7				○	○
24					○

Key: ● = complementing;
○ = noncomplementing.

Table 1253B   Complementation map

td 1	—————————————
td 2	————————
td 6	————————————
td 7	—————————————
td 24	———————————

Noncomplementing pairs of mutants are shown as overlapping segments and complementary pairs as nonoverlapping segments.

1254 In genetic studies of the *r*II region in phage T4 Benzer (1955) infected *Escherichia coli* strain K (on which mutants do not grow) with *r* mutants two at a time. The results are shown in diagram form. The two lines in each case represent the two mutant genomes in *E. coli* K as a consequence

Table 1254

Pairs of strains infecting *E. coli* strain K	Phenotype
1. ———•—————— 104	Normal number of phage produced; plaques formed
2. —•———————— 102	Normal number of phage produced; plaques formed
3. ———•—————— 104 / ——•———————— 47	Few, if any, phage produced; no plaques
4. ——•———————— 102 / —•—————————— 51	Few, if any, phage produced; no plaques
5. ———•—————— 104 or 47 / ——•———————— 102 or 51	Normal number of phage produced; plaques formed
6. —•———•—————— 104 47 / —•———•—————— 102 51	Normal number of phage produced; plaques formed

of mixed infection.   The dots represent the mutant sites (mutons) in the *r* strains but do not represent the positions of the mutations in the *r*II region. A *wild-type*, r^+, genome is represented by a line without dots.   In other experiments Benzer (1955) showed that *wild-type* recombinants occur in all pairwise crosses between these four mutants.   Recombination frequency is greatest between 47 and 102 and somewhat smaller between 104 and 51.

    a Using the terms cistron (gene), recon, muton, cis, and trans, explain these results, giving the number of genes (cistrons) involved and the sequence of the mutons.

    b State whether *r* or r^+ is dominant and why.

1255   You have 1,000 *r* mutants of T4 which map in the same region of the genetic map.   How many pairwise crosses would be required to map all the mutants?

1256   The *r*II region of phage T4 consists of two contiguous cistrons, A and B (Benzer, 1957, 1962).   The *r* mutants in this region are individually incapable of reproducing within (and hence of lysing) the *r*-immune strain K of *E. coli*, but mixed infection of certain of these mutants can infect and lyse.   The five naturally occurring mutations 47, 51, 101, 102, and 104 were recombined in pairs to form double-mutant (cis) strains referred to as r^1r^2.   The mutations are present in the same DNA strand; r^1 and r^2 represent any two mutations. Mixed infections of *E. coli* K with the five single-mutant strains, the various r^1r^2 strains, and *wild type*, r^+, two at a time, in all possible combinations were performed and studied for burst size, which indicates the number of phage particles developed per host cell.

    a If mutation 51 occurs in the B cistron, in which of the cistrons does each of the other mutations lie and why?

Table 1256

Strains used in mixed infection

r^+	r^1r^2	Burst size
		253
104	102	251
102	101	247
51	102	0*
51	101	256
47	104	0
47	101	0
51	47	253
47	102	249
51	104	264
101	104	0

* 0 refers to failure of phage development or, if some develop, defective phage that are incapable of lysing the host.

b All mutants revert to *wild type*, r^+, except 104.   What is the probable nature of this mutation?

1257   In phage, *r*II deletion mutants can be mapped by a spot test.   Mutants are added two at a time, to molten overlay agar seeded with a mixture of many K (λ) and a few B cells of *E. coli*.   The overlay is poured onto a plate and allowed to harden, and then drops of suspensions of the different *r*II mutants are placed on marked areas of the plate.   The plates are then incubated overnight at 37°C.   As a few B cells divide, they become infected with both kinds of phage mutants where the phage were deposited on the plate.   They lyse and liberate mostly *r*II phage and, if the deletions in the two mutants do not overlap, r^+ recombinant phage as well.   There are not enough B cells for the *r*II mutants to form visible plaques.   However, there are many K (λ) cells, and thus r^+ recombinants, which can grow on this strain (*r*II mutants cannot), and form visible plaques.   Thus the appearance of plaques where a phage mutant has been deposited indicates that the deletions of this mutant and the mutant in the overlay do not overlap.   On the other hand, absence of plaques indicates that the deletions do overlap.   Benzer (1959) tested hundreds of *r*II deletion mutants in phage T4 in this manner for their ability to produce r^+ recombinants.   The results for eight such mutants are given in the table.

Table 1257

	H88	B37	184	C51	782	C33	347	B138
H88	−	−	−	−	−	−	−	−
B37		−	+	+	+	+	+	+
184			−	−	+	−	+	+
C51				−	+	+	+	+
782					−	−	−	+
C33						−	+	+
347							−	+
B138								−

Key: + = recombinants formed; − = no recombinants.

a Draw a topological representation of these mutants.

b How would you test to determine whether the mutants are in the A or B gene (cistron) or both?

1258   **a**   The data in Table 1258*A* for pairwise crosses between *r*II mutants of T4 are from Benzer (1961).   When K (λ) *E. coli* bacteria, on which the mutants do not grow, are mixedly infected with pairs of these deletion mutations and with two other *r*II mutants 250 and 187, the following results in Table 1258*B* are obtained.   These data permit one to determine the sequence

Table 1258*A*

	1605	164	1589	196
1605	−	−	−	−
164		−	−	+
1589			−	−
196				−

Key: + = recombinants formed; − = no recombinants.

Table 1258*B*

Mutants used in infections		Consequences
1605 × 164,	1605 × 196	Few, if any
1589 × 164,	1589 × 196	phage produced
164 × 196,	164 × 187	Normal number
250 × 196,	250 × 187	of phage produced
250 × 1605,	187 × 1605	Few, if any
250 × 1589,	187 × 1589	phage produced

and extent of the deletions and to decide on the number of cistrons (genes) in this region.   Analyze these data and provide this information.

b The data in Table 1258*C* are the results of crosses between *r*II mutants in T4 phage.

(1) Is it possible to determine from these data whether the mutants in the horizontal or vertical column are deletion mutants?   Explain.

(2) Draw a topological representation of the mutations in the vertical column.

(3) What circumstantial evidence would indicate that these mutants are deletions?

Table 1258*C*

	1	2	3	4	5
A	−	+	+	+	+
B	−	−	+	+	+
C	−	−	−	+	+
D	−	−	−	−	+
E	−	−	−	−	−

Key: + = recombinants formed;
        − = no recombinants.

1259 Benzer (1961) tested seven mutants of *r*II region of phage T4 in all pairwise combinations by spot testing.   What conclusions can you draw from the data given regarding the kinds and extent of mutants involved?

Table 1259

	1272	1241	J3	PT1	PB242	A105	638
1272	—	—	—	—	—	—	—
1241		—	—	—	—	—	—
J3			—	—	—	—	—
PT1				—	—	—	—
PB242					—	—	—
A105						—	—
638							—

Key: + = recombinants formed; — = no recombinants.

1260 **a** Benzer (1961) tested numerous deletion mutants of the *r*II region of phage T4 using spot tests.   The results for eight such mutants tested in all pairwise combinations are given in Table 1260*A*.   Draw a topological representation of these mutations.

Table 1260*A*

	1272	1364	168	1993	184	1605	PT8	W8-33
1272	—	—	—	—	—	—	—	—
1364		—	—	—	+	+	+	+
168			—	—	+	+	+	+
1993				—	—	—	+	+
184					—	+	+	+
1605						—	—	+
PT8							—	—
W8-33								—

Key: + = recombinants formed; — = no recombinants.

b Eight point mutants, all producing recombinants with each of the other seven, were tested against the deletions studied in (a).   The results of the spot tests are given in Table 1260*B*.   Draw a map of the point mutations and draw lines below the map indicating the extent and end points of the deletions.

Table 1260B

	1272	1364	168	1993	184	1605	PT8	W8-33
A_1	−	+	+	+	+	+	+	+
A_2	−	−	−	+	+	+	+	+
A_3	−	+	+	−	−	+	+	+
B_{10}	−	+	+	+	+	+	−	+
A_5	−	+	+	−	+	−	+	+
A_6	−	+	+	+	+	−	−	+
B_5	−	+	+	+	+	−	−	+
B_8	−	+	+	+	+	+	−	−

Key: + = recombinants formed; − = no recombinants.

1261 *Amber* mutants of T4 phage multiply on strain CR63 of *E. çoli* K12, a permissive host, but not on the nonpermissive strain B. When these mutants infect B, the infection is abortive; i.e., reproduction is blocked at some stage in development. *E. coli* B is infected with four *amber* mutants two at a time (approximately three phage of each genotype per cell). The number of infected cells is measured, and the phage yield per plaque is assayed on strain CR63 and also on strain B. The burst size is then calculated.

$$\text{Burst size} = \frac{\text{phage yield}}{\text{number of cells infected}}$$

The normal burst size is 200 per cell. The results are as shown.

Table 1261

	Burst size on CR63				Burst size on B			
	am^1	am^2	am^3	am^4	am^1	am^2	am^3	am^4
am^1	0	2	200	200	0	1	4	3
am^2		0	200	200		0	5	4
am^3			0	2			0	1
am^4				0				0

a Construct a genetic map, enclosing markers in the same complementation group with parentheses.

b How many of the phage in the crosses $am_1 \times am_3$, $am_1 \times am_1$, and $am_3 \times am_4$ are recombinants? *Wild type* in genotype? Explain.

1262 *E. coli* K12 strain A cannot synthesize leucine, and strain B cannot synthesize leucine or arginine. The mutations, all of which are close enough to be cotransducible, are shown below, a_1 and a_2 being in the same complementation group:

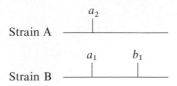

A three-factor cross is carried out by transduction to determine which of a_1 or a_2 is closer to b_1, using phage grown on strain A to infect B and vice versa. The recipient cells are then plated on minimal medium.   A hundredfold fewer transductant prototrophs $(a_1^+ a_2^+ b_1^+)$ are obtained when strain A is the donor than when strain B is the donor.   What is the order of the three markers a_1, a_2, and b_1?

1263   *Wild-type* λ phage, c^+, form *turbid* plaques; c^- mutants form *clear* plaques. Two *suppressor-sensitive* (*amber*) mutants of λ, c^- $Su1^-$ $Su2^+$ and c^+ $Su1^+$ $Su2^-$ are crossed at 37°C by growing them together in the permissive host strain of bacteria in which *amber* mutants (which carry the chain-termination codon UAG in the middle of the gene) can grow.   In the permissive host UAG is translated as sense; in the nonpermissive strain as nonsense.   The cells are incubated to lysis and the lysate diluted and plaque-assayed on the permissive and nonpermissive hosts.

Table 1263

	Permissive host (plaque type)	Nonpermissive host (plaque type and number)
Parental:		
c^- $Su1^-$ $Su2^+$	clear	none
c^- $Su1^+$ $Su2^-$	turbid	none
Recombinant:		
c^+ $Su1^-$ $Su2^+$	clear	none
c^- $Su1^+$ $Su2^+$	clear	clear 10
c^- $Su1^-$ $Su2^-$	clear	none
c^+ $Su1^-$ $Su2^+$	turbid	none
c^+ $Su1^+$ $Su2^-$	turbid	turbid 90
c^+ $Su1^-$ $Su2^-$	turbid	none

a What is the order of the three markers c, $Su1$, and $Su2$?
b Is there evidence of linkage?   Explain.

1264   In the fungus *Sordaria fimicola* crosses between *gray*-spored strains and *white-spored wild-type* strains, made by Kitani and Olive (1967), yielded asci with the types of linear spore illustrated.   Outside markers always showed regular

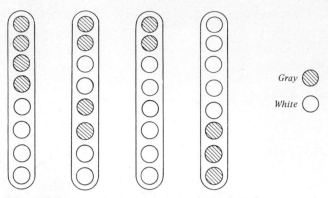

Figure 1264

4:4 segregations.   State the kind of recombination event that can account for each segregation pattern.   (See Holliday, 1964; Whitehouse, 1965; Whitehouse and Hastings, 1965.)

1265   In *Neurospora crassa* the loci *ad-1* (*adenineless*), *pan-2* (*pantothenicless*), and *trp-2* (*tryptophanless*) are in linkage group 6 and closely linked.   The sequence of *pan-2* mutations sites 23, 36, and 72 in relation to the other two genes and the centromere is

ad-centromere-23-36-72-*trp*

(The *ad*-centromere distance is very short.)   Ordered-spore analysis by Case and Giles (1964) showed that 13 of 1,457 asci from the cross $ad^- 23^- 72^+ 36^-$ $trp^+ \times ad^+ 23^+ 72^- 36^+ trp^-$ exhibited aberrant segregation patterns.   Four of these ascus patterns are shown.   With the aid of illustrations, describe the interallelic recombination events apparently responsible for each of the four types of asci.

Table 1265

Ascus 1	Ascus 2	Ascus 3	Ascus 4
$ad^+ 23^+ 72^- 36^+ trp^-$	$ad^+ 23^+ 72^- 36^+ trp^+$	$ad^- 23^+ 72^- 36^+ trp^+$	$ad^+ 23^+ 72^- 36^+ trp^-$
$ad^+ 23^+ 72^- 36^+ trp^-$	$ad^+ 23^+ 72^- 36^+ trp^+$	$ad^- 23^+ 72^- 36^+ trp^+$	$ad^+ 23^+ 72^- 36^+ trp^-$
$ad^+ 23^+ 72^- 36^+ trp^-$	$ad^+ 23^+ 72^- 36^+ trp^-$	$ad^- 23^- 72^+ 36^- trp^+$	$ad^+ 23^+ 72^+ 36^- trp^+$
$ad^+ 23^+ 72^- 36^+ trp^-$	$ad^+ 23^+ 72^- 36^- trp^-$	$ad^- 23^- 72^+ 36^- trp^+$	$ad^+ 23^+ 72^+ 36^- trp^+$
$ad^- 23^+ 72^+ 36^- trp^+$	$ad^- 23^- 72^- 36^+ trp^+$	$ad^+ 23^+ 72^- 36^+ trp^-$	$ad^- 23^+ 72^- 36^+ trp^-$
$ad^- 23^+ 72^+ 36^- trp^+$	$ad^- 23^- 72^+ 36^- trp^+$	$ad^+ 23^+ 72^- 36^+ trp^-$	$ad^- 23^+ 72^- 36^+ trp^-$
$ad^- 23^+ 72^+ 36^- trp^+$	$ad^- 23^- 72^+ 36^- trp^-$	$ad^+ 23^+ 72^- 36^+ trp^-$	$ad^- 23^- 72^+ 36^- trp^+$
$ad^- 23^+ 72^+ 36^- trp^+$	$ad^- 23^- 72^+ 36^- trp^-$	$ad^+ 23^+ 72^- 36^+ trp^-$	$ad^- 23^- 72^+ 36^- trp^+$

Key: − = mutated pattern; *+ = wild type.*

1266 In *Neurospora crassa*, *pdx* and *pdxp* are mutant forms of the *wild-type* gene *pdx*⁺ controlling pyridoxine synthesis. The gene locus is bounded closely on either side by the markers *pyr* (*pyrimidine requirement*) and *co* (*colony-type variant*). In crosses between $pyr^+ pdxp^- co^-$ and $pyr^- pdx^- co^+$ (Mitchell, 1955) the vast majority of tetrads showed the expected 2:2 ratio of *pyr*⁺ from *pyr*⁻, *pdx*⁻ from *pdxp*⁻, and *co*⁺ from *co*⁻, but four unusual asci contained *pdx*⁺ spores. Three of these are shown.

Table 1266

Spore pair	Ascus 1	Ascus 2	Ascus 3
1	$pyr^+ pdxp^- co^-$	$pyr^+ pdxp^- co^-$	$pyr^+ pdx^+ co^+$
2	$pyr^+ pdx^+ co^-$	$pyr^+ pdx^+ co^-$	$pyr^+ pdxp^- co^-$
3	$pyr^- pdxp^- co^+$	$pyr^- pdx^- co^+$	$pyr^- pdx^- co^+$
4	$pyr^- pdx^- co^+$	$pyr^- pdx^- co^+$	$pyr^- pdx^+ co^-$

In addition Mitchell showed that:

1 *pdx*⁺ spores were never found in the crosses $pdx^- \times pdx^-$ or $pdxp^- \times pdxp^-$.

2 Analysis of randomly collected *pdx*⁺ spores gave the following distribution of genotypes for the outside markers:

$pyr^- co^+$ 13
$pyr^- co^-$ 7
$pyr^+ co^+$ 7
$pyr^+ co^-$ 5

This violation of 2:2 segregation (referred to as *gene conversion* by Lindegren and Winkler) may be explained theoretically in more than one way. Show why these results cannot be explained (1) by mutation or (2) reciprocal crossing-over and offer an explanation that accounts for all the observations.

REFERENCES

Agarwal, K. L., et al. (1970), *Nature*, **227**:27.

Alikhanian, S. I., and V. Z. Pogosov (1969), *Genetics*, **61**:773.

Ames, B. N., B. Garry, and L. A. Herzenberg (1960), *J. Gen. Microbiol.*, **22**:369.

————, R. F. Goldberger, P. E. Hartman, R. G. Martin, and J. R. Roth (1967), in V. V. Koningsberger and L. Bosch (eds.), "Regulation of Nucleic Acid and Protein Biosynthesis," pp. 272–287, Elsevier, Amsterdam.

Armstrong, J. B., and J. Adler (1969), *Genetics*, **61**:66.

Barish, N., and A. S. Fox (1956), *Genetics*, **41**:45.

Beadle, G. W., and V. L. Coonradt (1944), *Genetics*, **29**:291.

Benzer, S. (1955), *Proc. Natl. Acad. Sci.*, **41**:344.

———— (1957), in W. D. McElroy and B. Glass (eds.), "The Chemical Basis of Heredity," pp. 70–93, Johns Hopkins, Baltimore.

———— (1959), *Proc. Natl. Acad. Sci.*, **45**:1607.

Benzer, S. (1961*a*), *Harvey Lect. Ser.* **56** (1960–1961).

—— (1961*b*), *Proc. Natl. Acad. Sci.*, **47**:403.

—— (1962), *Sci. Am.*, **206**(January):70.

Bernstein, H. (1964), *J. Theor. Biol.*, **6**:347.

——, R. S. Edgar, and G. H. Denhardt (1965), *Genetics*, **51**:987.

Bevan, E. A., and R. A. Woods (1962), *Heredity*, **17**:141.

Bonner, D. M., Y. Suyama, and J. A. Demoss (1960), *Fed. Proc.*, **19**:926.

Brenner, S. (1955), *Proc. Natl. Acad. Sci.*, **41**:862.

—— (1959), *Ciba Found. Symp. Biochem. Hum. Genet.*, p. 304.

Brookhaven Symp. Biol. **17** (1964), "Subunit Structure of Proteins: Biochemical and Genetic Aspects."

Carlson, E. A. (1959), *Genetics*, **44**:347.

Carlson, P. S. (1971), *Genet. Res.*, **17**:53.

Carlton, B. C. (1966), *J. Bacteriol.*, **91**:1795.

Case, M. E. (1958), *Proc. 10th Int. Congr. Genet. Montreal*, **2**:45.

—— and N. H. Giles (1958*a*), *Cold Spring Harbor Symp. Quant. Biol.*, **23**:119.

—— and —— (1958*b*), *Proc. Natl. Acad. Sci.*, **44**:378.

—— and —— (1964), *Genetics*, **49**:529.

Catcheside, D. G. (1960), *Proc. R. Soc. (Lond.)*, **B153**:179.

—— and A. Overton (1958), *Cold Spring Harbor Symp. Quant. Biol.*, **23**:137.

Chourey, P. S. (1971), *Genetics*, **68**:435.

Chovnick, A. (1961), *Genetics*, **46**:493.

—— (1966), *Proc. R. Soc. (Lond.)*, **B164**:198.

——, V. Finnerty, A. Schalet, and P. Duck (1969), *Genetics*, **62**:145.

——, A. Schalet, R. P. Kernaghan, and M. Krauss (1964), *Genetics*, **50**:1245.

——, ——, ——, and J. Talsma (1962), *Am. Nat.*, **96**:281.

Clowes, R. C. (1960), *Symp. Soc. Gen. Microbiol.*, **10**:92.

Coddington, A., and J. R. S. Fincham (1965), *J. Mol. Biol.*, **12**:152.

Coe, E. H. (1964), *Genetics*, **50**:571.

Crawford, I. P., and C. Yanofsky (1958), *Proc. Natl. Acad. Sci.*, **44**:1161.

——, S. Sikes, N. O. Belser, and L. Martinez (1970), *Genetics*, **65**:201.

Crick, F. H. C., and L. E. Orgel (1964), *J. Mol. Biol.*, **8**:161.

Cuénot, L. (1904), *Arch. Zool. Exp. Gen.*, (43)**2**: (notes et revue) 45.

Daniel, V., J. S. Beckmann, S. Sarid, J. L. Grimberg, M. Herzberg, and U. Z. Littauer (1971), *Proc. Natl. Acad. Sci.*, **68**:2117.

Demerec, M. (1964), *Proc. Natl. Acad. Sci.*, **51**:1057.

—— and Z. E. Demerec (1956), *Brookhaven Symp. Biol.*, **8**:75.

—— and P. E. Hartman (1959), *Annu. Rev. Microbiol.*, **13**:377.

—— and Z. Hartman (1956), *Carnegie Inst. Wash., D.C., Publ.* 612, p. 5.

—— and H. Ozeki (1959), *Genetics*, **44**:269.

de Serres, F. J. (1956), *Genetics*, **41**:668.

—— (1960), *Genetics*, **45**:555.

—— (1969), *Mutat. Res.*, **8**:43.

—— and H. E. Brockman (1968), *Genetics*, **58**:79.

Dorn, G. L., and A. B. Burdick (1962), *Genetics*, **47**:503.

Dubinin, N. P. (1933), *J. Genet.*, **27**:443.

—— and B. N. Sidorov (1934), *Am. Nat.*, **68**:377.

Edgar, R. S., G. H. Denhardt, and R. H. Epstein (1964), *Genetics*, **49**:635.

Enomoto, M. (1966), *Genetics*, **54**:715.

Epstein, R. H., A. Bolle, C. M. Steinberg, E. Kellenberger, E. Boy de la Tour, R. Chevalley, R. S. Edgar, M. Susman, G. H. Denhardt, and A. Lielausis (1963), *Cold Spring Harbor Symp. Quant. Biol.*, **28**:375.

Esposito, M. S. (1968), *Genetics*, **58**:507.

Fincham, J. R. S. (1959), *10th Int. Congr. Genet. Montreal, 1958*, **1**:335–363.

―――― (1962), *J. Mol. Biol.*, **4**:257.

―――― (1966), "Genetic Complementation," Benjamin, New York.

―――― and J. A. Pateman (1957), *Nature*, **179**:741.

Finnerty, V. G., and A. Chovnick (1970), *Genet. Res.*, **15**:351.

――――, P. Duck, and A. Chovnick (1970), *Proc. Natl. Acad. Sci.*, **65**:939.

Forrest, H. S., E. Glassman, and H. K. Mitchell (1956), *Science*, **124**:725.

Garen, A., and S. Garen (1963), *J. Mol. Biol.*, **7**:13.

Giles, N. H., C. W. H. Partridge, and N. J. Nelson (1957), *Proc. Natl. Acad. Sci.*, **43**:305.

Gillie, O. J. (1968), *Genetics*, **58**:543.

Glassman, E. (1962), *Proc. Natl. Acad. Sci.*, **48**:1491.

―――― (1965), *Fed. Proc.*, **24**:1243.

―――― and H. K. Mitchell (1959), *Genetics*, **44**:546.

―――― and ―――― (1959), *Genetics*, **44**:153.

Gross, J., and E. Englesberg (1959), *Virology*, **9**:314.

Gross, S. R. (1962), *Proc. Natl. Acad. Sci.*, **48**:922.

―――― and R. E. Webster (1963), *Cold Spring Harbor Symp. Quant. Biol.*, **28**:543.

Gussin, G. N. (1966), *J. Mol. Biol.*, **21**:435.

Hartman. M., and Z. Hartman (1956), *Carnegie Inst. Wash., D.C.*, Publ. 612, p. 5.

Hartman, P. E., Z. Hartman, and D. Serman (1960*a*), *J. Gen. Microbiol.*, **22**:354.

――――, J. C. Loper, and D. Serman (1960*b*), *J. Gen. Microbiol.*, **22**:323.

Hayes, H. (1968), "Genetics of Bacteria and Their Viruses," 2d ed., Blackwell, Oxford.

Hexter, W. M. (1958), *Proc. Natl. Acad. Sci.*, **44**:768.

Holliday, R. (1964), *Genet. Res.*, **5**:282.

Huang, P. C. (1964), *Genetics*, **49**:453.

Hubacek, J., and S. W. Glover (1970), *J. Mol. Biol.*, **50**:111.

Hubby, J. L. (1961), *Drosophila Inf. Serv.*, **35**:46.

Hutter, R., and J. A. De Moss (1967), *Genetics*, **55**:241.

Ishikawa, T. (1962), *Genetics*, **47**:1147.

Kakar, S. N. (1963), *Genetics*, **48**:957.

Kaplan, S., Y. Suyama, and D. M. Bonner (1964), *Genetics*, **49**:145.

Khorana, H. G. (1970), *IUPAC Symp. Chem. Nat. Prod.*, Rega, U.S.S.R.

Kitani, Y., and L. S. Olive (1967), *Genetics*, **57**:767.

――――, ――――, and A. S. El-Ani (1962), *Am. J. Bot.*, **49**:697.

Lacy, A. M., and D. M. Bonner (1961), *Proc. Natl. Acad. Sci.*, **47**:72.

Laughnan, J. R. (1955), *Am. Nat.*, **89**:91.

Lennox, E. S. (1955), *Virology*, **1**:190.

Lewis, E. B. (1942), *Genetics*, **27**:153.

―――― (1945), *Genetics*, **30**:137.

―――― (1951), *Cold Spring Harbor Symp. Quant. Biol.*, **16**:159.

―――― (1952), *Proc. Natl. Acad. Sci.*, **38**:953.

―――― (1967), in R. A. Brink (ed.), "Heritage from Mendel," p. 17, University of Wisconsin Press, Madison.

Lifschytz, E., and R. Falk (1969), *Mutat. Res.*, **8**:147.

Lindegren, C. C. (1953), *J. Genet.*, **51**:625.

Loper, J. C., M. Grabnar, R. C. Stahl, Z. Hartman, and P. E. Hartman (1964), *Brookhaven Symp. Biol.*, **17**:15.

McClintock, B. (1944), *Genetics*, **29**:478.

MacKendrick, M. E., and G. Pontecorvo (1952), *Experientia*, **8**:390.

Mangelsdorf, P. C., and G. S. Fraps (1931), *Science*, **73**:241.

Mitchell, M. B. (1955), *Proc. Natl. Acad. Sci.*, **41**:935.

Miyake, T., and M. Demerec (1960), *Genetics*, **45**:755.

Morgan, T. H. (1926), "The Theory of the Gene," Yale University Press, New Haven.

Morpurgo, G., and L. Volterra (1968), *Genetics*, **58**:529.

Murray, N. E. (1963), *Genetics*, **48**:1163.

Nadler, H. L., C. M. Chacko, and M. Rachmeler (1970), *Proc. Natl. Acad. Sci.*, **67**:976.

Nelson, O. E. (1962), *Genetics*, **47**:737.

Oliver, C. P. (1940), *Proc. Natl. Acad. Sci.*, **26**:452.

—— (1941), *Genetics*, **26**:163.

Ozeki, H. (1956), *Carnegie Inst. Wash., D.C., Publ.* 612, p. 97.

Pateman, J. A., and J. R. S. Fincham (1958), *Heredity*, **12**:317.

Pontecorvo, G. (1952), *Adv. Enzymol.*, **13**:121.

—— (1958), "Trends in Genetic Analysis," Columbia, New York.

Pritchard, R. H. (1955), *Heredity*, **9**:343.

Radford, A. (1970), *Mol. Gen. Genet.*, **109**:241.

Raffel, D., and H. J. Muller (1940), *Genetics*, **25**:541.

Reithel, F. J. (1963), *Adv. Protein Chem.*, **18**:123.

Roper, J. A. (1950), *Nature*, **166**:956.

—— (1953), in G. Pontecorvo, The Genetics of *Aspergillus nidulans*, *Adv. Genet.*, **5**:208.

—— and R. H. Pritchard (1955), *Nature*, **175**:639.

Schalet, A. (1971), *Mol. Gen. Genet.*, **110**:82.

——, R. P. Kernaghan, and A. Chovnick (1964), *Genetics*, **50**:1261.

Schlesinger, M. J., and C. Levinthal (1963), *J. Mol. Biol.*, **7**:1.

—— and —— (1965), *Annu. Rev. Microbiol.*, **19**:267.

Smith, D. W. E., and B. N. Ames (1964), *J. Biol. Chem.*, **239**:1848.

—— and —— (1965), *J. Biol. Chem.*, **240**:3056.

Smith, Oliver H. (1967), *Genetics*, **57**:95.

Stadler, D. R., and A. M. Towe (1963), *Genetics*, **48**:1323.

Stent, G. S. (1963), "Molecular Biology of Bacterial Viruses," Freeman, San Francisco.

—— (1971), "Molecular Genetics," Freeman, San Francisco.

Stephens, S. G. (1948), *Genetics*, **33**:191.

—— (1951), *Cold Spring Harbor Symp. Quant. Biol.*, **16**:131.

Stern, C. (1943), *Genetics*, **28**:441.

—— and E. W. Schaeffer (1943), *Proc. Natl. Acad. Sci.*, **29**:351.

Sturtevant, A. H. (1913), *J. Exp. Zool.* **14**:43.

—— and T. H. Morgan (1923), *Science*, **57**:746.

Sundaram, T. K., and J. R. S. Fincham (1967), *J. Mol. Biol.*, **29**:433.

Suyama, Y., A. M. Lacy, and D. M. Bonner (1964), *Genetics*, **49**:135.

Tessman, I. (1965), *Genetics*, **51**:63.

Visconti, N. (1953), *Genetics*, **138**:5.

Watson, J. D. (1970), "Molecular Biology of the Gene," Benjamin, New York.

Welshons, W. J., and E. S. Von Halle (1962), *Genetics*, **47**:743.

Whitehouse, H. L. K. (1965), *Sci. Prog.*, **53**:285.

—— and P. J. Hastings (1965), *Genet. Res.*, **6**:27.

Winkler, H. (1930), "Die Konversion der Gene," Jena, Gustav Fischer.

Woods, R. A., and E. A. Bevan (1966), *Heredity*, **21**:121.

Woodward, D. O. (1959), *Proc. Natl. Acad. Sci.*, **45**:846.

—— (1960), *Qt. Rev. Biol.*, **35**:313.

——, C. W. H. Partridge, and N. H. Giles (1958), *Proc. Natl. Acad. Sci.*, **44**:1237.

——, ——, and —— (1960), *Genetics*, **45**:535.

Wright, S. (1949), *Genetics*, **34**:245.

—— (1959), *Genetics*, **44**:1001.

Yanofsky, C. (1967), *Sci. Am.*, **216** (May):80.

—— and E. S. Lennox (1959), *Virology*, **8**:425.

Yen, T. T. T., and E. Glassman (1965), *Genetics*, **52**:977.

27
Biochemical Genetics

NOTATION

A list of biochemical abbreviations used in this and subsequent chapters is given in Appendix Table A-7.

QUESTIONS

1267 **a** In any chain of biochemical reactions how could you tell whether a mutation produced a block before or after a given step?

b Is a genetic block always complete? Why?

c "A mutant gene can change the rate of development by changing the efficiency of catalysis of a single enzyme." Explain.

1268 It has been observed that phenylalanine serves as a competitive inhibitor of the tyrosinase-catalyzed tyrosine system in vitro (Miyamoto and Fitzpatrick, 1957). It is also known that *phenylketonurics* tend to have lighter skin and hair pigmentation than their *normal* brothers and sisters. Suggest a reason for the latter observation in view of the former one.

1269 Present evidence for the one-gene–one-enzyme hypothesis and explain why it has now been modified to the form one gene, one polypeptide. Illustrate your discussion with examples.

1270 Several instances of single-gene mutations causing a double growth-factor requirement are known. In *Neurospora*, for example, a single mutation results in a nutritional requirement for the amino acids methionine and threonine. Show how these would be distinguished from two-gene mutations and discuss the possible mechanisms underlying the double requirement (see Wagner et al., 1958, 1960; Radhakrishnan et al., 1960).

1271 A strain of *Neurospora* isolated after x-ray treatment is unable to grow either on minimal medium or on minimal plus any one added compound. The addition of two specific separate compounds, however, satisfies the growth requirements of this strain. What different biochemical-genetic situations might account for your findings?

1272 Some traits in man are incompletely penetrant and show variable expressivity. For example, about 90 percent of individuals with the rare dominant mutant gene for *blue sclera* express the trait, and the degree of expression differs greatly among *affected* individuals. Suggest two features of biochemical mutants in other organisms, e.g., *Neurospora*, *Escherichia*, that would explain this condition as a metabolic error.

1273 Illustrate or describe, with the aid of hypothetical biochemical reactions, three different ways in which a gene mutation may produce an allele that is dominant to the *wild-type* allele.

1274 The majority of mutant proteins differ from *wild-type* (normal) proteins by having single amino acid substitutions at specific residue sites. Some of these substitutions have no effect on enzyme function, others affect enzyme activity slightly, and still others affect it severely or cause its complete cessation.   Offer an explanation to account for this variability of effect.

1275   Explain how it is possible to determine from an enzyme study whether alleles in a particular series of multiple alleles governing a certain metabolic step are acting in the direction of standard or not.

1276   Suggest the forms of gene-controlled metabolic pathways that could explain the following dihybrid interaction ratios:

    a 9:3:4   **b** 15:1   **c** 9:7

1277   **a** What are the primary consequences to the biochemical constitution of cells or tissues of a metabolic block?

    b Certain biochemical mutants accumulate only normal amounts of precursor behind the genetic block.   A few others accumulate large amounts of other substances apparently unrelated to the metabolic pathway in which the block occurs.   Suggest possible explanations for the latter.

1278   **a** In what respect can gene-caused diseases such as *diabetes mellitus* and *phenylketonuria* be "cured" (alleviated)?

    b What are some of the possible ways in which gene-caused diseases might be overcome?   Discuss, citing examples if possible.

    c Discuss the advantages and disadvantages of curing or alleviating inborn errors of metabolism.

    d Do such cured individuals have any moral obligation to society?   Discuss.

1279   In one strain of *Neurospora*, an enzyme catalyzing one of the steps in the chain of reactions leading to the formation of thiazole is lacking.   This strain is also unable to synthesize thiamine.   Explain how this is possible.

1280   What types of biochemical-genetic situations might provide possibilities for pleiotropic effects of a single mutation?

1281   Why is it that in the establishment of criteria for the detection of carriers of hereditary disease biochemical and physiological criteria are likely, for the most part, to be more useful than morphological criteria?

1282   Give some reasons why it would be a good idea for every medical practitioner to have training in genetics.

1283   In some cases a particular gene mutation results in the loss of a specific enzyme activity.   Does this mean that the enzyme is not produced?   Explain.

1284   Ribonuclease contains one polypeptide chain only (124 amino acids); insulin contains two: A, with 21 amino acids and B, with 30 amino acids.   Alkaline phosphatase has two chains of the same kind, whereas human adult hemoglobin contains four polypeptide chains of two kinds, α and β.   Which of these proteins will have a:

    a Primary structure?

    b Secondary structure?

c Tertiary structure?

d Quaternary structure?

Explain.

1285 T4 phage lysozyme consists of one polypeptide chain 160 amino acids long (Inouye and Tsugita, 1966). Read this paper or the one by Phillips (1966) and then answer the following:

a How many peptide fragments do you expect after treating the enzyme with trypsin? With chymotrypsin? With cyanogen bromide?

b How would you determine the amino acid sequence in the N-terminal tryptic peptide?

c Show how you can place tryptic and chymotryptic fragments in proper sequence if:

(1) The amino acid sequence of each is known.

(2) Only the amino acid content is known.

(3) Only N-terminal and C-terminal residues have been identified.

d Which amino acids may play a role in determining the tertiary structure of this enzyme, and how?

See Schroeder (1968) and Dickerson and Geis (1969) for methods for elucidating primary structure.

1286 The functional specificity of a protein is dependent on its final three-dimensional structure. This has been established for myoglobin (Kendrew, 1961, 1963) and T4 lysozyme (Phillips, 1966), which contain only one polypeptide chain and therefore have a tertiary structure, and human adult hemoglobin, a tetramer consisting of two α and two β chains and having a quaternary structure (Perutz, 1963, 1964).

a Outline the forces important in maintaining tertiary and quaternary structure. See the above papers for illuminating presentations of how these forces operate.

b Since the tertiary and quaternary structures are dependent on the amino acid sequence of the component polypeptide chains, with the aid of illustrations, show (1) the kinds of mutational substitutions which can affect tertiary and quaternary structures and their affinity for substrates and (2) those which cannot.

PROBLEMS

1287 a A tetrapeptide contains the amino acids histidine, tyrosine, leucine, and phenylalanine. Enzymatic degradation produces three kinds of dipeptides: Tyr-Leu, Tyr-His, and Phe-His. What is the amino acid sequence in this tetrapeptide?

b The α polypeptide (267 residues) of E. coli tryptophan synthetase, when treated with cyanogen bromide, breaks into five fragments. The N-terminal one, called the F-1 fragment, with residues 1 through 83, can be further fragmented by trypsin (which splits bonds at the carboxyl side of arginine and lysine) and chymotrypsin (which hydrolyzes bonds on the carboxyl side of phenylalanine, tyrosine, tryptophan, and leucine). The peptides containing the first 30 residues of this chain possess the amino acid sequences in Table 1287. In each peptide the N-terminal amino acid is presented to the left. Methionine is the N-terminal amino acid in the α chain (Guest et al., 1967; Drapeau and Yanofsky, 1967*a,b*). Determine the sequence of the 30 amino acids.

Table 1287

Peptide number	Amino acid sequence
	Tryptic peptides (TP)
9	*Met-Gln-Arg
14	Tyr-Glu-Ser-Leu-Phe-Ala-Gln-Leu-Lys
19	Glu-Arg
27	Met-Gln-Arg-Tyr-Glu-Ser-Leu-Phe
21	Lys
12	Glu-Gly-Ala-Phe-Val-Pro-Phe-Val-Thr-Leu-Gly-Asp-Pro-Gly-Ile
10	Lys-Glu-Gly-Ala-Phe-Val-Pro-Phe-Val-Thr-Leu-Gly-Asp-Pro-Gly-Ile
	Chymotryptic peptides (CP)
5	Met-Gln-Arg-Tyr
27	Met-Gln-Arg-Tyr-Glu-Ser-Leu-Phe
36	Lys-Glu-Arg-Lys-Glu-Gly-Ala-Phe
37	Lys-Glu-Arg-Lys-Glu-Gly-Ala-Phe-Val-Pro-Phe
8	Ala-Gln-Leu-Lys-Glu-Arg-Lys-Glu-Gly-Ala-Phe-Val-Pro-Phe

* N-terminal amino acid sequence.

c A 16–amino acid peptide from the human α chain contains one each of the amino acids lysine, aspartic acid, threonine, glutamic acid, proline, glycine, alanine, valine, leucine, and tyrosine and two each of histidine, serine, and phenylalanine. Konigsberg and Hill (1962) obtained two peptides following chymotrypsin digestion. These in turn were further hydrolyzed using pepsin and papain. Nine peptides (including the initial two) were isolated. Only N-terminal residues and several residues near the C terminus were identified. Amino acids whose sequence is not known are enclosed in parentheses, and a subscript 2 after an amino acid indicates that there are two in the peptide.

Thr-(His, Pro, Tyr, Phe$_2$)   Asp-Leu-Ser
(His, Asp, Ser$_2$, Gly, Ala, Leu)-Glu-Val-Lys   Phe-Pro-His-Phe
Thr-Tyr   Glu-Val-Lys
Asp-(His, Ser$_2$, Gly, Leu)-Ala   Ser-Ala
His-Gly

What is the sequence of amino acids in this hexadecapeptide?   Explain.

1288 a A solution of the peptide shown below was treated with the proteolytic enzyme trypsin, resulting in the formation of specific hydrolysis products.[1] The resulting digest was subjected to paper electrophoresis at pH 3 and the hydrolysis products were visualized by spraying the paper with ninhydrin, a compound which reacts with amino acids and peptides to produce a purple color.

Leu-Ile-Lys-Val-His-Lys-Glu-Asp-Lys-Glu-Asp-Asp
↑   ↑
Amino   Carboxy
terminal   terminal
residue   residue
NH$_2$ COOH

Note: (1) See Appendix Table A-4 for amino acid structures; see Mahler and Cordes (1966, p. 20), for description of peptide bond linkage.   (2) Carboxyl groups are polar and therefore hydrophilic.

At pH 3:

1   All free amino groups carry a single positive charge (NH$_3{}^+$).
2   The imidazole nitrogen of histidine carries a single positive charge.

$$
\begin{array}{c}
\text{HC}-\overset{\overset{\textstyle H^+}{|}}{N} \\
\text{HC}-N \\
\text{H}_2\text{C} \quad H \\
\text{H}_2\text{N}-\text{CH}-\text{CO}_2\text{H}
\end{array}
$$

3   Free carboxyl groups are uncharged (COOH).
4   The nitrogen atoms in the peptide bonds are uncharged.

(1) Indicate which peptide bonds are cleaved (hydrolyzed) by the enzyme trypsin and draw the structures of the resulting hydrolysis products.

(2) Indicate with a diagram how you would expect the hydrolysis products to move relative to the origin and the electrodes.   Explain your reasoning.

(3) Suppose that instead of spraying the paper after electrophoresis, it had been subjected to chromatography (in a hexane-water solvent system),

[1] For peptide bonds cleaved by trypsin see Bernhard (1968, pp. 137–138), White et al. (1964, p. 149), or Mahler and Cordes (1966, p. 76).

such that the direction of chromatography was perpendicular to the direction of electrophoresis. Would visualization of the hydrolysis products following the combined electrophoresis-chromatography operation reveal an improved resolution over that obtained by electrophoresis alone? Explain with the aid of a diagram.

b Aldolase has four identical N-terminal amino acid residues and four identical C-terminal ones. What kind of structure would this enzyme possess?

1289 In *Drosophila*, the recessive *bw* (on chromosome 2) causes a *dark-brown eye*, and the recessive *st* (on chromosome 3) causes a *bright-scarlet eye*. Flies homozygous for both recessives have *white eyes* (Wright, 1932).

Bw—St—	*red* (*wild type*)
bwbw St—	*brown*
Bw— stst	*scarlet*
bwbw stst	*white*

Outline a hypothetical biochemical pathway or pathways that would give this type of gene interaction and demonstrate why, according to your explanation, each genotype shows its specific phenotype.

1290 In *Drosophila*, the dominant genes *V* and *Cn* control sequential steps in a biochemical chain of reactions involved in the production of brown eye pigment (xanthommatin) as shown below.

$$
\begin{array}{ccc}
V & & Cn \\
\downarrow & & \downarrow
\end{array}
$$

Tryptophan → formylkynurenine → kynurenine → hydroxykynurenine → xanthommatin

    V substance   *Cn* substance   Brown pigment

The recessive alleles, *v* and *cn*, block the production of brown pigment at tryptophan → formylkynurenine and kynurenine → hydroxykynurenine, respectively, so that flies of either mutant strain have eyes with a scarlet hue rather than the deep-*red* characteristic of *wild-type* flies (Ziegler, 1961).

a What are the expected phenotypes and phenotypic ratios in the F_2 of a cross between *VV cncn* and *vv CnCn*?

b What is the term applied to the type of gene interaction that yields this kind of breeding behavior?

1291 Bateson, Saunders, and Punnett (1905) crossed a true-breeding *white*-flowered variety of the sage *Salvia horminum* with a true-breeding *pink*-flowered variety. The F_1's had *purple* flowers and when self-fertilized produced 255 *purple*-, 92 *pink*-, and 114 *white*-flowered F_2's. The same investigators obtained F_1's with *purple* flowers from a cross between true-breeding *white*-flowered varieties of sweet pea (*Lathyrus odoratus*). The F_1's when self-fertilized produced 651 plants, 382 with *purple* and 269 with *white* flowers. After stating the types of gene interaction involved, outline a gene-controlled reaction system to account for each of the two types of gene interaction.

1292 In terms of gene-controlled biochemical pathways explain the following dihybrid interactions:

1 Two true-breeding *white*-flowered sweet peas when crossed produced *purple*-flowered F_1's, which when self-fertilized produced 9 *purple* : 7 *white* F_2 progeny.

2 Two wheat plants from true-breeding *spring*-growth-habit varieties when crossed produced F_1's which also expressed *spring* habit. The F_2's of self-fertilized F_1's, however, segregated 15 *spring* : 1 *winter*.

3 A guinea pig with *brown* fur crossed with an *albino* one produced progeny with *black* fur. These when intercrossed produced progeny in the ratio 9 *black* : 3 *brown* : 4 *albino*.

1293 In corn, a cross-fertilizing species that can be self-fertilized, *wild-type* plants have *purple*-colored stems. Six true-breeding mutant lines are established, two with *brown* stems and four with *red* stems. The results of crosses among these mutants are shown.

Table 1293

Cross	F_1	F_2
1. *brown* 1 or 2 × *red* 1, 2, 3, or 4	*purple*	9 *purple* : 3 *red* : 3 *brown*: 1 *white*
2. *brown* 1 × *brown* 2	*brown*	All *brown*
3. *red* 1 or 2 × *red* 3 or 4	*purple*	9 *purple* : 7 *red*
4. *red* 1 × *red* 2	*reddish-blue*	1 *reddish-blue* : 1 *red*
5. *red* 3 × *red* 4	*blue*	5 *purple* : 4 *blue* : 7 *red*

a How many different *wild-type* genes (cistrons) mutated to give the six mutant strains?

b Are the genes linked or not?

c Enzymes specified by each of the genes were studied.

1 In cross 2 (*brown* 1 × *brown* 2) none of the F_1's or F_2's showed any enzyme B activity.

2 In cross 4 (*red* 1 × *red* 2) the F_1's showed some enzyme R activity; one-half the F_2's showed the same amount of enzyme R activity as the F_1's; others did not show any.

3 In cross 5 (*red* 3 × *red* 4) four different kinds of enzymes (*A, B, R, P*) were detected, one identical to that in *purple* plants. In the F_2 one-fourth of the individuals produced the same kinds of enzymes as the F_1's.

How many kinds of polypeptide chains do these enzymes possess, and how many of each kind of chain (one or more) are present?

d Show whether the genes act in the same or different biochemical pathways. (Provide an explanation for each of your answers.)

1294   The *red-lilac-blue* flower colors in higher plants are due to anthocyanins, sap-soluble molecules consisting of an anthocyanidin (a flavonoid) to which one or two sugar residues are attached through an oxygen.   If only one sugar is attached, it is always joined to the anthocyanidin at the 3 position.   If two sugars are attached, they are always present at the 3 and 5 positions.

Anthocyanidin

1   Wit (1937) working with the China aster (*Callistemma* [= *Callistephus*] *chinensis*) obtained the results shown in Table 1294*A* in crosses between true-breeding *blue*, *lilac*, and *deep-pink* varieties.   The anthocyanins in *blue*, *lilac*, and *deep-pink* varieties had *delphinins*, *cyanins*, and *pelargonins*, respectively.

Table 1294*A*

Cross	F$_1$	F$_2$
lilac × *deep-pink*	*almost-lilac*	294 *lilac* : 91 *deep-pink*
blue × *lilac*	*blue*	282 *blue* : 105 *lilac*
blue × *deep-pink*	*deep-pink*	195 *blue* : 64 *deep-pink*

Delphinin

Cyanin

Pelargonin

where R = sugar $(CHO)_n$

2   Further crosses by Wit between true-breeding lines gave the results shown in Table 1294*B*.

Table 1294 B

Cross	F_1	F_2
blue × salmon-pink	blue	135 blue
		43 deep-pink
		31 slaty-blue*
		16 salmon-pink*
lilac × salmon-pink	lilac	51 lilac
		13 deep-pink
		17 slaty-lilac
		4 salmon-pink

* *Salmon-pink* and *slaty-blue* are lighter in color than *pink* and *blue* respectively.

The anthocyanins in the *blue, slaty-blue, pink,* and *salmon-pink* F_2's were as follows:
Blue: Delphinin (see formula).
Slaty-blue:

Pink: Pelargonin (see formula).
Salmon-pink:

a What is the genetic basis for the difference among:
  (1) The *blue, lilac,* and *deep-pink* strains?
  (2) The phenotypes in each of the last two crosses?
b Account genetically and biochemically for the phenotype differences that occur in each of the first two crosses.
c What would be the formula of the anthocyanin molecule for *slaty-lilac*? Why?

1295 In most mammals tyrosinase is believed to catalyze the conversion of tyrosine to melanin. A *wild-type* and three *mutant* true-breeding strains of mice have the genotypes and phenotypes shown. Assume tyrosinase contains one kind of polypeptide chain present once; i.e., the enzyme is a monomer.

Table 1295

Genotype	Coat color
CC	black
$c^a c^a$	brown
$c^b c^b$	light yellow
$c^c c^c$	white (albino)

a What kinds of mutations in the *wild-type* allele C and what kinds of polypeptide changes (viz. substitutions, deletions of amino acids) could give rise to the phenotypes shown?   Illustrate.

b Classify each allele as fully morphic, hypomorphic (leaky), or amorphic (nonleaky).

1296 In *Neurospora*, methionine is an essential amino acid; i.e., it is required for growth.   It is synthesized via a chain of chemical reactions, each under the control of a single enzyme, as shown below (Teas et al., 1948; Fischer, 1957):

$$\text{gene }A \qquad \text{gene }B \qquad \text{gene }C \qquad \text{gene }D$$

$$\longrightarrow \text{Cysteine} \longrightarrow \text{cystathionine} \longrightarrow \text{homocysteine} \longrightarrow \text{methionine}$$

Methionineless strains, incapable of synthesizing methionine, result from mutation at any of the four loci.   Three mutant strains of independent origin showed the following characteristics:

1   The first mutant will grow if supplied either cystathionine, homocysteine, or methionine but not if supplied only cysteine.

2   The second mutant grows when either homocysteine or methionine is provided but not if supplied only cysteine or cystathionine or both.

3   The third mutant grows only if methionine is supplied.

a At which step does a metabolic block appear to occur in each of the mutant strains?   Explain.

b How could you determine experimentally whether a block caused by a mutation occurs before or after a given reaction?

1297 Srb and Horowitz (1944) obtained by x-ray or ultraviolet irradiation 15 mutants of *Neurospora crassa* which lacked the ability to synthesize arginine. Crosses between *wild type* and each mutant gave a 1:1 ratio of *wild type*: *arginineless*.   Eight of the mutants were found to be allelic with one of the seven mutants to be considered here.

1   Crosses between these seven mutants (A to G) taken two at a time produced *wild type* and *arginineless* progeny in a 1:3 ratio.

2   Strains A to D grew on a minimal medium if ornithine, citrulline, or arginine was added.   Strains E and F could not utilize ornithine but grew if the other two compounds were supplied.   Strain G grew only if arginine was added to the minimal medium.

With reasons for your answers, state the minimum number of genes involved in *arginine* synthesis, whether they act in the same or different pathways, and, if in the same pathway, the sequence of biochemical reactions.

1298   Using ultraviolet light, Horowitz et al. (1945) obtained four mutant strains of *Neurospora* unable to synthesize choline.   Crosses between each of the mutant strains and *wild type* produced asci whose spores segregated 1 *wild type* : 1 *cholineless*.   Subsequent studies with *cholineless* strains 1 and 2 revealed the following:

1   *Cholineless-2* requires much less choline to produce normal growth than *cholineless-1*. Whereas in unsupplemented minimal media *cholineless-1* produces almost no growth, *cholineless-2* produces approximately 20 percent of the dry weight attained on an optimum concentration of choline.
2   *Cholineless-1* can grow if either monomethylaminoethanol or dimethylaminoethanol is added to a minimal medium instead of choline.
3   *Cholineless-2* grows fairly well in dimethylaminoethanol but poorly on the monomethyl compound; however, it does accumulate the latter compound or a very similar one in the medium in which it grows.

a   Suggest a scheme for the synthesis of choline in *Neurospora*.
b   Offer an explanation to account for the fact that the second mutant strain grows, although poorly, on a minimal medium whereas the first one shows no growth on such a food supply.
c   Show the results you would expect in crosses between the two mutant strains.   Strains from some of the spores from this cross when crossed with *wild-type Neurospora* produced asci whose spores segregated 1 *wild-type* : 3 *mutant*. Of the *mutant* types, $\frac{1}{3}$ possessed the biochemical properties of *cholineless-1*, $\frac{1}{3}$ the properties of *cholineless-2*, and $\frac{1}{3}$ the properties of the parent *mutant* strain.   Are these results consistent with your hypothesis?   Explain.

1299   Horowitz (1947) isolated four mutant strains of *Neurospora* which were unable to synthesize methionine.   Each differed from *wild-type Neurospora* by a mutant gene at one locus.   Strain *me-4* grew if cysteine, cystathionine, homocysteine, or methionine was added to the minimal medium; *me-3* grew if any of the latter three compounds were added, whereas *me-2* and *me-1* grew on minimal medium only if homocysteine and methionine and methionine only, respectively, were added.   The chemical formulas of the compounds involved are

```
SH            CH₂—S—CH₂      SH         CH₃
|             |     |        |          |
CH₂           CHNH₂ CH₂      CH₂        S
|             |     |        |          |
CHNH₂         COOH  CHNH₂    CH₂        CH₂
|                   |        |          |
COOH                COOH     CHNH₂      CH₂
                             |          |
                             COOH       CH₂
                                        |
                                        CHNH₂
                                        |
                                        COOH
Cysteine      Cystathionine   Homocysteine Methionine
```

Crosses between any two of the four mutant strains always produced some spores that gave rise to *wild-type Neurospora*.   In the light of this fact explain these results genetically, including the biochemical mechanisms by which genes carry out the various conversions involved.

1300   In *Neurospora*, the last step in the synthesis of adenine is catalyzed by adenylosuccinase, which splits succinic acid off adenylosuccinic acid to yield adenine (Giles et al., 1957).   A large number of *adenineless* mutants, all requiring the enzyme adenylosuccinase, when crossed with *wild type* gave a 1 : 1 ratio of *adenine : adenineless* ascospores.

a Does this observation prove that the synthesis of adenylosuccinase is specified by alleles of one gene only?   Explain.

b How would you determine whether the *adenylosuccinase-requiring* mutants were due to mutations at the same locus?   Illustrate the results expected (1) if the mutants are allelic and (2) if they are not allelic.

1301   In a *wild-type* (*red-eyed*) strain of *Drosophila melanogaster* a *scarlet*-eye mutant appears.   Crossed with *wild-type* flies, it produced *wild-type* F_1's and F_2's in the ratio 3 *wild type* : 1 *scarlet*.   Two separate revertants to *wild-type* eyes subsequently appear in the true-breeding *scarlet* strain, and by subsequent breeding within this strain, a progeny strain of each is obtained that is true-breeding for the reversion.   The first, when crossed with *wild-type* flies, produces F_1's that breed true for *wild type*.   The second, similarly crossed, produces *wild-type* F_1's and F_2's that segregate 15 *wild type* : 1 *scarlet*.

a Why does the second and not the first revertant segregate for eye color in the F_2 ?

b Outline one possible mode of action of the gene or genes involved in the reversion to *wild type* observed in the second cross.

1302   The compound eye of *wild-type Drosophila melanogaster* is dull red due to the presence of two classes of pigment, the bright-red drosopterins and the brown xanthommatins. Biopterin and sepiapterin (and no doubt other pterins) are precursors of the drosopterins, apparently in the same biosynthetic pathway. *Wild-type* males accumulate the yellow compound sepiapterin in their testes. Flies homozygous for *sepia, se*, a chromosome-3 mutant, lack drosopterins in the eyes, but sepiapterin accumulates in the testes. *Drosophila* homozygous for *brown, bw*, a chromosome-2 mutant, do not accumulate drosopterin in the eyes and have colorless testes (Hadorn and Mitchell, 1951; Hubby and Throckmorton, 1960).

a Outline how the precursors are related to each other in the biochemical pathway leading to the synthesis of drosopterin and explain why you think the relationship is of this kind.

b Show the phenotypes and their expected proportions in the F_1 and F_2 of *sese BwBw* × *SeSe bwbw*.

c Do either of the genes act differently in different tissues?

1303 The compound eye of *Drosophila melanogaster* consists of many facets. Xanthommatins and drosopterins are usually bound to core granules in the facets, which are rich in protein and RNA (Ziegler, 1961). In *wild-type* flies precursors (which are carried to the eye by the hemolymph) are converted by these facet granules to the above pigments, but in flies carrying the recessive X-linked gene *w* (*white*-eye) they are incapable of doing so. The metabolic pathways for the synthesis of the two pigments are shown below.

$$
\text{Tryptophan} \xrightarrow{V} \text{formylkynurenine} \longrightarrow \text{kynurenine} \xrightarrow{Cn} \text{3-hydroxy-}
$$

$$
\text{kynurenine} \xrightarrow{\substack{St \\ or \\ Cd}} \xrightarrow{\substack{Cd \\ or \\ St}} \longrightarrow \text{brown pigment} \tag{1}
$$

Apparently

$$
\xrightarrow{Bw} \underset{\text{Yellow pigment}}{\text{biopterin}} \longrightarrow \text{sepiapterin} \xrightarrow{Se} \text{red pigment} \tag{2}
$$

Show the types and proportions of flies that would be obtained in the F_1 and testcross of the following matings.

a *ww BwBw stst* × *W/Y bwbw StSt* **b** *WW sese bwbw* × *w/Y SeSe BwBw*
c *ww vv CnCn* × *W/Y VV cncn* **d** *WW VV bwbw* × *w/Y vv BwBw*

1304 *Vermilion, v*, and *cinnabar, cn*, are recessives at different loci which produce similar phenotypic effects (*bright-red* eyes); both phenotypes lack the brown pigment xanthommatin. In 1936 Beadle and Ephrussi transplanted imaginal disks (larval tissues giving rise to the adult eye) from *Drosophila* donor larvae of one genotype into host larvae of a different genotype. Explain the results shown in the table, stating which eye primordia develop autonomously (by gene action in the eye transplant itself) and which develop non-autonomously (by gene action in the hemolymph).

Table 1304

Donor larvae		Host larvae		Color of eye transplants
Genotype	Phenotype	Genotype	Phenotype	
CnCn VV WW	*wild type*	*CnCn VV ww*	*white*	*wild type*
CnCn VV ww	*white*	*CnCn VV WW*	*wild type*	*white*
CnCn VV WW	*wild type*	*CnCn vv WW*	*vermilion*	*wild type*
CnCn VV WW	*wild type*	*cncn VV WW*	*cinnabar*	*wild type*
CnCn vv WW	*vermilion*	*CnCn VV WW*	*wild type*	*wild type*
cncn VV WW	*cinnabar*	*CnCn VV WW*	*wild type*	*wild type*
cncn VV WW	*cinnabar*	*cncn VV WW*	*cinnabar*	*wild type*

1305 In an ultraviolet-treated culture of *Neurospora* growing on inositolless medium an *inositolless* mutant appears.   After lengthy culture the mutant regains the ability to grow on inositolless medium. Possible explanations for this reversion to the *wild-type* traits are:

1   Reverse mutation of the mutant allele to the *wild-type* allele
2   Contamination of the mutant cultures with *wild-type Neurospora*
3   A mutation at another locus to an allele, which "suppresses" the effect of the mutant allele
4   Mutation at the same locus but at a site different from the original mutation
5   Physiological adaptation involving no genetic change

Outline a series of experiments to determine which of these explanations is the correct one.

1306 a In *Drosophila*, flies homozygous for either *v* (*vermilion*) or *cn* (*cinnabar*) produce no brown pigment and therefore have *bright-red* eyes. When imaginal disks of these flies are transplanted into *wild-type* hosts, they develop the *dull-red* color characteristic of the eyes of *wild type*. It has been shown that the *wild-type* host contributes a diffusable substance or substances required by the transplants to develop *wild-type* color. Three explanations are therefore possible for the failure of *cn* and *v* strains to develop *wild-type* color:

1   The *wild-type* alleles at the *v* and *cn* loci may be structurally and functionally identical, and the alleles *v* and *cn* may therefore fail to produce the same precursor of brown pigment.
2   They may function in different ways so that *v* fails to bring about formation of *V substance* and *cn* the formation of *Cn* substance, both substances being precursors of brown pigment.

a The genes *v* and *cn* may be acting dependently, i.e., in different series of chemical reactions, thus

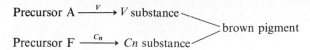

b The genes may be concerned with different steps in the same pathway so that the product of one acts as a precursor for the formation of the other; thus

Precursor ⟶ *Cn* substance ⟶ *V* substance ⟶ brown pigment

Outline the results expected in reciprocal transplants between *vv* and *cncn* flies under each of the three conditions, 1, 2*a*, and 2*b*, and give reasons for these results.

b When reciprocal transplants are made, it is found that *cinnabar* disks in *vermilion* hosts develop *cinnabar* eye color but *vermilion* disks develop *wild-type* color in *cinnabar* hosts. Which hypothesis in (a) do these results support? Using these facts, explain the results of reciprocal transplantation.

1307 The carotenoid pigments in the endosperm (3*n*) of corn are controlled by a single pair of alleles, *Yy*. The relationship between endosperm genotype, phenotype, and amount of vitamin A is as shown (Mangelsdorf and Fraps,

Table 1307

Endosperm phenotype	No. of *Y* alleles	Genotype	Average number of units of vitamin A per gram
white	0	*yyy*	0.05
pale yellow	1	*Yyy*	2.25
dilute yellow	2	*YYy*	5.00
deep yellow	3	*YYY*	7.50

1931). Describe the type of effect (amorphic = nonleaky or hypomorphic = leaky) each of the alleles appears to have on endosperm color and vitamin A content.

1308 **a** In 1944 Tatum, Bonner, and Beadle showed that two independently arising *tryptophanless Neurospora* mutants yielded *wild-type* recombinants when crossed. Strain A grew on minimal medium if indole or anthranilic acid was added; strain B grew only if indole was supplied and accumulated large amounts of anthranilic acid in the medium. The uptake of indole was greatly increased if serine was also present. Under these circumstances tryptophan appeared in the medium. In 1948 Mitchell and Lein discovered a third mutant strain C (*trp-3*), which required tryptophan for growth and could not utilize indole. Cell-free extracts of this mutant, unlike similar extracts prepared from *wild-type* strains, lacked the enzyme tryptophan synthetase and were unable to synthesize tryptophan even in the presence of indole and serine (Yanofsky, 1960). Strain C produced *wild-type* recombinants with each of the first two mutants. Discuss the implications of these findings with regard to the metabolic pathway of tryptophan synthesis in *Neurospora*, indicating how many gene loci there are and how they are involved in the process.

b Yanofsky and Bonner (1955) studied 25 *tryptophanless* mutants whose requirements cannot be satisfied by indole or anthranilic acid. All responded well to tryptophan. Each was unable to synthesize tryptophan synthetase or produces it in defective form. Crosses between any two of these mutants produced, with extremely rare exceptions, *tryptophanless* progeny only. Discuss the genetic and biochemical implications of these findings.

c Reversion to *tryptophan independence* occurred in many of the mutant strains. Several revertants in crosses with original *wild type* produced only *wild-type* (*tryptophan-independent*) progeny. Others produced asci segregating 6 *pseudo wild type* : 2 *mutant*. The *pseudo wild types* grew slower than *wild types* on medium lacking tryptophan and responded markedly to the addition of this amino acid to the medium. Account genetically for the two types of revertants and offer an explanation to account for the *pseudo-wild-type* phenotype.

1309 Whether it is possible to transplant tissues from one individual or line to another depends on the genotypes of the donor and host. Genes concerned with transplantation specificities are known as *histocompatibility loci* (Billingham and Silvers, 1962). Studies of skin grafting involving inbred strains A and B of the Syrian hamster and the F_1 and F_2 generations of the cross between them give the following results:

1   A → A, B → B, and F_1 (from A × B) → F_1 (from A × B) are accepted and incorporated into the recipient, but A → B or B → A transplants are not.

2   The F_1's accept skin grafts from A or B, but A and B reject grafts from the F_1's.

3   The F_2's obtained from a series of crosses between F_1 individuals fall into four phenotypic classes.

Table 1309

Accept graft from	No. in class
A inbred strain	103
B inbred strain	97
Both inbred strains	292
Neither inbred strain	38

a How many *histocompatibility loci* are involved in determining acceptance or rejection of grafts?

b Present in diagram form the genotypes of the parents and the F_1 and F_2 phenotypes that will account for the observed results.

c Suggest a mechanism to account for the acceptance or rejection of skin grafts (see Burnet, 1961; Medawar, 1961; and Hildemann, 1970).

1310 Whether one individual accepts or rejects tissues from another depends on the genotypes of the donor and host at each of several *histocompatibility*, *H*, loci. Each *histocompatibility* antigen is determined by a single, autosomal, codominant allele at an *H* locus. A recipient will accept a graft only if he carries all *H* genes present in the donor.

a For the cases in the table indicate whether the recipient will accept or reject skin from the donor.

Table 1310

Donor	Recipient
Inbred strain A	Inbred strain B
Inbred strain A	F_1 hybrid from inbred A × inbred B
F_1 hybrid	Either inbred strain A or B
F_2 hybrid	F_1 hybrid
Either inbred A or B or F_1 hybrid	F_2 hybrids

b Show why acceptance of grafts from either inbred parent strain by F_1's rules out the possibility that *H* antigens are determined by recessive genes.

c It is possible to establish a series of inbred lines each differing from the common inbred progenitor by a different single *H* allele at one *H* locus only. Otherwise these congenic lines are genetically identical. Results with four such lines are as follows:

Line 1 × line 2	F_1 rejects graft from common inbred parent
Line 1 × line 3	F_1 accepts graft from common inbred parent
Line 4 × line 3	F_1 accepts graft from common inbred parent
Line 1 or 2 × line 4	F_1 rejects graft from common inbred parent

How many *H* loci and how many alleles at each are involved in graft acceptance and rejection in these lines? Explain.

1311 Giles et al. (1957) studied 21 *adenine-requiring* mutants of independent origin in *Neurospora crassa*. All the *mutants* were found to be blocked in the terminal step in adenine biosynthesis, involving the splitting of adenosine monophosphate succinate (AMPS) to adenosine monophosphate (AMP), and to lack or have impaired activity for the AMPS-splitting enzyme, adenylosuccinase. Crosses of *mutants* in all possible combinations with rare exceptions yielded only auxotrophic progeny. *Wild type* × any *mutant* produce *wild types* and *mutants* in a 1:1 ratio.

a What is the apparent genetic basis for enzyme control?

b How many kinds of polypeptide chains does the enzyme possess?   Explain.

1312   In 1920 Sturtevant reported the occurrence of a bilateral gynandromorph in *Drosophila melanogaster* that developed from a zygote with the following sex-chromosome genotype:

$$\frac{sc\ W\ ec\ Rb\ ct\ v\ g\ f}{Sc\ w^e\ Ec\ rb\ \ Ct\ V\ G\ f}$$

The recessive alleles at the different loci were *sc* (*scute* bristles), w^e (*eosin* eyes), *ec* (*echinus* eyes), *rb* (*ruby* eyes), *ct* (*cut wings*), *v* (*vermilion* eyes), *g* (*garnet* eyes), and *f* (*forked* bristles).   The male side of the gynandromorph showed the recessive traits *scute*, *echinus*, *cut*, *garnet*, and *forked*.   The female side showed the *forked*, *f*, trait only.

a Explain (1) how the gynandromorph originated and (2) why the *vermilion* allele did not express its phenotype on the male side of the fly.

b State whether the results with *vermilion* were to be expected on the basis of Beadle and Ephrussi's results?   (See Prob. 1594.)

1313   The alleles C, c^k, c^d, c^r, c^a in the *albino* series of the *pink-eyed sepia* breed of guinea pig have the effects shown on the relative amounts of melanin pigmentation in the hairs (Wright, 1949, 1959).   Classify each allele

Table 1313

Genotype	Percent melanin pigment	Phenotype
C-	100	*Full*
$c^k c^k$	88	
$c^k c^c$	65	
$c^k c^r$	54	
$c^k c^a$	36	
$c^d c^d$	31	*Intermediate*
$c^d c^r$	19	
$c^d c^a$	14	
$c^r c^r$	12	
$c^r c^a$	3	
$c^a c^a$	0	*White*

as dominant, hypomorphic, or amorphic and give a biochemical explanation for the different effects of these alleles.

1314   Hsia et al. (1956) demonstrated that the levels of phenylalanine in the blood plasma at 1-, 2-, and 4-hour intervals after feeding a standard dose of phenylalanine were about twice as high in normal parents of *phenyl-ketonurics* as in persons with no history of *phenylketonuria* in their pedigrees. What do you conclude from these findings?

1315 Gross (1958), and Gross and Fein (1960) found that all *arom-1* mutants of *Neurospora crassa* require the addition of a mixture of *p*-aminobenzoic acid, phenylalanine, tyrosine, and tryptophan minimal medium for growth.   These mutants possess two enzymes, dehydroshikimic acid dehydrase and proto-catechuic acid oxidase (which act on dehydroshikimic acid and protocatechuic acid, respectively), not detected in extracts from *wild types*.   Moreover, the mutants lack the enzyme dehydroshikimic acid reductase, which is normally present in *wild type*, and shikimic acid, found in some other mutants.   Do the *arom-1* mutations appear to be pleiotropic in their effects?   Explain. What phenotypes and ratios are expected in crosses between *wild types* and *arom-1* mutants?

1316 Human adult hemoglobin is a tetrameric protein, made up of two dimers, each consisting of an identical pair of polypeptide chains.   One dimer contains two α chains; the other, two β chains.   *Normal* hemoglobin is symbolized as $\alpha_2{}^A \beta_2{}^A$ whether the individual tested is heterozygous or homozygous for hemoglobin genes; viz. in individuals with *sickle-cell* trait (*Hb-S*) molecules are $\alpha_2{}^A \beta_2{}^A$ and $\alpha_2{}^A \beta_2{}^S$; molecules with the polypeptide constitution $\alpha_2{}^A \beta^S \beta^A$ are never found (Itano and Robinson, 1959). Subscript 2 refers to the number of chains present.

The following notation is used for phenotypic descriptions:

A = normal α and β chains only

G = mutant α chain (α^G; asparagine at position 68 replaced by lysine), normal β

C = normal α chain, mutant β chain (β^C; glutamic acid at position 6 replaced by lysine)

X = mutant α chain; mutant β chain ($\alpha^G \beta^C$)

The symbols for phenotypes represent types of hemoglobin present (see Table 1316).

Table 1316

Phenotype symbol	Hemoglobins present
A	$Hb\text{-}A \ (\alpha_2{}^A \beta_2{}^A)$
AC	$Hb\text{-}A \ (\alpha_2{}^A \beta_2{}^A) + Hb\text{-}C \ (\alpha_2{}^A \beta_2{}^C)$
AG	$Hb\text{-}A \ (\alpha_2{}^A \beta_2{}^A) + Hb\text{-}G \ (\alpha_2{}^G \beta_2{}^A)$
$ACGX$	$Hb\text{-}A + Hb\text{-}C + Hb\text{-}G + Hb\text{-}X \ (\alpha_2{}^G \beta_2{}^C)$

The pedigree shows the distribution of mutant hemoglobins Atwater et al. (1960) found in a Negro family.   Analyze the pedigree to show the number of genes involved and whether there is any evidence of linkage.

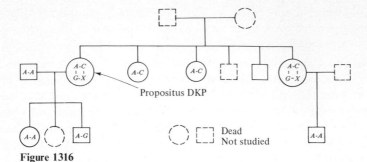

Propositus DKP

Dead
Not studied

Figure 1316

1317 Baglioni and Ingram (1961) showed that individuals who synthesize all four of the chains α^A, α^G, β^A, and β^C produced four types of adult hemoglobin in approximately equal numbers:

Hemoglobin $A = \alpha_2{}^A \beta_2{}^A$ Hemoglobin $C = \alpha_2{}^A \beta_2{}^C$
Hemoglobin $G = \alpha_2{}^G \beta_2{}^A$ Hemoglobin $X = \alpha_2{}^G \beta_2{}^C$

Note: No hybrid molecules, e.g., $\alpha_2{}^A \beta^A \beta^C$, were formed. Show, with illustrations where possible, what conclusions can be drawn from these data regarding the allelic relationships, chain synthesis, dimerization, and assembly of dimers into hemoglobin molecules.

1318 Hemoglobins S (*Hb-S*) and C (*Hb-C*) in man are specified by alleles of one gene (Hunt and Ingram, 1958). *Hb-S* and *Hb-C* each differ from *normal*, *Hb-A*, hemoglobin by one amino acid substitution in the β chain at position 6. Glutamic acid in the *Hb-A* β chain is replaced by valine and lysine in *Hb-S* and *Hb-C*, respectively. Hemoglobin *Hopkins-2* (*HO-2*) differs from *normal* hemoglobin by an amino acid substitution in the α chain. Lysine replaces leucine at position 65. Certain individuals with both *HO-2* and *Hb-S* chains are known who have one parent like themselves and the other *normal*. The sibs of these *HO-2*, *Hb-S* individuals may be of four types for the hemoglobins they produce (Smith and Torbert, 1958):

No Hb-S, no HO-2
No Hb-S, HO-2
Hb-S, no HO-2
Hb-S, HO-2

Explain, stating the bearing these results have on the one-gene–one-enzyme hypothesis.

1319 The enzyme tryptophan synthetase (TSase) catalyzes the following three reactions (Crawford and Yanofsky, 1958; Yanofsky and Rachmeler, 1958):

Indole + L-serine → L-tryptophan   (1)

Indoleglycerol phosphate ⇌ indole + triose phosphate   (2)

Indoleglycerol phosphate + L-serine → L-tryptophan + triose phosphate   (3)

Reaction (3) is the physiologically important reaction (viz. it carries out the major tryptophan synthesis in the organism). TSase in *Escherichia coli* consists of two different polypeptide chains, A and B. The normal B subunit can catalyze reaction 1 even in the absence of subunit A but does so very inefficiently. The normal A protein alone can catalyze reaction (2). Both subunits are required for the catalysis of any of the reactions at maximal rates. Reaction (3) occurs only in the presence of the AB complex (Crawford and Yanofsky, 1958).

In a study of numerous mutations affecting the formation of TSase in *E. coli*, Yanofsky and Crawford (1959) observed the following:

1 In certain mutants which do not revert and do not recombine with two or more other tryptophan synthetase mutants none of the three reactions occurs.

2 Some mutations affect the A protein only. The mutant A protein in these strains (in the presence of normal B protein) is effective in reaction (1) only (at the *wild-type* rate).

3 Other mutations affect the B protein only. These mutant B proteins (in the presence of normal A protein) catalyze reaction (2) only (at a *wild-type* rate).

4 The mutants producing defective A protein accumulate indoleglycerol phosphate. The mutants producing defective A proteins and defective B proteins revert naturally to tryptophan independence.

Interpret these results with respect to genetic control of TSase synthesis and activity.

1320 The molecular weights of the α and β chains of *Escherichia coli* tryptophan synthetase are 29,500 and 49,500. That of the entire enzyme is 159,000.
  a If the average molecular weight of each amino acid is 110, approximately how many amino acids does each chain contain?
  b How many chains does the enzyme contain? Explain.

1321 In *Escherichia coli* the two very closely linked nonsense mutants 5972 and 9778 affecting tryptophan synthesis have the following effects (Ito and Yanofsky, 1966):

1 Mutant 5972 cannot convert chorismic acid to anthranilic acid due to lack of anthranilate synthetase activity but can carry out all other reactions.

2 Mutant 9778 lacks not only anthranilate synthetase activity but also phosphoribosyl anthranilate transferase activity required for conversion of anthranilic acid to *N*-(5'-phosphoribosyl anthranilate (PRA).

3 When extracts from the two mutants are mixed, anthranilate synthetase activity is restored.

Did the two nonsense mutations occur in the same or different genes, and what is the nature (number of kinds of polypeptides) of each of the enzymes? Explain.

1322 a The genes, polypeptides, and enzymes involved in the five sequential enzymatic steps in the tryptophan pathway in *Escherichia coli* are as follows (after Ito and Yanofsky, 1966; Goldberg et al., 1966):

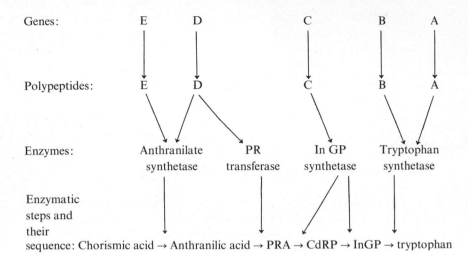

Genes: E   D   C   B   A

Polypeptides: E   D   C   B   A

Enzymes: Anthranilate   PR   In GP   Tryptophan
            synthetase   transferase synthetase synthetase

Enzymatic
steps and
their
sequence: Chorismic acid → Anthranilic acid → PRA → CdRP → InGP → tryptophan

Table 1322

Complementation group	Growth response on minimal medium		Accumulation	Other facts
	Anthranilic acid	Indole		
A	+	+	None	Lack anthranilate synthetase; no interallelic complementation
B	−	−	Indole glycerol; some anthranilic acid	Deficient in tryptophan synthetase; no inter-allelic complementation
C Few	+	+	None	Defective only in anthranilate synthetase
Most	−	+	Anthranilic acid	Do not complement anthranilic acid–utilizing mutants; lack anthranilate synthetase, PRA isomerase, and InGP synthetase activity
D	−	+	Anthranilic acid	Lack PR activity; no interallelic comple-mentation
E	+	+	None	Do not lack any of the enzymatic activities tested

Nonsense mutants a, b, c, d, e, are combined in the following merozygotic combinations:

c and d	d and d	a and b	e and d
c and b	d and a	b and b	e and c

Which of these merozygotes should show complementation (growth in minimal medium) and why?

b In *Aspergillus nidulans* the sequential metabolic pathway is the same as in *E. coli*. The manner of gene control of the enzymes is, however, somewhat different, as revealed by the mutant studies summarized in the table (data from Hutter and DeMoss, 1967; Roberts, 1967). Using a hypothetical set of polypeptides for these genes, show how the polypeptides participate in enzyme assembly and function in *A. nidulans*.

1323 a Lactose dehydrogenase (LDH) contains two types of polypeptide (or subunit) chains, A and B. It exists in five active forms, or *isozymes*, LDH-1 to LDH-5. LDH-1 and LDH-5 are pure isozymes, each with one kind of chain only, A and B, respectively. When their chains are mixed in equal proportions and are allowed to associate and reassemble spontaneously, the isozymes 1 to 5 appear in the respective proportions 1:4:6:4:1 (Markert, 1963). How many subunits does each isozyme contain and why? Using symbols A and B, describe each of the isozymes.

b The zone-electrophoresis LDH-isozymic patterns (= phenotype) on cylindrical starch gel from brain extracts of *Peromyscus maniculatus* individuals that are *homozygous normal*, *homozygous mutant*, and *heterozygous* are illustrated. Matings among the three types produced the results tabulated.

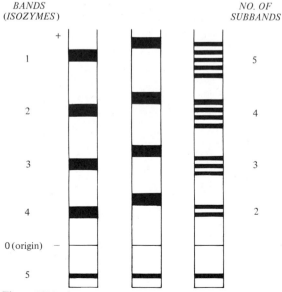

Figure 1323 (*From Shaw and Barto, 1963.*)

Table 1323

Parental phenotype	Offspring phenotype					
	Homozygous normal		Heterozygous		Homozygous mutant	
	♂	♀	♂	♀	♂	♀
homozygous normal × homozygous normal	13	11	0	0	0	0
homozygous normal × heterozygote	12	15	11	16	0	0
heterozygote × heterozygote	2	1	0	5	2	2

(1) How many allele pairs are responsible for the phenotypic differences?

(2) The A and B chains are specified by genes A and B, respectively. Of polypeptides A and B, do you expect one or both to exist in alternative forms and why?

(3) Bands 1 to 4 share common polypeptide (B) not present in band 5, which contains polypeptide A only. If only one polypeptide exists in alternative forms, which is it and why?

(4) Give the genotypes of the three types of individuals and describe (or give) the tetrameric-polypeptide constitution of each band and subband in *homozygous normal*, *homozygous mutant*, and *heterozygous* individuals. Give the relative proportions of the subunits in each isozyme in the heterozygotes.

1324 The mouse, like all mammals, produces glucose-6-phosphate dehydrogenase (G6PD) in isozyme forms. A male from a true-breeding strain with the *fast* form is crossed with a female from a true-breeding line with the *slow* form of the enzyme. The isozyme phenotypes of F_1 progeny are shown in the following zymogram:

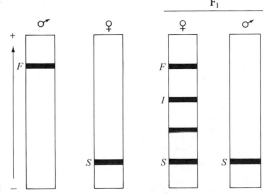

Figure 1324 F = fast band; S = slow band; I = intermediate band.

Is G6PD a monomer or a multimer? If the latter, specify the number of polypeptides per molecule. Explain your reasoning.

1325 Schwartz (1960) found a different form of a certain esterase in each of three true-breeding lines of *Zea mays*. Line *a* was characterized by a *slow* (S) migration rate to the negative pole at pH 8.6 during electrophoresis; lines *b* and *c* possessed *normal* (N) and *fast* (F) migration rates, respectively. Mixtures of enzymes from different lines showed two components each (d, e, f in table). F₁ hybrids from crosses between N and F lines and N and S lines form three components each. One of these in each F₁ is intermediate between the parental component in mobility. Zymograms (starch gels stained to reveal location of the various isozymes) of the various esterase types are illustrated. The results of similar tests on F₂'s are tabulated.

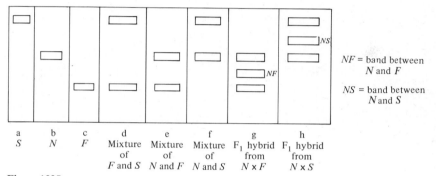

NF = band between N and F

NS = band between N and S

a	b	c	d	e	f	g	h
S	N	F	Mixture of F and S	Mixture of N and F	Mixture of N and S	F₁ hybrid from N × F	F₁ hybrid from N × S

Figure 1325

Table 1325

Cross	In F₁	F₂					
		Type	No.	Type	No.	Type	No.
N × F	N, NF, F	F only	141	F, NF, N	292	N only	146
N × S	N, NS, S	N only	21	N, NS, S	41	S only	20
F × S	F, FS,* S	F only	139	F, NS	314	S only	141

* FS band is in normal position.

Describe the mode of genetic control of this enzyme and state whether you consider the enzyme to be monomeric, dimeric, or trimeric, and why.

1326 By means of horizontal gel electrophoresis and the zymogram technique Scandalios (1967) studied the genetic control of alcohol dehydrogenase (ADH) isozymes (multiple molecular forms of an enzyme) in corn (*Zea mays*). Each homozygous strain studied showed two types of ADH isozyme zones:

one fast and faint, the other slower and deep-staining. Genetically different homozygous strains differed in their electrophoretic mobility in each of the two zones. One of the parents always had a faster-moving isozyme (represented by a band in the zymogram) than the other in each of the two zones. Hybrids from crosses between such strains showed only the parental isozymes in the fast-moving zone and both parental and an intermediate isozyme in the slow zone, as shown in the drawing. Crosses between strains

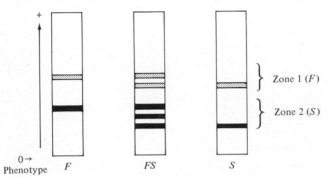

Figure 1326 Electrophoretic patterns obtained from extracts of single kernels. Parental types are shown to the left and right of the F_1 hybrid pattern. The arrow indicates the direction of migration at pH 8.5.

Table 1326

Parents		Progeny phenotypes (isozymes)					
		Zone 1			Zone 2		
♀	♂	F	FS	S	F	F + H* + S	S
F	F	80	0	0	80	0	0
S	S	0	0	80	0	0	80
S	F	0	90	0	0	90	0
F	FS	42	48	0	42	48	0
S	FS	0	42	38	0	42	38
FS	FS	22	49	19	22	49	19

Note: Reciprocals of third, fourth, and fifth crosses gave basically identical results.
* H = intermediate band.

showing *fast* (F) and *slow* (S) bands (isozymes) in each of the zones gave the results shown in the table.

a Are each of the two ADH zones of activity specified by one or two genes? Explain.

b Which of these ADH enzymes is a monomer and which a dimer and why?

1327 **a** What is somatic-cell genetics? Of what potential genetical importance are man-mouse somatic-cell hybrids? Explain.

b DNA synthesis from sugars and amino acids is blocked by aminopterin. An alternative synthetic pathway depends on previously formed nucleosides and the enzymes thymidine kinase (TdK) and hypoxanthine guanine phosphoribosyl transferase (HGPRT). Only cells with both enzymes can grow in Littlefield's (HAT) medium, containing aminopterin and nucleosides. A certain diploid strain of mouse cells and one of human cells lack TdK and HGPRT, respectively. From mixed cultures, hybrid cells containing the diploid mouse complement plus one human (group E) chromosome were isolated by Migeon and Miller (1968). All clones of these hybrid cells [40 mouse chromosomes + 1 human (group E) chromosome] grew on Littlefield's medium, which selects for TdK. When the clones were removed from the selective medium and transferred to a medium containing 5-bromodeoxyuridine, which kills cells containing TdK, none of the surviving cells contained this E-group chromosome. What conclusions can be drawn from these studies regarding the genetics of thymidine kinase in man, and why? See Ruddle et al. (1971) and Silvana Santachiara et al, (1970).

c The structural-gene loci specifying human glucose-6-phosphate dehydrogenase (G6PD) and HGPRT are X-linked. Miller et al. (1971) derived six interspecific somatic hybrid cell lines by fusing mouse cells able to synthesize G6PD but deficient in ability to produce HGPRT with human diploid cells with the *wild-type* (*normal*) genotype at both the G6PD and HGPRT loci. Cell fusion was mediated using ultraviolet-inactivated Sendai virus. Hybrid cells and their clones were grown on HAT medium. Human HGPRT and mouse G6PD were present in all six hybrids and the 105 clones derived from them. In 2 of the 6 hybrids and in 47 of the 105 clones from the other four hybrid lines, human G6PD was absent. Note the following points:

1 Reversion to HGPRT synthesis was not observed in control fusions within the mouse parental lines.

2 HGPRT produced by all mouse-human hybrid cells, with or without human G6PD, was always of the human type.

3 Inactivated Sendai virus can produce chromosome breaks.

4 After cloning, cells did not undergo phenotypic changes.

Which of the following explanations is the most plausible to account for the absence of human G6PD in 47 out of 105 clones, and why?

(1) Use of human cell donors with a slightly leaky or nonleaky mutant allele at the G6PD locus.

(2) Somatic mutation of *wild-type* to *mutant* allele at the G6PD locus.

(3) Repression of synthesis of human G6PD or inhibition of activity of the enzyme.

(4) Back mutation or virus-dependent induction at mouse HGPRT locus, resulting in synthesis of HGPRT indistinguishable from human HGPRT.

(5) Loss of the G6PD structural locus.

d In somatic-cell hybrids between human and mouse cells, which initially contain a complete diploid complement of chromosomes from both species, human chromosomes are preferentially lost more or less at random and clones can be obtained with different numbers and combinations of human chromosomes.   By determining positive correlations between human genes and chromosomes in hybrid cells linkage relationships can be established. Shows (1972) produced somatic-cell hybrids between human W1-38 diploid fibroblasts and LTP, a diploid mouse line deficient for thymidine kinase (TdK) and hypoxanthine-guanine phosphoribosyl transferase (HGPRT), both determined by X-linked genes.   The phenotypes of 10 human enzymes or enzyme subunits (polypeptides), each specified by one structural gene, in 13 independently formed hybrid clones is shown in the following table.

Table 1327

Clones	LDH-A	Es-A$_4$	G6PD	MDH	PGM$_1$	LDH-B	Pep-B	AK	Pep-A	Pep-C
WIL-1	−	−	+	−	−	+	+	−	+	−
WIL-2	−	−	+	−	−	+	+	−	−	−
WIL-3	+	+	+	−	+	−	−	−	+	+
WIL-5	−	−	+	−	−	−	−	−	+	−
WIL-6	+	+	+	+	−	−	−	−	−	−
WIL-7	+	+	+	+	−	−	−	−	+	−
WIL-8	+	+	+	+	+	+	+	−	+	+
WIL-9	+	+	+	+	+	+	+	−	+	+
WIL-10	+	+	+	+	−	+	+	−	+	+
WIL-12	+	+	+	+	−	−	−	−	+	−
WIL-14	−	−	+	−	+	−	−	−	−	−
WIL-15	+	+	+	−	−	+	+	−	+	−
WIL-19	+	+	+	+	−	−	−	−	+	−

Legend: (+)—enzyme activity; (−)—no enzyme activity: LDH—lactate dehydrogenase, A—one polypeptide, B—a different polypeptide; Est-A$_4$—esterase-A; Pep-A – peptidase A; Pep-B—peptidase-B; Pep-C—peptidase C; G6PD-glucose-6-phosphate dehydrogenase (X-linked); MDH—malate dehydrogenase; AK—adenylate kinase; PGM$_1$—phosphoglucomutase.

(1) Which human chromosome is common to all 13 clones?   Explain.

(2) Which genes show linkage?   Explain.

(3) How might you proceed to determine which chromosome or chromosomes carry the linkage group or groups detected?   Indicate the results expected if linkage exists and when it does not.   See Boone et al., 1972.

1328   It is possible to determine whether a mutation which has resulted in loss of enzyme activity has also resulted in the loss of ability to synthesize the enzymes.   This can be done by injecting animals (e.g., rabbits) with the enzyme from the *wild-type* strain to obtain antibodies against the enzyme.

Extracts of the mutant strains are then tested for reaction with these antibodies. Reaction indicates the presence of a protein immunologically identical to the *wild-type* enzyme. Such proteins are called CRM (cross-reacting material). Strains synthesizing CRM are termed CRM$^+$, and those not able to do so are called CRM$^-$ (Suskind, 1957; Suskind and Jordan, 1959). Mutations at structural-gene loci can result in the formation of CRM$^+$ enzymes which are less effective in catalyzing the reaction controlled by the *wild-type* enzyme. Moreover, quantitative differences occur among the enzymes specified by these mutant alleles with respect to this function. Other mutant alleles fail to produce CRM.

a Does failure to produce CRM indicate that no enzyme is formed? Explain.

b Would missense and/or nonsense changes be expected to produce each of these kinds of enzymes or no enzymes? Explain and illustrate.

1329 At least four separate X-linked mutations cause a deficiency of the enzyme glucose-6-phosphate dehydrogenase (G6PD), which plays a role in cellular respiration. Persons with the enzyme deficiency are *normal* except when they inhale pollen of the broad bean (*Vicia faba*), eat the raw bean itself (favism), or ingest certain drugs like naphthalene, sulfanilamide, and primaquine. What is the most likely genetic relationship of these mutations? How would you determine whether your explanation is correct?

1330 Yanofsky and Bonner (1955) studied 25 allelic, *td*, mutants unable to synthesize tryptophan which were deficient for or possessed a defective tryptophan synthetase. Many of the mutants reverted to *pseudo wild type* due to mutations at suppressor loci. Four suppressors were tested for their suppressive activities against various mutants. *Su 2* suppressed the effect of *td 2*; *Su 3* and *Su 24* suppressed both allele *td 3* and allele *td 24*; and *Su 6* suppressed alleles *td 2* and *td 6*. *Su 3* and *Su 24* are allelic with each other; the other two suppressors are unlinked to each other and *Su 3* and *Su 24*. None of the suppressors was capable of suppressing all *td* alleles. The suppressed mutants form less enzyme than *wild type*. Do the suppressors take over the function of the gene they suppress? Explain. If not, how do they appear to function?

1331 In man, glucose-6-phosphate dehydrogenase (G6PD) deficiency and hypoxanthine guanine phosphoribosyl transferase (HGPRT) deficiency are due to different X-linked recessive genes. Diploid ($2n = 46$) fibroblasts from a G6PD-deficiency male were hybridized by Siniscalco et al. (1969) with fibroblasts from a HGPRT-deficiency male. The $4n = 96$ mononucleate hybrid cells synthesized both G6PD and HGPRT. No sex chromatin was observed in these cells. It is not known whether a single cell produces both enzymes simultaneously.

a Account for the fact that both enzymes were synthesized by the hybrid cells.

b Do you think both enzymes are produced in the same cell (are both X's functional simultaneously)? See Silagi et al. (1969).

REFERENCES

Allison, A. C. (1959), *Am. Nat.*, **93**:5.

Anfinsen, C. B. (1963), "The Molecular Basis of Evolution," Wiley, New York.

Appella, E., and C. L. Markert (1961), *Biochem. Biophys. Res. Commun.*, **6**:171.

Armstrong, F-G. and H. Ishiwa (1971), *Genetics*, **67**:171.

Atwater, J. I., R. Schwartz, and L. M. Tocantins (1960), *Blood* (*J. Hematol.*), **15**:901.

Baglioni, C. (1963), in J. H. Taylor (ed.), "Molecular Genetics," pt. I, pp. 405–475, Academic, New York.

—— and V. M. Ingram (1961), *Nature*, **189**:465.

Bakay, B., and W. L. Nyhan (1971), *Biochem. Genet.*, **5**:81.

Bateson, W., E. R. Saunders, and R. C. Punnett (1905), *Rep. Evol. Comm. R. Soc.*, **2**:1–55, 80–99.

Beadle, G. W. (1937), *Genetics*, **22**:587.

—— (1945), *Physiol. Rev.*, **25**:643.

——, C. W. Clancy, and B. Ephrussi (1937), *Proc. R. Soc. (Lond.)*, **B112**:98.

—— and V. L. Coonradt (1944), *Genetics*, **29**:291.

—— and B. Ephrussi (1936), *Genetics*, **21**:225.

—— and —— (1937), *Genetics*, **22**:76.

—— and E. L. Tatum (1941), *Proc. Natl. Acad. Sci.*, **27**:499.

—— and —— (1945), *Am. J. Bot.*, **32**:678.

——, G. M. Robinson, R. Robinson, and R. Scott-Moncrieff (1939), *J. Genet.*, **37**:375.

Beet, E. A. (1949), *Ann. Eugen.*, **14**:279.

Benjamin, C. P. (1970), *Dev. Biol.*, **23**:62.

Bernhard, S. A. (1968), "The Structure and Function of Enzymes," Benjamin, New York.

Billingham, R. E., G. H. Sawchuk, and W. K. Silvers (1960), *Proc. Natl. Acad. Sci.*, **46**:1079.

—— and W. K. Silvers (1962), "Transplantation of Tissues and Cells," Wistar Institute, Philadelphia.

Bonner, D., E. L. Tatum, and G. W. Beadle (1943–1944), *Arch. Biochem.*, **3–4**:71.

Boone, C., T-R. Chen, and F. H. Ruddle (1972), *Proc. Natl. Acad. Sci.*, **69**:510.

Braunitzer, G., et al. (1961), *Z. Physiol. Chem.*, **325**:283.

Brewer, G. J., C. F. Sing, and E. R. Sears (1969), *Proc. Natl. Acad. Sci.*, **64**:1224.

Bryson, V., and A. J. Votel (eds.) (1965), "Evolving Genes and Proteins," Academic, New York.

Burnet, F. M. (1961), Immunological Recognition of Self, *Science*, **133**:307.

Butenandt, A., and G. Neubert (1955), *Z. Physiol. Chem. Hoppe-Selye*, **301**:109.

——, W. Weidel, and E. Becker (1940), *Naturwiss.*, **28**:63.

Carlton, B. C., J. R. Guest, and C. Yanofsky (1967), *J. Biol. Chem.*, **242**:5422.

Crawford, I. P., and C. Yanofsky (1958), *Proc. Natl. Acad. Sci.*, **44**:1161.

—— and —— (1959), *Proc. Natl. Acad. Sci.*, **45**:1280.

Dancis, J., et al. (1969), *Biochem. Genet.*, **3**:609.

Davis, R. H., and W. M. Thwaites (1963), *Genetics*, **48**:1551.

Dayton, T. O. (1956), *J. Genet.*, **56**:249.

Demerec, M., and Z. Hartman (1956), *Carnegie Inst. Wash., D.C., Publ.* 612, p. 5.

Dickerson, R. E., and I. Geis (1969), "The Structure and Action of Proteins," Harper & Row, New York.

Drapeau, G. R., and C. Yanofsky (1967*a*), *J. Biol. Chem.*, **242**:5413.

—— and —— (1967*b*), *J. Biol. Chem.*, **242**:5434.

Efron, Y. (1971), *Mol. Gen. Genet.*, **111**:97.

Ephrussi, B. (1942), *Q. Rev. Biol.*, **17**:327.

—— and M. C. Weiss (1968), *Sci. Am.*, **220** (April):26.

Fellous, M., and J. Dausset (1970), *Nature*, **225**:191.

Fincham, J. R. S., and P. R. Day (1965), "Fungal Genetics," Blackwell, Oxford.

Fischer, G. A. (1957), *Biochim. Biophys. Acta*, **25**:50.

Folling, A. (1934), *Z. Physiol. Chim.*, **227**:169.

Garrod, A. E. (1902), *Lancet*, **2**:1616.

―――― (1923), "Inborn Errors of Metabolism," Oxford University Press, London.

Giblett, E. R. (1969), "Genetic Markers in Human Blood," Blackwell, Oxford.

Giles, N. H., C. W. H. Partridge, and N. J. Nelson (1957), *Proc. Natl. Acad. Sci.*, **43**:305.

Glassman, E. (1956), *Genetics*, **41**:566.

Goldberg, M. E., T. E. Creighton, R. L. Baldwin, and C. Yanofsky (1966), *J. Mol. Biol.*, **21**:71.

Green, M. M. (1952), *Proc. Natl. Acad. Sci.*, **38**:300.

Gross, S. R. (1958), *J. Biol. Chem.*, **233**:1146.

―――― and A. Fein (1960), *Genetics*, **45**:885.

Grzeschik, K. H., P. W. Allderdice, A. Grzeschik, J. M. Opitz, O. J. Miller, and M. Siniscalco (1972), *Proc. Natl. Acad. Sci.*, **69**:69.

Guest, J. R., and C. Yanofsky (1966), *J. Biol. Chem.*, **241**:1.

―――――, G. R. Drapeau, B. C. Carlton, and C. Yanofsky (1967), *J. Biol. Biochem.*, **242**:5442.

Guidotti, G., R. J. Hill, and W. Konigsberg (1962), *J. Biol. Chem.*, **237**:2184.

Hadorn, E., and H. K. Mitchell (1951), *Proc. Natl. Acad. Sci.*, **37**:650.

Harris, H. (1966), "Human Biochemical Genetics," 2d ed., Cambridge University Press, Cambridge.

Harris, H. (1971), "The Principles of Human Biochemical Genetics," American Elsevier, New York.

Harris, J. I., F. Sanger, and M. A. Naughton (1956), *Arch. Biochem.*, **65**:427.

Hart, G. E. (1970), *Proc. Natl. Acad. Sci.*, **66**:1136.

Hartman, P. E., and S. R. Suskind (1969), "Gene Action," 2d ed., Prentice-Hall, Englewood Cliffs, N.J.

Hayes, W. (1968), "The Genetics of Bacteria and Their Viruses," 2d ed., Blackwell, Oxford.

Helinski, D. R., and C. Yanofsky (1966), in H. Neurath (ed.), "The Proteins," vol. IV, pp. 1–93, Academic, New York.

Hildemann, W. H. (1970), "Immunogenetics," Holden-Day, San Francisco.

Hill, R. J., and W. Konigsberg (1962), *J. Biol. Chem.*, **237**:3151.

Horowitz, N. H. (1947), *J. Biol. Chem.*, **171**:255.

―――――, D. Bonner, and M. B. Houlahan (1945), *J. Biol. Chem.*, **159**:145.

Hsia, D. Y. (1959), "Inborn Errors of Metabolism," Year Book, Chicago.

―――――, K. W. Driscoll, W. Troll, and W. E. Knox (1956), *Nature*, **178**:1239.

Hubby, J. L., and L. H. Throckmorton (1960), *Proc. Natl. Acad. Sci.*, **46**:65.

Hunt, J. A., and V. M. Ingram (1958), *Nature*, **181**:1062.

Hutter, R., and J. A. DeMoss (1967), *Genetics*, **55**:241.

Ingram, V. M. (1957), *Nature*, **180**:326.

―――― (1963), "The Hemoglobins in Genetics and Evolution," Columbia, New York.

―――― (1966), "The Biosynthesis of Macromolecules," Benjamin, New York.

Inouye, M., and A. Tsugita (1966), *J. Mol. Biol.*, **22**:193.

Itano, H. A., and E. A. Robinson (1959), *Nature*, **183**:1799.

―――― and ―――― (1960), *Proc. Natl. Acad. Sci.*, **46**:1492.

Ito, J., and C. Yanofsky (1966), *J. Biol. Chem.*, **241**:4112.

Jervis, G. A. (1938), *J. Biol. Chem.*, **126**:305.

Johnson, F. M. (1971), *Genetics*, **68**:77.

Jorgensen, E. C., and T. A. Geissman (1955), *Arch. Biochem. Biophys.*, **55**:389.

Kao, F-T., and T. T. Puck (1968), *Proc. Natl. Acad. Sci.*, **60**:1275.

―――― and ―――― (1970), *Nature*, **228**:329.

Kartha, G., J. Bello, D. Harker (1967), *Nature*, **213**:862.

Kendrew, J. C. (1961), *Sci. Am.*, **205**(December):96.

―――― (1963), *Science*, **193**:1259.

―――― et al. (1960), *Nature*, **185**:422.

Kiritani, T. M., and Y. Ikeda (1965), *Genetics*, **51**:341.
Klebe, R. J., T. R. Chen, and F. H. Ruddle (1970), *Proc. Natl. Acad. Sci.*, **66**:1220.
Knudson, A. G. (1969), *Annu. Rev. Genet.*, **3**:1.
Konigsberg, W., and R. J. Hill (1962), *J. Biol. Chem.*, **237**:2547.
Kusano, T., C. Long, and H. Green (1971), *Proc. Natl. Acad. Sci.*, **68**:82.
Lawrence, W. J. C. (1957), *Heredity*, **11**:337.
———— and J. R. Price (1940), *Biol. Rev.*, **15**:35.
Lein, J., H. K. Mitchell, and M. B. Houlahan (1948), *Proc. Natl. Acad. Sci.*, **34**:435.
Loper, J. C., M. Grabnar, R. C. Stahl, Z. Hartman, and P. E. Hartman (1964), *Brookhaven Symp. Biol.*, **17**:15.
MacIntyre, R. J. (1971), *Genetics*, **68**:483.
Mahler, H. R., and E. H. Cordes (1966), "Biological Chemistry," Harper & Row, New York.
Mangelsdorf, P. C., and G. S. Fraps (1931), *Science*, **73**:241.
Markert, C. L. (1963), *Science*, **140**:1329.
———— and F. Møller (1959), *Proc. Natl. Acad. Sci.*, **45**:753.
Medawar, P. B. (1961), *Science*, **133**:303.
Migeon, B. R., and C. G. Miller (1968), *Science*, **162**:1005.
————, S. W. Smith, and C. L. Leddy (1969), *Biochem. Genet.*, **3**:583.
Miller, D. J., et al. (1971), *Proc. Natl. Acad. Sci.*, **68**:116.
Minna, J., et al. (1971), *Proc. Natl. Acad. Sci.*, **68**:234.
Mitchell, H. K., and J. Lein (1948), *J. Biol. Chem.*, **175**:481.
Miyamoto, M., and T. B. Fitzpatrick (1957), *Nature*, **179**:199.
Murayama, M. (1966), *Science*, **153**:145.
Nabholz, M., V. Miggiano, and W. Bodmer (1969), *Nature*, **223**:358.
Neel, J. V. (1949), *Science*, **110**:64.
Neufeld, E. F., and J. C. Fratantoni (1970), *Science*, **169**:141.
Newmeyer, D. (1962), *J. Gen. Microbiol.*, **28**:215.
Nolte, D. J. (1959), *Heredity*, **13**:233.
Pauling, L., H. A. Itano, S. J. Singer, and I. C. Wells (1949), *Science*, **110**:543.
Penrose, L. S. (1935), *Lancet*, **229**:192.
Perutz, M. F. (1962), "Protein and Nucleic Acids: Structure and Function," American Elsevier, New York.
———— (1963), *Science*, **140**:863.
———— (1964), *Sci. Am.*, **211**(November):64.
Phillips, D. C. (1966), *Sci. Am.*, **215**(November):78.
Pontecorvo, G. (1971), *Nature*, **230**:367.
Race, R. R., and R. Sanger (1962), "Blood Groups in Men," 4th ed., Blackwell, Oxford.
Radhakrishnan, A. N., R. P. Wagner, and E. E. Snell (1960), *J. Biol. Chem.*, **235**:2322.
Roberts, C. F. (1967), *Genetics*, **55**:233.
Ruddle, F. H., V. M. Chapman, T. R. Chen, and R. J. Klebe (1971), *Nature*, **227**:251.
Sanderson, A. R., P. Cresswell, and K. I. Welsh (1971), *Nat. New Biol.*, **230**:8.
Sanger, F. (1955), *Symp. Soc. Exp. Biol.*, **9**:10.
Sansome, F. W., and J. Philp (1939), "Recent Advances in Plant Genetics," Churchill, London.
Scandalios, J. G. (1967), *Biochem. Genet.*, **1**:1.
———— (1969), *Biochem. Genet.*, **3**: 37.
Schroeder, W. A. (1968), "The Primary Structure of Proteins," Harper & Row, New York.
Schwartz, A. G., P. R. Cook, and H. Harris (1971), *Nat. New Biol.*, **230**:5.
Schwartz, D. (1960), *Proc. Natl. Acad. Sci.*, **46**:1210.
Scott-Moncrieff, R. (1936), *J. Genet.*, **32**:117.
———— (1939), *Ergeb. Enzymforsch.*, **8**:277.
Shaw, C. R. (1965), *Science*, **149**:936.
———— and E. Barto (1963), *Proc. Natl. Acad. Sci.*, **50**:211.
Shows, T. B. (1972), *Proc. Natl. Acad. Sci.*, **69**:348.

Silagi, S., G. Darlington, and S. A. Bruce (1969), *Proc. Natl. Acad. Sci.*, **62**:1085.

Silvana-Santachiara, A., M. Nabholz, V. Miggiano, A. J. Darlington, and W. Bodmer (1971), *Nature*, **227**:248.

Siniscalco, M., et al. (1969), *Proc. Natl. Acad. Sci.*, **62**:793.

Smith, E. W., and J. V. Torbert (1958), *Bull. Johns Hopkins Hosp.*, **102**:38.

Smithies, O. (1955), *Biochem. J.*, **61**:629.

———, G. E. Connell, and G. H. Dixon (1962), *Nature*, **196**:232.

Snell, G. D., and J. H. Stimpfling (1966), in E. L. Green (ed.), "Biology of the Laboratory Mouse," 2d ed., pp. 457–491, McGraw-Hill, New York.

Srb, A. M., and N. H. Horowitz (1944), *J. Biol. Chem.*, **154**:129.

Stanbury, J. B., J. B. Wyngaarden, and D. S. Fredrickson (eds.) (1966), "The Metabolic Basis of Inherited Diseases," 2d ed., McGraw-Hill, New York.

Sturtevant, A. H. (1920), *Proc. Soc. Exp. Biol. Med.*, **17**:70.

Suskind, S. R. (1957), *J. Bacteriol.*, **74**:308.

——— and E. Jordan (1959), *Science*, **129**:1614.

Tatum, E. L., and D. Bonner (1944), *Proc. Natl. Acad. Sci.*, **30**:30.

———, D. Bonner, and G. W. Beadle (1944), *Arch. Biochem.*, **3**:477.

Teas, H. J., N. H. Horowitz, and M. Fling (1948), *J. Biol. Chem.*, **172**:651.

Ursprung, H., and J. Leone (1965), *J. Exp. Zool.*, **160**:147.

Wagner, R. P., and H. K. Mitchell (1964), "Genetics and Metabolism," 2d ed., Wiley, New York.

———, A. N. Radhakrishnan, and E. E. Snell (1958), *Proc. Natl. Acad. Sci.*, **44**:1047.

———, C. Somers, and A. Berquist (1960), *Proc. Natl. Acad. Sci.*, **46**:708.

Weiss, M. C., and H. Green (1967), *Proc. Natl. Acad. Sci.*, **58**:1104.

White, A., P. Handler, and E. L. Smith (1964), "Principles of Biochemistry," 3d ed., McGraw-Hill, New York.

Wit, F. (1937), *Genetica*, **19**:1.

Wright, S. (1932), *Am. Nat.*, **66**:282.

——— (1949), *Genetics*, **34**:245.

——— (1959), *Genetics*, **44**:1001.

Yanofsky, C. (1960), *Bacteriol. Rev.*, **24**:221.

——— (1964), in I. C. Gunsalus and R. Y. Stanier (eds.), "The Bacteria," vol. V, pp. 373–417, Academic, New York.

——— (1967), *Annu. Rev. Genet.*, **1**:117.

——— and D. M. Bonner (1955), *Genetics*, **40**:761.

——— and I. P. Crawford (1959), *Proc. Natl. Acad. Sci.*, **45**:1006.

———, G. R. Drapeau, J. R. Guest, and B. C. Carlton (1967), *Proc. Natl. Acad. Sci.*, **57**:296.

——— and M. Rachmeler (1958), *Biochim. Biophys. Acta*, **28**:640.

Ziegler, I. (1961), *Genetics*, **10**:349.

28
Protein
Synthesis

Transcription and Translation

NOTATION

1 Biochemical abbreviations used in this and subsequent chapters are listed in Appendix Table A-7.

2 The structure and abbreviations of amino acids are shown in Appendix Table A-4.

3 Prefixes used with the metric system are explained in Appendix Table A-6.

QUESTIONS

1332 Discuss the function of each of the following in protein synthesis:

Aminoacyl-RNA synthetase	R_1, R_2, S
ATP	Ribosomal proteins
DNA	rRNA
F1, F2, F3 initiating factors	30S ribosomal subunit
GTP	50S ribosomal subunit
mRNA	tRNA
Peptide transferase	Transfer factors I and II
Polysome	

See *Cold Spring Harbor Symp. Quant. Biol.* (1969, 1970), and Watson (1970).

1333 **a** Discuss evidence that DNA is not directly involved in protein synthesis.

 b List and briefly discuss the circumstantial evidence implicating RNA as an intermediate in protein synthesis. See Chantrenne, 1961.

1334 **a** Compare the ribosomes of *Escherichia coli* and mammals with respect to structure and chemical content. (See Petermann, 1965; Osawa, 1968.)

 b What features of ribosome formation in animal cells distinguish the process from that occurring in bacteria? (See Maden, 1968.)

1335 **a** What genetic attributes does RNA share with DNA?

 b RNA is of three types in eukaryotes and bacteria. What are they? Where are they located in the cell? Where are they produced? What are their characteristics and functions?

 c Viruses form only two types of RNA. Which type do the viruses not synthesize and why?

 d Explain how an amino acid is activated and then attached to its specific tRNA.

 e What is the evidence that a mRNA may be long enough to code for several proteins?

 f Explain why polypeptides specified by a polycistronic message are not coupled.

 g The central dogma of molecular biology is now represented by the schematic diagram

$$DNA \underset{(2)}{\overset{(1)}{\rightleftarrows}} RNA \overset{(3)}{\longrightarrow} Protein$$

(1) Give the common descriptive name of an enzyme involved in each of the steps 1 to 5.

(2) Give examples of biological systems where you would expect to find each of the enzymes you name above.

h Present both theoretical and experimental evidence for the existence of polyribosomes. Discuss the initiation and termination of polypeptide synthesis and the translation of polycistronic messenger molecules using R17 as a model system.

1336 Show how the following statements are related to the current concepts of genetic transcription and translation.

a In bacteria and eukaryotes, tissues very active in protein synthesis contain large amounts of cytoplasmic RNA (Caspersson, 1941). RNA is also present in the nucleus and nucleolus (Brachet, 1957).

b There is a positive correlation between the rate of protein synthesis and the amount of cytoplasmic RNA (Kurnick, 1955).

c In 1943 Claude discovered particles called ribosomes (Roberts, 1958), which occur in large numbers in all cells (Palade, 1955).

d Radioactive amino acids are rapidly incorporated into proteins at the ribosomes and appear later in the protein fraction (Littlefield et al., 1955; McQuillen et al., 1959; Rabinowitz and Olson, 1956).

e The bulk (80 to 90 percent) of the RNA in a cell is contained in the ribosomes (Schackman et al., 1952). Ribosomes from different species are very uniform in size (Hall and Doty, 1959; Kurland, 1960). Moreover, the ratio of the bases in rRNA is remarkably uniform throughout nature (Belozersky and Spirin, 1960).

f If cells that are synthesizing proteins are treated with ribonuclease (which digests free RNA), protein synthesis stops immediately (Brachet, 1954).

g Bacteria cultured under different conditions grow at different rates; the amount of DNA per bacterium remains constant, but the amount of RNA and protein rises markedly with the growth rate (Maaløe, 1960).

h Under conditions of thymine starvation DNA synthesis stops, but RNA and protein synthesis does not (Barner and Cohen, 1957).

i Living cultures treated with either ribonuclease or uracil analogs (Slotnick et al., 1953) or deprived of uracil (Pardee, 1954) immediately stop producing proteins.

j RNA synthesis can continue at least temporarily in the absence of protein synthesis (Wisseman et al., 1954).

1337 *Acetabularia mediterranea* is a relatively large unicellular and uninucleate marine alga with a rhizoid, a stalk, and an umbrellalike cap which may be differently shaped in different strains. The nucleus (2n) remains in the rhizoid. The development of the cap, which is a late feature of growth, requires specific enzymes. Stalks may be severed near the base and grafted on to rhizoids from which the stalks have been removed. The grafted stalks develop to

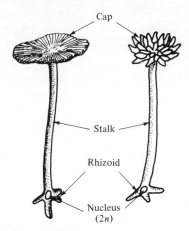

Cap

Stalk

Rhizoid

Nucleus
(2n)

Figure 1337

maturity, producing caps.   The following experiments were performed before the cap was formed.

1   When Hämmerling (1953) grafted enucleated stalks of one strain on to rhizoids of another strain, the morphology of the umbrella that developed corresponded to the strain of the nucleate rhizoid.

2   Hämmerling (1953) and Brachet and Chantrenne (1956) separated the stalks from the nucleated rhizoid and found that:

a   The stalks alone survive for 2 months and develop caps characteristic of their strain.

b   The stalks maintain protein synthesis for 2 weeks after separation, at the same rate as in nonseparated stalks (Baltus, 1959).

Discuss the implications of these findings for the statement that DNA acts directly as template for protein synthesis.

1338   According to the collinearity hypothesis, the specific base-pair content and sequence in a gene specifies the amino acid content and sequence in the corresponding polypeptide.

a Explain and illustrate how the sequence of base pairs (or bases, if the DNA or RNA is single-stranded) is *transcribed* into *m*RNA and how the *m*RNA is *translated* into the sequence of amino acids in the polypeptide chain. For discussion purposes use a hypothetical gene that codes for a hypothetical protein three amino acids long.

b State the functions of all of the components involved in the *translation* process.

1339   The schematic representation of the relationships between aminoacyl-*t*RNAs and *m*RNA at the ribosome is incorrect in some respects.   Redraw the illustration and indicate the errors you have corrected.

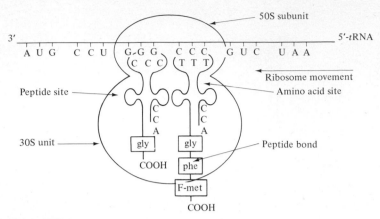

Figure 1339

1340 A *t*RNA with its attached amino acid is isolated, and subsequently the amino acid is changed into another one.   Such a *t*RNA with its changed amino acid is introduced into a cell-free, protein-synthesizing system.
  a Would you expect the altered amino acid to be incorporated into polypeptides?
  b Would it be at the residue site occupied by the amino acid from which it was derived?
  c What conclusion could you draw from this experiment?

1341 The first step in protein synthesis is the formation of an initiation complex (Nomura et al., 1967; Guthrie and Nomura, 1968).   What is the initiation complex, and what events follow to begin polypeptide synthesis?   See the papers by the above authors as well as those by Mangiarotti and Schlessinger (1966), Schlessinger et al. (1967), Nomura and Lowry (1967), Kaempfer et al. (1968), and Heywood (1970).

1342 **a** What is the role of codons UAA, UGA, UAG in protein synthesis?
  b What effect does each of the following have on polypeptide synthesis?
    (1) Nonsense mutation.
    (2) Missense mutation.
    (3) Frame-shift mutation.

1343 If AUG and GUG codons are polypeptide-initiating codons, why have many synthetic messenger RNAs without these codons (such as poly-U) nevertheless managed to direct in vitro polypeptide synthesis?

1344 The complete sequence is known for a number of *t*RNAs, e.g., for Ala-*t*RNA. Does this indicate that all the base sequence of its gene is also known? Why?

1345 Do you expect mutations in genes specifying aminoacyl-*t*RNA synthetases to be lethal to (1) haploid organisms, (2) diploid ones?

1346   **a** Discuss the lines of evidence that suggest that *mRNA* is single-stranded.
       b Since only one strand of DNA is transcribed into *mRNA* in vivo, what is or
       may be the function of the complementary strand?

1347   In vivo, DNA-dependent RNA polymerase transcribes one strand only.   In
       vitro, the same enzyme transcribes both strands.   Offer an explanation to
       account for this difference in behavior of the transcription enzyme based on
       the fact that template DNA is fragmented by the process of extraction.

1348   **a** The $A + T : G + C$ ratios in the DNA of cattle and rat are very similar.
       Would you expect the *tRNAs*, *rRNAs*, and *mRNAs* of the two species to
       be very similar?   Explain.
       b The base compositions of DNA of the tuberculosis bacillus and *E. coli* are
       shown in the table.   Do you expect differences between the two species in
       their:
       (1) *rRNAs*?
       (2) *tRNAs*?
       (3) *mRNAs*?
       Why?

Table 1348

	A	T	G	C
Tuberculosis bacillus	15.1	14.6	34.9	35.4
E. coli	26.1	23.9	24.9	25.1

1349   One cell makes one kind of protein only.   Others make a great variety of
       protein molecules.   In which of these two kinds of cells would you expect
       a broad distribution of polysome sizes and why?   Outline how you would
       proceed to support your expectation.

1350   In what respects (size, molecular weight, etc.) are *rRNAs* of different species
       alike?   In what ways are they different?   Is this also true for *tRNA*?   Explain.

1351   The *tRNA* from all sources have the same overall base composition
       $[(A + U)/(G + C) = 0.6]$, molecular weight (25,000), and sedimentation
       constants (4S).   They also function nonspecifically; viz. *tRNA* plus amino
       acid–activating enzymes from *E. coli* mixed with ribosomes plus *mRNA* from
       rabbit (*Oryctolagus*) reticulocytes synthesize rabbitlike hemoglobin.   Never-
       theless, they hybridize specifically; e.g., *E. coli tRNA* hybridizes only with *E. coli*
       DNA (Giacomoni and Spiegelman, 1962; Goodman and Rich, 1962).   Account
       as completely as possible for these paradoxical properties of *tRNA*.

1352   Approximately 0.3 percent of the DNA clustered in the nucleolus-organizer
       region in the genome of *Drosophila melanogaster* is concerned with ribosomal
       RNA (18S and 28S) synthesis (Ritossa and Spiegelman, 1965).   Ritossa and

Spiegelman favored the idea that the genes for rRNA are repeated consecutively along the chromosome 100 or more times to account for the large proportion of DNA set aside for this function.

a Suggest why such a large portion of the genome should be set aside for this purpose.

b Offer an alternative explanation to that of Ritossa and Spiegelman.

1353 Polypeptides polymerize by the stepwise addition of individual amino acids, beginning at the amino-terminal end.

a Should tRNA molecules be attached to the amino or carboxyl end of the amino acid? (See Zachau et al., 1958.)

b Would you expect a growing polypeptide chain to be terminated at its growing end by a tRNA molecule? Explain. (See Gilbert, 1963a,b; Takanami and Yan, 1965a.)

1354 At any given time each functioning ribosome is attached to only one growing polypeptide chain. Suggest why.

1355 If cell-free extract active in protein synthesis is centrifuged after incorporating amino acids, the growing protein chains sediment attached to 70S ribosomes which contain mRNA. When such extracts are treated with minute amounts of ribonuclease, the growing protein continues to sediment with the 70S particles. Is the nascent protein attached to mRNA or directly bound to ribosomes? Explain. (See Risebrough et al., 1962.)

1356 Outline an experiment that would indicate whether tRNA molecules are chromosomal (DNA) in origin.

1357 a The complete α and β chains of rabbit hemoglobin each contain an average of 17 leucine residues. Warner (see Rich, 1963; Warner and Rich, 1964a,b) found an average of 7.4 leucine residues per polypeptide per ribosome in an in vitro polysome system synthesizing hemoglobin. All ribosomes in the polysomal pool are active in protein synthesis. How can you explain these findings?

b In the reticulocyte of rabbit, which almost exclusively makes hemoglobin, polysomes are composed mostly of five ribosomes, occasionally four and five. In the human HeLa cell (a tumor cell) some of the polysomes contain 30 to 40 ribosomes, and mammalian cells infected by poliomyelitis virus polysomes consist of 50 to 70 ribosomes. To what do you attribute the differences in polysome sizes in the cases mentioned?

1358 The electron micrograph shows a thin fiber, from E. coli, with attached strings of granules (each about 200 Å in diameter). DNase treatment destroys the fiber but not the strings of granules, whereas RNase removes the granular strings from the fiber. Free strings with granules are never observed (Miller and B. R. Beatty, 1969; Miller et al., 1970).

a Which of these structures represents:

(1) A portion of the E. coli chromosome?

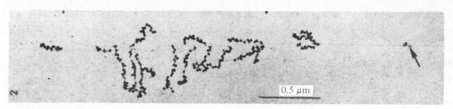

0.5 μm

Figure 1358

(2) Ribosomes?

(3) *m*RNA?

b Is translation coupled with transcription? Explain.

c May some of the granules be RNA polymerase molecules? If so, which and why?

1359 *Phenylketonuria* is due to an autosomal recessive gene, *p* (Penrose, 1935; Jervis, 1939), which results in the absence or a defective form of the enzyme phenylalanine hydroxylase from the liver, where it is synthesized after birth (Jervis, 1953). *Affected* individuals are unable to convert phenylalanine to tyrosine. As a consequence a series of pleiotropic effects occur, including excessive phenylalanine in the blood, urine, and spinal fluid, reduced skin pigmentation and lighter hair than in *normal* brothers and sisters, and frequently some degree of mental impairment (Harris, 1966). Miyamoto and Fitzpatrick (1957) have shown that phenylalanine is a competitive inhibitor of the tyrosine-tyrosinase system in vitro. Hsia et al. (1956) found that phenylalanine blood levels at 1-, 2-, and 4-hour intervals following standard dosage of phenylalanine were twice as high in the heterozygous, *Pp*, parents of *phenylketonurics*, *pp*, as in individuals homozygous for the *normal* allele *P*.

a Offer an explanation to account for the reduced pigmentation and lighter hair of *phenylketonurics*.

b What conclusions can be drawn from the findings of Hsia et al. (1956)? Do they have practical applications? Explain.

c Accumulation of phenylalanine might account for all symptoms in *phenylketonurics*. What kind of diet should *affected* individuals be restricted to, and when should it be started? Explain. (See Hsia et al., 1958.)

1360 State whether the following are true or false and why.

a Each of the 20 amino acids has its own specific activating enzyme.

b The ribosome is the site of protein synthesis.

c Polypeptide chains grow at their carboxyl end.

d The 23S RNA is not a dimer of 16S RNA.

e Any ribosome may serve as a "workbench" for the synthesis of more than one kind of polypeptide chain.

f Since nonribosomal RNA does not compete for the DNA, sites complementary to *r*RNA in hybridization experiments indicate that both types are synthesized on the DNA and at the same sites.

g That *t*RNA and associated amino acid–activating enzymes from one species when mixed with the ribosomes and attached *m*RNA from another species can bring about the synthesis of proteins of the kind directed by the *m*RNA proves that the code is universal.

h If the code is degenerate, only 20 different *t*RNA molecules should exist.

i The genetic code is universal or nearly so.

j Ribosomes of different species are comparatively indifferent to the origin of the genetic messages to which they respond.   They therefore must possess identical base compositions and sequences.

k The *t*RNA and *r*RNA, like *m*RNA, are formed on a DNA template.

l Nascent polypeptide chain is bound to the 30S subunit of the 70S ribosome.

m 30S and 50S ribosomal particles can couple to form 70S monomers regardless of whether *m*RNA, *t*RNA, potassium ions, and magnesium ions are present or not.

n Peptide-bond formation on the 70S ribosome involves at least two binding sites.

o All *t*RNAs carry amino acids.

p One messenger RNA is formed for an entire operon.

q There is only one kind of *t*RNA molecule capable of accepting any one amino acid.

r Methylation does not change the coding properties of *t*RNA.

PROBLEMS

1361 Tobacco mosaic virus (TMV) consists of a core of RNA surrounded by a hollow cylinder of protein; it causes the mosaic disease of *Nicotiana tabacum*. The various strains of this virus are characterized by the type of effect they have on their hosts.   *Normal* (N) produces a mottling of the leaves, and *Plantago* (P) produces a distinct ring pattern of diseased cells on host leaves.   In 1957 Fraenkel-Conrat and Singer combined the protein from *Normal* with RNA from *Plantago* and vice versa and obtained active viruses, which were

Table 1361

Experiment	Virus strain Protein	Virus strain RNA	Nature of disease	Percent weight of amino acids in protein of progeny* Ala	Arg	Met	Phe	Thr	Tyr
1	N	N	N	6.5	9.5	0	7.2	8.9	4.1
2	P	N	N	6.9	9.7	0	7.1	8.9	4.3
3	N	P	P	8.5	8.5	2.2	5.4	7.5	6.2
4	P	P	P	8.5	8.9	2.0	5.3	7.2	6.3

* The results shown are typical of those obtained with 13 other amino acids.

permitted to infect leaves of tobacco. The results of the infection study and the analysis of protein for the hybrid viruses and original strains are summarized in the table. Do these data indicate whether protein or RNA or both have the capacity to carry genetic information. Explain.

1362   When growing bacteria are infected with a virulent phage, e.g., T2, the synthesis of bacterial DNA stops immediately. Moreover net synthesis of RNA also stops, unlike that in growing cells, while synthesis of proteins continues at its preinfection rate. Using isotopic labeling, Volkin and Astrachan (1956, 1957) showed that immediately after *E. coli* cells are infected by T2, a very small amount of RNA is rapidly synthesized and just as rapidly destroyed (has a high turnover rate). The base compositions of this RNA, the total RNA of *E. coli*, and the DNA of the host and the T2 phage are tabulated.

Table 1362

| | RNA | | DNA | |
| | Total | High | | |
Base	*E. coli*	turnover	*E. coli*	T2
Adenine	23	31	25	32
Cytosine*	23	18	25	18
Guanine	31	22	25	18
Thymine (DNA) or uracil (RNA)	23	30	25	32

* 5-Hydroxymethylcytosine replaces cytosine in viral DNA.

a What appears to be the function of this new RNA and why?
b How does the T2 DNA appear to be related to this function? (See Hall and Spiegelman, 1961, for confirmation or refutation of your answer.)

1363   According to the one-gene–one-ribosome–one-enzyme hypothesis the RNA in the ribosomes is the template which determines protein structure. It implies that each gene determines the structure of one kind of ribosome, which in turn serves as a template for one protein. Discuss each of the following observations as evidence for or against this hypothesis:

1   The base ratios of DNA and *r*RNA bear the kind of relationship shown in the table.
2   The molecular weight of *r*RNA is uniform throughout nature.
3   When *E. coli* become infected with virulent T2 or T4 phage, net synthesis of RNA and bacterial protein is immediately and permanently arrested, and immediately thereafter phage protein is synthesized under the direction of phage DNA at about the rate of protein synthesis in uninfected bacteria.

See Jacob and Monod (1961) for a critical review of evidence bearing on this hypothesis.

Table 1363

Source	DNA				rRNA			
	A	T	G	C	A	U	G	C
E. coli	0.25	0.25	0.25	0.25	0.25	0.21	0.31	0.22
Yeast	0.31	0.33	0.19	0.17	0.26	0.26	0.28	0.20

1364 **a** Ehrenstein et al. (1963) synthesized rabbit hemoglobin molecules in a cell-free system containing ribosomes from rabbit reticulocytes, the amino acids cysteine and alanine, and *t*RNAs for these amino acids from *E. coli*. They studied the effects of modifying this system on peptide 13 in the α chain, which normally contains cysteine but not alanine, with the following results:

1   Cysteine attached to its *t*RNA was incorporated into peptide 13.
2   When cysteine was changed to alanine by Raney nickel while attached to its *t*RNA, only alanine was found present in peptide 13.
3   Alanine bonded to its own *t*RNA was not incorporated into this peptide.

Show how these observations answer the question: Is the *m*RNA code recognized by the *t*RNA or the amino acid?

b Dintzis (1961) studied hemoglobin synthesis in rabbit red blood cells to determine the direction of polypeptide synthesis. For the purposes of this problem make the following assumptions:

1   Six peptides of equal size are produced by treating hemoglobin with trypsin, and each peptide has only one lysine molecule at the carboxyl end of each peptide:

NH$_2$-	⋯ Lys	⋯ Lys	⋯ Lys	⋯ Lys	⋯ Lys	⋯ Lys	-COOH
Peptide	1	2	3	4	5	6	

2   It takes 60 minutes to synthesize a complete chain, and immediately upon completion of a chain it is released from the ribosome.
3   When the experiment is initiated (by the addition of ^{14}C-lysine to a culture of red blood cells), there are only five classes of chain lengths present on the ribosome. Each class terminates with a lysine residue, and no class is present in excess.

(1) What made the choice of reticulocytes a particularly appropriate one for this problem? Can you think of another system in which this problem could be solved?
(2) Assuming that synthesis proceeds from the amino to the carboxyl end and that only radioactivity incorporated into completed hemoglobin chains is examined, draw a graph indicating the relative content of labeled lysine (^{14}C radioactivity per total lysine) in each peptide fragment (*y* axis) as a function of the peptide number (*x* axis). On this

graph show two curves, one indicating the results expected after 30 minutes of labeling and the second those expected after 1 hour of labeling.

(3) Repeat (2) for the situation in which the relative content of labeled lysine in incomplete chains (still attached to the ribosomes) is examined.

(4) Compare the results of (2) and (3) with those of Dintzis (see Prob. 1371) to determine the direction of polypeptide synthesis.

1365 **a** Somatic cells of the toad *Xenopus laevis* ($2n = 36$) have two nucleoli produced at corresponding nucleolus-organizer regions (segments of DNA) of a pair of homologous chromosomes. Elsdale et al. (1958) discovered a mutant female ($2n = 36$) with somatic nuclei containing only one nucleolus. This characteristic was transmitted to its progeny. When toads with only one nucleolus per somatic nucleus were interbred, they produced *binucleolar*, *uninucleolar*, and *anucleolar* progeny in a 1:2:1 ratio. The *uninucleolar* toads lacked a secondary constriction on one of the two homologs. *Anucleolar* toads, which showed no secondary constrictions, died as tadpoles. Brown and Gurdon (1964) showed that *anucleolar* embryos were unable to synthesize either the 18S or the 28S molecules of the ribosomes, which are known to differ in RNA base composition but to be synthesized in a coordinated manner (same time, same place in the cell). The survival of the *anucleolar* toads to the tadpole stage is made possible by the large pool of maternal ribosomes from the egg.

(1) Offer a hypothesis that will explain the inheritance, the lethality, and the biochemical data.

(2) Offer two plausible genetic explanations to account for the phenotypic distributions in the cross of two *uninucleolar* individuals. Discuss the apparent significance of these results.

b Wallace and Birnstiel (1966) found that the DNA of *anucleolar* tadpoles failed to hybridize with *wild-type X. laevis r*RNA whereas that of *uninucleolar* tadpoles hybridized with a quantity of *r*RNA intermediate to that annealed by the DNA of *binucleolar* and *anucleolar* tadpoles. The 18S and 28S *r*RNAs hybridize with 0.04 and 0.07 percent of homologous *wild-type X. laevis* DNA respectively. The 18S and 28S components differ in base sequences, annealing to different stretches of DNA, which together occupy approximately 1 percent of the genome. The potential number of nucleoli, the number of secondary constrictions, and the portion of the genome complementary in base sequence to *r*RNA all show a linear reduction in proportion to the dosage of the mutation. Do these results support your previous conclusion and how? If not, what is the nature of the mutant, where are ribosomal cistrons for 18S and 28S located, and why?

1366 The nucleolus-organizer (NO) region of *Drosophila melanogaster* is present on both the X and Y chromosomes. Strains containing one to four doses

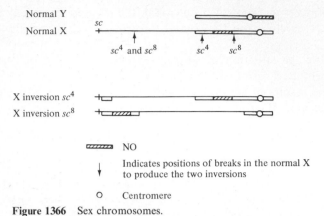

Figure 1366 Sex chromosomes.

of NO, as duplications and deletions, were derived from X-chromosome inversion stocks. DNAs extracted from these stocks were hybridized with isotopically labeled *r*RNA from *wild-type* XX or XY flies (Ritossa and Spiegelman, 1965). The kinds of sex chromosomes used to derive the four strains and the chromosome constitution and RNA-DNA hybridization data of the flies from the strains that were studied are shown.

Table 1366

Strain	Chromosome constitution	No. of NO per genome	Percent RNA hybridized with DNA
1		1	0.135
2		2	
		2	0.270
3		3	0.405
4		4	0.540

Percentages of hybridization are at saturation levels of DNA. The *wild-type* genome (DNA) saturates at 0.27 percent of total DNA.

a Set up breeding stocks carrying the four types of sex chromosomes and show how they can be used to produce the types of flies in the RNA-DNA hybridization experiments.

b Show what conclusions can be made regarding:
(1) The function of the DNA in the NO segment.
(2) The number of *r*RNA cistrons.
(3) Clustered vs. scattered distribution of *r*RNA cistrons.

c How would you determine whether 5S *r*RNA genes are in the NO region? (See Tartof and Perry, 1970; Wimber and Steffensen, 1970.)

1367 **a** In *Drosophila melanogaster* the approximate molecular weight of DNA per genome is 1.2×10^{11} (8×10^{11} nucleotide pairs). Moreover, 0.27 percent of the *wild-type* genome, specifically in the nucleolus-organizer region, hybridizes with *r*RNA.
   (1) Why must this molecular weight be divided by 2 to estimate the number of cistrons controlling *r*RNA synthesis?
   (2) Approximately how many cistrons are there per diploid set?

b DNA-RNA hybridization studies in the yeast *Saccharomyces cerevisiae* by Schweizer et al. (1969) have indicated that 0.064 to 0.08, 0.8, and 1.6 percent of the nuclear genome (DNA) hybridizes with 4S *t*RNA and 18S and 26S *r*RNA, respectively. The amount of DNA per haploid genome is 1.25×10^{10} daltons.

Table 1367

Type of RNA	Molecular weight	DNA hybridized	
		Percent of total DNA	Weight, daltons
4S	2.5×10^4	0.064–0.08	2×10^7 (max. value)
18S	0.7×10^4	0.8	2.0×10^8
26S	1.4×10^6	1.6	4.0×10^8

   (1) How many cistrons are there per genome for:
      (*a*) 18S *r*RNA?
      (*b*) 26S *r*RNA?
      (*c*) All *t*RNAs?
   (2) Assuming there are approximately 60 different species of *t*RNA on the average, approximately how many cistrons are there for each *t*RNA species?
      Note: The cistrons for different RNAs differ in size.   See Schweizer et al. (1969).

1368 By 1962 it had been established that each structural gene, a specific segment of a DNA (or RNA) molecule, directs the synthesis of a specific polypeptide chain.   Moreover, excellent circumstantial evidence indicated that DNA does not act directly as a template, that function being performed by mRNA (Brenner et al., 1961; Gros et al., 1961).   According to the adaptor hypothesis (Crick, 1958), each of the 20 amino acids is carried by an adaptor molecule to the template RNA; each adaptor recognizes the appropriate codon in the mRNA, thus correctly positioning the amino acid in the polypeptide specified by the gene.   The following experiments concern the role of tRNA in genetic coding.

1   Amino acids bonded to tRNA molecules were obtained from cell-free extracts.   Each amino acid was joined to a specific tRNA (Hoagland et al., 1957).
2   Chapeville et al. (1962), using a synthetic mRNA containing only uracil and guanine in an in vitro protein-synthesizing system from E. coli, found that this mRNA (poly-UG) templated the incorporation of phenylalanine and cysteine into polypeptides when these amino acids were attached to their own tRNA species.   When cysteine was reduced to alanine by Raney nickel after it had become attached to its tRNA, poly-UG stimulated the synthesis of a polypeptide containing phenylalanine and alanine.   When alanine was attached to its own tRNA, it was not incorporated into polypeptides synthesized on poly-UG.

a Show how these data support Crick's adaptor hypothesis.
b Show what results would have been expected if the code word in the mRNA were recognized by the amino acid itself (the tRNAs being merely carriers, bringing the amino acids to the ribosomes).
c If tRNA$_{Cys}$ remains attached to an amino acid that has been catalytically converted from cysteine to alanine, what suggestions can be made regarding the specific manner in which this tRNA becomes attached in the first place?

1369   In *Drosophila melanogaster* the approximately 130 or more identical genes for each of the rRNA species 18S and 28S are located in the DNA (rDNA) of the nucleolus-organizer (NO) regions, which are in corresponding positions of the X and Y chromosomes (Ritossa and Spiegelman, 1965).   This NO DNA, which is about 0.27 percent of the total *wild-type* genotype, is at the

Table 1369

	$\frac{rRNA}{DNA}$, % in hybrid (1)	$\frac{28S\ rRNA}{DNA}$, % in hybrid (2)	$\frac{18S\ rRNA}{DNA}$, % in hybrid (3)	$\frac{(1)}{(3)}$	$\frac{(1)}{(2)}$	$\frac{(2)}{(3)}$
Wild type	0.373	0.247	0.122	3.05	1.51	2.02
yw bb S1	0.165	0.111	0.053	3.11	1.48	2.09
yw bb S2	0.205	0.138	0.066	3.10	1.48	2.09
car bb S1	0.254	0.170	0.082	3.09	1.49	2.07
car bb S2	0.168	0.109	0.053	3.17	1.54	2.05
U C03 bb S1	0.185	0.123	0.060	3.08	1.50	2.05

bobbed, *bb*, locus where many hypomorphic alleles occur, each due to a partial deletion of the *bobbed* (locus) which cause small bristles and poor development. The severity of the mutant phenotype depends on the extent of the *bb* deletion. After denaturing DNA and using the technique of DNA-RNA hybridization along with RNase to degrade any RNA that did not hybridize, Quagliarotti and Ritossa (1968) studied flies homozygous for different hypomorphic alleles and *wild types* to determine the sequence and ratio of genes for 18S and 28S *r*RNA. The ratio of DNA complementary to 18S and 28S in *wild-type* and different *mutant* flies is shown.

a Suggest the ratio between the number of genes for 18S *r*RNA and that for 28S *r*RNA and explain it.

b What is the ratio between the amount of DNA complementary to 28S *r*RNA and that complementary to 18S *r*RNA?

c The genes for the two *r*RNAs may be arranged as follows:

(1) Genes for 28S *r*RNA are clustered in one block, and those for 18S *r*RNA are clustered in another block, the two blocks being adjacent to each other.

(2) The two kinds of genes are interspersed in the *bb* locus, e.g., in alternating sequence 18S-28S-18S-28S $\cdots$.

Which arrangement do these data support and why?

1370 Brenner et al. (1961) grew cultures of *E. coli* in a medium containing heavy nitrogen, ^{15}N, and heavy carbon, ^{13}C, so that all the nitrogen (in the RNA) and carbon (in the protein and RNA) of the ribosomes would be heavy. The cells were then infected with T4 phage and immediately transferred to a medium containing normal (light) nitrogen, ^{14}N, and carbon, ^{12}C. Any new ribosomes formed after infection would incorporate ^{14}N and ^{12}C only and would be light.[1]

1 In one experiment, from the second to the seventh minute after transfer to ^{14}N-^{12}C medium bacteria were fed ^{32}P to label any RNA formed after infection. These were mixed with a fiftyfold excess of cells grown on ^{14}N-^{12}C-^{31}P medium.

2 In the second experiment, for the first 2 minutes after transfer to ^{14}N-^{12}C medium the bacteria were given ^{35}S to label any protein formed after infection.

3 Subsequently in both experiments cells were ruptured, ribosomes purified and centrifuged and the amount of radioactivity associated with heavy (^{15}N-^{13}C) and light (^{14}N-^{12}C) bands (= ribosomes) determined. The results are illustrated.

a Are new ribosomes formed after infection? Discuss evidence bearing on your answer.

b Do these results refute or support the one-gene–one-ribosome–one-protein hypothesis, according to which each gene specifies the synthesis of one kind of specialized ribosome (specifically RNA) which in turn directs the synthesis of the corresponding protein? If they refute this hypothesis, what appears to be the function of ribosomes?

---

[1] Heavy and light ribosomes can be distinguished by centrifuging purified ribosomes from disrupted cells in a cesium chloride solution, in which they form separate, distinct bands.

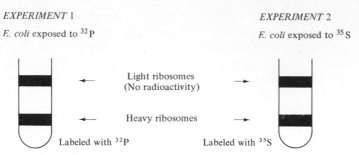

Figure 1370

c The RNA formed after infection is short-lived.   What is or appears to be its function?   See Gros et al. (1961) for experimental observations bearing on these questions.

1371 Rabbit reticulocytes synthesize α and β polypeptide chains of hemoglobin. In each cell, both finished chains (separated from ribosomes) and incomplete chains (at different stages of formation on ribosomes; time t in the figure)

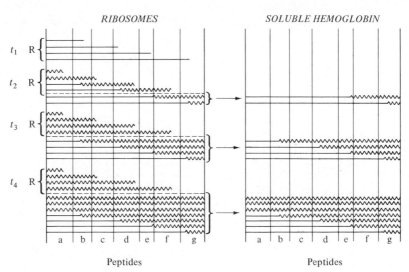

Figure 1371 Analysis of peptide fragments for radioactivity.   Straight lines represent unlabeled chains or peptide fragments.   Zigzag lines represent radioactive chains or peptide fragments formed after addition of ^{3}H-leucine at time t_1.   The groups of peptides designated R are unfinished and attached to the ribosomes.   The rest of the polypeptides, having been completed, are present in the soluble hemoglobin.   In the ribosomes at time t_2, the top two lines represent peptide chains formed completely from amino acids during the time interval between t_1 and t_2.   The middle two lines represent incomplete chains synthesized during the time interval which are still attached to the ribosomes.   The bottom two lines represent complete chains, separated from the ribosomes and mixed with other complete chains of soluble hemoglobin. (*Redrawn from H. M. Dintzis and P. M. Knopf, in H. J. Vogel et al. (eds.), "Informational Macromolecules," p. 376, Academic Press Inc., New York, 1963.*)

are present at any one time.   Since results obtained with both chains are basically identical, consider the results (modified after Dintzis, 1961) obtained with the α chain, which contains approximately 150 amino acids.   From the known composition of this chain and the fact that trypsin cleaves polypeptides on the carboxyl side of arginine and lysine residues, it is possible to fragment the complete chains into seven peptides a, b, c, d, e, f, g.   The peptides, each with one leucine residue, occur in the above sequence from the amino to the carboxyl end.   A culture of reticulocytes is exposed to ^{3}H-labeled leucine for a short time.   Samples of cells are withdrawn after various times of exposure (time t_2, time t_3, and time t_4), the complete and incomplete chains are separated from each other and subjected to tryptic digestion (the vertical lines in the figure indicate places where trypsin fragments the chains).   The peptide fragments are then characterized and analyzed for radioactivity with the results shown in figure 1371.   Which of the following methods of chain formation do the data support and why?

1   A number of separate peptides are assembled first and are combined to form a long chain.

2   The chain is formed by sequential addition of amino acids, starting at either the amino or the carboxyl end.

3   Chain formation may begin internally and proceed to both ends.   See Bishop et al. (1960) and Goldstein and Brown (1961).

1372   Rabbit reticulocytes, which synthesize α and β polypeptide chains of hemoglobin, contain both finished and incomplete chains at any one time.   A dose of labeled amino acids is injected into unlabeled cells.   Shortly after, finished hemoglobin molecules are isolated and chains separated and analyzed for location of the radioactive label.   Where do you expect to find the label in the polypeptides and why?

1373   The 16S and 23S RNAs of the ribosomes are similar in base composition. Moreover, their molecular weights are almost exactly in the relation of 2 : 1. Discuss the bearing of Yankofsky and Spiegelman's (1963) DNA-RNA hybridization experiments with *Bacillus megaterium* on the questions:
   a Is 23S RNA a dimer of 16S RNA?
   b Do the two RNAs have a common origin?
   c What percentage of total DNA is complementary to 23S RNA?   To 16S RNA?

1   A fixed amount of heat-denatured DNA was hybridized with increasing amounts of 16S and 23S ribosomal RNA to determine the ratio of each kind of RNA to DNA in the hybrid at saturation (when no more of that RNA, 16S or 23S, can hybridize with DNA).   The proportions of ribonuclease-resistant DNA-RNA hybrids are shown in Fig. 1373*A*.   Saturation curves indicate the fraction of the DNA molecules specifying the 16S and 23S ribosomal RNAs.

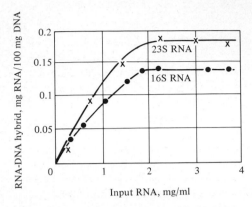

Figure 1373*A* (*Redrawn from S. Spiegelman, Hybrid Nucleic Acids, Sci. Am., May, 1964, p. 54.*)

2 Hybridization mixtures containing fixed amounts of heat-denatured DNA and saturating concentrations of 23S RNA labeled with ^{32}P were prepared. To these mixtures, increasing amounts of 16S RNA, labeled with ^{3}H, were added until no more hybridization occurred. The relative amounts of ^{32}P and ^{3}H in the hybrids were determined. The results are illustrated in Fig. 1373*B*.

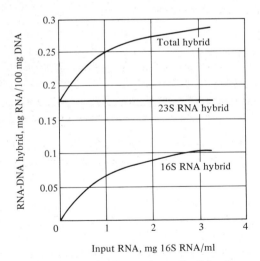

Figure 1373*B* (*Redrawn from S. Spiegelman, Hybrid Nucleic Acids, Sci. Am., May, 1964, p. 54.*)

1374 **a** It has been shown (Weiss and Nakamoto, 1961; Hurwitz and Furth, 1962; Hurwitz and Anders, 1963; Hurwitz et al., 1960, 1961) that the RNA polymerases of *E. coli* and mammals require all four ribonucleoside 5'-triphosphates (UTP, ATP, GTP, and CTP) simultaneously for in vitro synthesis of RNA. Deoxyribonuclease inhibits and DNA addition stimulates RNA formation. The base ratios of DNA and RNA in T2 phage, *E. coli,* and calf thymus were found by Hurwitz et al. (1961) to be as given in Table 1374*A*. Two synthetic DNA polymers were added to

Table 1374*A*

DNA	$\dfrac{A + T}{G + C}$	Nucleotide incorporation in RNA, nanomoles				$\dfrac{A + U}{C + G}$
		AMP	UMP	GMP	CMP	
T2	1.86	0.54	0.59	0.31	0.30	1.85
Thymus	1.35	3.10	3.30	2.00	2.20	1.52
E. coli	1.00	2.70	2.74	2.90	2.94	0.93

mixtures containing all four bases and RNA polymerase. One contained thymine (poly-T), the other contained adenine and thymine in alternating sequence (poly-AT). The synthetic RNA formed when poly-T was template contained adenine, whereas that produced when poly-AT was template contained uracil and adenine in alternating sequence. What appear to be the functions of DNA and of RNA polymerase in RNA synthesis (viz. what kind of substrate does the enzyme use, and how is templating performed)?

b It has been shown (Nirenberg and Matthaei, 1961) that the synthetic *m*RNA containing uracil yields polyphenylalanine only as the product of in vitro protein synthesis, whereas a mixture of polyuracil and polyadenine chains yields no polypeptide. Does this information help to elucidate the role of RNA polymerase more clearly? If so, how?

c 1 The base compositions of single-stranded ϕX174 DNA and the enzymatically synthesized RNA using this DNA as template are given in Table 1374*B*.

Table 1374*B*

	A	T(U)	G	C
ϕX174 DNA	0.25	0.33	0.24	0.18
RNA	0.32	0.25	0.20	0.23

2 With rare exceptions, in all RNA molecules in all species, the amounts of A and U differ, and the same is true for G and C.

3 When 3′-deoxyadenosine is added to cells synthesizing RNA chains, this inhibitor is first phosphorylated to 3′-deoxyadenosine-P $\sim$ P $\sim$ P and then joined to the 3′ end of RNA molecules. The latter event terminates RNA synthesis.

Is RNA single- or double-stranded? Does synthesis of RNA chains occur in the 3′-to-5′ or the reverse direction?

d 1   RNA polymerase contains five different polypeptide chains α, β, β^1, ω, and σ; two σ chains are present, along with one of each of the other chains (Travers and Burgess, 1969).   The complete enzyme (holoenzyme) is easily dissociated into two subunits, a *core* polymerase ($\alpha_2 \beta\beta^1\omega$) and σ (Burgess et al., 1969; Berg et al., 1969).

2   Core enzyme can synthesize RNA chains starting anywhere along either strand of a gene (Sugiura et al., 1970; Bautz et al., 1969).

3   By itself σ has no catalytic function, but the holoenzyme ($\sigma + \alpha_2 \beta\beta^1\omega$) synthesizes RNA chains complementary to only one strand of each gene and begins synthesis at specific points in the DNA complex (Travers and Burgess, 1969).

4   Synthesis of all RNA chains appears to start with either A or G (Bremer and Bruner, 1968).

5   Stop signals (specific base sequences) exist to terminate RNA synthesis at specific points along the DNA.   When the ρ factor (a specific protein) is absent from cells, chain elongation is not terminated at stop signals (J. W. Roberts, 1969).

6   Once a gene has been transcribed, hydrogen bonds re-form between the complementary chains.

(1) What is the function of (a) the core enzyme, (b) σ, and (c) ρ factor?

(2) Illustrate the transcription of an RNA chain using a hypothetical DNA representing three consecutive genes.

See papers on transcription by many authors in *Cold Spring Harbor Symp. Quant. Biol.* (1970).

1375   A short-lived species is formed shortly after T2 infection of *E. coli*.   A few minutes later T2 proteins are synthesized.   Hall and Spiegelman (1961) labeled T2 DNA with radioactive hydrogen, ^{3}H, and the RNA formed after infection with radioactive phosphorus, ^{32}P.   The β particles emitted by these isotopes have different energies and so can be distinguished from each other using a scintillation spectrometer.   Prior to mixing the labeled RNA and labeled DNA, the DNA was extracted and heated to 95°C for 15 minutes to separate the strands (denaturation) and then rapidly cooled to keep the strands separate.   RNA was also extracted.   Subsequently the following experiments were performed:

1   In experiment 1, the labeled DNA and RNA were mixed at 65°C and allowed to cool slowly for 30 hours (permits hybridization).   Thereafter the samples were subjected to density-gradient centrifugation.

2   In experiment 2, labeled denatured DNA and labeled RNA were mixed but not heated and immediately centrifuged.

3   In experiment 3, denatured and labeled DNAs from T5 (same overall base composition as T2), *E. coli*, and *Pseudomonas aeruginosa* were separately mixed with the T2-specific RNA (^{32}P) and allowed to cool slowly.

4   In the control experiment, undenatured (native) labeled T2 DNA and labeled RNA were slowly cooled and then centrifuged.[1]

---

[1] Since RNA has a higher density than DNA, the two molecules can be separated by adding them to a cesium chloride solution and centrifuging.   After centrifugation is completed, the sample tube is punctured at the bottom, the contents removed drop by drop, and radioactivity of the various fractions determined.

The results of these experiments are illustrated.

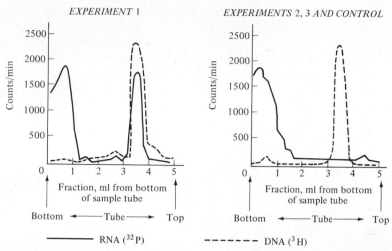

Figure 1375   (*Redrawn from S. Spiegelman, Hybrid Nucleic Acids, Sci. Am., May, 1964, p. 55.*)

a What are the requirements for effective hybridization?
b Interpret the results shown in each of the two graphs.
c What do these results indicate regarding the relationship of the RNA and DNA of T2, T5, *P. aeruginosa*, and *E. coli*?
d How might you confirm your deductions in (c)?
See Hayashi and Spiegelman (1961) for interpretation of these experiments.

1376   To determine whether 23S ribosomal RNA is formed on a DNA template or not, the following procedure was followed (Spiegelman, 1964).

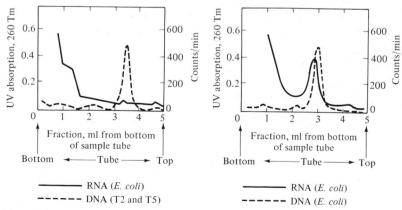

Figure 1376   (*Redrawn from S. Spiegelman, Hybrid Nucleic Acids, Sci. Am., May, 1964, pp. 51 and 52.*)

1 Organisms were chosen with a DNA base ratio greatly different from that of *E. coli.*
2 Using a suitable isotope, e.g., ^{32}P, all the RNA in a culture of *E. coli* was labeled. The cells were then transferred to a nonradioactive medium to permit the unstable labeled messenger RNA to disappear.
3 The labeled *E. coli rRNA* was mixed with single-stranded DNA from *E. coli* and T2 and T5 phages, heated, and slowly cooled. Next the mixtures were treated with ribonuclease and centrifuged in a cesium chloride solution. Sample tubes were punctured at the bottom, the solution was removed drop by drop, and the fractions were analyzed for the presence of RNA and DNA. The results are illustrated.

a Explain why each procedure was followed.
b Is ribosomal RNA formed on a DNA template? Explain.

1377 Genes can be located not only by well-established cytogenetic methods but also cytologically utilizing RNA-DNA hybridization. Pardue et al. (1970) prepared slides of cells from salivary glands of *Drosophila hydei* as follows. Glands were treated with RNase and then with sodium hydroxide to denature the DNA of the preparation. Radioactive (^3H) *rRNA* from *Xenopus laevis* was then added to salivary-gland preparations; the slides were covered with a coverslip and incubated at 66°C for 14 hours to permit RNA-DNA hybridization. After hybridization, the slides were coated with Kodak NTB-2 emulsion, stored for 94 days, and then developed and stained with Giemsa.

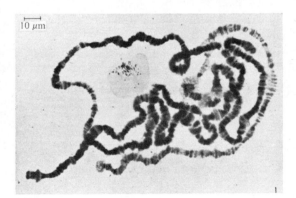

Figure 1377 [*From M. L. Pardue, S. A. Gerbi, R. A. Eckhardt, and J. G. Gall, Cytological Localization of DNA Complementary to Ribosomal RNA in Polytene Chromosomes of Diptera, Chromosoma,* **29**: *272 (1970).*]

The figure shows an autoradiograph of a salivary-gland cell of *D. hydei* (with an entire 2*n* complement) treated as described above. Black dots (silver grains) are due to emissions of β particles by ^3H of radioactive *rRNA*. Identify the various chromosomal components and indicate where the genes for *rRNA* are located and why.

1378 *t*RNA molecules are approximately 80 nucleotides long. Each nucleotide has an average molecular weight of 330 and each *t*RNA molecule an average molecular weight of 2.5×10^4. In 1962 Giacomoni and Spiegelman obtained the results shown in base-composition and DNA-RNA hybridization experiments in *E. coli.*

1   Base composition of RNA hybridization with DNA.

Table 1378

RNA hybridized	Base composition in moles, %				Percent G + C	Purine/ pyrimidine
	C	A	U	G		
Total *t*RNA	27.2	20.6	18.2	34.0	61.2	1.23
RNase-resistant *t*RNA	27.2	18.6	19.0	35.2	62.4	1.20
*r*RNA	24.3	25.0	19.7	31.0	55.3	1.27
*m*RNA	24.7	24.1	23.5	27.7	52.4	1.07

2   *E. coli* DNA was saturated with *t*RNA at 0.025 percent of total DNA.

3   In one DNA-RNA hybridization experiment *t*RNA from *E. coli* (^{32}P) and *Bacillus megaterium* (^{3}H) were incubated with DNA from *E. coli.* In another, the *t*RNA from *B. megaterium* was incubated with DNA from the same species and *E. coli.* The hybridization results are shown in the illustration.

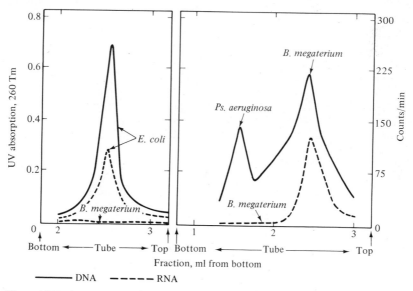

Figure 1378   (*Redrawn from S. Spiegelman, Hybrid Nucleic Acids, Sci. Am., May, 1964, p. 53.*)

a Indicate why these results show that *t*RNA molecules are specified by segments of DNA.

b The *E. coli* genome has a molecular weight of 4×10^9. With a non-degenerate code what portion of total DNA should hybridize with *t*RNA? Is degeneracy indicated? Explain.

1379 In *Diplococcus pneumoniae* either of the two DNA strands, as well as the duplex, can bring about transformation. Moreover, one of the two chains of DNA contains sufficiently more of the heavier purine guanine and the heavier pyrimidine thymine than the other, so that the two chains, after denaturation and density-gradient centrifuging, form separate bands. Using DNA from a *novobiocin-resistant* strain, Guild and Robison (1963) obtained two fractions by this means. They found that the fraction containing the heavier chain required about 45 minutes at 25°C to begin transforming *novobiocin-sensitive* cells to resistant ones. The lighter fraction modified the phenotype of transformed cells almost immediately. The cell-generation time (time for one DNA replication) at 25°C is approximately 40 minutes. Analyze this information to show that *m*RNA is copied from only one of the strands of DNA although either can transform the cell's genotype.

1380 In the mature bacterial virus ϕX174, DNA is single-stranded and circular. After the virus infects a bacterial cell, this strand serves as a template for the synthesis of the complementary strand, resulting in the double-stranded replicating form (RF), also in a circle. Hayashi et al. (1963) conducted hybridization tests between ^{3}H-labeled *m*RNA, transcribed from RF, and the DNA of the mature virus and RF (both labeled with ^{32}P). Although the single-stranded DNA of the mature virus, heated or unheated, did not hybridize with *m*RNA, the heat-denatured (chains-separated) RF DNA did. The base composition of *m*RNA, the DNA of ϕX174, and strand of RF complementary to original strand were found to be as given in the table.

Table 1380

Nucleic acid	Base ratio			
	C	A	U (T)	G
*m*RNA	18.5	23.8	33.1	23
ϕX174, DNA	19	25	33	23
ϕX174 DNA, complementary strand	23	33	25	19

Are transcribed messenger RNAs in vivo complementary to:

a Both chains of DNA?

b One chain only?

c Both chains but only one at any one time?

Illustrate and explain your answer.   See Hayashi et al. (1964) and Hayashi and Hayashi (1966) for a complete picture of genetic transcription in vivo and how and why it may differ in vitro.

1381   The two strands of the DNA of the phage SP8, virulent on *Bacillus subtilis*, have different densities and therefore can be separated by centrifugation. The heavier (*H*) strand is richer in pyrimidines and the lighter (*L*) strand is richer in purines.   Only the *H* strand forms hybrids with *m*RNA isolated from SP8-infected bacteria, but both strands anneal with RNA synthesized in vitro using SP8 DNA as primer (Marmur and Greenspan, 1963).

a Why does *m*RNA formed in vivo hybridize with the *H* strand only?

b How does the RNA formed in vitro differ from that synthesized in vivo?

1382   a Warner et al. (1963) gently extracted material, including ribosomes, from rabbit reticulocytes after 45-second exposure to ^{14}C-labeled amino acids. The cell contents were placed in a sucrose gradient and centrifuged, and fractions in the tube were removed in sequence from bottom to top and studied by photometric determination and radioactivity counts.   Ribosomes contain large amounts of RNA, which strongly absorb ultraviolet radiation at a wavelength of 2600 Å.   Therefore, the optical density (amount of absorption) was read at this wavelength to reveal the presence of these organelles.   The radioactivity of each fraction was measured using a radiation counter.   Because the measurement indicates presence of labeled amino acids, it also indicates which fractions contain growing polypeptide chains.

1   The normal distribution of ribosomes and the pattern of distribution of newly synthesized polypeptides are shown in Fig. 1382*A*.

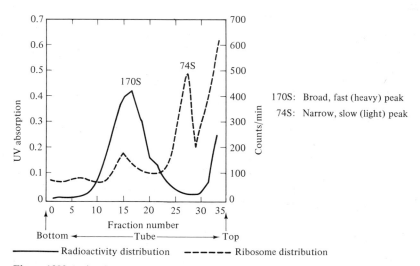

Figure 1382*A*   (*Redrawn from A. Rich, Polyribosomes, Sci. Am., Dec. 1963, p. 48.*)

2   When a cell-free medium containing the ribosomes from cells incubated and fed ^{14}C amino acids was treated with ribonuclease before centrifugation, no fast (heavy) peak appeared.   Only the slow (light) peak occurred, and it contained all the radioactivity.

3   The gently extracted material subjected to gentle grinding and centrifugation produced five peaks instead of two; i.e., two original ones (170S and 76S) and three new ones (154S, 134S, and 108S)   (Fig. 1382B).

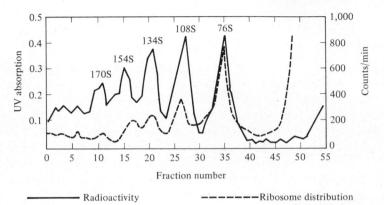

——————— Radioactivity   ——————Ribosome distribution

Figure 1382B   (*Redrawn from A. Rich, Polyribosomes, Sci. Am., Dec., 1963, p. 49.*)

4   Electron micrographs by Warner et al. (1962, 1963) and Slayter et al. (1963) showed the results tabulated.   Ribosomes in the first four fractions are connected by a thin thread 10 to 15 Å in diameter.

Table 1382

Fraction	Number of ribosomes
170S	5
154S	4
134S	3
108S	2
76S	1

With reasons, answer each of the following questions:
(1) Is hemoglobin synthesis carried out on groups of ribosomes or single ribosomes?
(2) What cellular component holds the ribosomes together in the multiple ribosomal structures?
(3) What is the predominant number of ribosomes per multiple structure?

(4) Is the *m*RNA used more than once in polypeptide synthesis?

(5) Can more than one polypeptide chain be associated with each active ribosome?   (See Warner and Rich, 1964*a,b*.)

b If the distance between ribosomes (230 Å in diameter) on a polyribosome averages 90 Å and the distance between the bases in *m*RNA is 3.4 Å, calculate the length of messenger required and the average number of ribosomes per polyribosome to synthesize a *β*-hemoglobin chain of approximately 150 amino acids.

1383 In T4 phage, *r*⁺ (*wild type*) produce small progeny plaques on both *E. coli* strains B and K.   The *r* mutants in the *r*II region, which consists of two contiguous genes *A* and *B*, produce large plaques on B but no plaques (no progeny) on K when they infect the hosts individually.   When K is simultaneously infected with mutants *r101* and *r103*, no progeny are produced; when infected with *r103* and *r104*, large numbers of plaques are formed.   The genomes of mutants *r103* and *r104* together produce the same phenotype as *r*⁺.   Explain the noncomplementation in the first cross and the complementation in the second one in terms of transcription and translation.

1384 Isolated 50S ribosomal subunits can be induced to catalyze peptide-bond formation between peptidyl-*t*RNA and aminoacyl-*t*RNA in the absence of template and 30S subunits (Monro, 1969).   Is peptidyl transferase activity associated with the 30S, 50S, or both subunits?

1385 The illustration shows a model of polysome function in protein synthesis.

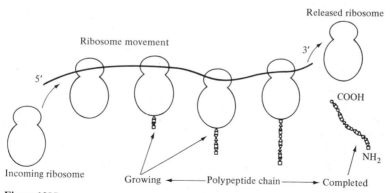

Figure 1385

The model predicts that under protein-synthesizing conditions:

1   It should be possible to attach single ribosomes (70S) to polysomes.
2   Single ribosomes should be released from polysomes.
3   Polypeptide chains should be released from polysomes.

Discuss the bearing of each of the following experiments on these predictions:

a Extracts of living cells in the process of protein synthesis incubated by Rich et al. (1963) for varying periods and subjected to sucrose-gradient centrifugation showed at the beginning a large polysome peak and a smaller single-ribosome peak. As incubation proceeded, the number and size of polysomes decreased and the number of single ribosomes increased. At the end of 90 minutes of incubation most of the polysomes disappeared; only single ribosomes were present.

b Goodman and Rich (1963) incubated a suspension of living HeLa cells for $1\frac{1}{2}$ minutes with ^{14}C amino acids. The cells were then chilled to stop protein synthesis. This process loaded the polysomes with ^{14}C amino acids that were joined into growing polypeptide chains. Next the cells were broken and the ribosomes and polysomes were isolated, using appropriate procedures, and resuspended in a fresh cell extract identical with that removed except that it contained normal (nonradioactive) amino acids. The suspension was incubated and radioactivity measured in the polysome fraction and soluble protein fraction (free of polysomes). As incubation proceeded, radioactivity decreased in the former fraction and increased in the latter.

c A mixture of ^3H-labeled single ribosomes from HeLa cells and an unlabeled extract of polyribosomes plus single ribosomes from the same source was incubated briefly and subjected to sucrose-gradient centrifugation. Some of the tritium-labeled ribosomes had become attached to polysomes, as indicated by a test for radioactivity. Twice as many single labeled ribosomes were attached to polysomes composed of five ribosomes as to polysomes composed of ten ribosomes when the total number of ribosomes in each fraction was equal (Goodman and Rich, 1963).

1386 In amphibians the DNA specifying the synthesis of 18S and 28S *r*RNAs is in the nucleolus-organizer (NO) region which forms the nucleolus. This DNA is highly redundant, containing large numbers of tandemly arranged *r*RNA genes, each capable of transcribing a 45S *r*RNA precursor, which yields the 18S and 28S *r*RNA molecules, and a further RNA message that is subsequently degraded. In each gene the DNA regions specifying one 18S RNA and one 28S RNA are separated by a degradable high G-C stretch that does not specify *r*RNA. Miller and Beatty (1969) and Miller et al. (1970) were the first to observe the structure of individual genes and associated

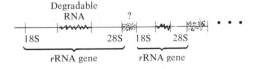

Figure 1386A   Short stretch of DNA of NO region.

transcription products whose function is known.   Using an electron microscope, they observed the tandemly arranged genes for 45S *r*RNA precursors in extrachromosomal nucleoli of the amphibian *Triturus viridescens*. Visualization of these genes is possible because many precursor molecules are simultaneously synthesized on each gene and associated with it.

During pachytene-diplotene of amphibian oocytes the NO DNA is multiplied, and the nucleolus disperses to produce about 1,000 extrachromosomal nucleoli within each nucleus. These, like chromosomal nucleoli, synthesize *r*RNA precursors.   In thin sections of fixed oocytes each extrachromosomal nucleolus shows a fibrous core, which contains DNA, RNA, and protein, surrounded by a granular cortex, which contains only RNA and protein (Fig. 1386*B*).   When the contents of an oocyte nucleus are placed in deionized water, the nucleolar cores and cortices separate, and the core unwinds.   The unwound core consists of a circular axial fiber, 100 to 300 Å in diameter and 20 to 1,000 μm in length, which is coated discontinuously with matrix material.   Matrix-coated segments (units) are separated by matrix-free segments (Figs. 1386*C* and 1386*D*).   Treatment of the circular axial fiber with DNase breaks the fiber axes; trypsin treatment of axial fibers reduces them to a diameter of about 30 Å but does not break them.   A matrix unit consists of about 100 fibrils each connected by one end to the core axis (Figs. 1386*D* and 1386*E*).   These increase in length from the thin to the thick end of the unit.   Matrix fibrils can be removed from the core axis by RNase, pepsin, or trypsin but not by DNase.   When RNA of intact oocytes is labeled, silver grains appear only over the matrix units,

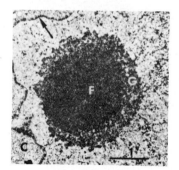

Figure 1386*B*   Thin section of extrachromosomal nucleolus from *Triturus viridescens* oocyte.   Fibrous core (F) surrounded by a granular cortex (G). Portions of nuclear membranes (arrow) at cytoplasm (C) are visible.   [*From O. L. Miller, Jr., and B. R. Beatty, Visualization of Nucleolar Genes, Science:* **164**: *955 (1969).*]

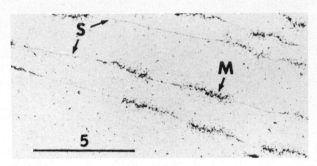

Figure 1386C Portion of nucleolar core from *T. viridescens* oocyte, showing matrix units (M) separated by matrix-free segments (S) of the core axis. [*From O. L. Miller, Jr., and B. R. Beatty, Visualization of Nucleolar Genes, Science,* **164**: *956 (1969).*]

Figure 1386D A segment of the nucleolar core from *T. viridescens* showing a tandemly arranged series of matrix units. (*From J. D. Watson, "Molecular Biology of the Gene." W. A. Benjamin, Inc., 1970.*) *Courtesy of O. L. Miller, Jr., and Barbara Beatty, Biology Division, Oak Ridge National Laboratory.*

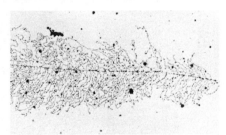

Figure 1386E A matrix unit showing a gradient of increasingly long fibrils each attached at one end to the core axis. [*From O. L. Miller, Jr., and B. R. Beatty, Visualization of Nucleolar Genes, Science,* **164**: *956 (1969).*]

not in matrix-free segments.   RNA polymerases are associated with matrix fibrils.

a What are the chemical components of (1) fiber axis, (2) fibrils?   Explain.

b Is each matrix unit a nongenic region, a gene, two genes, or more genes? Explain.   Answer the same question for the nonmatrix region.

c Is each fibril an 18S *r*RNA, a 28S *r*RNA, or a 45S *r*RNA precursor molecule?   Explain.

d Why are fibrils in a matrix unit of different length?

e Is each gene transcribed once, twice, or many times? Provide evidence for your answer.

f Each oocyte contains hundreds of *r*RNA genes. Suggest a reason why so many are required.

g Suggest how the circular DNA molecules in the extrachromosomal nucleoli are synthesized.

1387 **a** The *β* chain of hemoglobin in man is 146 amino acids long. What would be the length of the *m*RNA molecule required to direct the synthesis of this chain?

b An enzyme consists of one polypeptide chain only and has a molecular weight of 300,000. Assuming the average weight of an amino acid is 100, how many nucleotides are there in the gene controlling it?

c The single-stranded DNA of bacteriophage ϕX174 has 4.5×10^3 nucleotides. How many proteins of molecular weight 30,000 could be coded by this genome?

d The DNA of *Aspergillus* has 4×10^7 nucleotide pairs. If an average gene contains 1,500 nucleotide pairs, how many genes does *Aspergillus* possess?

1388 When $5'$-AAA $\cdots$ AC[(Ap)$_n$ C]-$3'$ was used as a messenger in the rabbit reticulocyte cell-free system, the polypeptide product was NH$_2$-Lys-Lys $\cdots$ Lys-Asn-COOH. The coding triplet for lysine is AAA and for asparagine AAC (Lamfrom et al., 1966). Is this *m*RNA read from the $3'$ or the $5'$ end? Explain. Does the NH$_2$ end or the COOH end of the polypeptide correspond to the $3'$ end of the *m*RNA?

1389 Thach et al. (1965) showed that the peptide synthesized under the direction of the messenger $5'$-AAAUUU-$3'$ is NH$_2$-Lys-Phe-COOH and not NH$_2$-Phe-Lys-COOH. In what direction does messenger translation proceed along the polynucleotide chain?

1390 Kossel et al. (1967), using a synthetic messenger RNA with the repeating sequence $5'$-(UAUC)$_n$ $\cdots$ $3'$ in a cell-free amino acid-incorporating system from *E. coli* B, found that the messenger directed the synthesis of a polypeptide with the repeating tetrapeptide sequence Tyr-Leu-Ser-Ile. Chymotrypsin digestion (breaks peptide bonds on carboxyl side of tyrosine) produced tetrapeptides, NH$_2$-Tyr-Leu-Ser-Ile-COOH. Show how those results established that the *m*RNA is read $5'$ to $3'$.

1391 In vitro, as well as in vivo, the coat-protein gene of f2 virus (single-stranded RNA) directs the synthesis of its polypeptide. The N-terminal amino acid sequence of this protein specified by the *wild-type* and a *nonsense*, *Su4A*, allele of the gene were found by Webster et al. (1966) to be as shown.

a Why is F-methionine present in coat protein produced in vitro but not in that produced in vivo?

b Why are different results obtained with the mutant in hosts Su^+ and Su^-?

Table 1391

Virus allele	Genotype of E. coli used as host or source of extract	System	N-terminal residues
wild type	Su^+	in vivo	Ala-Ser-Asp-Phe-Thr-Gln-Phe-Val
wild type	Su^+	in vitro	F-Met-Ala-Ser-Asp-Phe-Thr-Gln-Phe-Val
nonsense	Su^-	in vivo	Ala-Ser-Asp-Phe-Thr
nonsense	Su^+	in vivo	Ala-Ser-Asp-Phe-Thr-Ser-Phe-Val
	Su^-	in vitro	F-Met-Ala-Ser-Asp-Phe-Thr

c What does Su^+ do that Su^- cannot?

d What is the base-pair sequence at the beginning of the coat-protein gene?

1392 When an amino acid enters a growing polypeptide chain, it is linked to the chain by a peptide bond formed between its amino group and the carboxyl group in the last amino acid in the chain. One of the two methionine tRNAs in E. coli carries methionine to the messenger RNA in the formylated form (F-Met-tRNA$_F$). The formyl group, CHO, replacing H in NH$_2$, prevents this amino group from forming a peptide bond. tRNAs for amino acids other than methionine are not formylated (Marcker, 1965; Clarke and Marcker, 1966a,b, 1968).

a Show how a peptide bond is formed.

b Does formylated methionine appear to occupy a special position in the polypeptide chain? If so, what is it and why? Illustrate.

1393 Methionine tRNA of E. coli exists in two forms only, one of which can have its methionine formylated. In a cell-free system from E. coli, only two of a large number of synthetic messengers (random poly-UG and random poly-UAG) were capable of incorporating methionine into polypeptide chains. When only the formylated Met-tRNA was present, both random poly-UAG and random poly-UG directed the synthesis of polypeptides which contained methionine, but only in the start position (NH$_2$ end of chain). The same results were obtained even when the formyl group was not attached to formylatable Met-tRNA. The nonformylated Met-tRNA responded only to random poly-UAG. The polypeptide so produced has methionine only in internal positions (Clarke and Marcker, 1966a, 1968). Binding of Met-tRNA$_M$ and F-Met-tRNA$_F$ to ribosomes in the presence of various trinucleoside diphosphates gave the results tabulated. Synthetic messengers poly-$(GUG)_n$ and poly-$(AUG)_n$, in which the same sequences were repeated over and over, lead to formation of polypeptides with formylmethionine in the start position. The rest of poly-$(GUG)_n$-directed chains contained valine, whereas the rest of the poly-$(AUG)_n$ polypeptides contained methionine.

Table 1393

Trinucleotide	Met-tRNA$_M$	F-Met-tRNA$_F$ or Met-tRNA$_M$
UAG	−	−
AUU	−	−
AAG	−	−
AUG	+	+
UUG	−	−
GUG	−	+
UGU	−	−

Key: + = binding; − = no binding.

a What are the apparent functions in protein synthesis of each of the two species of tRNA that carry methionine?

b What may be the role of the formyl group in protein synthesis?

c Which are the chain initiating codons? Do they perform different functions within the gene? Explain.

1394 The codons for polypeptide-chain initiation are AUG and GUG. Moreover, mRNA from the RNA phage f2 can direct the synthesis in a cell-free system of phage coat protein of which the in vitro N-terminal amino acid sequence is F-Met-Ala-Ser-Asp-Phe-Thr. In vitro experiments by Nomura and Lowry (1967) revealed the following:

1 Addition of f2 RNA to 30S ribosome particles stimulates binding of F-Met-tRNA$_F$ to these particles. Such stimulation does not occur with 70S ribosomal particles, 30S-50S complexes, or 50S subunits alone.

2 f2 mRNA does not direct the binding of any other tRNA to 30S particles.

3 f2 mRNA directs binding of all tRNAs (other than tRNA$_F$) to 70S or 30S-50S complexes (formed by addition of 50S to 30S subunits).

4 When the synthetic mRNA AUG (1:1:1) is used in random sequence, which contains the initiator codon AUG and also valine codons GUA, GUU, and GUG in addition to others, the 30S particles bind F-Met-tRNA$_F$ only. Addition of 50S particles results in the 30S-50S complex binding Val-tRNA and Met-tRNA$_M$ but not F-Met-tRNA$_F$.

a The most widely accepted hypothesis of the mechanism of polypeptide-chain initiation until 1968 was that 70S ribosomes bind to the initiator codon in mRNA and that the resultant mRNA-ribosome complex binds F-Met-tRNA$_F$. Do these results indicate that this hypothesis has to be modified? If so, how?

b Since ribosomes of polysomes are known to be 70S, what events appear to follow the initiation steps to permit protein synthesis to continue? (See Nomura et al., 1967; Guthrie and Nomura, 1968.)

1395 AUG codes for F-methionine if it is in the initiator (first) position of a gene and for nonformylatable methionine when it is located elsewhere in the gene. The synthetic messenger poly-AUG, which contains the three bases in repeating trinucleotide sequence, directs the synthesis of polymethionine only in the presence of both F-Met-tRNA$_F$ and Met-tRNA$_M$ at low (apparently in vivo) magnesium-ion concentration. In in vitro experiments by Ghosh and Khorana (1967) at low magnesium-ion concentration the significant findings were as follows:

1 The 30S ribosome particles can bind F-Met-tRNA$_F$ in the presence of poly-AUG at low magnesium-ion concentration.

2 The tRNA$_M$ (carrying methionine) could be bound only after addition of 50S particles to the 30S particles.

3 The F-Met-tRNA$_F$–30S–poly-AUG complex, when supplemented with 50S particles and Met-tRNA$_M$, resulted in the formation of 70S ribosomes containing both F-Met-tRNA$_F$ and Met-tRNA$_M$. This 70S complex resulted in the synthesis of the dipeptide F-Met-Met.

Most molecular biologists believe that peptide-bond formation on the 70S ribosomes involves two binding sites (Gilbert, 1963a,b; Warner and Rich, 1964b; Suzuka et al., 1966). If these results support this belief, what may be the function of the two binding sites? (See Bretscher and Marcker, 1966.)

1396 70S ribosomes consist of two subunits, 30S and 50S. According to the most widely accepted view of translation (at least until recently), the 70S particle attaches to the mRNA at the initiator codon and then reads the mRNA by moving, one codon at a time, to the terminal (nonsense) codon of the gene, at which the polypeptide chain synthesis is terminated and the 70S ribosome released. Kaempfer et al. (1968) labeled ribosomes in growing bacteria with heavy isotopes (^{13}C, ^{15}N) and then transferred the bacteria to a medium containing light isotopes (^{12}C, ^{14}N), where growth continued. As growth progressed, heavy ribosomes were progressively replaced by two species of hybrid 70S particles, heavy 50S and light 30S and vice versa. Are 70S subunits permanently associated in vivo? Explain. See Kaempfer (1968), who also discusses other aspects of ribosome behavior pertinent to protein synthesis; Schlessinger et al. (1967); and Davis (1971).

1397 Esterification of amino acids to tRNAs is catalyzed by aminoacyl-tRNA synthetases, each kind being specific for one amino acid. The Ser-tRNA synthetase attaches serine to two different Ser-tRNA species. One responds to codons AGU and AGC; the other responds to codons UCA and UCG (Sundharadas et al., 1968). Does the anticodon serve as the recognition site for the synthetase? Explain.

1398 The following data are based on results of Capecchi (1967), Scolnick et al. (1968), Caskey et al. (1968), and Milman et al. (1969). In an in vitro system containing the messenger AUGAAAUUUUAA, ribosomes, tRNAs, amino acids, and initiating and elongation factors, a polypeptide containing F-Met-Lys-Phe is synthesized but not released from the ribosome. When the protein factors R_1, R_2, and S are added, the results shown in the table are obtained.

Table 1398

Factor	Polypeptide chain
R_1 with or without S	Released from ribosome
R_2 with or without S	Released from ribosome
S alone	Not released

a What appears to be the function of each of these factors?

b Do the data imply there is a *t*RNA species for chain-terminating codons in nonsuppressor-containing, *Su*⁻, hosts?

1399 Electrophoresis in polyacrylamide gel columns by Leboy et al. (1964) showed that 1 of the 19 or 20 ribosomal proteins in the 30S subunit (the K character) is present only in *E. coli* strain K12. Phage P1kc grown on *streptomycin-resistant, str-r*, K12 was used to infect *streptomycin-sensitive, str-s*, strain C. All the 34 *str-r* transductants selected possessed the K character of the donor. What do these results imply regarding the location of the cistron for this protein? See O'Neil et al. (1969).

1400 a Malkin and Rich (1967) showed that proteolytic enzymes acting on the growing polypeptide chain on reticulocyte ribosomes are unable to digest the 30 to 35 amino acids proximal to the peptidyl *t*RNA. What do these results suggest regarding the location of the *t*RNA bearing the nascent polypeptide chain?

b Takanami et al. (1965*b*) showed that the fragment of *m*RNA bound to ribosomes is resistant to RNase digestion. Suggest why bound *m*RNA might be resistant to RNase.

c Kuechler and Rich (1970) found that if an initiation complex is formed involving *E. coli* ribosomes, R17 *wild-type* RNA, and F-Met-*t*RNA_F, and if the unprotected *m*RNA is digested with RNase, the remaining complex is still able to synthesize the N-terminal pentapeptide (F-Met-Ala-Ser-Asn-Phe) of the coat protein attached to ribosome-bound *t*RNA. How many codons is the initiator site from the point of entry of *m*RNA on the ribosome? Explain.

d When an *amber* mutant of R17 virus carrying an *amber* codon in the seventh position of the coat-protein gene is used in an amino acid–incorporating system, a free hexapeptide (F-Met-Ala-Ser-Asn-Phe-Thr) is produced. In a system deficient in asparagine a tripeptide (F-Met-Ala-Ser) is formed which is attached to *t*RNA bound to the ribosome (in this complex the *amber* codon is four codons from the peptidyl site occupied by *t*RNA containing the tripeptide). After RNase digestion of unprotected *m*RNA the ribosomes were added to an in vitro system with all amino acids. A hexapeptide (N-terminal, see above) is formed and released. Is the peptidyl site the same distance from the point of entry of *m*RNA on the ribosome as the initiator site?

e The portion of the R17 coat-protein gene *m*RNA protected by the ribosome from RNase digestion was isolated and sequenced by Steitz (1969) with the following results:

5′-AGAGCCUAACCGGGGUUUGAAGGAUGGCUUCUAACUUU-3′

Are the results in agreement with those of Kuechler and Rich? Illustrate by assigning amino acids to codons in the sequenced messenger.

1401 Initiation of protein synthesis involves the formation of an initiation complex in which the interaction of *m*RNA, F-Met-*t*RNA$_F$, and initiation factors (F1, F2, and probably F3) occurs on the 30S subunit, the complex being subsequently joined by a 50S particle to form the active 70S ribosome that reads the rest of the genetic message. Mukundan et al. (1968) investigated the

Table 1401

Experiment	In vitro reaction mixture	Amount of F-Met-*t*RNA$_F$ bound,* picomoles	
		Plus AUG	Minus AUG
1*a.* Derived 30S	Complete†	4.50	0.32
	Minus GTP	0.26	0.31
	Minus factors	0.25	0.33
	Minus factors and GTP	0.30	0.31
b. Native 30S	Complete	2.94	0.65
	Minus GTP	0.99	
	Minus factors	0.82	
2. Native 30S‡	Complete	14.1	1.9
	Minus GTP	3.2	1.6
	Minus GTP plus GMP-PCP	11.9	1.6
3*a.* Binding of F-Met-*t*RNA$_F$ to 30S	Complete (GTP)	3.5	0.3
	Complete (GMP-PCP)	3.2	0.3
b. F-Met-puro synthesis	Complete (GTP)	1.3	0.94
	Plus 50S	5.1	1.00
	Complete (GMP-PCP)	1.0	0.98
	Plus 50S	1.0	1.02

* Experiment 3*b* refers to picomoles of F-Met-puro formed.

† Complete reaction mixture contains 30S particles, initiation factors, F-Met-*t*RNA$_F$, GTP, AUG, magnesium ions, NH^4, tris buffer, and DTT.

‡ Basically identical results were obtained with derived 30S.

requirements for formation of the initiation complex and obtained the results shown.

1 Experiment 1 shows the binding of F-Met-$tRNA_F$ to native and derived 30S particles.
2 Experiment 2 compares the binding of F-Met-$tRNA_F$ to native and derived 30S particles when GTP and GMP-PCP are used.
3 Experiment 3 indicates the reaction between F-Met-$tRNA_F$ prebound to 30S particles and puromycin (puro) in the presence and absence of 50S particles.

a Are GTP and initiation factors (F1 and F2) required for binding of F-Met-$tRNA_F$ to ribosomes? Are the sites of action on the 30S or 50S subunits?

b F-methionine forms a peptide bond with puromycin. Do peptide bonds form on 30S, 50S, or both particles? Explain. See Monro (1967), Hille et al. (1967).

c Does GTP function in peptide-bond formation? What other functions does GTP perform? Which of these can GMP-PCP carry out? See Revel et al., (1968), Hershey et al. (1969), Kolakofsky et al. (1969).

1402 The chain-initiating amino acid is formylmethionine. Livingstone and Leder (1969) found an enzyme that removes the formyl group from the amino acid. Do you think the enzyme acts on N-F-Met-$tRNA_F$ before or after methionine forms the initial peptide bond with the second amino acid?

1403 a When streptomycin is added to a cell-free system containing ribosomes from a *streptomycin-sensitive*, *str-s*, bacterium, protein synthesis is inhibited and amino acids other than those dictated by the genetic code in the messenger are incorporated. In a similar system containing ribosomes from a *streptomycin-resistant*, *str-r*, strain, protein synthesis proceeds normally, and normal proteins are produced. The ribosome consists of two subunits; one, the larger, sediments on centrifugation as 50S, and the smaller one sediments as 30S. Normal ribosomes can be reconstituted by mixing the subunits under appropriate conditions, and thus "hybrid" ribosomes, containing 30S particles from one strain of bacteria and 50S from another, can be produced. The effects of streptomycin on hybrid ribosomes from *streptomycin-resistant* and -*sensitive* bacteria are shown in Table 1403A. What roles do the 30S particles appear to play in translation?

Table 1403A

	50S subunit	
30S subunit	From *str-s*	From *str-r*
From *str-s*	Inhibition; abnormal proteins	Inhibition; abnormal proteins
From *str-r*	No inhibition; normal proteins	No inhibition; normal proteins

b Ribosomal proteins and *r*RNAs can also be separated from either subunit and can be similarly reconstituted to re-form the original particles (30S subunits contain about 20 proteins and 20 *r*RNAs).   The data for the effect of streptomycin on reconstituted ribosomes prepared in this way are shown in Table 1403*B*.   Is it the *r*RNA or the *r*-protein that is involved in the inhibition and translational error caused by streptomycin?

Table 1403*B*

	From *resistant* strain	From *sensitive* strain
Protein of 30S	No inhibition; normal proteins	Inhibition; abnormal proteins
*r*RNA of 30S	No inhibition; normal proteins	No inhibition; normal proteins

c The P-10 protein of 30S subunits, derived from a *sensitive* strain, causes inhibition and translation error with streptomycin; that from *resistant* strains does not.   Moreover, systems using synthetic messengers are not affected (it is known that these messengers do not require initiating sequences such as that for formylmethionine).   What do you think the P-10 protein does in translation?   See Nomura (1969) and Ozaki et al. (1969).

REFERENCES

Abelson, J. N., et al. (1970), *J. Mol. Biol.*, **47**:15.

Adams, J. M., and M. R. Capecchi (1966), *Genetics*, **55**:147.

Aloni, Y., L. E. Hatlen, and G. Attardi (1971), *J. Mol. Biol.*, **56**:555.

Ames, B. N., and P. E. Hartman (1963), *Cold Spring Harbor Symp. Quant. Biol.*, **28**:349.

Apirion, D., S. L. Phillips, and D. Schlessinger (1969), *Cold Spring Harbor Symp. Quant. Biol.*, **34**:117.

—— and D. Schlessinger (1968), *J. Bacteriol.*, **96**:768.

Attardi, G., and F. Amaldi (1970), *Annu. Rev. Biochem.*, **39**:183.

Baglioni, C. (1962), *Proc. Natl. Acad. Sci.*, **44**:1880.

Baltus, E. (1959), *Biochim. Biophys. Acta*, **33**:337.

Barner, H. D., and S. S. Cohen (1954), *J. Bacteriol.*, **68**:80.

—— and —— (1957), *J. Bacteriol.*, **74**:350.

Bautz, E. K. F., and B. D. Hall (1962), *Proc. Natl. Acad. Sci.*, **48**:400.

——, F. A. Bautz, and J. L. Dunn (1969), *Nature*, **223**:1022.

Beaudet, A. L., and C. T. Caskey (1971), *Proc. Natl. Acad. Sci.*, **68**:619.

Belozersky, A. N., and A. S. Spirin (1960), in E. Chargaff and J. N. Davidson (eds.), "The Nucleic Acids," vol. III, pp. 147–185, Academic, New York.

Berg, D., K. Barrett, D. Hinkle, J. McGrath, and M. Chamberlin (1969), *Fed. Proc.*, **28**:659.

Berg, P., and E. J. Ofengand (1958), *Proc. Natl. Acad. Sci.*, **44**:78.

Birnbaum, L. S., and S. Kaplan (1971), *Proc. Natl. Acad. Sci.*, **68**:925.

Birnstiel, M., et al. (1968), *Nature*, **219**:454.

Bishop, J. O., J. Leahy, and R. S. Schweet (1960), *Proc. Natl. Acad. Sci.*, **46**:1030.

Brachet, J. (1954), *Nature*, **174**:876.

——— (1957), "Biochemical Cytology," Academic, New York.

——— and H. Chantrenne (1956), *Cold Spring Harbor Symp. Quant. Biol.*, **21**:329.

Bremer, H., and R. Bruner (1968), *Mol. Gen. Genet.*, **107**:6.

———, M. W. Konrad, K. Gaines, and G. S. Stent (1965), *J. Mol. Biol.*, **13**:540.

Brenner, S., F. Jacob, and M. Meselson (1961), *Nature*, **190**:576.

Bretscher, M. (1966), *Cold Spring Harbor Symp. Quant. Biol.*, **31**:289.

——— (1968), *J. Mol. Biol.*, **34**:131.

——— (1969), *Cold Spring Harbor Symp. Quant. Biol.*, **34**:651.

———, and K. Marcker (1966), *Nature*, **211**:380.

Brown, D. D., and I. B. Dawid (1968), *Science*, **160**:272.

——— and J. B. Gurdon (1964), *Proc. Natl. Acad. Sci.*, **51**:139.

Bryson, V., and H. J. Vogel (eds.) (1965), "Evolving Genes and Proteins," Academic, New York.

Burgess, R. R. (1969), *J. Biol. Chem.*, **244**:6168.

———, A. Travers, J. Dunn, and E. K. F. Bautz (1969), *Nature*, **221**:43.

Caffier, H., et al. (1971), *Nat. New Biol.*, **229**:239.

Capecchi, M. R. (1967), *Proc. Natl. Acad. Sci.*, **58**:1144.

Caskey, T., E. Scolnick, R. Tompkins, J. Goldstein, and G. Milman (1969), *Cold Spring Harbor Symp. Quant. Biol.*, **34**:479.

———, R. Tompkins, E. Scolnick, T. Caryk, and M. Nerenberg (1968), *Science*, **162**:135.

Caspersson, T. (1941), *Naturwiss.*, **29**:33.

Chamberlin, M., and P. Berg (1962), *Proc. Natl. Acad. Sci.*, **48**:81.

Chantrenne, H. (1961), "The Biosynthesis of Proteins," Pergamon, New York.

Chapeville, F., F. Lipmann, G. von Ehrenstein, B. Weisblum, W. J. Ray, Jr., and S. Benzer (1962), *Proc. Natl. Acad. Sci.*, **48**:1086.

Clark, B. F. C., and K. A. Marcker (1966*a*), *J. Mol. Biol.*, **17**:394.

——— and ——— (1966*b*), *Nature*, **211**:378.

——— and ——— (1968), *Sci. Am.*, **218**(January):36.

Claude, A. (1943), *Biol. Symp.*, **10**:111.

Cohen, S. S. (1948), *J. Biol. Chem.*, **174**:281.

Cold Spring Harbor Symp. Quant. Biol., **26**(1961), "Cellular Regulatory Mechanisms."

Cold Spring Harbor Symp. Quant. Biol., **28** (1963), "Synthesis and Structure of Macromolecules."

Cold Spring Harbor Symp. Quant. Biol., **34** (1969), "The Mechanism of Protein Synthesis."

Cold Spring Harbor Symp. Quant. Biol., **35** (1970), "Transcription of Genetic Material."

Colli, W., I. Smith, and M. Oishi (1971), *J. Mol. Biol.*, **56**:117.

Craig, N., and R. P. Perry (1971), *Nat. New Biol.*, **229**:75.

Craven, G. R., R. Gavin, and T. Fanning (1969), *Cold Spring Harbor Symp. Quant. Biol.*, **34**:129.

Crick, F. H. C. (1958), *Symp. Soc. Exp. Biol.*, **12**:138.

Davern, C. I., and M. Meselson (1960), *J. Mol. Biol.*, **2**:153.

Davies, J. (1964), *Proc. Natl. Acad. Sci.*, **51**:659.

Davis, B. D. (1971), *Nature*, **231**:153.

Dekio, S., R. Takata, and S. Osawa (1970), *Mol. Gen. Genet.*, **109**:131.

Deusser, E., G. Stöffler, H. G. Wittmann, and D. Apirion (1970), *Mol. Gen. Genet.*, **109**:298.

Dintzis, H. M. (1961), *Proc. Natl. Acad. Sci.*, **47**:247.

Doctor, B. P., J. Loebel, M. Sodd, and D. Winter (1969), *Science*, **163**:693.

Dubnoff, J. S., and U. Maitra (1969), *Cold Spring Harbor Symp. Quant. Biol.*, **34**:301.

Dunn, J. L., F. A. Bautz, and E. K. F. Bautz (1971), *Nat. New Biol.*, **230**:94.

Egawa, K., Y. C. Choi, and H. Busch (1971), *J. Mol. Biol.*, **56**:565.

von Ehrenstein, G., B. Weisblum, and S. Benzer (1963), *Proc. Natl. Acad. Sci.*, **49**:669.

Eisenstadt, J. M., and G. Brawerman (1967), *Proc. Natl. Acad. Sci.*, **58**:1560.

Elsdale, T. R., M. Fischberg, and S. Smith (1958), *Exp. Cell Res.*, **14**:642.

Fincham, J. R. S., and P. R. Day (1971), "Fungal Genetics," 3d ed., Blackwell, Oxford.

Flaks, J., P. Leboy, E. Berge, and C. G. Kurland (1966), *Cold Spring Harbor Symp. Quant. Biol.*, **31**:623.

Fogel, S., and P. S. Sypherd (1968), *Proc. Natl. Acad. Sci.*, **59**:1329.

Fraenkel-Conrat, H., and B. Singer (1957), *Biochim. Biophys. Acta*, **24**:540.

Gall, J. G. (1968), *Proc. Natl. Acad. Sci.*, **60**:553.

Garen, A. (1968), *Science*, **160**:149.

Gefter, M. L., and R. Russell (1969), *J. Mol. Biol.*, **39**:145.

Geiduschek, E., and R. Haselkorn (1969), *Annu. Rev. Biochem.*, **38**:647.

Ghosh, H. P., and H. G. Khorana (1967), *Proc. Natl. Acad. Sci.*, **58**:2455.

——, D. Soll, and H. G. Khorana (1967), *J. Mol. Biol.*, **25**:275.

Giacomoni, D., and S. Spiegelman (1962), *Science*, **138**:1328.

Gierer, A. (1963), *J. Mol. Biol.*, **6**:148.

Gilbert, W. (1963a), *J. Mol. Biol.*, **6**:374.

—— (1963b), *J. Mol. Biol.*, **6**:389.

Goldstein, A., and B. J. Brown (1961), *Biochim. Biophys. Acta*, **53**:438.

Goldstein, J., et al. (1970), *Proc. Natl. Acad. Sci.*, **65**:430.

Goodman, H. M., et al. (1968), *Nature*, **217**:1019.

—— and A. Rich (1962), *Proc. Natl. Acad. Sci.*, **48**:2101.

—— and —— (1963), *Nature*, **199**:318.

Grau, O., A. Guha, E. Geiduschek, and W. Szybalski (1969), *Nature*, **224**:1105.

Gros, F., H. Hiatt, W. Gilbert, C. G. Kurland, R. W. Risebrough, and J. D. Watson (1961), *Nature*, **190**:581.

Guild, W. R., and M. Robison (1963), *Proc. Natl. Acad. Sci.*, **50**:106.

Gupta, N. K., et al. (1968), *Proc. Natl. Acad. Sci.*, **60**:1338.

Guthrie, C., H. Nashimoto, and M. Nomura (1969), *Proc. Natl. Acad. Sci.*, **63**:384.

—— and M. Nomura (1968), *Nature*, **219**:232.

Hall, B. D., and P. Doty (1959), *J. Mol. Biol.*, **1**:111.

—— and S. Spiegelman (1961), *Proc. Natl. Acad. Sci.*, **47**:137.

Hämmerling, J. (1953), *Int. Rev. Cytol.*, **2**:475.

Hardy, J., C. Kurland, P. Voynew, and G. Mora (1969), *Biochemistry*, **8**:2897.

Harris, H. (1966), "Human Biochemical Genetics," Cambridge University Press, Cambridge.

Hartman, P. E., and S. R. Suskind (1969), "Gene Action," Prentice-Hall, Englewood Cliffs, N.J.

Hatlen, L., and G. Attardi (1971), *J. Mol. Biol.*, **56**:535.

Hayashi, M., M. N. Hayashi, and S. Spiegelman (1963), *Proc. Natl. Acad. Sci.*, **50**:664.

——, ——, and —— (1964), *Proc. Natl. Acad. Sci.*, **51**:351.

—— and S. Spiegelman (1961), *Proc. Natl. Acad. Sci.*, **47**:1564.

Hayashi, M. N., and M. Hayashi (1966), *Proc. Natl. Acad. Sci.*, **55**:635.

Hayes, W. (1968), "The Genetics of Bacteria and Their Viruses," 2d ed., Blackwell, Oxford.

Hecht, L. I., M. L. Stephenson, and P. C. Zamecnik (1959), *Proc. Natl. Acad. Sci.*, **45**:505.

Hershey, J. W. B., K. F. Dewey, and R. E. Thach (1969), *Nature*, **222**:944.

Heywood, S. M. (1970), *Nature*, **225**:696.

Hill, W. E., G. P. Rossetti, and K. E. van Holde (1969), *J. Mol. Biol.*, **44**:263.

Hille, M. B., M. J. Miller, K. Iwasaki, and A. J. Wahba (1967), *Proc. Natl. Acad. Sci.*, **58**:1652.

Hoagland, M. B., M. L. Stephenson, J. F. Scott, R. J. Hecht, and P. C. Zamecnik (1958), *J. Biol. Chem.*, **231**:241.

——, P. C. Zamecnik, and M. L. Stephenson (1957), *Biochim. Biophys. Acta*, **24**:215.

Holley, R. W. (1966), *Sci. Am.*, **214**(February):30.

—— et al. (1965), *Science*, **147**:1462.

Hsia, D. Y. Y., K. W. Driscoll, W. Troll, and W. E. Knox (1956), *Nature*, **178**:1239.

——, W. E. Knox, K. V. Quin, and R. S. Paine (1958), *Pediatrics*, **21**:178.

Hurwitz, J., and J. T. August (1963), *Prog. Nucleic Acid Res.*, **1**:59.

——, ——, and R. Diringer (1960), *Biochem. Biophys. Res. Commun.*, **3**:15.

Hurwitz, J., and J. J. Furth (1962), *Sci. Am.*, **206**(February):41.

——, ——, M. Anders, P. J. Oritz, and J. T. August (1961), *Cold Spring Harbor Symp. Quant. Biol.*, **26**:91.

Imamoto, F., T. Yamane, and N. Sueoka (1965), *Proc. Natl. Acad. Sci.*, **53**:1456.

Ingram, V. M. (1965), "The Biosynthesis of Macromolecules," Benjamin, New York.

Ishitsuka, H., and A. Kaji (1970), *Proc. Natl. Acad. Sci.*, **66**:168.

Itoh, T., E. Otaka, and S. Osawa (1968), *J. Mol. Biol.*, **33**:109.

Jacob, F., and J. Monod (1961), *J. Mol. Biol.*, **3**:318.

Jervis, G. A. (1939), *J. Ment. Sci.*, **85**:719.

—— (1953), *Proc. Soc. Exp. Biol. Med.*, **82**:514.

Jordan, B. R., B. G. Forget, and R. Monier (1971), *J. Mol. Biol.*, **55**:407.

Kaempfer, R. (1968), *Proc. Natl. Acad. Sci.*, **61**:106.

—— and M. Meselson (1968), *J. Mol. Biol.*, **34**:703.

——, ——, and H. J. Raskas (1968), *J. Mol. Biol.*, **31**:277.

Kaji, A., K. Igarashi, and H. Ishitsuka (1969), *Cold Spring Harbor Symp. Quant. Biol.*, **34**:167.

Kaji, H., I. Suzuki, and A. Kaji (1966), *J. Biol. Chem.*, **241**:1251.

Kaltschmidt, E., M. Dzionaya, D. Donner, and H. G. Wittmann (1967), *Mol. Gen. Genet.*, **100**:364.

——, ——, and H. G. Wittmann (1970), *Mol. Gen. Genet.*, **109**:292.

—— and H. G. Wittmann (1970), *Proc. Natl. Acad. Sci.*, **67**:1276.

Knight, E., Jr., and J. E. Darnell (1967), *J. Mol. Biol.*, **28**:491.

Knowland, J., and L. Miller (1970), *J. Mol. Biol.*, **53**:321.

Kolakofsky, D., K. F. Dewey, J. W. B. Hershey, and R. E. Thach (1968a), *Proc. Natl. Acad. Sci.*, **61**:1066.

——, ——, and R. Thach (1969), *Nature*, **223**:694.

——, T. Ohta, and R. Thach (1968b), *Nature*, **220**:244.

Kossel, H., A. Morgan, and H. Khorana (1967), *J. Mol. Biol.*, **26**:449.

Kreider, G., and B. L. Brownstein (1971), *J. Mol. Biol.*, **61**:135.

Kuechler, E., and A. Rich (1970), *Nature*, **225**:920.

Kurland, C. G. (1960), *J. Mol. Biol.*, **2**:83.

Kurnick, N. B. (1955), *J. Histochem. Cytochem.*, **3**:290.

Lamfrom, H., C. S. McAlughlin, and A. Sarabhai (1966), *J. Mol. Biol.*, **22**:355.

Lampen, J. O., and V. Bryson (eds.) (1968), "Organizational Biosynthesis," Academic, New York.

Leboy, P. S., E. C. Cox, and J. G. Flaks (1964), *Proc. Natl. Acad. Sci.*, **52**:1367.

Levitt, M. (1969), *Nature*, **224**:759.

Lewis, J. B., and P. Doty (1970), *Nature*, **225**:510.

Lipmann, F. (1969), *Science*, **164**:1024.

Littlefield, J. W., E. B. Keller, G. Gross, and P. Zamecnik (1955), *J. Biol. Chem.*, **217**:111.

Livingston, D. M., and P. Leder (1969), *Biochemistry*, **8**:435.

Lodish, H. F., and H. D. Robertson (1969), *Cold Spring Harbor Symp. Quant. Biol.*, **34**:655.

Lu, P., and A. Rich (1971), *J. Mol. Biol.*, **58**:513.

Lucas-Lenard, J., and A.-L. Haenni (1969), *Proc. Natl. Acad. Sci.*, **63**:93.

——, P. Tao, and A.-L. Haenni (1969), *Cold Spring Harbor Symp. Quant. Biol.*, **34**:455.

McCarthy, B. J., and J. J. Holland (1965), *Proc. Natl. Acad. Sci.*, **54**:880.

McConkey, E. H., and J. W. Hopkins (1964), *Proc. Natl. Acad. Sci.*, **51**:1197.

McQuillen, K., R. B. Roberts, and R. J. Britten (1959), *Proc. Natl. Acad. Sci.*, **45**:1437.

Maaløe, O. (1960), *Symp. Soc. Gen. Microbiol.*, **10**:272.

Maden, B. E. H. (1968), *Nature*, **219**:685.

Madison, J. T. (1968), *Annu. Rev. Biochem.*, **37**:131.

——, G. Everett, and H. Kung (1966), *Cold Spring Harbor Symp. Quant. Biol.*, **31**:409.

Maitra, U., and J. Hurwitz (1965), *Proc. Natl. Acad. Sci.*, **54**:815.

Malkin, L. I., and A. Rich (1967), *J. Mol. Biol.*, **26**:329.

Mangiarotti, G. (1969), *Nature*, **222**:947.
—— and D. Schlessinger (1966), *J. Mol. Biol.*, **20**:123.
Marcker, K. (1965), *J. Mol. Biol.*, **14**:63.
Marmur, J., and C. M. Greenspan (1963), *Science*, **142**:387.
—— and D. Lane (1960), *Proc. Natl. Acad. Sci.*, **46**:453.
Matsuura, S., Y. Tashiro, S. Osawa, and E. Otaka (1970), *J. Mol. Biol.*, **47**:383.
Miller, L., and J. Knowland (1970), *J. Mol. Biol.*, **53**:329.
Miller, O. L., Jr., et al. (1970), *Cold Spring Harbor Symp. Quant. Biol.*, **35**:505.
—— and B. R. Beatty (1969), *Science*, **164**:955.
——, B. A. Hamkalo, and C. A. Thomas, Jr. (1970), *Science*, **169**:392.
Milman, G., J. Goldstein, E. Scolnick, and T. Caskey (1969), *Proc. Natl. Acad. Sci.*, **63**:183.
Miyamoto, M., and T. B. Fitzpatrick (1957), *Nature*, **179**:199.
Modolell, J., D. Vazquez, and R. E. Monro (1971), *Nat. New Biol.*, **230**:109.
Möller, W., and J. Widdowson (1967), *J. Mol. Biol.*, **24**:367.
Monier, R., et al. (1969), *Cold Spring Harbor Symp. Quant. Biol.*, **34**:139.
Monro, R. E. (1967), *J. Mol. Biol.*, **26**:147.
—— (1969), *Nature*, **223**:903.
——, T. Staehelin, M. L. Celma, and D. Vazquez (1969), *Cold Spring Harbor Symp. Quant. Biol.*, **34**:357.
Moore, P. B., R. R. Traut, H. Noller, P. Pearson, and H. Delivs (1968), *J. Mol. Biol.*, **31**:441.
Morell, P., J. Smith, D. Dubanau, and J. Marmur (1966), *Biochemistry*, **6**:258.
Mukundan, M. A., J. W. B. Hershey, K. F. Dewey, and R. E. Thach (1968), *Nature*, **217**:1013.
Nakamoto, T., and D. Kolakofsky (1966), *Proc. Natl. Acad. Sci.*, **55**:606.
Nichols, J. L. (1970), *Nature*, **225**:147.
Nirenberg, M., and J. H. Matthaei (1961), *Proc. Natl. Acad. Sci.*, **47**:1588.
Noll, H., T. Staehelin, and F. O. Wettstein (1963), *Nature*, **198**:632.
Nomura, M. (1969), *Sci. Am.*, **221**(October):28.
—— and C. V. Lowry (1967), *Proc. Natl. Acad. Sci.*, **58**:946.
——, ——, and C. Guthrie (1967), *Proc. Natl. Acad. Sci.*, **58**:1487.
——, S. Mizushima, M. Ozaki, P. Traut, and C. V. Lowry (1969), *Cold Spring Harbor Symp. Quant. Biol.*, **34**:49.
Ohta, T., S. Sarkar, and R. E. Thach (1967), *Proc. Natl. Acad. Sci.*, **58**:1638.
—— and R. E. Thach (1968), *Nature*, **219**:238.
Oishi, M., and N. Sueoka (1965), *Proc. Natl. Acad. Sci.*, **54**:483.
O'Neil, D. M., L. S. Baron, and P. S. Sypherd (1969), *J. Bacteriol.*, **99**:242.
Osawa, S. (1968), *Annu. Rev. Biochem.*, **37**:109.
——, E. Otaka, T. Itoh, and T. Fukui (1969), *J. Mol. Biol.*, **40**:321.
Otaka, E., et al. (1970), *J. Mol. Biol.*, **48**:499.
Ozaki, M., S. Mizushima, and M. Nomura (1969), *Nature*, **222**:333.
Palade, G. E. (1955), *J. Biophys. Biochem. Cytol.*, **1**:59.
Pardee, A. B. (1954), *Proc. Natl. Acad. Sci.*, **40**:263.
Pardue, M. L., et al. (1970), *Chromosoma*, **29**:268.
Parenti-Rosina, R., A. Eisenstadt, and J. Eisenstadt (1969), *Nature*, **221**:363.
Penrose, L. S. (1935), *Lancet*, **229**(2):192.
Perani, A., O. Tiboni, and O. Ciferri (1971), *J. Mol. Biol.*, **55**:107.
Perkowska, E., H. C. Macgregor, and M. L. Birnstiel (1968), *Nature*, **217**:649.
Perry, R. P. (1967), *Prog. Nucleic Acid Res.*, **6**:219.
Petermann, M. L. (1965), "The Physical and Chemical Properties of Ribosomes," American Elsevier, New York.
—— and A. Pavlovec (1966), *Biochim. Biophys. Acta*, **114**:264.
Prescott, D. M. (1964), *Prog. Nucleic Acid Res.*, **3**:35.
Quagliarotti, G., and F. M. Ritossa (1968), *J. Mol. Biol.*, **36**:57.
Rabinowitz, M., and M. E. Olson (1956), *Exp. Cell Res.*, **10**:747.

Revel, M., M. Herzberg, A. Becarevic, and F. Gros (1968), *J. Mol. Biol.*, **33**:231.

————, ————, and H. Greenshpan (1969), *Cold Spring Harbor Symp. Quant. Biol.*, **34**:261.

————, J. C. Lelong, G. Brawerman, and F. Gros (1968), *Nature*, **219**:1016.

Rich, A. (1963), *Sci. Am.*, **209**(December):44.

————, J. R. Warner, and H. M. Goodman (1963), *Cold Spring Harbor Symp. Quant. Biol.*, **28**:269.

Riley, M., A. B. Pardee, F. Jacob, and J. Monod (1960), *J. Mol. Biol.*, **2**:216.

Risebrough, R. W., A. Tissières, and J. D. Watson (1962), *Proc. Natl. Acad. Sci.*, **48**:430.

Ritossa, F. M., K. C. Atwood, and S. Spiegelman (1966), *Genetics*, **54**:819.

———— and S. Spiegelman (1965), *Proc. Natl. Acad. Sci.*, **53**:737.

Roberts, G. C. K., et al. (1969), *Proc. Natl. Acad. Sci.*, **62**:1151.

Roberts, J. W. (1969), *Nature*, **224**:1168.

Roberts, R. B. (1958), in R. B. Roberts (ed.), "Microsomal Particles and Protein Synthesis," introduction, Pergamon, Washington.

Rossett, R., and R. Monier (1963), *Biochim. Biophys. Acta*, **68**:653.

Russell, R. L., et al. (1970), *J. Mol. Biol.*, **47**:1.

Salas, M., M. Hille, J. Last, A. J. Wahba, and S. Ochoa (1967a), *Proc. Natl. Acad. Sci.*, **57**:387.

————, M. J. Miller, A. J. Wahba, and S. Ochoa (1967b), *Proc. Natl. Acad. Sci.*, **57**:1865.

Schackman, H. K., A. B. Pardee, and R. J. Stanier (1952), *Arch. Biochem. Biophys.*, **38**:245.

Schlessinger, D., and D. Apirion (1969), *Annu. Rev. Microbiol.*, **23**:387.

————, G. Mangiarotti, and D. Apirion (1967), *Proc. Natl. Acad. Sci.*, **58**:1782.

Schmidt, D. A., et al. (1970), *Nature*, **225**:1012.

Schreier, M. H., and H. Noll (1971), *Proc. Natl. Acad. Sci.*, **68**:805.

Schweizer, E., C. MacKechnie, and H. O. Halvorson (1969), *J. Mol. Biol.*, **40**:261.

Scolnick, E., G. Milman, M. Rosman, and T. Caskey (1969), *Proc. Natl. Acad. Sci.*, **64**:1235.

————, R. Tompkins, T. Caskey, and M. Nerenberg (1968), *Proc. Natl. Acad. Sci.*, **61**:768.

Sirlin, J. L., J. Jacob, and K. I. Kato (1962), *Exp. Cell Res.*, **27**:355.

Slayter, H. S., J. R. Warner, A. Rich, and C. E. Hall (1963), *J. Mol. Biol.*, **7**:652.

Slotnick, C. J., D. W. Visser, and S. C. Rittenberg (1953), *J. Biol. Chem.*, **203**:647.

Smith, A. E., K. A. Marcker, and M. B. Mathews (1970), *Nature*, **225**:184.

Smith, I., D. Dubnau, P. Morell, and J. Marmur (1968), *J. Mol. Biol.*, **33**:123.

Spiegelman, S. (1964), *Sci. Am.*, **210**(May):48.

Spirin, A. S., and L. P. Gavrilova (1969), "The Ribosome," Springer-Verlag, Berlin.

Staehelin, T., D. Maglott, and R. E. Monro (1969), *Cold Spring Harbor Symp. Quant. Biol.*, **34**:39.

————, F. O. Wettstein, H. Oura, and H. Noll (1964), *Nature*, **201**:264.

Stanley, W. M., Jr., M. Salas, A. Wahba, and S. Ochoa (1966), *Proc. Natl. Acad. Sci.*, **56**:290.

Steitz, J. (1969), *Nature*, **224**:957.

Stevens, A. (1960), *Biochem. Biophys. Res. Commun.*, **3**:92.

Stretton, A., S. Kaplan, and S. Brenner (1966), *Cold Spring Harbor Symp. Quant. Biol.*, **31**:173.

Stutz, E., and H. Noll (1967), *Proc. Natl. Acad. Sci.*, **57**:774.

Sueoka, N., and T. Yamane (1962), *Proc. Natl. Acad. Sci.*, **48**:1454.

Sugiura, M., T. Okamoto, and M. Takanami (1970), *Nature*, **225**:598.

Sugiyama, T. (1969), *Cold Spring Harbor Symp. Quant. Biol.*, **34**:687.

Sundharadas, G., et al. (1968), *Proc. Natl. Acad. Sci.*, **61**:693.

Suzuka, I., H. Kaji, and A. Kaji (1966), *Proc. Natl. Acad. Sci.*, **55**:1483.

Sypherd, P. S. (1971), *J. Mol. Biol.*, **56**:1971.

————, D. M. O'Neill, and M. M. Taylor (1969), *Cold Spring Harbor Symp. Quant. Biol.*, **34**:77.

Takanami, M., and M. Yan (1965a), *Proc. Natl. Acad. Sci.*, **54**:1450.

————, Y. Yan, and T. H. Jukes (1965b), *J. Mol. Biol.*, **12**:761.

Takata, R., et al. (1970), *Mol. Gen. Genet.*, **109**:123.

Tartof, K. D., and R. P. Perry (1970), *J. Mol. Biol.*, **51**:171.

Thach, R. E., et al. (1969), *Cold Spring Harbor Symp. Quant. Biol.*, **34**:277.

Thach, R. E., M. A. Cecere, T. A. Sundararajan, and P. Doty (1965), *Proc. Natl. Acad. Sci.*, **54**:1167.

Thach, S. S., and R. E. Thach (1971), *Nat. New Biol.*, **229**:219.

Tissières, A., D. Schlessinger, and F. Gros (1960), *Proc. Natl. Acad. Sci.*, **46**:1450.

——, J. D. Watson, D. Schlessinger, and B. R. Hollingsworth (1959), *J. Mol. Biol.*, **1**:221.

Tompkins, R. K., E. M. Scolnick, and C. T. Caskey (1970), *Proc. Natl. Acad. Sci.*, **65**:702.

Traub, P., and M. Nomura (1968a), *Proc. Natl. Acad. Sci.*, **59**:777.

—— and —— (1968b), *Science*, **160**:198.

—— and —— (1969a), *Cold Spring Harbor Symp. Quant. Biol.*, **34**:63.

—— and —— (1969b), *J. Mol. Biol.*, **40**:391.

Traut, R. R., et al. (1967), *Proc. Natl. Acad. Sci.*, **57**:1294.

Travers, A. A. (1969), *Nature*, **223**:1107.

—— (1970), *Nature*, **225**:1009.

—— (1971), *Nat. New Biol.*, **229**:69.

—— and R. R. Burgess (1969), *Nature*, **222**:537.

Vogel, H. J., V. Bryson, and J. O. Lampen (eds.) (1963), "Informational Macromolecules," Academic, New York.

Volkin, E., and L. Astrachan (1956), *Virology*, **2**:149.

—— and —— (1957), in W. E. McElroy (ed.), "The Chemical Basis of Heredity," pp. 686–695, Johns Hopkins, Baltimore.

Wagner, R. P., and H. K. Mitchell (1964), "Genetics and Metabolism," 2d ed., Wiley, New York.

Wahba, A. J., et al. (1969), *Cold Spring Harbor Symp. Quant. Biol.*, **34**:291.

Wallace, H., and M. L. Birnstiel (1966), *Biochim. Biophys. Acta*, **114**:296.

Waller, J. P. (1964), *J. Mol. Biol.*, **10**:319.

Warner, J. R., P. M. Knopf, and A. Rich (1963), *Proc. Natl. Acad. Sci.*, **49**:122.

—— and A. Rich (1964a), *Proc. Natl. Acad. Sci.*, **51**:1134.

—— and —— (1964b), *J. Mol. Biol.*, **10**:202.

——, ——, and C. E. Hall (1962), *Science*, **138**:1399.

Watson, J. D. (1963), *Science*, **140**:17.

—— (1970), "Molecular Biology of the Gene," Benjamin, New York.

Webster, R. E., D. L. Engelhardt, N. D. Norton, and W. Konigsberg (1967), *J. Mol. Biol.*, **29**:27.

——, ——, and N. D. Zinder (1966), *Genetics*, **55**:155.

Weeks, C. O., and S. R. Gross (1971), *Biochem. Genet.*, **5**:505.

Weisblum, B., S. Benzer, and R. W. Holley (1962), *Proc. Natl. Acad. Sci.*, **48**:1449.

Weiss, J., R. Pearson, and A. Kelmers (1968), *Biochemistry*, **7**:3479.

Weiss, S. B. (1960), *Proc. Natl. Acad. Sci.*, **46**:1020.

—— and T. Nakamoto (1961), *Proc. Natl. Acad. Sci.*, **47**:694.

Wettstein, F. O., and N. Noll (1965), *J. Mol. Biol.*, **11**:35.

Wilhelm, J. M., and R. Haselkorn (1969), *Cold Spring Harbor Symp. Quant. Biol.*, **34**:793.

Wimber, D. E., and D. M. Steffensen (1970), *Science*, **170**:639.

Wisseman, C. L., et al. (1954), *J. Bacteriol.*, **67**:662.

Yamane, T., and N. Sueoka (1963), *Proc. Natl. Acad. Sci.*, **50**:1093.

Yankofsky, S. A., and S. Spiegelman (1962a), *Proc. Natl. Acad. Sci.*, **48**:1069.

—— and —— (1962b), *Proc. Natl. Acad. Sci.*, **48**:1466.

—— and —— (1963), *Proc. Natl. Acad. Sci.*, **49**:538.

Zachau, H. G., G. Acs, and F. Lipmann (1958), *Proc. Natl. Acad. Sci.*, **44**:885.

——, D. Dutting, H. Feldman, F. Mecheres, and W. Darau (1966), *Cold Spring Harbor Symp. Quant. Biol.*, **31**:417.

Zamecnik, P. C., E. B. Keller, J. W. Littlefield, M. B. Hoagland, and R. B. Loftfield (1956), *J. Cell Comp. Physiol.*, **47**:81.

——, M. L. Stephenson, and L. I. Hecht (1958), *Proc. Natl. Acad. Sci.*, **44**:73.

Zubay, G. (1962), *J. Mol. Biol.*, **4**:347.

29
Coding,
Collinearity,
and
Suppressors

29
Coding,
Collinearity,
and
Suppressors

NOTATION

1 Biochemical abbreviations used in this and subsequent chapters are listed in Appendix Table A-7.

2 A coding dictionary appears in Appendix Table A-5.

3 Prefixes used with the metric system are explained in Appendix Table A-6.

QUESTIONS

1404 Gamow (1954) pointed out that since the genetic language contains only four letters, A, U (= T), C, and G, if all code words are of the same size, codons must be at least three bases long.

a Show why codons cannot consist of one or of two bases.

b Discuss two lines of research that indicate that codons are three bases long.

c Since codons are three bases long, 64 different triplets can exist. Illustrate these using the branching method.

d How many of the 64 triplets will contain (1) no adenine, (2) at least one adenine?

1405 Answer each of the following as briefly as possible:

a Explain what is meant by a degenerate code and illustrate your answer to show degeneracy in translation.

b Discuss the various lines of evidence indicating that the code is (1) nonoverlapping and (2) degenerate.

c Why are many mutational sites expected within a gene?

d Explain how single base-pair substitutions in DNA are reflected in phenotypic changes.

e How does the linear sequence of four kinds of nucleotides in DNA of a gene determine the linear sequence of the 20 amino acids in the corresponding polypeptide chain?

f Would you expect nonsense mutants to have polar effects?

1406 A basic concept in molecular biology is the collinearity of gene, DNA, RNA, and protein. Discuss the various lines of evidence supporting or proving this concept.

1407 **a** Which are the nonsense triplets and why are they so termed?

b Are nonsense mutations identical with stop codons?

c Explain why polypeptides specified by a polycistronic message are not coupled.

d What is characteristic of polypeptide chains specified by genes carrying nonsense mutations? Do missense mutations have the same effect? Answer using any known structural genes or a hypothetical one and its protein product.

e How would you demonstrate that a mutant owed its phenotype to a nonsense triplet in one of its genes?

f What kinds of suppressor mutation would you expect to act on nonsense triplets?

1408   a Of the two molecules 5-bromouracil and proflavine which is more likely to produce leaky mutants and why?

b Do you expect *amber* mutants in the same gene to complement each other? Why?

c Why is UAA more likely than UAG to be the regular chain-terminating triplet?

d What is the direction of polynucleotide-chain growth in RNA transcription from a DNA template?

e If the *m*RNA code word for valine is GUA, what are the corresponding DNA and *t*RNA sequences?

f Do codons have the same meaning in vitro as in vivo?

1409   A polysome contains 10 ribosomes 150 Å apart, held together by a mono-cistronic *m*RNA. Approximately how many amino acids does this messenger specify in the corresponding polypeptide chain?

1410   With regard to the genetic code, discuss the significance of each of the following findings:

a Cell-free systems containing ribosomes and *m*RNA from rabbit reticulocytes and charged *t*RNAs from *Escherichia coli* synthesize rabbit hemoglobin (Ehrenstein and Lipmann, 1961).

b Studies on protein synthesis in rat liver by Staehelin et al. (1964) indicated that for every piece of *m*RNA 90 nucleotides long a hemoglobin polypeptide fragment was synthesized containing 30 amino acids.

1411   Are the chain-terminating codons UAG, UGA, and UAA read by *t*RNAs or by specific proteins?

1412   a In formylmethionine, the formyl, CHO, group replaces one hydrogen atom in the amino group, NH_2, of the methionine molecule. Since the formyl group prevents peptide-bond formation, is formylmethionine likely to be present at one of the ends of the polypeptide chain or internally? If at one end, which and why?

b Is there a special *t*RNA involved in initiation of synthesis of the polypeptide chain? If so, which one is it, and how does it function?

1413   a Does most of the degeneracy in the code involve the first, second, or third base of a codon?

b Amino acids like tyrosine and histidine are coded by two triplets only. Can the *t*RNA have inosine at the 5′ end of the anticodon?

c A highly purified species of *E. coli* alanine-*t*RNA can recognize three alanine codons, GCU, GCC, and GCA (Söll et al., 1965). Explain this on the basis of the wobble hypothesis (see Prob. 1476).

1414   Would you expect mitochondrial aminoacyl-RNA synthetases and their respective *t*RNAs to be exclusively associated with mitochondrial-protein synthesis? Explain. (See Barnett et al., 1967.)

1415   Streptomycin can probably act as a suppressor by distorting the structure of the 30S subunit of the ribosome.   *Wild-type E. coli* have glutamine at a certain position in the polypeptide chain specified by the *B* gene.   A mutant strain, induced by 5-bromouracil, produces a fragment of this polypeptide which is one-half the normal length.   Addition of streptomycin to the minimal medium seeded with this mutant causes the mutant to grow at a slow rate and to form some complete B proteins, with the *wild-type* amino acid sequence.   Explain how streptomycin might cause this effect.

1416   A *wild-type* phage T4 strain has tyrosine at position 30 in the head protein. A mutant strain has serine in place of tyrosine; otherwise the two proteins are identical.   A revertant strain arises, due to mutation at a locus different from that producing the mutant strain, which is phenotypically *wild type* and has tyrosine at position 30, as in the original strain.   Outline two mechanisms by which the mutant strain could revert to *wild type* or *pseudo wild type*.

1417   How would you determine in *Salmonella typhimurium* whether a revertant has a suppressor mutation at a different locus, a suppressor mutation at a second site in the same gene, or a true reversion of the original mutation site?

1418   Show how a mutation which changes the base sequence of a *t*RNA molecule might act as a suppressor mutation.

1419   Reversions of missense mutations can arise as a result of mutation outside the codon in which the original mutation is located, whereas reversions of nonsense mutations must result from mutations within the nonsense codon. Why should this difference exist?

1420   Suppressed strains produce two types of proteins: one like that of the un-suppressed mutant strain and another with physical (although not always) and enzymatic properties characteristic of the *wild-type* protein.   For example, in the missense *E. coli A*-gene mutant *A36* A-type polypeptides have arginine in place of glycine at a specific position in peptide CP2.   When *A36* carries a suppressor, *Su36*$^+$, some A polypeptides have arginine and others have glycine (as in *wild-type* A polypeptides) at this site (Brody and Yanofsky, 1963). Explain how this is possible.

1421   **a** Why can a given suppressor gene suppress mutations in a number of different genes?

   b Suppressor genes misread nonsense codons, and each inserts a specific amino acid into the polypeptide chain at the position of the nonsense triplet.   In *E. coli*, the suppressor genes *Su1*$^+$, *Su2*$^+$, and *Su3*$^+$ suppress only the *amber* (UAG) codon, inserting serine, glutamine, and tyrosine, respectively, at the nonsense position.   *Su4*$^+$ and *Su5*$^+$ suppress both *amber* and *ochre* (UAA) codons by inserting tyrosine and a basic amino acid (probably lysine) respectively.   Account for the difference in behavior of the two classes of suppressors in the light of the wobble hypothesis (see Prob. 1476).

   c UGA was the last nonsense codon to be discovered, and it was found that none of the suppressors *Su1*$^+$, . . . , *Su5*$^+$ could suppress this codon.

(1) Explain, with a diagram if you like, how a nonsense codon interferes with the normal functioning of a gene. Make sure you point out whether nonsense codons interrupt gene function at the level of transcription or translation.

(2) Indicate why neither the *amber* (UAG) nor the *ochre* (UAA) suppressors could suppress the UGA nonsense codon.

(3) Suggest the amino acids one might expect the suppressor of the UGA codon to have inserted and indicate your reasoning.

1422 **a** Are operons transcribed into polycistronic *m*RNA molecules?

b The position of a nonsense mutation in a gene strongly influences the degree of polarity. Which mutants would be more polar, those closest to the operator or those farthest from it?

c Mutations are called polar when in addition to affecting the gene in which they occur they also lower the output of other genes farther from the operator. Three such mutations are those giving rise to the nonsense codons UAG, UGA, and UAA. Suggest how these mutations might produce their polar effects.

PROBLEMS

1423 One of Benzer's *r*II deletion mutants in phage T4, 1589, extends from within the A cistron into the left end of the B cistron, as illustrated.

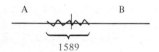

Figure 1423

a This mutant complements with B but not A mutants. What does this tell you concerning (1) the activities of the gene products produced by translation of the portions of the A and B cistrons present in *r*1589 and (2) the left end region of the B cistron?

b Nonsense mutants, mapping to the left of the *r*1589 deletion, when crossed with *r*1589 produce double mutants (*r* nonsense, *r*1589). Describe how these double mutants can be obtained and isolated. (Note that all *r*II mutants grow in *E. coli* B but not in *E. coli* K.)

c The double mutants do not complement with D mutants, whereas *r*1589 by itself does. Explain this in terms of the direction of translation of the *m*RNA code and the effect of the nonsense mutation on the size (complete vs. incomplete) of the polypeptide product.

d The nonsense mutant (by itself) complements B mutants. Account for this in terms of punctuation between the A and B cistrons.

    e A deletion mutant missing four bases and mapping in the A cistron to the left of the *r*1589 segment is crossed with *r*1589 to produce a double mutant.

        (1) Would the double mutant have B activity (complement with B mutants)? *Note:* The deletion mutant by itself has B activity.

        (2) What name has been given to mutations consisting of the insertion and deletion of a number of bases not divisible by 3?

        (3) Suppose that the double mutant has B activity; how would you explain these results in terms of generation of nonsense codons by the deletion of four bases?

    f Mutant 1589 shows no A activity but almost *wild-type* B activity in *E. coli* B. The addition or deletion of a single base pair in the A cistron suppresses B activity.   What hypothesis do these results support?

1424 Explain why a frame-shift mutation can be suppressed by a second frame-shift mutation if:

    a The region between the two frame shifts is nonessential.

    b The first frame shift in the direction of translation of *m*RNA does not create a nonsense codon before the second opposite frame shift.

Now explain whether it is possible for a deletion mutant to revert.   Would you expect a *wild-type* or *pseudo-wild-type* revertant and why?

1425 *Wild-type*, r^+, T4 phage grow on *E. coli* strains B and K and form small plaques on both.   Mutants, *r*, in the *r*II region, which consists of two contiguous cistrons, A and B, grow on B (form large plaques) but not on K.

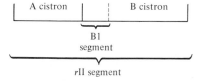

Figure 1425*A*

Proflavine is known to induce base-pair additions $(+)$ and deletions $(-)$. Crick et al. (1961) induced an *r* mutant (FC-O) in the nonessential B1 region by proflavine.   They characterized it as $+$, assuming the addition of a base pair.   Natural revertants (suppressed mutants), assumed to carry single-base deletions and hence characterized as $-$, were shown to involve sites near the FC-O site and were *pseudo wild type*.   Subsequently suppressors of suppressors of FC-O were isolated, as were suppressors of suppressors of suppressors of FC-O.   All were located in the B1 segment.   By recombination, various groups of deletions and additions were obtained.   The phenotypes of some of these combinations of suppressors are shown.

Table 1425

Hypothetical possibilities
for FC-O in combination with
successive reversions*

(1)	(2)	B activity (plaque type on *E. coli* B)
$+^1$	$-^1$	*r*
$+^1 -^2$	$-^1 +^2$	*pseudo wild type*
$+^1 +^2$	$-^1 -^2$	*r*
$+^1 +^2 +^3$	$-^1 -^2 -^3$	*pseudo wild type*

* FC-O is designated $+^1$ if it is due to an addition of a base pair;
$-^1$ if due to a deletion of a basic pair.   First suppressive revertants:
$+^2$ if addition; $-^2$ if deletion.   Second suppressive revertants:
$+^3$ if addition; $-^3$ if deletion.

a Using the above data, present arguments for the following features of the
genetic code:
(1) Codons are triplets (three consecutive bases long).
(2) Initiation of translation in a cistron occurs from a fixed point.
Normally substitution, addition, or deletion of base pairs in the A cistron
does not affect the expression of the B cistron and vice versa.   Deletion 1589,
comprising the right end of the A cistron and a small part (B1) of the left end of
the B cistron, alters this situation.   Using deletion 1589 in combination with
additions and deletions in the A cistron to the left of 1589 Crick et al. observed
the results shown in Fig. 1425*B*.

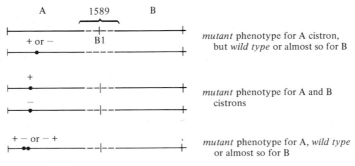

mutant phenotype for A cistron,
but *wild type* or almost so for B

mutant phenotype for A and B
cistrons

mutant phenotype for A, *wild type*
or almost so for B

Figure 1425*B*

b Using these data, present an argument or arguments to show that the genetic
code:
(1) Has punctuation marks between cistrons.
(2) Is degenerate.
(3) Is read from a fixed point.

c What bearing, if any, do these data have on the following aspects of the genetic code?
(1) Collinearity.
(2) Overlapping vs. nonoverlapping nature of the code.
(3) Its commaless nature.

d Describe how you would determine that the region B1 in the B cistron covered by 1589 is nonessential to the function of the B polypeptide.

e Although the region B1 is in fact nonessential to B function, one can find spontaneous rII mutants, deficient in B function, located in B1.   What can you conclude about the nature of these spontaneous mutants?

f Why couldn't you use 1589 as a deletion standard in a spot test to localize unknown B cistron mutants either inside or outside B1?

g Outline the experiments which proved that the genetic code in the rII region is read in a discontinuous manner; i.e., the region between the A and B cistrons contains a stop signal for the reading of the A gene and a start signal for the reading of the B gene.

1426 Crick et al. (1961) found that most of the double mutants ($+$ $-$ or $-$ $+$) in the B1-2 (nonessential) region of the B gene in the rII region of phage T4 have a *wild-type* or *pseudo-wild-type* phenotype.   Some such double mutants in this segment, even though the additions and deletions were close together, possessed a mutant phenotype (no B activity).   See Prob. 1425 for meaning of $+$ and $-$.

a Why do these results suggest that (1) the newly generated triplets in the region between the addition of a base pair and a deletion of a base pair are sense and (2) that the code is highly degenerate?

b Explain in molecular terms why the latter double mutants did not show any B activity.   Outline how you would test your explanation.

1427 Tsugita and Fraenkel-Conrat (1962) studied alterations by nitrous acid in the 158–amino acid coat protein of tobacco mosaic virus.   This chemical is known to effect base substitutions but not base additions or deletions. Usually only one amino acid at a time is changed.   Only in rare cases were two amino acids altered in a mutant, and these were never at adjacent sites. Some of the amino acid substitutions in the mutant polypeptides are shown.

Table 1427

Strain	Amino acid position				
	1	11	20	81	156
Wild type	NH_2-Ser	Val	Pro	Thr	Gly
Mutant 1	NH_2-Ser	Met	Pro	Thr	Gly
Mutant 2	NH_2-Ser	Val	Thr	Thr	Gly
Mutant 3	NH_2-Ser	Val	Thr	Thr	Leu

On which of the following aspects of coding do these results have a bearing? What kind, and why?

a Codons consist of three consecutive bases.

b The code is degenerate.

c The code is nonoverlapping.

d The codons are read from a fixed starting point.

1428 Serine occupies position 234 in the *wild-type* A polypeptide of *E. coli* tryptophan synthetase. A 2-aminopurine–induced *mutant* has glycine in position 234; treatment of this *mutant* with 2-aminopurine gives a *mutant* with aspartic acid at this position. Finally treatment of the second *mutant* with the same mutagen produces a *revertant*, with serine replacing aspartic acid at this site. What information do these results provide about the fundamental properties of the genetic code? Explain.

1429 The *E. coli* suppressor gene *Su6*⁺ causes the *amber* triplet UAG to code for leucine. Gopinathan and Garen (1970) in experiments with fractionated *t*RNA from *Su6*⁻ and *Su6*⁺ strains showed that in *Su6*⁻ strains there were two species of leucyl-*t*RNA which bind to ribosomes in the presence of leucine codon UUG. In *Su6*⁺ strains only one of these *t*RNA species was present; the other was replaced by a homologous leucyl-*t*RNA which binds to ribosomes only in the presence of UAG.

a What appears to have been the consequence of the suppressor mutation *Su6*⁻ to *Su6*⁺?

b What is the most plausible mechanism for this transformation?

c Why is a suppressor mutation that alters the codon recognition of a *t*RNA potentially lethal? Why is *Su6*⁺ not lethal?

d The normal amino acid sequence in a certain region of the lysozyme molecule produced by *wild-type* T4 phage during the late stages of infection of *E. coli* is

···Lys-Ser-Pro-Ser-Leu-Asn-Ala···

A frame-shift mutation was induced in the viral gene for lysozyme by treating T4 phage with an acridine dye. The mutation involved the deletion of a base in the DNA codon for the serine residue which is nearest the amino-terminal end of the fragment shown. A reversion of the *mutant* to the *wild-type* phenotype was obtained by treating the mutant with acridines and causing addition of a base in the alanine codon. When the amino acid sequence in the lysozyme molecule produced by the *revertant* was examined, it was found to be

···Lys-Val-His-His-Leu-Met-Ala···

(1) Because of degeneracy, there are a large number of theoretically possible sequences for the *wild-type* *m*RNA which would give rise to the amino acid sequence in the *wild-type* polypeptide fragment shown. Calculate how many.

(2) In fact only one of these sequences was present in the T4 lysozyme *mRNA* molecule. The unique sequence can be determined, since upon deletion of a base in the serine codon and addition of a base in the alanine codon it must have given rise to an *mRNA* sequence in the *revertant*, which coded for the altered lysozyme sequence shown. Determine the *wild-type mRNA* sequence and indicate which base must have been deleted from the serine codon and which must have been added to the alanine codon to produce the revertant *mRNA*.

1430   The satellite tobacco necrosis virus (STNV) is a very small nucleoprotein. Its RNA genome consists of only 1,200 nucleotides. Not counting tryptophan (which was not determined), its protein coat contains 372 amino acid residues (Reichmann, 1964). Discuss the implications of the above data with regard to (1) the probable number of genes in the genome and (2) the coding ratio of base pairs per amino acid.

1431   Both poly-UC (3:1) and poly-UG (3:1) stimulate the incorporation of leucine into polypeptides but to one-third the extent of phenylalanine, which is incorporated to the greatest extent. By countercurrent distribution Weisblum et al. (1962) separated *E. coli tRNA* into two samples. Each sample was charged with ^{14}C leucine and the response of each to poly-UC and poly-UG was determined. Leucine attached to *tRNA* in sample 1 was incorporated into polypeptides by poly-UC, whereas leucine attached to *tRNA* from sample 2 was incorporated into polypeptides only when poly-UG was the messenger.

  a These results clearly indicate the role of *tRNA* in protein synthesis. What is it and why?

  b Is coding degeneracy involved? If so, provide an explanation of the mechanism involved.

1432   **a** Nirenberg and Matthaei (1961) included a different ^{14}C-labeled amino acid in each of 20 mixtures containing all amino acids. These were added to cell-free translation systems containing ribosomes, *tRNAs*, enzymes, synthetic messenger containing uracil only, and energy source. The only mixture in which polypeptides were labeled was the one containing labeled phenylalanine. When poly-U was omitted from the system or paired with an adenine chain, no protein containing this amino acid was synthesized. Discuss the significance of this discovery from the coding point of view.

  b Synthetic *mRNA* containing only adenine templates the in vitro synthesis of a polypeptide containing lysine only (Gardner et al., 1962), and that containing only uracil leads to a polypeptide composed of phenylalanine only (Nirenberg and Matthaei, 1961). Since adenine and uracil (thymine) are complementary, what do these results imply regarding the transcription process?

1433   **a** Salas et al. (1965) found that in a cell-free *E. coli–Lactobacillus arabinosus* system oligonucleotides of type AAA $\cdots$ AAC, with the AAC codon at the 3' end of the chain, directed the synthesis of Lys-Lys-Lys $\cdots$ Lys-Asn polypeptides with NH_2-terminal lysine and COOH-terminal asparagine. What is the direction of translation of the genetic code and why?

b In a similar system, the 5′AAAAACA ⋯ AAA-3′ oligonucleotide promotes the incorporation of lysine, asparagine, and minute amounts of threonine, with asparagine largely in the NH_2-terminal position (M. A. Smith et al., 1966). Do these results support the conclusion drawn in (a)? Suggest a reason why asparagine was not always present in the NH_2-terminal position.

1434 a The following three synthetic messengers have bases incorporated at random:
(1) 1 U : 5 C   (2) 1 A : 1 C : 4 U   (3) 1 A : 1 C : 1 G : 1 U
Calculate the frequencies of the different codons for each messenger.

b Two synthetic *mRNAs* contain cytosine and guanine. When cytosine is in excess, more alanine is incorporated into polypeptides than valine, and vice versa when guanine is in excess.

(1) Show the codons containing guanine and cytosine which could possibly code for valine and alanine.

(2) A synthetic messenger contains 80 percent guanine and 20 percent cytosine incorporated at random. Using Appendix Table A-5, determine the percentage of the various amino acids expected to be incorporated into polypeptides.

(3) How much more valine than alanine is expected to be incorporated into polypeptides?

c Synthetic messengers containing adenine and cytosine in a 4:1 ratio direct in vitro synthesis of polypeptides containing 4 times as many lysine as asparagine residues, 16 times as many asparagine as proline residues, and 16 times as many lysine as threonine residues. What codons specify each of these amino acids?

1435 Using a synthetic *mRNA* containing adenine and cytosine in a ratio of 3:7 incorporated at random, Jones and Nirenberg (1962) found that the proteins synthesized contained the following frequencies of amino acids: aspartic acid 3.2 percent, glutamic acid 3.5 percent, histidine 11.7 percent, lysine 3.0 percent, proline 68.0 percent, and threonine 10.5 percent. Do the observed frequencies of amino acids approximate the frequencies expected on the basis of random incorporation of bases into the messenger?

1436 Beginning with Nirenberg and Matthaei's cracking of the code in 1961, in vitro experiments using synthetic messengers have been used to determine code words specifying the various amino acids. Where the synthetic messenger consists of more than one kind of nucleotide (base), the initial concentration of nucleotides determines the base ratio in the polymer. The nucleotides are incorporated at random. The initial experiments identified the code words but did not permit the determination of the sequence of bases in the code words. One of these experiments was reported on by Wahba et al. (1963), who incorporated the bases adenine (A) and cytosine (C) into the synthetic messenger in a 5:1 ratio.   They obtained the results tabulated.   After showing the types of code words that are possible with A and C and the probabilities of these in a synthetic messenger, assign code words to the amino acids. Explain your assignment.

Table 1436

Amino acid incorporated	Amino acid incorporation*
Asparagine	24.2
Glutamine	23.7
Histidine	6.5
Lysine	100
Proline	7.2
Threonine	26.5

* The incorporation of an amino acid is given as percent incorporation of that amino acid whose incorporation is promoted to the greatest extent by the polymer.   For example, incorporation of lysine is promoted to the greatest extent; it is therefore given a value of 100 percent.   Other incorporations are relative to this.

1437   Position 210 in the *wild-type* A protein of *E. coli* tryptophan synthetase is occupied by glycine.   In mutants *A46* and *A47* this glycine is replaced by glutamic acid and valine respectively.   Both replacements are associated with loss of enzyme activity.   No *wild-type* recombinants (glycine at 210) are produced in crosses between these mutants (Guest and Yanofsky, 1966; Helinski and Yanofsky, 1962).   Crosses between mutants *A46* and *A23* (arginine replaces glycine at 210) produce 0.002 percent *wild-type* recombinants (Henning and Yanofsky, 1962*a*).   Treatment of *A23* with mutagens produces new mutant strains.   Those with glycine or serine at 210 produce a fully active enzyme, whereas those with threonine in the same position produce a partially active enzyme (Henning and Yanofsky, 1962*b*; Guest and Yanofsky, 1966).

a What is the minimum number of (1) alternative forms that a mutable site can exist in, (2) adjacent mutable sites that specify a single amino acid?

b Is a unique amino acid sequence required for enzyme activity?
Justify both answers.

1438   Wahba et al. (1963) used synthetic polynucleotides A-C (5:1), A-C (1:5), and U-C (5:1) as messengers in in vitro protein-synthesizing systems.   (The nucleotides are incorporated at random.)   The percent amino acid incorporations were as shown in each of the experiments.   Do these data indicate that the code is degenerate?   If so, for which amino acid or amino acids?

Table 1438

Amino acid incorporated	A-C		U-C (5:1)
	5:1	1:5	
Asparagine	24.2	5.2	
Glutamine	23.7	5.3	
Histidine	6.5	23.4	
Leucine			22.2
Lysine	100.0*	1.0	
Phenylalanine			100.0*
Proline	7.2	100.0*	5.1
Serine			23.6
Threonine	26.5	20.8	

Percent amino acid incorporation

* 100.0 is standard; see Table 1436.

1439 Using randomly ordered synthetic polyribonucleotides as messenger, it has been shown that the codons for valine, leucine, and cysteine contain two U's and one G.   Using the ribosome-binding technique (see Prob. 1442), Leder and Nirenberg (1964*b*) obtained the results shown.   What is the base sequence of

Table 1439

Cell-free system plus	Radioactive aminoacyl-*t*RNA bound to ribosomes, picomoles	
	^{35}S-Cys-*t*RNA	^{14}C-Leu-*t*RNA
No messenger	0.29	0.76
UGU	1.46	0.78
UUG	0.32	1.74
GUU	0.34	0.92
UUU	0.32	
UG	0.21	0.92
GU	0.34	0.86
UU	0.26	0.88

a The codon for valine and why?

b Val-*t*RNA anticodon when read from 5' to 3'?

1440 In in vitro systems with high magnesium-ion content the synthetic messengers 5'-AAAAACAA···AAA-3' end and 5'-AAAACAA···AAA-3' direct the synthesis of lysine polypeptides with asparagine and threonine, respectively, at the NH$_2$-terminal ends.

a What are the codons for lysine, asparagine, and threonine?

b What is the direction of translation of the *m*RNAs?

c Which codon initiates the formation of each of the polypeptides?

d Would you expect these messengers to direct synthesis if magnesium-ion concentration was low?

Explain each of your answers.

1441 The base sequence in the DNA strand complementary to *m*RNA is 5'-TAC TAA CTT AGC CTC GCA TCA ⋯ 3'.

a What amino acids are coded by this sequence?

b An adenine is inserted in this strand after the first guanine from the left. The resulting polypeptide is six amino acids long. Where is the newly produced nonsense codon located, and what is the amino acid sequence in the fragment?

1442 Nirenberg and Leder (1964) showed that if a trinucleotide messenger is added to a cell-free system containing ribosomes and the corresponding aminoacyl-*t*RNA, the charged-*t*RNA binds to the ribosome (pairs with the *m*RNA) and the resulting complex is retained on nitrocellulose membranes. Free aminoacyl-*t*RNAs do not adsorb. The above procedure was repeated for each trinucleotide in 20 different media, all carrying the full complement of 20 amino acids but with a different ^{14}C-labeled amino acid in each. After each experiment the amount of ^{14}C radioactivity associated with the membrane was

Table 1442

Experiment	Type of RNA used as template	Activity of templates for ^{14}C-aminoacyl-*t*RNA picomoles of ^{14}C-aminoacyl-*t*RNA bound to ribosomes		
		^{14}C-Val-*t*RNA	^{14}C-Phe-*t*RNA	^{14}C-Leu-*t*RNA
1	None*	0.38	0.22	0.37
	Poly-U	0.23	4.73	0.27
	Poly-UG	2.65	1.93	0.24
2	None	0.40	0.22	0.62
	GUU†	1.11	0.27	0.56
	UGU	0.40	0.25	0.53
	UUG	0.37	0.25	0.44
3	None	0.18	0.22	0.69
	GU‡	0.20	0.22	0.69
	UG	0.19	0.21	0.71
	UU	0.18	0.20	0.67

* Values represent background binding of ^{14}C-aminoacyl-*t*RNA to ribosomes in the absence of template RNA.

† Triplets of trinucleoside diphosphate type.

‡ Doublets of dinucleoside monophosphate type.

determined.   The results of Leder and Nirenberg (1964*a*) using trinucleotides and other types of templates are given in the table.

a Can doublets as well as triplets code for amino acids?   Explain.

b Do the tested triplets and poly messengers code for either of the three amino acids?   If so, which codons specify each amino acid?

1443 Using the ribosome-binding technique (see Prob. 1442 for details), Nirenberg et al. (1965) tested the template activity (ability to bind charged *t*RNA to ribosomes) of 26 trinucleotides (trinucleoside diphosphate type).   The binding results between these 26 triplets and 8 of the amino acids are tabulated.

Table 1443

Trinucleotide	Activity of templates for ^{14}C-aminoacyl-*t*RNA, picomoles of ^{14}C-aminoacyl-*t*RNA bound to ribosomes								
	Ala	Asp	Asn	Glu	Gln	His	Met	Pro	Ser
ACU	−0.16	−0.08	−0.03	−0.02	−0.03	−0.02	−0.03	−0.02	−0.05
ACC	−0.15	−0.03	−0.05	−0.02	−0.22	0.01	0.00	−0.01	0.03
ACA	−0.04	0.05	0.05	0.01	−0.03	0.00	−0.04	0.05	−0.07
ACG	−0.15	0.02	−0.03	−0.04	0.03	0.03	0.02	−0.01	−0.15
GCU	0.71	−0.08	−0.05	0.01	−0.13	−0.02	−0.18	−0.02	−0.18
CCU	−0.15	−0.03	0.00	0.00	0.04	0.01	0.03	0.04	−0.08
UCG	−0.20	0.00	0.07	0.05	−0.23	0.01	−0.11	0.06	1.09
GAU	−0.01	1.29	0.33	0.00	−0.02	−0.03	−0.09	−0.03	0.01
GAC	−0.05	1.32	0.19	0.62	−0.10	0.02	−0.10	−0.01	0.01
GAA	−0.07	0.01	0.00	−0.02	−0.32	−0.04	0.00	−0.07	−0.23
CAU	0.01	−0.01	−0.03	−0.02	0.00	0.52	−0.04	−0.01	0.02
CAC	−0.02	−0.01	0.04	−0.02	0.14	0.26	−0.01	0.01	−0.03
CAA	0.02	−0.06	−0.07	0.05	2.05	0.02	−0.12	−0.01	−0.02
CAG	−0.13	0.02	−0.03	0.02	2.60	−0.03	−0.02	−0.01	−0.09
UAA	−0.02	−0.01	−0.07	0.04	−0.30	−0.08	−0.09	0.02	0.02
UAG	−0.07	0.01	0.12	0.00	0.00	−0.01	−0.02	0.00	0.00
AGU	−0.07	0.00	0.04	−0.02	−0.03	0.03	−0.05	−0.04	0.27
AGC	0.43	0.03	0.30	0.19	0.00	0.03	−0.05	−0.01	0.17
AGA	−0.06	0.01	0.00	0.04	0.00	0.00	−0.06	0.01	0.03
GGU	−0.02	0.01	−0.02	−0.01	−0.21	−0.03	−0.08	−0.07	−0.07
CGC	0.14	0.02	−0.13	0.07	−0.07	−0.03	0.01	−0.05	−0.12
CGA	−0.20	−0.01	−0.01	0.04	−0.05	−0.03	−0.17	−0.12	−0.22
UGC	0.28	0.05	0.12	0.10	−0.05	−0.02	−0.18	0.02	0.02
UGA	−0.12	0.07	0.14	0.01	−0.13	−0.03	0.00	0.00	0.00
AUG	−0.01	−0.06	0.02	−0.01	−0.14	−0.02	1.00	0.00	−0.05
CUG	−0.14	−0.01	0.04	0.01	0.11	0.03	0.10	0.03	0.03
Background binding*	0.50	0.21	0.21	0.12	1.65	0.25	0.40	0.14	0.58

* Binding of ^{14}C-aminoacyl-*t*RNA to ribosomes in the absence of trinucleotides.   All the values were obtained by subtracting background binding of ^{14}C-aminoacyl-*t*RNA from binding upon addition of a trinucleotide.

a With reasons for your answers state which triplet(s) specify each of the amino acids.

b What other information do these data reveal about the code?

1444 In 1966 Jones, Nishimura, and Khorana showed that the long-chain synthetic messenger poly-AC containing the bases adenine and cytosine in alternating sequence [ACACAC $\cdots$ (AC)$_n$] stimulated the synthesis of polypeptides containing threonine and histidine in alternating sequence (Thr-His-Thr-His $\cdots$).

a These results prove that codons possess an odd number of bases. How?

b A synthetic messenger containing 5 A : 1 C, randomly incorporated, directed the synthesis of a polypeptide containing the amino acids shown in Prob. 1435 (Wahba et al., 1963). Using these data and the information from the 1966 experiment of Jones et al., assign codons to threonine and histidine.

1445 The synthetic messenger poly-AAG (with the repeating sequence AAG $\cdots$ AAG) directs synthesis of three homopolypeptides, polylysine, polyglutamic acid, and polyarginine. The synthetic messengers poly-UC, poly-AC, poly-UG, poly-AG (each with two nucleotides in alternating sequence) direct the synthesis of the polypeptides shown, each with the amino acids in alternating sequence (Khorana et al., 1966).

Table 1445

Messenger	Polypeptide
(UC)$_n$	Ser-Leu
(UG)$_n$	Val-Cys
(AC)$_n$	Thr-His
(AG)$_n$	Glu-Arg

a Why does poly-AAG direct the synthesis of three kinds of homopolypeptides?

b Show why these results establish the triplet and nonoverlapping properties of the genetic code.

1446 Khorana and his colleagues have synthesized long RNA molecules with various repeating sequences of bases. Some of these, together with the amino acid sequence in the polypeptide chain or chains synthesized in vitro, are shown (from Nishimura et al., 1965a,b; Jones et al., 1966; Khorana, 1966; Khorana et al., 1966; Kossel et al., 1967).

a Determine from the above results what the in vitro codons for each amino acid are and compare your results with the codons assigned by Nirenberg, Khorana et al.

b Assign (with reasons) each of the following sequences to its correct amino acid: CUU, UCU, UUC, CUC, ACA, CAC.

c Why do (GUA)$_n$ and (GAU)$_n$ code for only two rather than three homopolypeptides?

d Why do $(GAUA)_n$ and $(GUAA)_n$ fail to stimulate synthesis of polypeptides? (Short peptides, two to three amino acids long, are formed to a small extent.)

Table 1446

mRNA base sequence	Amino acid sequence in polypeptide or polypeptides
$(UC)_n$	Ser-Leu
$(UG)_n$	Val-Cys
$(AC)_n$	Thr-His
$(AG)_n$	Arg-Glu
$(UUC)_n$	(Phe-Phe), (Ser-Ser), (Leu-Leu)
$(UUG)_n$	(Leu-Leu), (Cys-Cys), (Val-Val)
$(AAG)_n$	(Lys-Lys), (Arg-Arg), (Glu-Glu)
$(CAA)_n$	(Gln-Gln), (Asn-Asn), (Thr-Thr)
$(UAC)_n$	(Tyr-Tyr), (Thr-Thr), (Leu-Leu)
$(AUC)_n$	(Ile-Ile), (Ser-Ser), (His-His)
$(GUA)_n$	(Val-Val), (Ser-Ser)
$(GAU)_n$	(Asp-Asp), (Met-Met)
$(UAUC)_n$	(Try-Leu-Ser-Ile)
$(UUAC)_n$	(Leu-Leu-Thr-Tyr)
$(GAUA)_n$	None
$(GUAA)_n$	None

1447   Kossel, Morgan, and Khorana (1967) showed that the synthetic messenger $(UAUC)_n$ with U at the 5' end and C at the 3' end of the chain directed the synthesis of the repeating tetranucleotide sequence Tyr-Leu-Ser-Ile with tyrosine at the NH_2 end of the polypeptide and isoleucine at the carboxyl end.   The messenger $(UUAC)_n$ with U and C at the 5' and 3' ends, respectively, directed the incorporation of the repeating amino acid sequence Leu-Leu-Thr-Tyr with leucine at the amino and tyrosine at the carboxyl end of the chain.

Do these results prove whether:
(1) The direction of translation of mRNA is 5' to 3' or the reverse?
(2) The code is triplet and nonoverlapping?
Give reasons for your answers.

1448   The six codons for arginine fall into two groups: (1) AGA, AGG and (2) CGA, CGC, CGU, CGG (Brimacombe et al., 1965; Morgan et al., 1966).   Söll et al. (1966*b*) separated two arginine tRNA species from yeast by countercurrent distribution.   The purified tRNA fractions were charged with radioactive (^{14}C) amino acid and tested for binding to ribosomes in the presence of trinucleotides with the above codons.   Arg-tRNA I binds to *E. coli* ribosomes in the presence of CGU, CGA, and CGC, whereas Arg-tRNA II binds to ribosomes in the presence of AGA and AGG.   Weisblum et al. (1967) studied the transfer of arginine into the *wild-type* α chain of rabbit hemoglobin which has three arginine residues at positions 31, 92, and 141. The other two arginine residues are in the β chain. *t*RNA I transfers its arginine to position 141; *t*RNA II transfers its arginine to position 31; neither species of *t*RNA transfers its arginine to position 92.

a Is there a discrete *t*RNA species for the recognition of each codon, or can one *t*RNA species recognize more than one codon? If the latter, how can this be done? (See Crick, 1966a,b.)

b Which codons occur at positions 31, 92, and 141 in the α chain of rabbit hemoglobin? Explain. In some mutant α chains arginine at 31 is replaced by lysine, arginine at 141 by histidine, and arginine at 92 by leucine or glutamine. Does this information help to confirm your codon assignments? Explain.

c Which base occurs at the third position in the anticodon of *t*RNA that reads the three codons CGU, CGA, CGC?

d Do these results provide support for the wobble hypothesis (see Prob. 1476 for details)? Explain.

1449 Genetic and biochemical studies by Hartwell and McLaughlin (1968) revealed the following.

1 Two *temperature-sensitive* mutants (ts^-341 and ts^-443) of the yeast *Saccharomyces cerevisiae* grow at a *wild-type* rate at 23°C (permissive temperature) but very slowly at 36°C (nonpermissive temperature), at which protein synthesis is rapidly inhibited. Tetrad analysis of crosses between these mutants and a *wild-type* (ts^+) strain gave the results in Table 1449*A*. By random ascospore analyses, approximately 2×10^{-5} of the spores from $ts^-341 \times ts^-443$ were ts^+.

Table 1449*A*

	No. of tetrads with a $ts^-:ts^+$ segregation pattern of				
Cross	4:0	3:1	2:2	1:3	0:4
$ts^-341 \times ts^+$	0	7	87	2	0
$ts^-443 \times ts^+$	0	3	16	2	0
$ts^-341 \times ts^-443$	27	0	0	0	0

2 Cell-free extracts from ts^-341 and ts^+ were assayed for their ability to catalyze the transfer of ^{14}C amino acids onto *t*RNA. Aminoacyl-*t*RNA synthetase activities at 25°C, after incubation at 36°C for $\frac{1}{2}$ hour (to permit inactivation of temperature-sensitive enzymes), are shown in Table 1449*B* for five amino acids. The results with other amino acids showed the same results as four of these five.

3 The rate of formation of Ile-*t*RNA by extracts of the mutants is the same as that of ts^+ at 25°C but decreases as the temperature increases.

Table 1449*B*

Amino acid	Nanomoles of aminoacyl-*t*RNA formed per milligram of protein per hour at 25°C	
	ts^+	ts^-341
Alanine	2.70	2.78
Arginine	4.05	3.69
Histidine	1.48	1.75
Isoleucine	3.35	0.45
Proline	1.54	1.35

4   The segregation patterns of *high* vs. *low* enzyme (Ile-*t*RNA synthetase) activity and ts^- vs. ts^+ in the cross $ts^-341 \times ts^+$ are shown in Table 1449*C*.

Table 1449*C*

Tetrad	*ts*	Enzyme activity (counts per minute of ^{14}C-Ile-*t*RNA formed per 30-minute incubation)	
		25°C	50°C
I	+	1,300	390
	−	2,300	60
	−	2,400	40
	+	1,950	620
III	+	3,200	800
	−	1,400	20
	−	1,900	20
	+	2,400	750
II	+	2,500	450
	−	3,200	20
	+	1,000	400
	−	1,750	20

Discuss the genetics and biochemistry of the two mutants as completely as possible.   Include references to data relevant to a specific conclusion. Be sure to state whether the mutant genes are (1) structural, (2) allelic or not and why, and (3) if structural, what their role in protein synthesis is.

1450 **a** Gene 34 in T4 phage is the structural gene for tail-fiber antigen, absent in lysates of all *amber* mutants but present in all *temperature-sensitive* mutants for this gene.   What molecular explanations can you give for the two types of mutants?

b Nonsense mutants in gene 23 produce incomplete head protein in *E. coli* strain B and complete head proteins when they infect *E. coli* strain CR63. Why does the mutant produce some complete proteins on CR63 but not on B?

1451 In T4 phage *amber*, *am*, mutants grow on *E. coli* strain CR63 but not on *E. coli* strain B.   *Temperature-sensitive*, *ts*, mutants grow on both strains at 25°C but on neither at 42°C.   Intragenic complementation occurs between many *ts* mutations in the same gene but not between *am* or *ts* and *am* mutations in the same gene (Epstein et al., 1963; Edgar et al., 1964).

   a What is or appears to be the molecular basis for the two types of mutation?

   b How would you test your explanation, assuming that the protein controlled by the gene or genes being studied has been identified and is easily analyzable?

1452 *E. coli* mutant *A46* for the A polypeptide of tryptophan synthetase is enzymatically inactive.   Glycine at position 210 (of *wild type*) is replaced by glutamic acid.   The analysis of A polypeptides in occasional revertants to *pseudo wild type* shows that glutamic acid still occupies position 210 and that a mutation has occurred at position 174, lying between position 210 and the amino end of the A chain, as shown in the table.   A change at one

Table 1452

	A protein	
Strain	174	210
Wild type	Tyr	Gly
A46	Tyr	Glu
Revertant (*A47 PR8*)	Cys	Glu

amino acid position which is detrimental to enzyme activity could be "cured" by a change at a different position in the same polypeptide chain.   Suggest an explanation for these findings.

1453 Terzaghi et al. (1966) compared lysozyme from a *wild-type*, e^+, phage T4 with that from a *pseudo-wild-type* revertant strain, *eJ42 eJ44*, induced by proflavine. All the peptides in the lysozymes from the two strains were identical except peptide 10, which is eight amino acids long.   The following sequences were found:

Wild-type lysozyme NH_2-Thr-Lys-Ser-Pro-Ser-Leu-Asn-Ala-COOH
Pseudo wild type NH_2-Thr-Lys-Val-His-His-Leu-Met-Ala-COOH

The mutations involved in producing the original *mutant eJ42* and the *pseudo-wild-type* strain were changes involving single base pairs.

   a How many mutations occurred to produce the original mutant *lysozymeless* strain?   To produce the revertant strain from the original mutant?

   b What kinds of changes were they and why?   Use the codons from Appendix Table A-5 to verify your answer.

1454   The amino acid sequence of a segment of the A polypeptide specified by the
E. coli tryptophan synthetase A gene in *wild type* and in a *revertant*, induced
by a mutagenic chemical ICR-13 in the ultraviolet-induced mutant 9813, is
given in the table (Brammar et al., 1967).   The two polypeptides are otherwise identical.

Table 1454

Strain	Amino acid sequence (NH$_2$ $\longrightarrow$ COOH)
Wild type	Thr-Tyr-Leu-Leu-Ser-Arg-Ala-Gly-Val
ICR-13–induced revertant	Thr-Phe-Cys-Cys-His-Glu-Gln-Gly-Val

a What kinds of mutation (substitution, addition, deletion) did ultraviolet
light and ICR-13 induce?   Explain, using the codon assignments in
Appendix Table A-5.
b Show how these results confirm the conclusion of Crick et al. (1961) that
the genetic message is arranged in nucleotide triplets and translated
sequentially from a fixed starting point.

1455   Mutations in the *e* gene of phage T4 affect the structure of the enzyme
lysozyme which breaks down the bacterial cell wall, permitting lysis and
plaque formation.   Terzaghi et al. (1966) and Okada et al. (1966) isolated
three proflavine-induced *e* mutants, *eJ17*, *eJ42*, and *eJ44*.   Crosses involving
these mutants and their recombinants produced the results shown in Table

Table 1455*A*

Cross	Recombinants in the progeny
eJ17 × *eJ44*	*wild type* (plaques with large holes)
	pseudo wild type 1 (plaques with small holes)
pseudo wild type 1 × *wild type*	*eJ17*; *eJ44*
eJ42 × *eJ44*	*wild type*
	pseudo wild type 2
pseudo wild type 2 × *wild type*	*eJ42*; *eJ44*

1455*A*.   The amino acid sequence in the *wild-type* and the two *pseudo-wild-
type* strains were identical except for the corresponding peptides (called A),
which showed the sequences given in Table 1455*B*.
a Are the *pseudo wild types* single or double mutants?   Explain.
b What kinds of changes were induced in the DNA of the *e*$^+$ gene by
proflavine to produce each of the mutants *eJ17*, *eJ42*, and *eJ44*?   Show
why the changes can only be of this type and no other.

Table 1455*B*

Strain	Amino acid sequence in peptide A
e^+	NH$_2$-Lys-Ser-Pro-Ser-Leu-Asn-Ala-COOH
pseudo wild type 1	NH$_2$-Lys-Val-His-His-Leu-Met-Ala-COOH
pseudo wild type 2	NH$_2$-Lys-Ser-Val-His-His-Leu-Met-Ala-COOH

1456   The *A*-gene mutants 1, 2, 3, and 4 of *E. coli* have the amino acid substitutions shown in Table 1456*A* in the A polypeptide of tryptophan synthetase.   These

Table 1456*A*

Mutant	Substitution
1	Val for Glu at position 48
2	Arg for Leu at position 176
3	Ile for Thr at position 182
4	Val for Gly at position 212

point mutants were tested in transduction experiments against four deletion mutants, all with one end of their deletion in the T1 locus (see Fig. 1459 for location of T1 relative to A).   The results in Table 1456*B* were obtained.

Table 1456*B*

Point mutant	Deletion mutant			
	A	B	C	D
1	+	+	+	−
2	+	+	−	−
3	+	−	−	−
4	−	−	−	−

Key:
+ = *wild-type* (*trp*⁺) *recombinants formed*; − = *no such recombinants.*

Explain these recombination data in terms of the relative positions of point mutations and deletion termini in the A cistron.

1457   The amino acid sequence of a portion of the *wild-type* A polypeptide of *E. coli* tryptophan synthetase and of the corresponding region in the polypeptide specified by a double-frame-shift mutant A9813PR8 due to a base-pair deletion and a base-pair addition 13 base pairs apart is shown (Brammar, Berger, and Yanofsky, 1967).

Table 1457

Polypeptide	Amino acid residue
	173 174 175 176 177 178 179
Wild type	NH₂-Thr-Tyr-Leu-Leu-Ser-Arg-Ala-COOH
Double mutant	NH₂-Thr-Phe-Cys-Cys-His-Gly-Ala-COOH

a Using Appendix Table A-5, determine the base sequence of *m*RNA for both the *wild-type* and the *double-mutant* polypeptides.

b Which base was deleted and where? Which was added and where?

1458 Three *A*-gene mutations in *E. coli* have the following effects on the primary structure of the A polypeptide of tryptophan synthetase:

1   *A3* results in substitution of valine for glutamine at position 48.
2   *A446* results in substitution of cysteine for tyrosine at position 174.
3   *A223* results in substitution of isoleucine for threonine at position 182 (Yanofsky, 1967).

a Which of these amino acid sites is closest to the amino end of the protein?

b State whether the recombination rate should be higher or lower between *A446* and *A223* than between *A3* and *A446* and why.

1459 The tryptophan operon of *E. coli* consists of five structural genes and an operator. The *A* gene directs synthesis of the A polypeptide of tryptophan

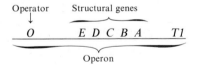

Figure 1459

synthetase; the *E* (*anth*) gene directs anthranilic acid synthetase formation; *T1* (outside the operon) determines *resistance* vs. *susceptibility* to *T1* phage.

1   Yanofsky (1967) crossed seven-point mutations in *A* with a set of five deletion mutants, in which the deletions extended from within *A* to within *T1*. The data are reproduced in Table 1459*A*.

2   In three-point transduction crosses involving *A*-gene point mutations Yanofsky et al. (1964) and Guest and Yanofsky (1966) classified tryptophan-independent recombinants as *anth*⁺ vs. *anth*⁻ with the results shown in Table 1459*B*.

3   Two-point transduction crosses between *A*-gene point mutations revealed the recombination frequencies in Table 1459*C*.

4   Studies of the primary structure of the A polypeptide (consisting of 267 amino acids) revealed that the *wild-type* amino acid sequence and the amino acid substitutions in the various point mutants were as shown in Table 1459*D*. Each mutant protein differed from *wild-type* protein by one amino acid substitution only.

Table 1459A

Point mutant	Deletion mutant				
	6	*50*	*70*	*898*	*229*
3	−	+	+	+	+
446	−	−	+	+	+
223	−	−	+	+	+
23	−	−	−	+	+
46	−	−	−	+	+
58	−	−	−	−	+
78	−	−	−	−	+

Key: + = tryptophan-independent recombinants;
− = no such recombinants.

Table 1459B

Donor	Recipient	Trp^+ recombinants	
		$anth^+$	$anth^-$
$anth^+$ 46	$anth^-$ 223	14	112
$anth^-$ 46	$anth^+$ 23	3	14
$anth^-$ 78	$anth^+$ 58	13	4

Table 1459C

Cross	Approximate percent recombination
446 × 223	0.34
223 × 23	0.93
46 × 58	0.56
58 × 78	0.001
23 × 46	0.001

Table 1459D

Mutant	Amino acid site	Amino acid present	
		In *wild type*	In *mutant*
223	182	Thr	Ile
23	210	Gly	Arg
46	210	Gly	Glu
78	233	Gly	Cys
58	233	Gly	Asp

a Derive the relative locations of these point mutations with respect to the E gene.

b Explain how the data of any one of the crosses in (b) prove that the sites of the base-pair substitutions in the two A mutants concerned are different.

c Compare the genetic map with the positions of the amino acid substitutions in each of the point mutant strains and state the conclusion(s) you draw from this comparison.

1460 The *E. coli* mutants *A23* and *A46* in the *A* gene replace glycine at position 210 with arginine and glutamic acid, respectively, in the A polypeptide or tryptophan synthetase. In three-factor transduction crosses involving the *anth* locus at the operator end of the tryptophan operon, Guest and Yanofsky (1966) found that of 17 tryptophan-independent recombinants in the cross *anth⁻ A46* (donor) × *anth⁺ A23* (recipient), 14 were *anth⁻* and 3 *anth⁺*.

a Did the mutations producing the mutant strains *A23* and *A46* occur in the same or different base pairs and why?

b After assigning codons to glycine (in the *wild-type* and recombinant strains), arginine, and glutamic acid, orient the bases in the codons relative to the *anth* locus and peptide chain.

c What is the molecular nature of the mutations producing *A23* and *A46*?

1461 Sarabhai et al. (1964) found that 10 base-analog-induced mutants of phage T4 affecting the protein coat of the head of the virus grew on *E. coli* strain CR63

Table 1461A

Cross	Recombination, %	Cross	Recombination, %
B17 × H11	0.80	*C137 × B278*	0.33
B17 × H11 + B278	0.14	*C137 × B272 + B278*	0.28
		C137 × B278 + A489	0.06
B278 × A489	1.30	*H36 × B278*	0.46
B278 × B17 + A489	0.34	*H36 × B272 + B278*	0.40
		H36 × B278 + A489	0.06
B272 × B278	2.20		
B272 × H11 + B278	0.32	*H36 × C137*	0.22
		H36 × C137 + A489	0.05
B272 × B17	1.30	*C137 × H36 + A489*	0.20
B272 × B17 + A489	0.24		
		C208 × A489	0.34
C140 × B17	0.27	*C208 × B17*	4.0
C140 × H11 + B17	0.04	*C208 × B17 + A489*	0.33
C140 × B17 + B272	0.20	*C208 × B278 + A489*	0.29
H32 × B272	0.5		
H32 × B17 + B272	0.36		
H32 × B272 + B278	0.05		

but not strain B.   No progeny were produced when *E. coli* B was infected with all possible combinations of these mutants, two at a time.   Two- and three-factor crosses were performed by infecting strain CR63 with a mixture of the parental phages in a 1:1 ratio.   In each cross the proportion of *wild-type* recombinants to total progeny was determined and the percent recombinations calculated.   (*Note:* Designations *B17*, *H11*, and so on, refer to point mutations.   Therefore, *B17* × *H11* is a two-factor (point) cross and *B17* × *B278* is a three-factor cross.)   The results of these crosses are shown in Table 1461*A*.   The defective proteins produced by each of the six mutants *B17*, *B272*, *B278*, *A489*, *C137*, and *H32* and the protein produced by *wild-type* T4, upon infection of *E. coli* B, were as shown in Table 1461*B*.

Table 1461*B*

Strain	Proteins produced
B17	NH_2
B272	NH_2-Cys*
H32	NH_2-Cys-HisT7C
B278	NH_2-Cys-HisT7C-TyrC12b
C137	NH_2-Cys-HisT7C-TyrC12b-TrpT6
A489	NH_2-Cys-HisT7C-TyrC12b-TrpT6-ProT2a-TrpT2
Wild-type T4	NH_2-Cys-HisT7C-TyrC12b-TrpT6-ProT2a-TrpT2-TyrC2-HisC6

* Cys, HisT7C, etc., designate the peptides in the protein.

a Did the base analogs induce missense or nonsense mutations?   Explain.

b Why did the mutants grow on *E. coli* CR63 but not on *E. coli* B?

c In how many genes did these mutations occur?   Explain.

d Map the six mutants, whose head proteins have been sequenced with respect to peptides, and discuss the bearing of these results on the *sequence hypothesis*, which states the amino acid sequence of a protein is specified by the nucleotide sequence of the gene determining that protein.

1462   The A polypeptide of tryptophan synthetase in *E. coli* is specified by gene *A*, one of five genes in the tryptophan operon (see Prob. 1458).   Gene *A* is distal to the operator and gene *E* (= *anth*), which specifies anthranilate synthetase (converts chorismic acid to anthranilic acid).   In the *wild-type* A protein, glycine (specified by codon GGA)[1] occupies position 210.   In mutant *A23*, arginine (AGA) replaces glycine, and in mutant *A46* glutamic acid (GAA) replaces glycine.   Because *A23-A46* mutational events affect different base pairs within the glycine codon, Guest and Yanofsky (1966) performed the cross shown, by transduction, to determine the relative positions of the base changes in the codons specifying the mutant proteins *A23* and *A46*.   *Wild-type* recombinants (GGA) were selected and scored for

[1] Base sequences are those of the DNA strand corresponding to *m*RNA.

Table 1462

Cross		Wild-type (trp^+ cys^+) recombinants	
Donor	Recipient	$anth^+$	$anth^-$
cys^+ $anth^-$ A46 × cys^- $anth^+$ A23		3	14

the nonselective marker ($anth^+$ vs. $anth^-$). From previous studies it is known that polypeptide synthesis begins at the amino, NH_2, terminal and moves toward the carboxyl, COOH, end; the mRNA is translated from the 5′ to the 3′ end; the direction of growth of mRNA during transcription from DNA is also from the 5′ to the 3′ end; the gene and protein are collinear.

a Is the glycine codon at position 210 oriented

Explain.

b What are the positions of the base changes in the codons specifying the mutant proteins relative to the 5′ and 3′ ends of the mRNA?

c Show the orientations of the codons relative to the order of the mutant sites.

d Show the relative orientations of the genetic map, gene (DNA), mRNA, and polypeptide chain for the trytophan synthetase gene.

1463 In phage T4 a series of mutant strains induced by hydroxylamine have amino acid substitutions at position 30 in lysozyme. Mutant A has serine; mutant B, derived from A, has lysine; and mutant C, derived from B, has phenylalanine. Only the first strain can revert naturally to *wild type*. None of the crosses A × B, A × C, or B × C produce *wild-type* recombinants.

a What amino acid is present at position 30 in the *wild-type* lysozyme? Explain.

b Give the base sequences of the sense and antisense strands.

1464 Adult human hemoglobin consists of two α chains and two β chains, 141 and 146 amino acid residues long respectively. Since 1957 many amino acid substitutions have been found in both the α and β chains of adult human hemoglobin. The partial data shown, from Ingram (1963), are typical.

a Are changes in the two chains determined by allelic or nonallelic genes?

b Would you expect the mutant sites for β chains to be arranged at random or in some specific sequence? Explain with the aid of illustrations.

c Which of the intermutant crosses are not expected to produce recombinants and why?

Table 1464

Chain strain	Position in chain								
α	1	2	16	30	57	58	68	116	141
wild type (Hb-A)	Val	Leu	Lys	Glu	Gly	His	Asn	Glu	Arg
Hb-I	Val	Leu	Asp	Glu	Gly	His	Asn	Glu	Arg
Hb-G$_{Honolulu}$	Val	Leu	Lys	Gln	Gly	His	Asn	Glu	Arg
Hb-N$_{Norfolk}$	Val	Leu	Lys	Glu	Asp	His	Asn	Glu	Arg
Hb-M$_{Boston}$	Val	Leu	Lys	Glu	Gly	Tyr	Asn	Glu	Arg
Hb-G$_{Philadelphia}$	Val	Leu	Lys	Glu	Gly	His	Lys	Glu	Arg

β	1	2	6	7	26	63	121	146	
wild type (Hb-A)	Val	His	Glu	Glu	Glu	His	Glu	His	
Hb-S	Val	His	Val	Glu	Glu	His	Glu	His	
Hb-C	Val	His	Lys	Glu	Glu	His	Glu	His	
Hb-G$_{San Jose}$	Val	His	Glu	Gly	Glu	His	Glu	His	
Hb-M$_{Saskatoon}$	Val	His	Glu	Glu	Glu	Tyr	Glu	His	
Hb-E	Val	His	Glu	Glu	Lys	His	Glu	His	
Hb-O$_{Arabia}$	Val	His	Glu	Glu	Glu	His	Lys	His	

1465 Three *m* (very small plaques) mutants of independent origin in the m^+ gene in phage T2 gave the results shown.

Table 1465

Cross	Recombinant (m^+) progeny
$m_1^- \times m_2^-$	0
$m_1^- \times m_3^-$	0
$m_2^- \times m_3^-$	a few

a What is the sequence of the three mutants? Explain.

b What is the molecular nature (viz. inversion, substitution, addition) of base pairs of the m_1^- mutant?

c Is it possible to obtain *pseudo-wild-type* revertants of the m_1^- mutant? Why?

d The mutants m_2^- and m_3^- synthesized polypeptide chains of normal length (assume 150 amino acids). In m_2^- arginine was substituted for histidine at position 7; in m_3^- proline was substituted for leucine at position 85.

e In another mutant, m_4^-, valine replaced leucine at position 85. *Wild types* occurred with rare frequency in crosses between m_4^- and m_3^-.
 (1) Account for this result on a molecular basis.
 (2) Are the mutations transitions or transversions? Explain.

1466 a In *Neurospora crassa, wild-type, td$^+$,* strains synthesize tryptophan synthetase. Three mutant strains arise at the *td* locus, each with a defective
enzyme showing no activity. Crosses involving *fl$^+$fl$^-$* (*fluffy* vs. *nonfluffy*)
and *Aa* (*black* vs. *albino* spores), which are 2 and 1 map units to the left
and right of the *td* locus, respectively, gave the results in Table 1466*A*.

Table 1466*A*

Cross	Parents	Recombinant progeny	Frequency per 50,000 ascospores
1	*fl$^+$ td$_1^-$ A × fl$^-$ td$_2^-$ a*	*fl$^-$ td$^+$ A*	8
2	*fl$^+$ td$_1^-$ A × fl$^-$ td$_3^-$ a*	*fl$^+$ td$^+$ a*	2
3	*fl$^-$ td$_2^-$ A × fl$^+$ td$_3^-$ a*	*fl$^-$ td$^+$ a*	13

Show the relative positions of the mutation sites in the three mutant strains
with respect to *fl* and *A* on a linkage map.
  b Assume that the amino acids present at certain sites in the *wild-type* and
mutant polypeptides were as given in Table 1466*B*. Show whether these
results are consistent with the collinearity hypothesis.

Table 1466*B*

Allele	\multicolumn{4}{Amino acid position}

Allele	1	7	12	30
td$^+$	Arg	Lys	Glu	Pro
td$_1^-$	Arg	Lys	Ala	Pro
td$_2^-$	Arg	Lys	Glu	Leu
td$_3^-$	Arg	Thr	Glu	Pro

1467 Ingram and others have established the primary structure of both the α and β
chain of human adult hemoglobin. Hunt and Ingram (1958) showed that the
amino acid sequence in peptide 4 (at the NH$_2$ end of the β chain) of

Table 1467

Hemoglobin type	Amino acid position							
	1	2	3	4	5	6	7	8
A	Val-His-Leu-Thr-Pro-Glu-Glu-Lys
S	Val-His-Leu-Thr-Pro-Val-Glu-Lys
C	Val-His-Leu-Thr-Pro-Lys-Glu-Lys

normal (*A*) hemoglobin and of hemoglobins *S* and *C* in patients with *sickle-cell anemia* and *mild anemia*, respectively, were as tabulated. With respect to all other peptides the three molecules are identical.

a Consult Appendix Table A-5 and present comparative linear sequences of bases and amino acids to show how the structure of DNA may be related to the structures of the hemoglobins *A*, *S*, and *C*.

b Could some recombinants from crosses between *S* and *C* have an *A* hemoglobin?

1468 Position 210 in the *wild-type* A protein of *E. coli* tryptophan synthetase is occupied by glycine. Henning and Yanofsky (1962*a*) showed that the following sequences of amino acid substitutions and restitutions occur at this position:

ile $\rightleftharpoons$ arg $\rightleftharpoons$ gly $\rightleftharpoons$ glu

a Show the codons and one-letter alterations that will account for the changes shown.

b Which of the crosses *arg* × *glu*, *arg* × *ile*, and *glu* × *ile* would you expect to produce *wild-type* recombinants by crossing-over within the position-210 codons and why? State where DNA exchange would have to occur in the codon to produce the *wild-type* codon.

c Which substitutions are due to base-pair transitions and which to transversions?

1469 The differences between the insulins of most mammalian species lie in the α polypeptide, 20 amino acid residues long, at positions 8, 9, and 10. Only a few differ in the β, and at position 30 only. The partial data tabulated are from Harris et al. (1956) and Mahler and Cordes (1966).

Table 1469

Source	α8	α9	α10	β30
Beef	Ala	Ser	Val	Ala
Pig and sperm whale	Thr	Ser	Ile	Ala
Sheep	Ala	Gly	Val	Ala
Horse	Thr	Gly	Ile	Ala
Man	Thr	Ser	Ile	Thr
Dog	Thr	Ser	Ile	Ala
Rabbit	Thr	Ser	Ile	Ser

a Would you conclude that the 8, 9, 10, and 30 regions are critical with respect to insulin activity?

b Using Appendix Table A-5, develop a hypothesis suggesting how these differences could have arisen.

c Would single base-pair substitutions account for all amino acid substitutions?

1470 In a certain operon with three genes the sequence is operator (*O*)-*A*-*B*-*C*. The antisense strand of the DNA of gene *A* (analogous to its *m*RNA) has the codon GGA (glycine) at position 48 and UAU (tyrosine) at position 174. Since the genetic map, DNA, *m*RNA, and protein are collinear, show in Table 1470 which end (1) of the DNA sense strand, (2) of the DNA antisense strand, and (3) of *m*RNA, is the 5′ and which the 3′ and (4) which end of the protein would have the amino and which the carboxyl group.

Table 1470

	Genetic map of gene *A*	
	Position 48	Position 174
DNA: Sense strand		
Antisense strand		
*m*RNA		
A protein		

1471 **a** Tsugita (1962*a,b*) and Fraenkel-Conrat (1964) found that treatment of tobacco mosaic virus (TMV) (single-stranded RNA) with nitrous acid brought about the substitution of proline by leucine and serine by phenylalanine at several positions in the coat protein but never the reverse. Explain.

b The *wild-type* coat protein of TMV contains proline at position 20. Treatment with nitrous acid produced mutant variants with amino acid substitutions at this position according to the scheme. Treatment of the final

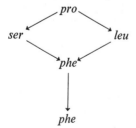

phe mutant with nitrous acid cannot induce further amino acid substitutions at this position.

(1) What are the codons for each of the amino acids shown? (Give correct base sequences.)

(2) Do these results indicate degeneracy? If so, for which of the codons shown?

1472 The protein coat of tobacco mosaic virus (TMV) which surrounds the single-stranded RNA contains 158 amino acids (Tsugita et al., 1960). Treatment of *wild-type* TMV with nitrous acid causes deamination of cytosine to uracil and adenine to hypoxanthine and results in the formation of mutant strains each with a single amino acid substitution. The amino acid substitutions and their positions in the protein in five mutant strains are shown in Table 1472

Table 1472

Mutant strain	Position in protein	Amino acid	
		In *wild-type* protein	In *mutant* protein
1	148	Ser	Phe
2	59	Arg	Gly
3	103	Glu	Gly
4	20	Pro	Leu
5	129	Ile	Val

(Tsugita, 1962*a*; Fraenkel-Conrat, 1964). Consulting Appendix Table A-5, indicate the codons in the *wild-type* TMV at each of these sites and the nucleotide change responsible for each of the mutants.

1473 Some rare (mutant) forms of human adult hemoglobin show single amino acid substitutions in the β chain, caused by different alleles of a single gene. Four of these types are shown.

Table 1473

Type of hemoglobin	Amino acid position	Change involved in amino acid substitutions, *wild-type* → *mutant*
S	6	Glu → Val
C	6	Glu → Lys
E	26	Glu → Lys
$G_{\text{San Jose}}$	7	Glu → Gly

a If the codons for lysine, valine, and glycine are AAG, GUG, and GGG respectively, and a mutation involving a single base-pair change causes each of these rare forms, what is the codon for glutamic acid?

b Which of these changes represent transitions and which represent transversions?

c In the α chain, hemoglobins *I* and *Norfolk*, respectively, contain aspartic acid at positions 16 and 57 in the place of lysine and glycine at those positions.
 (1) What is the codon for aspartic acid?
 (2) Did both changes occur via the same base-pair substitution mechanisms?

1474   The *t*RNA of *E. coli* contains two different methionine-accepting *t*RNAs, one (Met-*t*RNA$_F$) of which can have a hydrogen atom in the amino group of its methionine replaced by a formyl, CHO, group.   The methionine in the other *t*RNA species (Met-*t*RNA$_M$) is, like the other amino acids on *t*RNAs, nonformylatable (Clark and Marcker, 1966*a*).   Of the trinucleotides tested for ability to bind amino acids to ribosomes they found the following.

1   Only AUG bound the methionine carried by Met-*t*RNA$_M$, whereas both AUG and GUG bound that carried by Met-*t*RNA$_F$.   GUG also bound valine.   Of the synthetic messengers tested for their ability to bring about synthesis of polypeptides containing methionine (in a cell-free *E. coli* system) only poly-UAG and poly-UG gave positive results (Clark and Marcker, 1966*a,b*, 1968).

2   Using Met-*t*RNA$_M$, methionine was incorporated only in the presence of poly-UAG, and only internally.   Using Met-*t*RNA$_F$, methionine was incorporated into polypeptides with both synthetic messengers but only in the starting position; this was true regardless of whether the methionine was formylated or not.   Formylation increased the rate of protein synthesis.

Note: *t*RNAs for other amino acids are not formylatable.   Subsequent studies by Ghosh et al. (1967*b*) revealed that if poly-(UG)$_n$ with a repeating sequence of uracil and guanine was used as messenger, polypeptide chains were synthesized with the amino acid sequence F-Met-(Cys-Val)$_n$.   Formyl-methionine was in the N-terminal position only.   The synthesis was dependent on the addition of Met-*t*RNA$_F$ (whether formylated or not).   Poly-(AUG)$_n$ directed the synthesis of polymethionine, which required both Met-*t*RNA$_F$ and Met-*t*RNA$_M$.

a With reasons for your answers state:
(1) Which is the initial codon or which are the initiating codons.
(2) The role of *t*RNA$_F$ and *t*RNA$_M$.
(3) The apparent role of the formyl group in protein synthesis.

b Describe the system whereby polypeptide translation is initiated, including the role of codons and *t*RNAs, and suggest a role for formylation in poly-peptide synthesis.

1475   The N-terminal amino sequence in the coat protein of phage R17 synthesized in vivo is Ala-Ser-Asn-Phe-Thr.   To test the involvements of F-Met-*t*RNA in protein synthesis Adams and Capecchi (1966) added this compound to an in vitro *E. coli* extract with phage R17 RNA as messenger and found that the in vitro coat protein had the amino acid sequence F-Met-Ala-Ser-Asn-Phe-Thr.   Suggest why methionine should be the first amino acid in the in vitro product and not in the in vivo one.

1476   In 1966*b* Crick suggested that during translation, while standard base pairing (A with U and G with C) occurs in the first two positions of the triplet, there may be some play (or wobble) in the pairing at the third base site so that a particular species of *t*RNA may pair with more than one codon; thus the base I at the 5′ end of the anticodon can pair with the bases A, U, or

C at the 3′ end of a codon, G at the wobble position can pair with the U or C, while U can pair with A or G. Base C at the 5′ end of the anticodon would pair with G and A only with U.

a Show what codons can pair with the anticodons $tRNA_{Phe(GAA)}$ and $tRNA_{Ala(IGC)}$.

b There are two kinds of serine $tRNAs$, $tRNA_{SerI(IGC)}$ and $tRNA_{SerII(UGA)}$. Consulting Appendix Table A-5, list the codons recognized by each of these $tRNAs$.

c The anticodons of six $tRNAs$ are tabulated. Predict the codons these $tRNAs$ would pair with according to the wobble hypothesis and check whether these are in agreement with the code (see Appendix Table A-5).

Table 1476

Species and amino acid	Anticodon in $tRNA$ $3' \rightarrow 5'$
Alanine	CGI
Valine	GAI
Phenylalanine	AAG
Tyrosine	AUG, AUC
Formylmethionine	UAC

See Söll and RajBhandary (1967) and Söll, Cherayil, and Bock (1967) for convincing evidence for multiple recognition of codons by one kind of $tRNA$.

d Explain how one $tRNA$ can recognize several codons provided they differ in the last base in the codon.

e If an amino acid is coded for by all four bases in the third position, i.e., proline (CCU, CCC, CCA, CCG), according to the wobble theory at least how many $tRNAs$ must there be for this amino acid? *Note:* Codon-anticodon pairing occurs in an antiparallel fashion.

1477 The very closely related phages f2, MS2, and R17 have a single strand of RNA about 3,300 to 3,500 nucleotides long which contains three genes that specify the synthesis of RNA synthetase, the coat, and A proteins (Jeppesen et al., 1970). The results of nucleotide and amino acid sequencing of the RNA and proteins respectively are detailed below.

1 The 5′ end of MS2, R17, and probably f2 contains an identical untranslated sequence of 129 nucleotides followed by the nucleotide sequence of the first of the three genes (Adams and Cory, 1970; De Wachter et al., 1971; Ling, 1971; Steitz, 1969).

2 The 5′-terminal sequence of MS2, R17, and f2 RNA is

GGGUGGGACC(CCUUU)CGGGGUCCUGCUCAACUUCCUGUCGAGCUAAU-
GCCAUUUUUAAUGU(CUUU)AGCGAGACGCUACCAUG(CUAUCGGCUGU-
AGGUAGCCG)GAAUUCCAUUCCUAGGAGGUUUGACCU
·AUG·CGA·GCU·UUU·AGU·G···

3   The nucleotide sequences (5′ to 3′) of (*a*) three short segments, each including a region just before and extending into each R17 gene (Steitz, 1969), (*b*) a 72-nucleotide segment, including the latter part of the coat-protein gene, an untranslated region of 36 nucleotides between the coat protein and synthetase genes, and the beginning of the synthetase gene (Nichols, 1970), and (*c*) two fragments from the coat-protein gene of f2 and R17 encompassing 100 nucleotides (Nichols and Robertson, 1971) are as follows:

a   Three short segments including the beginning of each R17 gene

Coat protein	UAACCGGGGUUUGAAGCAUGGCUUCUAACUUU
Synthetase	AAACAUGAGGAUUACCCAUGUCGAAGACAACAAAG···
A protein	CCUAGGAGGUUUGACCUAUGCGAGCUUUUAGUG···

b   The 72-nucleotide segment is

(G)CA AAC UCC GGU AUC UAC UAA UAG AUG CCG GCC AUU CAA ACA UGA GGA UUA CCC AUG UCG AAG <u>ACA</u> <u>ACA</u> <u>AAG</u>

c   Two coat-protein fragments

4   The amino acid sequence of (*a*) the 129 residues in the R17 coat protein, which is identical to that of f2 except for a single amino acid difference at position 88 (R17 contains methionine, f2, leucine) (Weber, 1967) and (*b*) the N-terminal regions of the A protein and RNA synthetase (Steitz, 1969) are given below.

a   Amino acid sequence of R17 coat protein:

Ala-Ser-Asn-Phe-Thr-Gln-Phe-Val-Leu-Val-Asn-Asp-Gly-Gly-Thr-Gly-Asn-
　　　　　　　　　　　5　　　　　　　　　　10　　　　　　　　　　15

Val-Thr-Val-Ala-Pro-Ser-Asn-Phe-Ala-Asn-Gly-Val-Ala-Glu-Trp-Ile-Ser-Ser-Asn-
　20　　　　　　　　　25　　　　　　　　　30　　　　　　　　　35

Ser-Arg-Ser-Gln-Ala-Tyr-Lys-Val-Thr-CMCys-Ser-Val-Arg-Gln-Ser-Ser-Ala-
　　　　　40　　　　　　　　　45　　　　　　　　　50

Gln-Asn Arg-Lys-Tyr-Thr-Ile-Lys-Val-Glu-Val-Pro-Lys-Val-Ala-Thr-Gln-Thr-
　　　55　　　　　　　　60　　　　　　　　65　　　　　　　　70

Val-Gly-Gly-Val-Glu-Leu-Pro-Val-Ala-Ala-Trp-Arg-Ser-Tyr-Leu-Asn-Met-
　　　　　75　　　　　　　　80　　　　　　　　85

Glu-Leu-Thr-Ile-Pro- Ile-Phe-Ala-Thr-Asn-Ser-Asp-CMCys-Glu-Leu-Ile-Val-
　　90　　　　　　　95　　　　　　　100　　　　　　　105

Lys-Ala-Met-Gln-Gly-Leu-Leu-Lys-Asp-Gly-Asn-Pro-Ile-Pro-Ser-Ala-Ile-Ala-
　　　　110　　　　　　　115　　　　　　　120

Ala-Asn-Ser-Gly-Ile-Tyr
　　　125　　　　　　　129

b   N-terminal amino acid sequence:

A protein	F-Met-Arg-Ala-Phe-Ser →
Synthetase	F-Met-Ser-Lys-Thr-Thr-Lys →

Note: F-Met, the first amino acid in all three proteins, is removed after the polypeptides are synthesized.

a Indicate one possible function the initial 129-nucleotide sequence may perform.

b At what position does the first initiating codon start, and which one is it?

c Which of the three genes is closest to the 5′ end of the genome?   Explain, showing the latter sequence of the 5′ untranslated region and the first five codons of this gene and the amino acids they specify.

d What is the sequence of the other two genes relative to each other and the one nearest the 5′ end?   Explain in the same manner as in (c).

e With respect to the 72-nucleotide sequence:

(1) Which region contains the 36 nucleotides between the coat protein and synthetase genes?

(2) How many and which amino acids in the coat protein and synthetase are specified by the 27-nucleotide sequence?

(3) How many stop codons are there at the end of the coat-protein gene? Do you think this number is a common occurrence?   Is this number always required?   Explain.   Why may this or a greater number of such successive codons have come into existence?

f One of the two f2 and R17 fragments that Nichols and Robertson compared codes for amino acids 81 to 99 of the coat protein; the other codes for amino acids 124 to 129 and also contains an untranslated region of eight codons beyond the 3′ terminus of the coat-protein gene.

(1) Four nucleotide differences occur between f2 and R17 in these two fragments.   Where are they, and which changes do they involve?

(2) Which one accounts for the single amino acid difference between the two proteins?   Why do the other three changes not alter the message?

1478 Söll et al. (1967) showed that the specific binding of yeast phenylalanine-*t*RNA to yeast of *E. coli* ribosomes was stimulated by the two trinucleotides UUU and UUC, both of which are codons for phenylalanine.   Söll and RajBhandary (1967), using an *E. coli* cell-free protein-synthesizing system with only one purified species of yeast phenylalanine-*t*RNA, found that phenylalanine was incorporated into polypeptides by either poly-U or poly-(UUC)$_n$ as messengers.   The primary structure of this *t*RNA contains no AAA sequence but does contain a 2′OMeGAA sequence [2′OMeG is 2′-(*O*-methyl) guanosine]. Do these data support the codon-anticodon pairing as proposed in the wobble hypothesis (Crick, 1966*b*)?

1479 In *wild-type E. coli* glycine occurs at position 210 in the A polypeptide of tryptophan synthetase.   In the mutant *A46*, which does not grow on minimal medium, glycine is replaced by glutamic acid.   A revertant of *A46*, designated *A46 PR8*, grows on minimal medium but much more slowly than *wild type*. When this revertant is crossed with *wild type*, some of the recombinants are *A46* and others *PR8* (do not grow on minimal medium).   The cross

$T^+ PR8^-$ × $T^- PR8^+ A46^-$ produced 101 $PR8^+ A46^+$, of which 81 were T^- and 20 were T^+, and 85 $PR8^- A46^-$, of which 20 were T^- and 65 were T^+. Note that T is to the left of the A gene, in which the two mutant sites PR and $A46$ occur. The amino acid sequences at the relevant positions in the *wild-type* A protein and in the three mutant A chains were found to be as shown. At all other positions the proteins were identical and

Table 1479

Strain	Amino acid positions	
	174	210
Wild type	Tyr	Gly
A46	Tyr	Glu
A46 PR8	Cys	Glu
PR8	Cys	Gly

wild type in constitution (Helinski and Yanofsky, 1963). Map the mutations responsible for *A46* and *PR8* and discuss the significance of the fact that the mutant *A46 PR8* is phenotypically *pseudo wild type*.

1480 a A cross is made between a *wild-type* revertant from a particular *r*II mutant of phage T4 and *wild type*. About 2 percent of the phage particles in the progeny are of *r*II-mutant phenotype. When these mutants were crossed with the original mutant, about half of them bred true (producing no *wild type*); the other half produced *wild types* and *mutants* in approximately equal proportions. Was the revertant due to a mutation at the same site as the original mutation? Explain by diagraming the crosses, indicating the genotype and phenotype of all phage involved.

b (1) The extreme left end of the *r*IIB cistron of phage T4 is not essential for its activity. However, Sarabhai and Brenner (1967a) found that (1) an *ochre* mutant *oc* and (2) a frame-shift mutation *fs*(+), both mapping in that region, could separately eliminate the B function. When a double mutant *oc fs*(+) was constructed, still no B activity was detected. Surprisingly, treatment of the double mutant with amino-purine resulted in the restoration of B function by a third mutation *m*. The relative locations of *oc*, *fs*(+), and *m* are (the arrow indicates the direction of reading):

Can you propose a satisfactory explanation for these findings? What would be the expected phenotype of *m*, *oc m*, and *m fs*(+)?

(2) A new drug has been found to possess interesting mutagenic properties. *E. coli* mutants, *met⁻*, induced by the drug revert, *met⁻* → *met⁺*, both by nitrosoguanidine (believed to cause substitution mutations) and by ICR-191 (substituted acridine, believed to induce frame-shift mutations). The *met⁻* mutants do not undergo phenotypic curing by antibiotics such as streptomycin or neomycin.   Some of the "revertants" seem to be due to external suppressors, which, however, fail to suppress nonsense mutants of T4.   From these observations, can you suggest a mechanism for the mutagenic action of this drug?

(3) A spontaneous *lac⁻* mutant of *E. coli* fails to revert by any of the known mutagens or mutators.   However, it maps as a point mutation and reverts spontaneously at a low but appreciable frequency.   What could be the probable nature of this mutational event?

1481 Benzer and Champe (1962) and Garen and Siddiqi (1962) noted that *amber* mutants are due to the formation of nonsense triplets within the gene causing premature polypeptide-chain termination at the residue specified by the nonsense codon.   The same has been found to hold true for *ochre* mutants (Brenner and Beckwith, 1965).   To determine the base composition and sequence of the *amber* and *ochre* mutants Brenner et al. (1965) studied the production and reversion of *r*II *amber* and *ochre* mutants in phage T4 and also investigated head-protein *amber* mutants to define the amino acids connected to the *amber* triplet.   Their experiments showed that:

1   No *amber* or *ochre* mutants reverted to *wild type* when treated with hydroxylamine, NH_2OH, which reacts with cytosine and changes G-C base pairs to A-T pairs.

2   *Ochre* mutants can be converted into *amber* mutants by treating with 2-aminopurine, which causes (A-T ↔ G-C changes). *Ochre* mutants cannot be induced to mutate to *amber* with NH_2OH.

3   Both *amber* and *ochre* mutants can be induced by NH_2OH from *wild type*.

These data indicated that the *amber* and *ochre* triplets (codons) on DNA have at least one A-T pair in common and that *amber* contains a G-C pair which corresponds to an A-T pair in *ochre*.

a Analyze the data to show that this is so.

b Is it possible to establish the orientation of these bases with respect to the sense and antisense strands of DNA?   Why?

From previous studies it was known that *r*II genes express their functions before the onset of DNA synthesis.   Therefore only changes in the sense strand of DNA will register a phenotypic effect in the first cycle of growth. Changes (mutations) in the antisense strand would not produce their mutant effects until the second cycle of growth.   Brenner et al. treated *wild-type*, *r⁺*, T4 phage with NH_2OH and studied *amber* and *ochre* mutants in the *r*II region.   Some of these mutants produced their effects before the first round of DNA replication.   Others did not do so until after the first round of DNA replication.

c These data permit a definite decision regarding the number of A-T pairs in the *amber* and *ochre* triplet and the orientation of these base pairs with respect to the sense and antisense strands of DNA.

(1) Analyze the data and give this information.

(2) Show the possible *amber* and *ochre* triplets in the *m*RNA.

Note: A C → U change in sense strand will produce a defective *m*RNA before replication. Such a change in the antistrand requires one replication to effect a change in the *m*RNA.

d *Amber* codons can arise from codons for glutamine or tryptophan which in *m*RNA are CAG and UGG respectively. What is the base composition of *m*RNA *amber* and *ochre* triplets? Explain.

1482 *Phosphate-negative* nonsense mutants in *E. coli* are incapable of synthesizing functional alkaline phosphatase. In nonsense mutants, only protein fragments are formed. The size of the fragment depends on the position of nonsense codons in the gene. Weigert and Garen (1965*b*) established the base sequence in one of the nonsense codons by comparing the amino acid substitutions in 21 revertant strains arising from the nonsense mutant *H12*. The revertants, due to mutation in the nonsense codon responsible for *H12*, either arose spontaneously or were induced by nitrous acid, ethyl methanesulfonate, or *N*-methyl-*N'*-nitro-*N*-nitrosoguanidine. In each of these revertants, tryptophan at a specific site in peptide 1 in *wild-type* alkaline phosphatase was replaced by one of the following six amino acids: glutamine, glutamic acid, tyrosine, serine, leucine, or lysine. The RNA codons for tryptophan and these six amino acids are shown. Thus the nonsense triplet must have

Table 1482

Amino acid	RNA codon
Tryptophan	UGG
Glutamine	CAA, CAG
Glutamic acid	GAA, GAG
Tyrosine	UAU, UAC
Serine	UCU, UCC, UCA, UCG, AGU, AGC
Leucine	UUA, UUG, CUU, CUA, CUC, CUG
Lysine	AAA, AAG

arisen from the tryptophan triplet by a single-base change and must also by a single-base change provide one of the possible codons for each of these amino acids. Only one triplet satisfies all the conditions. Which is it and why?

1483 Weigert and Garen (1965*a*) showed that three nonsense mutants, 12, 45, and G-5, at different sites in the alkaline phosphatase structural gene in *E. coli* could arise from either a codon for tryptophan or a codon for glutamine. All three mutants were suppressed by the same suppressor (*Su1*), which replaced tryptophan with serine in suppressed mutant 12, whereas in the other two *Su1* caused the substitution of serine for glutamine. Account for the fact that the original codon in which a mutation occurs to produce a nonsense codon can vary but the amino acid inserted by a specific suppressor gene, e.g., *Su1* is in all cases the same, e.g., serine.

1484   Two of the three nonsense codons are UAG and UAA. They respond differently to suppressors, and either can be converted into the other. *Wild-type* alkaline phosphatase in *E. coli* has glutamine at a specific position in peptide C40. A nonsense *phosphatase-negative* mutant *U28(N2)*, due to mutation in the glutamine codon of *wild-type E. coli*, was transformed by a mutation, *U28(N1)*. From both mutants *phosphatase-positive* revertants were isolated which were due to mutations in the nonsense codons. The amino acids that were substituted for glutamine (present in *wild-type* enzyme) in 29 revertants of *U28(N2)* and 24 from *U28(N1)* are shown in Table 1484*A* (Weigert et al., 1967*a*).

Table 1484*A*

Original nonsense mutant	Amino acid substitutions in revertants	Total no. of occurrences	Proportion of tryptophan substitutions
U28(N2)	Glu → Glu	8	0 in 29
	Glu → Gln	11	
	Glu → Ser	3	
	Glu → Leu	2	
	Glu → Lys	5	
U28(N1)	Glu → Glu	2	6 in 24
	Glu → Gln	12	
	Glu → Tyr	2	
	Glu → Lys	2	
	Glu → Trp	6	

a The nine possible triplets that are related to UAG by single-base substitutions are shown in Table 1484*B*. Show the nine triplets related to UAA by single-base substitutions. Using Appendix Table A-5, assign amino acids to the triplets.

Table 1484*B*

Nonsense codon	Related codons			Nonsense codon	Related codons		
UAG	AAG	UCG	UAC	UAA	——	——	——
	CAG	UUG	UAU		——	——	——
	GAG	UGG	UAA		——	——	——

b Compare the amino acids that can be coded by triplets for ability to distinguish between UAG and UAA. Provide the kinds of substitutions that were obtained in revertants of the mutants *U28(N2)* and *U28(N1)* and state, with reasons, which is UAG and which UAA.

1485 Zipser and Newton (1967) demonstrated that:

1 The nonsense mutant (*NG813*) in the *z* (*β*-galactosidase) gene of *E. coli* is due to a single-base change in a *wild-type* codon. This same codon also mutates to form the *amber* (UAG) mutant *NG1012*.

2 *NG813* is convertible to UAA by a single-base change. The possible codons in *NG813* that can be changed by single-mutational steps to the *ochre* codon UAA and the possible codons that the *wild-type* codon, which gives rise to *NG813* and the *amber* (UAG) codon *NH1012*, can have are shown.

Table 1485

Possible mutant codons which are one mutational step away from *ochre* (UAA)	Possible *wild-type* codons which are one mutational step away from *amber* (UAG)
UAU (Tyr)	UAC (Tyr)
UAC (Tyr)	UAU (Tyr)
UUA (Leu)	UUG (Leu)
UCA (Ser)	UCG (Ser)
UGA (—)	UGG (Trp)
AAA (Lys)	AAG (Lys)
CAA (Gln)	CAG (Gln)
GAA (Glu)	GAG (Glu)

a Which is the nonsense codon in *NG813*? Explain your decision.

b Which codon in the *wild-type z* gene can mutate to form both *NG813* and *NG1012*? Explain.

1486 The RNA of phage R17 in a certain mutant strain is known to carry a nonsense mutation in the coat-protein gene. This RNA directs the synthesis of some complete coat-protein molecules in a cell-free protein-synthesizing system derived completely from the permissive, Su^+, host strain S26RIE of *E. coli*. Serine replaces glutamine, which occurs in *wild-type* head proteins. No complete coat proteins are formed when the cell-free system is derived from the isogenic nonpermissive, Su^-, strain S26. When *t*RNAs from the Su^+ strain of *E. coli* were added singly to an otherwise nonpermissive, Su^-, system, complete head proteins were produced only when serine-accepting *t*RNA were added (Capecchi and Gussin, 1965).

a Which cellular component is responsible for suppressor activity?

b What effect might mutation of Su^- to Su^+ have on the component to cause some of the proteins in the suppressed strain to have a complete complement of amino acids identical to *wild-type* protein except, in this case, with serine in place of glutamine?

1487 **a** J. D. Smith et al. (1966), using a transducing phage $\phi80$ carrying the *E. coli* nonsense suppressor $Su3^+$, found that when the phage and its suppressor replicate, an increased synthesis of tyrosine *t*RNA occurs. This tyrosine *t*RNA recognizes UAG and appears not to recognize the two normal tyrosine codons UAU and UAC. Replication of a $\phi80$ strain carrying the *wild-type* $Su3^-$ allele leads to a similar increase of tyrosine *t*RNA, which recognizes the normal tyrosine codons UAU and UAC but not UAG. What do these results suggest regarding the function of the *Su* gene?

b Andoh and Ozeki (1968) found, using phage $\phi80$ $Su3^+$, that the DNA from this phage hybridizes with the tyrosine *t*RNA from both the Su^+ and Su^- species of *E. coli*. One segment per DNA molecule hybridizes with one *t*RNA molecule. Do these results confirm your answer in (a)? Explain.

c What part of the *t*RNA might mutation of $Su3^-$ to $Su3^+$ change? Explain.

1488 The head proteins produced by *wild-type* T4 phage and by the mutant H36 on two different *E. coli* strains have the N-terminal amino acid sequences shown (Stretton and Brenner, 1965).

Table 1488

Strain	N-terminal amino acid sequences
Wild type	Ala-Gly-Val-Phe-Asp-Phe-Gln-Asp-Pro-Ile-Asp-Ile-Arg $\cdots$
H36 on *E. coli* strain 1	Ala-Gly-Val-Phe-Asp-Phe
H36 on *E. coli* strain 2	Ala-Gly-Val-Phe-Asp-Phe and
	Ala-Gly-Val-Phe-Asp-Phe-Ser-Asp-Pro-Ile-Asp-Ile-Arg $\cdots$

a Was the mutation from *wild type* to *H36* a substitution, deletion, or addition of a base pair?

b Why does *H36* produce an incomplete protein on strain 1 of *E. coli* and both complete and incomplete types on strain 2?

c Why is glutamine replaced by serine in *H36* proteins on strain 2?

d Would you expect the phenotype of *H36* on strain 2 to be *mutant*, *wild type*, or *pseudo wild type*? Explain.

1489 In *E. coli*, an *amber* codon (UAG) in the β-galactosidase gene causes chain termination in $Su3^-$ strains. In $Su3^+$ strains the chain-terminating triplet is read as if it spelled tyrosine, and both NH_2-terminal fragments and whole β-galactosidase proteins are produced. Goodman et al. (1968) showed that:

1 $Su3^+$ originated as a spontaneous mutant from an $Su3^-$ strain due to a single base-pair change.

2 Infection of *E. coli* $Su3^-$ lac^- *amber* with $\phi80$ carrying $Su3^+$ results in suppression, associated with increase in the amount of tyrosine *t*RNA in the cells, which binds only

to UAG in ribosome-binding experiments.   The corresponding *t*RNA in cells infected with ϕ80 *Su3*⁻ recognizes the normal tyrosine codons UAU and UAC.

3   The nucleotide sequences in the *t*RNAs specified by the allelic genes *Su3*⁺ and *Su3*⁻ were identical except for base change in oligonucleotide 17 (a short segment):[1]

Sequence in *Su3*⁺ strains   ACUCUAA*AψCUG ⋯
Sequence in *Su3*⁻ strains   ACUGUAA*AψCUG ⋯

a Suppressor gene *Su3* specifies a species of *t*RNA.   What is the nature of the change *Su3*⁻ → *Su3*⁺, and what portion of the *t*RNA does it affect? Explain.

b Does the base sequence in the *Su3*⁻ and *Su3*⁺ strains accord with coding properties noted for the *Su*⁻ and *Su*⁺ *t*RNAs?

c Would you expect more than one species of tyrosine *t*RNA in *E. coli*? Would these be specified by genes other than *Su3* and why?   See Goodman et al. (1968).

1490 a Nonsense mutants cause premature termination of polypeptide chain growth.   It is known that addition (+) or deletion (−) of a base pair in the B1 segment of the *B* gene in the *r*II region of phage T4 results in a mutant phenotype (no B activity).   If both an addition and a deletion are present in this region, a *pseudo-wild-type* phenotype is restored.   Brenner and Stretton (1965) found that double and triple mutants had the phenotypes shown when grown on *E. coli* K12 *Su*⁻ (a strain on which nonsense mutants do not grow).   Do nonsense mutants exert their effects at the level of translation or at the level of transcription?   Explain.

Table 1490

Type	Mutant	Phenotype
double	*amber*, + or *amber*, −	no growth
double	*ochre*, + or *ochre*, −	no growth
triple	+, *amber*, −	near *wild-type* growth
triple	+, *ochre*, −	near *wild-type* growth

b Although most double (+ −) mutants in the B1 segment grow on *E. coli* K12 *Su*⁻, one (*FC73*, *FC23*) does not.   It does, however, grow on *E. coli* K12.   Offer an explanation for these results.

1491 The following data are based on a series of elegant experiments by Gorini and Kataja (1964), Davies et al. (1964), Gorini (1966), and others.

[1] ψ stands for pseudouridylic acid, which can form the same base pairs as the base uracil.   A* is a modified form of A.

1   An arginineless *E. coli* strain unable to synthesize ornithine transcarbamylase (OTC) because of a single mutant gene and sensitive to streptomycin grows if arginine or citrulline is added to minimal medium but does not produce OTC under these conditions.   Streptomycin can act as a substitute for the growth factors.   When added to minimal medium, the mutant grows, although slowly, and produces small amounts of OTC.   The same results were obtained with other mutants.   This reaction to streptomycin is determined by a mutant gene nonallelic with the OTC gene.

2   In a cell-free system containing all amino acids and poly-U as RNA if the ribosomes, *t*RNA, and enzymes came from an *E. coli streptomycin-resistant* strain, only phenylalanine was incorporated into the polypeptides, as expected, regardless of whether streptomycin was added or not.   However, if the ribosomes, *t*RNA, and enzymes came from the *streptomycin-sensitive E. coli* strain, the results depended on whether streptomycin was added or not.   When streptomycin was not added, only phenylalanine was incorporated; but when streptomycin was added, it decreased the incorporation of phenylalanine and caused the misincorporation of isoleucine.

3   By mixing 30S and 50S subunits from *streptomycin-sensitive* and *streptomycin-resistant* strains, hybrid and reconstituted parental ribosomes were formed.   The effect of streptomycin on phenylalanine incorporation by these ribosomes in the in vitro poly-U system is shown in the table.

Table 1491   Percent inhibition of phenylalanine incorporation by stremptomycin

Constitution of hybrids		Experiment			
30S	50S	1	2	3	4
Sensitive	*Sensitive*	50	40	78	78
Sensitive	*Resistant*	84	37	47	73
Resistant	*Resistant*	14	3	7	0
Resistant	*Sensitive*	20	10	12	17

a Discuss these data fully, being sure to state the apparent effect of the second mutant gene, why the *streptomycin-sensitive* strain produces OTC and grows on minimal when streptomycin is added, and why this antibiotic has little effect on the *streptomycin-resistant* strain.

b Would you expect molecules of OTC produced in the presence of streptomycin to have a different amino acid sequence from that synthesized without the antibiotic?   Explain.

See Brownstein and Lewandowski (1967) and Apirion (1966) for more recent information on the effect of suppressors on ribosome structure.

REFERENCES

Abelson, J. M., M. L. Gefter, L. Barnett, A. Landy, R. L. Russell, and J. D. Smith (1970), *J. Mol. Biol.*, **47**:15.

Adams, J. M., and M. R. Capecchi (1966), *Proc. Natl. Acad. Sci.*, **55**:147.

—— and S. Cory (1970), *Nature* **227**:570.

Altman, S., S. Brenner, and J. D. Smith (1971), *J. Mol. Biol.*, **56**:195.

Andoh, T., and H. Ozeki (1968), *Proc. Natl. Acad. Sci.*, **59**:792.

Apirion, D. (1966), *J. Mol. Biol.*, **16**:285.

—— and D. Schlessinger (1967), *Proc. Natl. Acad. Sci.*, **58**:206.

Baglioni, C. (1963), in Taylor (1963), pp. 405–475.

Barnett, W. E., D. H. Brown, and J. L. Epler (1967), *Proc. Natl. Acad. Sci.*, **57**:1775.

Basilio, C., A. J. Wahba, P. Lengyel, J. F. Speyer, and S. Ochoa (1962), *Proc. Natl. Acad. Sci.*, **48**:613.

Beale, D., and H. Lehmann (1965), *Nature*, **207**:259.

Benzer, S., and S. P. Champe (1961), *Proc. Natl. Acad. Sci.*, **47**:1025.

—— and —— (1962), *Proc. Natl. Acad. Sci.*, **48**:1114.

Berger, H., W. J. Brammar, and C. Yanofsky (1968), *J. Mol. Biol.*, **34**:219.

—— and A. W. Kozinski (1969), *Proc. Natl. Acad. Sci.*, **64**:897.

—— and C. Yanofsky (1967), *Science*, **156**:394.

Birge, E. A., and C. G. Kurland (1970), *Mol. Gen. Genet.*, **109**:356.

Brammar, W. J., H. Berger, and C. Yanofsky (1967), *Proc. Natl. Acad. Sci.*, **58**:1499.

Bremer, H., M. W. Konrad, K. Gaines, and G. S. Stent (1965), *J. Mol. Biol.*, **13**:540.

Brenner, S. (1957), *Proc. Natl. Acad. Sci.*, **43**:687.

——, L. Barnett, E. R. Katz, and F. H. C. Crick (1967), *Nature*, **213**:449.

—— and J. R. Beckwith (1965), *J. Mol. Biol.*, **13**:629.

—— and A. O. W. Stretton (1965), *J. Mol. Biol.*, **13**:944.

——, ——, and S. Kaplan (1965), *Nature*, **206**:994.

Bretscher, M. S. (1966), *Cold Spring Harbor Symp. Quant. Biol.*, **31**:289.

Brimacombe, R., J. Trupin, M. Nirenberg, P. Leder, M. Bernfield, and T. Jaouni (1965), *Proc. Natl. Acad. Sci.*, **54**:954.

Brody, S., and C. Yanofsky (1963), *Proc. Natl. Acad. Sci.*, **50**:9.

Brown, J. C., and A. E. Smith (1970), *Nature*, **226**:610.

Brownstein, B. L., and L. J. Lewandowski (1967), *J. Mol. Biol.*, **25**:99.

Capecchi, M. R. (1966), *Proc. Natl. Acad. Sci.*, **55**:1517.

—— and G. N. Gussin (1965), *Science*, **149**:417.

Carbon, J., C. Squires, and C. W. Hill (1970), *J. Mol. Biol.*, **52**:571.

Carlton, B. C., and C. Yanofsky (1963), *J. Biol. Chem.*, **238**:2390.

Caskey, C. T., et al. (1968), *Science*, **162**:125.

Casselton, L. A. (1971), *Sci. Prog.*, **59**:143.

Champe, S. P., and S. Benzer (1962), *Proc. Natl. Acad. Sci.*, **48**:523.

Chan, T. S., R. E. Webster, and N. D. Zinder (1971), *J. Mol. Biol.*, **56**:101.

Chapeville, F., F. Lipman, G. von Ehrenstein, B. Weisblum, W. J. Roy, and S. Benzer (1962), *Proc. Natl. Acad. Sci.*, **48**:1086.

Clark, B. F. C., S. K. Dube, and K. A. Marcker (1968), *Nature*, **219**:484.

—— and K. A. Marcker (1965), *Nature*, **207**:1038.

—— and —— (1966a), *J. Mol. Biol.*, **17**:394.

—— and —— (1966b), *Nature*, **211**:378.

—— and —— (1968), *Sci. Am.*, **218**(January):36.

Cold Spring Harbor Symp. Quant. Biol., **28**(1963), "Synthesis and Structure of Macromolecules."

Cold Spring Harbor Symp. Quant. Biol., **31**(1966), "The Genetic Code."

Cold Spring Harbor Symp. Quant. Biol., **34**(1969), "The Mechanism of Protein Synthesis."
Crick, F. H. C. (1962), *Sci. Am.*, **207**(October):66.
——— (1963), *Prog. Nucleic Acid Res.*, **1**:163.
——— (1966*a*), *Sci. Am.*, **215**(October):55.
——— (1966*b*), *J. Mol. Biol.*, **19**:548.
——— (1967), *Proc. R. Soc. (Lond.)*, **B167**:331.
———, L. Barnett, S. Brenner, and R. J. Watts-Tobin (1961), *Nature*, **192**:1227.
——— and S. Brenner (1967), *J. Mol. Biol.*, **26**:361.
———, J. S. Griffith, and L. E. Orgel (1957), *Proc. Natl. Acad. Sci.*, **43**:416.
Davies, J., W. Gilbert, and L. Gorini (1964), *Proc. Natl. Acad. Sci.*, **51**:883.
———, D. S. Jones, and H. G. Khorana (1966), *J. Mol. Biol.*, **18**:48.
De Wachter, R., et al. (1971), *Proc. Natl. Acad. Sci.*, **68**:585.
Edgar, R. S., G. H. Denhardt, and R. H. Epstein (1964), *Genetics*, **49**:635.
Ehrenstein, G. von, and F. Lipmann (1961), *Proc. Natl. Acad. Sci.*, **47**:941.
Engelhardt, D. L., R. E. Webster, R. C. Wilhelm, and N. D. Zinder (1965), *Proc. Natl. Acad. Sci.*, **54**:1791.
Epstein, R. H., et al. (1963), *Cold Spring Harbor Symp. Quant. Biol.*, **28**:375.
Fitch, W. M., and E. Margoliash (1967), *Biochem. Genet.*, **1**:65.
Flaks, J. G., P. S. Leboy, E. A. Birge, and C. G. Kurland (1966), *Cold Spring Harbor Symp. Quant. Biol.*, **31**:623.
Fraenkel-Conrat, H. (1964), *Sci. Am.*, **211**(October):47.
Gallucci, E., and A. Garen (1966), *J. Mol. Biol.*, **15**:193.
Gamow, G. (1954), *Nature*, **173**:318.
Gardner, R. S., et al. (1962), *Proc. Natl. Acad. Sci.*, **48**:2087.
Garen, A. (1968), *Science*, **160**:149.
———, S. Garen, and R. C. Wilhelm (1965), *J. Mol. Biol.*, **14**:167.
——— and O. Siddiqi (1962), *Proc. Natl. Acad. Sci.*, **48**:1121.
Gesteland, R. F., W. Salser, and A. Bolle (1967), *Proc. Natl. Acad. Sci.*, **58**:2036.
Ghosh, H. P., and H. G. Khorana (1967*a*), *Proc. Natl. Acad. Sci.*, **58**:2455.
———, D. Söll, and H. G. Khorana (1967*b*), *J. Mol. Biol.*, **25**:275.
Goldstein, J., T. P. Bennett, and L. C. Craig (1964), *Proc. Natl. Acad. Sci.*, **51**:119.
Goodman, H. M., et al. (1968), *Nature*, **217**:1019.
Gopinathan, K. P., and A. Garen (1970), *J. Mol. Biol.*, **47**:393.
Gorini, L. (1966), *Sci. Am.*, **214**(April):102.
——— (1969), *Cold Spring Harbor Symp. Quant. Biol.*, **34**:101.
——— and J. R. Beckwith (1966), *Annu. Rev. Microbiol.*, **20**:401.
——— and E. Kataja (1964), *Proc. Natl. Acad. Sci.*, **3**:487.
Guest, J. R., and C. Yanofsky (1966), *Nature*, **210**:799.
Gupta, N. K., et al. (1968), *Proc. Natl. Acad. Sci.*, **60**:1338.
Harris, J. T., F. Sanger, and M. A. Naughton (1956), *Arch. Biochem. Biophys.*, **65**:427.
Hartwell, L. H., and C. S. McLaughlin (1968), *Proc. Natl. Acad. Sci.*, **59**:422.
Hawthorne, D. C. (1969), *J. Mol. Biol.*, **43**:71.
Hayes, W. (1968), "The Genetics of Bacteria and Their Viruses," 2d ed., Blackwell, Oxford.
Helinski, D. R., and C. Yanofsky (1962), *Proc. Natl. Acad. Sci.*, **48**:173.
——— and ——— (1963), *J. Biol. Chem.*, **238**:1043.
Henning, U., and C. Yanofsky (1962*a*), *Proc. Natl. Acad. Sci.*, **48**:183.
——— and ——— (1962*b*), *Proc. Natl. Acad. Sci.*, **48**:1497.
——— and ——— (1963), *J. Mol. Biol.*, **6**:16.
Hirsch, D. (1971), *J. Mol. Biol.*, **58**:439.
——— and L. Gold (1971), *J. Mol. Biol.*, **58**:459.
Holley, R. W. (1966), *Sci. Am.*, **214**(February):30.
———, J. Apgar, G. A. Everett, J. T. Madison, M. Marquisee, S. H. Merrill, J. R. Penswick, and A. Zamir (1965), *Science*, **147**:1462.

Hunt, J. A., and V. M. Ingram (1958), *Nature*, **181**:1062.
Imamoto, F., J. Ito, and C. Yanofsky (1966), *Cold Spring Harbor Symp. Quant. Biol.*, **31**:235.
⸻ and C. Yanofsky (1967), *J. Mol. Biol.*, **28**:1.
Ingram, V. M. (1957), *Nature*, **180**:326.
⸻ (1963), "The Hemoglobins in Genetics and Evolution," Columbia, New York.
Jacobson, K. B. (1971), *Nat. New Biol.*, **231**:17.
Jeppesen, P. G. N., et al. (1970), *Nature*, **226**:230.
Jones, D. S., S. Nishimura, and H. G. Khorana (1966), *J. Mol. Biol.*, **16**:454.
Jones, O. W., and M. W. Nirenberg (1962), *Proc. Natl. Acad. Sci.*, **48**:2115.
⸻ and ⸻ (1966), *Biochem. Biophys. Acta*, **119**:400.
Kaplan, S., A. O. W. Stretton, and S. Brenner (1965), *J. Mol. Biol.*, **14**:528.
Khorana, H. G. (1966), *J. Mol. Biol.*, **18**:48.
⸻ et al. (1966), *Cold Spring Harbor Symp. Quant. Biol.*, **31**:39.
Kossel, H., A. R. Morgan, and H. G. Khorana (1967), *J. Mol. Biol.*, **26**:449.
Landy, A., J. Abelson, H. M. Goodman, and J. D. Smith (1967), *J. Mol. Biol.*, **29**:457.
Leder, P., E. L. Skogerson, and D. J. Roufa (1969), *Proc. Natl. Acad. Sci.*, **62**:928.
⸻ and M. W. Nirenberg (1964a), *Proc. Natl. Acad. Sci.*, **52**:420.
⸻ and ⸻ (1964b), *Proc. Natl. Acad. Sci.*, **52**:1521.
Lengyel, P. (1967), in Taylor (1967), pp. 192–212.
⸻, J. F. Speyer, and S. Ochoa (1961), *Proc. Natl. Acad. Sci.*, **47**:1936.
⸻, ⸻, C. Basilio, and S. Ochoa (1962), *Proc. Natl. Acad. Sci.*, **48**:282.
Leonard, N. J., H. Iwamura, and J. Eisinger (1969), *Proc. Natl. Acad. Sci.*, **64**:352.
Ling, V. (1971), *Biochem. Biophys. Res. Commun.*, **42**:82.
Lodish, H. F. (1968), *J. Mol. Biol.*, **32**:47.
Lu, P., and A. Rich (1971), *J. Mol. Biol.*, **58**:513.
Madison, J. T., G. A. Everett, and H. Kung (1966), *Science*, **153**:531.
Mahler, H. R., and E. H. Cordes (1966), "Biological Chemistry," Harper & Row, New York.
Marcker, K. A., B. F. C. Clark, and J. S. Anderson (1966), *Cold Spring Harbor Symp. Quant. Biol.*, **31**:279.
⸻ and F. Sanger (1964), *J. Mol. Biol.*, **8**:835.
Margoliash, E. (1963), *Proc. Natl. Acad. Sci.*, **50**:672.
Marshall, R. E., C. T. Caskey, and M. Nirenberg (1967), *Science*, **155**:820.
Martin, R. G., J. H. Matthaei, O. W. Jones, and M. W. Nirenberg (1962), *Biochem. Biophys. Res. Commun.*, **6**:410.
Matthaei, J. H., et al. (1966), *Cold Spring Harbor Symp. Quant. Biol.*, **31**:25.
Miller, R. C., et al. (1971), *J. Mol. Biol.*, **56**:363.
Model, P., R. E. Webster, and N. D. Zinder (1969), *J. Mol. Biol.*, **43**:177.
Morgan, A. R., R. P. Wells, and H. G. Khorana (1966), *Proc. Natl. Acad. Sci.*, **56**:1899.
Newton, W. A., J. R. Beckwith, D. Zipser, and S. Brenner (1965), *J. Mol. Biol.*, **14**:290.
Nichols, J. L. (1970), *Nature*, **225**:147.
⸻ and H. D. Robertson (1971), *Biochem. Biophys. Acta*, **228**:676.
Ninio, J. (1971), *J. Mol. Biol.*, **56**:63.
Nirenberg, M. W. (1963), *Sci. Am.*, **208**(March):80.
⸻ and P. Leder (1964), *Science*, **145**:1399.
⸻, ⸻, M. Bernfield, R. Brimacombe, J. Trupin, F. Rottman, and C. O'Neal (1965), *Proc. Natl. Acad. Sci.*, **53**:1161.
⸻ and J. H. Matthaei (1961), *Proc. Natl. Acad. Sci.*, **47**:1588.
⸻, ⸻, and O. W. Jones (1962), *Proc. Natl. Acad. Sci.*, **48**:104.
Nishimura, S., D. S. Jones, and H. G. Khorana (1965a), *J. Mol. Biol.*, **13**:302.
⸻, ⸻, E. Ohtsuka, H. Hayatsu, T. M. Jacob, and H. G. Khorana (1965b), *J. Mol. Biol.*, **13**:283.
Ohlsson, B. M., P. F. Strigini, and J. R. Beckwith (1968), *J. Mol. Biol.*, **36**:209.

Okada, Y., E. Terzaghi, G, Streisinger, J. Emrich, M. Inouye, and A. Tsugita (1966), *Proc. Natl. Acad. Sci.*, **56**:1692.

Ozaki, M., S. Mizushima, and M. Nomura (1970), *Nature*, **222**:333.

Pestka, S., R. Marshall, and M. Nirenberg (1965), *Proc. Natl. Acad. Sci.*, **53**:639.

Reeves, R. H., and J. R. Roth (1971), *J. Mol. Biol.*, **56**:523.

Reichmann, M. E. (1964), *Proc. Natl. Acad. Sci.*, **52**:1009.

Rosen, B., F. Rothman, and M. G. Weigert (1969), *J. Mol. Biol.*, **44**:363.

Rosset, R., and L. Gorini (1969), *J. Mol. Biol.*, **39**:95.

Russell, R. L., J. N. Abelson, A. Landy, M. L. Gefter, S. Brenner, and J. D. Smith (1970), *J. Mol. Biol.*, **47**:1.

Salas, M., M. A. Smith, W. M. Stanley, A. J. Wahba, and S. Ochoa (1965), *J. Biol. Chem.*, **240**:3988.

Sambrook, J. F., D. P. Fan, and S. Brenner (1967), *Nature*, **214**:452.

Sarabhai, A., and S. Brenner (1967a), *J. Mol. Biol.*, **26**:141.

——— and ——— (1967b), *J. Mol. Biol.*, **27**:145.

———, A. O. W. Stretton, S. Brenner, and A. Bolle (1964), *Nature*, **201**:13.

Scolnick, E., R. Tompkins, C. T. Caskey, and M. Nirenberg (1968), *Proc. Natl. Acad. Sci.*, **61**:768.

Signer, E. R., et al. (1965), *J. Mol. Biol.*, **14**:153.

Smith, A. E., and K. A. Marcker (1970), *Nature*, **226**:607.

Smith, E. L. (1962), *Proc. Natl. Acad. Sci.*, **48**:859.

Smith, J. D., J. B. Abelson, B. F. Clark, H. M. Goodman, and S. Brenner (1966), *Cold Spring Harbor Symp. Quant. Biol.*, **31**:479.

Smith, M. A., M. Salas, W. M. Stanley, A. J. Wahba, and S. Ochoa (1966), *Proc. Natl. Acad. Sci.*, **55**:141.

Söll, D., J. D. Cherayil, and R. M. Bock (1967), *J. Mol. Biol.*, **29**:97.

———, ———, D. S. Jones, R. D. Faulkner, A. Hampel, R. M. Bock, and H. G. Khorana (1966a), *Cold Spring Harbor Symp. Quant. Biol.*, **31**:51.

———, D. S. Jones, E. Ohtsuka, R. D. Faulkner, R. Lohrmann, H. Hayatsu, H. G. Khorana, J. D. Cherayil, A. Hampel, and R. M. Bock (1966b), *J. Mol. Biol.*, **19**:556.

———, E. Ohtsuka, D. J. Jones, R. Lohrmann, H. Hayatsu, S. Nishimura, and H. G. Khorana (1965), *Proc. Natl. Acad. Sci.*, **54**:1378.

——— and U. L. RajBhandary (1967), *J. Mol. Biol.*, **29**:113.

Soll, L., and P. Berg (1969), *Nature*, **223**:1340.

Speyer, J. F. (1967), in Taylor (1967), pp. 137–191.

———, P. Lengyel, C. Basilio, and S. Ochoa (1962a), *Proc. Natl. Acad. Sci.*, **48**:63.

———, ———, ———, and ——— (1962b), *Proc. Natl. Acad. Sci.*, **48**:441.

Staehelin, T., F. O. Wettstein, H. Oura, and H. Noll (1964), *Nature*, **201**:264.

Stanley, W. M., M. Salas, A. J. Wahba, and S. Ochoa (1966), *Proc. Natl. Acad. Sci.*, **56**:290.

Steffensen, D. M., and D. E. Wimber (1971), *Genetics*, **69**:163.

Steitz, J. A. (1969), *Nature*, **224**:957.

Streisinger, G., J. Emrich, Y. Okada, A. Tsugita, and M. Inouye (1968), *J. Mol. Biol.*, **31**:607.

———, Y. Okada, J. Emrich, J. Newton, A. Tsugita, E. Terzaghi, and M. Inouye (1966), *Cold Spring Harbor Symp. Quant. Biol.*, **31**:77.

Stretton, A. O. W., and S. Brenner (1965), *J. Mol. Biol.*, **12**:456.

——— and ——— (1967), *J. Mol. Biol.*, **26**:137.

Strigini, P., and L. C. Gorini (1970), *J. Mol. Biol.*, **47**:517.

Sueoka, N., and T. Yamane (1963), in H. J. Vogel, V. Bryson, and J. O. Lampen (eds.), "Informational Macromolecules," pp. 205–227, Academic, New York.

Sundararajan, T. A., and R. E. Thach (1966), *J. Mol. Biol.*, **19**:74.

Takanami, M., and Y. Yen (1965), *Proc. Natl. Acad. Sci.*, **54**:1450.

Taylor, J. H. (ed.), "Molecular Genetics," Academic, New York, pt. I (1963), pt. II (1967).

Terzaghi, E., Y. Okada, G. Streisinger, J. Emrich, M. Inouye, and A. Tsugita (1966), *Proc. Natl. Acad. Sci.*, **56**:500.

Tsugita, A. (1962*a*), *J. Mol. Biol.*, **5**:284.

—— (1962*b*), *J. Mol. Biol.*, **5**:293.

——, et al. (1969), *J. Mol. Biol.*, **41**:349.

—— and H. Fraenkel-Conrat (1960), *Proc. Natl. Acad. Sci.*, **46**:636.

—— and —— (1962), *J. Mol. Biol.*, **4**:73.

——, D. T. Gish, J. Young, H. Fraenkel-Conrat, C. A. Knight, and W. M. Stanley (1960), *Proc. Natl. Acad. Sci.*, **46**:1463.

Twardzik, D. R., E. H. Grell, and K. B. Jacobson (1971), *J. Mol. Biol.*, **57**:231.

Wahba, A. J., R. S. Miller, C. Basilio, R. S. Gardner, P. Lengyel, and J. R. Speyer (1963), *Proc. Natl. Acad. Sci.*, **49**:880.

Weber, K. (1967), *Biochemistry*, **6**:3144.

Webster, R. E., D. L. Engelhardt, and N. D. Zinder (1966), *Proc. Natl. Acad. Sci.*, **55**:155.

——, ——, ——, and W. Konigsberg (1967), *J. Mol. Biol.*, **29**:27.

Weigert, M. G., and A. Garen (1965*a*), *J. Mol. Biol.*, **12**:448.

—— and —— (1965*b*), *Nature*, **206**:992.

——, E. Lanka, and A. Garen (1967*a*), *J. Mol. Biol.*, **23**:391.

——, ——, and —— (1967*b*), *J. Mol. Biol.*, **23**:401.

Weisblum, B., S. Benzer, and R. W. Holley (1962), *Proc. Natl. Acad. Sci.*, **48**:1449.

——, J. D. Cherayil, R. M. Bock, and D. Söll (1967), *J. Mol. Biol.*, **28**:275.

—— and J. Davies (1968), *Bacteriol. Rev.*, **32**:493.

Wells, R. D. (1967), *J. Mol. Biol.*, **27**:273.

Wittmann, H. G. (1960), *Virology*, **12**:609.

—— and B. Wittmann-Liebold (1963), *Cold Spring Harbor Symp. Quant. Biol.*, **28**:589.

Woese, C. R. (1968), *Proc. Natl. Acad. Sci.*, **59**:110.

Yamane, T., T-Y. Ching, and N. Sueoka (1963), *Cold Spring Harbor Symp. Quant. Biol.*, **28**:569.

Yanofsky, C. (1960), *Bacteriol. Rev.*, **24**:221.

—— (1963*a*), *Proc. Natl. Acad. Sci.*, **50**:9.

—— (1963*b*), *Genetics*, **48**:1065.

—— (1963*c*), *Cold Spring Harbor Symp. Quant. Biol.*, **28**:581.

—— (1967), *Sci. Am.*, **216**(May):80.

——, V. Horn, and D. Thorpe (1964), *Science*, **146**:1593.

Ycas, M. (1969), "The Biological Code," Wiley, New York.

Yourno, J., and S. Heath (1969), *J. Bacteriol.*, **100**:460.

Zimmermann, R. A., R. T. Garvin, and L. Gorini (1971), *Proc. Natl. Acad. Sci.*, **68**:2263.

Zipser, D., and A. Newton (1967), *J. Mol. Biol.*, **25**:567.

Zubay, G., L. Cheong, and M. Gefter (1971), *Proc. Natl. Acad. Sci.*, **68**:2195.

30
Development
and
Regulation

QUESTIONS

1492 Theoretically all cells in a multicellular organism have the same genotype. If so, why do different cells differentiate at different times?

1493 **a** Development is not a matter of simple enlargement of a preformed germ but is epigenetic. Discuss experimental findings that support this view.

OR

b According to a certain hypothesis, the chromosomes of the egg contain determiners for the eyes, ears, and other body organs, and as cell division takes place, the determiners for each organ are sorted out. Explain how the chromosomes divide at cell division and why this method of division of the fertilized egg cannot account for this assortment.

1494 What influence, if any, do embryonic nuclei have on the differentiation of an embryo during cleavage?

1495 Explain how induction causes differentiation during early embryonic growth, giving an example of the process.

1496 Discuss the following effects briefly, using the terms *organizing substance*, *competence*, and *target tissue* where relevant.
a In Danforth's *short-tail* mice, *Sdsd*, it has been shown that failure of the ureter to reach a particular region of the fetus during embryonic growth results in failure of a kidney to develop at that place.
b Male fowl, *hh*, are *cock-feathered* because of a genetic effect on the response of the skin to the sex hormones.

1497 Describe briefly an example to illustrate the following:
a Gene control of an organizer's effectiveness.
b Gene control of a target tissue's competence.
c Synthesis of cytoplasmic substances under nuclear control.
d The limitation of the period of action (expression) for some genes controlling development.

1498 Would you say that the discovery that the dorsal lip is an organizer in the newt embryo disproves that genes and chromosomes are the physical basis of inheritance? Tell why or why not. Is the dorsal lip itself present in the fertilized egg of the newt? Would you say then that it is to be classified as germ plasm or as a character?

1499 Merogonic hybrids are produced by the introduction of a sperm nucleus of one species into the enucleated egg of another. The hybrid zygotes thus produced do not usually complete their embryonic development, the period before termination differing with the species involved. In general, the closer the relationship between two species, the longer this period will be. Suggest an explanation for these observations.

1500 Can the direction of cell division be genetically controlled? If so, give an example.

1501 Are the observed nuclear changes in the nuclear-transplantation work in amphibia of a reversible nature, or are they stable?   How would you test your hypothesis?   See Briggs and King (1959), Gurdon (1963).

1502 Answer each of the following briefly:

a List and discuss the criteria that might be used to distinguish V-type position effects from gene mutations.

b What is the *spreading effect* associated with V-type position effects? Advance a hypothesis that could explain it.

c Compare the structure of polytene and lampbrush chromosomes and the functions of puffs and loops.

d Discuss the structure and function of the nucleolus-organizing region.

e The *tortoise-shell* pattern appears only in female cats and the rare male of constitution XXY.   Why?

f Offer an explanation for the fact that some nuclei in human female (XX) somatic cells do not contain a Barr body.

g Would you expect incompletely sex-linked genes like *bobbed*, *bb*, in *Drosophila* to show dosage compensation?   Explain.

h Would you expect all females heterozygous for the X-linked mutant gene for *hemophilia* to have the same degree of bleeding?   Discuss.

i If *a* represents the number of X chromosomes and *b* the number of sets of autosomes, develop a formula for the maximum number of Barr (sex-chromatin) bodies per somatic-cell nucleus.

j What is the basis for the deduction that it is the heteropycnotic (condensed) X rather than the isopycnotic X that is functionally inactive in *normal* female (XX) mammals?

k Why do heterozygous carriers of X-linked recessive disorders have a much greater phenotypic variability than heterozygous carriers of autosomal recessive disorders?

1503 Variegation (mosaicism) is fairly common among both plants and animals. Briefly discuss three possible causes of this phenomenon and outline one or more experiments to distinguish among the alternatives.

1504 **a** Define stable and variegated types of position effects and explain the differences between them.   Explain how each can be used in studies of gene expression.

b What facts suggest that position-effect variegation is operating at the transcriptional (or possibly the translational) level of gene expression?

c Could you detect a V-type position effect if the gene product were diffusible? Explain.

1505 Position-effect rearrangements in the mouse are always associated with rearrangements of the X, the chromosome known to become heteropycnotic in one of the X's of an XX somatic cell.   What does this indicate regarding the kind of chromatin present in mouse autosomes?   Why?

1506 Ohno and Cattanach (1962) showed that a segment of an autosome inserted into the X behaved like an integral part of the X cytologically and functionally. Would you expect a piece of an X translocated to an autosome (with centromere) to lose its ambivalent nature and behave like an integral part of the autosome? Explain. See Ohno and Lyon (1965).

1507 When a segment of euchromatin is moved close to or within heterochromatin, it becomes heterochromatinized and vice versa. Explain how these changes may occur and the effects such cytological changes may have on gene action.

1508 **a** Why are lampbrush chromosomes important in genetics?

b What inherent features of chromosome puffs and lampbrush chromosomes expedite rapid protein synthesis?

c What materials seem to be involved in the puffs of giant salivary-gland chromosomes?

d Are puffing patterns specific and, if so, of what significance might this be from the standpoint of developmental gene action?

e Why are lampbrush chromosomes not observed in amphibian male meiocytes or amphibian somatic cells?

f In salivary-gland chromosomes puffs are surrounded by RNA. How would you show that this RNA is (1) single-stranded, (2) transcribed from the puff DNA and not RNA that was synthesized elsewhere in the genome and accumulated on the puff?

1509 Lampbrush chromosomes in growing oocytes of animals ranging from mollusks to mammals synthesize large quantities of RNA. Davidson et al. (1966) showed that most of this RNA is ribosomal and is conserved at least until gastrulation. Further $rRNA$ synthesis does not occur in the toad *Xenopus laevis* at least until gastrulation (Brown and Gurdon, 1964). Elucidate the apparent role of this excessive amount of $rRNA$ in embryogenesis.

1510 In unfertilized sea urchin eggs, concentrations of actinomycin D, sufficient to completely inhibit $mRNA$ synthesis, reduce but do not completely inhibit protein synthesis. Explain.

1511 Why is there little or no transcription during mitosis?

1512 The activation of a gene (getting the gene to produce its $mRNA$ and subsequently its specific protein or enzyme) is probably dependent upon the sequential operation of at least two mechanisms in eukaryotes. One operates at the chromosomal level and the other in a highly special way at the individual gene level.

a *Briefly* discuss the process or processes at the chromosomal level. Include in your discussion reference to recent experiments that support your statements.

b Outline *briefly* the sequence of events, following the chromosomal process, that result in a structural gene producing its specific polypeptide product.

1513   According to the Lyon hypothesis (Lyon, 1961; Beutler et al., 1962) in man, mouse, and probably all mammals all X's in excess of one (e.g., in females, and individuals such as XXY's, XXX's, etc.) are inactivated in all the cells early in embryonic development.   It is a matter of chance which one of the two or more X's remains active in a given cell.   Discuss the cytological and genetic evidence for this hypothesis, from either man or mouse.

1514   Outline how you would proceed to obtain evidence for or against the Lyon hypothesis with the use of a sex-linked biochemical mutation.

1515   Compare the phenotypic consequences of the Lyon hypothesis for X-linked traits (Prob. 1513):

a In females and in males.

b Among different heterozygous females.

c In different tissues of heterozygous females.

d Among female and male monozygotic twins for the expression of polygenic traits to whose genetic background the X chromosome contributes.

1516   In male mammals the single X in somatic cells always behaves euchromatically; in female mammals one X remains euchromatic, the other manifests positive heteropycnosis (is heterochromatic) along its entire length.   Why does this difference in cytological behavior of X's exist between the two sexes?   What is the functional consequence of this behavior?

1517   Compare and contrast the dosage-compensation mechanism in mammals with the mechanism in *Drosophila.*

1518   Histones suppress gene action (DNA-dependent RNA synthesis) in eukaryote chromosomes.   Suggest a possible mechanism.

1519   Read the papers by Huang and Bonner (1962), Allfrey et al. (1963), Izawa et al. (1963*a*), and Littau et al. (1964), which are concerned with regulation of genetic activity at the chromosome level, and briefly summarize the salient conclusions in a logical sequence.

1520   When lactose is added to a growing culture of *Escherichia coli,* the cells begin to make enzymes necessary for lactose utilization.   In contrast, when tryptophan is added to the culture medium, synthesis of the enzymes of the tryptophan pathway ceases.   Contrast the two systems, explaining why they act in reverse directions.

1521   **a** A mutant arises in *Salmonella* that is unable to synthesize the ten enzymes mediating the conversion of phosphoribosyl pyrophosphate and ATP to histidine.   Offer two explanations to account for the behavior of this mutant and outline how you would proceed to distinguish between these two alternatives.

b Compare and contrast unregulated (constitutive) and regulated (inducible or repressible) enzyme synthesis.

c Why would one expect mutations in a promoter to alter the potential for operon expression?   Can operator mutants affect the rate of operon expression?

d Gilbert and Müller-Hill (1966) isolated a protein that binds to isopropyl thiogalactoside (IPTG), which, like lactose, is an inducer of the *lac* operon. There was no binding between IPTG and proteins from i^- deletion and some i^- (revertible) mutants. Moreover, differences in binding of the protein isolated from *wild type* and certain i^- mutants and IPTG were also detected. What is the significance of these facts?

1522 The *his S* gene is the structural gene for histidyl-*t*RNA synthetase in *Salmonella* (Roth and Ames, 1966). Would you expect mutations in this gene to result in derepression of the histidine operon and why?

1523 Mutant alleles (due to nonsense mutations) of genes that code for a *t*RNA molecule or provide a binding site for a repressor molecule would not be detected with suppressor mutations. Why?

1524 Human beings with X-linked *agammaglobulinemia* are unable to synthesize three different immunoglobulins (Fudenberg and Hirschhorn, 1964). Suggest two possible explanations.

1525 What role does feedback play in the molecular control of genetic activity?

1526 Do you expect operons to occur in eukaryotes? Explain. See Giles et al. (1967*a,b*) and Rines, Case, and Giles (1969).

PROBLEMS

1527 **a** In snails, the two cells formed at the end of the first division of the zygote, when separated, develop differently and abnormally. However, in the sea urchin, any one of the four cells formed after the second division of the fertilized eggs gives rise to a *normal* individual (Barth, 1949). If initial segregation of cytoplasmic materials is the basis of differentiation, contrast the sea urchin with the snail in regard to the way in which cytoplasmic organization may be related to cleavage.

b Monozygotic twins in man are derived from the same fertilized egg. With regard to its initial developmental organization, does the human egg resemble that of the sea urchin or the snail?

c It is possible to separate the cells of two-, four-, and eight-cell embryos of a number of animal species and culture them individually. In the sea urchin, each of the eight cells of an eight-cell embryo so treated develops into a complete individual. In the newt and frog, only the cells of two-cell embryos develop complete embryos. In snails, the first two cells if separated develop only into half-embryos, which disintegrate. Account for the difference in the totipotency of the early blastomeres of these species.

d In armadillos the 4 or 5 offspring that arise at any one birth are identical. This being so, what is the earliest cleavage division in which the critical separation of cytoplasmic materials essential to the formation of a *normal* individual can occur in armadillos?

1528   The newt has no organ that corresponds to the sucker of a tadpole, and the frog has no structure corresponding to the balancer of the newt. However, a newt embryo into which tadpole ectoderm has been transplanted can develop a sucker if the transplantation site corresponds with the site of sucker development in the tadpole. In the reciprocal transplant, newt ectoderm forms a balancer in the place where it arises in the newt (Spemann and Schotte, 1932). Discuss this phenomenon with regard to intergeneric organizer competence and tissue lability.

1529   In mice, growth rate of *dwarfs, dwdw*, falls behind that of *normals, Dw—*, shortly after birth, and by the seventeenth day it stops. They then begin to lose weight, and some die. Later the survivors resume a slow growth and eventually reach about a quarter of the *normal* adult weight. At maturity their tails, snouts, and ears are unusually short, and their thyroid, thymus, pituitary, and adrenal glands are small (Grüneberg, 1952; Boettiger and Osborn, 1938). *Normal* young with anterior pituitaries removed grew and developed exactly like *dwarfs*. *Dwarf* young implanted with pituitary glands from *normal* young approached the *normal* growth rate, size, glandular, and morphological development, and fertility (Smith and MacDowell, 1930). Discuss these effects briefly, using appropriate terms from the following list: competence, organizer or organizing tissue; pleiotropy, phenocopy, and target tissue.

1530   *Tfm (testicular feminization)* vs. *tfm (normal sexual development)* is an X-linked allele pair in mice. Testosterone, administered in vivo to *tfm/tfm* female, or castrated *tfm*/Y male mice, causes:

1   An immediate translation of preexisting kidney *m*RNAs
2   A hundredfold increase in kidney alcohol dehydrogenase and glucuronidase
3   A very marked increase in the transcription of kidney *r*RNA genes
4   An increase in an RNA polymerase, tentatively identified as RNA polymerase I, which is specific for the transcription of *r*RNA genes

Tfm/Y mice showed neither response 2 nor 3 and gave no evidence of response 1 or 4 (Ohno, 1971). Compare this type of control mechanism with that of the *lac* operon. See Ohno (1971) for an operonlike hypothesis of hormone action.

1531   a   In the domestic fowl, birds heterozygous for the *Creeper* allele, *Cr*, have very short legs and wings. In homozygotes this allele is lethal before hatching. In homozygous embryos, ossification and marrow development of the bones do not take place. The spleen, the only remaining source of red blood cells, becomes greatly enlarged but fails to offset the ensuing anemia. The lack of oxygen stimulates and greatly enlarges the heart. The following data are taken from a series of extensive developmental studies of this phenomenon (Fell and Landauer, 1935; Hamburger, 1941).

    1   Transplantation of limb buds to growth-restricting media (lacking certain essential

metabolites) showed that ossification failure occurred in primordia from *noncreeper*, *crcr*, embryos only if transplanted before a certain stage in development (when birds were less than 3.8 mm long).

2   Limb buds from *CrCr* embryos taken as early as 35 hours after being laid, when limb buds first appear, and implanted in *normal* host embryos developed as typical *creeper* limbs, with ossification failure.

3   Optic vesicles transplanted in the eye region of host embryos at about the same time developed as indicated in the table.

Show whether the *creeper* trait represents true or spurious pleiotropy.

Table 1531

Implant	Host	Development
CrCr vesicle	*crcr*	*normal*
crcr vesicle	*crcr*	*normal*
crcr vesicle	*CrCr*	*coloboma**

* Absence of iris, a characteristic *creeper* defect.

b Rudiments of chick limb tissue from *crcr* birds grow well in vitro in a medium containing an extract from *noncreeper*, *crcr*, embryos.   The growth of this tissue is reduced if the culture medium contains an extract from *CrCr* embryos.   When *CrCr* and *crcr* limb tissues are cultured simultaneously in the same medium containing *noncreeper* embryo extract, growth of the normal limb is inhibited (Elmer and Pierro, 1964, 1966).   Explain.

1532   In 1936 Danforth discovered a *short-tail* mouse mutant whose *short-tail* descendants when crossed by Dunn et al. (1940) gave the following results: 159 *normal*, 365 *short-tail*, and 153 *tailless* (die within 24 hours after birth). Embryological studies by Gluecksohn-Schoenheimer (1943) revealed that *tailless* mice lack kidneys, urethra, several of the posterior vertebrae, and the rectal and anal openings.   *Short-tail* mice almost always have reduced kidneys and ureters, and one or both kidneys may be absent.   The variability of anomalies of kidney and ureter structure in 109 *short-tails* is sketched in 14 classes whose frequencies are indicated.   *Note:* Kidneys do not hang by themselves as adrenals may.   Wherever a kidney occurs, a complete ureter extends to it.   The ureter arises as a bud from the mesonephros (which functions as a kidney in fish and amphibia).   With the use of appropriate terms explain the genetics and embryological observations of *short-tail* and *tailless* mice (see Glueksohn-Waelsch and Rota, 1963).

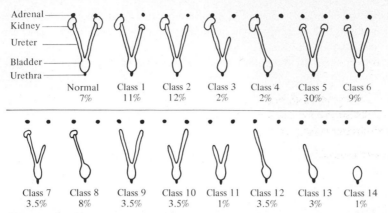

| Adrenal | Kidney | Ureter | Bladder | Urethra |

| Normal | Class 1 | Class 2 | Class 3 | Class 4 | Class 5 | Class 6 |
| 7% | 11% | 12% | 2% | 2% | 30% | 9% |

| Class 7 | Class 8 | Class 9 | Class 10 | Class 11 | Class 12 | Class 13 | Class 14 |
| 3.5% | 8% | 3.5% | 3.5% | 1% | 3.5% | 3% | 1% |

Figure 1532 (*Redrawn from Adrian M. Srb, Ray D. Owen, and Robert S. Edgar,*
"*General Genetics,*" *2d ed., p. 386, W. H. Freeman and Company, Copyright © 1965;
after S. Gluecksohn-Schoenheimer in H. Grüneberg, "Genetics of the Mouse,"
Cambridge University Press, 1943.*)

Hen feathering vs. *cock feathering* in fowl (males ZZ, females ZW) is determined
by a single pair of autosomal alleles. *HH* and *Hh* males are *hen-feathered;*
hh males are *cock-feathered.* Females of any genotype are *hen-feathered.* In
Sebright Bantams both sexes are *hen-feathered.* In White Leghorns males
are *cock-feathered;* females, *hen-feathered.* Work done in the 1920s and 1930s
significant in elucidating the control of these traits may be summarized as
follows:

1   Removal of testes in *hen-feathered* males, *H—*, or of the ovary in females (regardless
of genotype) results in *cock feathering* (Roxas, 1926; Eliot, 1928).
2   A castrated Sebright Bantam cock, *HH*, that had developed *cock feathering* became
hen-feathered again after implantation of a Leghorn testes, *hh*. In the reciprocal transplantation the castrated Leghorn male continued to express *cock feathering* (Roxas, 1926).
3   Danforth (1930) and Danforth and Foster (1929) showed that transplantation of
potentially *cock-feathered* skin from an *hh* male to a female of any genotype, *H—* or *hh*,
results in *hen feathering.* Transplantation of skin from *hh* females to *cock-feathered*, *hh*,
males results in *cock feathering.* Transplants of skin from *hen-feathered*, *H—*, males
and females to *cock-feathered* males, *hh*, continue to develop *hen feathers.* Finally,
transplants of skin from females to males in breeds in which both sexes are *hen-feathered*
do not alter feathering type. Note that within breeds like the Leghorn, transplants
acquire the trait of their host.

a Which is the organizing tissue and which the target tissue in the expression
of feather type? Explain.
b On what are the sexual differences in feathering based? Explain.
c Is the initial effect of the *H* and *h* alleles in the organizing or target
tissues? Explain.
d Summarize the control of feather type as briefly as possible.

1534 The thyroid produces four thyroxin hormones, which stimulate the rate of oxidative metabolism and regulate growth and development of all body cells and tissues. Insufficient secretion of the hormones causes *dwarfism.* The anterior lobe of the pituitary gland secretes six distinct hormones; one of these is the thyrotropic (thyroid-stimulating) hormone, which regulates various aspects of thyroid function, including development and maintenance; another is somatropin, the growth hormone. The general functional relationship of these glands and their hormones is

Pituitary-gland cells	thyrotropin + → somatropin	thyroid-gland cells	thyroxin → hormones	body → cells and tissues	regulation of → growth and development

Newborn mice homozygous for the autosomal recessive gene *dw* implanted with pituitaries from *normal, Dw—,* young grow almost normally and reach nearly normal size and normal fertility (Smith and MacDowell, 1930, 1931). *Dwarf* beef cattle homozygous for a recessive autosomal trait controlled by one or a few genes have subnormally functioning thyroids (Carroll et al., 1951).

a At what step in the above sequence might each recessive gene be acting and why?

b Outline how you might proceed to determine whether your answer to (a) is correct or not.

1535 In certain breeds of poultry, like the Rhode Island Reds, males are *cock-feathered* and females *hen-feathered;* in other breeds like the Sebright Bantams, both sexes are *hen-feathered.* How could you determine by grafting procedures whether the difference between the two kinds of breed is due to the fact that the alleles act in the skin (target tissue)?

1536 Plants in certain varieties of Chinese primrose (*Primula sinensis*) homozygous or heterozygous for the autosomal dominant gene *R* bear *red* flowers regardless of temperature during flower development. In *rr* varieties, flowers are *white* when the temperature during the critical period during early floral development is above 86°F. When temperatures are lower than 86°F during this period, the flowers produced are *red.* Since different flowers develop at different times, it is possible, by making the temperature fluctuate above and below 86°F during inflorescence development, to obtain both *red* and *white* flowers on a single plant. Assuming that a single enzyme is involved, suggest a mechanism or mechanisms to explain these observations.

1537 In *Drosophila melanogaster,* the recessive X-linked gene *fused* wing vein, *fu,* has pleiotropic effects causing partial sterility and ovarian tumors in *fufu* females and a reduced number of ocelli in both *fufu* females and *fu/Y* males. *Fufu* females produce offspring of the expected genotypes and in the expected proportions:

Fufu ♀ × *fu/Y* ♂ → 1 *fused* ♀ : 1 *nonfused* ♀ : 1 *fused* ♂ : 1 *nonfused* ♂

Fused (*fufu*) females × *Fu*/Y males produce females only; the male progeny die in early embryogenesis. Embryonic development is normal in all viable progeny. The cross *fufu* females × *fu*/Y males is sterile; no progeny are produced (Lynch, 1919; Counce, 1956). Zygotes from the latter cross develop normally for 5 hours after fertilization. Thereafter development in the germ layers of the ovary is aberrant and leads to the pleiotropic effects mentioned above. These embryos do not hatch but remain alive far beyond the hatching time of *wild-type* embryos (Counce, 1956). When *fufu* ovaries are transplanted into a genetically normal environment, they behave autonomously, the eggs being as inviable as they are in *fufu* mothers (Clancy and Beadle, 1937).

a Why do these results indicate that *Fu* and not *fu* is required for viability?

b When does the *Fu* allele act (produce *Fu* substance)? Explain.

c What is the function of the *Fu* substance, and where is it probably produced and stored?

1538 The courtship and mating activities of males of the platyfish (*Xiphophorus* [= *Platypoecilus*] *maculatus*) and the swordtail (*X. helleri*) are basically similar. Nevertheless both qualitative and quantitative differences have been demonstrated for 10 male sexual patterns by Clark et al. (1954). One of these is *pecking*, a characteristic of the sexually active male platyfish but not of the male swordtail which consists of a series of biting movements on the bottom gravel or sides of the aquarium. Another courtship trait is *retiring*, which refers to the male's sudden backing away from the female with his body limp and his dorsal and caudal fins folded. Whereas the male platyfish exhibits this trait, the male swordtail does not. The male behavior patterns in parental, F_1, and F_2 generations and backcrosses to both parents are tabulated. How is each of these behavioral characters inherited?

Table 1538

Strain or cross	Pecking, %	Retiring, %
Platyfish	81.8	72.8
Swordtail	0.0	0.0
F_1 (platyfish × swordtail)	0.0	20.0
F_2 ($F_1 \times F_1$)	39.3	23.0
F_1 × platyfish	20.0	10.0
F_1 × swordtail	0.0	0.0

1539 In domestic fowl, the mutant autosomal gene *F* (*frizzle*) changes the structure and morphology of the feathers; the change being mild (curly feathers) in heterozygotes, *Ff*, and extreme in homozygotes, *FF*. In *extreme-frizzle* birds the defective feathers are curly, brittle, and easily broken, leading to near nakedness. Various abnormalities result from this primary aberration. Reduced insulation results in rapid heat loss, lowers body temperature, and impairs the ability to adjust to changing environmental temperatures. Consequently the metabolic system of these birds suffers stresses, causing increased heart rate and food consumption, enlarged spleen, crop, and gizzard, and many other abnormalities (Landauer, 1942*a,b*). Krimm (1960) showed that feather keratin (a protein) in *FF* birds is much more poorly organized than that of *normal, ff,* birds and that the amino acid content of the mutant protein differs from that of normal keratin. Outline in chronological sequence the cause-effect relationship between the DNA, keratin, and the deleterious pleiotropic effects in *extreme-frizzle* fowl.

1540 The eggs of the silkworm moth are either *oval* or *spindle-shaped,* and the larvae *striped* or *nonstriped.* The shape of the egg depends solely on the strain to which the mother belongs. However, the larvae from crosses between *striped* and *nonstriped* strains are always *striped,* and those from crosses between these hybrids segregate 3 *striped* : 1 *nonstriped.* Contrast gene action, in the control of these two characters, with respect to (1) autonomy of the mature egg, (2) the stage of the sexual process at which induction occurs.

1541 In *Drosophila melanogaster,* the recessive autosomal mutant gene *lozenge-clawless,* lz^{cl}, causes many abnormalities of the eyes, tarsals, and female genitalia (Anders, 1947). *Lozenge-clawless* females lack sperm-storage organs and ovarian glands and are sterile. Mutant, $lz^{cl}lz^{cl}$, males have normal genitalia and are fertile. Anders (1947, 1955) found that XX flies homozygous for both lz^{cl} and *tra* (which renders XX flies phenotypically male) were phenotypically male with normal male genitalia. Which of the genes, lz^{cl} or *tra,* acts first in development? Explain.

1542 In *Chironomus,* the special cells of the salivary gland which contain permanently paired polytenic homologous chromosomes form a small lobe next to the general excretory duct. In *pallidivittatus* their excretion contains SZ granules, which are lacking in *tentans.* Interspecific hybrids are fertile. Their special cells produce SZ granules in amounts reduced compared to the quantities in *pallidivittatus.* Of the offspring of hybrids $\frac{1}{4}$ resemble *pallidivittatus,* $\frac{1}{2}$ F_1 hybrids, and $\frac{1}{4}$ *tentans* in granular content. The secretory type and chromosome constitution of 25 F_2's are tabulated. In *pallidivittatus,* fourth chromosomes from nuclei of specialized cells show a puff close to their centromeres. These chromosomes in other cells of the gland (which produce clear secretions) have no corresponding puff. In *tentans,* chromosome 4 does not form a puff in the corresponding position in any of the salivary-gland cells. In hybrids, only one of the two fourth chromosomes in nuclei of special cells possess the puff. None of the other hybrid cells possess this puff. Although the two fourth chromosomes differ in that one has a puff and the other does not, the number of chromosome bands in the fourth chromosomes of the two species is the same (Beermann, 1961).

a Is the formation of SZ granules specified by one, two, or more genes? Explain.

Table 1542 Chromosomes and chromosome arms

F$_2$ fly	2L	2R	3L	3R	4	SZ
1	T	H	P	P	H	+
2	T	P	P	P	P	+
3	H	H	P	P	P	+
4	P	H	T	T	H	+
5	H	H	H	H	P	+
6	H	H	P	P	P	+
7	H	P	P	P	H	+
8	H	P	H	H	P	+
9	T	H	P	P	P	+
10	H	T	H	H	P	+
11	T	H	H	H	H	+
12	T	H	H	H	T	−
13	T	T	P	H	H	+
14	T	H	T	T	H	+
15	T	H	P	P	H	+
16	H	H	T	T	H	+
17	T	H	H	H	H	+
18	T	H	H	H	T	−
19	H	T	T	T	H	+
20	H	H	P	P	H	+
21	T	H	H	H	H	+
22	P	P	T	T	H	+
23	P	P	H	H	P	+
24	H	H	H	H	P	+
25	H	P	H	P	T	−

Key: T = both homologs from *tentans*; P = both homologs from *pallidivittatus*, H = one homolog from *tentans*, one from *pallidivittatus*; fourth chromosome is telocentric; + = present; − = absent.

b In what way do puffs, granules, and the gene (or genes) appear to be related?

c If only one gene is involved, is it more likely to be a structural or an operator type? Explain.

1543 The puffing pattern of any region of any polytene chromosome varies from tissue to tissue, as shown in the table for region 14 of chromosome 4 of *Chironomus tentans* (data from Beermann, 1956). Using autoradiography, Pelling (1959) showed that ^{3}H-thymidine injected into *Chironomus* larvae was distributed evenly throughout the polytene chromosomes. When ^{3}H-uridine was injected, the label was first located only in puffs, Balbiani rings, and nucleoli. With longer incubation the cytoplasm also became labeled. Injecting actinomycin D (inhibitor of DNA-dependent RNA synthesis) first and then ^{3}H-uridine resulted in cessation of RNA synthesis and shrinkage of all puffs.

Table 1543

Tissue	Locus					
	1	2	3	4	5	6
Salivary gland	−	−	+	−	−	+
Malpighian tubule	−	−	+	+	+	+
Rectum	+	+	−	+	+	+
Midgut	−	−	−	−	−	−

Key: + = puffing; − = no puffing.

a What appears to be the function of puffs?

b Suggest an explanation for (1) the puffing sequence in any one tissue and (2) the differences in puffing sequence between tissues.

1544 Somatic nuclei of *wild-type Chironomus* (2*n*) have nucleoli at corresponding positions on a pair of homologs. Beermann (1960) established a balanced lethal strain carrying only one nucleolus per somatic nucleus. This strain always produced about 25 percent progeny which lacked nucleoli. These flies develop normally to the gastrula stage, when aberrant features appear, followed by death. Offer a hypothesis, in molecular terms, which will account for these results.

1545 Multicellular organisms begin embryonic development, with almost all nuclear genes turned off, under the direction of gene products (*r*RNA, *t*RNA, certain *m*RNAs, and a few enzymes) synthesized during diplotene in the oocyte. Moreover, the genes first activated in embryonic nuclei are apparently the same ones which contributed to storage during oogenesis. Ohno et al. (1968) studied allele activation at the autosomal locus for 6-phosphogluconate dehydrogenase (PGD) in the Japanese quail (*Coturnix coturnix japonica*) with the following results:

1 PGD (a dimer) is found in growing ova, in embryos at all stages of development, and in every adult tissue.

2 PGD stored in fertilized-egg cytoplasm is exhausted in embryos by 24 hours of incubation.

3 The four allelic homozygote embryos, *aa*, a^1a^1, a^2a^2, a^3a^3, formed single *very slow*, *slow*, *fast*, and *very fast* electrophoretic bands, respectively. Each 36- to 48-hour heterozygous embryo formed three bands, two like those of the parents and one of intermediate mobility.

a Is phenotypic expression by early embryos entirely under the control of maternal genes?

b Suggest why the genes functioning in the oocyte are among the first to begin activity in the embryo.

1546 Many enzymes exist in two or more molecular forms, called *isozymes*, varying in relative concentration from tissue to tissue, changing during embryogenesis. Lactic dehydrogenase (LDH), for example, consists of two kinds of polypeptides, A and B. Isozyme LDH-1 contains only A, and LDH-5 contains only B. When these subunits are isolated and then mixed in equal proportions to permit reassembly, five LDH isozymes are found in the ratio 1 LDH-1 : 4 LDH-2 : 6 LDH-3 : 4 LDH-4 : 1 LDH-5 (Markert and Møller, 1959).

a How many polypeptides does each isozyme contain? Explain.

b LDH-4 is dissociated and then reaggregated. What isozymes should appear and in what proportions?

c Suggest a possible function for these variations in isozyme ratios (1) from one tissue to another at a given time and (2) at different times in development in the same tissue.

1547 In the squash (*Cucurbita pepo*) two pairs of alleles affect fruit shape. Plants that are *A—B—* are *disk*-shaped (very wide), those *A—bb* or *aaB—* *spherical* (width equals length), and *aabb elongate* (sausage-shaped) (Sinnott, 1927). If growth is entirely dependent on cell division, suggest how these genes might control shape.

1548 Evidences of a functional role of histones, which, together with some other protein and DNA, constitute the major components of eukaryote chromosomes, have been obtained by Huang and Bonner (1962) from in vitro studies of DNA-dependent RNA synthesis. Their results were as follows:

1 The DNA as it is present in chromatin is less effective in supporting DNA-dependent RNA synthesis than an equal amount of pure DNA.

2 A comparison of chromatin as template with and without the histones removed gave the results shown in Table 1548*A*.

Table 1548*A*

Reaction mixtures (system)	RNA synthesized*
Crude chromatin (histone:DNA ratio = 1:1.1)	400
Solubilized chromatin, histone removed	2,090

* RNA synthesized per milligram of original crude chromatin per picometer of nucleotide incorporated per 10 minutes.

3 Further studies indicated that the addition of pure histones to pure DNA to produce reconstituted 2,4-dinitrophenol (DNP) or the addition of pure histones to a DNA-containing reaction mixture immediately before the addition of RNA polymerase consistently produces the results in Table 1548*B*.

Table 1548B

Template used	RNA synthesized*
None	720
DNA	1,440
Reconstituted DNP (histone:DNA ratio 1.04:1)	576
DNA + histone	0

* RNA synthesized per 10 minutes per picometer of nucleotide incorporated.

4   DNA of DNP after removal of histones is still able to support RNA synthesis.

a Discuss the apparent role of histones in RNA synthesis.

b In view of your answer to (a), why should whole (native) chromatin as isolated from pea embryo possess approximately 20 percent ability to support DNA-dependent RNA synthesis?
See Allfrey et al. (1963), Marushige and Bonner (1966), Marushige and Ozaki (1967), and Dahmus and Bonner (1970).

1549   Pea (*Pisum sativum*) cotyledons synthesize a specific pea-seed reserve globulin, which is not produced in other pea tissues, e.g., flowers and buds.   Chromatin isolated from cotyledons and buds in an in vitro system produced the tabulated results (Bonner et al., 1963).

Table 1549

Source of chromatin	Pea-seed reserve globulin
Pea cotyledon	Synthesized
Pea buds	Not synthesized
Pea bud chromatin minus the histone	Synthesized

a What substance is involved in repression of gene activity?
b How may it function in gene-action suppression?

1550   Offer an explanation for the fact that protamines replace histones in sperm chromosomes of certain animal species.

1551   The *wild-type* alleles of the euchromatic X-chromosome genes in *Drosophila melanogaster* in their normal position do not express a *variegated* (mottled) phenotype in either the homozygous or heterozygous state:

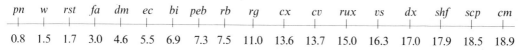

pn	w	rst	fa	dm	ec	bi	peb	rb	rg	cx	cv	rux	vs	dx	shf	scp	cm
0.8	1.5	1.7	3.0	4.6	5.5	6.9	7.3	7.5	11.0	13.6	13.7	15.0	16.3	17.0	17.9	18.5	18.9

Demerec (1940) found one female with *notched* wings and *variegated* eyes (white eyes mottled with red spots) among offspring of a cross between *y sc w* females and x-rayed *wild-type* males. It carried an insertion (= translocation) of a euchromatic segment of the X chromosome (which included all genes beginning with *w* and ending with *vs*) in its normal order in the heterochromatin of the fourth chromosome between segments 101F and 102A. Crosses between *notch* females heterozygous at all loci in the insertion and males carrying recessive alleles at these loci produced *notch* females with one normal X chromosome (with the recessive alleles) and another X involved in the translocation (with *wild-type* alleles at all loci). These females expressed a *variegated* phenotype at the *w*, *rst*, *fa*, *dm* loci in the distal end of the insertion and at the *rg*, *cx*, *rux*, and *vs* loci in the proximal end and a *normal* (*wild-type*) phenotype at the *ec*, *bi*, *peb*, and *rb* loci situated between the proximal and distal groups. Demerec also found a female with an insertion of the X-chromosome segment possessing the loci *w*, *rst*, *fa*, and *dm* at one end of the heterochromatin (next to the centromere) in the left arm of chromosome 3. In flies heterozygous at these four loci a *variegated* phenotype was expressed in characters controlled by genes at the *w*, *rst*, and *fa* loci. The expression at the *dm* locus was *wild type*.

a Is gene mutation during ontogeny a plausible explanation for the *variegated* phenotypes among the progeny of the two females? Explain.

b If not, offer an alternative mechanism for the mottling in terms of euchromatin and heterochromatin.

c Account for the fact that in the first translocation genes in the distal and proximal groups, but not those in the center of the insertion, show a *variegated* phenotype whereas in the second translocation the genes closest to heterochromatin express mottling and the one farthest from the heterochromatin does not.

1552 a In the mouse (females XX, males XY) the *b* locus (*B* = *wild type* and *b* = *brown* coat color) is on chromosome 8. A *variegated* mutant female arose among the offspring of irradiated *wild-type* males mated to *brown* females. Results of matings involving this female and her offspring are shown in Table 1552*A* (Russell and Bangham, 1961). *Note:* (1) All *variegated* animals are *partially sterile* females. They have small *brown* areas irregularly interspersed with *wild-type* ones. The size of their litters ranges from 3.0 to 4.5; that of *fully fertile* females ranges from 5.7 to 11.0. (2) Matings among F_2's, involving crosses identical to crosses 2, 3, and 4 among F_1's, gave results similar to those in crosses 2, 3, and 4 respectively.

(1) Suggest a hypothesis to account for these facts.

(2) Indicate what cytological observations would confirm your hypothesis.

b Matings between *variegated* females and *brown* males produced some *partially sterile*, *brown* females which in crosses with *wild-type* males produced *partially sterile* females, all *wild type*. Such females in crosses with

Table 1552*A*

Cross	Gener-ation	Offspring					
		Variegated		Wild type		Brown	
		♂	♀	♂	♀	♂	♀
1. *Variegated* mutant ♀ × *brown* ♂	F$_1$	0	9	8	1	5	3
2. *Brown* F$_1$ ♀ × *brown* F$_1$ ♂	F$_2$	0	0	0	0	$\frac{1}{2}$	$\frac{1}{2}$
3. *Wild-type* F$_1$ ♀ × *brown* F$_1$ ♂	F$_2$	0	0	$\frac{1}{4}$	$\frac{1}{4}$	$\frac{1}{4}$	$\frac{1}{4}$
4. *Variegated* F$_1$ ♀ × *brown* F$_1$ ♂	F$_2$	0	275	278*	45†	336‡	352§

* Most sterile. † Most fertile. ‡ Most fertile, few sterile. § Most fertile, few partially sterile.

brown males produced some *variegated* females, all *partially sterile*. Show whether these results agree with your hypothesis.

c The allele pair *Tata* is X-linked. *TaTa* or *Ta*/Y show a *tabby* coat, *tata* or *ta*/Y show *wild type* and *TaTa* have a *mosaic* coat for the two patterns (called *heterozygous tabby*). *Brown variegated, partially sterile, hetero-zygous-tabby* females were produced from matings between the *brown variegated, partially sterile, wild-type* females and *bb*/*Ta*/Y males. When these females were test-mated with *brown, tabby* (*bb Ta*/Y) males, they produced the progeny shown in Table 1552*B*.

Table 1552*B*

Coat color	Tabby genotypes			
	♀		♂	
	TaTa	*Tata*	*Ta*/Y	*ta*/Y
Brown	78	14	51	16
Variegated	0	68		
Wild type	11	0	16	57

These data permit a definite decision whether your hypothesis is correct or not. State your conclusions.

d Outline experiments to test your hypothesis. See Ohno and Cattanach (1962), Ohno (1967), Russell (1963), and Russell and Montgomery (1969) for the correct interpretation of these and similar results.

1553 a The gene loci *p* (*pink* eye) and *c^{ch}* (*chinchilla* coat color) in the mouse *Mus musculus* are located in the central segment of chromosome 1, an autosome. From the mating of *pp c^{ch}c^{ch}* females with *wild-type*, *PP CC*, males treated with triethylenemelamine, Cattanach (1961) obtained one female with a *variegated* phenotype in which the central segment of chromosome 1 (about one-third of the autosome) carrying the *wild-type* alleles P and C was inserted into the X as shown. In animals with the two

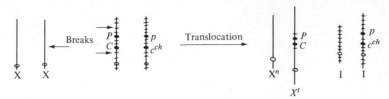

Figure 1553

chromosome 1s after translocation (hemizygous for p and c^{ch}):

(1) $X^t X^n$ females and $X^t X^n Y$ males expressed *variegation* (patches of *wild-type* coat color) and *white* (pc^{ch} coat color).

(2) $X^t Y$ males and $X^t O$ females express a *wild-type* coat color.

State whether these results are due to V-type position effect or dosage compensation and why.

b Ohno and Cattanach (1962) showed that in *variegated* $X^t X^n$ females and $X^t X^n Y$ males the somatic cells in *wild-type* patches contained a condensed (tightly coiled) X^n and a normal (isopycnotic) X^t (behaving in the same manner as the euchromatic autosomes). pc^{ch} patches contain cells with a condensed X^t and a normal X^n. No condensed chromosomes were found in somatic cells of $X^t Y$ and $X^t O$ animals. The normal X always behaves isopycnotically in somatic cells. Are these results in accord with your explanation in (a)? Discuss briefly.

Note: X^n = normal chromosome

X^t = translocated chromosome

1554 Suppose you have a population of mice some of which are heterozygous for a reciprocal translocation between chromosome 8 and the X chromosome and also heterozygous for the alleles Bb (*wild type* vs. *brown* coat) on chromosome 8. Explain how you would test the Lyon hypothesis (see Question 1513) autoradiographically.

1555 Assume that in the mouse two true-breeding strains possess the phenotypes shown in Table 1555A. The F_1 and F_2 results from the cross *black, normal* female × *yellow, ocular-albinism* male are given in Table 1555B.

Table 1555A

Strain	Phenotype
1	*black* coat, retina completely pigmented (*normal*)
2	*yellow* coat, retina lacking pigment (*ocular albinism*)

Table 1555B

F_1 ♀ *mosaic* coat color,* *irregular* retina pigmentation†
    ♂ *black, normal*

F_2 Phenotype	♀	♂
mosaic, irregular	150	1
black, normal	150	148
black, irregular	24	0
mosaic, normal	26	0
yellow, ocular albinism	0	152
black, ocular albinism	0	23
yellow, normal	0	27

* Some patches black, others yellow.

† Some patches of cells pigmented, others lack pigment.

a Suggest a cytogenetic explanation for these results. Include in your explanation the number of genes involved, their location, linkage, and percent recombination, if any, and why (with the exception of one individual) all *mosaic irregularly pigmented* individuals are females.

b Account for the one *mosaic, irregularly pigmented* male.

c Four sets of female monozygotic twins and four sets of male monozygotic twins were studied in the F_1 of the cross outlined. The intrapair difference in size and weight for the female twins was significantly different from the intrapair difference for male twins. Is this expected on the basis of your hypothesis? Explain.

1556 In *Drosophila melanogaster* the genes *Y* (*gray* body), *W* (*dull red* eyes), and *Spl* (*normal* wing) are normally located near the tip of the euchromatic left arm of the X chromosome. Müller (1930), Lewis (1950), and Judd (1955) obtained reciprocal translocations between the X and the fourth chromosome in which a segment containing these genes was moved to a heterochromatic region near the centromere of the right arm of chromosome 4, as shown in Fig. 1556A. In translocation heterozygotes of the type shown (Fig. 1556B), a *variegated* phenotype was expressed for eye color (all flies had white eyes with dull-red speckles) and wing type (some flies had a normal and others a split wing).

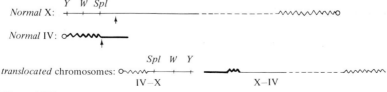

Figure 1556A

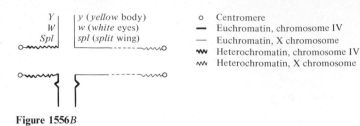

Y | | y (*yellow* body)
W | | w (*white* eyes)
Spl | | spl (*split* wing)

o   Centromere
—   Euchromatin, chromosome IV
—   Euchromatin, X chromosome
⋙   Heterochromatin, chromosome IV
⋘   Heterochromatin, X chromosome

Figure 1556*B*

Matings between such females and *y w spl* males produced *y w Spl* females from which a homozygous translocation strain true-breeding for *y w Spl* was obtained.   Such females, crossed with *y W Spl* males, produced *yy Ww* flies with *wild-type* eyes.   These flies were all translocation heterozygotes with breaks at the same points as in the original 4-X translocation.   The same results were obtained when *wᵃ* (*apricot* eyes) was used in place of *W*; *w* is an amorph (nonactive).

a Explain why the *mottled* phenotype is not due to gene mutation and indicate how heterochromatin may be related to such *variegated* phenotypes.

b Offer an explanation for the fact that in the first translocation heterozygote variegation was expressed for the traits controlled by *Splspl* and *Ww* but not for those specified by *Yy*.

1557 Lyon (1962) reported on mosaic F_1 mice doubly heterozygous in the repulsion phase for mutant alleles of two X genes, one affecting coat color and the other hair structure.   Patches of fur were either *mutant* for one character and *wild type* for the other, or vice versa, and never *mutant* or *wild type* for both.   State whether or not this supports the Lyon hypothesis and why.

1558 The structural gene *Pgd* for the synthesis of the enzyme 6-phosphogluconate dehydrogenase (PGD), a dimer, is at approximately 0.9 on the X chromosome of *Drosophila melanogaster*.   Kazazian et al. (1965), using starch-gel electrophoresis, showed that certain strains have a single PGD band (phenotype A) whose electrophoretic mobility is greater than that of the single band of other strains (phenotype B).   In all females produced by mating PGD A with PGD B strains the cells containing the enzyme produce three different molecules and therefore bands of this enzyme: PGD A, PGD B, and PGD AB (intermediate mobility).   The AB molecule was dissociated and reassociated to form the same entity.   Explain how each cell that synthesizes PGD can form three types of molecules A, B, and AB and discuss the bearing of these data on the mechanism of dosage compensation in this organism.   (See Young, 1968, and Steele et al., 1968.)

1559 *Primary hyperuricemia* is the (pathological) expression of the mutant allele *jh* of a normal X-linked gene *Jh*.   The primary biochemical consequence is a deficiency of the enzyme hypoxanthine guanine phosphoribosyl transferase (HGPRT) which converts the bases hypoxanthine and guanine into their ribonucleotides. Salzmann et al. (1968) demonstrated the existence of the enzyme in fibroblasts cultured in medium containing ^{3}H-hypoxanthine and studied by autoradiography. The cells were treated with RNase before microscopic examination. The autoradiographic phenotypes of cells in uncloned populations appeared in the frequencies shown in Table 1559A.

Table 1559A

Individuals studied	Cells without label ($-$)	Cells with strong label ($+$)	Cells with intermediate label ($+$ $-$)
Affected ♂	3,342	0	0
Heterozygous ♀ 1	82	788	46
Heterozygous ♀ 2	39	941	50
Normal ♂	0	2,079	0
Normal ♀	0	2,027	0

Single-cell fibroblast clones from heterozygous and *normal* females gave the results in Table 1559B.

Table 1559B

Type of ♀	Clones without label ($-$)	Clones with label ($+$)
Heterozygote 1	61	78
Heterozygote 2	99	124
Normal	0	221

Key: $+$ = *wild-type* clones with normal enzyme activity;
$-$ = *mutant* clones with little or no enzyme activity.

a The authors conclude that these results support the single-active-X hypothesis. State whether you agree and why.

b A low but definite degree of labeling occurs in cells from mutant males and ($-$) clones from heterozygous females. This may be accounted for by the presence of utilizable radioactive contaminant in the hypoxanthine. What other reason is possible? How would you test whether this is occurring?

See Felix and De Mars (1969), Dancis et al. (1969), for the biochemical block involved, identification of heterozygotes and *affected* individuals before birth, and therapy of *affected* individuals to avert the neurological and developmental symptoms which constitute the *Lesch-Nyhan syndrome*.

1560 Ohno and Makino (1961) have shown that in somatic cells of human females one of the X chromosomes, like the autosomes, remains in an extended state during interphase and prophase, while the other becomes heavily condensed (heterochromatic), forming the Barr body. In male somatic cells the single X never manifests positive heteropycnosis at these stages but remains euchromatic. Offer an explanation for these phenomena and discuss the functional significance of this behavior.

1561   The Duchenne type of *muscular dystrophy*, a rare human trait with onset usually before the seventh year of life and death by age fifteen, can be detected by increased levels of creatine phosphokinase (CPK) and histological changes in the muscle. The test results of Pearson et al. (1963) on one family, typical of many studied, are presented in the pedigree.

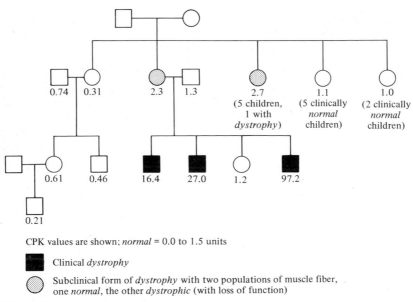

CPK values are shown; *normal* = 0.0 to 1.5 units

■   Clinical *dystrophy*

◯   Subclinical form of *dystrophy* with two populations of muscle fiber, one *normal*, the other *dystrophic* (with loss of function)

Figure 1561

a Why does this form of *muscular dystrophy* occur almost exclusively in boys?

b On the basis of one gene with two alleles, suggest genotypes and dominance relationships of individuals marked with a star.

c A few females have been reported to be clinically *affected*. Among subclinically *affected* females dystrophic muscle fibers range from few to many, and clinical symptoms appear only after about 50 percent of functional muscle is lost. What does this suggest regarding the dynamic relationship between the two kinds of X's with regard to gene inactivation according to the Lyon hypothesis?

d A genetic counselor is asked to indicate the likelihood of a clinically *affected* child in a family in which the female parent is ◯ and the male parent is ☐. What should he tell them?

1562 In man, the gene specifying the synthesis of glucose-6-phosphate dehydrogenase (G6PD) is on the X chromosome; the allele *G* specifies normal production and *g* no synthesis of the enzyme. Beutler et al. (1962) found that individually tested red blood cells of *normal* males, $X^G Y$, and *normal* females, $X^G X^G$, show the same amount of G6PD activity. Some cells of heterozygous females, $X^G X^g$, were normal and others deficient for G6PD, but none showed intermediate activity. *Affected males*, $X^g Y$, and *affected females*, $X^g X^g$, show no enzyme activity. In vitro studies by Davidson et al. (1963) of clones from single skin cells of $X^G X^g$ females also showed that some had *normal* and others little or no enzyme activity.

a (1) Why do erythrocytes of *normal* females not produce twice as much enzyme as those of *normal* males?

(2) Why do heterozygous, $X^G X^g$, females produce half the amount produced by *normal* males?

b XO, XY, XX, XXX, XXXY, and XXXX cells that are diploid for all autosomes and carry only the normal allele *G* all synthesize the same amount of G6PD (Grumbach et al., 1962; Harris et al., 1963). Relate this to your hypothesis in (a).

1563 Roberts (1929) described three women who had patches of smooth skin (without sweat glands) and patches of normal skin and some of whose sons were affected with *anhidrotic ectodermal dysplasia* (absence of sweat glands). Offer an explanation for this phenotypic mosaicism.

1564 At least one of the genes determining a certain polygenic character is on the X chromosome. Would you expect greater intrapair differences in male than in female monozygotic twins and why? (See Vandenberg et al., 1962.)

1565 The mule is a hybrid between a female horse (*Equus caballus*; $2n = 64$) and a male donkey (*E. asinus*; $2n = 62$). In the mule karyotype (somatic chromosome number 63) the paternal and maternal chromosomes, including the two X's in females, can easily be distinguished cytologically. Mukherjee and Sinha (1964) exposed cultured female mule leukocytes to tritiated thymidine for 5 to 6 hours and then studied chromosome duplication by autoradiography. In 17 of 33 cells at metaphase, the heavily labeled chromosome was the X from the donkey. In the remaining 16 it had the size and morphology of the X of the horse. What significant fact do these observations reveal, and what bearing do they have on the Lyon hypothesis?

1566 Deficiency of hypoxanthine guanine phosphoribosyl transferase (HGPRT) due to an X-linked recessive gene causes *Lesch-Nyhan syndrome* with a distressing type of mental deficiency and many other abnormalities. Carriers of the recessive gene can be detected by biochemical tests on cultured skin fibroblasts. A pregnant woman diagnosed as a carrier had a sample of the fetus's cells removed from her uterus. Some were normal, others abnormal (Fujimoto et al., 1968). Is the baby a male or female, would it be born normal or abnormal, and would it be a carrier or not? Explain.

1567   Epstein (1969) measured oocyte activity of G6PD and LDH in XO and XX female mice and found that LDH activity of both types of oocytes was the same but that the G6PD activity in XX oocytes was double that in XO oocytes.   What is the probable cause of the difference in activity of these two enzymes?

1568   In mice, the X chromosome carries a histocompatibility locus.   Bailey (1963) obtained hybrids between two appropriate highly inbred strains of mice.   When skin of an F_1 female was grafted to her brother, it was rejected in patches.   Interpret these results in the light of the dosage-compensation mechanism in mammals.

1569   In kangaroos (females XX; males XY) one of the two X chromosomes in females is late-replicating, genetically inactive, and heterochromatic.   The X (X_e) of euros (*Macropus robustus erubescens*), which possess a *slow*-moving electrophoretic form (*S*) of glucose-6-phosphate dehydrogenase (G6PD), is about $1\frac{1}{2}$ times larger than the X (X_w) of wallaroos (*M. r. robustus*), which produce a *fast*-moving (*F*) G6PD.   The karyotypes of the two subspecies, which interbreed freely and produce viable offspring, are otherwise identical. In F_1 females from euro female × wallaroo male, the X_w always replicates later than X_e.   The reverse is true in F_1 females from wallaroo female × euro male crosses (Sharman, 1971).   The difference between *fast* and *slow* electrophoretic variants of G6PD is due to different alleles of one X-linked gene. Reciprocal crosses between euros and wallaroos produced the F_1 results shown (Richardson et al., 1971).   Is the mode of dosage compensation for X chromo-

Table 1569

	Parents		F_1	
♀		♂	♀	♂
1. Euro (*S*)		Wallaroo (*F*)	*S*	*S*
2. Wallaroo (*F*)		Euro (*S*)	*F*	*F*

somes in kangaroos the same as in eutherian mammals (random X inactivation)?   If not, indicate how eutherian X inactivation may have evolved from the marsupial type of X inactivation.   See Cooper (1971).

1570   The female mule has one X (X_H) from her horse mother and the other X (X_D) from her donkey father.   The two X's are cytologically different. Giannelli et al. (1969) found that the late-replicating X is X_D in 90 percent of somatic cells studied.   However, electrophoretic studies of G6PD (X-linked) of these females indicate that almost all of it is of the horse type.   Is the X which replicates late and forms the sex-chromatin body the genetically inactive one?

1571   The lactose region in *E. coli* consists of the regulator gene, *i*, the operator, *o*, and three structural genes *z* (makes β-galactosidase, an intracellular enzyme which hydrolyzes lactose into glucose and galactose), *y* (specifies permease, which aids in getting lactose into the cell), and *a* (specifies acetylase, required for lactose utilization). The operon consists of the operator and the three structural genes (consider the first two only). The sequence of the genes in the lactose region is shown. Jacob et al. (1960) and Jacob and Monod

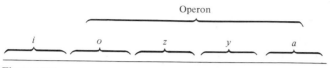

Figure 1571

(1961*a*, 1961*b*) found the functions of the alleles at the first four loci to be as shown in Table 1571*A*.

a What is the expected order of dominance of the alleles at the regulator locus? At the operator locus? Explain.

b Jacob et al. (1960) and Jacob and Monod (1961*a*, 1961*b*) showed that the phenotypes of some of the haploid genotypes were those given in Table 1571*B*.

Table 1571*A*

Gene type	Function

Regulator:

i^+ *Wild-type* allele for a diffusible repressor (protein) which inhibits synthesis of proteins specified by the structural genes by binding with o^+ operator DNA; lactose inactivates *wild-type* repressor

i^- Constitutive allele which results in no or an inactive repressor

i^s Superrepressor allele; results in a modified repressor protein which is unable to combine with the inducer lactose; the protein, being unbound, can bind to o^+ DNA and inactivate it

i^d Constitutive allele, inactivates i^+ product by aggregation

i^q Produces more repressor than i^+

Operator:

o^+ *Wild-type* operator DNA, which turns on or permits synthesis of *m*RNA of structural genes; this operator is sensitive to, and inactivated by, i^+ and i^s repressors; i.e., these repressors prevent synthesis of operon *m*RNA

o^c Constitutive allele which causes operator DNA to be insensitive to i^+ and i^s repressors, permitting permanent synthesis of operon *m*RNA

β-Galactosidase:

z^+ Specifies synthesis of *wild-type* β-galactosidase

z^- Determines synthesis of no β-galactosidase or its mutant form, *Cz* (assume former)

Permease:

y^+ Specifies *wild-type* permease

y^- Determines no or defective permease (assume former)

Table 1571B

Genotype	Inducer present		Inducer absent	
	β-galactosidase	Permease	β-galactosidase	Permease
1. $i^+ o^+ z^+ y^+$	+	+	−	−
2. $i^+ o^+ z^- y^+$	−	+	−	−
3. $i^+ o^+ z^+ y^-$	+	−	−	−
4. $i^- o^+ z^+ y^+$	+	+	+	+
5. $i^s o^+ z^+ y^+$	−	−	−	−
6. $i^+ o^c z^+ y^+$	+	+	+	+
7. $i^s o^c z^+ y^+$	+	+	+	+
8. $i^+ o^c z^+ y^-$	+	−	+	−

Key: + = synthesis; − = no synthesis of enzyme.

Do the phenotypes of these genotypes verify the functions of the alleles at the i and o loci? Indicate the genotypes that verify a particular function.

c Phenotypes of various diploid genotypes were shown by the above authors to be those given in Table 1571C.

Table 1571C

Genotype	Inducer present		Inducer absent	
	β-galactosidase	Permease	β-galactosidase	Permease
1. $i^+ o^+ z^+ y^+/i^+ o^+ z^- y^-$	+	+	−	−
2. $i^- o^+ z^+ y^+/i^+ o^+ z^- y^-$	+	+	−	−
3. $i^- o^+ z^- y^+/i^+ o^+ z^+ y^+$	+	+	−	−
4. $i^+ o^+ z^+ y^+/i^- o^+ z^- y^-$	+	+	−	−
5. $i^s o^+ z^+ y^+/i^+ o^+ z^+ y^+$	−	−	−	−
6. $i^s o^+ z^+ y^+/i^- o^+ z^+ y^+$	−	−	−	−
7. $i^+ o^c z^- y^+/i^+ o^+ z^+ y^-$	+	+	−	+
8. $i^+ o^c z^+ y^-/i^+ o^+ z^- y^+$	+	+	+	−
9. $i^+ o^+ z^+ y^-/i^- o^c z^- y^+$	+	+	−	+
10. $i^- o^c z^+ y^-/i^+ o^+ z^- y^+$	+	+	+	−
11. $i^s o^+ z^+ y^+/i^+ o^c z^+ y^+$	+	+	+	+

(1) Determine the dominance relationship between z^+ and z^-; y^+ and y^-.

(2) Are the allelic relationships at the i locus outlined in (a) verified by the phenotypes of genotypes 2 to 6? Answer the same question for the relationships at the operator locus by analyzing the results of genotypes 7 to 11.

(3) Does the cis-trans position of the alleles at the regulator locus affect their functions? Is the same true of the alleles at the operator locus? Explain.

d For each of the following genotypes indicate whether β-galactosidase and permease synthesis is constitutive or inducible or both. All bacteria are $z^+ y^+$.

(1) $i^- o^+ / i^- o^c$ (2) $i^+ o^+ / i^- o^c$ (3) $i^+ o^c / i^- o^+$
(4) $i^+ o^+ / i^+ o^+$ (5) $i^+ o^+ / i^+ o^c$ (6) $i^- o^+ / i^- o^+$
(7) $i^+ o^+ / i^- o^-$ (8) $i^+ o^c / i^- o^c$ (9) $i^s o^+ / i^- o^+$
(10) $i^s o^- / i^+ o^-$ (11) $i^- o^+ / i^s o^+$ (12) i^- / i^s
(13) i^s / i^d (14) i^q / i^d (15) i^- / i^d

e For each of the following genotypes, which are also *wild types* at the z and y loci, indicate whether β-galactosidase and permease would be produced (1) if inducer is present and (2) if inducer is absent.

(1) $i^- o^+$ (2) $i^s o^c$ (3) $i^+ o^c$
(4) $i^+ o^+$ (5) $i^s o^+$ (6) $i^s o^-$
(7) $i^- o^c$ (8) $i^+ o^c$

1572 In an inducible bacterial system a is the regulator gene, b the operator, and c and d the operon structural genes whose *wild-type* alleles specify enzymes C and D and whose mutant alleles specify enzymes C^- and D^-. *Mutant allele a^- permits constitutive synthesis of the enzymes in either of two ways: by producing an inducer, thus eliminating the need for such a compound (a^+ would be an amorph not producing inducer), or by not producing a repressor specified by the a^+ allele, which in the absence of inducer combines with *wild-type, b^+, operator and prevents operon *m*RNA formation. Inducer inactivates repressor, allowing transcription of operon *m*RNA. The phenotypes produced by various diploid genotypes when inducer is present and when it is absent are shown in the table.

a What is the mechanism of action of alleles at the a locus? Explain.

b If the second alternative is correct, do the a alleles act only on loci in cis positions? If not, what is the probable nature (diffusible vs. nondiffusible) of the repressor? Explain.

Table 1572

	Inducer present				Inducer absent			
Genotype	C	C^-	D	D^-	C	C^-	D	D^-
$a^+ b^+ c^- d^- / a^- b^+ c^+ d^+$	+	+	+	+	−	−	−	−
$a^- b^+ c^+ d^+ / a^- b^+ c^- d^-$	+	+	−	+	+	+	−	+
$a^+ b^+ c^+ d^- / a^- b^+ c^- d^+$	+	+	+	+	−	−	−	−
$a^- b^+ c^- d^- / a^+ b^+ c^+ d^+$	+	+	+	+	−	−	−	−
$a^- b^+ c^- d^- / a^- b^+ c^+ d^+$	+	+	+	+	+	+	+	+

1573   Which of the following *E. coli* can hydrolyze lactose?   Do they do so constitutively or inducibly?

a $i^+ \, p^+ \, o^+ \, z^+ / i^+ \, p^+ \, o^c \, z^+$ **b** $i^- \, p^- \, o^+ \, z^+ / i^+ \, p^+ \, o^+ \, z^-$

c $i^+ \, p^+ \, o^+ \, z^+ / i^+ \, p^- \, o^c \, z^+$ **d** $i^+ \, p^+ \, o^+ \, z^+ / i^+ \, p^+ \, o^c \, z^-$

e $i^- \, p^+ \, o^+ \, z^+ / i^+ \, p^- \, o^c \, z^-$ **f** $i^+ \, p^+ \, o^+ \, z^- / i^+ \, p^- \, o^c \, z^+$

Note: *p* is the promoter located between *i* and *o*.

1574   Genetic experiments show that structural genes in an operon may be switched off by the repressor, a protein produced by the regulator gene interacting with the operator.   The repression may be due to the repressor binding to a receptor site on DNA, i.e., the operator, thus directly preventing transcription, or the repressor may interact with *m*RNA or *t*RNA to block translation.   To determine the mechanism of repression Ptashne (1967*b*) isolated and labeled the λ-phage repressor produced by the C_1 gene of the mutant *ind*, mixed it with λ DNA in one experiment and with λimm^{434} phage DNA (contains almost all the genes of λ except that the operator and repressor genes are derived from 434) in another, and sedimented the mixtures through a sucrose gradient to determine whether λ repressor would bind with λ DNA or λimm^{434} DNA or both.

a Does λ repressor bind to λ DNA?   To λimm^{434} DNA?   Explain.

b Does the repressor act at the transcription or the translation level?   Explain.

c What conclusion can you draw from the results of binding experiments with denatured DNA?

See Gilbert and Müller-Hill (1967), Ptashne and Gilbert (1970).

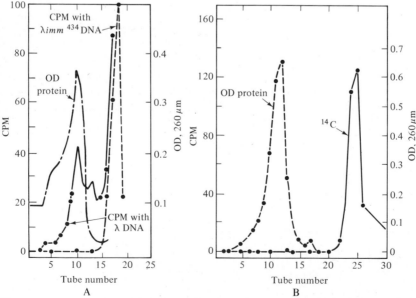

Figure 1574   A. Results of binding experiments to determine whether λ repressor binds with λ DNA or λimm^{434} DNA or both or neither.   B. Results of binding experiments with denatured DNA.

1575 The linked structural genes in the arabinose operon in *E. coli* are shown, along with a contiguously linked region marked *X* and a gene *C*. The enzymes specified by the three structural genes and the reactions the enzymes control are also indicated. The unlinked gene *E* specifies L-arabinose permease, which is concerned with active transport of arabinose.

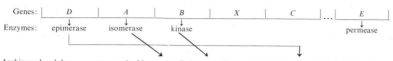

Arabinose breakdown sequence: L-Arabinose ⇌ L-ribulose → L-ribulose 5-phosphate ⇌ D-xylulose-5-phosphate

Englesberg et al. (1965) and Sheppard and Englesberg (1967) obtained the results given.

1   Haploids are shown in Table 1575*A*.

Table 1575*A*

Genotype	Arabinose present	Arabinose absent
$E^+ D^+ A^+ B^+ C^+$	All four enzymes produced to same extent	Deficient in all four enzymes
$E^+ D^+ A^+ B^+ C^c$		
$E^+ D^+ A^+ B^+ C^-$	Deficient in all four enzymes	

2   Merodiploids are shown in Table 1575*B* for isomerases only (essentially the same results have been obtained for enzymes specified by the other three genes).

Table 1575*B*

Genotype	Arabinose	Enzyme activity, units	
		Diploid	Control*
$A^- C^+/A^+ C^-$	Present	3.4	0.7
$A^- C^c/A^+ C^-$	Absent	1.8	0.06
$A^- C^c/A^+ C^+$	Absent	0.1	0.1

* Males and females alone were tested.

3   Deletion of *C* results in no permease synthesis.

a Is *C* an operator or regulator? Discuss data that support your answer.
b If *C* is a regulator, does it function negatively, like i^+ in the lactose system, by repressing enzyme synthesis or positively by inducing enzyme formation? Explain.
c What is the allelic relationship between C^+, C^c, and C^-? Explain.

d Indicate whether you expect isomerase and kinase synthesis and activity in the following stable merodiploids when arabinose is and is not present:

$F'A^+ B^+ C^-/A^- B^+ C^+$ $F'A^- B^+ C^+/A^+ B^+ C^-$

$F'A^+ B^+ C^-/A^- B^+ C^c$ $F' A^+ B^- C^+/A^- B^+ C^c$

e Deletion 709 covers gene C, the region between C and B, and part of B. Induced isomerase and kinase enzyme levels in this mutant in a haploid and merodiploid were as shown in Table 1575*C* (Sheppard and Englesberg,

Table 1575*C*

	Isomerase	Kinase
Deletion 709 ($D^- A^+$)	≤ 1	0.1
$F'A^- B^+ C^+/D^- A^+$ *deletion*	≤ 1	16.2

1967).   What is the probable function of the region X, and why?   See Englesberg, Squires, and Meronk (1969) and Englesberg et al. (1969) for an answer to this question and a detailed insight into this operon.

1576 The *R1* and *R2* regulator genes control the formation of a repressor for alkaline phosphatase.   Garen and Garen (1963) found that certain constitutive mutations in these regulator genes respond to an external suppressor. Why do these results suggest that the products of the regulator genes (repressor for alkaline phosphatase) are protein molecules?   See Garen and Otsuji (1964); also Morse and Yanofsky (1969*a,b*) for similar studies with the *E. coli* tryptophan regulatory gene.

1577 In *E. coli*, the regulator gene is closely linked to an operon consisting of two structural genes (consider only one) and an operator.   In the table the

Table 1577

Genotype	Phenotype	
	Inducer absent	Inducer present
$a^- b^+ c^+$	S	S
$a^+ b^+ c^-$	S	S
$a^+ b^- c^-$	s	s
$a^+ b^- c^+/a^- b^+ c^-$	S	S
$a^+ b^| c^+/a^- b^- c^-$	s	S
$a^+ b^+ c^-/a^- b^- c^+$	s	S
$a^- b^+ c^+/a^+ b^- c^-$	S	S

Key: S = enzyme synthesized in normal quantities; s = little or no synthesis; + = *normal* allele; − = *mutant*.

regulator, operator, and structural genes are listed in correct sequence. The phenotypes of the genotypes under induced and noninduced conditions are as shown. Which of these genes is a regulator? An operator? The structural gene? Explain.

1578 Jacob, Ullman, and Monod (1964) showed that certain mutations in the operator region which affect the rate at which the lactose operon is expressed cannot be in the operator DNA because operator-constitutive, o^c, mutations which destroy function of the operator so that it has no affinity for repressor do not alter the potential for operon expression. They merely permit constitutive *m*RNA formation. This site for initiation of operon expression, the promoter, must be between either *i* and *o* or *o* and *z*. Ippen et al. (1968) isolated ultraviolet-induced mutants L8 and L37 (point mutations) which coordinately reduce the rate of expression of the structural genes but do not alter either the operator or the *i* gene. These, like the o^c mutants, are cis-dominant, in that they only reduce the rate of expression of the operon on the same chromosome. Ippen et al. also isolated a series of deletion mutants which always include the distal end (*a* gene) of the operon and extend varying distances toward and past the operator. These deletions include all the genetic material between the *T1* locus and the determined

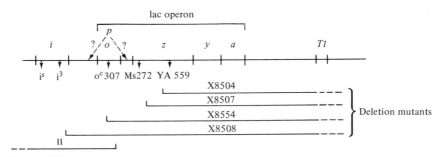

Figure 1578

deletion end, as shown. The frequency of recombination between the deletion mutants and point mutations in the operon *i* gene and L8 and L37 are tabulated.

a What is the map position of the promoter mutants? Explain.

b Deletion mutant II impairs operon expression. Do these results support your answer to (a)?

c None of the point mutations of the operator, o^c, are suppressed by *amber* or *ochre* suppressors.

(1) Is the operator translated into proteins?

(2) Are the promoter mutants defective in transcription or translation? Explain.

(3) If the latter, what is the probable function of the promoter? How might it act? See Ippen et al. (1968).

Table 1578

Mutant	Deletion			
	X8504	X8507	X8554	X8508
z^- YA559	0	0	0	0
z^1 M5272	+	+	0	0
o^c 307	n.t.	n.t.	0	n.t.
i^3	n.t.	n.t.	0.19	0.069
i^s	+	+	0.24	+
L8	+	+	0.065	0
L37	0.79	0.34	0.086	0

Key: + = recombinants obtained; − = no recombinants obtained; n.t. = not tested.

1579 The structural genes of the tryptophan operon of *Escherichia coli*, the enzymes they specify, and the reactions they control are as follows:

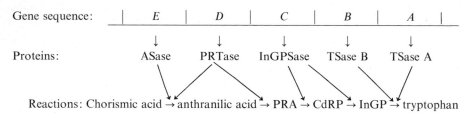

Gene sequence: | *E* | *D* | *C* | *B* | *A* |

Proteins: ASase   PRTase   InGPSase   TSase B   TSase A

Reactions: Chorismic acid → anthranilic acid → PRA → CdRP → InGP → tryptophan

where   ASase = anthranilate synthetase
     PRTase = phosphoribosyl anthranilate transferase
     InGPSase = indoleglycerolphosphate synthetase
     TSase = tryptophan synthetase
     InGP = indole-3-glycerol phosphate
     PRA = N-(5′-phosphoribosyl) anthranilate
     CdRP = 1-(*O*-carboxyphenylamino)-1-deoxyribulose-5-phosphate

a Deletions at the *E* end lead to nonrepressible transcription of *trp* genes. At which end of the operon is the operator located, and at which end of the operon would you expect *m*RNA synthesis to begin?

b The formation of the tryptophan enzymes is coordinate. Nonsense mutants 1 and 2 occur in gene *E* (with 2 closer to *D* than 1) and nonsense mutant 3 in gene *B*.

(1) On which side would you expect these mutations to reduce the relative rates of synthesis of genes, the side away from or toward the operator?

(2) Which nonsense mutant, 1 or 2, would have a more drastic effect on enzyme synthesis?   Answer the same question for mutants 1 and 3.

1580 In the repressible trytophan operon (see Prob. 1579 for details), the repressor becomes active after reacting with tryptophan, binding to the operator and preventing transcription of operon *m*RNA.   Matsushiro et al. (1965) assayed a number of *tryptophanless* mutant strains as well as *wild types* for enzymes of the tryptophan pathway under derepressed conditions.   The results for *wild type*, two E-gene mutants, and one B-gene mutant are shown.

Table 1580   Levels of tryptophan enzymes

	Gene and enzyme				
	E ↓ ASase	D ↓ PRTase	C ↓ InGPSase	B ↓ TSase B	A ↓ TSase A
wild type	100	100	100	73	60
E mutant PA309	0	5	5.5	5	11
E mutant T15N	0	12	25.5	19	20
B mutant W4627	100	120	114.5	3.8	<1

No *A*-gene mutations or deletions involving *A* but not *E* alter capacity to produce ASase.   Deletions of *E* lead to nonrepressible transcription of tryptophan genes.

a Where is the operator located relative to the structural genes in the operon? Explain.   See Hiraga (1969).

b $\phi 80$ *dt* carries all tryptophan genes, whereas $\phi 80$ *dt anth*^{del} carries tryptophan operon with a deleted *E* gene and segment next to it.   In $\phi 80$ *dt* heterogenotes $A^- B^+ anth^+/A^+ B^+ anth^+$ or $A^+ B^- anth^+/A^+ B^+ anth^+$, synthesis of all tryptophan enzymes is completely repressed by tryptophan. In $\phi 80$ *dt anth*^{del} heterogenotes $A^- B^+ anth^+/A^+ B^+ anth^{del}$ synthesis of TSase B is partially repressed by tryptophan while synthesis of A is not repressed.   The reverse is true of $\phi 80$ *anth*^{del} heterozygotes $A^+ B^- anth^+/A^+ B^+ anth^{del}$.   In both cases ASase is repressed normally.

(1) To which of the operator, *o*, mutants in the lactose system is this deletion comparable?

(2) Does it produce its effect in cis or trans or both positions?

1581 Under a variety of conditions the synthesis of the tryptophan biosynthetic enzymes in *wild-type E. coli* is coordinate (see Prob. 1579 for enzymes specified by genes in the tryptophan operon).   Yanofsky and Ito (1966) found that enzyme production by two groups of mutants under repressed conditions were as shown (R^- is a regulator mutation producing a repressor which is only partially efficient and therefore permits elevated production of tryptophan enzymes under repression conditions).   These workers also showed that within a gene, mutations in the second group on the side close to the operator

Table 1581

Mutant	Class*	Percentage of wild-type value†				
		ASase	PRTase	InGPSase	TSase B	TSase A
R^- A9952	1	100	104	114	100	90
R^- B9763	1	100	—	108	0	103
R^- C3404	1	100	—	0	115	111
R^- D9885	1	100	0	91	117	107
R^- E9547	1	0	—	—	96	97
R^- A9796	2	100	95	100	49	0
R^- B40	2	100	111	83	0	3
R^- C9905	2	100	82	0	26	33
R^- D10242	2	0	0	8	9.3	10
R^- E9851	2	0	2	3	4	4

* Mutants in class 1 form CRM$^+$; those in class 2 do not.

† ASase value set at 100 percent in each case.

(E-gene) end had a more pronounced effect on protein synthesis specified by more distal genes in the operon than changes toward the other end of the gene.

a Mutants in group 2, but not 1, responded to certain suppressors. What is the probable nature of mutations in the two groups?

b What is the difference in effect of mutations in the two groups on the relative rates of polypeptide synthesis?

c Offer an explanation to account for the tabular facts and the additional stated facts.

1582 In *Salmonella typhimurium* the structural genes specifying the synthesis of the ten enzymes in the histidine biosynthetic pathway are clustered (in juxtaposition) in one region of the genome (Loper et al., 1964). Synthesis of all these enzymes is repressed simultaneously and to the same extent by excess histidine. Most mutations in this region result in the modification of structure and function of only one enzyme. However, mutations that occur at one end of this region result in nonsynthesis of all ten enzymes, although it is known that the structural genes for at least nine of them are *wild type*. Why should the first kind of mutation cause structural alteration or loss of only one enzyme whereas the latter affects the synthesis of all enzymes?

1583 Polarity mutations are changes in structural genes in a multigenic operon which lead to a relative decrease in the levels of all enzymes specified by genes on the operator-distal side of the mutated gene (Jacob and Monod, 1961*a*; Newton et al., 1965; Yanofsky and Ito, 1966; Martin et al., 1966*a,b*). To decide whether a nonsense codon causes polarity by reducing translation of the *m*RNA region beyond the nonsense codon, or by affecting the relative representation of different regions of the operon in the *m*RNA population, or by both means, Imamoto and Yanofsky (1967*a,b*) isolated tryptophan-operon *m*RNA of *E. coli* from cultures of *wild type* and various nonsense mutants following a shift from repression to derepression conditions. The messenger was pulse-labeled with ³H-uridine shortly after initiation of derepression and studied in two ways: the total amount of *trp-m*RNA in each strain was determined, and *trp-m*RNA segments corresponding to different segments of the operon were measured by hybridizing with DNA from $\phi 80$ phage carrying different segments of the operon:

1 Strong polar mutants (nonsense mutations in the initial segment of the gene) contained less total *trp-m*RNA than *wild-type* strains.

2 Using $\phi 80$ (no *trp* genes), $\phi 80$ pt *AB* (carrying *A* and *B* genes), $\phi 80$ pt *A–C* (carry genes *A, B, C*), $\phi 80$ pt *C–E* (carry genes *C, D, E*), and $\phi 80$ pt *A–E* (all *trp* structural genes) as *trp*-DNA sources in DNA-RNA hybridization experiments, Imamoto and Yanofsky found that:

   a   15 percent *m*RNA of *wild type* is hybridizable with pt *AB* DNA, 55 percent is hybridizable with pt *A–C* DNA, and 83 percent is hybridizable with pt *C–E* DNA.

   b   *trp-m*RNA of strong-polarity mutants was deficient in *m*RNA regions corresponding to the operon genes on the operator-distal side of the mutated gene.

3 Sucrose-gradient sedimentation profiles of *trp-m*RNA of strong-polarity mutants indicated that most of the *trp-m*RNA molecules of each mutant were smaller than normal.

4 The size of the main type of *trp-m*RNA species in different mutants corresponded to the location of the nonsense mutation relative to the operator: the farther from the operator the longer the *m*RNA.

5 The same number of *trp-m*RNAs are produced by *wild-type* and polar mutants. Relative number of normal vs. short *trp-m*RNA molecules in different polar mutants was correlated with the severity of polarity.

6 Polarity mutations may cause premature termination of transcription in the vicinity of a nonsense mutation.

a Suggest a mechanism which will explain:
   (1) Reduction in the amount of polypeptide synthesis for genes distal to the one with the nonsense codon.
   (2) The lower level of total operon *m*RNA.
   (3) The reduction in size of at least some operon-messenger molecules in mutants with nonsense codons.
   (4) The effect of the location of the nonsense codon in the gene on the synthesis of *m*RNA and polypeptides of distal genes.
b Webster and Zinder (1969) showed that ribosomes detach from RNA phage messenger, in vitro, after reading strong polar or nonpolar nonsense codons. Moreover, ribosomes can attach to internal regions of RNA phage messenger in vitro (Lodish, 1968; Webster and Zinder, 1969). In view of the above findings, can polarity be explained solely by *threading* (a mechanism proposed by Martin et al., 1966; Yanofsky and Ito, 1966), i.e., attachment of

ribosomes only at the first gene of a polycistronic *m*RNA and detachment of ribosomes from the messenger after encountering the first codon?

c In addition to confirming that polarity in *E. coli* is associated with a diminished amount of detectable messenger from regions distal to nonsense codons, Morse and Yanofsky (1969*b*) showed that the kinetics of messenger synthesis in *wild type* and in polar and missense mutants is the same (i.e., RNA polymerase transcribed the entire operon disregarding defective codons) but the amount of messenger in polar mutants distal to the nonsense codon was greatly reduced.

(1) Do these results support or refute the 1967 Imamoto and Yanofsky mechanism for polarity of coupling transcription with translation, with the possibility of transcription termination beyond a nonsense codon?

(2) If not, outline a plausible mechanism involving rapid degradation of *m*RNA to account for all these facts and see how it compares with the apparently correct explanation for polarity by Morse and Yanofsky (1969*b*).

1584 Regulation of the histidine operon in *Salmonella typhimurium* involves at least six genes (Antón, 1968). Mutations at any one of these loci destroy the cell's ability to fully repress the histidine biosynthetic enzymes. Fink and Roth (1968) studied dominance relationships between *mutant* and *wild-type* alleles at each of these loci using merodiploids containing an F' episome with the *wild-type* allele of the gene involved and the chromosome carrying the *mutant* allele. The levels of histidine biosynthetic enzymes, histidinol phosphate phosphatase (specified by *hisB*), and histidinol dehydrogenase (specified by *hisD*) were assayed as an index of derepression of the operon. Results for

Table 1584

Line	Regulatory mutation	Genotype	Specific activity of*	
			B enzyme	D enzyme
	wild type	$T^+ W^+$	1	1
	hisT	T	26.3	
		$F'\,T^+/T^-$	3.5	
	hisW$^-$	W	16.1	
		$F'\,W^+/W^-$	3.6	
1	*wild type*	$O^+ B^+ D^+$	1	1
2	*hisO$^-$*	$O^- B^- D^+$	0	12.4
3	*hisO$^-$*	$O^- B^+ D^-$	12.0	0
4	*hisO$^-$*	$O^- B^+ D^+$	11.2	15.1
5	*hisO$^-$*	$F'\,O^+ B^+ D^+/O^- B^- D^+$	1.7	15.0
6	*hisO$^-$*	$F'\,O^+ B^+ D^+/O^- B^+ D^-$	16.5	1.2
7	*hisO$^-$*	$F'\,O^+ B^+ D^-/O^- B^+ D^+$	9.5	15.3

* All enzyme activities are expressed relative to *wild type* taken as 1.

three genes (*hisI*, *W*, and *D*) are presented.   The relative specific activities are given first for each *mutant* and then for the diploid (same strain carrying the episome).

a For each of these loci state whether the *mutant* or *wild-type* allele is dominant and why, and if dominant, whether in both cis and trans.

b Where would you expect *hisO* to map relative to the histidine operon structural genes and why?

1585 Kiho and Rich (1965) measured the relative size of polyribosomes showing β-galactosidase activity in *E. coli*.   Strains with deletions in the *y-a* (permease-acetylase) part of the lactose operon had smaller polyribosomes than *wild type*.   *Amber* mutants in the *z*, *y*, and *a* genes also had smaller polyribosomes. *Wild-type* revertants of these had the larger *wild-type* polyribosomes.   Do these results indicate that there is one *m*RNA for each gene or one messenger per operon?   Explain.

1586 The position of a nonsense mutation in a gene strongly influences the degree of polarity; operator-proximal mutants allow less efficient expression of the distal genes in the operon than operator-distal mutants.   Thus, there is a gradient of polarity along a gene (Newton et al., 1965; Yanofsky and Ito, 1966; Martin et al., 1966; Fink and Martin, 1967).

a Suggest a mechanism to explain this phenomenon (see Morse and Yanofsky, 1969*b*).

b Why do nonsense but not missense mutations have polar effects?

c Do nonsense mutations in operon structural genes produce their effects at the transcription or translation level?   Explain.

1587 Attardi et al. (1963) found that lactose operon *m*RNA synthesis in *E. coli* increased with addition of inducer.   Do inducers act at the translation or transcription level?

REFERENCES

Adelberg, E. A., and H. E. Umbarger (1953), *J. Biol. Chem.*, **205**:475.
Allfrey, V. G., R. Faulker, and A. E. Mirsky (1964), *Proc. Natl. Acad. Sci.*, **51**:786.
———, V. C. Littau, and A. E. Mirsky (1963), *Proc. Natl. Acad. Sci.*, **49**:414.
Ames, B. N., and B. Garry (1959), *Proc. Natl. Acad. Sci.*, **45**:1453.
———, R. F. Goldberger, P. E. Hartmen, R. G. Martin, and J. R. Roth (1967), in V. V. Koningsberger and L. Bosch (eds.), "Regulation of Nucleic Acid and Protein Synthesis," pp. 272–287, Elsevier, Amsterdam.
Anders, G. (1947), *Rev. Suisse Zool.*, **54**:269.
——— (1955), *Z. Vererbungsl.*, **87**:113.
Antón, D. N. (1968), *J. Mol. Biol.*, **33**:533.
Attardi, G., et al. (1963), *Cold Spring Harbor Symp. Quant. Biol.*, **28**:363.
Bahn, E. (1971), *Hereditas*, **67**:79.
Bailey, D. W. (1963), *Science*, **141**:631.
Baker, W. K. (1963), *Am. Zool.*, **3**:57.
——— (1968), *Adv. Genet.*, **14**:133.
——— (1971), *Proc. Natl. Acad. Sci.*, **68**:2472.
Barth, L. G. (1949), "Embryology," Dryden, New York.

Becker, H. J. (1962), *Chromosoma*, **13**:341.

Beckwith, J. R. (1964), *J. Mol. Biol.*, **8**:427.

―――― and D. Zipser (eds.) (1970), "The Lactose Operon," Cold Spring Harbor Laboratory, Cold Spring Harbor, N.Y.

Beermann, W. (1956), *Cold Spring Harbor Symp. Quant. Biol.*, **21**:217.

―――― (1960), *Chromosoma*, **11**:263.

―――― (1961), *Chromosoma*, **12**:1.

―――― and U. Clever (1964), *Sci. Am.*, **210**(April):58.

Berendes, H. D. (1966), *Chromosoma*, **20**:32.

Beutler, E., M. Yeh, and V. F. Fairbanks (1962), *Proc. Natl. Acad. Sci.*, **48**:9.

Boettiger, E. G., and C. M. Osborn (1938), *Endocrinology*, **22**:447.

Bonner, J., et al. (1968), *Science*, **159**:47.

――――, R. C. C. Huang, and R. V. Gilden (1963), *Proc. Natl. Acad. Sci.*, **50**:893.

Bonnevie, K. (1934), *J. Exp. Zool.*, **67**:443.

Briggs, R., and T. J. King (1959), in J. Brachet and A. E. Mirsky (eds.), "The Cell," vol. 1, pp. 537–617, Academic, New York.

Brown, D. D., and I. B. Dawid (1968), *Science*, **160**:272.

―――― and ―――― (1969), *Annu. Rev. Genet.*, **3**:127.

―――― and J. B. Gurdon (1964), *Proc. Natl. Acad. Sci.*, **51**:139.

Brown, S. W. (1966), *Science*, **151**:417.

Brumbough, J. A. (1971), *Dev. Biol.*, **24**:392.

Callahan, R., A. J. Blume, and E. Balbinder (1970), *J. Mol. Biol.*, **51**:709.

Carroll, F. D., P. W. Gregory, and W. C. Rollins (1951), *J. Anim. Sci.*, **10**:16.

Case, M. E., and N. H. Giles (1971), *Proc. Natl. Acad. Sci.*, **68**:58.

Cattanach, B. M. (1961), *Z. Vererbungsl.*, **92**:165.

――――, J. N. Perez, and C. E. Pollard (1970), *Genet. Res.*, **15**:183.

Chandley, A. C. (1969), *Nature*, **221**:70.

Changeaux, J. P. (1961), *Cold Spring Harbor Symp. Quant. Biol.*, **26**:313.

Clancy, C. W., and G. W. Beadle (1937), *Biol. Bull.*, **72**:47.

Clark, E., L. R. Aronson, and M. Gordon (1954), *Bull. Am. Mus. Nat. Hist.*, **103**:139.

Clever, U. (1963), *Chromosoma*, **14**:651.

―――― (1965), *Chromosoma*, **17**:309.

―――― (1968), *Annu. Rev. Genet.*, **2**:11.

Cohen, G., and F. Jacob (1959), *C. R. Acad. Sci.*, **248**:3490.

Cohen, M. M., and M. C. Rattazzi (1971), *Proc. Natl. Acad. Sci.*, **68**:544.

Cooper, D. W. (1971), *Nature*, **230**:292.

Counce, S. J. (1956), *Z. Vererbungsl.*, **87**:462.

Crandall, B. F., and R. S. Sparkes (1971), *Biochem. Genet.*, **5**:451.

Crippa, M., and G. P. Tocchini-Valentine (1971), *Proc. Natl. Acad. Sci.*, **68**:2769.

Dahmus, M. E., and J. Bonner (1970), *Fed. Proc.*, **29**:1255.

Dancis, J., et al. (1969), *Biochem. Genet.*, **3**:609.

Danforth, C. H. (1930), *Biol. Gen.*, **6**:99.

―――― and F. Foster (1929), *J. Exp. Zool.*, **52**:443.

Datta, R. K., and A. S. Mukherjee (1971), *Genetics*, **68**:269.

Davidson, E. H. (1969), "Gene Activity in Early Development," Academic, New York.

―――― and R. J. Britten (1971), *J. Theor. Biol.*, **32**:123.

―――― and B. R. Hough (1970), *J. Exp. Zool.*, **172**:25.

―――― et al. (1966), *Proc. Natl. Acad. Sci.*, **56**:856.

Davidson, R. G., H. M. Nitowsky, and B. Childs (1963), *Proc. Natl. Acad. Sci.*, **50**:481.

Demerec, M. (1940), *Genetics*, **25**:618.

Denis, H. (1968), *Adv. Morphog.*, **7**:115.

de Reuck, A. V. S., and J. Knight (eds.) (1966), "Histones: Their Role in the Transfer of Genetic Information," Little, Brown, Boston.

Dubinin, N. P., and B. V. Sidorov (1935), *Biol. Zh.*, **4**:555.
Dunn, L. C., S. Gluecksohn-Schoenheimer, and V. Bryson (1940), *J. Hered.*, **31**:343.
Ebert, J. (1970), "Interacting Systems in Development," 2d ed., Holt, New York.
Eliot, T. S. (1928), *Physiol. Zool.*, **1**:286.
Elmer, W. A. (1968), *Dev. Biol.*, **18**:76.
―――― and L. J. Pierro (1964), *Am. Zool.*, **4**:381.
―――― and ―――― (1966), *Am. Zool.*, **6**:510.
Englesberg, E., J. Irr, J. Power, and N. Lee (1965), *J. Bacteriol.*, **90**:946.
Englesberg, E., D. Sheppard, C. Squires, and F. Meronk (1969), *J. Mol. Biol.*, **43**:281.
――――, C. Squires, and F. Meronk (1969), *Proc. Natl. Acad. Sci.*, **62**:1100.
Epstein, C. J. (1969), *Science*, **163**:1078.
Fan, H., and S. Penman (1970), *J. Mol. Biol.*, **50**:655.
Felix, J. S., and R. De Mars (1969), *Proc. Natl. Acad. Sci.*, **62**:536.
Fell, H. B., and W. Landauer (1935), *Proc. R. Soc. (Lond.)*, **B118**:133.
Fink, G. R., and R. G. Martin (1967), *J. Mol. Biol.*, **30**:97.
―――― and J. R. Roth (1968), *J. Mol. Biol.*, **33**:547.
Frenster, J. H. (1965), *Nature*, **206**:680.
Fudenberg, H. H., and K. Hirschhorn (1964), *Science*, **145**:611.
Fujimoto, W. Y., et al. (1968), *Lancet*, **2**:511.
Fuller, J. L., and W. R. Thompson (1960), "Behavior Genetics," Wiley, New York.
Gabrusewycz-Garcia, N. (1971), *Chromosoma*, **33**:421.
Gall, J. G. (1968), *Proc. Natl. Acad. Sci.*, **60**:553.
―――― and H. G. Callen (1962), *Proc. Natl. Acad. Sci.*, **48**:562.
Galton, M., and S. F. Holt (1964), *Cytogenetics*, **3**:97.
Garen, A., and S. Garen (1963), *J. Mol. Biol.*, **6**:433.
―――― and N. Otsuji (1964), *J. Mol. Biol.*, **8**:841.
Gartler, S. M., et al. (1971), *Science*, **172**:572.
Gerhart, J. C., and A. B. Pardee (1963), *Cold Spring Harbor Symp. Quant. Biol.*, **28**:491.
Giannelli, F., et al. (1969), *Heredity*, **24**:175.
Gilbert, W., and B. Müller-Hill (1966), *Proc. Natl. Acad. Sci.*, **56**:1891.
―――― and ―――― (1967), *Proc. Natl. Acad. Sci.*, **58**:2415.
Giles, N. H., et al. (1967a), *Proc. Natl. Acad. Sci.*, **58**:1453.
―――― et al. (1967b), *Proc. Natl. Acad. Sci.*, **58**:1930.
Gluecksohn-Schoenheimer, S. (1943), *Genetics*, **28**:341.
Gluecksohn-Waelsch, S. (1951), *Adv. Genet.*, **4**:1.
―――― and T. R. Rota (1963), *Dev. Biol.*, **7**:432.
Greenaway, P. J., and K. Murray (1971), *Nat. New Biol.*, **229**:223.
Gross, P. R. (1967), *Curr. Top. Dev. Biol.*, **2**:1.
―――― (1968), *Annu. Rev. Biochem.*, **37**:631.
Grumbach, M. M., P. A. Marks, and A. Morishima (1962), *Lancet*, **1**:1330.
Grüneberg, H. (1948), *Symp. Soc. Exp. Biol. Camb.*, **2**:155.
―――― (1952), "The Genetics of the Mouse," 2d ed., M. Nyhoff, The Hague.
Guha, A., Y. Saturen, and W. Szybalski (1971), *J. Mol. Biol.*, **56**:53.
Gurdon, J. B. (1963), *Qt. Rev. Biol.*, **38**:54.
―――― (1968), *Sci. Am.*, **219**(December):24.
Gustafson, T. (1971), *Am. Sci.*, **59**:452.
Hadorn, E. (1951), *Adv. Genet.*, **4**:53.
―――― (1961), "Developmental Genetics and Lethal Factors," Wiley, New York.
Hamburger, V. (1941), *Physiol. Zool.*, **14**:355.
Hamerton, J. L., et al. (1969), *Nature*, **222**:1277.
―――― et al. (1971), *Nature*, **232**:312.
Harris, H. (1970), "Cell Fusion," Harvard University Press, Cambridge, Mass.
――――, D. A. Hopkinson, and N. Spencer (1963), *Ann. Hum. Genet.*, **27**:59.

Hiraga, S. (1969), *J. Mol. Biol.*, **39**:159.

Huang, R. C., and J. Bonner (1962), *Proc. Natl. Acad. Sci.*, **48**:1216.

Humphrey, R. R. (1960), *Dev. Biol.*, **2**:105.

Hutton, J. J. (1971), *Biochem. Genet.*, **5**:315.

Ikeda, K., and W. D. Kaplan (1970), *Proc. Natl. Acad. Sci.*, **67**:1480.

Imamoto, F., and C. Yanofsky (1967a), *J. Mol. Biol.*, **28**:1.

―― and ―― (1967b), *J. Mol. Biol.*, **28**:25.

Ippen, K., J. H. Miller, J. Scaife, and J. Beckwith (1968), *Nature*, **217**:825.

Izawa, M., V. G. Allfrey, and A. E. Mirsky (1963a), *Proc. Natl. Acad. Sci.*, **49**:544.

――, ――, and ―― (1963b), *Proc. Natl. Acad. Sci.*, **49**:811.

Jacob, F., and J. Monod (1959), *C. R. Acad. Sci.*, **249**:1282.

―― and ―― (1961a), *J. Mol. Biol.*, **3**:318.

―― and ―― (1961b), *Cold Spring Harbor Symp. Quant. Biol.*, **26**:193.

――, D. Perrin, C. Sanchez, and J. Monod (1960), *C. R. Acad. Sci.*, **250**:1727.

――, A. Ullman, and J. Monod (1964), *C. R. Acad. Sci.*, **258**:3125.

Judd, B. H. (1955), *Genetics*, **40**:739.

Kazazian, H. H., W. J. Young, and B. Childs (1965), *Science*, **150**:1601.

Kedes, L. H., and M. L. Birnstiel (1971), *Nat. New Biol.*, **230**:165.

Kiho, Y., and A. Rich (1965), *Proc. Natl. Acad. Sci.*, **54**:1751.

Klose, J., and U. Wolf (1970), *Biochem. Genet.*, **4**:87.

Korge, G. (1970), *Nature*, **225**:386.

Krimm, S. (1960), *J. Mol. Biol.*, **2**:247.

Landauer, W. (1942a), *Am. Nat.*, **76**:1.

―― (1942b), *Biol. Symp.*, **6**:127.

Lewis, E. B. (1950), *Adv. Genet.*, **3**:73.

Lieberman, M. M., and A. Markovitz (1970), *J. Bacteriol.*, **101**:965.

Lifschytz, E., and D. L. Lindsley (1972), *Proc. Natl. Acad. Sci.*, **69**:182.

Littau, V. C., et al. (1964), *Proc. Natl. Acad. Sci.*, **52**:93.

Lodish, H. (1968), *J. Mol. Biol.*, **32**:681.

Loper, J. C., et al. (1964), *Brookhaven Symp. Biol.*, **17**:15.

Lynch, C. J. (1919), *Genetics*, **4**:501.

Lyon, M. F. (1961), *Nature*, **190**:372.

―― (1962), *Am. J. Hum. Genet.*, **14**:135.

―― (1966), *Genet. Res.*, **8**:197.

―― (1970), *Sci. Prog.*, **58**:117.

―― (1971), *Nat. New Biol.*, **232**:229.

Markert, C. L., and F. Møller (1959), *Proc. Natl. Acad. Sci.*, **45**:753.

Martin, R. G. (1969), *Annu. Rev. Genet.*, **3**:181.

―― et al. (1966a), *Cold Spring Harbor Symp. Quant. Biol.*, **31**:215.

――, D. F. Silbert, D. W. E. Smith, and H. J. Whitfield, Jr. (1966b), *J. Mol. Biol.*, **21**:357.

Marushige, K., and J. Bonner (1966), *J. Mol. Biol.*, **15**:160.

―― and H. Ozaki (1967), *Dev. Biol.*, **16**:474.

Matsushiro, A., K. Sato, J. Ito, S. Kida, and F. Imamoto (1965), *J. Mol. Biol.*, **11**:54.

McKusick, V. A. (1964), "On the X-Chromosome of Man," American Institute of Biological Sciences, Washington.

Miller, G., L. Berlowitz, and W. Regelson (1971), *Chromosoma*, **32**:251.

Miller, L., and D. D. Brown (1969), *Chromosoma*, **28**:430.

Miller, O. L., Jr., and B. R. Beatty (1969a), *Science*, **164**:955.

―― and ―― (1969b), *J. Cell. Physiol.*, **74**(suppl. 1):225.

―― et al. (1970), *Cold Spring Harbor Symp. Quant. Biol.*, **35**:505.

Monod, J., J. P. Changeaux, and F. Jacob (1963), *J. Mol. Biol.*, **6**:306.

Morse, D. E. (1971), *J. Mol. Biol.*, **55**:113.

――, R. D. Mosteller, and C. Yanofsky (1969), *Cold Spring Harbor Symp. Quant. Biol.*, **34**:729.

Morse, D. E., and C. Yanofsky (1969a), *J. Mol. Biol.*, **41**:317.
—— and —— (1969b), *Nature*, **224**:329.
Mukherjee, A. S., and W. Beermann (1965), *Nature*, **207**:785.
Mukherjee, B. B., and R. G. Milet (1972), *Proc. Natl. Acad. Sci.*, **69**:37.
—— and A. K. Sinha (1964), *Proc. Natl. Acad. Sci.*, **51**:252.
Muller, H. J. (1930), *J. Genet.*, **22**:299.
—— and W. D. Kaplan (1966), *Genet. Res.*, **8**:41.
Newton, W. A., J. R. Beckwith, D. Zipser, and S. Brenner (1965), *J. Mol. Biol.*, **14**:290.
Nuclear Physiology and Differentiation (1969), *Genetics*, **61**(suppl.): 1.
Ohno, S. (1967), "Sex Chromosomes and Sex-linked Genes," Springer-Verlag, New York.
—— (1971), *Nature*, **234**:134.
—— and B. M. Cattanach (1962), *Cytogenetics (Basel)*, **1**:129.
——, R. Dofuku, and U. Tettenborn (1971), *Clin. Genet.*, **2**:1.
—— and M. F. Lyon (1965), *Chromosoma*, **16**:90.
—— and M. E. Lyon (1970), *Clin. Genet.*, **1**:121.
—— and S. Makino (1961), *Lancet*, **1**:78.
——, C. Stenius, L. C. Christian, and C. Haris (1968), *Biochem. Genet.*, **2**:197.
Pearson, C. M., W. M. Fowler, and S. W. Wright (1963), *Proc. Natl. Acad. Sci.*, **50**:24.
Pelling, C. (1959), *Nature*, **184**:655.
Pogo, B. G. T., V. G. Allfrey, and A. E. Mirsky (1966), *Proc. Natl. Acad. Sci.*, **55**:805.
Ptashne, M. (1967a), *Proc. Natl. Acad. Sci.*, **57**:306.
—— (1967b), *Nature*, **214**:232.
—— and W. Gilbert (1970), *Sci. Am.*, **222**(June):36.
Richardson, B. J., A. B. Czuppon, and G. B. Sharman (1971), *Nat. New Biol.*, **230**:154.
Rines, H. W., M. E. Case, and N. H. Giles (1969), *Genetics*, **61**:789.
Roberts, E. (1929), *J. Am. Med. Assoc.*, **93**:277.
Roth, J. R., and B. N. Ames (1966), *J. Mol. Biol.*, **22**:325.
—— et al. (1966), *Cold Spring Harbor Symp. Quant. Biol.*, **31**:383.
Roxas, H. A. (1926), *J. Exp. Zool.*, **46**:63.
Ruddle, F. H., et al. (1971), *Nat. New Biol.*, **232**:69.
Russell, L. B. (1963), *Science*, **140**:976.
—— and J. W. Bangham (1961), *Genetics*, **46**:509.
—— and C. S. Montgomery (1969), *Genetics*, **63**:103.
Salzmann, J., R. De Mars, and P. Benke (1968), *Proc. Natl. Acad. Sci.*, **60**:545.
Scott, J. P., and J. L. Fuller (1965), "Genetics and the Social Behavior of the Dog," University of Chicago Press, Chicago.
Sharman, G. B. (1971), *Nature*, **230**:231.
Sheppard, D. E., and E. Englesberg (1967), *J. Mol. Biol.*, **25**:443.
Sinnott, E. W. (1927), *Am. Nat.*, **61**:333.
Smith, P. E., and E. C. MacDowell (1930), *Anat. Rec.*, **46**:249.
—— and —— (1931), *Anat. Rec.*, **50**:85.
Spemann, H., and O. Schotte (1932), *Naturwiss.*, **20**:463.
Steele, M. W., W. J. Young, and B. Childs (1968), *Biochem. Genet.*, **2**:159.
——, ——, and —— (1969), *Biochem. Genet.*, **3**:359.
Steinberg, R. A., and M. Ptashne (1971), *Nat. New Biol.*, **230**:76.
Stern, C. (1954), *Am. Sci.*, **42**:213.
—— (1960), *Genet. Cytol.*, **2**:105.
Steward, F. C., M. O. Mapes, A. E. Kent, and R. D. Holsten (1964), *Science*, **143**:20.
Tobler, J., J. T. Bowman, and J. R. Simmons (1971), *Biochem. Genet.*, **5**:111.
Tompkins, G. M., et al. (1969), *Science*, **166**:1474.
——, and D. W. Martin, Jr. (1970), *Annu. Rev. Genet.*, **4**:91.
Tompkins, R. (1970), *Dev. Biol.*, **22**:59.
Umbarger, H. E. (1956), *Science*, **123**:848.

Umbarger, H. E. (1962), *Cold Spring Harbor Symp. Quant. Biol.*, **26**:301.

Ursprung, H. (1963), *Am. Zool.*, **3**:71.

Vandenberg, S. G., V. A. McKusick, and A. B. McKusick (1962), *Nature*, **194**:505.

Vogel, H. J. (1957*a*), *Proc. Natl. Acad. Sci.*, **43**:491.

—— (1957*b*), in W. D. McElroy and B. Glass (eds.), "The Chemical Bases of Heredity,"
pp. 276–289, Johns Hopkins, Baltimore.

Waddington, C. H. (1962), "New Patterns in Genetics and Development," Columbia University
Press, New York.

Wallace, H., J. Morray, and W. H. R. Langridge (1971), *Nat. New Biol.*, **230**:20.

Webster, R. E., and N. D. Zinder (1969), *J. Mol. Biol.*, **42**:425.

Westerveld, A., et al. (1971), *Nat. New Biol.*, **234**:20.

Yanofsky, C., and J. Ito (1966), *J. Mol. Biol.*, **21**:313.

—— and —— (1967), *J. Mol. Biol.*, **24**:143.

Young, W. J. (1968), *J. Hered.*, **57**:58.

Zuckerkandl, E. (1964), *J. Mol. Biol.*, **8**:128.

31
Inbreeding, Outbreeding, and Heterosis

31
Inbreeding, Outbreeding, and Heterosis

QUESTIONS

1588 Wright (1921*a,b,c*) classified mating systems into five basic types:

1 Random
2 Genetic assortative (inbreeding)
3 Phenotypic assortative
4 Genetic disassortative (outbreeding)
5 Phenotypic disassortative

Explain what type of mating occurs in each of these systems and which system would have the greatest effect per generation in changing the genetic and phenotypic composition of a cross-fertilizing population of an animal or plant species.

1589 **a** Some people believe that inbreeding per se favors an increase in the frequency of recessive alleles in a population. Demonstrate that this is incorrect and that inbreeding affects only the distribution of the alleles among the genotypes.

b Describe the conditions necessary for inbreeding to cause harmful effects, in man and other animals, and show why without these conditions it cannot have harmful effects.

1590 In 1862 Charles Darwin stated that nature "abhors perpetual self-fertilization." Discuss this statement, citing present-day evidence for and against it.

1591 **a** (1) Describe what is meant by *degree of inbreeding*.

(2) Describe, with the help of an illustration, the genetic effects of inbreeding and how they are related to the degree of inbreeding and to the viability of offspring. Are the genetic effects the same in both cross- and self-fertilizing organisms?

(3) State which kind of organism (cross-fertilizing or self-fertilizing) is expected to suffer the harmful effects of inbreeding more severely and why.

(4) State what the ultimate effect of inbreeding in a population will be and show what effects would accrue by it to a habitually outbreeding population.

b (1) Describe what is meant by outbreeding.

(2) Describe the genetic effects of outbreeding and their relation to the viability of offspring.

(3) State what the ultimate effect of continued outbreeding in a population will be and show what effects an occasional round of inbreeding would have on such a population.

1592 The results of self- and cross-fertilization in many cross-fertilizing organisms are basically the same. Shull's (1910) conclusions in corn, listed below, are representative of results obtained in cross-fertilizing animals and plants.

a The progenies of self-fertilized plants are of inferior size, vigor, and productiveness compared with those of crossbred plants derived from the same source.   This is true whether the parents are above average or below it.

b The decrease in size and vigor which accompanies self-fertilization is greatest in the first generation and becomes less so in each succeeding generation until there is no further loss of vigor.

c Self-fertilized families of common origin differ from one another in definite hereditary morphological characters.

d A cross between sibs within a self-fertilized family shows little or no improvement over self-fertilization in the same family.

e The progeny of a cross between plants of self-fertilized families has the same vigor, size, and productiveness as families which have never been self-fertilized.

f Reciprocal crosses between distinct self-fertilized families are equivalent, possessing the characters of the original corn.

g Hybrids between certain families isolated by self-fertilization show yields superior to yields obtained from the original crossbred stock.

h The yield and the quality of the crop produced are functions of the particular combination of self-fertilized parental types, and these qualities remain the same whenever the cross is repeated.

i The F_1 hybrids are no more variable than the pure strains which enter into them.

j The F_2 shows much greater variation than the F_1.

k The yield per acre of the F_2 is less than that of the F_1.

Explain each of the observed facts.   See Wright (1922).

1593 **a** What effect does inbreeding have on (1) allele frequency and (2) heterozygosity?

b Self-fertilization results in a reduction in vigor in one species of plants but not in another.   What conclusion could you draw regarding the form of breeding natural to the two species?

c Under what circumstances would inbreeding not have deleterious consequences?

d Discuss the relatively different roles of inbreeding and outbreeding in the development of a well-adapted race of a cross-fertilizing species.

e Why is selection within a pure line futile?

f Can a population simultaneously mate at random with respect to one pair of alleles and assortatively with respect to others?   Explain.

g How is heterosis related to effects of inbreeding?

PROBLEMS

1594   Show, by means of diagrams of inheritance, which of the following systems of mating would be most efficient in achieving homozygosis:

a Random mating   **b** Brother-sister mating

c Self-fertilization   **d** First-cousin mating

1595 **a** A plant is heterozygous at two loci. How many of these will be heterozygous after (1) one, (2) two, (3) eight, (4) one hundred generations of self-fertilization?

b Two parental lines in a plant species differ by four allele pairs. They are crossed, and the offspring are allowed to self-pollinate for four generations (giving rise to an F_5). What percentage of the F_5 may be expected to be homozygous at all four loci?

1596 Demonstrate that the following rates of reduction in heterozygosity per generation are correct:

a One-half for self-fertilization.

b One-fourth for brother-sister matings.

c One-eighth for half-brother–half-sister matings.

d One-sixteenth for cousin matings.

1597 A single *tall*, *Tt*, pea plant and its progeny in subsequent generations are self-fertilized. How many generations would it take to attain approximately 94 percent homozygosity?

1598 Three corn plants are selected, one heterozygous for *starchy* endosperm, *Wxwx*, and two homozygous for *waxy* endosperm, *wxwx*. If all plants produce the same number of progeny, what percentage of the population will be *Wxwx* after four generations of self-fertilization?

1599 **a** Demonstrate that offspring from a first-cousin marriage have 1 chance in 64 of being homozygous for an autosomal recessive allele for which one of the common great-grandparents is heterozygous.

b The incidence of *galactosemia*, determined by a recessive autosomal allele, *g*, is about 1 in 40,000 in the United States.

(1) What is the expected incidence of the disease among the offspring of first-cousin marriages in the United States?

(2) If a heterozygous man marries his first cousin, what is the chance that their first child will be *galactosemic*?

(3) If the man marries an unrelated woman, what is the chance that their first child will be *affected*? How many times more likely is the conception of a *gg* child if the man marries a first cousin than if he marries an unrelated woman?

1600 The frequency of the recessive allele *p* causing *phenylketonuria* is about 1 in 200 in most North American populations.

a What is the risk of a child having the disease when the parents are unrelated and mating at random?

b How much is this risk enhanced if the parents are first cousins?

c If a *normal* man marries his niece, determine the probability of their first child being *affected* (assume that his sister's husband is homozygous *normal*). Would the chance of producing an *affected* child be greater or smaller if the man married his first cousin?

1601 Sjögren (1931) showed that 1 percent of all marriages in Sweden are between first cousins. This 1 percent was responsible for 15 percent of all children with *juvenile amaurotic idiocy* (autosomal recessive). The frequency of this disease in the population as a whole was 1 in 30,000. What is the frequency of the allele responsible for this trait?

1602 In the pedigree shown, in which the recessive genes for *Duchenne muscular dystrophy* (X-linked) and *infantile amaurotic idiocy* (autosomal) are segregating, a marriage occurs between first cousins, as indicated by the double line.

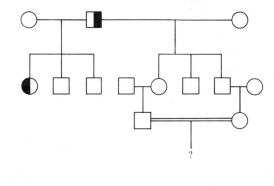

Muscular dystrophy   Infantile amaurotic idiocy   Unaffected

Figure 1602

What is the probability that the offspring of the first-cousin marriage will be *affected*:
a With *infantile amaurotic idiocy*?
b With *Duchenne muscular dystrophy*?
c With both conditions?
d With neither?

1603 Two first cousins are planning to marry and raise a family. A common uncle died in childhood of autosomal recessive *juvenile amaurotic idiocy*. They therefore wish a realistic estimate of the chance that any child they produce will be *affected*, so that they can decide whether to have their own children or to adopt.
a What estimate should you give?
b If they had had no knowledge of any occurrence of *juvenile amaurotic idiocy* in their pedigree, what would your estimate have been if the frequency of heterozygous carriers of the allele is 0.003?
c How would this compare with estimates for unrelated couples?

1604 **a** Two grandchildren of a woman with autosomal recessive *albinism* marry.
(1) What is the probability that both husband and wife carry the allele for *albinism*?
(2) If both are carriers, what is the probability that a successful pregnancy will give an *affected* child?

(3) Answer the above questions for the sex-linked recessive form.

b A marriage involves first cousins. What is the probability that a child will be affected with *alcaptonuria* (rare recessive autosomal trait, incidence 1 in 40,000):

(1) If the common grandmother is *affected*?

(2) If the wife's father is *affected*?

(3) If none of the relatives are *affected*?

1605 Two clones of a wild plant species are subjected to continuous yearly reproduction for 10 years as follows:

1 One clone is vegetatively reproduced annually.

2 The other is reproduced annually by self-fertilization.

The numbers of offspring produced per generation are, on the average, the same in the two groups. State which of these two resulting lines is more likely to carry a fairly large number of deleterious recessive genes and why.

1606 Morton et al. (1956) discuss methods by which data from first-cousin marriages can be used to estimate the number of recessive lethals causing death before reproduction (adulthood) carried by phenotypically *normal* individuals in a human population. Basically, the method derives from the observation that for any heterozygous individual (X), the probability of any grandchild (Y) by a cousin marriage being homozygous for a given gene is $\frac{1}{16}$.

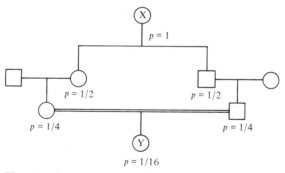

Figure 1606

For *n* such genes (carried by X) the probability of death of such a grandchild would then be $\frac{1}{16} + \frac{1}{16} + \cdots$ to *n* terms $= n/16$. Sutter (1958) obtained the tabulated results from church records in rural France. Estimate the average

Table 1606   Deaths before adulthood (including stillbirths)

Parents first cousins	0.25
Parents unrelated	0.12
Excess in deaths from first-cousin marriages	0.13

number of recessive lethal equivalents per person (a recessive lethal equivalent is either one gene that causes death of homozygotes before reproduction or two genes each causing death of half the homozygotes before adulthood).

1607 The frequency of autosomal alleles *K* and *k* in a large randomly mating population of snails is 0.6 and 0.4.

  a What will be the frequency of *KK*, *Kk*, *kk* genotypes after:
  (1) One generation of self-fertilization?
  (2) Two generations of self-fertilization?
  (3) One generation of brother-sister mating?
  (4) One generation of first-cousin mating?
  b What genotypic frequencies would you expect if the inbreeding coefficient was 0.3?
  c If the snail population also carries the allelic genes *M* and *m* in frequencies 0.99*M* and 0.01*m*, what frequencies of *kk* and *mm* individuals are expected if only first cousins mate?
  d Compare these frequencies with those expected upon random mating.

1608 Two different plant species are brought under cultivation. One is self-pollinating, the other cross-pollinating. Which one is more likely to respond to selection and why?

1609 When animals from natural populations are inbred, various deleterious recessive traits appear in later generations. Moreover, the frequency of death in utero and from birth to reproductive age increases sharply.

  a Why have these alleles not been eliminated from the natural populations by natural selection?
  b Why do the traits show up more frequently upon inbreeding than in natural populations?

1610 Would a newly arisen viable mutation have a greater chance of establishment in a self-fertilizing or a cross-fertilizing organism?

1611 A species of organism which is normally cross-fertilizing and in which mating has been at random for certain loci is suddenly subjected to enforced self-fertilization of all individuals continuously for several generations. What effects will this have, if any, on allelic, genotypic, and phenotypic frequencies for any one of these loci?

1612 Allard (1960) has described some of the factors accounting for differences among plants in the effect of forced inbreeding and summarizes the type of effect by saying, "Inbreeding is regularly deleterious in some species (alfalfa, maize), sometimes deleterious in others (onions, sunflowers), and has little or no ill effects in still others (cucurbits and the self-pollinated species)." Describe three genetic factors that may account for these differences.

1613 Offer an explanation for each of the following statements:
  a The hybrid from crossing two distinct varieties of a species is usually more vigorous than either parent.

b Forced inbreeding usually leads to a decline in vigor in many species that are normally cross-fertilizing.

c The majority of the offspring produced by F_1 hybrids from crosses between inbred lines in cross-fertilizing species show a reduction in vigor.

d Hybrids between true-breeding varieties or species are usually no more variable in phenotype than their parents.

e Continued self-fertilization has no effect on vigor in self-fertilizing species.

1614 Using the self-fertilizing garden bean (*Phaseolus vulgaris*) Johannsen (1903, 1926) grew many plants of the Princess variety from a mixture of seeds from many plants which varied in weight from very heavy to very light.   From his population he chose 19 plants for further propagation, keeping the seed progeny from each plant separate.   From each seed progeny he selected the heaviest and the lightest seed for growing.   This practice was followed for six generations, with the following results.

1 In the first planting, the plants from heavier seeds bore seeds of greater average weight than those grown from lighter seeds.

2 Seeds from any one of the 19 plants could be shown to vary considerably in weight.

3 The results of one of his selection experiments, for his line 19, typical of those of all 19 lines, were as shown.

Table 1614

Harvest year	Average weight of seeds, cg			
	Selected parent seeds		Progeny seeds	
	Lighter	Heavier	Lighter	Heavier
1902	30	40	36	35
1903	25	42	40	41
1904	31	43	31	33
1905	27	39	38	39
1906	30	46	38	40
1907	24	47	37	37

Data from Adrian M. Srb, Ray D. Owen, and Robert S. Edgar, "General Genetics," 2d ed., table 14.10, W. H. Freeman Company, copyright © 1965; after Johannsen, "Elemente der exacten Erblichkeitslehre," Gustav Fischer, Jena, 1926.

These experiments led to the *pure-line hypothesis*.   Describe this hypothesis and show how the data support it.

1615 Define coefficient of inbreeding.   Calculate the coefficients for the progeny of the following marriages:

a Uncle and niece.

b First cousins.

    c Second cousins.

    d Half first cousins.

    e Half brother and half sister.

1616 Most pure breeds of dogs are more susceptible to diseases than mongrels. What genetic explanations can you offer for this?

1617 Why may persistent inbreeding in a number of pure lines, followed later by crosses between them, result in more vigorous individuals than are produced by crosses among a similar number of lines that have not been inbred?

1618 In the pedigree shown what are the coefficients of inbreeding for the sibships in which one or more individuals are deficient in *pyruvate kinase* (a red blood-cell autosomal recessive trait)?

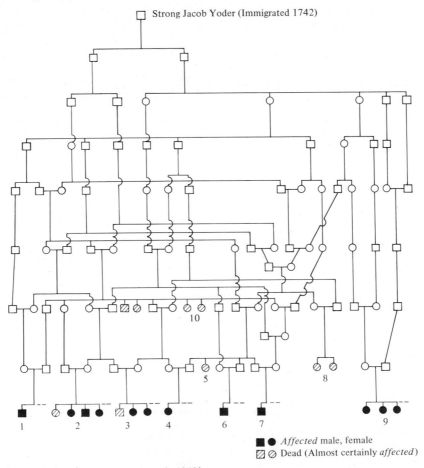

Figure 1618 (*From Bowman et al., 1965.*)

1619 Determine the coefficient of inbreeding of the bull Domino from the pedigree given. Assume that Lamplighter has an inbreeding coefficient of $\frac{1}{8}$.

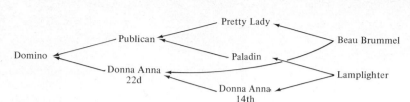

Figure 1619

1620   If 40 percent of the plants in a plot of wheat are heterozygous at a given locus, what proportion would be expected to be homozygous at this locus after two generations of self-fertilization assuming all plants produce the same number of seeds all of which grow to maturity?

1621   **a** Why is the heterotic vigor of hybrid plants lost when they are bred together for a number of generations?

    b Why may selection for vigor in inbred lines delay the attainment of homozygosity?

1622   **a** Briefly explain the dominance hypothesis (Jones, 1917) and the over-dominance hypothesis (East, 1908; Shull, 1909; Hull, 1945) formulated to explain heterosis. Which would you favor and why? In which is hetero-zygosity ascribed an incidental rather than a causal role in hybrid vigor?

    b List two ways in which hybrid vigor could be fixed in a population.

    c Thus far it has been impossible to isolate true-breeding lines of corn showing the high degree of vigor and uniformity found in hybrid corn. Suggest two of the possible reasons for this failure.

1623   Davenport (1908), Bruce (1910), and Keeble and Pellew (1910) proposed the dominance hypothesis to explain the depression in vigor on inbreeding and the heterosis of F_1 resulting from crossing inbred lines. Many investigators have shown that the F_2's from F_1's showing hybrid vigor are symmetrically distributed for heterotic characters. Moreover, it has not been possible thus far to obtain lines breeding true for F_1-hybrid vigor.

    a With these facts as a basis, what are the two main objections to this hypothesis?

    b How did Jones (1917) reconcile these apparent discrepancies?

    c See Collins (1921) for his contribution clarifying the explanation.

1624   With four inbred lines, six single crosses and three double crosses are possible. Anderson (1938) crossed the four inbred lines of corn designated 23, 24, 26, and 27 in all possible combinations and obtained the following single-cross yields (in bushels per acre):

23×24 41.7
23×26 62.6
23×27 70.8
24×26 65.6
24×27 72.1
26×27 64.2

Which of the double crosses, (23 × 24) × (26 × 27), (23 × 26) × (24 × 27), or (23 × 27) × (24 × 26), would show the highest yield and why?

a In corn, the best method of predicting the yield of the double-cross hybrid (A × B) × (C × D) is to average the yields of the four single crosses A × C, A × D, B × C, and B × D (Jenkins, 1934; Anderson, 1938).   Why?

b If A and B are inbred lines from one variety and C and D are single lines from another variety, which two of the six possible single crosses would you mate to obtain maximum heterosis and why?

See Eckhardt and Bryan (1940) for data and discussion of results bearing on the question.

REFERENCES

Allard, R. W. (1960), "Principles of Plant Breeding," Wiley, New York.
Anderson, D. C. (1938), *J. Am. Soc. Agron.*, **30**:209.
Bowman, H. S., V. A. McKusick, and K. R. Dronamraju (1965), *Am. J. Hum. Genet.*, **17**:1.
Bruce, A. B. (1910), *Science*, **32**:627.
Chai, C. K. (1970), *J. Hered.*, **61**:3.
Collins, G. N. (1921), *Am. Nat.*, **55**:116.
Crow, J. F. (1952), in C. W. Gowen (ed.), "Heterosis," pp. 282–297, Iowa State College Press, Ames.
Davenport, C. B. (1908), *Science*, **28**:454.
Dobzhansky, T. (1955), "Evolution, Genetics and Man," Wiley, New York.
Dronamraju, K. R. (1964), *Cold Spring Harbor Symp. Quant. Biol.*, **29**:81.
East, E. M. (1908), *Rept. Conn. Agric. Exp. Stn. for 1907*, pp. 419–428.
———— and H. K. Hayes (1912), *USDA Bur. Plant Ind. Bull.* 243, p. 1.
———— and D. F. Jones (1919), "Inbreeding and Outbreeding," Lippincott, Philadelphia.
Eckhardt, R. C., and A. A. Bryan (1940), *J. Am. Soc. Agron.*, **32**:347.
Ellerstrom, S., and A. Hagberg (1967), *Hereditas*, **57**:319.
Falconer, D. S. (1960), "An Introduction to Quantitative Genetics," Oliver & Boyd, Edinburgh.
Falk, C. T. (1971), *Heredity*, **27**:125.
Felsenstein, J. (1971), *Genetics*, **68**:581.
Fisher, R. A. (1965), "The Theory of Inbreeding," 2d ed., Oliver & Boyd, Edinburgh.
Freire-Maia, N. (1964), *Cold Spring Harbor Symp. Quant. Biol.*, **29**:31.
Gajdusek, D. C. (1964), *Cold Spring Harbor Symp. Quant. Biol.*, **29**:121.
Gowen, J. W. (ed.) (1952), "Heterosis," Iowa State College Press, Ames.
Hayes, H. K., F. H. Immer, and D. C. Smith (1955), "Methods of Plant Breeding," McGraw-Hill, New York.
Hull, F. H. (1945), *J. Am. Soc. Agron.*, **37**:134.
Jenkins, M. T. (1934), *J. Am. Soc. Agron.*, **26**:199.
Johannsen, W. (1903), "Elemente der exacten Erblichkeitslehre," Gustav Fischer, Jena; reprinted in J. A. Peters (ed.), "Classic Papers in Genetics," pp. 20–26, Prentice-Hall, Englewood Cliffs, N.J., 1959.
Jones, D. F. (1917), *Genetics*, **2**:466.
———— (1924), *Genetics*, **9**:405.
Keeble, J., and C. Pellew (1910), *J. Genet.*, **1**:47.
Krieger, H., N. Freire-Maia, and J. B. C. Azevedo (1971), *Am. J. Hum. Genet.*, **23**:8.
Lerner, I. M. (1958), "The Genetical Basis of Selection," Wiley, New York.
Li, C. C. (1955), "Population Genetics," University of Chicago Press, Chicago.
McDaniel, R. G., and I. V. Sarkissian (1968), *Genetics*, **59**:465.

Maruyama, T. (1971), *Genetics*, **67**:437.

Morton, N. E. (1958), *Am. J. Hum. Genet.*, **10**:344.

—— (1961), *Prog. Med. Genet.*, **1**:261.

——, J. F. Crow, and H. J. Muller (1956), *Proc. Natl. Acad. Sci.*, **42**:855.

Mukai, T., I. Yoshikawa, and K. Sano (1966), *Genetics*, **53**:513.

Richey, F. D. (1946), *J. Am. Soc. Agron.*, **38**:833.

Sarkissian, I. V., and H. K. Srivastava (1969), *Proc. Natl. Acad. Sci.*, **63**:302.

Schull, W. J. (1959), *Eugen. Q.*, **6**:102.

—— et al. (1970), *Am. J. Hum. Genet.*, **22**:239.

—— and J. V. Neel (1964), "The Effect of Consanguinity on Japanese Children," Harper & Row, New York.

Shull, G. H. (1909), *Rep. Am. Breeders' Assoc.*, **4**:296.

—— (1910), *Am. Breeders' Mag.*, **1**:98.

Sjögren, T. (1931), *Hereditas*, **14**:197.

Spofford, J. B. (1969), *Am. Nat.*, **103**:407.

Sutter, J. (1958), *Biol. Med.*, **47**:563.

—— and J. M. Goux (1964), *Cold Spring Harbor Symp. Quant. Biol.*, **29**:41.

Wright, S. (1921a), *Genetics*, **6**:111.

—— (1921b), *Genetics*, **6**:144.

—— (1921c), *Genetics*, **6**:167.

—— (1922), *Am. Nat.*, **56**:330.

—— (1923), *J. Hered.*, **14**:339.

Yamaguchi, M., et al. (1970), *Am. J. Hum. Genet.*, **32**:145.

Zirkle, C. (1952), in Gowen (1952), pp. 1–13.

32
Population
Genetics

QUESTIONS

1625 **a** What is Hardy-Weinberg equilibrium?

    b What are the conditions necessary for the maintenance of this equilibrium in any population? Discuss.

    c What evidence is required before concluding that an allele pair is in Hardy-Weinberg equilibrium?

    d What kinds of sexually reproducing species are likely to show this type of equilibrium?

    e For a facultative cross-fertilizing species (i.e., one that may cross-fertilize or self-fertilize) describe the general nature of the balance between self- and cross-fertilization that would satisfy the definition of panmixis.

1626 **a** Discuss the basic differences, from the point of view of the population geneticist, that exist between controlled or laboratory populations and natural populations.

    b Clearly distinguish between allelic, genotypic, and Hardy-Weinberg equilibrium, using a specific example to illustrate your answer.

1627 **a** Show what is fundamentally wrong with the following statement: If *brown* eyes in man is due to a dominant allele at a single locus, the frequency of individuals showing the trait may be expected to increase in the population until a frequency of about 3 in 4 is obtained.

    b State whether you agree with the following statement and why, referring to concrete examples to support your answer: Traits controlled by dominant alleles are always more frequent in a population than those controlled by recessive alleles.

    c State whether you agree with the following statement: Deleterious recessives are not at equilibrium in present-day human populations. Explain.

    d The following is a statement from an article entitled Sickle Cell Anemia in *Scientific American*, August 1951: "About one in forty of those who have the *sickle-cell trait* are homozygotes and therefore possess it in the exaggerated degree which leads to *sickle-cell anemia*. Because the trait is hereditary, this number is expected to increase." Do you agree with this statement and why?

1628 In *Biston betularia*, the allele which produces the *carbonaria* form of the moth is dominant. A century ago, however, the *carbonaria* heterozygote was much lighter in color than it is now, and the allele was not dominant. How is this accounted for?

1629 Answer each of the following questions as briefly as possible.

    a Do you think the frequency of dominant lethal alleles changes from generation to generation? Explain.

    b What frequencies of p and q give the greatest proportion of heterozygotes in Hardy-Weinberg equilibrium?

    c Explain why it becomes increasingly difficult to eliminate a recessive allele from a population as its frequency becomes lower.

d What is one type of genetic change that contributes to evolutionary change in a nonadaptive fashion?

e Which would you expect to be the more vulnerable to selection, a sex-linked recessive or an autosomal recessive? Why?

f What are the conditions under which the frequency of individuals homozygous for an allele equals the rate of mutation to that allele?

g Why is the frequency of a recessive allele that causes sterility in a population at equilibrium equal to the square root of the rate of mutation to it?

h A dominant allele of one gene and a recessive allele of another have identical adaptive values. Against which type of allele will selection be more effective?

i Does a disturbance of allelic equilibrium necessarily lead to a disturbance of genotypic equilibrium? Explain.

j Does a disturbance of genotypic equilibrium necessarily lead to a change in the frequencies of alleles?

1630 State whether you agree with the following statements and why.

a The prevalence of a mutant in a population is inversely related to its selection coefficient.

b Mutant alleles which are the least detrimental to reproductive potential cause the greatest prevalence of affliction.

1631 In the genus *Drosophila* the heaviest genetic loads appear to occur in common, ecologically versatile species, whereas the lightest loads appear to be found in rare, specialized species. Offer an explanation for this.

1632 Explain why in cross-fertilizing species mutants appearing in natural populations at frequencies that remain constant from generation to generation frequently appear to be deleterious when cultured in the laboratory.

1633 **a** Certain cave-dwelling species of animals are blind. Can you account for this in terms of population genetics?

b Mutations that reduce wing size are deleterious in the bees, wasps, and flies. Under certain circumstances, however, such reductions might be advantageous to an insect. What are these circumstances?

1634 **a** Distinguish between polymorphism, transient polymorphism, and balanced polymorphism.

b What are the criteria for balanced polymorphism?

c Aside from heterozygote superiority, there are a number of other ways in which a balanced polymorphism may be maintained. Discuss briefly, referring where possible to actual studies.

d Discuss the biological significance of balanced polymorphism.

PROBLEMS

1635 **a** In sheep *white* vs. *yellow* fat is controlled by a single pair of autosomal alleles (Mohr, 1934). In a large randomly mating flock 23 percent of the sheep have *white* fat and 77 percent have *yellow* fat.

(1) Can you tell from these data whether *white* is recessive or dominant? Explain.

(2) If not, show how you would determine which allele is dominant.

(3) Explain how you would determine whether the alleles are in Hardy-Weinberg equilibrium.

b A sheep rancher in Iceland finds that the recessive allele *y* for *yellow* fat has become established in his flock of 1,024 and that about 1 out of every 256 sheep expresses the trait.

(1) The rancher wishes to know how many of the *normal* sheep carry the recessive allele. Assuming the population is randomly mating for this gene and all genotypes have the same reproductive fitness, what is this proportion?

(2) How many of the 1,020 *white* animals can be expected to be homozygous?

(3) Since only *white* individuals are selected for breeding, why is it that the recessive allele has not been completely eliminated from the population? How would you proceed to accomplish this? (Assume methods are now available for detecting heterozygotes.)

1636 In 1958 Matsunaga and Itoh reported the following *MN* blood-typing data (number observed) from the mining town of Ashibetsu in Hokkaido, Japan:

$L^M L^M$ 406
$L^M L^N$ 744
$L^N L^N$ 332

This population consisted of 741 married couples. The six types of mating and their frequencies were as follows:

$L^M L^M \times L^M L^M$ 58
$L^M L^M \times L^M L^N$ 202
$L^M L^N \times L^M L^N$ 190
$L^M L^M \times L^N L^N$ 88
$L^M L^N \times L^N L^N$ 162
$L^N L^N \times L^N L^N$ 41

a Show whether this population is at Hardy-Weinberg equilibrium.

b Derive expected frequencies for the various possible kinds of mating and state from inspection whether the data indicate that mating was at random with respect to the $L^M L^N$ locus.

1637 The human serum protein haptoglobins *haptoglobin-1*, *Hp-1*, and *haptoglobin-3*, *Hp-3*, are specified by a single pair of autosomal codominant alleles, *Hp-1* and *H p-3*, respectively (Smithies and Walker, 1955). Kamel and Hammoud (1966) tested 219 Egyptians for the presence of each haptoglobin in their blood. Of these,

9 showed only *haptoglobin-1* (presumed *Hp-1,Hp-1*)

75 showed only *haptoglobin-3* (presumed *Hp-3,Hp-3*)

135 showed both haptoglobins (presumed *Hp-1,Hp-3*)

a What are the frequencies of the two alleles in the gene pool?

b Are these results consistent with those expected of a large population mating at random?

1638 In 1956 Grubb and Laurell discovered two human gamma globulin types, $Gm(a^+)$ and $Gm(a^-)$. The phenotypic frequencies among the offspring of different kinds of matings in the 28 Swedish families they studied are shown.

Table 1638

Matings		Children	
Parents	Frequency	$Gm(a^+)$	$Gm(a^-)$
1. $Gm(a^+) \times Gm(a^+)$	9	30	3
2. $Gm(a^+) \times Gm(a^-)$	14	25	12
3. $Gm(a^-) \times Gm(a^-)$	5	0	24
Total	28	55	39

a What is the probable mode of inheritance of this pair of traits and why?

b What are the allele frequencies in this population sample? (Show your method of derivation.)

c Derive expected frequencies for the three kinds of matings and test them against the observed frequencies to show whether there is any evidence of a departure from random mating.

1639 You are permitted to take samples of certain species once a year from a large natural park. The species you sample shows variation for a known pair of alleles lacking dominance, typical samples taken being as shown.

Table 1639

Year	AA	AA^1	A^1A^1
1961	22	76	102
1962	24	72	84

What is the simplest interpretation of the data if the organism concerned is:

a A mammal?

b A bisexual (monoecious) plant?

c A unisexual (dioecious) plant?

1640 The gamma globulin of human blood serum exists in two forms, $Gm(a^+)$ and $Gm(a^-)$, specified respectively by an autosomal dominant gene $Gm(a^+)$ and its recessive allele $Gm(a^-)$. Broman et al. (1963) recorded the tabulated

Table 1640

	No. tested	Phenotype, %	
Region		$Gm(a^+)$	$Gm(a^-)$
Norrbotten county	139	55.40	44.60
Stockholm city and rural district	509	57.76	42.24
Malmöhus and Kristianstad counties	293	54.95	45.05

phenotypic frequencies in three Swedish populations. Assuming the populations were at Hardy-Weinberg equilibrium, calculate the frequency of heterozygotes in each population.

1641 *Phenylketonuria*, a lethal in early life, occurs in Caucasoids with a frequency of about 1 in 40,000 individuals (Stern, 1960).

a What is the probability that a Canadian of Caucasoid origin is heterozygous for the recessive autosomal allele causing this disease?

b What are the allele frequencies *p* and *q* for the *normal* and *phenylketonuric* alleles respectively?

1642 a Show briefly how and to what extent each of the following, operating singly, is expected to affect the distribution of a relatively rare allele in a natural population:

(1) Mutation.

(2) Selection.

(3) Genetic drift.

(4) Migration.

(5) Mating pattern.

b Assuming that (1) to (4), plus a given mating pattern, all operate in a certain population, show how a change in intensity of one of them, e.g., selection, would affect this distribution.

1643 Large populations of a certain rodent are placed on each of two rodent-free islands for a population study. Each population consists of 10 percent *colored* animals and 90 percent *albinos*, the difference is due to different alleles of one gene and is uniform for a blood type *Kk*, due to the combined effect of a pair of codominant alleles.

a The animals placed on the first island are all known to be homozygous for one or the other of the coat-type alleles.

(1) Give the expected allelic, genotypic, and phenotypic frequencies for each pair of alleles when this population reaches Hardy-Weinberg equilibrium if the traits are neutral with regard to selection.

(2) Show the expected frequencies of the six phenotypic classes representing the possible combination of phenotypes for these loci at equilibrium.

b Of the animals placed on the second island, only the phenotypes for coat type are known. If, at equilibrium, these phenotypes are found to be distributed as 31 percent *colored* and 69 percent *albino*, what were the genotypic distribution frequencies of the original populations?

1644 Among 3,000 Shorthorn cattle, 260 are *white*, 1,430 are *red*, and 1,310 are *roan*. Is this consistent with the assumption that the traits are controlled by a single pair of autosomal alleles and that mating has been at random for this allele pair?

1645 Calculate the proportion of heterozygotes that are the progeny of heterozygote × heterozygote matings in a panmictic population, assuming there is no selective advantage for any phenotype.

1646 In mice *short* vs. *long* hair is determined by a single pair of alleles with S for *short* dominant (Grüneberg, 1952). You find that an isolated wild population of mice contains 99 percent *short-haired* individuals. If the population is large and panmictic, and if the genotypes are binomially distributed in the same proportions in all generations, what is the expected frequency of s?

1647 Assume *blue eyes* is determined by a recessive allele of an autosomal gene. In a certain panmictic population 16 percent have *blue eyes*.
a What portion of the *brown-eyed* persons are heterozygous for the recessive allele for *blue eyes*?
b In what percentage of marriages would both parents be heterozygous?
c What is the expected frequency of *blue-eyed* children in families where both the mother and the father are *brown-eyed*?
d In a large group of families in which one parent is *brown-eyed* and the other *blue-eyed*, what is the expected frequency of *brown-eyed* children?

1648 *Sickle-cell anemia*, a severe deformation or sickling of the red blood cells associated with anemia, mental impairment, and death before the age of reproduction, is determined by an autosomal allele, Hb^S, in the homozygous condition. *Sickle-cell trait*, a very mild form of the disease difficult to detect, is the result of heterozygosity for Hb^S ($Hb^A Hb^S$) (Beet, 1949; Neel, 1949). In North American Negroes *sickle-cell trait* appears in approximately 9 out of every 100 individuals.
a Calculate the frequencies of the Hb^A and Hb^S alleles in this population.
b Which type(s) of matings can give *anemic* offspring? If marriages occur at random with respect to the allele pair, what is the probability for such matings and that for an *anemic* child?
c Compare the probability for *anemic* children from (b) with the frequency of *anemics* in the population as derived by the Hardy-Weinberg formula and explain your results.
d Derive the proportion of North American Negroes suffering from *sickle-cell anemia* 50 generations from now by means of the formula

$$q_n = \frac{1}{q_n} - \frac{1}{q_0}$$

e If *anemics* die before reproductive age, how did the allele become so common in this group?

1649 In human beings, any one of several alleles at the *Rh* locus produces the D antigen, and individuals that carry such alleles and therefore produce the antigen are classed as Rh^+. Other alleles at this locus fail to produce the D antigen. Individuals containing only the latter alleles are classed as Rh^- (Race and Sanger, 1968). Landsteiner and Wiener (1941) found that 15 percent of the people tested in New York City were Rh^-. Considering all alleles producing *D* as *R*, and those not as *r*, and assuming for each of your calculations that the above value is 16 percent, answer the following:

a If the large population is mating at random, what are the frequencies of the genotypes *RR*, *Rr*, and *rr*?

b If the sex ratio is 1:1, in what proportion of all marriages is there a probability of having Rh^+ babies?

c A man and his wife are told, when they have their first child, that their serological genotypes are *R—* and *rr* respectively and that there is hence a rather high risk of further children being *erythroblastotic*. On investigation you learn that, out of a relatively large sample of 1,000 second births to Rh^+ father–Rh^- mother matings studied in New York, the frequency of *erythroblastotic* children is 100. On the assumption that the statistics given for New York are characteristic of North American populations as a whole, what is the chance that a second child born to this couple would be *erythroblastotic*?

1650 Of a population of 235 true-breeding *brown-eyed* individuals on a South Sea island only 18 remain after a disease epidemic: 10 young girls, 2 young men, and 6 very old men. Then 8 Canadian males, of whom 4 are homozygous *blue-eyed*, 2 heterozygous *brown-eyed*, and 1 homozygous *brown-eyed*, settle on this island. Assuming that *brown* vs. *blue* eyes is controlled by a single pair of autosomal alleles with *B* (for *brown*) dominant, that intermarriage is at random with respect to eye color, and that all families average 4 children:

a What are the expected allelic frequencies in this parental population? Is it at allelic equilibrium?

b What are the expected genotypic and phenotypic frequencies after the population has reached Hardy-Weinberg equilibrium?

c Show the types of marriages that would occur in the parental generation and the types and frequencies of genotypes and phenotypes among their progeny. Show whether or not the population is at Hardy-Weinberg equilibrium the first generation after random mating.

d Would Hardy-Weinberg equilibrium be achieved in the next generation?

e Suppose that in the fifth generation a study is made of the distribution of eye colors and that only 4.5 percent of the residents are *blue-eyed*. Of the alternative causal factors—mutation, selection, genetic drift, and migration—which is most likely to have effected the allele-frequency change and why?

1651   On the basis of allele-frequency analysis of data from a randomly mating population Snyder (1934) concluded that the *ability* vs. *inability* to taste phenylthiocarbamide (PTC) is determined by a single pair of autosomal alleles, of which *T* for *taster* is dominant to *t* for *nontaster*.   Of the 3,643 Caucasians (whites) tested, 70 percent were *tasters* and 30 percent *nontasters*. Assume the population satisfies the conditions of Hardy-Weinberg equilibrium.

    a Calculate the frequencies of the alleles *T* and *t* and the frequencies of the genotypes *TT*, *Tt*, and *tt*.

    b Determine the probability of a *nontaster* child from a *taster* × *taster* mating.

    c Determine the probability of a *taster* child from *taster* × *nontaster* mating.

1652   In man, the ability to taste phenylthiocarbamide (PTC) is determined by a pair of alleles, *T* for *taster* being dominant to *t* for *nontaster*.

    a In one large randomly mating population the frequency of *nontasters* is 0.04; in another it is 0.64.   What is the frequency of heterozygotes, *Tt*, in each of these populations?

    b Snyder (1934) tested a random sample of 800 United States families for *ability* vs. *inability* to taste PTC, taken from a population in which the frequency of *t* was 0.537.   The results were as shown.

Table 1652

Parents	No. of couples	Offspring		Average family size
		Tasters	*Nontasters*	
taster × *taster*	425	929	130	2.5
taster × *nontaster*	289	483	278	2.6
nontaster × *nontaster*	86	5	218	2.6

    (1) Assuming that the numbers of offspring per family do not deviate from the listed averages, calculate the frequency of parental genotypes as closely as possible and compare them with expected frequencies to derive an answer to the question: Are the genotypes for this pair of traits binomially distributed?

    (2) Discuss possible explanations for the five *tasters* from *nontaster* × *nontaster* matings.

1653   The *MN* blood-group frequencies (in percent) in a sample of 1,279 English people are *M* (homozygous, $L^M L^M$) = 28.38, *MN* (heterozygous, $L^M L^N$) = 49.57, *N* (homozygous, $L^N L^N$) = 22.05 (Race and Sanger, 1968).

    a Calculate the frequencies of the alleles L^M and L^N and determine by the use of the Hardy-Weinberg formula whether or not this population is at genetic equilibrium.

    b You are studying an isolated community and find that the proportion of individuals with *N* blood type is approximately 4 percent.

(1) What are the allele frequencies $p(L^M)$ and $q(L^N)$?

(2) What is the probable reason for this great difference in allele frequencies between this community and the general population?

c If the first population, which consists of 2,400 people, mated at random with the isolated community whose size is twice as large, what will the expected frequencies of phenotypes in the new population be?

1654 *Baldness* vs. *nonbaldness* in man is a sex-influenced pair of traits determined by a single pair of autosomal alleles: *B* for *baldness* is dominant in males, and *b* for *nonbaldness* is dominant in females. It may be assumed that human populations are at Hardy-Weinberg equilibrium for these alleles. Data from a certain large population indicate that 1 percent of the women are *bald*.

a How many women are heterozygous?

b How many men are *bald*?

c If mating occurs at random for this allele pair, what proportion of the marriages will be between a *bald* woman and a *nonbald* man?

d A *nonbald* couple have one son. What is the probability he will become *bald*?

e The husband of a *bald* woman died very young, and therefore it is impossible to determine whether he would have been *bald* or not. What is the probability that the daughter of this couple will become *bald*?

In another population approximately 50 percent of all men are *bald*.

f How many women are *bald*?

g What is the frequency of marriages between *bald* men and *nonbald* women?

1655 In a large randomly mating population the frequencies of the I^A, I^B, and i alleles controlling A, B, and O blood groups antigens are 0.6, 0.3, and 0.1, respectively. What are the expected frequencies for the blood groups A, B, AB, and O?

1656 Calculate the following, assuming Hardy-Weinberg equilibrium:

a The proportion of *Kk*'s if *kk*'s are normal in fertility and viability and have a frequency of 1 percent.

b The frequency of *K* if the only other allele at this locus, *k*, is homozygous in 49 percent of the population.

c The frequencies of *K* and *k* if 50 percent of the people in a population are heterozygotes.

1657 **a** Alleles S_1 and S_2 occur at a single locus on the X chromosome. The proportion of the S_1 allele is 0.60 in the female half of the population and 0.40 in the male half. What are the expected proportions among males and females in the two succeeding generations?

b If *color blindness* is due to a recessive allele at a certain locus on the X chromosome and 18 women in 20,000 are *color-blind* in a particular population, what is the expected frequency of *color-blind* men in this population?

c The frequency of a certain X-linked affliction in men is 1 in 20,000 and that of heterozygous women is 1 in 9,000. If *affected* individuals of a generation are prevented from mating, what are the expected frequencies of *affected* males and of heterozygous females in the next generation?

1658 In cats, the genotypes *BB* and *B/Y* are *black*, *bb* and *b/y yellow*, and *Bb tortoiseshell*. In a sample of 281 Boston cats Todd (1964) found the phenotypic distribution shown. Determine whether the population was at Hardy-Weinberg equilibrium.

Table 1658

Sex	Black	Tortoiseshell	Yellow
♀	102	48	4
♂	99	0	28

1659 *Color blindness* in man is controlled by an X-linked recessive allele.
a In a certain population, it occurs 20 times more frequently in males than in females.
(1) What is the frequency of the allele for *color blindness*?
(2) What is the frequency of heterozygous females?
b If mating is at random, what frequency of *color blindness* would you predict among women in a population at Hardy-Weinberg equilibrium in which 9 percent of the men are *color-blind*?

1660 Why is the mutation rate for dominant lethal mutations in diploid organisms (e.g., for *retinoblastoma* in man) equal to one-half the frequency of *affected* individuals?

1661 It is difficult to arrive at a reliable estimate of mutation rates to recessive alleles in human populations. What are some of the reasons, and how might reliable estimates be obtained?

1662 The recessive allele *a* for *juvenile amaurotic idiocy*, a condition that is lethal at a very early age, has a frequency of 0.003 in North American whites (Neel and Schull, 1958). Suppose the mutation rate of *A* to *a* is doubled; what proportion of the gametes will carry the allele after a new equilibrium has been established?

1663 In a large, isolated, randomly mating population in which the genotypes *AA*, *Aa*, and *aa* have the same reproductive fitness, the ratio of forward ($A \rightarrow a$) to back mutation ($a \rightarrow A$) is 5:1. What would be the frequencies of alleles *A* and *a* when equilibrium was reached?

1664 Mørch (1941) has shown that the reproductive fitness of *chondrodystrophic dwarfs* is 20 percent that of *normals*. Of 94,075 births in a Copenhagen hospital 10 were *dwarfs*, 8 to *normal* parents, 2 to parents of whom one was a *dwarf*. From this information calculate:
a The mutation rate of A_1 (*normal*) $\rightarrow A_2$ (*dwarfs*).
b The frequency of A_1 at genotypic equilibrium.

1665 Among 30,000 births in a particular hospital, 30 children had *achondroplastic dwarfism*, a trait caused by an autosomal dominant allele. Of these *affected* children 20 had one parent *affected*; the remaining 10 were all of *normal* parents. What is the mutation rate of the recessive to the dominant allele?

1666 A certain large human population is at equilibrium for the autosomal recessive lethal allele a for *juvenile amaurotic idiocy*, which causes death before reproductive age. The mutation rate $A \rightarrow a$ is 1 in 490,000 (1 gamete in 490,000 carries a instead of A). Ignore reverse mutation, the effect of which is, for all purposes, nil.
a What is the frequency of a?
b What is the frequency of heterozygotes?
c State the proportion of the population that would die because of this allele.

1667 In a large randomly mating population at equilibrium 1 child in 90,000 is born with *cystic fibrosis*, a condition caused by a recessive autosomal lethal, causing death before sexual maturity. Derive:
a The rate of mutation, u, for C to c.
b The proportion of the gene pool that contains c.
c The proportion of heterozygotes in the population.

1668 In a human population the equilibrium frequency of *congenital total color blindness*, which is caused by a recessive autosomal allele, is 1 in 80,000. *Afflicted* individuals have extremely poor vision and may as a result have a lowered reproductive fitness. If their reproductive fitness is 0.5, what is the mutation rate necessary to maintain this frequency of the trait in the population?

1669 The frequency of individuals homozygous for a recessive autosomal allele for *sterility* affecting males and females alike is 1 in 1,500. If the rate of mutation of this allele is doubled, what will be the frequency of *steriles* when a new equilibrium is attained?

1670 If the adaptive value of a recessive allele is 1.0, the homozygotes have a frequency of 0.01, and the mutation rate of the dominant to the recessive allele is 10^{-5} at equilibrium, what is the reverse mutation rate?

1671 *Pseudohypertrophic muscular dystrophy* is a sex-linked recessive trait which occurs only in males, who cannot reproduce. Of 63,000 males born in Utah during a period of 10 years, 18 were *afflicted* (Neel and Schull, 1958). What is the mutation rate?

1672 A recessive lethal allele causes the death of 1 person in every 20 homozygous for it before the age of reproduction ($s = 0.05$). The mutation rate from the dominant to the mutant allele is 1 in 200,000. The population is at equilibrium for this allele.
a What proportion of homozygous recessives die before reproductive age?
b How much more frequent would homozygotes be if the recessive lethal allele were fully penetrant ($s = 1$)?
c What would be the frequency of the recessive allele if $s = 1$?

1673　Assuming that the gene *A* for *achondroplasia* in dogs mutates to *a* (*normal*) at the rate of 10^{-5} and that back mutation does not occur, what will be the frequency of the dominant allele after:

   a 10 generations?
   b 100 generations?
   c 1,000 generations?
   d 5,000 generations?

1674　*Kuru*, a fatal paralysis, occurs only in the Fore, a people native to New Guinea, where it is very frequent. The disease is at present responsible for the death of about half of the females and a tenth of the males in the tribe. The affliction, for which there is no known cure, leads to death a few months from the time of onset. Bennett et al. (1959) have postulated that an autosomal allele *Ku* is responsible for the disease. Homozygous, *KuKu*, individuals, both male and female, show early onset. Heterozygous females are late-onset victims, *affected* as adults; and heterozygous males and *kuku* individuals of both sexes are *unaffected*. Explain how such a highly undesirable allele can become widespread in this one area and suggest measures to reduce its frequency and prevent its spread to other parts of the world.

1675　**a** In Europe, up to about 1848, in many species of moths the vast majority of individuals were *light-colored*; e.g., in *Biston betularia* at least 99 percent of the population was estimated to be *light-colored*. With industrialization the frequency of *melanic* (dark-colored) variants in industrial areas increased until they now are the predominant forms, comprising 95 to 99 percent of many species. In most of the known cases *melanism* is caused by a single autosomal dominant allele (Kettlewell, 1961).

   (1) Propose an explanation for these changes and suggest how you might test your hypothesis experimentally.

   (2) There is some evidence that the *melanic* alleles were originally recessive. How might such changes in dominance relationship have come about? What evolutionary significance could such changes have?

   b Equal numbers of *melanic* and *light* forms of *B. betularia* were released into an unpolluted wood in Dorset, England, a relatively unindustrialized area. Five species of birds were observed to eat 190 *betularia*, of which 164 were *melanic* and only 26 were *light*. In a polluted wood near Birmingham, an industrial area, the two types were again released in a 1:1 ratio of *melanic* to *light*. Redstarts ate 15 *melanics* and 43 *light*. A series of release and recapture experiments supported the visual-predation hypothesis. For example, near Birmingham 154 *melanics* and 73 *light* were marked and released; of 98 moths recaptured 82 were *melanic* and 16 *light* (Kettlewell, 1958). Do these results support your hypothesis and why?

1676　What data are required to distinguish between allele-frequency changes due to genetic drift and those caused by natural selection? (See Dobzhansky and Pavlovsky, 1957.)

1677 Oysters reproduce at a prolific rate. About 1915 the *Malpique disease*, fatal to *affected* individuals of this species, appeared in Prince County, Prince Edward Island, which includes Malpeque Bay in the Gulf of St. Lawrence. The graph, taken from a study reported by Needler and Logie (1947), shows oyster yields in this area for the 27 years following the appearance of the disease. (The area sampled included some districts not affected by the disease.)

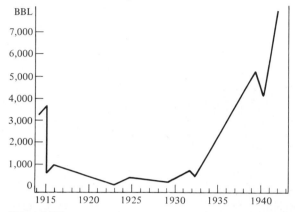

Figure 1677

Explain the rise and fall in yields of oysters from 1915 to 1942.

1678 In a given population, a certain condition is caused by a dominant allele, *B*, which has a penetrance of 0.5. Half the population are *Bb*, and the other half *bb*; only *Bb* × *bb* matings occur. In another population the condition is caused by a recessive allele, *b*, which is fully penetrant, and for which the population is at equilibrium; the frequency of *bb*'s is 0.16. Which population will have the higher frequency of *affected* individuals:
a After one generation of complete selection against the condition?
b After ten generations of such selection?

1679 It is sometimes suggested as a eugenic measure that those suffering from serious inherited defects should be prevented from reproducing, in order to reduce the frequency of the trait in future generations. Suppose that such a defect were present with a frequency of 1 in 40,000.
a How long would it take to reduce the frequency to 1 in 100,000?
b What assumption did you make in answering this question?

1680 A corn plant heterozygous for a recessive allele, *w*, which causes *albinism*, is self-fertilized. The offspring consist of 275 *green* and 85 *white* seedlings. The latter die within about 3 weeks of their appearance.
a The 275 *green* plants are permitted to cross-fertilize, and the seed is sown the next spring. The number of seeds is not determined, and the investigator, due to illness, is unable to look at the progeny until 6 weeks

after seeding, at which time he finds 8,000 *green* plants. How many *albino* seedlings may be expected to have germinated and died? What is the frequency of *w* in this generation?

b The population in the next generation consists of 5,000 plants. How many can be expected to be *green* and why?

c How many generations would it take to reduce the frequency of *w* to 1 in 20?

1681 You wish by artificial selection to reduce the frequency of a recessive trait in a large randomly mating population in which the frequency of the recessive allele is 0.5.

a Show the types and proportions of the different phenotypes in this population before complete selection ($s = 1$) against the recessive trait.

b Determine the frequency of the allele in the population after one, two, and three generations of such selection.

1682 *Galactosemia*, lethal early in life, is due to a recessive autosomal allele *g* (Holzel and Komrower, 1955). It is present with a frequency of 0.001 in a certain equilibrium population. Radiation induces a mutation from *G* to *g* in a gamete that effects fertilization. If the population is mating at random, what is the chance of a *galactosemic* child being produced as a result of this event? How many generations on the average would it take for the newly arisen allele for *galactosemia* to combine with another to produce a *galactosemic*?

1683 *Phenylketonuria*, a condition that is normally lethal in early childhood, is due to a recessive autosomal allele *p*. Assume the mutation for *P* to *p* has been found to be 1 in 360,000. Due to recent advances in medicine, however, it has been possible for several years to alleviate the condition and permit the alleviated *pp* to live normally. Assume that 90 percent of all potential *phenylketonurics* are now "cured" and that they have the same reproductive fitness as *normal* individuals.

a What will be the ratio for *phenylketonuric* deaths before vs. those after the change?

b What was the initial frequency of *p*? What would be its frequency among the generation of children born after the change?

1684 Chung et al. (1959) found that *deaf-mutism* as it occurs in Northern Ireland is in some families an autosomal dominant trait and in other families a non-allelic autosomal recessive. The dominants are caused by mutation arising with a frequency of 5×10^{-5} and the recessives with a frequency of 3×10^{-5} per gamete. Assume that reproductive fitness of dominant mutant homo-zygotes and heterozygotes is 25 percent that of *normal* homozygotes and the same is true for recessive mutant homozygotes in relation to *normal* homo-zygotes and heterozygotes. What is the expected frequency of each of the two mutant alleles when equilibrium is attained?

1685 **a** Assuming (1) that *feeblemindedness* is due to a single autosomal recessive allele, (2) that it occurs with an average frequency of 0.0049, (3) that the rate of mutation is negligible, and (4) that all genotypes have equal reproductive fitness:

(1) Calculate the effect of six generations of sterilization of all *feebleminded* individuals on (*a*) the frequency of the allele and (*b*) the incidence of the trait.

(2) How many generations of sterilization are necessary to reduce the frequency of *feeblemindedness* from 0.0049 to 0.0017?

(3) Show what effect assortative mating would have on a program designed to reduce the frequency of *feeblemindedness*.

b If the trait were a result of the interaction of recessive alleles at five loci, would sterilization cause a slower or a faster rate of reduction than that you have described in (a)?

1686 What do you think would happen to the frequency of a recessive autosomal allele if its adaptive value changed from 0 to 0.5?   Explain.

1687 An autosomal and a sex-linked allele each have an *s* value (selection coefficient) of 0.5 and arise by mutation at the same rate.   Mutant phenotypes for the autosomal recessive, however, appear more frequently than those controlled by the sex-linked recessive.   Explain.

1688 In a certain population at equilibrium, $p(A) = q(a) = 0.5$, for a gene governing color differences, selection is introduced by the appearance of a new type of predator in the area which operates against the recessive phenotype.   For the *aa* genotype the *s* values are 1.0, 0.70, 0.10, and 0.01.   Calculate the corresponding values of *q* and show graphically the relation between *s* and *q*:
a For the population after one generation of selection.
b For the population when it has attained a new equilibrium.

1689 a What is meant by the terms *genetic load* and *genetic death*?
b Explain why, in a population at equilibrium, different deleterious genes, although possessing different adaptive values and therefore different selection coefficients, e.g., the genes for *juvenile amaurotic idiocy* ($s = 1$), *retinoblastoma* ($s = 0.9$), and *chondrodystrophy* ($s = 0.8$), nevertheless may be expected to produce genetic deaths equal to their respective mutation rates.
c In a population the proportion of individuals that carry a particular allele depends on both the mutation rate and on the degree of selection against the allele.   State whether or not you agree with the statement and why. Reconcile this statement with (b), illustrating your remarks by an example in which a deleterious recessive allele *a* arises and has an *s* value of 0.5.
d Compare the effects of mutation and selection on genetic loads in Mendelian populations.

1690 Since individuals with dominant autosomal *retinoblastoma* and recessive autosomal *juvenile amaurotic idiocy* almost never reach the age of reproduction, both are lethal traits.   Assume the genes occur in separate populations.

a If both defective alleles arise by mutation of the *normal* allele at the rate of 3×10^{-5} per gamete, what are the expected frequencies of each when equilibrium is reached?

b Which population would have a higher proportion of homozygous *normals*?

1691 In a population of 50,000 people, 20,000 suffer from a genetic disease caused by an autosomal recessive *a* in the homozygous condition. If these individuals are prevented from reproducing and the population size remains the same, how many *afflicted* individuals will occur in the next generation:

a If the mutation rates for $A \rightarrow a \rightarrow A$ are so low as to be negligible.

b If the mutation rate from $A \rightarrow a$ is approximately 1 in 1,200?

1692 An interesting instance of a stable but disturbed equilibrium has been brilliantly investigated by Allison (1954, 1955). *Sickle-cell anemia* may be regarded as a recessive trait since the sickling process and its attendant severe anemia are suppressed in heterozygotes. These can be distinguished from *normals* by electrophoretic detection of sickle-cell hemoglobin in their blood or by a blood test in which the red blood cells become sickle-shaped when the sample is deoxygenated. Recessive homozygotes (*sickle-cell anemics*) do not usually survive long enough to reproduce. (It is generally believed that about one-fourth of them survive to reproduce, but in view of the probability that the families of these would also be greatly reduced in size, it is assumed here that their reproductive capacity is so small that it can be considered to be zero.) It is expected that with such a strong reduction ($s = 1$) against the *sickle-cell* allele over many generations its frequency in any present-day population would be extremely low. This is true in most human populations. In some East African tribes, however, its frequency is abnormally high. This is entirely due to a selective advantage of heterozygotes over homozygous *normals*, occasioned by the superior ability of the former to recover from the severe malarial infections that occur in the region, so that the frequency of heterozygotes is actually higher than that of *normals*.

a Assuming that the frequency $p \, (Hb^S) = 0.8$ and $q \, (Hb^A) = 0.2$, what would be the expected genotypic equilibrium frequencies for $Hb^S Hb^S$, $Hb^S Hb^A$, and $Hb^A Hb^A$?

b After one generation, with complete selection against homozygotes, what would the gametic frequencies be? What would the frequency of heterozygotes be?

c To offset this reduction in frequency of heterozygotes, what would the selective advantage for heterozygotes have to be?

1693 Individuals who are homozygous for the sickling allele Hb^S are called *sickle-cell anemics*; they usually die from acute anemia (hemoglobinemia) before adolescence. Those heterozygous for the allele are not affected or show only slight anemia; they can be detected by a blood test in which the red blood cells become sickle-shaped when the sample is deoxygenated. They are resistant to malaria. Individuals homozygous for the normal allele Hb^A

Table 1693

Population	Location	Frequency Hb^SHb^S	Hb^SHb^A
1	America	0.0895	0.0005
2	Malaria-infested region of Africa	0.012	0.388

show neither anemia nor sickling; they are susceptible to malaria. The table summarizes some data arising from studies of this trait in large samples from adult Negro populations.

a What are the allele frequencies in these populations?

b Demonstrate that these populations are not in Hardy-Weinberg equilibrium.

c Describe the causes of departure for the two situations.

d For population 2, assuming that the reproductive fitness of Hb^AHb^A individuals is one-quarter that of Hb^SHb^A individuals (viz. the former have, on the average, 4 times as many children per individual as the latter) use the relationships

$$p\,(Hb^S) = \frac{S_2}{S_1 + S_2} \quad \text{and} \quad q\,(Hb^A) = \frac{S_1}{S_1 + S_2}$$

to find the value of S_1.

Note: S_1 is the coefficient of selection against Hb^AHb^A's relative to Hb^SHb^A's.

1694 Migration of genetically distinct populations may take many forms, including the following:

a A large number of individuals may move from a particular locale in one direction into an adjacent inhabited area.

b A number of small groups may emigrate into areas that are separate, isolated, and uninhabited.

c Movement may take place in all possible directions between different populations.

Contrast the genetic consequences of these three migration patterns.

1695 Each of two highly detrimental recessive genes *a* and *b* when homozygous causes death during infancy. In one large population *a* has a frequency of 0.04, and *b* is absent. In another population of the same size, *b* has a frequency of 0.03, and *a* is absent. Mating is at random with respect to these loci.

a These two populations are unified by massive intermigration and subsequently come to constitute a single large Mendelian population. Compare the frequencies of *affected* individuals in the new population with those of the older ones for each of the two diseases.

b Does miscegenation under these circumstances have a deleterious or a beneficial effect?

1696 The Old Order Amish is a religious sect.   Members of Amish communities rarely marry outside the group.   In the Amish of Lancaster County, Pennsylvania, a form of dwarfism described as *Ellis–van Creveld syndrome* occurs (short limbs, disproportionate dwarfism, polydactyly, dysplasia of the finger nails, knockknees, abnormally short upper lip).   About 5 per 1,000 births and about 2 per 1,000 among the 8,000 living members of the group are characterized by the syndrome, which is absent from other Amish communities, such as those of Ohio and Indiana.   It is probable that the ancestry of all *afflicted* persons in the Pennsylvania community trace back to a Samuel King and his wife, who immigrated to that area in 1774, although this couple are known not to have been *affected*.   All *affected* individuals are known to have had *normal* parents (McKusick et al., 1964).

   a State the most likely mode of inheritance of the syndrome and derive the frequency of the causal allele and of heterozygotes (with selection and under the assumption that the population is large and randomly mating).

   b Provide an explanation to account for the high frequency of the syndrome in the Pennsylvania Amish, together with its complete absence among the other Amish groups.

1697 In a certain localized geographic population of a certain species of beetle all parents die before their offspring reproduce.   There is sufficient food in the area for 1,000 individuals to reach maturity.   The original population is homozygous for a recessive autosomal allele causing *spotted* wings.   In each generation there is an average migration of 100 beetles homozygous for the dominant allele for *nonspotted* into this population.   (These individuals are from a nearby population in which only the *nonspotted* condition occurs.)   Show the proportions of *spotted* vs. *nonspotted* beetles (1) after one and two generations of migration and (2) after a very long period.

1698 An isolated large human population is at equilibrium with regard to the recessive allele for *albinism*, which is present with a frequency of 0.1.   Another isolated population of the same size has never produced any *albinos*.   A prolonged period of extensive intermigration occurs between these two populations.   If random mating is a continuous characteristic of both populations, what is the expected frequency of *albinos* in each of these populations afterward?

1699 Glass (1956) studied the *ABO* blood groups of a very small religious sect in Franklin County, Pennsylvania, known as the Dunkers.   The ancestors of this sect consisted of 27 families who in the early eighteenth century came from the Rhineland region of Germany, near Krefeld.   These people have remained relatively isolated, both sexually and culturally, from other people in the United States.   Comparisons of present-day *ABO* blood-group proportions are given in the table.

   a What is the most likely explanation for the shift in allele frequency in the Dunker isolate?

Table 1699

Group	No. of people	Percent of *ABO* blood groups			
		O	*A*	*B*	*AB*
Dunkers	228	35.5	59.3	3.1	2.2
Rhineland Germans	3,036	40.7	44.6	10.0	4.7
United States	30,000	45.2	39.5	11.2	4.2

b Suggest how this study might have been extended so as to confirm or contradict your answer.
See Cavalli-Sforza (1969).

1700 In a survey of *albinos* among the Indian populations of Arizona and New Mexico, Woolf (1965) noted that in the majority of these populations no *albinos* were present.   However, their frequency was high in three populations: 1 in 227 among Indians of Arizona, 1 in 140 among the Jemez Indians of New Mexico, and 1 in 247 among the Zuni of New Mexico.   All three populations are culturally but not linguistically related.   Other Indian tribes contain *albinos* at a much lower frequency (not as low, however, as that of the Caucasian population in the United States, which is 1 in 20,000).   Explanations for the high frequency of *albinos* in these Indian populations are:

1   Cultural selection in past generations
2   Gene flow from one population to another
3   Genetic drift
4   Selection for the heterozygote

a Which of these seems to be the most likely cause?
b Suggest how this survey might be extended to confirm or contradict your answer.

1701 The water snake *Natrix sipedon* exhibits four major types of banding pattern: *A, no bands*; *B, slight banding*; *C, intermediate banding*; *D, complete banding*.   All four types occur on the islands in western Lake Erie.   Almost all snakes from the mainland surrounding Lake Erie are of type *D*.   Moreover, except in the area of western Lake Erie and one area in Tennessee, all known populations of this species are of type *D*.   The distribution of the banding types in large samples of adult snakes and litters from pregnant females from the islands is illustrated.[1]   The following processes might be responsible for the significant difference in pattern-type frequencies between adult and litter populations:
a Mutation   **b** Genetic drift   **c** Migration   **d** Selection
Which of these factors seems most likely to be responsible and why?

---

[1] The percentage of relatively *unbanded* individuals is higher in adult than in young (litter) populations.   Snakes in the laboratory show no evidence of pattern changes during development.

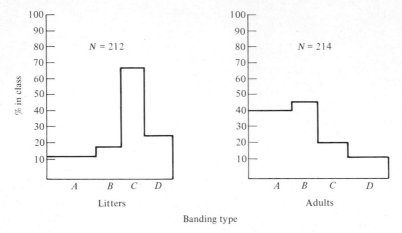

Figure 1701 Banding-type distribution. [*From J. H. Camin and P. R. Ehrlich, Natural Selection in Water Snakes (Natrix sipedon L. on Islands in Lake Eire, Evolution,* **12**:*508 (1958).*]

1702 The following represents a summary of data presented by Boyd (1950):

1 Among American Indians the tribal values for the frequency of *s* (secretor) vary from zero to 0.01; in Japanese, a nontribal people, the frequency is 0.24.

2 In American Indians the *Rh* allele is absent; in different Asiatic populations it varies in frequency from 0.6 to 0.015.

3 In the majority of American Indian tribes the *B* blood-group allele is absent; the remainder show values ranging from 0.16 to 0.024; in Asiatics the frequencies (tribal or otherwise) are never lower than 0.27.

4 Among American Indians the tribal values for the *N* blood-group allele vary from 0.01 to 0.08; in Eastern Asiatics the frequencies are 0.18 for Chinese and 0.20 for Japanese.

It is a commonly accepted belief that the American Indians are the descendants of Asiatics who crossed the Bering Strait from Siberia. Assuming that at some place in Siberia these peoples "rested" for some considerable period, show how these observed differences in gene frequencies may be explained:

a Between Asiatics and American Indians.

b Between tribes of American Indians.

1703 The frequency of blood group *B* is 20 to 25 percent in all populations from Asia. It drops to 15 percent in a belt containing the Arabian peninsula, Eastern Russia, and Finland and to 10 percent in Continental Europe; in certain fringe areas of Europe (toward the end of the Iberian and Italian peninsulas), it drops considerably below this value (Boyd, 1950). What does this suggest regarding the prehistory and history of the peoples of Eurasia?

1704 Different human populations vary markedly in the frequencies of the various blood groups, tabulated for the *A, B, O* series.

Table 1704

Population	Phenotype, %			
	O	*A*	*B*	*AB*
Swiss (Berne)	39.79	47.25	8.64	4.31
Scottish (Stornoway)	50.00	32.00	14.50	3.50
Swedes (Uppsala)	35.80	48.07	9.50	5.64
Hindus (Bombay)	31.76	29.20	28.25	10.79
Navaho Indians (New Mexico)	75.26	24.74	00.00	00.00

a Suggest a likely explanation for these differences.
b Can you determine from these data whether alleles have different selective values in different environments?

1705 In man the *Rh-negative* blood antigen is controlled by the Rh^- allele. This allele does not occur in American Indians but is present with a frequency of 15 percent in most European populations. To what degree would the fact that the Rh^- allele does not occur in American Indians allow you to identify an individual as belonging to this race?

1706 The *MN, Rh*, and *ABO* blood-group frequencies vary from one human population to another and from one geographic region to another. Propose a hypothesis to account for this variation and show by means of a simple mathematical model the process by which they may originate.

1707 In a certain diploid species in a restricted geographic region, a harmful recessive mutation $(A \rightarrow a)$ occurs in about 1 out of 5,000 individuals in every generation. A large sample of this population is introduced into an environment in which the mutant allele produces a beneficial effect. What would probably happen to the frequency of the allele in this new environment? Do you think that its relation to the other allele, *A*, at the locus (i.e., its recessiveness) would change? Explain.

1708 The numbered questions are to be answered for (a) *or* (b).
a *Pseudohypertrophic muscular dystrophy* in man (gradual muscular degeneration) is controlled by a recessive X-linked allele, *p*. It usually is lethal in the early teens, and no *affected* males have been known to reproduce. The mutation rate u $(P \rightarrow p)$ is close to 10^{-4} (Neel and Schull, 1958); the reverse rate V $(p \rightarrow P)$ cannot be determined because of the rareness of the *p* allele, but reverse rates for deleterious traits in general are probably lower than forward rates. For the sake of the arguments to be presented in this question, assume that V is actually higher (10^{-3}) than *u*. The frequency of the *p* allele is very low (probably close to 0.001).

b *Infantile amaurotic idiocy*, a lethal at ages two to three years, is controlled by an autosomal recessive, *p*, allele which arises from the *normal* allele, *P*, at the rate of 1 in 100,000. The back-mutation rate is 1 in 10^6.

(1) What would be the frequencies of *P* and *p* in the population at equilibrium due to mutation pressure only?

(2) At this allele frequency level, if a population consisted of 500,000 individuals, how many of the alleles would be *P* and how many would be *p*? How many of each allele would mutate to the alternative one in any generation?

(3) The actual frequencies of *P* and *p* are the reverse of those expected under mutation pressure alone. Moreover, *p* is extremely infrequent. Why doesn't the recessive allele increase in frequency if the above mutation rates are correct?

(4) What are the equilibrium frequencies of *P*, *p*, *PP*, and *Pp*?

(5) You are asked how often the trait would appear in any generation at equilibrium. What would your answer be?

(6) Suppose that the biochemical error causing this form of disorder is identified and it is now possible to alleviate or "cure" the *afflicted* individuals. What effect would this have on the frequency of the *p* allele in subsequent generations? Discuss.

1709 The phenotypes *black*, *yellow*, and *tortoiseshell* are controlled by an X-linked pair of alleles *Bb* in which there is no dominance. Heterozygotes express the *tortoiseshell* phenotype. The table shows phenotypic distributions reported for London, Singapore, and Boston cat populations.

Table 1709

Location	Sex	Black	Tortoise-shell	Yellow	Total	Reference
London	♀	277	54	7	338	Searle, 1949
	♂	311	1	42	353	
Singapore	♀	63	55	12	130	Searle, 1959
	♂	74	0	38	108	
Boston	♀	102	48	4	154	Todd, 1964
	♂	99	0	28	127	

a Determine the frequencies of the *B* and *b* alleles and state whether or not mating was at random with respect to these alleles for each of the three populations. Are all populations at Hardy-Weinberg equilibrium?

b How would you interpret the difference in allele frequencies that exist between the three populations?

c *Kinky* tail occurs in 69 percent of the Singapore cats but is absent from the London and Boston populations. Give a genetic explanation for this difference.

d A *black* tom and a *yellow* female from the Boston population are introduced into an isolated community that has been without any cats; a *yellow* tom and a *black* female of the same origin are introduced into another community, also isolated and also without cats.

(1) Show the expected frequencies for *B* and *b* for the two populations on the assumption that the alleles are neutral in selection and that members of each generation mate only with the same generation.

(2) Suggest how this situation can be related to genetic drift.

1710 What long-range effects on the human species can be predicted to accrue from the relaxation of the intensity of selection against human diseases such as *phenylketonuria* and *galactosemia* occasioned by developments in modern medical genetics?

1711 Do you think that all effective eugenic programs must be based on detailed knowledge of the genetics of the trait being selected? Discuss.

1712 If a human trait is already present with a very low frequency in the population, would institution of a program discouraging reproduction by *affected* persons be likely to cause significant changes in the frequency of the trait? Discuss fully.

1713 Dobzhansky and Pavlovsky (1955) studied a population of *Drosophila tropicalis* in the vicinity of Lancetilla, Honduras, and found that it consisted chiefly of heterozygotes for an inversion in the second chromosome. Slightly more than 50 percent of the fertilized eggs developed to the adult stage. Most of the mortality occurred in the egg stage, but some larval and a small amount of pupal mortality also occurred. Offer an explanation of these observations.

1714 You observe color variation in an aggregate of birds. Show how you would determine whether or not a true genetic polymorphism existed, assuming you have adequate facilities.

1715 In a certain population of beetles, *three-spot* vs. *four-spot* wings is controlled by an autosomal pair of alleles. The allele *S* for *three-spot* has a frequency of 0.44 and *s* for *four-spot* a frequency of 0.55. What are the possible explanations for the intermediate frequencies of the two alleles?

1716 In the South American species *Drosophila polymorpha*, body color is mono-genically controlled, the body being *dark*, *EE*, *intermediate*, *Ee*, or *light*, *ee*. In natural populations, the frequencies of the three traits remain constant for many generations. Da Cunha (1949) found that a laboratory population of F_2's from a cross between *dark*, *EE*, and *light*, *ee*, consisted of 1,605 *dark*, 3,767 *intermediate*, and 1,310 *light* individuals. Determine the relative adaptive values of the three genotypes and explain why the frequencies of the traits remain constant from generation to generation.

1717 A certain natural population of animals is at genotypic equilibrium. The reproductive fitness of the three genotypes at the *A* locus are as shown.

Table 1717

Genotype	A_1A_1	A_1A_2	A_2A_2
Fitness	0.5	1.0	0

a How do you account for this genotypic equilibrium?

b Derive the frequencies of A_1 and A_2 at this equilibrium.

1718 *Drosophila melanogaster* homozygous for the *Arrowhead* inversion are placed together with flies homozygous for the *Chiricuhua* inversion in a population cage. Sampling after one generation of mating reveals that the two kinds of inverted chromosomes are present at frequencies of 0.8 (*Arrowhead*) and 0.2 (*Chiricuhua*). After 15 generations, these frequencies are found to have changed to 0.6 and 0.4 respectively; the latter values remain unchanged in generations later than the fifteenth. State the most probable explanation for these results and describe how the change and subsequent stabilization of ratios comes about.

1719 In *Drosophila melanogaster*, *Cncn* (*red* vs. *cinnabar* eyes), *Bb* (*gray* vs. *black* body), and *Byby* (*normal* vs. *blistery* wing) are autosomal pairs of alleles. Samples of three large natural adult populations, each classified for a different pair of traits, are found to have the following genotypes:

Population A	31 *cncn*	171 *Cncn*	60 *CnCn*
Population B	182 *BB*	391 *Bb*	152 *bb*
Population C	100 *ByBy*	372 *Byby*	40 *byby*

Compare these distributions with those expected for a population at Hardy-Weinberg equilibrium. Account for any differences.

1720 a What is meant by transient polymorphism?

b Populations of *Biston betularia* in industrial areas of England are now almost entirely *melanic*. The transition took about 50 years. In other species the transition has taken place much more rapidly. For example, the change to *melanism* at Hamburg, Germany, in a species of the genus *Tethea* took place in the period 1904–1912. What factors can you think of that may affect the rate of displacement of one allele by another?

1721 Is most of the genetic load a mutation load or a balanced one? Discuss. See Morton (1960), Crow (1963), and Hershowitz (1960) for their analysis of the situation.

REFERENCES

Allison, A. C. (1954), *Br. Med. J.*, **1**:290.
——— (1955), *Cold Spring Harbor Symp. Quant. Biol.*, **20**:239.
——— (1956), *Sci. Am.*, **195**(August):87.
Ayala, F. J., J. R. Powell, and T. Dobzhansky (1971), *Proc. Natl. Acad. Sci.*, **68**:2480.
Beet, E. A. (1949), *Ann. Eugen.*, **14**:279.
Bennett, J. H., F. A. Rhodes, and H. N. Robson (1959), *Am. J. Hum. Genet.*, **11**:169.

Berg, K. (1969), *Hum. Hered.*, **19**:239.

Blatherwick, C., and C. Wills (1971), *Genetics*, **68**:547.

Blumberg, B. S., and J. E. Hesser (1971), *Proc. Natl. Acad. Sci.*, **68**:2554.

Boyd, W. C. (1950), "Genetics and the Races of Man," Little, Brown, Boston.

Broman, B., A. Heiken, and J. Hirschfeld (1963), *Acta Genet. Stat. Med.*, **13**:132.

Brues, A. M. (1964), *Evolution*, **18**:379.

Cain, A. J., and P. M. Sheppard (1954), *Am. Nat.*, **88**:321.

Camin, J. H., and P. R. Ehrlich (1958), *Evolution*, **12**:504.

Candela, P. B. (1942), *Hum. Biol.*, **14**:413.

Carson, H. L., and J. E. Seto (1969), *Evolution*, **23**:493.

Cavalli-Sforza, L. L. (1969), *Sci. Am.*, **221**(August):30.

——, T. Barrai, and W. F. Edwards (1964), *Cold Spring Harbor Symp. Quant. Biol.*, **29**:9.

——, and W. F. Bodmer (1971), "The Genetics of Human Populations," Freeman, San Francisco.

Chung, C. S., O. W. Robison, and N. E. Morton (1959), *Ann. Hum. Genet.*, **23**:357.

Cold Spring Harbor Symp. Quant. Biol., **15** (1950), "Origin and Evolution of Man."

Cold Spring Harbor Symp. Quant. Biol., **20** (1955), "Population Genetics: The Nature and Causes of Genetic Variability in Populations."

Cold Spring Harbor Symp. Quant. Biol., **24** (1959), "Genetics and Twentieth Century Darwinism."

Cold Spring Harbor Symp. Quant. Biol., **29** (1964), "Human Genetics."

Crow, J. F. (1963), *Am. J. Hum. Genet.*, **15**:310.

——, and M. Kimura (1970), "An Introduction to Population Genetics Theory," Harper & Row, New York.

——, and N. E. Morton (1960), *Am. Nat.*, **94**:413.

Da Cunha, A. B. (1949), *Evolution*, **3**:239.

Darwin, C. (1956), "The Origin of Species," 6th ed., Oxford University Press, Oxford.

—— (1958), "The Voyage of the Beagle," Bantam Books, New York.

Dobzhansky, T. (1950), *Sci. Am.*, **182**(September):32.

—— (1951), "Genetics and the Origin of Species," 3d ed., Columbia, New York.

—— (1970), "Genetics of the Evolutionary Process," Columbia, New York.

——, and O. Pavlovsky (1955), *Proc. Natl. Acad. Sci.*, **41**:289.

——, and —— (1957), *Evolution*, **11**:311.

——, B. Spassky, and T. Tidwell (1963), *Genetics*, **48**:361.

Ehrlich, P. R., and R. W. Holm (1963), "The Process of Evolution," McGraw-Hill, New York.

Falconer, D. S. (1961), "Introduction to Quantitative Genetics," Ronald, New York.

Felsenstein, J. (1971), *Am. Nat.*, **105**:1.

Ford, E. B. (1964), "Ecological Genetics," Wiley, New York.

Fujino, K., and T. King (1968), *Genetics*, **59**:79.

Glass, B. (1953), *Sci. Am.*, **189**(August):76.

—— (1954), *Adv. Genet.*, **6**:95.

—— (1956), *Am. J. Phys. Anthrop.*, **14**:541.

Glass, H. B., and C. C. Li (1953), *Am. J. Hum. Genet.*, **5**:1.

Grant, V. (1964), "The Architecture of the Germplasm," Wiley, New York.

Grubb, R., and A. B. Laurell (1956), *Acta Path. Microbiol. Scandia*, **39**:390.

Grüneberg, H. (1952), *Bibliog. Genet.*, **15**:1.

Haldane, J. B. S. (1931), "The Causes of Evolution," Harper, New York.

—— (1960), *J. Genet.*, **57**:351.

Hardy, G. H. (1908), *Science*, **28**:49.

Hershowitz, T. H. (1960), *Nuclear Inform.*, **3**(2):1.

Holzel, A., and G. M. Komrower (1955), *Arch. Dis. Child.*, **301**:155.

Hopkinson, D. A., N. Spencer, and H. Harris (1963), *Nature*, **199**:969.

Johnson, F. M., et al. (1966), *Proc. Natl. Acad. Sci.*, **56**:119.

Jones, D. A. (1967), *Sci. Prog.*, **55**:379.
Kamel, N. H., and E. I. Hammoud (1966), *J. Med. Genet.*, **3**:279.
Kettlewell, H. B. D. (1958), *Heredity*, **12**:51.
———— (1959), *Sci. Am.*, **200**(March):48.
———— (1961), *Annu. Rev. Entomol.*, **6**:245.
Kimura, M., and T. Ohta (1971), *Nature*, **229**:467.
King, J. C. (1971), "The Biology of Race," Harcourt Brace Jovanovich, New York.
King, J. L. (1967), *Genetics*, **55**:483.
Kitzmiller, J. B., and H. Laven (1959), *Cold Spring Harbor Symp. Quant. Biol.*, **24**:173.
Koehn, R. K., J. E. Perez, and R. B. Merritt (1971), *Am. Nat.*, **105**:51.
Kojima, K. (1970), "Mathematical Topics in Population Genetics," Springer-Verlag, New York.
————, and K. N. Yarborough (1967), *Proc. Natl. Acad. Sci.*, **57**:645.
Lai, L., S. Nevo, and A. G. Steinberg (1964), *Science*, **145**:1187.
Lamotte, M. (1959), *Cold Spring Harbor Symp. Quant. Biol.*, **24**:65.
Landsteiner, K., and A. S. Wiener (1941), *J. Exp. Med.*, **74**:309.
Lerner, I. M. (1958), "The Genetic Basis of Selection," Wiley, New York.
Lewontin, R. C. (1971), *Proc. Natl. Acad. Sci.*, **68**:984.
————, and J. L. Hubby (1966), *Genetics*, **54**:595.
Li, C. C. (1961), "Human Genetics," McGraw-Hill, New York.
Lin, C. C., et al. (1969), *Biochem. Genet.*, **3**:609.
McKusick, V. A., et al. (1964), *Bull. Johns Hopkins Hosp.*, **114–115**:306.
McLaughlin, P. J., and M. O. Dayhoff (1970), *Science*, **168**:1469.
Mandel, S. P. H. (1959), *Nature*, **183**:1347.
Mather, K. (1953), *Symp. Soc. Exp. Biol.*, **7**:66.
Matsunaga, E., and S. Itoh (1958), *Ann. Hum. Genet.*, **22**:111.
Mayo, O. (1970), *Ann. Hum. Genet.*, **33**:307.
Mettler, L. E., and T. G. Gregg (1969), " Population Genetics and Evolution," Prentice-Hall,
    Englewood Cliffs, N.J.
Milkman, R. D. (1967), *Genetics*, **35**:493.
Mohr, O. L. (1934), *J. Hered.*, **25**:246.
Mørch, E. T. (1941), *Opera Domo Biol. Hered. Hum. Univ. Hafniensis*, **3**:1.
Morton, N. E. (1960), *Am. J. Human Genet.*, **12**:348.
————, J. Crow, and H. J. Muller (1956), *Proc. Natl. Acad. Sci.*, **42**:855.
Mourant, A. E. (1954), "The Distribution of the Human Blood Groups," Thomas, Springfield, Ill.
Mukai, T., and T. Yamazaki (1968), *Genetics*, **59**:513.
Needler, A. W. H., and R. R. Logie (1947), *Trans. Soc. Can.*, 3d ser. sect. V, **41**:73.
Neel, J. V. (1949), *Science*, **110**:64.
———— and H. F. Falls (1957), *Science*, **114**:419.
————, and W. J. Schull (1958), "Human Heredity," University of Chicago Press, Chicago.
Nei, M., and Y. Imaizumi (1966), *Heredity*, **21**:461.
Race, R. R., and R. Sanger (1968), "Blood Groups of Man," 5th ed., Blackwell, Oxford.
Rendel, J. M. (1959), *Evolution*, **13**:425.
Searle, A. G. (1949), *J. Genet.*, **49**:214.
———— (1959), *J. Genet.*, **56**:111.
Shapiro, H. L. (1936), "The Heritage of the Bounty," Simon & Schuster, New York.
Sheppard, P. M. (1958), "Natural Selection and Heredity," Hutchinson, London.
Smithies, O., and N. F. Walker (1955), *Nature*, **176**:1265.
Snyder, L. H. (1934), *Genetics*, **19**:1.
———— (1947), *J. Biol. Med.*, **19**:817.
————, and P. R. David (1957), "The Principles of Heredity," Heath, Boston.
Spencer, W. P. (1947), *Am. Nat.*, **81**:237.
Spiess, E. B. (ed.) (1962), "Papers on Animal Population Genetics," Little, Brown, Boston.

Stebbins, G. L. (1966), "Processes of Organic Evolution," Prentice-Hall, Englewood Cliffs, N.J.

Stern, C. (1960), "Principles of Human Genetics," Freeman, San Francisco.

Sutton, H. E., et al. (1959), *Ann. Hum. Genet.*, **23**:175.

Thoday, J. M., and J. B. Gibson (1962), *Nature*, **198**:1164.

Todd, N. B. (1964), *Heredity*, **19**:47.

Turner, J. R. G. (1970), *Science Prog.*, **58**:219.

Wallace, A. R. (1871), "Contributions to the Theory of Natural Selection," Macmillan, New York.

Wallace, B. (1963), *Proc. Natl. Acad. Sci.*, **49**:801.

Weinberg, W. (1908), *Jahresh. Ver. Vaterl. Natuurkd. Wuerttemb. Stuttg.*, **64**:368; English trans. in S. H. Boyer (ed.), "Papers on Human Genetics," pp. 4–15, Prentice-Hall, Englewood Cliffs, N.J., 1963.

Wills, C. (1970), *Sci. Am.*, **222**(March):98.

Woolf, C. M. (1965), *Am. J. Hum. Genet.*, **17**:23.

———, and R. B. Grant (1962), *Am. J. Hum. Genet.*, **14**:391.

Wright, S. (1921), *Genetics*, **6**:167.

——— (1931), *Genetics*, **16**:97.

——— (1951), *Ann. Eugen.*, **15**:323.

——— (1968–1969), "Evolution and the Genetics of Populations," 3 vols., University of Chicago Press, Chicago.

Zuckerkandl, E., and L. Pauling (1962), "Horizons in Biochemistry," Academic, New York.

33
The Genetics
of Race
and
Species Formation

QUESTIONS

1722 What is the largest and most inclusive taxonomic category that falls within the definition of a Mendelian population?

1723 Different species of animals may in some instances be more closely similar in appearance than different breeds within a species are. For example, several species of bears appear more alike than certain breeds of dogs, e.g., the Labrador retriever and the Chihuahua. Why do we designate the former as belonging to different species while grouping all dogs in the same species?

1724 **a** What other mechanisms besides gene mutation may be considered as providing raw material for evolution?

b State what basic features of evolution theory apply to sexually and asexually reproducing organisms alike and what basic differences may exist in the evolutionary process with respect to these two kinds of organisms.

1725 **a** Define *race* and discuss the genetic differences that exist between two such entities, illustrating your answer from man.

b Discuss existing fallacious concepts of race.

c "Webster's New Collegiate Dictionary" defines race as "A division of mankind possessing constant traits that are transmissible by descent and sufficient to characterize it as a distinct human type." In what respects is this inaccurate and inadequate as a genetic definition?

1726 **a** Are racial differences real? If so, why is the number of races arbitrary?

b Explain why genetic studies of populations are a prerequisite to classifying a species into races.

c Is it easier to classify organisms into species than into races? Explain.

d Although both species and races are natural biological entities, the lines of demarcation between the former are more distinct than between the latter. Why?

e What is the important change that transforms races into species?

f Do you think that the number of human races will increase in the future? Explain.

g Why is it impossible to decide the race to which an individual belongs by comparing his phenotype with a racial average?

h People with *O*, *A*, *B*, or *AB* blood groups do not belong to distinct races. Why?

1727 Explain how species relationships can be measured at (1) the protein level and (2) DNA level through in vitro hybridization.

1728 Some people contend that interracial marriages are biologically undesirable, others contend the reverse. Which do you believe and why?

1729 **a** What is the most essential genetical criterion for designating two isolated populations of a certain kind of organism:

(1) As distinct races?

(2) As distinct species?

b By the criterion in (2) does an allotetraploid (e.g., cultivated tobacco) qualify as a separate species, distinct from each of the diploid species from which it arose?

1730 Show how the concepts of *race*, *breed*, and *variety* are similar and how they differ.   Can any of these terms apply to asexually reproducing species?

1731 State which of the following you would consider to be a *race* and why:
a A breed of dog.
b A variety of common wheat.
c A local population of an asexually reproducing organism inhabiting a certain geographic region.

1732 What are the mechanisms by which new species may arise suddenly?   Illustrate your answer with examples.

1733 Dobzhansky's (1951) definition of a species is: Species are, accordingly, groups of populations the gene exchange between which is limited or prevented in nature by one, or by a combination of several, reproductive mechanisms. Can this be applied to organisms that reproduce only by asexual means? Explain.

1734 What reasons can you give for including all human beings in the one species, *Homo sapiens*?

1735 Domesticated species of plants and animals are more variable than the corresponding wild species.   How do you account for this?

1736 Would you expect two species to be separated by one reproductive barrier or by more than one?   Explain.

1737 Is morphological specificity alone adequate for defining a species?   Explain, citing examples.   If not, what other criteria should be used?   Which of these criteria can and which cannot be used to distinguish species among asexual organisms?

1738 The increase in chromosome number, occurring in polyploids, is in itself not important in speciation.   However, when it is combined with interspecific hybridization, it becomes an important factor in speciation.   Discuss.

1739 Explain how sympatric races may become reproductively isolated.

1740 Do you think that populations of cross-fertilizing species must be broken up geographically into different isolated subpopulations before they can diverge into new and different species?   Give reasons for your decision.

1741 Certain species occur only in restricted geographic regions.   What genetic explanation can you offer to account for this?

1742 The Monterey and Bishop pines in California are different sympatric species which shed their pollen at different times.   Do you think that hybrids between them would have a chance of becoming established in nature?

1743 In cross-fertilizing organisms most races are allopatric, but this is not necessarily true in asexually reproducing or self-fertilizing forms.   Why?

1744 What are the different kinds of barriers causing reproductive isolation? Discuss each and cite an example where possible.

1745 Certain evolutionists consider geographic isolation to be necessary before a population can become divided into two or more by reproductive isolating mechanisms.
a What are some of the reasons for this?
b Cite evidence indicating that this is so, at least in some cases.

1746 The three primary processes governing the rate and direction of evolution are mutation, recombination, and selection. Is any one of these more important than the others and why?

1747 Can natural selection promote the establishment of reproductive isolating mechanisms? If so, in what ways?

1748 Using two parental species of different chromosome number, illustrate the process of speciation by amphidiploidy.

1749 In allotetraploid organisms like cultivated tobacco and the macaroni wheats, a fairly large number of characters are controlled by duplicate genes. In the allohexaploid bread wheat certain characters are controlled by triplicate and others by duplicate genes. Account for the origin of these systems.

1750 Discuss the role of heterochromatin in karyotype evolution involving changes in chromosome number, referring to specific examples where possible.

1751 Explain why there is little or no correlation between rates of mutation and rates of evolution.

1752 According to King and Jukes (1969), "Most evolutionary change in proteins may be due to neutral mutations and genetic drift." Do you agree? See Richmond (1970).

PROBLEMS

1753 In the perennial cinquefoil plant species (*Potentilla glandulosa*) in California, three allopatric races inhabit the *coast* ranges, the *foothills*, and *alpine* regions, respectively, of the Sierra Nevada. In the coastal habitat, where temperatures rarely fall below the freezing point, plants of this species grow throughout the year; in contrast, the habitats of the other regions are characterized by cold winters with heavy snowfall. When reciprocally transplanted, the *coastal* and *foothills* races do not thrive as well as they do in their native habitats. Nevertheless, the *foothills* race shows vegetative growth during the winter on the coast, and the *coastal* race becomes dormant in the winter in the *foothills*. Both these races are usually killed within a year when transplanted to the *alpine* zone. The *alpine* race, growing as a dwarf in its own environment, remains so when planted on the coast. Hybrids between the races are vigorous and fertile. The species was once very uniform genotypically and phenotypically (Clausen, Keck, and Hiesey, 1940). Describe the probable manner in which the adaptation of geographic races to specific habitats arose.

1754 Sturtevant (1920, 1929) performed reciprocal crosses between *Drosophila melanogaster* and *D. simulans*, both of which have X-Y sex determination, and obtained sterile progeny with rudimentary gonads. The results of his studies on the inheritance in these offspring of traits known to be X-linked in the parental species are shown.

Table 1754

Parents		Offspring	
♀	♂	♀	♂
melanogaster (XX) × *simulans*		regular	none
melanogaster (XXY) × *simulans*		regular*	exceptional*
simulans (XX) × *melanogaster*		regular	none
simulans (XXY) × *melanogaster*		regular	exceptional

* *Regular* offspring inherit X-linked traits (viz. males inherit them only from female parent, and females from either parent); *exceptional* offspring inherit these traits abnormally (viz. males inherit them from the male parent only, and females inherit them from the female parent only).

a Suggest a hypothesis to account for this type of interspecific hybrid sterility.

b Suggest experiments that might have been made to test this hypothesis had the offspring been fertile and outline the results expected.

1755 Giving reasons for your decision, state which of the following accounts represent genic and which represent chromosomal sterility.

a The diploid hybrid between *Primula verticillata* and *P. floribunda* shows mainly bivalents at meiosis and yet is almost completely sterile. The allotetraploid *P. kewensis*, formed by doubling the chromosome number of the F_1 hybrid, has bivalents with only the occasional quadrivalent and is fertile (Newton and Pellew, 1929; Upcott, 1939).

b Two strains of *Sorghum vulgare* from India, morphologically similar to other known strains of the same species, produce hybrids that are sterile when crossed with these latter. Meiosis in the hybrids is normal, but the male gametophytes abort (Hadjinov, 1937).

c In the hybrid between radish (*Raphanus sativus*; $2n = 18$) and cabbage (*Brassica oleracea*; $2n = 18$) no chromosome pairing takes place, and the plants are almost completely sterile. When the chromosome number is doubled, the resulting plants ($2n = 36$) have a regular meiosis; 18 bivalents are formed, and they are fully fertile (Karpechenko, 1928).

1756 Two species can give rise to a new species after interspecific hybridization by allopolyploidy (amphidiploidy), by introgression (natural selection of recombinant progeny) of the interspecific hybrid crossed with its parents, or by the occurrence of recombinants that are reproductively isolated from the parental species among its own progeny.   The following is an account of the work of Lewis and Epling (1946) and Epling (1947) in the larkspur (*Delphinium* spp.).   *D. gypsophilum* is intermediate in morphology between *D. hesperium* and *D. recurvatum* and occupies a new habitat which is intermediate between that of the other two species.   All three species are diploid with $2n = 16$.   The F_1 hybrid between *D. recurvatum* and *D. hesperium* resembles *D. gypsophilum*.   The progeny from the cross (F_1 hybrid × *D. gypsophilum*) are more fertile than those obtained by backcrossing the F_1 hybrid to either parent or by crossing *gypsophilum* with either of the other two species.   What is the phylogenetic relationship among these three species, and which of the above speciating methods is responsible for it?   Give reasons for your decisions.

1757   Moore (1946) collected living samples of frogs (*Rana*) from different geographic regions along the east coast of the United States differing considerably in habitat.   These populations were intermated, and the F_1 hybrids were studied for development.   The development of F_1 hybrids from the Vermont, New Jersey, central Florida, and southern Florida populations is shown (modified after Moore, 1946).   The average seasonal temperatures for egg-to-adult development are Vermont, 45°F; New Jersey, 40°F; central Florida, 29°F; and southern Florida, 27°F.

Table 1757   Rates of development and mature head size in hybrids

♀	♂			
	Vermont	New Jersey	Central Florida	Southern Florida
Vermont	Normal	Normal or very slight acceleration; head normal or very slightly enlarged	Moderate retardation; head considerably enlarged	Marked retardation; head extremely enlarged
New Jersey		Normal	Very slight retardation; head very slightly enlarged	Slight retardation; head moderately enlarged
Central Florida		Slight retardation; head slightly enlarged	Normal	Normal; head normal
Southern Florida	Marked retardation; head considerably reduced			Normal

Note: (1) Vermont–southern Florida hybrids are so inviable few would survive in nature; all other hybrids develop into *normal* adults. (2) The greater the distance between populations the greater the retardation in rate of development, the expression of the defects, and the proportion of inviable offspring.

a Would you classify these four populations as different races of the same species? Explain.

b If so, how would you account for:

(1) The gradual manner in which reproductive (genetic) isolation becomes established?

(2) Failure of these populations to reach the status of distinct species?

1758 In general, the process of speciation has been considered to take place only in the presence of geographic isolation (allopatric speciation). Evidence for sympatric speciation has been found in the studies of Darwin's finches on the Galapagos Islands, where different species developed from the ancestral mainland finch on different islands, although extensive intermigration between islands is possible. Similar evidence is provided by the "species flocks" of shrimps in Lake Baikel, where 300 species of gammarid shrimps exist (more than are known in the rest of the world) (Lack, 1947; Brooks, 1950). Explain briefly how habitat isolation might lead to speciation under these conditions.

1759 A plant different from all others appears in a population of a cross-fertilizing species. How would you determine whether the plant is an unusual genetic recombinant within the species or a new species?

1760 The hybrids in the table produced from crosses between two species of hawksbeards (*Crepis*) are described (Hollingshead, 1930). Crosses between

Table 1760

Parents	Offspring
C. capillaris × *C. tectorum* strain 1498	1 *viable* : 1 *inviable**
C. capillaris × *C. tectorum* plant 8†	All *inviable*
C. capillaris × *C. tectorum* plant 10†	All *viable*
C. capillaris × *C. tectorum* plant 13†	1 *viable* : 1 *inviable*
C. capillaris × *C. tectorum* strain 1066	All *inviable*
C. capillaris × *C. tectorum* strain 1700	All *viable*

* Seedlings die in the cotyledon stage.

† Plants 8, 10 and 13 of *C. tectorum* are self-fertilized progeny of strain 1498.

different strains of *C. tectorum* produce no inviable offspring. In crosses between *C. tectorum* 1498 and *C. leontodontoides* all hybrids are inviable, and half the F$_1$ seedlings die in the cross *C. tectorum* × *C. bursifolia*. When *tectorum* is crossed with *setosa* or *taraxacifolia*, all offspring are viable.

Analyze these results to show the genetic basis of the mechanism of reproductive isolation. Under what conditions would reproductive isolation be
complete?

1761 What requirement(s) must be satisfied for a single mutation to give rise to
a new species?

1762 *Drosophila pachea*, found only in the Sonora Desert, breeds exclusively in the
stems of the senita cactus (*Lophocereus schottii*), which synthesize the sterol
Δ^7-stigmasten-3β-ol. It does not reproduce in the laboratory unless a piece
of cactus stem or the sterol is added to the medium. No other *Drosophila*
species utilize the stem of this cactus for breeding although two other
species in this area breed in the fruits of this cactus. The sterol and several
other alkaloids of this cactus are lethal to these other species (Heed and
Kircher, 1965). *D. pachea* is ecologically isolated from other sympatric
Drosophila species. Does it necessarily follow that the species arose by
divergence of sympatric races?

1763 Polyploidy has been relatively unimportant in the evolution of animals.
Polyploid animals that do occur are restricted to certain of the lower forms
(insects, crustacean, etc.). This cannot be said for chromosomal aberrations
such as reciprocal translocations, deletions, and inversions.
a Discuss some of the probable reasons for the apparent unimportance of
polyploidy in evolution in the animal kingdom.
b Cite evidence that supports the latter statement and explain why
aberrations have been more important than polyploidy in animal evolution.

1764 Grant (1966) crossed *Gilia malior* ($2n = 36$) × *G. modocensis* ($2n = 36$) and
obtained a highly sterile F_1 hybrid, intermediate in phenotype between the
parents, with an average of six bivalents per meiocyte. By intercrossing and
selecting for fertility, *fully fertile* ($2n = 36$) F_{10} plants were obtained with a
new combination of morphological characters. These were *sterile* with both
parents.
a How does this form of speciation differ from amphidiploidy?
b Suggest how the F_{10}'s were derived and what their chromosome constitution
might be like.

1765 Two groups of phenotypically similar and phylogenetically closely related
birds inhabit different nearby islands. The chromosomes of both groups are
the same in number, size, and morphology. Crosses made between the two
groups produce only sterile hybrids. What is the most probable cause of the
sterility?

1766 Discuss the evolutionary importance, if any, of (1) deletions, (2) duplications,
(3) inversions, and (4) translocations.

1767   Additional questions on the genetics of raciation and speciation are listed under polyploidy, aneuploidy, and chromosome aberrations as follows:

Chapter 20		Chapter 22	
855	858	1019	1024
856	899	1020	1025
857	900	1021	1027
		1022	1029

1768   In groups of closely related species and genera some have gained or lost large quantities of DNA relative to others, their apparent progenitors (Keyl, 1965; McCarthy, 1967; Rees and Jones, 1967; Stebbins, 1966). What methods could be used to determine whether the differences in DNA content are due to redundant (duplicate) genes or to unique genes?   See Britten and Kohne (1970), Hoyer et al. (1964, 1965), Ingram (1963), McCarthy (1967), Marmur et al. (1963), Zuckerkandl (1965), and Zuckerkandl and Pauling (1965).

REFERENCES

Anderson, E. (1949). "Introgressive Hybridization," Wiley, New York.
Blair, W. F. (1958), *Am. Nat.*, **92**:27.
Boyd, W. C. (1950), "Genetics and the Races of Man," Little, Brown, Boston.
Britten, R. J., and D. E. Kohne (1968), *Science*, **161**:529.
———, and ——— (1970), *Sci. Am.*, **222**(April):24
Brooks, J. L. (1950), *Q. Rev. Biol.*, **25**:131.
Bryson, V., and H. J. Vogel (eds.) (1965), "Evolving Genes and Proteins," Academic, New York.
Buettner-Janusch, J. (ed.) (1963–1964), "Evolutionary and Genetic Biology of Primates," vols. I and II, Academic, New York.
Carson, H. L. (1959), *Cold Spring Harbor Symp. Quant. Biol.*, **24**:87.
——— (1970), *Science*, **168**:1414.
———, F. E. Clayton, and H. D. Stalker (1967), *Proc. Natl. Acad. Sci.*, **57**:1280.
Chen, T. R., and F. H. Ruddle (1970), *Chromosoma*, **29**:255.
Clausen, J., D. D. Keck, and W. M. Hiesey (1940), *Carnegie Inst. Wash., D.C., Publ.* 520, p. 1.
———, ———, and ——— (1947), *Am. Nat.*, **81**:114.
Cold Spring Harbor Symp. Quant. Biol., **24** (1959), "Genetics and Twentieth Century Darwinism."
Cold Spring Harbor Symp. Quant. Biol. Pop. Genet., **20** (1955), "The Nature and Causes of Genetic Variability in Populations."
Cold Spring Harbor Symp. Quant. Biol., **29** (1964), "Human Genetics."
Darwin, C. (1959), "The Origin of Species," 6th ed., Murray, London.
——— and A. R. Wallace (1859), *J. Linn. Soc.*, **3**:45.
de Wet, J. M. J., and J. R. Harlan (1970), *Evolution*, **24**:270.
Dobzhansky, T. (1933), *Proc. Natl. Acad. Sci.*, **19**:397.
——— (1950), *Sci. Am.*, **182**(September):32.
——— (1951), "Genetics and the Origin of Species," 3d ed., Columbia, New York.
——— (1962), "Mankind Evolving," Yale University Press, New Haven, Conn.
——— (1970), "Genetics of the Evolutionary Process," Columbia, New York.
———, L. Ehrman, O. Pavlovsky, and B. Spassky (1964), *Proc. Natl. Acad. Sci.*, **51**:3.
———, and O. Pavlovsky (1971), *Nature* **230**:289.
———, M. K. Hecht, and W. C. Steere (eds.) (1968), "Evolutionary Biology," vol. 2, Appleton-Century-Crofts, New York.

Dutta, S. K., N. Rochman, V. W. Woodward, and M. Mandel (1968), *Genetics*, **57**:719.
Ehrlich, P. R., and R. W. Holm (1963), "The Process of Evolution," McGraw-Hill, New York.
Ehrman, L. (1960), *Evolution*, **14**:212.
Epling, C. (1947), *Am. Nat.*, **81**:104.
———, and H. Lewis (1946), *Am. J. Bot.*, **33**:20s.
Ford, E. B. (1964), "Ecological Genetics," Wiley, New York.
Grant, V. (1963), "The Origin of Adaptations," Columbia, New York.
——— (1966), *Genetics*, **54**:1189.
Hadjinov, M. J. (1937), *Bull. Appl. Bot. Gen. Plant Breed.*, (2)**7**:417.
Hayman, D. L., et al. (1971), *Nature*, **231**:194.
Heed, W. B., and H. W. Kircher (1965), *Science*, **149**:758.
Hollingshead, L. (1930), *Genetics*, **15**:114.
Hoyer, B. H., E. T. Bolton, B. J. McCarthy, and R. B. Roberts (1965), in Bryson and Vogel (1956), pp. 581–590.
———, B. J. McCarthy, and E. T. Bolton (1964), *Science*, **144**:959.
———, ———, and ——— (1968), *Genet. Res.*, **12**:117.
Ingram, V. M. (1963), "The Hemoglobins in Genetics and Evolution," Columbia, New York.
Jukes, T. H. (1966), "Molecules and Evolution," Columbia, New York.
Karpechenko, G. D. (1928), *Z. Indukt. Abstamm.-Vererbungsl.*, **48**:1.
Kettlewell, H. B. D. (1959), *Sci. Am.*, **200**(March):48.
——— (1961), *Annu. Rev. Entomol.*, **6**:245.
Keyl, H. G. (1965), *Experientia*, **21**:191.
King, J. L., and T. H. Jukes (1969), *Science*, **164**:788.
Kitzmiller, J. B., and H. Laven (1959), *Cold Spring Harbor Symp. Quant. Biol.*, **24**:173.
Koopman, K. F. (1950), *Evolution*, **4**:135.
Lack, D. (1947), "Darwin's Finches," Cambridge University Press, Cambridge.
Laird, C. D., B. L. McConaughy, and B. J. McCarthy (1969), *Nature*, **224**:149.
Laven, H. (1959), *Cold Spring Harbor Symp. Quant. Biol.*, **24**:166.
Lewis, H., and C. Epling (1946), *Am. J. Bot.*, **33**:21s.
Lewontin, R. C., and M. J. D. White (1960), *Evolution*, **14**:116.
McCarthy, B. J. (1966), *Prog. Nucleic Acid Res.*, **4**:129.
——— (1967), *Bacteriol. Rev.*, **31**:215.
Mangelsdorf, P. C. (1958), *Proc. Am. Philos. Soc.*, **102**:454.
Marmur, J., S. Falkow, and M. Mandel (1963), *Annu. Rev. Microbiol.*, **17**:329.
Mayr, E. (1963), "Animal Species and Evolution," Harvard University Press, Cambridge, Mass.
Mertens, T. R. (1971), *Bioscience*, **21**:420.
Mettler, L. E., and T. G. Gregg (1969), "Population Genetics and Evolution," Prentice-Hall, Englewood Cliffs, N.J.
Moore, J. A. (1942), *Biol. Symp.*, **6**:189.
——— (1946), *Genetics*, **31**:304.
——— (1949), *Evolution*, **3**:1.
Morton, N. E. (1962), *Eugen. Q.*, **9**:23.
Newton, W. C. F., and C. Pellew (1929), *J. Genet.*, **20**:405.
Nolan, C., and E. Margoliash (1968), *Annu. Rev. Biochem.*, **37**:727.
Patterson, J. T., and W. S. Stone (1952), "Evolution in the Genus *Drosophila*," Macmillan, New York.
Rees, H., and R. N. Jones (1967), *Nature*, **216**:825.
Rensch, B. (1960), "Evolution above the Species Level," Columbia, New York.
Richmond, R. C. (1970), *Nature*, **225**:1025.
Sheppard, P. M. (1961), *Adv. Genet.*, **10**:165.
Simpson, G. G. (1964), *Science*, **146**:1535.
Sokal, R. R., and T. J. Crovello (1970), *Am. Nat.*, **104**:127.
Stalker, H. D. (1966), *Genetics*, **53**:327.

Stebbins, G. L. (1950), "Variation and Evolution in Plants," Columbia, New York.

———— (1966), "Processes of Organic Evolution," Prentice-Hall, Englewood Cliffs, N.J.

———— (1966), *Science*, **152**:1463.

Stone, W. S. (1962), *Univ. Tex. Publ.* 6204, p. 507.

Sturtevant, A. H. (1920), *Genetics*, **5**:488.

———— (1929), *Carnegie Inst. Wash., D.C., Publ.* 399, p. 1.

————, and T. Dobzhansky (1936), *Proc. Natl. Acad. Sci.*, **22**:448.

Thoday, J. M., and J. B. Gibson (1962), *Nature*, **193**:1164.

Upcott, M. (1939), *J. Genet.*, **39**:79.

Urbain, J. (1969), *Biochem. Genet.*, **3**:249.

Volpe, E. P., D. Duplantier, and E. M. Earley (1970), *Cytogenetics*, **9**:161.

West, J. L. (1970), *Evolution*, **24**:378.

White, G. B. (1971), *Nature*, **231**:184.

White, M. J. D. (1954), "Animal Cytology and Evolution," 2d ed., Cambridge University Press, Cambridge.

———— (1968), *Science*, **159**:1065.

Wright, S. (1932), *Proc. 6th Int. Congr. Genet.*, **1**:356.

———— (1955), *Cold Spring Harbor Symp. Quant. Biol.*, **20**:16.

Zuckerkandl, E. (1965), *Sci. Am.*, **212**(May):110.

————, and L. Pauling (1965), in Bryson and Vogel (1965), pp. 97–166.

————, and W. A. Schroeder (1961), *Nature*, **192**:984.

Appendix

Appendix

Table A-1 Product-moment formulas (ratio of products) and their associated recombination fractions

Type of cross	Repulsion $(AA\,bb \times aa\,BB)$	Coupling $(AA\,BB \times aa\,bb)$
Ratio of products	$\dfrac{ad}{bc}$	$\dfrac{bc}{ad}$

Recombination fraction:		
0.00	0.000000	0.000000
0.05	0.005031	0.003629
0.10	0.02051	0.01586
0.15	0.04763	0.03915
0.20	0.08854	0.07671
0.25	0.1467	0.1328
0.30	0.2271	0.2132
0.35	0.3377	0.3259
0.40	0.4898	0.4821
0.45	0.7013	0.6985
0.50	1.0000	1.0000
0.55	1.4317	1.4260

Note: a, b, c, and d are the numbers of individuals in the respective phenotype classes $A\_\_B\_\_$, $A\_\_bb$, $aa\,B\_\_$, and $aa\,bb$ in the F_2 of a dihybrid cross.

Table A-2   Structure of DNA bases, base analogs, and tautomers

DNA base	Tautomers		DNA base analog	Tautomers	
	Common	Rare		Common	Rare

Adenine

Guanine

Thymine

Cytosine

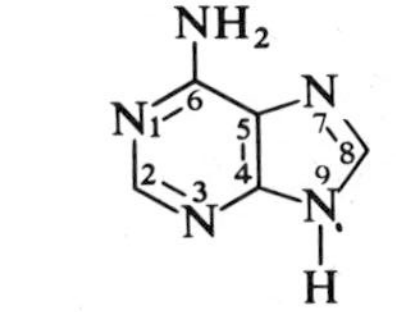

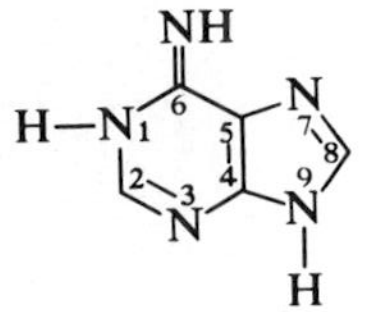

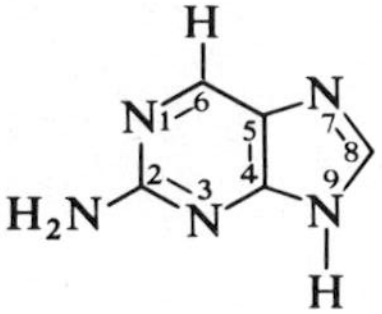

2-Aminopurine — Amino state

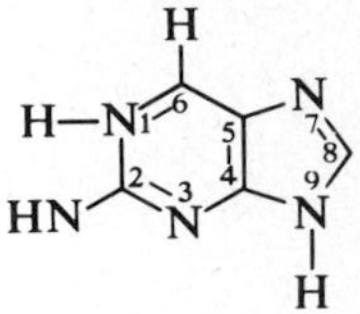

Xanthine

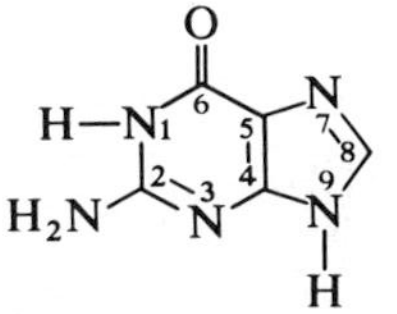

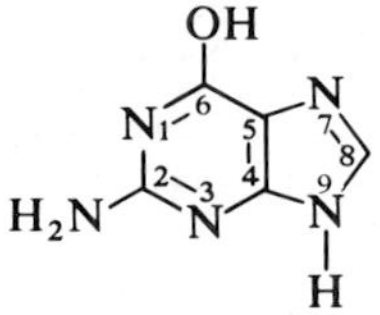

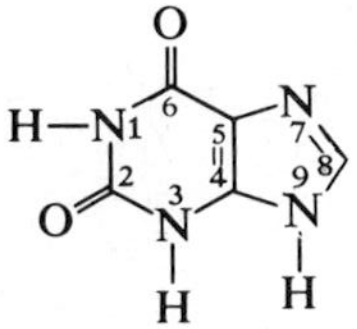

Hypoxanthine

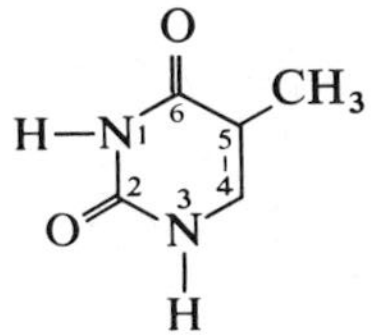

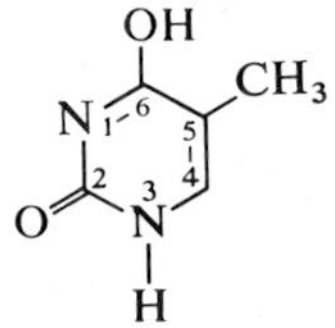

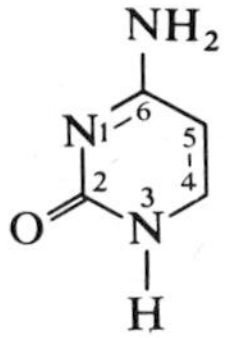

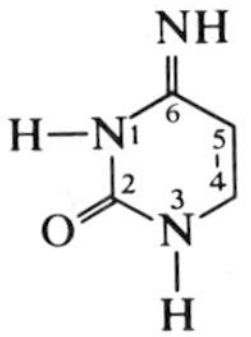

5-Bromouracil — Keto state / Enol state

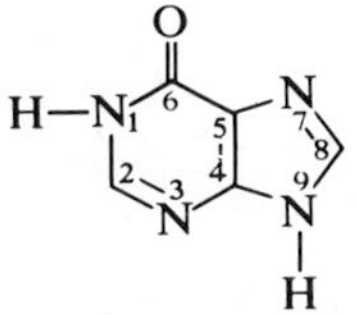

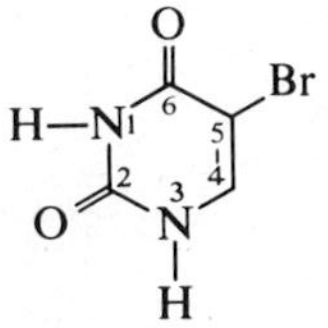

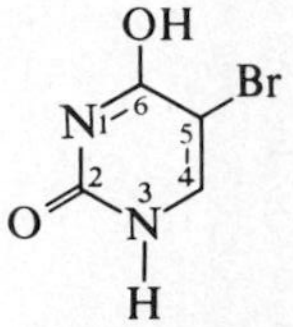

Uracil

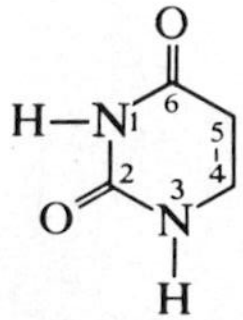

Imino state

Table A-3   Building blocks of DNA

Bases	Sugar

Pyrimidines

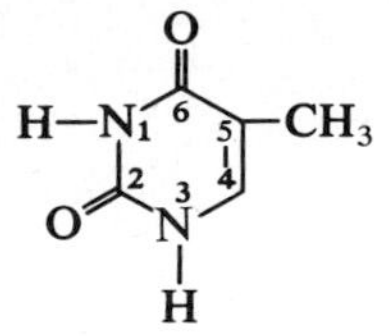

Thymine

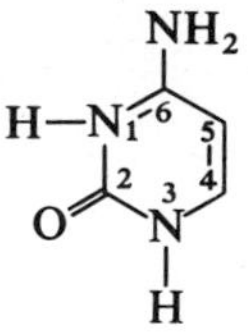

Cytosine

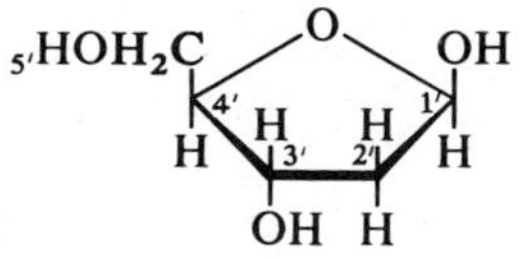

Deoxyribose

Purines

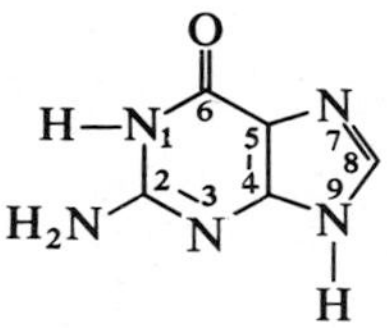

Guanine

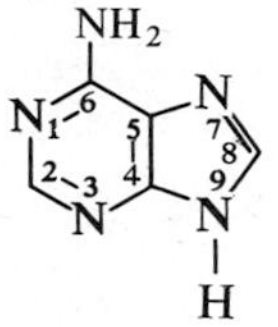

Adenine

Phosphate

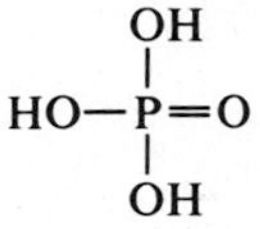

Table A-4 Structure and abbreviations of amino acids

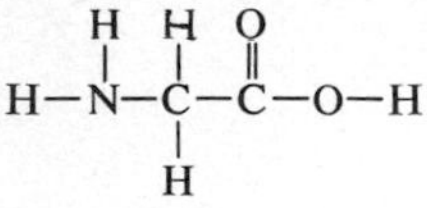

Glycine (Gly)

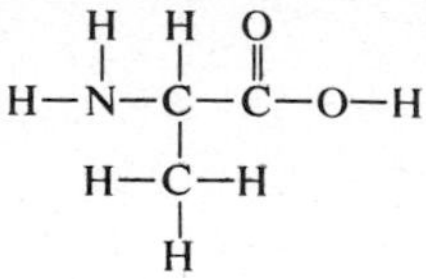

Alanine (Ala)

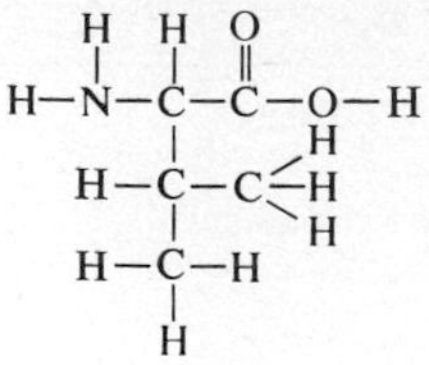

Valine (Val)

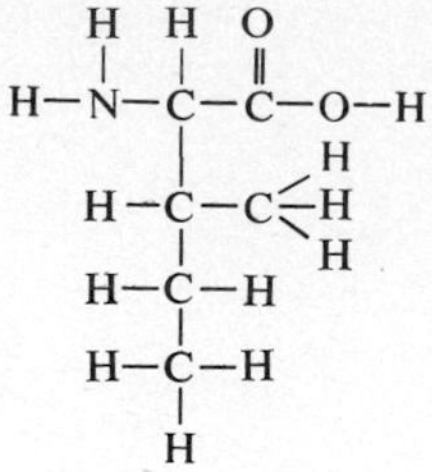

Isoleucine (Ile)

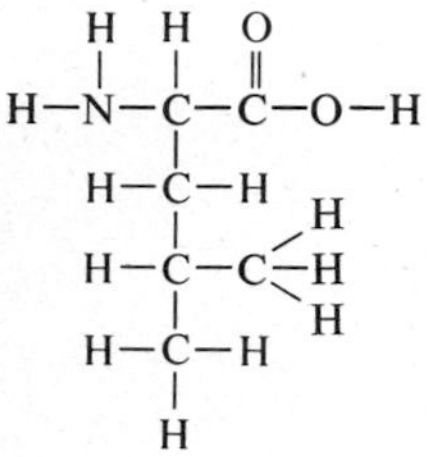

Leucine (Leu)

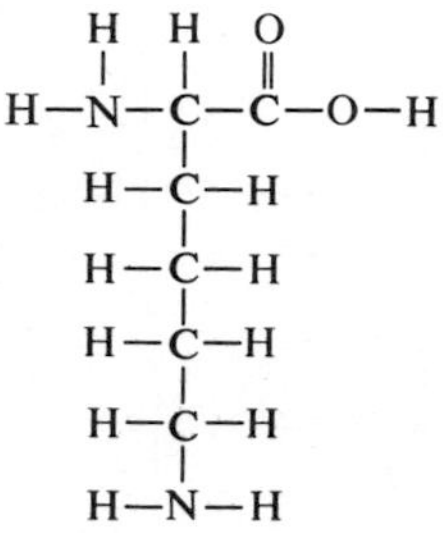

Lysine (Lys)

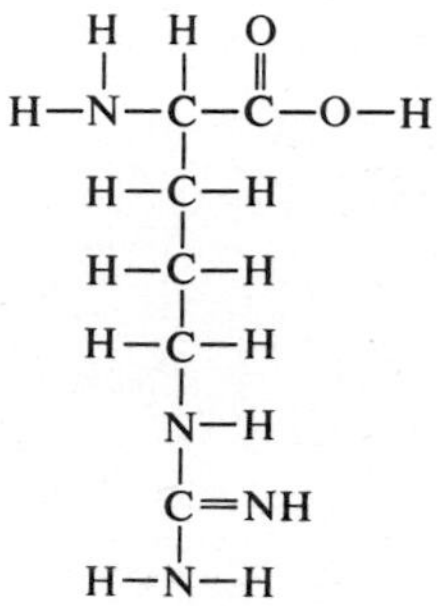

Arginine (Arg)

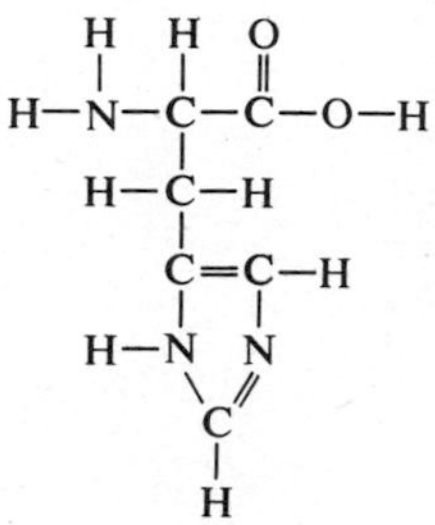

Histidine (His)

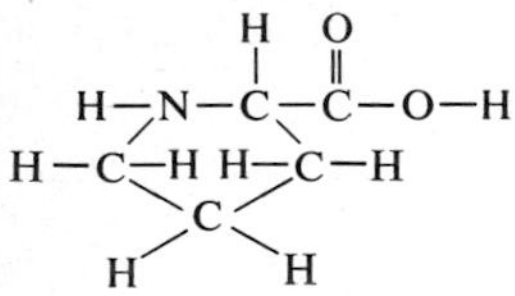

Proline (Pro)

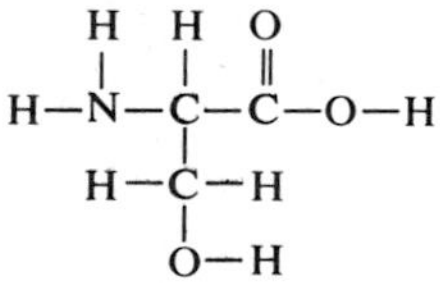

Serine (Ser)

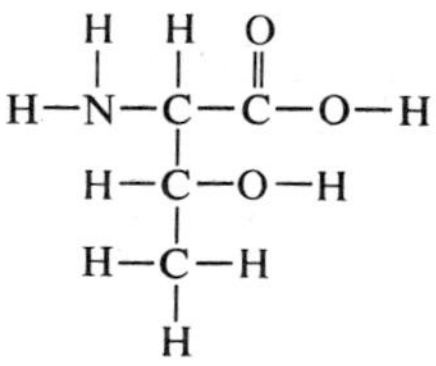

Threonine (Thr)

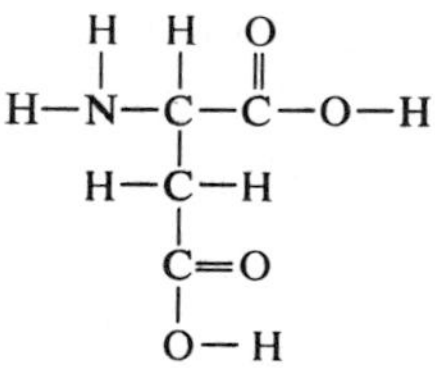

Aspartic acid (Asp)

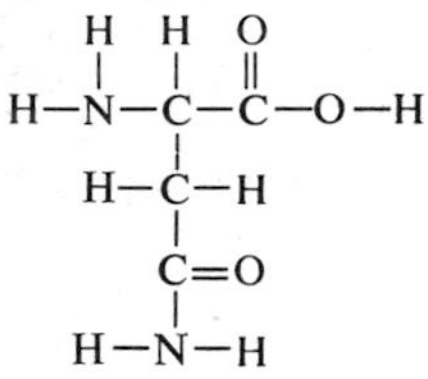

Asparagine (Asn)

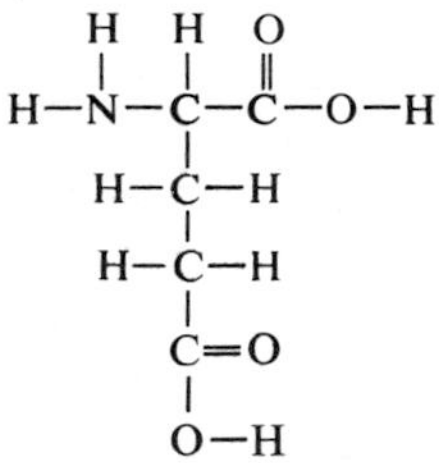

Glutamic acid (Glu)

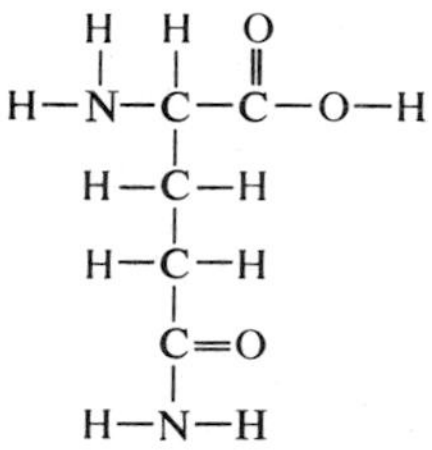

Glutamine (Gln)

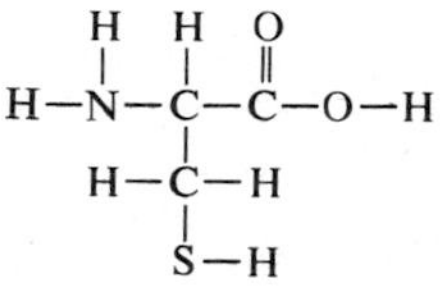

Cysteine (Cys)

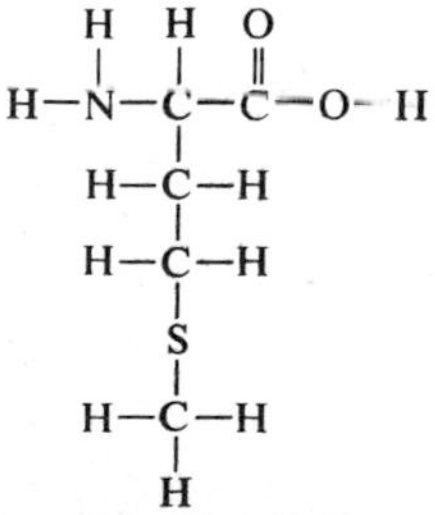

Methionine (Met)

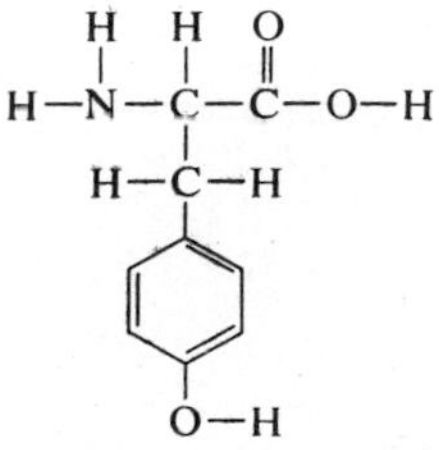

Tyrosine (Tyr)

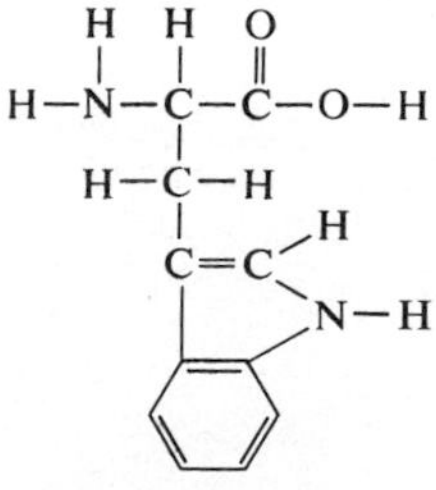

Tryptophan (Trp)

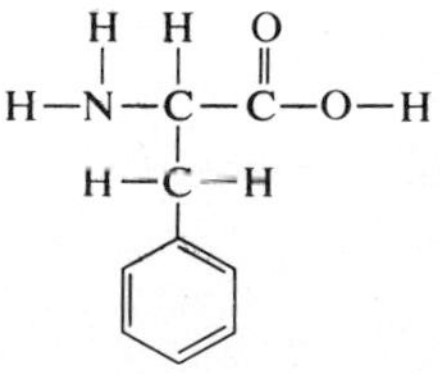

Phenylalanine (Phe)

Table A-5   Coding dictionary
*m*RNA codons (5′ → 3′) for the 20 amino acids

First letter	Second letter				Third letter
	U	C	A	G	
U	UUU ⎤ Phe UUC ⎦ UUA ⎤ Leu UUG ⎦	UCU ⎤ UCC ⎥ Ser UCA ⎥ UCG ⎦	UAU ⎤ Tyr UAC ⎦ UAA ⎤ Nonsense* UAG ⎦	UGU ⎤ Cys UGC ⎦ UGA Nonsense* UGG Trp	U C A G
C	CUU ⎤ CUC ⎥ Leu CUA ⎥ CUG ⎦	CCU ⎤ CCC ⎥ Pro CCA ⎥ CCG ⎦	CAU ⎤ His CAC ⎦ CAA ⎤ Gln CAG ⎦	CGU ⎤ CGC ⎥ Arg CGA ⎥ CGG ⎦	U C A G
A	AUU ⎤ Ileu AUC ⎥ AUA ⎦ AUG Met	ACU ⎤ ACC ⎥ Thr ACA ⎥ ACG ⎦	AAU ⎤ Asn AAC ⎦ AAA ⎤ Lys AAG ⎦	AGU ⎤ Ser AGC ⎦ AGA ⎤ Arg AGG ⎦	U C A G
G	GUU ⎤ GUC ⎥ Val GUA ⎥ GUG ⎦	GCU ⎤ GCC ⎥ Ala GCA ⎥ GCG ⎦	GAU ⎤ Asp GAC ⎦ GAA ⎤ Glu GAG ⎦	GGU ⎤ GGC ⎥ Gly GGA ⎥ GGG ⎦	U C A G

* Chain-terminating codons.

Table A-6   Metric equivalents

Metric prefixes

Prefix	Abbreviation	Meaning	Prefix	Abbreviation	Meaning
deci	d	10^{-1}	micro	μ	10^{-6}
centi	c	10^{-2}	nano	n	10^{-9}
milli	m	10^{-3}	pico	p	10^{-12}

Conversion factors for length

	Meter (m)	Micrometer* (μm)	Angstrom (Å)
1 meter (m)	1	10^6	10^{10}
1 centimeter (cm)	10^{-2}	10^4	10^8
1 millimeter (mm)	10^{-3}	10^3	10^7
1 micrometer* (μm)	10^{-6}	1	10^4
1 nanometer* (nm)	10^{-9}	10^{-3}	10
1 picometer* (pm)	10^{-12}	10^{-6}	10^{-2}
1 inch	2.54×10^{-2}	2.54×10^4	2.54×10^8

* The unit formerly called the micron (μ), i.e., 1×10^{-6} m, has been renamed the micrometer, and the prefixes millimicro- and micromicro- have been replaced as shown.

Some constants

Average molecular weight* of nucleotide = 330
Average molecular weight of amino acid = 110
Distance between nucleotides = 3.4 Å
Distance between two amino acids on a
  polypeptide chain = 3.8 Å
Diameter of DNA helix = 20 Å

* Sum of atomic weights of all atoms in molecule.

Approximate lower limits of resolution

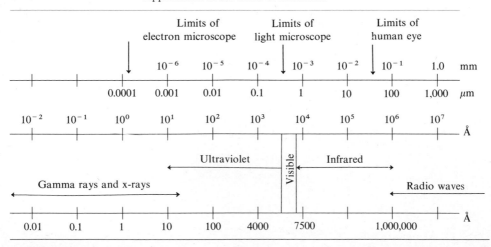

Table A-7 Biochemical abbreviations*

A	Adenine	HMC	5-Hydroxymethylcytosine
ADH	Alcohol dehydrogenase	HX	Hypoxanthine
AMP	Adenosine monophosphate	I	Inosine
AMPS	Adenosine monophosphate succinate	IMP	Inosine monophosphate
AP	2-Aminopurine	InGP	Indole 3-glycerol phosphate
ASase	Anthranilic synthetase	InGPSase	Indole glycerol phosphate synthetase
ATP	Adenosine triphosphate	Met-*t*RNA	Methionine-*t*RNA
BD	5-Bromodeoxyuridine	*m*RNA	Messenger RNA
BU	5-Bromouracil	NA	Nitrous acid
C	Cytosine	2'OMG	2'-(*O*-methyl) guanosine
CdRP	1-(*O*-carboxyphenylamino)-1-deoxyribulose-5-phosphate	OTC	Ornithine transcarbamylase
		p	Single phosphate residue
CMP	Cytidine monophosphate	P	Proflavine
CNBr	Cyanogen bromide	PAB	*p*-Aminobenzoic acid
CP	Chymotryptic peptide	PGD	6-Phosphogluconate dehydrogenase
CPK	Creatine phosphokinase		
CRM	Cross-reacting material	PR	Phosphoribosyl anthranilate transferase
CTP	Cytidine triphosphate		
DNase	Deoxyribonuclease	PRA	*N*-(5'-phosphoribosyl) anthranilate
DNP	2,4-Dinitrophenol		
DOPA	3,4-Dihydroxyphenylalanine	PTC	Phenylthiocarbamide
DTT	Dithiothreitol	RF	Replicating form
EES	Ethyl ethanesulfonate	RNA	Ribonucleic acid
EMS	Ethyl methanesulfonate	RNase	Ribonuclease
F-Met	Formylmethionine	*r*RNA	Ribosomal RNA
F-Met-*t*RNA	Formylmethionine-*t*RNA	T	Thymine
G	Guanine	TdK	Thymidine kinase
GMP-PCP	5'-Guanylmethylene diphosphonate (a GTP analog)	TP	Tryptic peptide
		*t*RNA	Transfer RNA
		*t*RNA$_F$	Carries formylmethionine
GP	Glyceraldehyde 3-phosphate	*t*RNA$_M$	Carries methionine
G6PD	Glucose-6-phosphate dehydrogenase	TSase	Tryptophan synthetase
		U	Uracil
GTP	Guanosine triphosphate	UTP	Uridine triphosphate
HA	Hydroxylamine	X	Xanthine
HAT medium	Hypoxanthine, aminopterin, and thymidine	XDH	Xanthine dehydrogenase
		ψ	Pseudouridylic acid
HGPRT	Hypoxanthine guanine phosphoribosyl transferase		

* Abbreviations of amino acids are explained in Appendix Table A-4.

Table A-8   Chi square (χ^2)

Probability of a larger value of χ^2

df	0.995	0.990	0.975	0.950	0.900	0.750	0.500	0.250	0.100	0.050	0.025	0.010	0.005
1	0.0^4393	0.0^3157	0.0^3982	0.0^2393	0.0158	0.102	0.455	1.32	2.71	3.84	5.02	6.63	7.88
2	0.0100	0.0201	0.0506	0.103	0.211	0.575	1.39	2.77	4.61	5.99	7.38	9.21	10.6
3	0.0717	0.115	0.216	0.352	0.584	1.21	2.37	4.11	6.25	7.81	9.35	11.3	12.8
4	0.207	0.297	0.484	0.711	1.06	1.92	3.36	5.39	7.78	9.49	11.1	13.3	14.9
5	0.412	0.554	0.831	1.15	1.61	2.67	4.35	6.63	9.24	11.1	12.8	15.1	16.7
6	0.676	0.872	1.24	1.64	2.20	3.45	5.35	7.84	10.6	12.6	14.4	16.8	18.5
7	0.989	1.24	1.69	2.17	2.83	4.25	6.35	9.04	12.0	14.1	16.0	18.5	20.3
8	1.34	1.65	2.18	2.73	3.49	5.07	7.34	10.2	13.4	15.5	17.5	20.1	22.0
9	1.73	2.09	2.70	3.33	4.17	5.90	8.34	11.4	14.7	16.9	19.0	21.7	23.6
10	2.16	2.56	3.25	3.94	4.87	6.74	9.34	12.5	16.0	18.3	20.5	23.2	25.2
11	2.60	3.05	3.82	4.57	5.58	7.58	10.3	13.7	17.3	19.7	21.9	24.7	26.8
12	3.07	3.57	4.40	5.23	6.30	8.44	11.3	14.8	18.5	21.0	23.3	26.2	28.3
13	3.57	4.11	5.01	5.89	7.04	9.30	12.3	16.0	19.8	22.4	24.7	27.7	29.8
14	4.07	4.66	5.63	6.57	7.79	10.2	13.3	17.1	21.1	23.7	26.1	29.1	31.3
15	4.60	5.23	6.26	7.26	8.55	11.0	14.3	18.2	22.3	25.0	27.5	30.6	32.8
16	5.14	5.81	6.91	7.96	9.31	11.9	15.3	19.4	23.5	26.3	28.8	32.0	34.3
17	5.70	6.41	7.56	8.67	10.1	12.8	16.3	20.5	24.8	27.6	30.2	33.4	35.7
18	6.26	7.01	8.23	9.39	10.9	13.7	17.3	21.6	26.0	28.9	31.5	34.8	37.2
19	6.84	7.63	8.91	10.1	11.7	14.6	18.3	22.7	27.2	30.1	32.9	36.2	38.6
20	7.43	8.26	9.59	10.9	12.4	15.5	19.3	23.8	28.4	31.4	34.2	37.6	40.0
21	8.03	8.90	10.3	11.6	13.2	16.3	20.3	24.9	29.6	32.7	35.5	38.9	41.4
22	8.64	9.54	11.0	12.3	14.0	17.2	21.3	26.0	30.8	33.9	36.8	40.3	42.8
23	9.26	10.2	11.7	13.1	14.8	18.1	22.3	27.1	32.0	35.2	38.1	41.6	44.2
24	9.89	10.9	12.4	13.8	15.7	19.0	23.3	28.2	33.2	36.4	39.4	43.0	45.6
25	10.5	11.5	13.1	14.6	16.5	19.9	24.3	29.3	34.4	37.7	40.6	44.3	46.9
26	11.2	12.2	13.8	15.4	17.3	20.8	25.3	30.4	35.6	38.9	41.9	45.6	48.3
27	11.8	12.9	14.6	16.2	18.1	21.7	26.3	31.5	36.7	40.1	43.2	47.0	49.6
28	12.5	13.6	15.3	16.9	18.9	22.7	27.3	32.6	37.9	41.3	44.5	48.3	51.0
29	13.1	14.3	16.0	17.7	19.8	23.6	28.3	33.7	39.1	42.6	45.7	49.6	52.3
30	13.8	15.0	16.8	18.5	20.6	24.5	29.3	34.8	40.3	43.8	47.0	50.9	53.7
40	20.7	22.2	24.4	26.5	29.1	33.7	39.3	45.6	51.8	55.8	59.3	63.7	66.8
50	28.0	29.7	32.4	34.8	37.7	42.9	49.3	56.3	63.2	67.5	71.4	76.2	79.5
60	35.5	37.5	40.5	43.2	46.5	52.3	59.3	67.0	74.4	79.1	83.3	88.4	92.0

Catherine M. Thompson.   Abridged from Table of Percentage Points of the χ^2 Distribution, *Biometrika*, **32**:188–189 (1941), by permission of the author and the editor of *Biometrika*.